MODERN
CONTROL THEORY

MODERN CONTROL THEORY

Second Edition

William L. Brogan, Ph.D.

Professor of Electrical Engineering
University of Nebraska

PRENTICE-HALL, INC.

Englewood Cliffs, New Jersey 07632

Library of Congress Cataloging in Publication Data

Brogan, William L. (date)
Modern control theory.

Bibliography: p.
Includes index.
1. Control theory. I. Title.
QA402.3.B76 1985 629.8′312 84-24818
ISBN 0-13-590316-5

Editorial/production supervision: *Raeia Maes*
Manufacturing buyer: *Anthony Caruso*

Printed in the United States of America

10 9 8 7 6 5 4 3 2

ISBN 0-13-590316-5 01

Prentice-Hall International, Inc., *London*
Prentice-Hall of Australia Pty., Limited, *Sydney*
Editora Prentice-Hall do Brasil, Ltda., *Rio de Janeiro*
Prentice-Hall Canada Inc., *Toronto*
Prentice-Hall Hispanoamerica, S.A., *Mexico*
Prentice-Hall of India Private Limited, *New Delhi*
Prentice-Hall of Japan, Inc., *Tokyo*
Prentice-Hall of Southeast Asia Pte. Ltd., *Singapore*
Whitehall Books Limited, *Wellington, New Zealand*

CONTENTS

3

FUNDAMENTALS OF MATRIX ALGEBRA / 70

4

VECTORS AND LINEAR VECTOR SPACES / 92

5

SIMULTANEOUS LINEAR ALGEBRAIC EQUATIONS / 118

6

TRANSFORMATIONS / 152

7

EIGENVALUES AND EIGENVECTORS / 173

8

FUNCTIONS OF SQUARE MATRICES AND THE CAYLEY-HAMILTON THEOREM / 201

9

STATE VARIABLES AND THE STATE SPACE DESCRIPTION OF DYNAMIC SYSTEMS / 225

10

ANALYSIS OF THE STATE EQUATIONS FOR CONTINUOUS-TIME LINEAR SYSTEMS / 256

PREFACE

This text presents an introduction to a range of topics which fall within the domain of modern control theory. Modern control theory provides a fundamental and unified method of approaching a wide variety of problems. It builds on and complements the classical methods of Nyquist, Bode, Evans, and Nichols. Because of this, an introductory course in classical linear control theory is a desirable, although not essential, prerequisite for this text. In the typical university curriculum, sufficient preparation is obtained by the junior year.

This text is suitable for advanced juniors, seniors, and beginning graduate students. It attempts to fill the gap between a first course in classical control theory and theoretically oriented graduate courses. However it is not just a stepping stone towards graduate study. The practitioner who has no interest in pursuing a program of graduate study will find many immediately useful results and techniques fully developed and illustrated. There is ample material for a two-semester sequence. The text can also be used for a more advanced one-semester course by skipping some of the earlier chapters and by selectively omitting one or more of the topics found in Chapters 14 through 17. This book should also be a useful supplement to other less problem oriented texts. Finally, a serious attempt has been made to make this material suitable for a program of self-study. This has been done by including numerous illustrative examples within the chapters and a large number of solved problems at the end of each chapter. Wherever possible, an intuitive approach and ample verbal discussion are provided rather than the more concise, but often cryptic, theorem-proof method of presentation.

The material presented herein divides into four general segments. Chapters 1 and 2 are introductory and, for the most part, would have been covered in prerequisite courses. An overview of the methods and goals of classical control theory is presented, without getting bogged down in minute details. The advantages of modern control theory can be more fully appreciated when viewed as a natural outgrowth of classical control theory.

Modern control theory depends heavily on matrix theory and linear algebra. Since this is not true in the classical approach, some of the necessary mathematical tools will be unfamiliar to some readers. The second segment of this text, Chapters 3 through 8, develops these topics. Chapters 3 and 4 develop the language of modern control theory, namely, matrix theory and linear algebra. Chapter 5 applies them to several technically important problem areas. These include linear programming and least squares problems, which are extremely useful in their own right. They also serve to sharpen the manipulative skills presented in Chapters 3 and 4. Chapter 6 is a slightly more abstract extension of the previous chapters. It is an introduction to some of the notions which are encountered in functional analysis treatments of modern control. This chapter could be omitted in an undergraduate class where it is not necessary to prove all of the results presented in later chapters. Chapters 7 and 8 present a concise but fairly complete treatment of eigenvalue-eigenvector problems and their principal applications to control theory.

The third segment of the text comprises Chapters 9 through 12. The notions of state and state space are defined and discussed in Chapter 9. The continuous-time and discrete-time linear state equations are analyzed in Chapters 10 and 11 respectively. By combining the discrete state equation and recursive least squares techniques, a deterministic development of the Kalman filter is given in the problems of Chapter 11. Intrinsic system properties of controllability and observability are discussed in Chapter 12.

Chapters 13 through 17 constitute the fourth segment of the text material. These chapters provide introductions to several of the diverse directions of specialization possible within the field of modern control theory. Chapter 13 explores some of the fundamental techniques for extending methods of linear analysis to nonlinear systems. Chapter 14 deals with stability analysis of linear and nonlinear systems, including the method of Lyapunov. The extension of classical techniques to multivariable systems, and the relationships between classical and modern techniques are pursued in Chapters 15 and 16. Chapter 17 introduces optimal control theory by means of Dynamic Programming and Pontryagin's Maximum Principle. The intention is to convey an understanding of the techniques while leaving the more rigorous development and proofs for advanced graduate courses. Finally, the large number of references and suggestions for further study form an important method of extending the coverage provided in these 17 chapters.

Major changes have been incorporated into this second edition. More emphasis is placed on digital systems (discrete-time, sampled data). The use of computers in control loops and in the design and analysis of control systems is now more commonplace. This fact is reflected throughout this edition, but especially in the many new examples and problems. Increased emphasis is given to engineering design of control systems, especially in the final two chapters. Ten years' experience with the first edition has brought reactions and comments from hundreds of students and other readers. These form the basis for many other improvements scattered throughout the book.

Since this book is a direct outgrowth of the earlier edition, all the people credited there still merit my appreciation. It is not possible to individually mention all those people whose feedback helped shape this edition. More direct contributions were made by former graduate students Dr. A. John Boye, Timothy H. Conway, Dr.

K. H. Gurubasavaraj, and Albert N. Hinrichs. Undergraduates Dave C. Ganow and Christopher C. Smith helped develop some useful programs. James T. Pritchard of Westinghouse Electric provided me with copies of several useful algorithms from the literature. Finally, the staff at Prentice-Hall are kindly acknowledged. Without their prodding this second edition would not have been started. Without their skillful assistance it would never have been finished.

William L. Brogan

MODERN CONTROL THEORY

Chapter

1

BACKGROUND AND PREVIEW

1.1 INTRODUCTION

Control theory is often regarded as a branch of the general, and somewhat more abstract, subject of systems theory [117].* The boundaries between these disciplines are often unclear, so a brief section is included to delineate the point of view of this book.

In order to put control theory into practice, a bridge must be built between the real world and the mathematical theory. This bridge is the process of modeling, and a summary review of modeling is included in this chapter [15, 109].

Control theory can be approached from a number of directions. The first systematic method of dealing with what is now called control theory began to emerge in the 1930s. Transfer functions and frequency domain techniques were predominant in these "classical" approaches to control theory. Starting in the late 1950s and early 1960s a time-domain approach using state variable descriptions came into prominence. This is what this book refers to as "modern" control theory. At the present time there is a blurring of the boundaries and a merging of these methods, along with new developments, into a unified theory. However, distinctions still exist, including the types of systems that can be dealt with efficiently, the forms of the system models, the methods of analysis, and even the philosophies that underlie the methods of design and analysis. Some of these distinctions are discussed in this chapter. A very brief look at the basic parts of classical theory will be reviewed in the next chapter. After that, the great majority of this text will be aimed at the state variable method. Bridges between the two will be built at several points throughout the book as seems appropriate.

*Reference citations are given numerically in the text in brackets. The references are in a single section at the end of the book.

1.2 SYSTEMS, SYSTEMS THEORY, AND CONTROL THEORY

According to the *Encyclopedia Americana*, a system is ". . . an aggregation or assemblage of things so combined by nature or man as to form an integral and complex whole . . . ." Mathematical systems theory is the study of the interactions and behavior of such an assemblage of "things" when subjected to certain conditions or inputs. The abstract nature of systems theory is due to the fact that it is concerned with mathematical properties rather than the physical form of the constituent parts.

Control theory is more often concerned with physical applications. A control system is considered to be any system which exists for the purpose of regulating or controlling the flow of energy, information, money, or other quantities in some desired fashion. In more general terms, a control system is an interconnection of many components or functional units in such a way as to produce a desired result. In this book control theory is assumed to encompass all questions related to design and analysis of control systems.

Figure 1.1 is a general representation of an *open-loop* control system. The input or control $\mathbf{u}(t)$ is selected based on the goals for the system and all available a priori knowledge about the system. The input is in no way influenced by the output of the system, represented by $\mathbf{y}(t)$. If unexpected disturbances act upon an open-loop system, or if its behavior is not completely understood, then the output will not behave precisely as expected.

Another general class of control systems is the *closed-loop* or *feedback* control system, as illustrated in Fig. 1.2. In the closed-loop system, the control $\mathbf{u}(t)$ is modified in some way by information about the behavior of the system output. A feedback system is often better able to cope with unexpected disturbances and uncertainties about the system's dynamic behavior. However, it need not be true that closed-loop control is always superior to open-loop control. When the measured outputs have errors which are sufficiently large, and when unexpected disturbances are relatively unimportant, closed-loop control can have a performance which is inferior to open-loop control.

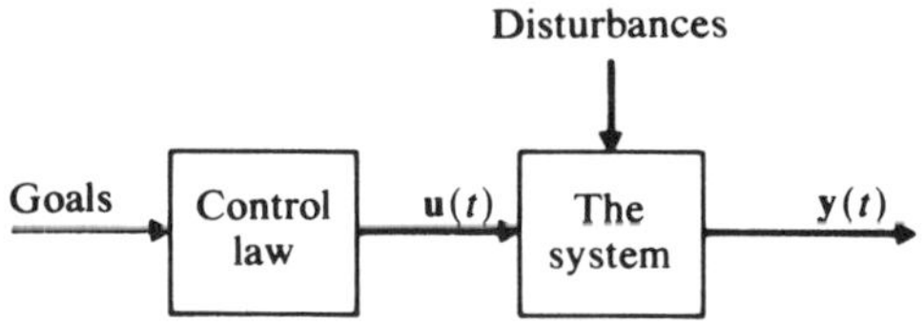

Figure 1.1 An open-loop control system.

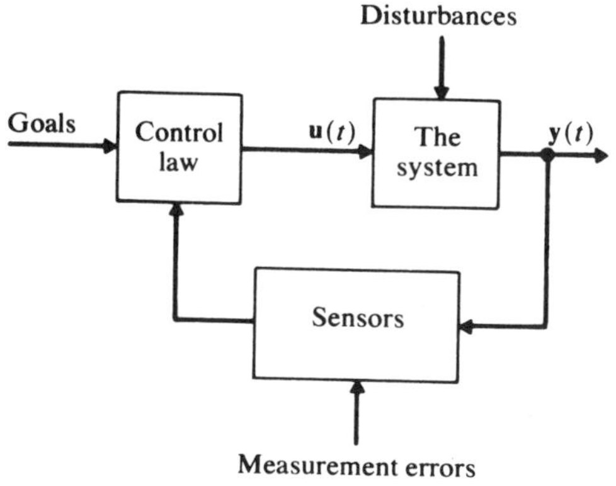

Figure 1.2 A closed-loop control system.

Example 1.1

A man's goal is to provide as much financial security as possible for his retirement years. He decides to have an extra \$300.00 per month deducted from his paycheck and deposited in a tax sheltered annuity. His "input" each month is $u(t)$ and the system output $y(t)$ is the accrued value in his account. Since $u(t)$ is in no way affected by the current economic climate or by $y(t)$, this is an open-loop system. ■

Example 1.2

Another man has the same goal of achieving financial security. He decides to directly control his investments in the stock market. His input $u(t)$ at any given time is influenced by his perception of the market conditions, how well he has done so far, and so forth. This is a feedback or closed-loop system. ■

Example 1.3

A typical industrial control system involves components from several engineering disciplines. The automatic control of a machine shown in Fig. 1.3 illustrates this. In this example, the desired time history of the carriage motion is patterned into the shape of the cam. As the cam-follower rises and falls, the potentiometer pick-off voltage is proportional to the desired carriage position. This signal is compared with the actual position, as sensed by another potentiometer. This difference, perhaps modified by a tachometer-generated rate signal, gives rise to an error signal at the output of the differential amplifier. The power level of this signal is usually low and must be amplified by a second amplifier before it can be used for corrective action by an electric motor or a servo valve and a hydraulic motor or some other prime mover. The prime mover output would usually be modified by a precise gear train, a lead screw, a chain and sprocket, or some other mechanism. Clearly, mechanical, electrical, electronic, and hydraulic components play important roles in such a system. ■

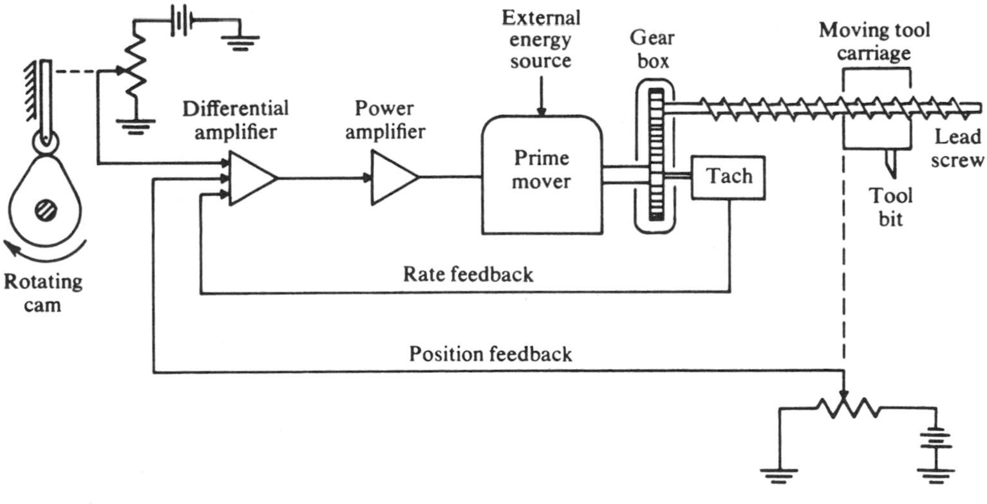

Figure 1.3

Example 1.4

The same ultimate purpose of controlling a machine tool could be approached somewhat differently using a small computer in the loop. The continuous-time or analog signals for position and velocity must still be controlled. Measurements of these quantities would probably be made directly in the digital domain using some sort of optical pulse counting circuitry. If analog measurements are made, then an analog-to-digital (A/D) conversion is necessary. The desired position and velocity data would be available to the computer in numeric form. The digital measurements would be compared and the differences would constitute inputs into a corrective control algorithm. At the output of the computer a digital-to-analog (D/A) conversion could be performed to obtain the control inputs to the same prime mover as in Example 1.3. Alternately, a stepper motor may be selected because it can be directly driven by a series of pulses from the computer. Figure 1.4 shows a typical control system with a computer in the loop. ■

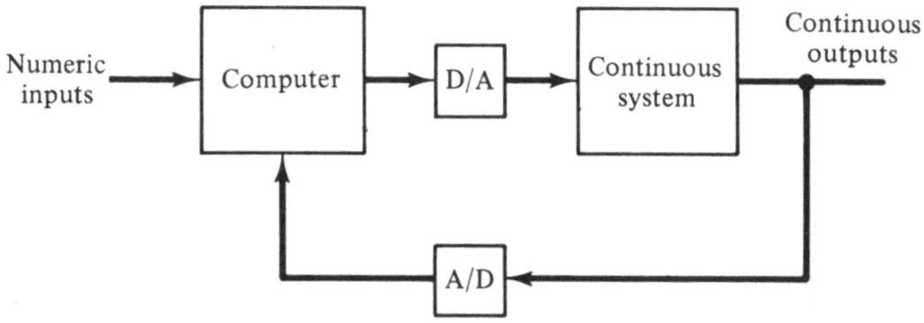

Figure 1.4

1.3 MODELING

Engineers and scientists are frequently confronted with the task of analyzing problems in the real world, synthesizing solutions to these problems, or developing theories to explain them. One of the first steps in any such task is the development of a mathematical model of the phenomenon being studied. This model must not be oversimplified, or conclusions drawn from it will not be valid in the real world. The model should not be so complex as to complicate unnecessarily the analysis.

System models can be developed by two distinct methods. *Analytical modeling* consists of a systematic application of basic physical laws to system components and the interconnection of these components. *Experimental modeling*, or modeling by synthesis, is the selection of mathematical relationships which seem to fit observed input-output data. Analytical modeling is emphasized here. Some aspects of the other approach are presented in Chapters 5, 11, and 13 (least-squares data fitting).

An outline of the analytical approach to modeling is presented in Fig. 1.5. The steps in this outline are discussed in the following paragraphs.

1. *The intended purposes of the model must be clearly specified.* There is no single model of a complicated system which is appropriate for all purposes. If the purpose is a detailed study of an individual machine tool, the model would be very

different from one used to study the dynamics of work flow through an entire factory.

2. *The system boundary is a real or imagined separation of the part of the real world under study, called the system, and the rest of the real world, referred to as the environment*. The system boundary must enclose all components or subsystems of primary interest, such as subsystems *A*, *B*, and *C* in Fig. 1.5*a*.

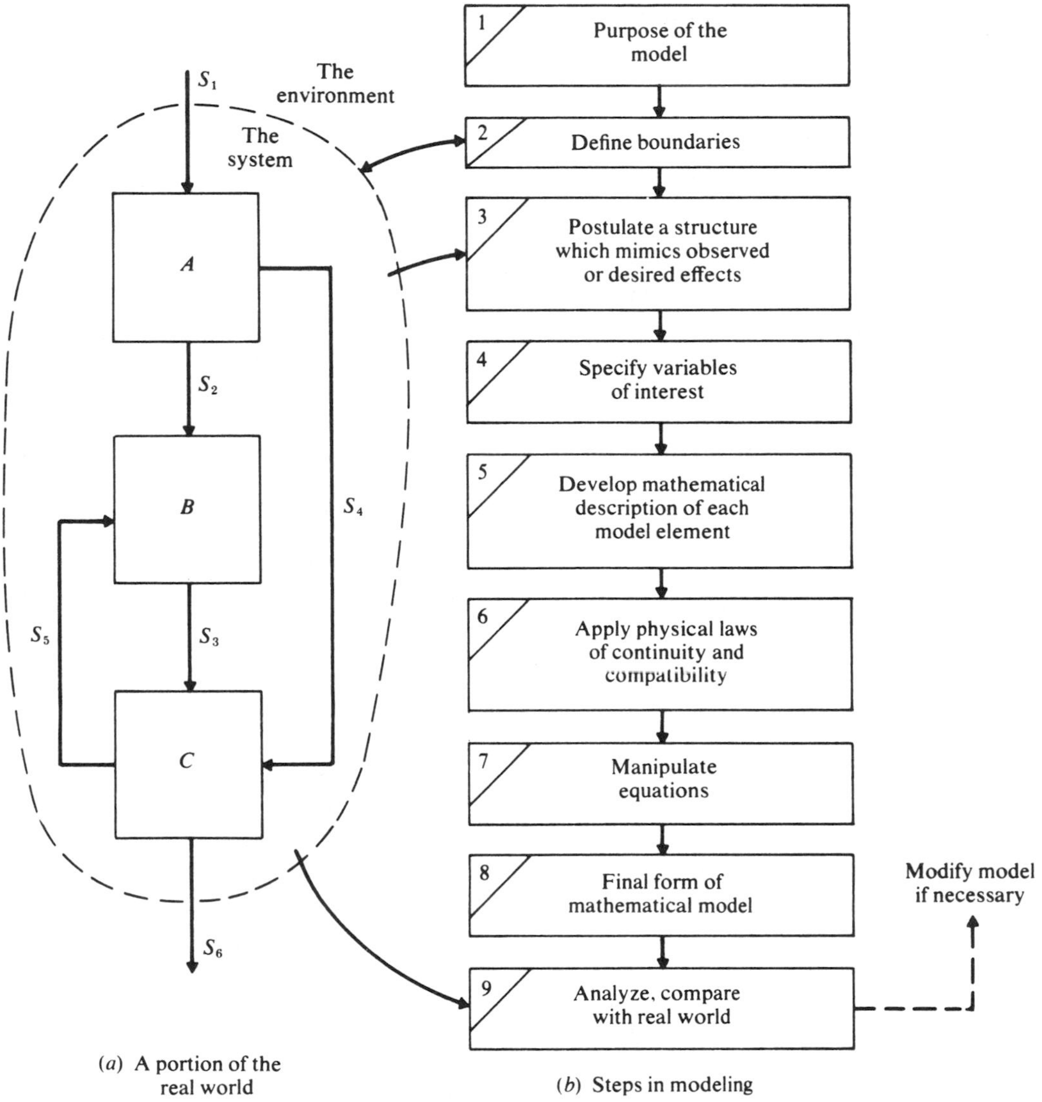

Figure 1.5 Modeling considerations.

A second requirement on the selection of the boundary is that all causative actions or effects (called signals) crossing the boundary be more or less one-way interactions. The environment can affect the system, and this is represented by the input signal S_1. The system output, represented by the signal S_6, should not affect the environment, at least not to the extent that it would modify S_1. If there is no interest in subsystem A, then a boundary enclosing B and C, and with inputs S_2 and S_4, could be used. Subsystem C should not be selected as an isolated system because one of its outputs S_5 modifies its input S_3 through subsystem B. The requirement is that all inputs are known, or can be assumed known for the purpose of the study, or can be controlled independently of the internal status of the system.

Example 1.5

The purpose of the models of Fig. 1.6 is to study the flow of work and information within a production system due to an input rate of orders. These orders could be an input from the environment, as in Model I. If the purpose is to study the effects of an advertising campaign, then orders are determined, at least in part, by a major system variable. In this case the rate of orders should be an internal variable, as in Model II. ■

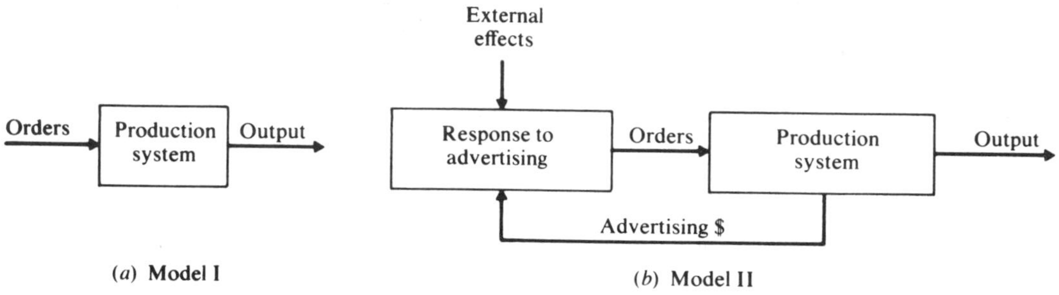

Figure 1.6

3. *All physical systems, whether they are of an electrical, mechanical, fluid, or thermal nature, have mechanisms for storing, dissipating, or transferring energy, or transforming energy from one form to another.* The third step in modeling is one of reducing the actual system to an interconnection of simple, idealized elements which preserve the character of these operations on the various kinds of energy. An electric circuit diagram illustrates such an idealization, with ideal sources representing inputs. In mechanical systems, idealized connections of point masses, springs, and dashpots are often used. In thermal or fluid systems, and to a certain extent in economic, political, and social systems, similar idealizations are possible. This process is referred to as *physical modeling*. The level of detail required depends on the type of information expected from the model.
4. *If the physical model is properly selected, it will exhibit the same major characteristics as the real system.* In order to proceed with development of a mathematical model, variables must be assigned to all attributes of interest. If a quantity of interest does not yet exist and thus cannot be labeled, a modification will be

required in Step 3 so as to include it. The classification of system types is discussed in the next section. This book deals mainly with *deterministic lumped-parameter systems*. In all lumped-parameter systems there are basically just two types of variables. They are *through variables* (sometimes called path variables or rate variables) and *across variables* (sometimes called point variables or level variables). Through variables flow through two-terminal elements and have the same value at both terminals. Examples are electric current, force or torque, heat flow rate, fluid flow rate, and rate of work flow through a production element. Across variables have different values at the two terminals of a device. Examples are voltage, velocity, temperature, pressure, and inventory level.

5. *Each two-terminal element in the idealized physical model will have one through and one across variable associated with it.* Multiterminal devices such as transformers or controlled sources will have more. In every device, mathematical relationships will exist between the two types of variables. These relationships, called *elemental equations*, must be specified for each element in the model. This step could uncover additional variables that need to be introduced. This would mean a modification of Step 4. Common examples of elemental equations are the current-voltage relationships for resistors, capacitors, and inductors. The form of these relations may be algebraic, differential, or integral expressions, linear or nonlinear, constant or time-varying.
6. *After a system has been reduced to an interconnection of idealized elements, with known elemental equations, equations must be developed to describe the interconnection effects.* Regardless of the physical type of the system, there are just two types of physical laws that are needed for this purpose. The first is a statement of *conservation* or *continuity* of the through variables at each node where two or more elements connect. Examples of this basic law are Kirchhoff's node equations, D'Alembert's version of Newton's second law, conservation of mass in fluid flow problems, and heat balance equations. The second major law is a *compatibility* condition relating across variables. Kirchhoff's voltage law around any closed loop is but one example. Similar laws regarding relative velocities, pressure drops, and temperature drops must also hold. Both of these laws yield linear equations in through or across variables, regardless of whether the elemental equations are linear or nonlinear. This fact is responsible for the name given to linear graphs, an extremely useful tool in applying these two laws.

Example 1.6

Consider the system with six elements, including a source v_0, shown in Fig. 1.7. Each element is represented as a branch of the linear graph, and the interconnection points are nodes. Each node is identified by an across variable v_i, and each branch has a through variable, called f_i, with the arrow establishing the sign convention for positive flow. Let

b(number of branches) $= 6$
s(number of sources) $= 1$
n(number of nodes) $= 4$

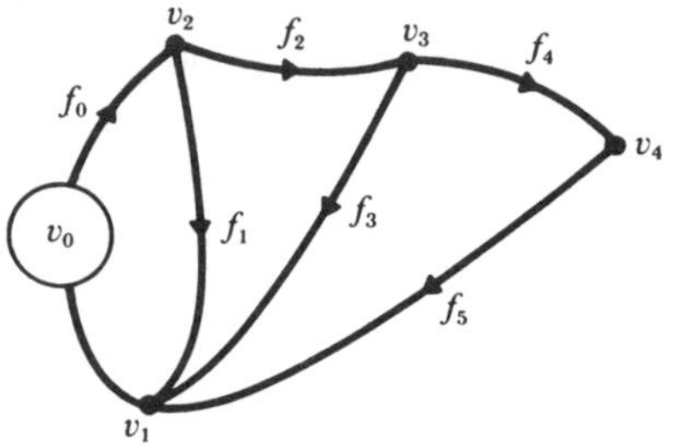

Figure 1.7

Two unknowns exist for each branch, except source branches have a single unknown. Thus there are $2b - s = 11$ unknowns, and eleven equations are needed. They are $b - s = 5$ elemental equations, $n - 1 = 3$ continuity equations (node 1 is used as a reference and is redundant)

$$f_0 - f_1 - f_2 = 0, \qquad f_2 - f_3 - f_4 = 0, \qquad f_4 - f_5 = 0$$

and $b - (n - 1) = 3$ compatibility (loop) equations. Letting $v_{ab} = v_a - v_b$, these are

$$v_{21} - v_0 = 0, \qquad v_{23} + v_{31} + v_{12} = 0, \qquad v_{34} + v_{41} + v_{13} = 0$$

These $2b - s$ equations can be used to determine all unknowns. ■

7. *This step consists of manipulating the elemental, continuity, and compatibility equations into a desired final form.* A further discussion of this step is presented in Sec. 1.5.
8. *Item 8 of Fig. 1.5 is the end result of the modeling process.* It may be arrived at by a process of iteration, as mentioned in Step 9.
9. *The model developed in the preceding steps should never be confused with the real world system being studied.* Whenever possible, the results produced by the real system should be compared with model results under similar conditions. If unacceptable discrepancies exist, the model is inadequate and should be modified.

1.4 CLASSIFICATION OF SYSTEMS

As a result of the modeling procedure of Sec. 1.3, it can be seen that the types of equations required to describe a system depend entirely on the types of elemental equations and the types of inputs from the environment. System models are classified according to the types of equations used to describe them. The family tree shown in Fig. 1.8 illustrates the major system classifications. Combinations of these classes can also occur. The most significant combination is the continuous-time system, digital controller of the type mentioned in Example 1.4. The digital signals are discrete-time in nature, that is, they only change at discrete time points. The most common approach to these problems is to represent the continuous-time part of the system by a discrete-time approximate model and then proceed with a totally discrete problem.

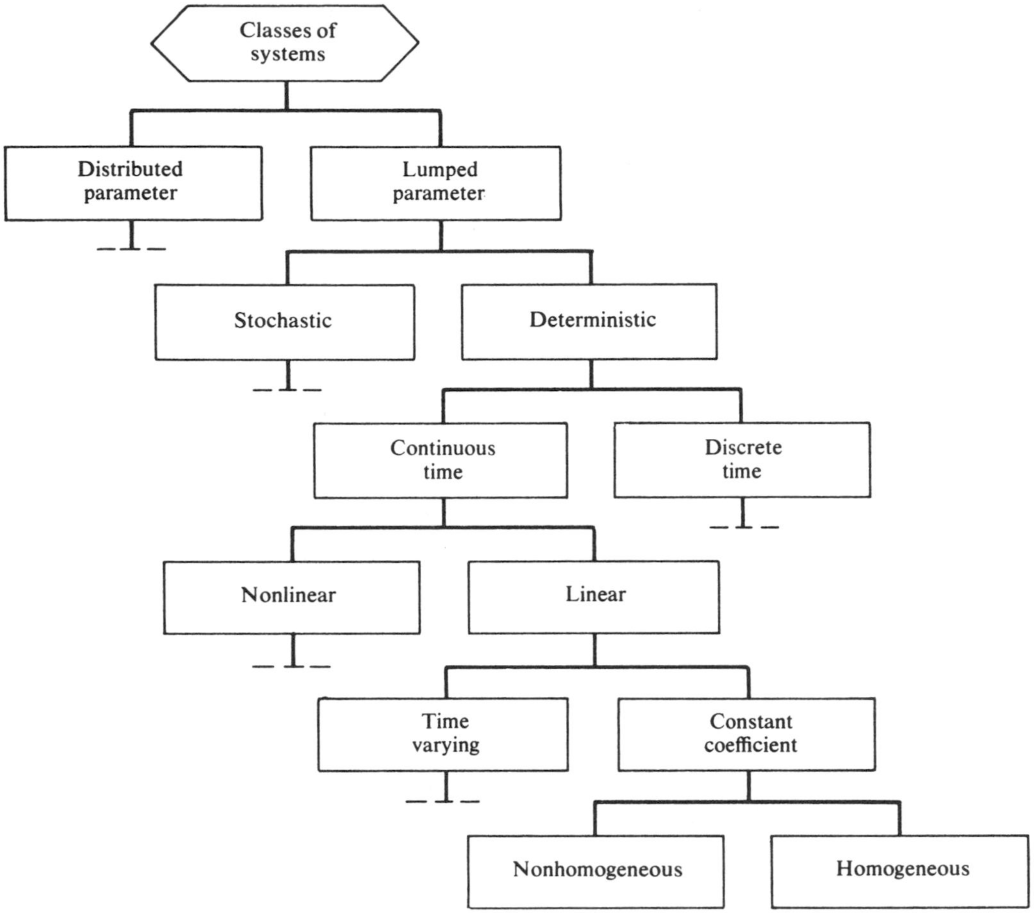

Figure 1.8 Major classes of system equations.

In Fig. 1.8 dashed lines indicate the existence of subdivisions similar to the others shown on the same level.

Distributed parameter systems require partial differential equations [10] for their description, for example, as in the description of currents and voltages at every spatial point along a transmission line. These will not be considered further, but can often be approximated by lumped-parameter models. Lumped-parameter systems are those for which all energy storage or dissipation can be lumped into a finite number of discrete spatial locations. They are described by ordinary difference equations, or in some cases by purely algebraic equations. Discrete component electric circuits fall into this category.

Systems which contain parameters or signals (including inputs) which can only be described in a probabilistic fashion (due to ignorance or actual random behavior) are called stochastic or random systems. Because random process theory [87] is not an assumed prerequisite for this text, emphasis will be on deterministic (nonrandom) lumped-parameter systems. There will be a few occasions, such as in the discussion of noisy measurements, where the random nature of certain error signals cannot be ignored.

If all elemental equations are defined for all time, then the system is a continuous-time system. If, as in sampling or digital systems, some elemental equations are defined or used only at discrete points in time, a discrete-time system is the result. Continuous-time systems are described by differential equations, discrete-time systems by difference equations.

If all elemental equations are linear, so is the system. If one or more elemental equations are nonlinear, as is the case for a diode, then the overall system is nonlinear. When all elemental equations can be described by a set of constant parameter values, as in the familiar *RLC* circuit, the system is said to be stationary or time-invariant or constant coefficient. If one or more parameters, or the very form of an elemental equation, vary in a known fashion with time, the system is said to be time-varying. Finally, if there are no external inputs and the system behavior is determined entirely by its initial conditions, the system is said to be homogeneous or unforced. With forcing functions acting, a nonhomogeneous system must be considered.

One additional distinction not shown in Fig. 1.8 could be made between large-scale (many variables) systems and small-scale systems. The degree of difficulty in analysis varies greatly among these system classifications. These differences have motivated different methods of approach. Modern control theory provides one of the most general approaches.

1.5 MATHEMATICAL REPRESENTATIONS OF SYSTEMS

During the modeling process, equations are developed which describe the behavior of each individual system element and also equations which describe the interconnections of these elements. These equations, or for that matter the corresponding linear graph, could be taken as the mathematical representation of the system. Normally, however, additional manipulations will be performed before the mathematical representation of the model is in final form.

Many forms are possible, but generally they divide into one of two categories: (*a*) input-output equations, and (*b*) equations which reveal the internal behavior of the system as well as input-output terminal characteristics.

Input-output equations are derived by a process of elimination of all system variables except those constituting inputs and those considered as outputs. For the system shown in Fig. 1.5*a*, this would mean expressing signals S_2, S_3, S_4, and S_5 in terms of the input signals S_1 and output signals S_6. This is done using the known dynamic relations of the subsystems *A*, *B*, and *C*. For example, S_2 and S_4 are related to S_1 through the dynamics of subsystem *A*. The input-output equations could constitute one or more differential or difference equations in any of the classes shown in Fig. 1.8. The independent variable is usually time, the dependent variables are the system outputs, and the inputs act as forcing functions.

When the constituent equations are linear with constant coefficients, Laplace or Z-transforms can be used to define input-output *transfer functions*. When more than one input and more than one output must be treated, matrix notation and the concepts of *transfer matrices* are convenient. It is assumed that the Laplace transform [4] is a tool familiar to most readers. The Z-transform [39, 69, 103] may be less familiar, and so a bare-bones minimum introduction to it is contained in the problems. Although time-domain methods will be stressed in this book, transforms and transfer functions will at times be useful. A reader with no prior exposure should also consult the references.

Integral forms of the input-output equations, using the system *weighting function*, are also widely used. The weighting function, or system impulse response, is obtained from the inverse Laplace or Z-transform of the input-output transfer function.

Equations which reveal the internal behavior of the system, and not just the terminal characteristics, can take several forms. Loop equations, involving all unknown loop currents (through variables), or node equations, involving all unknown nodal voltages (across variables), can be written. The choice between alternatives may depend on the number of loops versus the number of nodes. If everything is linear, transform techniques lead to the concepts of impedance and admittance matrices. Hybrid combinations of voltage and current equations are also used.

The *state space* approach provides another type of system representation [117]. Modern control theory is almost synonymous with the state space approach. This will be developed extensively in the remainder of this book. At this time it is sufficient to say that state variables consist of some minimum set of variables which are essential for completely describing the internal status, i.e., state of the system.

1.6 INTRODUCTION TO MODERN CONTROL THEORY

Several factors provided the stimulus for the development of modern control theory:

(a) The necessity of dealing with more realistic models of systems.
(b) The shift in emphasis towards optimal control and optimal system design.
(c) The continuing developments in digital computer technology.
(d) The shortcomings of previous approaches.
(e) A recognition of the applicability of well-known methods in other fields of knowledge.

The transition from simple approximate models, which are easy to work with, to more realistic models produces two effects. First, a larger number of variables must be included in the model. Second, a more realistic model is more likely to contain nonlinearities and time-varying parameters. Previously ignored aspects of the system, such as interactions with and feedback through the environment, are more likely to be included.

With an advancing technological society, there is a trend towards more ambitious goals. This also means dealing with complex systems with a larger number of interacting components. The need for greater accuracy and efficiency has changed the emphasis on control system performance. The classical specifications in terms of percent overshoot, settling time, bandwidth, etc., have in many cases given way to optimal criteria such as minimum energy, minimum cost, and minimum time operation. Optimization of these criteria makes it even more difficult to avoid dealing with unpleasant nonlinearities. Optimal control theory often dictates that nonlinear time-varying control laws be used, even if the basic system is linear and time-invariant.

The continuing advances in computer technology have had three principal effects on the controls field. One of these relates to the gigantic supercomputers. The size and class of problems that can now be modeled, analyzed, and controlled are considerably larger than they were when the first edition of this book was written.

The second impact of computer technology has to do with the proliferation and wide availability of microcomputers in homes and in the work place. Classical control theory was dominated by graphical methods because at the time that was the only way to solve certain problems. Now every control designer has easy access to powerful computer packages for systems analysis and design [49]. The old graphical methods have not yet disappeared, but have been automated. They survive because of the insight and intuition that they can provide. However, some different techniques are often better suited to a computer. Although a computer can be used to carry out the classical transform-inverse transform methods, it is usually more efficient for a computer to integrate differential equations directly.

The third major impact of computers is that they are now so commonly used as just another component in the control system. Their cost, size, and reliability make it possible to use them routinely in many systems. This means that the discrete-time and digital system control now deserves much more attention than it did in the past.

Modern control theory is well suited to the above trends because its time-domain techniques and its mathematical language (matrices, linear vector spaces, etc.) are ideal when dealing with a computer. Computers are a major reason for the existence of state variable methods.

Most classical control techniques were developed for linear constant coefficient systems with one input and one output (perhaps a few inputs and outputs). The language of classical techniques is the Laplace or Z-transform and transfer functions. When nonlinearities and time variations are present, the very basis for these classical techniques is removed. Some successful techniques such as phase-plane methods, describing functions, and other ad hoc methods, have been developed to alleviate this shortcoming. However, the greatest success has been limited to low-order systems. The state variable approach of modern control theory provides a uniform and powerful method of representing systems of arbitrary order, linear or nonlinear, with time-varying or constant coefficients. It provides an ideal formulation for computer implementation and is responsible for much of the progress in optimization theory.

Modern control theory is a recent development in the field of control. Therefore, the name is justified at least as a descriptive title. However, the foundations of modern control theory are to be found in other well-established fields. Representing a system in terms of state variables is equivalent to the approach of Hamiltonian mechanics, using generalized coordinates and generalized momenta. The advantages of this approach have been well-known in classical physics for many years. The advantages of using matrices when dealing with simultaneous equations of various kinds have long been appreciated in applied mathematics. The field of linear algebra also contributes heavily to modern control theory. This is due to the concise notation, the generality of the results, and the economy of thought that linear algebra provides.

ILLUSTRATIVE PROBLEMS

1.1 Explain why it would be inappropriate to consider a single branch of an electric network a system, even if the only variable of interest is the current through that branch.

The current in one branch affects the currents and voltages in other parts of the network. This, in turn, affects the current in the first branch. Because of the two-way coupling, the network must be considered as a whole, and must be solved using simultaneous equations.

1.2 Develop an electromechanical model of the fixed field, armature controlled dc motor. Consider the voltage supplied to the armature as the input and account for the observed dissipation of electrical energy and mechanical energy.

The dissipation of electrical energy can be accounted for by lumping all armature resistance into a resistor R. A noticeable phase shift between the supply voltage and the current through the armature windings can be accounted for by a lumped inductor L. The load and all rotating parts can be represented as a lumped inertia element J. Mechanical energy losses are accounted for by adding an ideal damper b between the rotating load and some fixed reference. The connection between the electrical and the mechanical aspects is obtained from Maxwell's equations. A moving charge in a magnetic field has a force exerted upon it, so that the armature torque T is a function of the armature current i_a. Likewise, a conductor moving in a magnetic field has a voltage induced in it, the back emf e_b. The model is shown schematically in Fig. 1.9.

The torque and the back emf are often approximated by the linear relationships

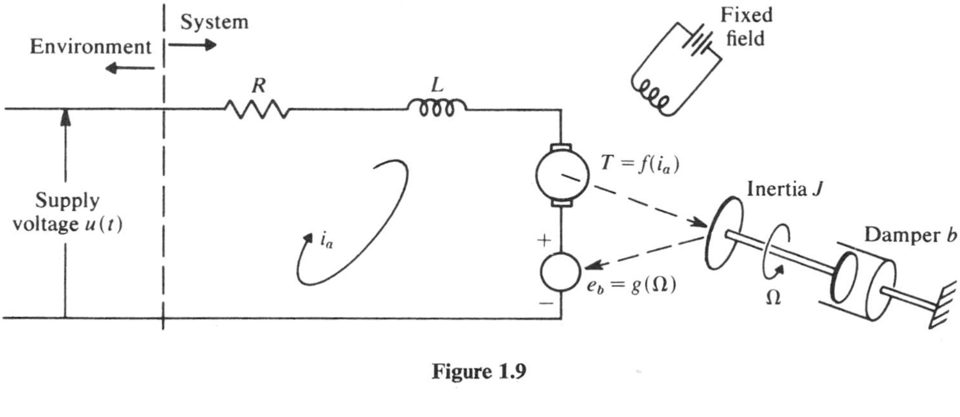

Figure 1.9

$T = Ki_a$ and $e_b = K\Omega$, where K is a constant for a particular motor. For the linear case the transfer function between the input voltage $u(t)$ and the output angle $y(t) = \int \Omega\, dt$ is derived as follows. The loop equation for the electrical circuit is $u(t) = L(di_a/dt) + Ri_a + e_b$. The mechanical torque balance is $T(t) = J\ddot{y} + b\dot{y}$. The Laplace transforms of these equations are $u(s) = (Ls + R)i_a(s) + e_b(s)$ and $T(s) = (Js^2 + bs)y(s)$. Solving gives $i_a(s) = [u(s) - e_b(s)]/(R + Ls)$. The electromechanical conversion equation then gives $T(s) = K[u(s) - e_b(s)]/(R + Ls)$. Equating the two forms for $T(s)$ and using $e_b(s) = Ksy(s)$ gives $(Js^2 + bs)y(s) = K[u(s) - Ksy(s)]/(R + Ls)$. The input-output transfer function is found from this,

$$y(s)/u(s) = K/[(Js^2 + bs)(R + Ls) + K^2 s]$$

1.3 Some electronic test gear (Fig. 1.10) is mounted near a large tank of liquid gas at -350°F. Develop a simple model which would be useful in estimating the coldest temperature at which the electronic equipment will need to operate.

Because of the insulation material, heat is allowed to flow only in one direction, from the 70° air through the electronic package to the -350° liquid gas. The environment, consisting of the two constant temperatures of 70 and -350, is represented by two ideal sources. There is a single unknown temperature T (across variable), that of the electronic package interior. The package has some thermal capacitance C and its end walls and the tank wall present thermal resistance, R_p and R_t, to heat flow Q. The linear graph is shown in Fig. 1.11.

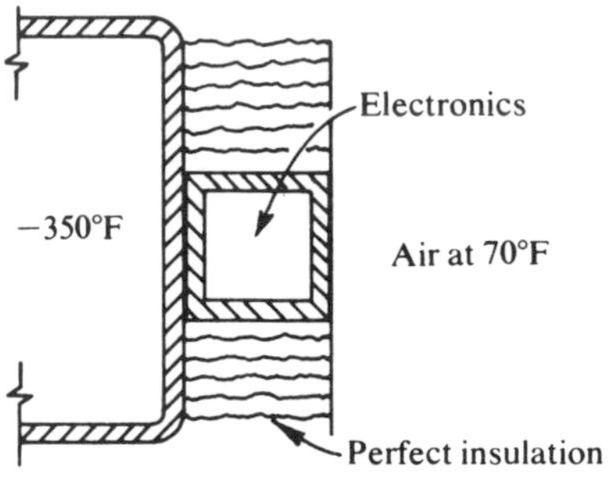

Figure 1.10

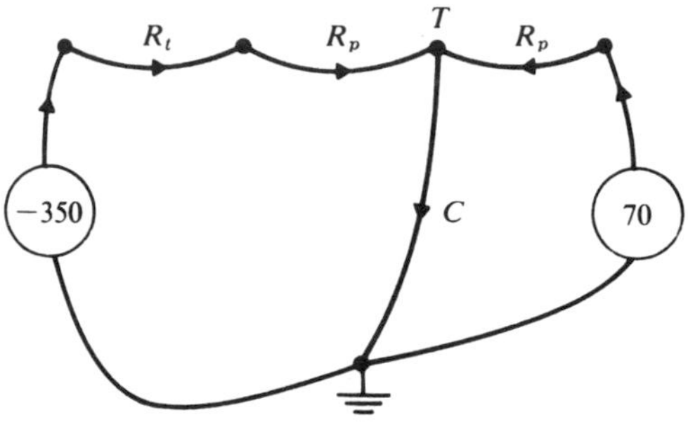

Figure 1.11

In steady state no heat flows in the branch representing the capacitance. Thus $70 - T = QR_p$ and $T - (-350) = Q(R_p + R_t)$. Eliminating the heat flow Q gives

$$\frac{70 - T}{R_p} = \frac{T + 350}{R_p + R_t} \quad \text{or} \quad T = \frac{70(R_p + R_t) - 350R_p}{2R_p + R_t}$$

If the thermal resistivity R_p of each end of the electronic package is 1°F sec/Btu and if the thermal resistivity R_t of the adjacent area of the tank is 2°F sec/Btu, then $T = [70(3) - 350(1)]/4 = -35$°F.

1.4 In many ways the flow of work through a factory is similar to fluid flow in a piping network. Figure 1.12 shows such a network. Two pumps deliver fluid at constant pressure P_1 and P_2, respectively. Six lumped approximations for fluid resistance R_i are indicated. They account for the pressure drop in each segment of pipe proportional to the flow Q through the segment. Three fluid capacitances C_i are indicated. The pressure at their base is proportional to the height of the standing fluid, that is, proportional to the integral of the flow into them. Two ideal elements, called fluid iner-

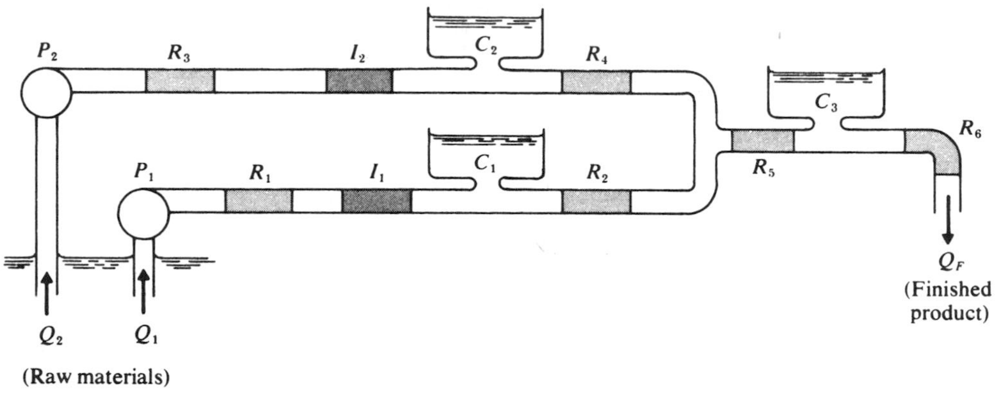

Figure 1.12

tances I_1 and I_2, are included to account for inertia effects. They cause a pressure drop proportional to the rate of change of flow. a. Draw the linear graph, label all variables; b. write the elemental equations; c. write the continuity equations; d. write the compatibility equations. Neglect all changes in height except in the capacitances, and use atmospheric pressure as the reference node.

a. Each distinct pressure (across variable) will form a system node. Between each pair of nodes a branch will represent the ideal element that accounts for the pressure change. This allows the construction of the linear graph of Fig. 1.13.

b. There are $n = 9$ nodes, $b = 13$ branches, and $s = 2$ sources. The $b - s = 11$ elemental equations are

$$P_{23} = R_3 Q_2 \qquad P_{14} = R_1 Q_1 \qquad P_{46} = I_1 \frac{dQ_1}{dt} \qquad P_{45} = I_2 \frac{dQ_2}{dt}$$

$$C_2 \frac{dP_{50}}{dt} = Q_4 \qquad C_1 \frac{dP_{60}}{dt} = Q_3 \qquad P_{57} = R_4 Q_6 \qquad P_{67} = R_2 Q_5$$

$$P_{78} = R_5 Q_7 \qquad C_3 \frac{dP_{80}}{dt} = Q_8 \qquad P_{80} = R_6 Q_F$$

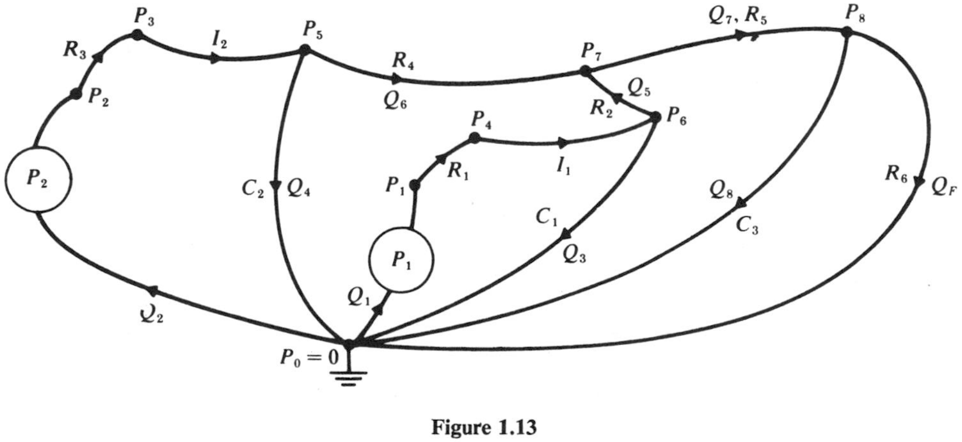

Figure 1.13

c. The $n - 1 = 8$ continuity equations are

$Q_1 = Q_1$ (twice) since Q_1 flows in three separate branches

$Q_2 = Q_2$ (twice) since Q_2 flows in three separate branches

$Q_1 - Q_3 - Q_5 = 0 \qquad Q_2 - Q_4 - Q_6 = 0 \qquad Q_5 + Q_6 - Q_7 = 0$

$Q_7 - Q_8 - Q_F = 0$

d. The $b - (n - 1) = 5$ compatibility equations are

$P_1 = P_{14} + P_{46} + P_{60} \qquad P_2 = P_{23} + P_{35} + P_{50}$

$P_{50} = P_{57} + P_{76} + P_{60} \qquad P_{60} = P_{67} + P_{78} + P_{80}$

$P_{08} + P_{80} = 0$

By eliminating variables in various ways, a set of differential equations involving only flow rates Q_i, or only nodal pressures P_i, or a combination of both could be obtained.

1.5 a. Write equations describing the lumped-parameter approximate model for the transmission line shown in Fig. 1.14. b. Find the input-output transfer function $y(s)/u(s)$. The input $u(t)$ is the source voltage v_s, and the output $y(t)$ is the load voltage v_L.

a. For simplicity, the line is segmented into three equal lengths as shown. The leakage conductance G is the reciprocal of the leakage resistance. More segments could be used in the same manner.

The values of R and L are obviously 1/3 that for the entire line, while G and C each have values equal to 1/2 that for the entire line. Since the source voltage is known, there are six unknowns: i_0, i_1, i_2, v_1, v_2 and v_L.

b. Writing loop equations, using the Laplace transforms of the elemental equations, gives

$v_s = [R + Ls]i_0 + v_1 \qquad v_1 = [R + Ls]i_1 + v_2$

$v_2 = [R + Ls]i_2 + v_L \qquad v_L = R_L i_2$

The nodal equations are $i_0 = (G + Cs)v_1 + i_1$ and $i_1 = (G + Cs)v_2 + i_2$. Letting $A = R + Ls$ and $B = G + Cs$ gives

$$v_s = Ai_0 + v_1 = (AB + 1)v_1 + Ai_1 = (AB + 1)v_2 + (A^2B + 2A)i_1$$

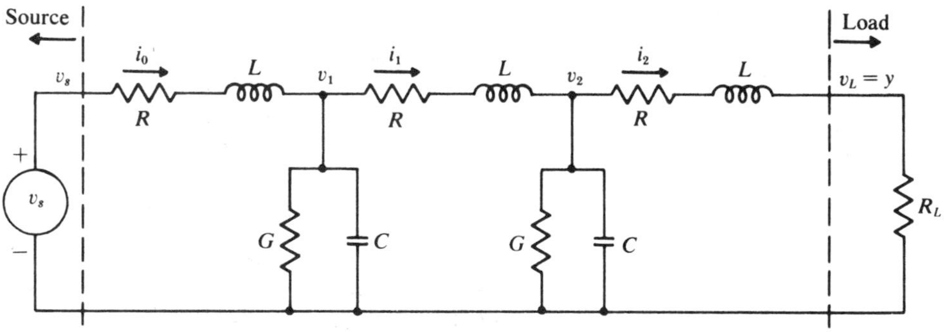

Figure 1.14

By continuing this process of substitution, a final expression containing only $v_s = u$ and $v_L = y$ is obtained:

$$\frac{y(s)}{u(s)} = \frac{R_L}{(A^3B^2 + 4A^2B + 3A) + R_L(A^2B + 3AB + 1)}$$

1.6 Derive a difference equation for the purely resistive ladder network shown in Fig. 1.15 (perhaps a dc version of the lumped approximation for a transmission line).

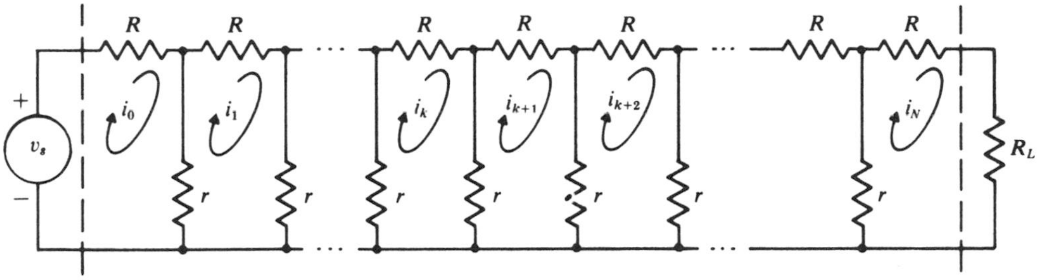

Figure 1.15

The difference equation for a typical $(k + 1)$th loop is obtained by writing a loop equation

$$(2r + R)i_{k+1} - ri_k - ri_{k+2} = 0$$

This holds for $1 \leq k + 1 \leq N - 1$. It is a second-order difference equation, and two boundary conditions are needed in order to uniquely specify the solution. The first and the last loops, which do not satisfy the general equation, provide the two necessary conditions:

$$v_s = (R + r)i_0 - ri_1$$

$$0 = (r + R + R_L)i_N - ri_{N-1}$$

1.7 A pair of dams in a flood control project is shown in Fig. 1.16. The water level at dam 1 at a given time t_k is $x_1(k)$, and $x_2(k)$ is the height at dam 2 at the same time. The amount of run-off water collected in reservoir 1 between times t_k and t_{k+1} is $Q_0(k)$. The water released from dams 1 and 2 during this period is denoted by $Q_1(k)$ and $Q_2(k)$. Develop a discrete-time model for this system.

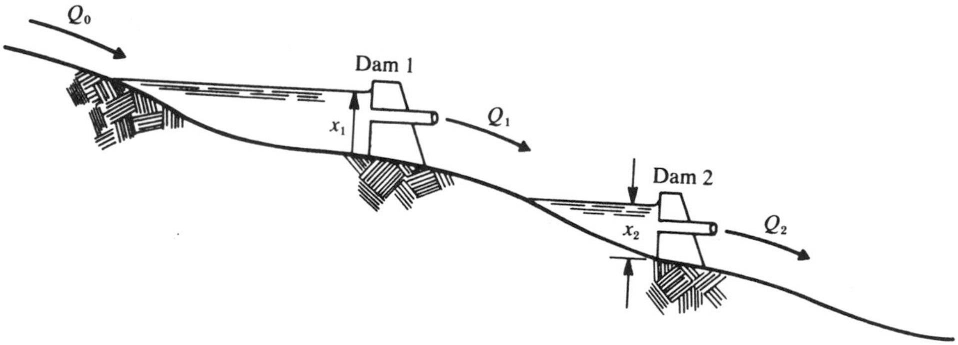

Figure 1.16

Conservation of flow requires

$$x_1(k+1) = x_1(k) + \alpha[Q_0(k) - Q_1(k)]$$
$$x_2(k+1) = x_2(k) + \beta[Q_1(k) - Q_2(k)]$$

These represent lumped-parameter discrete-time equations. If the amount of controlled spillages Q_1 and Q_2 are selected as functions of the water heights x_1 and x_2, a discrete feedback control system obviously results. Z-transform theory could be used to analyze such a system.

1.8 Draw the linear graph for the ideal transformer circuit of Fig. 1.17 noting that the transformer is a four-terminal element.

The linear graph is shown in Fig. 1.18*a* with the transformer represented as shown in Fig. 1.18*b*.

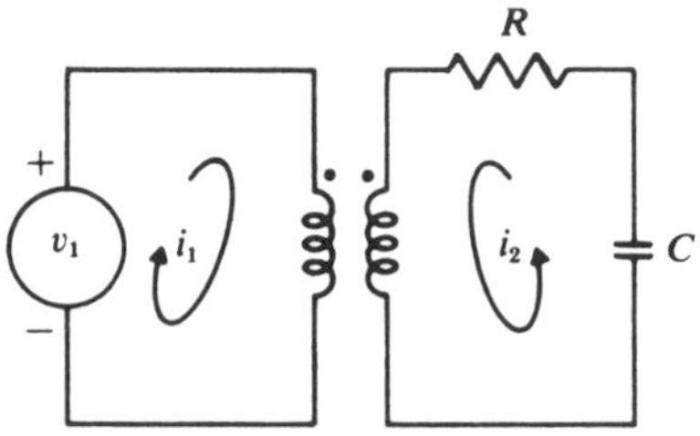

Turns ratio = N

Figure 1.17

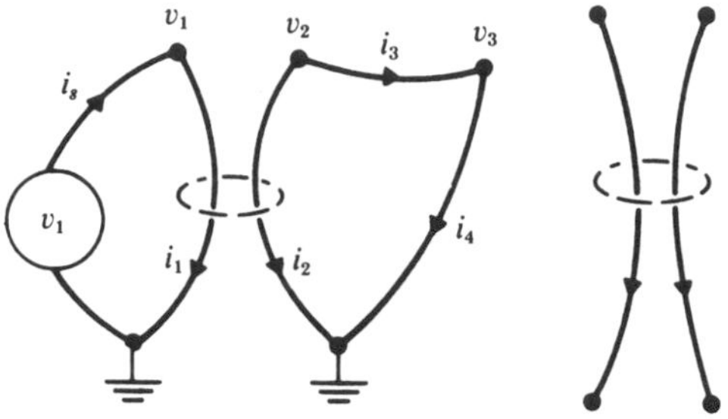

(*a*) (*b*)

Figure 1.18

The equations for this circuit are

Node equations: $i_s = i_1, \quad i_3 = -i_2, \quad i_3 = i_4$

Elemental euqations: $v_2 = Nv_1, \quad i_1 = -Ni_2$ (transformer)

$v_2 - v_3 = i_3 R$ (resistance)

$C\dot{v}_3 = i_4$ (capacitance)

Note that the transformer requires two equations for its specification. Similar multi-terminal elements are required whenever energy is transformed from one form to another. Transformers, transducers, and gyrators all have similar representation. Note also that an ideal transformer has zero instantaneous power flow into it, i.e.

$$v_1 i_1 + v_2 i_2 = v_1(-Ni_2) + (Nv_1)i_2 = 0$$

1.9 Develop a model of an automobile which would be appropriate for studying the effectiveness of the suspension system, tire characteristics, and seat design on passenger comfort.

For simplicity, lateral rolling motions are ignored. An idealized model might be represented as shown in Fig. 1.19. The displacements x_1 and x_2 are inputs from the environment (road surface). Masses m_1 and m_2 represent the wheels, while M and J represent the mass and pitching inertia of the main car body. The seat and passenger mass are represented by m_p. The elasticity and energy dissipation properties of the tires are represented by k_1, k_2, b_1, and b_2. The suspension system is represented by k_3, k_4,

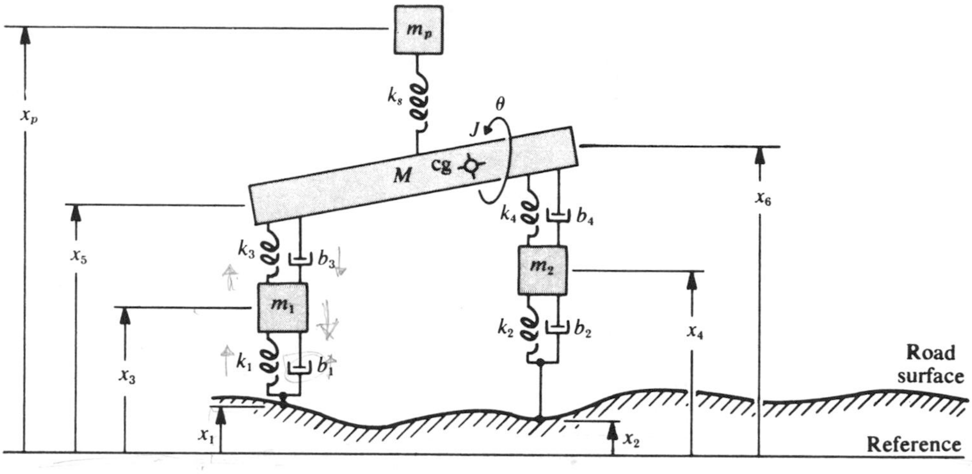

Figure 1.19

b_3, and b_4. The seat characteristics are represented by k_s. Newton's second law is applied to the wheels, giving

$$m_1\ddot{x}_3 = k_1(x_1 - x_3) + b_1(\dot{x}_1 - \dot{x}_3) + k_3(x_5 - x_3) + b_3(\dot{x}_5 - \dot{x}_3)$$
$$m_2\ddot{x}_4 = k_2(x_2 - x_4) + b_2(\dot{x}_2 - \dot{x}_4) + k_4(x_6 - x_4) + b_4(\dot{x}_6 - \dot{x}_4)$$

Letting l_1 be the distance from the left end to the center of gravity cg and letting l_2 be the distance to the seat mount, the following geometric relations can be obtained. Assuming small angles,

$$x_{cg} = x_5 + \frac{l_1}{l}(x_6 - x_5)$$

$$x_s - x_5 + \frac{l_2}{l}(x_6 - x_5) \qquad \text{and} \qquad \theta = \frac{x_6 - x_5}{l}$$

where l is the total length (wheel base). Summing forces on M gives

$$M\ddot{x}_{cg} = k_3(x_3 - x_5) + k_4(x_4 - x_6) + k_s(x_p - x_s) + b_3(\dot{x}_3 - \dot{x}_5) + b_4(\dot{x}_4 - \dot{x}_6)$$

Summing torques gives

$$J\ddot{\theta} = -l_1k_3(x_3 - x_5) + (l - l_1)k_4(x_4 - x_6) - (l_1 - l_2)k_5(x_p - x_s)$$
$$-l_1b_3(\dot{x}_3 - \dot{x}_5) + (l - l_1)b_4(\dot{x}_4 - \dot{x}_6)$$

Finally, summing forces on m_p gives $m_p\ddot{x}_p = k_s(x_s - x_p)$. This set of five coupled second-order differential equations, along with the geometric constraints, constitutes an approximate model for this system.

1.10 A typical common base amplifier circuit, using a *pnp* transistor, is shown in Fig. 1.20*a*. The *h*-parameter equivalent circuit for small signals within the amplifier mid-band frequency range is given in Fig. 1.20*b*. Draw the linear graph for the amplifier.

The input signal voltage is replaced by an ideal source v_s in series with the source resistance R_s. The three-terminal transistor device is described by the four hybrid parameters h_{ib}, h_{rb}, h_{fb}, and h_{ob}, which are straight line approximations to the various nonlinear device characteristics in the vicinity of the operating point.

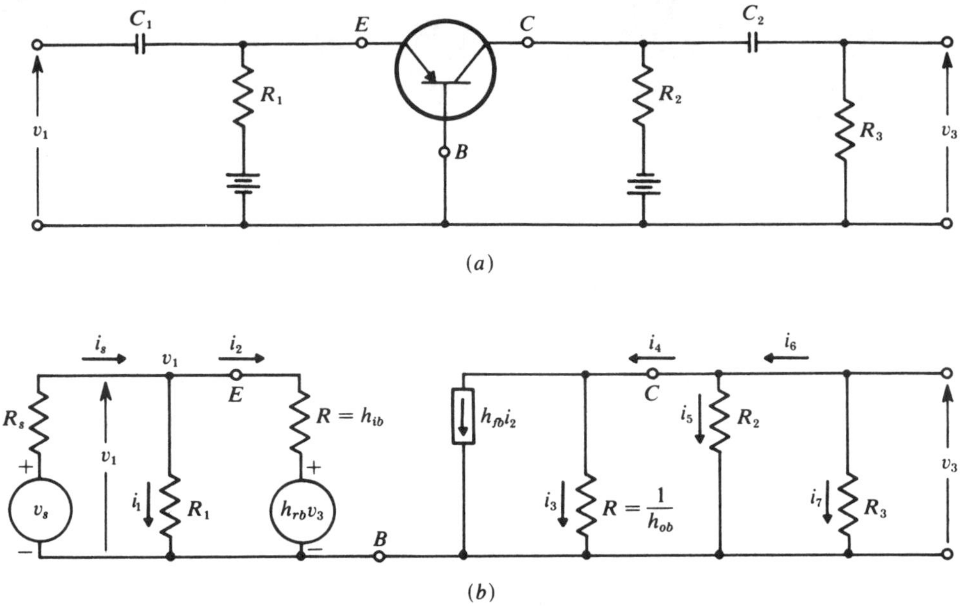

Figure 1.20

In this example there are two dependent or controlled sources, described by $v_d = h_{rb}v_3$ and $i_d = h_{fb}i_2$. Using the equations implied by the linear graph, Fig. 1.21

$$v_1 = i_2 h_{ib} + h_{rb} v_3 \qquad (E\text{–}B \text{ loop equation}) \qquad (1)$$

$$h_{fb} i_2 + i_3 + i_5 + i_7 = 0 \qquad (C \text{ node equation}) \qquad (2)$$

or

$$h_{fb} i_2 + v_3 h_{ob} + v_3/R_2 + v_3/R_3 = 0 \qquad \text{(using elemental equations)} \qquad (3)$$

From equation (3),

$$i_2 = \frac{-v_3}{h_{fb}}\left(h_{ob} + \frac{1}{R_2} + \frac{1}{R_3}\right) \qquad (4)$$

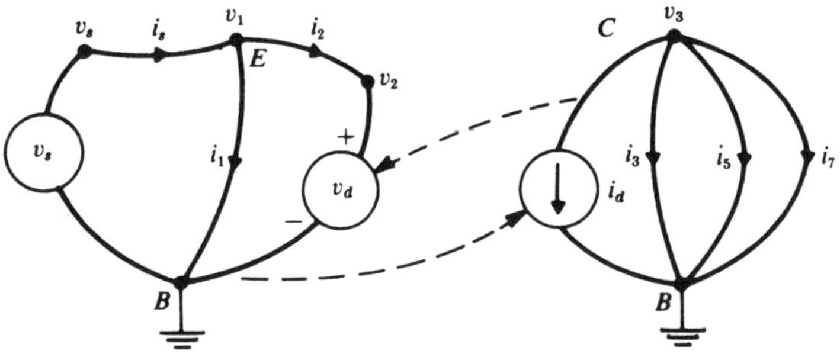

Figure 1.21

Using equation (4) in equation (1) gives the voltage input-output equation:

$$v_3 = \frac{-h_{fb}v_1}{h_{ib}(h_{ob} + 1/R_2 + 1/R_3) - h_{rb}h_{fb}}$$

1.11 Draw a block diagram for the motor system of Problem 1.2, preserving the individual identity of the electrical, mechanical, and conversion aspects, and illustrating the feedback nature of this system.

The block diagram of Fig. 1.22 is drawn directly from the constituent equations of Problem 1.2.

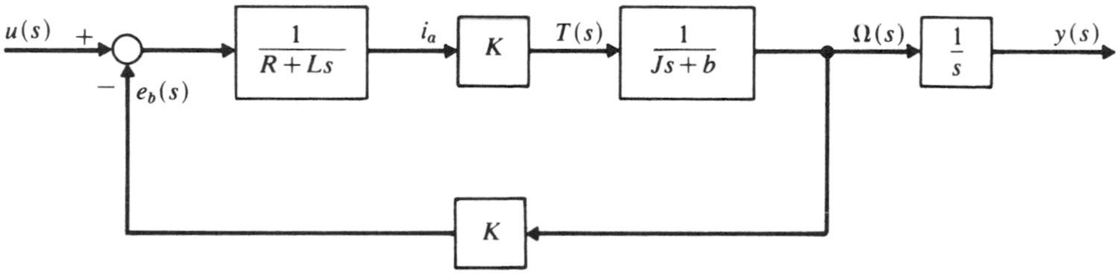

Figure 1.22

1.12 Classify the systems described by the following equations:

a. $\ddot{y} + t^2\dot{y} - 6y = u(t)$ b. $\dddot{y} + \ddot{y}y + 4y = 0$

c. $\dfrac{\partial^2 y}{\partial t^2} = a\dfrac{\partial^2 y}{\partial x^2}$ d. $\ddot{y} + a\dot{y} + by^2 = u(t)$

e. $\dot{y} + ay = u(t)$ if $t < t_1$ f. $\dot{y} + a \max(0, y) = 0$
$\dot{y} + by = u(t)$ if $t \geq t_1$

a. Lumped parameter, linear, continuous time, time variable coefficient (t^2), nonhomogeneous.
b. Lumped parameter, nonlinear (due to $\ddot{y}y$ term), continuous time.
c. Distributed parameter.
d. Lumped parameter, nonlinear (due to y^2 term).
e. Lumped parameter, linear, time variable.
f. Lumped parameter, nonlinear.

Changing of system characteristics, through switching at predetermined times, does not make a system nonlinear. However, if the switching depends on the magnitude of the dependent variable y, the system is nonlinear.

1.13 Use the concepts of Fig. 1.5, page 5, to discuss the development of a mathematical model for a rocket vehicle.

1. With such a vague problem statement, many purposes for this model could be considered, such as the structural adequacy of the design, or the temperature history of a component within the vehicle. Suppose that the purpose is to study the trajectory of the vehicle.
2. The boundary of the system is the physical envelope of the vehicle. The inputs from the external environment consist of atmospheric and gravitational effects, as well as a thrust force caused by the gases being expelled across the system boundary. Additional inputs are the mission data which specify key characteristics that the

trajectory should possess. Outputs are the components of position and velocity along the resulting trajectory.

3. The structure of this system consists of several subsystems. One of these is the vehicle dynamics subsystem, which relates forces and torques to the vehicle acceleration. Another is the kinematic subsystem, which relates accelerations to vehicle positions and velocities. A third subsystem is the navigation subsystem, which takes measurements of position, velocity, or acceleration and provides useful signals containing present position and velocity information. A fourth subsystem, the guidance system, accepts the position-velocity data, compares it with mission goals, and computes guidance commands. The final subsystem is the control subsystem. It accepts guidance commands as inputs, and its outputs are commanded body attitude angles or angular rates which will cause the vehicle to steer to the desired trajectory. The control system also turns the thrust on and off. The overall system is shown in Fig. 1.23.

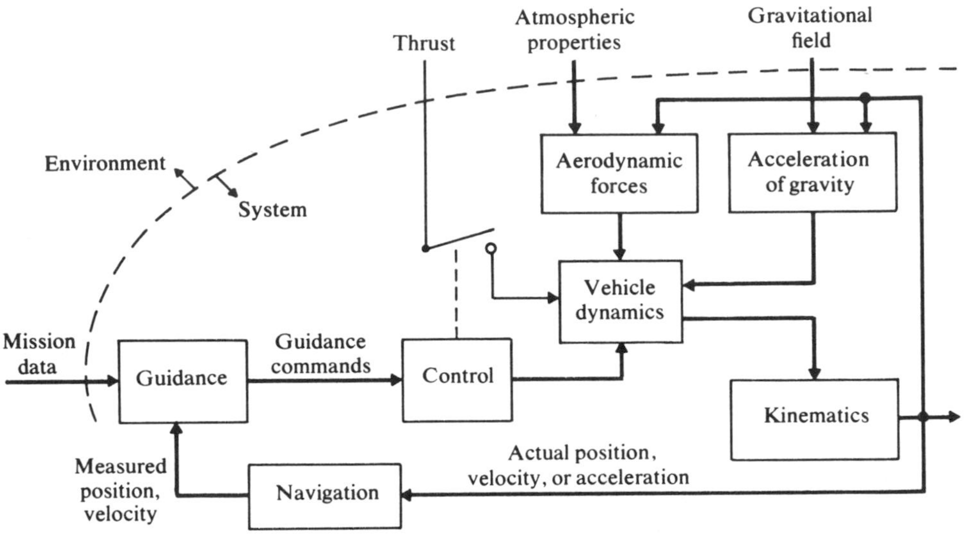

Figure 1.23

4. A wide diversity of models could be developed. If only position and velocity are of interest, the vehicle may be represented as a point mass and the attitude control system might be assumed perfect. If angular attitude information is desired, detailed equations for rotational motion may be required. The description could include elastic vehicle bending, spurious torques due to fuel sloshing, the dynamics of hydraulic control actuators, etc. Each of the other subsystems could be broken down into very fine detail, if required.
5. The remaining steps in Fig. 1.5 are relatively straightforward if the preceding steps have been carried out correctly. The resulting equations will be nonlinear and involve many variables, coordinate transformations, etc. Except in certain simple cases, computer simulation will be required.

1.14 Discuss the various levels at which control techniques can be applied in industrial systems.

Harold Chestnut [20] gives the four-level representation to control activities in the business environment shown in Fig. 1.24.

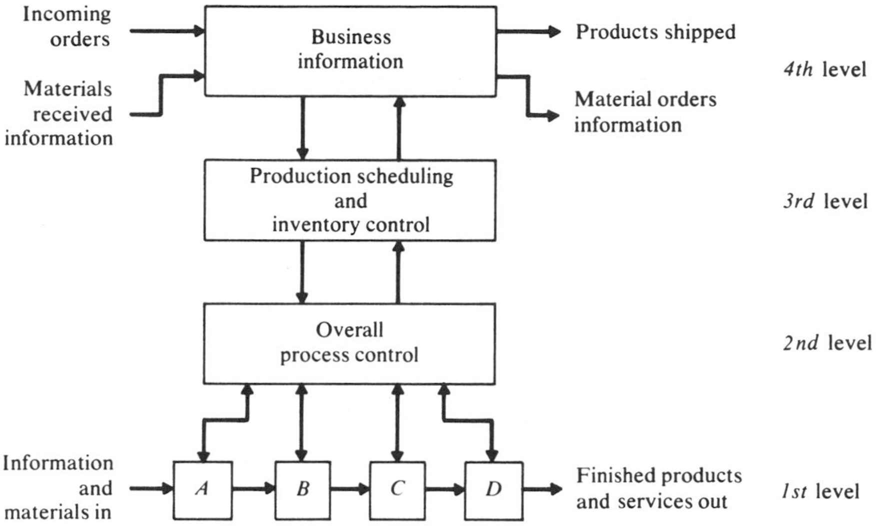

Figure 1.24

Individual operations *A*, *B*, *C*, and *D* might represent automatic machine tools, as in Example 1.3. At this level, a very complete model is required to study detailed behavior. The overall sequence *A*, *B*, *C*, *D* could represent a process such as an automated steel mill or chemical plant. The general characteristics of the sequence might be modeled in a way similar to the system of Problem 1.4. The characteristics of the interacting sequence would be described by the level 2 model, but details of individual elements need no longer be apparent. At level 3 a still broader view is taken. The model might be used to determine how work schedules should be set in order to efficiently use the system capability while maintaining optimum inventory, avoiding premium overtime pay, and meeting delivery schedules. At the 4th level, the broadest view is taken. A complete production line might be viewed as a simple time delay. Broader questions regarding market forecasts, new product development, plant expansion, and customer relations become dominant.

According to Mr. Chestnut, "Traditionally the automatic control engineers have focused their attention on the first and second levels of control, which are those associated with the fast control functions in the energy and materials ends of the industrial spectrum. With the current emphasis being developed in the systems aspect of the overall industrial process, more attention is being given to the third and fourth levels of control where significant economies in time and money and resources can and are being realized and can be more readily brought to the attention of the customers." Some of these "big picture" economic systems models are now being facilitated by the various spread sheet and data base management programs which are widely available on management's personal microcomputers.

1.15 Define the Z-transform of a continuous-time signal $y(t)$.

The Z-transform of $y(t)$, written $Z\{y(t)\} = Y(z)$, is defined as the result of a three-step operation:

i. Modulate $y(t)$ with a periodic train of Dirac delta functions, i.e.

$$y_s(t) = y(t) \sum_{n=-\infty}^{\infty} \delta(t - nT)$$

where T is the sample period. Since $\delta(\ )$ is zero, except when its argument is zero,

$$y_s(t) = \sum_{n=0}^{\infty} y(nT)\,\delta(t - nT)$$

assuming $y(t) = 0$ for $t < 0$.

ii. Laplace transform the impulse-modulated signal

$$Y^*(s) \triangleq \mathcal{L}\{y_s(t)\} = \sum_{n=0}^{\infty} y(nT)e^{-nTs}$$

Note that $y(nT)$ is no longer a function, but just a set of sample values. These act as constants as far as $\mathcal{L}$ is concerned.

iii. Make a change of variables $z = e^{Ts}$. Thus

$$Y(z) = Y^*(s)|_{z=e^{Ts}} = \sum_{n=0}^{\infty} y(nT)z^{-n}$$

The reason for the change of variables is to allow working with polynomials in z rather than transcendental functions in s.

1.16 What is the significance of the Z-transform as expressed in the previous problem?

The Z-transform of a function can be written in various other forms such as ratios of polynomials in z or z^{-1}. However, if a function's Z-transform can be manipulated into an infinite series in z^{-1}, then we can pick off the function's value at time $t = nT$ as the coefficient multiplying z^{-n}. This series form can be found by long division or by using knowledge of some standard infinite series results.

If $y(t)$ is a unit step, then all $y(nT) = 1$, so

$$Y(z) = \sum_{n=0}^{\infty} z^{-n} = \frac{1}{1 - z^{-1}}$$

Conversely, if $Y(z) = \dfrac{z}{z - 0.5}$, then long division gives

$$Y(z) = 1 + 0.5z^{-1} + 0.25z^{-2} + \cdots$$

From this it is immediately known that

$$y(0) = 1,\ y(T) = 0.5,\ y(2T) = 0.25, \ldots, \text{etc.}$$

1.17 What are some other methods of determining the inverse Z-transform?

i. There are extensive tables of transform pairs available [69]. It should be pointed out that the a function has unique Z-transform, but the inverse transform is not unique. Many functions have the same sample values, but are different between samples.

ii. A complicated transform expression can often be written as the sum of several simple terms, using a variation of partial fraction expansion. Then each simple term can be inverted.

iii. The formal definition of the inverse transform is

$$y(nT) = \frac{1}{2\pi j} \oint Y(z)z^{n-1}\,dz$$

The contour integral is around a closed path which encloses all singularities of $Y(z)$. This integral can be evaluated using Cauchy's residue theory of complex variables.

1.18 Why are Z-transforms useful when dealing with constant coefficient difference equations and sampled-data signals?

The Z-transform possesses all the advantages for these systems as does the Laplace transform with differential equations. It allows much of the solution effort to be carried out with only algebraic manipulations in z. After the transform of the desired output variable $Y(z)$ is isolated algebraically, then its inverse can be calculated to give $y(nT)$.

1.19 Analyze the difference equation

$$y(t_k) + a_2 y(t_{k-1}) + a_1 y(t_{k-2}) + a_0 y(t_{k-3}) = b_0 x(t_k) + b_1 x(t_{k-1})$$

$x(t_k)$ is a known input sequence.

Let $Y(z)$ and $X(z)$ be the Z-transforms of y and x, respectively. Since the Z-transform is a linear operator, it can be applied to each individual term in the sum, giving

$$Y(z) + a_2 z^{-1} Y(z) + a_1 z^{-2} Y(z) + a_0 z^{-3} Y(z) = b_0 X(z) + b_1 z^{-1} X(z)$$

The "delay operator" nature of z^{-1} has been used here. As should be apparent from Problems 1.15 and 1.16, a shift of n sample periods in the time domain is achieved by multiplying by z^{-n} in the Z-domain. Thus

$$Y(z) = \left[\frac{b_0 + b_1 z^{-1}}{1 + a_2 z^{-1} + a_1 z^{-2} + a z^{-3}}\right] X(z)$$

The output transform $Y(z)$ is the input transform $X(z)$ multiplied by a rational function of z^{-1} (or z). This rational function is the Z-domain transfer function $H(z)$.

1.20 What is the significance of the poles and zeros of $H(z)$?

Just as in the Laplace s-domain, the behavior of the system depends very heavily on the roots of the denominator of $H(z)$, i.e. the poles. A stable system must not have any s-plane roots with positive real parts. Since $z = e^{Ts}$ this means that in the z-plane all poles of a stable system must be inside the unit circle.

The zeros of the transfer function affect the magnitude of the various terms in the time-domain output. That is, the poles determine the system modes and the zeros help determine how strongly the modes will contribute to the total response. This is evident if $H(z)$ is expanded in partial fraction form.

PROBLEMS

1.21 Derive the elemental equation for the fluid storage tank of Fig. 1.25 and show that it is analogous to an electric capacitance. Let Q be the volume flow rate, P the pressure at the base of the tank of cross sectional area A, and h the height of the fluid.

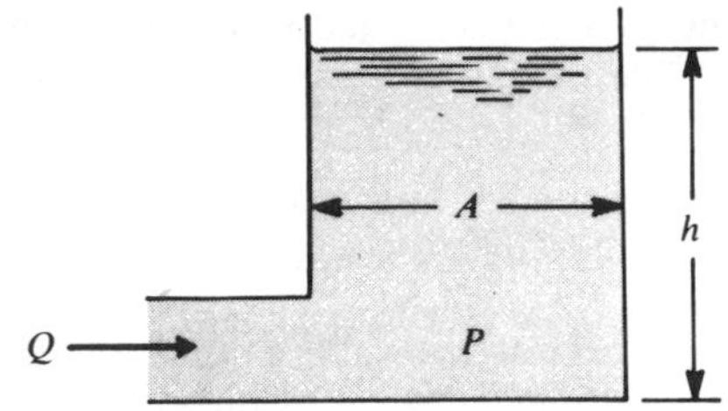

Figure 1.25

1.22 Show that an inventory storage unit can be modeled by an elemental equation analogous to an electric capacitance. Let the net flow of goods into inventory be Q items per unit time, and let the number of items in inventory at time t be $v(t)$.

1.23 If branch 1 of Fig. 1.26 contains an ideal capacitance C, branch 2 an ideal inductance L, and branch 3 an ideal resistance R, find the input-output equation relating v_0 and f_3.

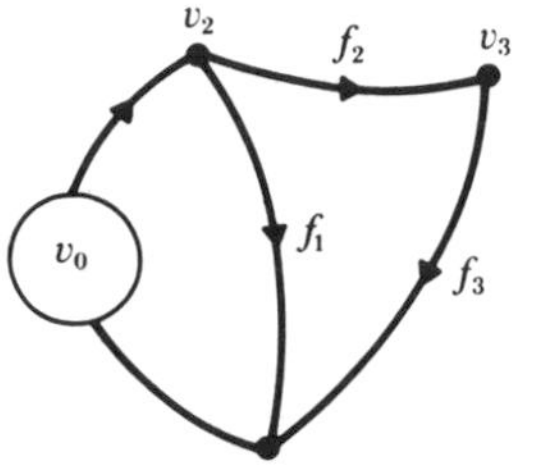

Figure 1.26

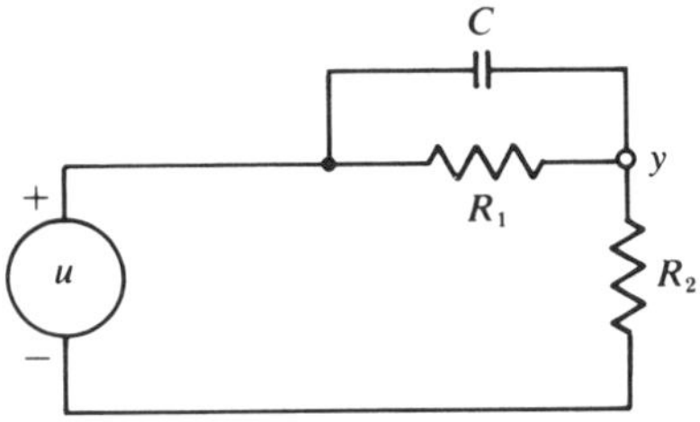

Figure 1.27

1.24 Derive the input-output differential equation for the network of Fig. 1.27. Treat $u(t)$ as the input voltage and $y(t)$ as the output voltage. Also give the input-output transfer function $y(s)/u(s)$.

1.25 A government agency would like a model for studying the effectiveness of its air pollution monitoring and control program. Discuss the factors involved in such a model.

1.26 Derive the Z-transform of $y(t) = e^{-0.1t}$. For $t < 0$, $y(t) = 0$. Use a sample time of $T = 2.0$ seconds.

1.27 The following values apply to the system in Problem 1.19.

$$a_0 = -0.064,\ a_1 = 0.56,\ a_2 = -1.4,\ b_0 = 10.0 \text{ and } b_1 = 5.0$$

Find the poles and zeros of $H(z)$. Does this represent a stable system?

1.28 For the same system, find $y(nT)$ if the input $x(nT)$ is the sampled version of a unit step function starting at $t = 0$.

Chapter

2

HIGHLIGHTS OF CLASSICAL CONTROL THEORY

2.1 INTRODUCTION

Classical control theory, at the introductory level, deals primarily with linear, constant coefficient systems. Few real systems are exactly linear over their whole operating range, and few systems have parameter values that are precisely constant forever. But many systems approximately satisfy these conditions over a sufficiently narrow operating range. This chapter reviews the classical methods which are applicable to linear, constant coefficient systems. More extensive discussions are in References 30, 31, 82, and 103.

2.2 SYSTEM REPRESENTATION

The consideration of linear, stationary systems is greatly simplified by the use of transform techniques and frequency domain methods. For continuous-time systems this means Laplace transforms (or sometimes Fourier transforms) [4]. Z-transforms provide equivalent advantages for discrete-time systems [39, 69, 103]. These methods are basic in classical control systems analysis. Thus algebraic equations in the transformed variables are dealt with rather than the system's differential or difference equations. Manipulation of the algebraic cause and effect relations is facilitated by the use of transfer functions and block diagrams or signal flow graphs [30].

2.3 FEEDBACK

Most systems considered in classical control theory are feedback control systems. A typical single-input, single-output continuous-time (totally analog) system is shown in Fig. 2.1*a*. Figure 2.1*b* shows a typical feedback arrangement for controlling a

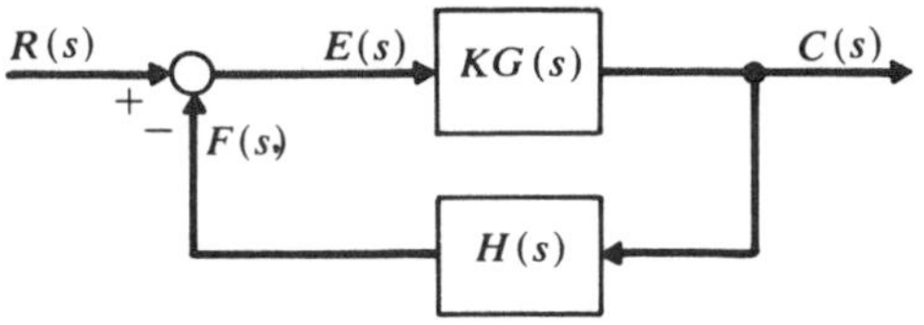

Figure 2.1a Elementary feedback control system.

continuous-time process using a digital controller. Notation commonly used in classical control theory will be used in this chapter. In both cases the input or reference signal is R, the output or controlled signal is C, the actuating or error signal is E, and the feedback signal is F. This should cause no confusion with the parameters R and C used in resistance and capacitance networks.

While there are many similarities between these two types of systems, the differences are significant enough to merit a brief separate discussion of each.

Continuous-Time Systems

The forward *transfer function is* $KG(s)$, where K is an adjustable gain. The forward transfer function often consists of two factors $G(s) = G_c(s)G_p(s)$, where $G_p(s)$ is fixed by the nature of the plant or process to be controlled. $G_c(s)$ is a compensation or controller transfer function which the designer can specify (within certain limits) to achieve desired system behavior. The *feedback transfer function* is $H(s)$. This often represents the dynamics of the instrumentation used to form the feedback signals, but it can also include signal conditioning or compensation networks. The designer may be able to at least partially specify $H(s)$ in some cases, and in other cases it may be totally fixed, or even just unity. In any case, the *open-loop transfer function* is $KG(s)H(s)$. It represents the transfer function around the loop, say from E to F, when the feedback signal is disconnected from the summing junction.

In the feedback system of Fig. 2.1*a* and *b* the actuating signal is determined by comparing the feedback signal with the input signal. When $H(s) = 1$, the unity feedback case, the comparison is directly between the output and the input. Then the difference E is truly an error signal.

A major part of classical control theory for continuous-time systems is devoted to the analysis of feedback systems like the one shown in Fig. 2.1*a*. Multiple input-output systems and multi-loop systems can also be considered using transfer function techniques (see Problems 3.2 through 3.7), although most of this book is devoted to a state variable approach instead. It is beneficial to have a thorough understanding of single-input, single-output systems before the multivariable case is considered. This chapter provides a review of the methods used in studying the behavior of C and E as influenced by R.

Example 2.1

Relations, in the Laplace transform domain, between the input R and the output C and between R and the error E are derived algebraically as follows. At the summing junction, $R - HC = E$. The relation between E and C is $KGE = C$. Elimination of E gives $KGR - KGHC = C$ so that $C = KGR/(1 + KGH)$. Using $E = C/KG$ gives $E = R/(1 + KGH)$. ∎

The system of Fig. 2.1*a* is the prototype for all continuous system discussions in this chapter. The following terminology will be used frequently. In general, $G(s) = g_n(s)/g_d(s)$ and $H(s) = h_n(s)/h_d(s)$ will be ratios of polynomials in s. The values of s which are roots of the numerator are called *zeros*. Roots of the denominator are called *poles*. In particular, the *open-loop zeros* are values of s which are roots of the numerator of the open-loop transfer function $KG(s)H(s) = Kg_n(s)h_n(s)/[g_d(s)h_d(s)]$. The *open-loop poles* are roots of the denominator of $KG(s)H(s)$. Since the closed-loop transfer function is $C(s)/R(s) = KG(s)/[1 + KG(s)H(s)] = Kg_n(s)h_d(s)/[g_d(s)h_d(s) + Kg_n(s)h_n(s)]$, the *closed-loop zeros* are all the roots of $g_n(s)h_d(s)$. The *closed-loop poles* are roots of $1 + KG(s)H(s) = 0$ or equivalently, roots of $g_d(s)h_d(s) + Kg_n(s)h_n(s) = 0$.

Discrete-Time Systems

There is a richer variety of possibilities when dealing with the digital control of continuous systems. The points of conversion from continuous-time signals to discrete-time signals (A/D) and back again (D/A) can vary from one application to the next. An analog feedback sensor could be used and then its output sampled, or a direct digital measurement may be used. The reference input R could be a continuous-time signal that needs to be sampled before being sent to the control computer, or it might be a direct digital input. Figure 2.1*b* is just one possible arrangement. Other con-

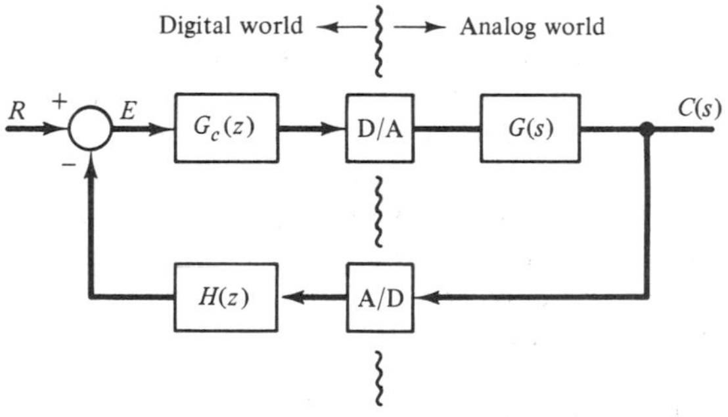

Figure 2.1*b*

figurations can be analyzed in a similar way. Before proceeding with the analysis, models of the A/D and D/A conversion processes are required. These conversions are also referred to as sampling and desampling or signal reconstruction, respectively.

Sampling. Assume a periodic sampler with period T. A convenient model of the A/D conversion is an impulse modulator, usually shown symbolically as a switch like the one of Fig. 2.2, where a general signal $y(t)$ is being sampled. Impulse modulation is not really what physically occurs, since no infinite amplitude signals such as $y^*(t_k)$ actually exist in the system. This series of impulse functions have infinite amplitude at the sample times, but it is their areas or strengths that represent the real signal amplitudes $y(t_k)$ mathematically. This artificial representation is used because:

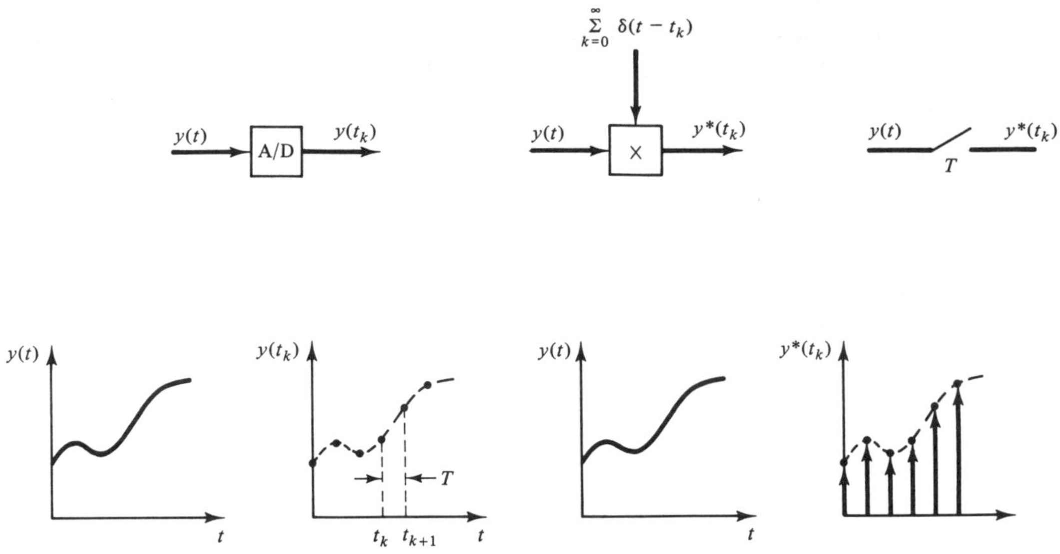

(a) Reality

(b) Impulse modulation model

Figure 2.2

1. It allows the use of Z-transforms, which simplify much of the analysis.
2. The correct answers are obtained (except for quantization effects) as long as it is understood that within the digital portion it is the strengths of the impulses, not their amplitudes, that describe the signals.
3. The correct effect on the continuous-time part of the system is obtained provided some sort of "hold" circuit is used on the impulse train before the signal reenters the analog world. There are various versions of hold devices. One function common to them all is an integration, which eliminates the impulses and once again gives a finite amplitude physical signal. This is the desampling function of the D/A.

Desampling. The only model of the D/A process to be considered here is another sampler (perfect time synchronization assumed), followed by a zero order hold (ZOH). The zero order hold integrates the difference between two consecutive impulses in the periodic impulse train shown in Fig. 2.2 and repeated in Fig. 2.3. Therefore the output is a piecewise constant signal whose value between t_k and t_{k+1}

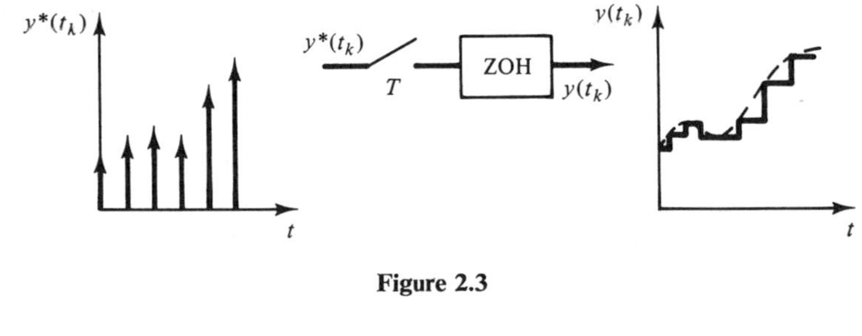

Figure 2.3

is clamped at $y(t_k)$ (again, ignoring quantization errors). If the computer made no modification to the signal between the A/D and D/A, the end-to-end effect of this sampling-desampling operation would be to create a piecewise constant approximation to the continuous input signal. From Chapter 1, the Laplace transform of an impulse-modulated signal is (after a change of variable) the Z-transform of the signal. Any linear operation that the computer algorithm performs on the signal samples between the A/D and D/A can be represented by a Z-transform domain transfer function, sometimes called a pulse transfer function. $G_c(z)$ and $H(z)$ in Fig. 2.1b are examples of this.

Extensive tables of Z-transforms are available. Use of these tables, plus a few simple rules, will allow systems like Fig. 2.1b to be analyzed almost as easily as, and with a great deal of similarity to, those of Fig. 2.1a. Of course, a complete understanding and appreciation will require a more thorough treatment, as can be found in the references. Some key rules of manipulation are:

1. *As a signal passes through the "switch," it is Z-transformed.*

2. *The transform of the signal out of a transfer function block is the product of that transfer function and the transform of the input signal.*

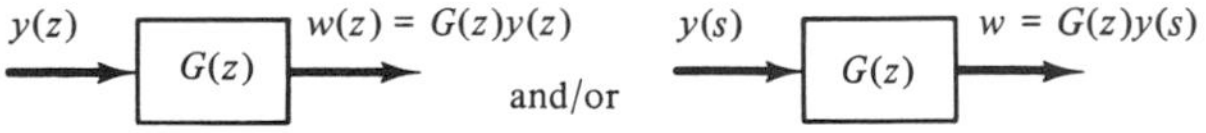

3. *Sampling a signal which is already sampled does not change it.*

or $Z\{y(z)\} = y(z)$

4. *Pulsed or Z-transformed signals (and transfer functions) and s-domain signals (and transfer functions) will appear together in the same expression at times.* The action of a sampler (or a Z-transform) on these mixed signals is illustrated below.

$$Z\{G(z)y(s)\} = G(z)y(z)$$

5. *The Z-transform operator is not associative for products.* The placement of "switches" in a block diagram is important.

$$w_1(z) = Z\{y(s)G_1(s)G_2(s)\} \neq w_2(z) = Z\{G_1(s)y(s)\}G_2(z)$$

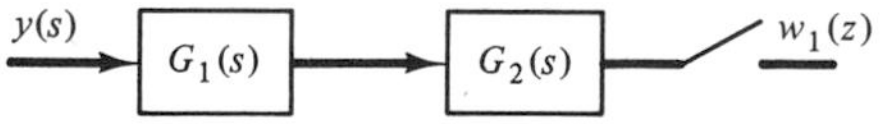

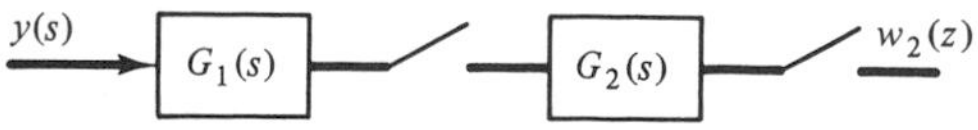

6. *When working with closed-loop systems like Fig. 2.1b, it is generally best to follow a two-step process.* First, algebraically solve for the variable at the input of a sampler in terms of external inputs and/or outputs of that sampler. Second, "close the loop" by passing through the sampler, i.e., taking the Z-transform. This will give a result which is entirely in the Z-domain and can be used to solve for the system output sequence at the sample times. If the sampling period is small enough compared to the rate of change of the signal, this approximation may be all that is needed to describe the continuous system output.

Example 2.2

The system of Fig. 2.1*b* is redrawn in Fig. 2.4, using symbolism just introduced for A/D and D/A operations. The transfer function for a ZOH is also used, $G_0 = (1 - z^{-1})/s$.

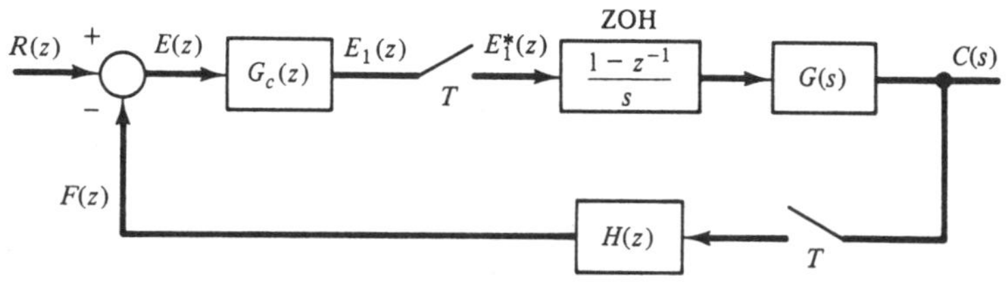

Figure 2.4

The input to the forward path sampler is first isolated:

$$E_1(z) = G_c(z)[R(z) - F(z)]$$
$$= G_c(z)[R(z) - E_1^*(z)Z\{(1 - z^{-1})G(s)/s\}H(z)]$$

Define the Z-transform of $G(s)/s$ times $(1 - z^{-1})$ as $G'(z)$. Then

$$E_1(z) = G_c(z)[R(z) - E_1(z)G'(z)H(z)]$$

Solving gives

$$E_1(z) = G_c(z)R(z)/[1 + G_c(z)G'(z)H(z)]$$

In this case the input to the selected sampler was already sampled, so the second step of passing through the sampler has no effect, i.e. $E_1^*(z) = E_1(z)$. The full two-step process is better illustrated by selecting the feedback sampler instead.

Before doing that, note that

$$C(s) = E_1(z)[(1 - z^{-1})G(s)/s)]$$

and that

$$C(z) = E_1(z)G'(z)$$
$$= G_c(z)G'(z)R(z)/[1 + G_c(z)G'(z)H(z)],$$

an expression very similar to the continuous system result.

The same example is reworked by isolating the input to the feedback sampler first.

$$C(s) = [R(z) - H(z)C(z)]G_c(z)(1 - z^{-1})G(s)/s$$

Then the traverse around the loop is completed by passing through the sampler to obtain

$$C(z) = [R(z) - H(z)C(z)]G_c(z)G'(z)$$

Solving this for $C(z)$ gives the same result as above.

In order to dispel the idea that a discrete system result can always be written from the continuous system result by substituting all the individual Z-transfer functions for their Laplace transform counterparts, the reader is urged to work through Problems 2.38, 2.39, and 2.41, which have different sampling arrangements. ■

Once $C(z)$ is found, the values of the output $C(t)$ can be determined at the sample times t_k by finding the inverse Z-transform. While this will not give the values of $C(t)$ between sample points, it often gives a sufficiently accurate representation of the continuous signal.

Control systems like those of Fig. 2.1a and b are used extensively because of the advantages that can be obtained by using feedback. The advantages of feedback control are:

1. The system output can be made to follow or track the specified input function in an automatic fashion. The name automatic control theory is frequently used for this reason.
2. System performance is less sensitive to variations of parameter values (see Problem 2.1).
3. System performance is less sensitive to unwanted disturbances (see Problem 2.4).
4. Use of feedback makes it easier to achieve the desired transient and steady-state response (see Problem 2.5).

The advantages of feedback are gained at the expense of certain disadvantages, the principal ones being:

1. The possibility of instability is introduced and stability becomes a major design concern. Actually, feedback can either stabilize or destabilize a system.
2. There is a loss of system gain, and additional stages of amplification may be required to compensate for this.
3. Additional components of high precision are usually required to provide the feedback signals (see Problems 2.2 and 2.3).

Once the system models are specified in terms of transfer functions and block diagrams, classical control theory is devoted to answering three general questions.

(a) What are appropriate measures of system performance that can be easily applied to feedback control systems?

(b) How can a feedback control system be easily analyzed in terms of these performance measures?

(c) How should the system be modified if its performance is not satisfactory?

2.4 MEASURES OF PERFORMANCE AND METHODS OF ANALYSIS IN CLASSICAL CONTROL THEORY

If the complete solutions for the system output $C(t)$ were available in analytical form for every conceivable input, system performance could be assessed. To obtain an analytical expression for $C(t)$, the inverse Laplace transform of

$$C(s) = \frac{KG(s)R(s)}{1 + KG(s)H(s)} \tag{2.1a}$$

is required. To determine $C(t_k)$ in the discrete-time case, or in the mixed discrete-continuous case, the inverse Z-transform of

$$C(z) = \frac{G_c(z)G'(z)R(z)}{1 + G_c(z)G'(z)H(z)} \tag{2.1b}$$

(or a similar expression) is needed. In both cases, if the denominators can be factored, partial fraction expansion can be used to obtain a sum of easily invertible terms. However, the denominator of equation (*2.1*) may be a high-degree polynomial. Also, an infinite number of possible inputs $R(s)$ or $R(z)$ could be considered. Rather than seek complete analytical solutions, classical control theory uses only certain desirable features, which C should possess, in order to evaluate performance. Methods of classical control theory were developed before the wide-spread availability of digital computers. As a result, all the techniques seek as much information as possible about the behavior of $C(t)$ or $C(t_k)$ without actually solving for them. The methods have been developed for ease of application and stress graphical techniques. They are still useful methods of analysis and design because of the insight they provide.

The problem of infinite variety for possible inputs is dealt with by considering important aperiodic and periodic signals as test inputs. Step functions, ramps, and sinusoids are common examples.

The general characteristics which a well-designed control system should possess are (1) stability, (2) steady-state accuracy, (3) satisfactory transient response, and/or (4) satisfactory frequency response.

1. *Stability means that $C(t)$ or $C(t_k)$ must not grow without bound due to a bounded input, initial condition, or unwanted disturbance.* (This intuitive definition is expanded upon in Chapter 14.) For linear constant coefficient systems, stability depends only on the locations of the roots of the closed-loop characteristic equation. The continuous system's characteristic equation is the denominator of equation (*2.1a*) set to zero. The discrete case uses the denominator of equation (*2.1b*). Both of these are of the form

$$1 + KF(\zeta) = 0 \tag{2.2}$$

with the complex variable ζ being either s or z. The principal difference is the stability region. The roots must be in the left-half s-plane for a stable continuous system. Since $z = e^{Ts}$, the entire left-half s-plane maps into the interior of the unit circle in the Z-plane. A stable discrete system must have all roots of its characteristic equation inside the unit circle. Methods of determining stability are discussed below.

a. *Routh's criterion* determines how many roots have positive real parts directly from the coefficients of the characteristic polynomial. The actual root locations are not found. The number of unstable roots of a continuous system are obtained directly. Routh's criterion can also be applied to a discrete system, but first a bilinear transformation $z = (w + 1)/(w - 1)$ is used to map the inside of the unit circle in the Z-plane into the left half of a new complex w-plane. This converts the characteristic equation into a polynomial in w, to which Routh's criterion can be applied.

b. *Root locus* is a graphical means of factoring the characteristic equation (or any algebraic polynomial of similar form). Both continuous and discrete systems can be described simultaneously by using equation (*2.2*) as the characteristic equation. The essence of the method is to consider $KF(\zeta) = -1$. Since $F(\zeta)$ is a complex number with a magnitude and a phase angle, this implies two conditions which are considered separately. They are, assuming the gain is positive and real,

$$\angle F(\zeta) = (1 + 2m)180^\circ \qquad \text{for any integer } m \tag{2.3a}$$

and

$$K\,|F(\zeta)| = 1 \tag{2.3b}$$

Thus root locus determines the closed-loop roots (and therefore stability) by working with the open-loop transfer function $KF(\zeta)$, which is normally available in factored form.

c. *Bode plots* are another graphical method which provides stability information for *minimum phase systems* (systems with no open-loop poles or zeros in the unstable region). Magnitude and phase angle are considered separately, as in equations (*2.3a*) and (*2.3b*), but the only values of ζ considered are on the stability boundary. This technique is widely used with continuous systems, in which case the stability boundary is defined by $s = j\omega$. This corresponds to the consideration of sinusoidal input functions with frequencies ω. This method is greatly simplified by using decibel units for magnitude and a logarithmic frequency scale for plotting. This allows for rapid construction of straight line asymptotic approximations of the magnitude plot. The critical point for stability, -1, becomes the point of 0 db and -180° phase shift. Bode techniques can also be applied to discrete systems by first using the same bilinear transformation as was mentioned under Routh's criterion. The stability boundary in the w-plane can also be characterized by the purely imaginary values $w = j\omega_w$. This "transformed" frequency ω_w is generally badly distorted from the true sinusoidal frequency, so intuition is of less value in this approach, insofar as stability margins, bandwidths, and similar concepts are concerned. For this reason, Bode methods are probably used less often with discrete systems, and they will not be pursued here. The same is more or less true for the following two frequency domain methods as well, so they are only discussed for continuous systems.

d. *Polar plots and Nyquist's stability criterion.* Polar plots convey much the same information as Bode plots, but the term $KG(j\omega)H(j\omega)$ is plotted as a locus of phasors with ω as the parameter. The critical point is again -1. Nyquist's stability criterion, which applies to nonminimum phase systems as well, states that the number of unstable closed-loop poles is $Z_R = P_R - N$, where N is the number of encirclements of the critical point -1 made by the locus of phasors. Counterclockwise encirclements are considered positive. P_R is the number of open-loop poles in the right-half plane.

e. *Log magnitude versus angle plots* are sometimes used for stability analysis. They contain the same information as Bode plots, but magnitude and angle are combined on a single graph with ω as a parameter.

2. *Steady-state accuracy requires that the signal $E(t)$, which is often an error signal, approach a sufficiently small value for large values of time.* The final value theorem facilitates analyzing the requirement without actually finding inverse transforms. That is, for continuous systems

$$\lim_{t\to\infty} \{E(t)\} = \lim_{s\to 0} \{sE(s)\} \tag{2.4a}$$

For discrete systems, the Z-transform version of the final value theorem is used,

$$\lim_{k\to\infty} \{E(t_k)\} = \lim_{(z\to 1)} \{(z-1)E(z)\} \tag{2.4b}$$

Both versions of the final value theorem are only valid when the indicated limits exist. By considering step, ramp, and parabolic test inputs, the useful parameters called *position*, *velocity*, and *acceleration* (or step, ramp, and parabolic) *error constants* are developed. These provide direct indications of steady-state accuracy (see Problem 2.8).

3. *Satisfactory transient response means there is no excessive overshoot for abrupt inputs, an acceptable level of oscillation in an acceptable frequency range, and satisfactory speed of response and settling time, among other things.* These are actually questions of relative stability, and depend upon the location of the closed-loop poles in the s-plane or Z-plane and their proximity to the stability boundary. Questions regarding transient response are best studied using root locus, since it is the only classical method which actually determines closed-loop pole locations. Bode, Nyquist, and log-magnitude plot methods also give information regarding transient response, at least indirectly. *Gain margin GM* is a measure of additional gain a system can tolerate with no change in phase, while remaining stable. *Phase margin PM* is the additional phase shift that can be tolerated, with no gain change, while remaining stable. Experience has shown

that acceptable transient response will usually require stability margins on the order of

$$PM > 30°, \qquad GM > 6 \text{ db}$$

These frequency domain stability margins can often be used to draw conclusions regarding transient performance, because many control systems have their response characteristics dominated by a pair of underdamped complex poles. For this case known correlations exist between frequency domain and time domain characteristics. A few approximate rules of thumb are [103]:

damping ratio $\cong 0.01\, PM$ (in degrees)
% overshoot $+ PM \cong 75$
(rise time)(closed-loop bandwidth in rad/sec) $\cong 0.45\,(2\pi)$.

Other response times have similar inverse relationships with bandwidth. The frequency of 0 db magnitude for the open-loop KGH term has an effect similar to bandwidth. Increasing this cross-over frequency increases bandwidth and decreases response times.

4. *Satisfactory frequency response implies such things as satisfactory bandwidth, limits on maximum input-to-output magnification, frequency at which this magnification occurs, as well as gain and phase margin specifications.* Bode, Nyquist, and log-magnitude-angle plots all are frequency response methods, and they deal with the open-loop transfer function. If the closed-loop characteristics, such as closed-loop bandwidth, must be determined, then the *Nichol's chart* [30] can be used. The Nichol's chart is a graphical conversion from open-loop magnitude-phase characteristics to closed-loop characteristics. Normally, one of the open-loop graphical methods is first used and the results are then transferred to a Nichol's chart. From this, the closed-loop frequency response characteristics can be read off directly.

2.5 METHODS OF IMPROVING SYSTEM PERFORMANCE

Whenever the performance of a feedback control system is not satisfactory, the following possible approaches should be considered.

1. A simple adjustment of the gain parameter K. This could be considered by using any of the analysis methods mentioned in the preceding paragraphs. From a consideration of the system's root locus, it is obvious that gain adjustment can only shift the closed-loop poles along well-defined loci. Perhaps no points on these loci give satisfactory results.

2. Minor changes in the system's structure, such as adding additional measurements to be used as feedback signals. Addition of minor feedback loops can alter the loci

of possible pole locations as K is varied. The inclusion of a rate feedback loop, using a tachometer for example, is a common means of improving stability.

3. Major changes in the system's structure or components. A hydraulic motor may perform better than an electric motor in some cases. A higher capacity pump or a more streamlined aerodynamic shape may be the answer in other cases.
4. Addition of compensating networks—i.e., $G_c(s)$ or $H(s)$—or digital algorithms—i.e., $G_c(z)$ or $H(z)$—to alter the root locus or to change the magnitude and phase characteristics in a critical frequency range.

Of these four techniques for improvement, only the second and fourth constitute what are usually referred to as *compensation techniques*. The advantages of root locus, Bode, and Nyquist methods of analysis are that compensating changes in the open-loop transfer function can be rapidly taken into account. The modifications may be made in order to reshape the locus, or improve gain or phase margins or increase the error constants. The classical methods thus constitute design techniques as well as analysis techniques. A process of design by analysis is usually used. That is, a compensating network is selected and then analyzed. However, a little experience gives great insight into the kinds of compensation that are needed. If the major problem is to improve relative stability with less concern for error constants, lead compensation networks are usually tried. If the system has acceptable stability margins, but poor steady-state accuracy, lag compensation networks will usually be appropriate. If a combination of both improvements is needed, a lag-lead network may give the desired results. More complicated networks, such as the bridged-T network, Butterworth filters, and so on, can be used to effectively cancel undesirable left-half-plane poles and replace them with more favorable ones. Cancellation compensation should never be used to eliminate unstable poles, because parameter tolerances will preclude exact cancellation. Even an infinitesimal error in cancellation will leave an unstable closed-loop pole. The form of the desired specifications and the personal preference of the designer will influence the choice of the analysis method. Extra insight can usually be gained by looking at a compensation problem from both the root locus and one of the frequency domain techniques.

An alternative method of design, through synthesis rather than analysis, is also possible. In this approach, the design specifications are translated into a desired closed-loop transfer function which satisfies them. Let the closed-loop transfer function be $M(z)$. This can be related to the compensator $G_c(z)$. For example, the system of Example 2.2 has

$$M(z) = G_c(z)G'(z)/[1 + G_c(z)G'(z)H(z)]$$

which can be solved to give

$$G_c(z) = \frac{M(z)}{[1 - M(z)H(z)]G'(z)} \tag{2.5}$$

Because of this result it is clear that certain restrictions must be imposed on $M(z)$ if the resulting compensator is to be physically realizable. This is discussed in Problems

2.23 ad 2.24. More details on this method can be found in References 39 and 69. Since discrete system compensators are just computer algorithms, there is no concern about synthesizing the results in terms of passive electrical components R, C, and maybe an occasional L that dominated classical control compensation in the early years. It is perhaps for this reason that the algebraic synthesis methods seem to be more widely used in discrete system design, although the continuous system version was worked out years earlier by Truxal [69]. Even in the continuous system domain the definition of what is practical now is quite different from the early years because of progress in technology, such as operational amplifiers and integrated circuit technology in general.

One final design parameter in discrete systems is the sampling period T. It can have a profound effect on system performance. Nyquist's sampling theorem tells us that a signal must be sampled at least twice per cycle of the highest frequency present in order to avoid losing information about the signal. The highest frequency present is often interepreted as the highest frequency of interest. This in turn can be related to system bandwidth since that is what determines which frequencies the system is capable of passing or responding to. The "twice" is strictly a theoretical limit based on an unachievable ideal low pass filter which would be needed to reconstruct the original signal from its sampled version. In reality, a cushion is provided by sampling at a considerably higher rate if possible. The reason for the sampling in the first place might be because of time-shared or multiplexed equipment, so possible T values may be restricted in many cases. In closed-loop systems T has another effect beyond the sampling theorem considerations. The value of T interacts with the loop gain K (and, of course, pole-zero locations, too) to determine system stability.

Example 2.3

Investigate the system of Fig. 2.5 for stability.

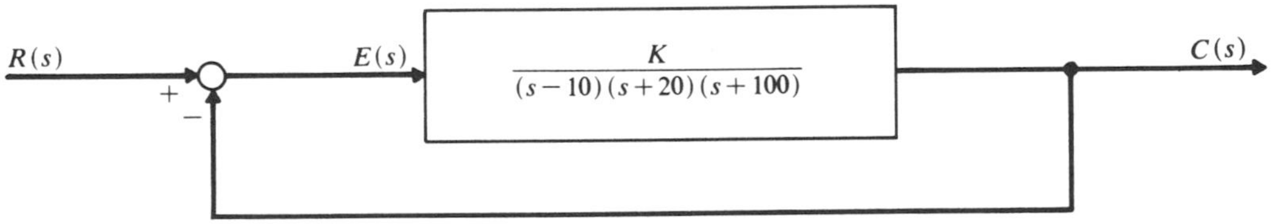

Figure 2.5

The characteristic equation is

$$1 + \frac{K}{(s-10)(s+20)(s+100)} = 0 \quad \text{or} \quad s^3 + 110s^2 + 800s - 20{,}000 + K = 0$$

The Routhian array is a table with one more row than the highest power of s in the characteristic equation. The first two rows are filled in a saw-tooth pattern with the coefficients of the characteristic equation. Each succeeding row is computed from terms in the two rows just above it. The pattern for the computed rows is as follows. Suppose two typical rows with a_i and b_i coefficients are available as in Table 2.1a. Then the c_i terms are given by

$$c_1 = (b_1a_2 - a_1b_2)/b_1, \qquad c_2 = (b_1a_3 - a_1b_3)/b_1, \qquad c_3 = (b_1a_4 - a_1b_4)/b_1$$

Each row is filled in from left to right until all remaining terms are zero. In this case c_4 and all higher c_i terms are zero because of blanks in the a_i and b_i rows.

TABLE 2.1a

a_1	a_2	a_3	a_4
b_1	b_2	b_3	b_4
c_1	c_2	c_3	c_4

TABLE 2.1b

s^3	1	800
s^2	110	$K - 20{,}000$
s	$\dfrac{108{,}000 - K}{110}$	
s^0	$K - 20{,}000$	

Table 2.1*b* gives the array for the system of Fig. 2.5. Routh's criterion states that the number of sign *changes* in the first column is equal to the number of roots in the right-half plane. For stability the first column must have all entries positive. Therefore, the system is stable if $20{,}000 \leq K \leq 108{,}000$. If $K < 20{,}000$, one sign *change* exists in column one and there will be one unstable root. If $K > 108{,}000$, there are two sign changes and therefore two unstable roots. If $K = 108{,}000$, the s row is zero. Whenever an entire row is zero, the coefficients of the preceding row are used to define the *auxiliary equation.* Roots of the auxiliary equation are also roots of the original characteristic equation. With $K = 108{,}000$, the auxiliary equation is $110s^2 + 88{,}000 = 0$, indicating poles at $s = \pm j\sqrt{800}$. With K at this maximum value, the system oscillates at $\omega = \sqrt{800}$ rad/sec. ■

Example 2.4

Investigate the steady-state following error for the system of Fig. 2.1 if $K = 100{,}000$ and the input is a unit step.

The error is

$$E(s) = \frac{1/s}{1 + \dfrac{K}{(s - 10)(s + 20)(s + 100)}}$$

Using the final value theorem, the steady-state error is

$$E(t)|_{ss} = \frac{1}{1 + \dfrac{K}{(-10)(20)(100)}} = -0.25$$

Since the input is unity, the steady-state output has a 25% error. ■

Example 2.5

Add compensation to the previous system in order to achieve a steady-state error of less than 10%. The oscillatory poles should have a damping ratio of $0.7 < \zeta < 0.9$ and a damped natural frequency of about 10 to 20 rad/sec.

In order to meet the steady-state error specifications, a gain increase by a factor of 2.2 is required. This would give an unstable system and then the final value theorem cannot be used. Compensation is required, and it will be added in the forward loop. Because of the form of the specifications, root locus will be used. First, the locus of points satisfying equation (*2.2*) is found. The following rules greatly simplify this procedure.

1. The number of branches of the root locus equals the number of open-loop poles. One closed-loop pole will exist on each branch.
2. One branch of the locus starts at each open-loop pole. One branch terminates at each open-loop zero and the remaining branches approach infinity.
3. The part of the locus on the real axis lies to the left of an odd number of poles plus zeros.
4. With $K = 0$, open and closed-loop poles coincide. As K increases, the closed-loop poles move along the loci. As $K \longrightarrow \infty$ each closed-loop pole approaches either an open-loop zero or infinity.
5. Branches that go to infinity do so along asymptotes with angles given by $\phi_i = 180°(1 + 2k)/(n - m)$ for $k = 0, \pm 1, \pm 2, \ldots$, and where n and m are the number of open-loop poles and zeros respectively.
6. The asymptotes emanate from the *center of gravity* given by cg = [(sum of real parts of all open-loop poles) − (sum of real parts of all open-loop zeros)]/$(n - m)$.
7. The loci are symmetric with respect to the real axis.

By using these rules and by testing the angle criterion at a few additional points off the real axis, the uncompensated root locus of Fig. 2.6 is obtained. It is obvious that the locus

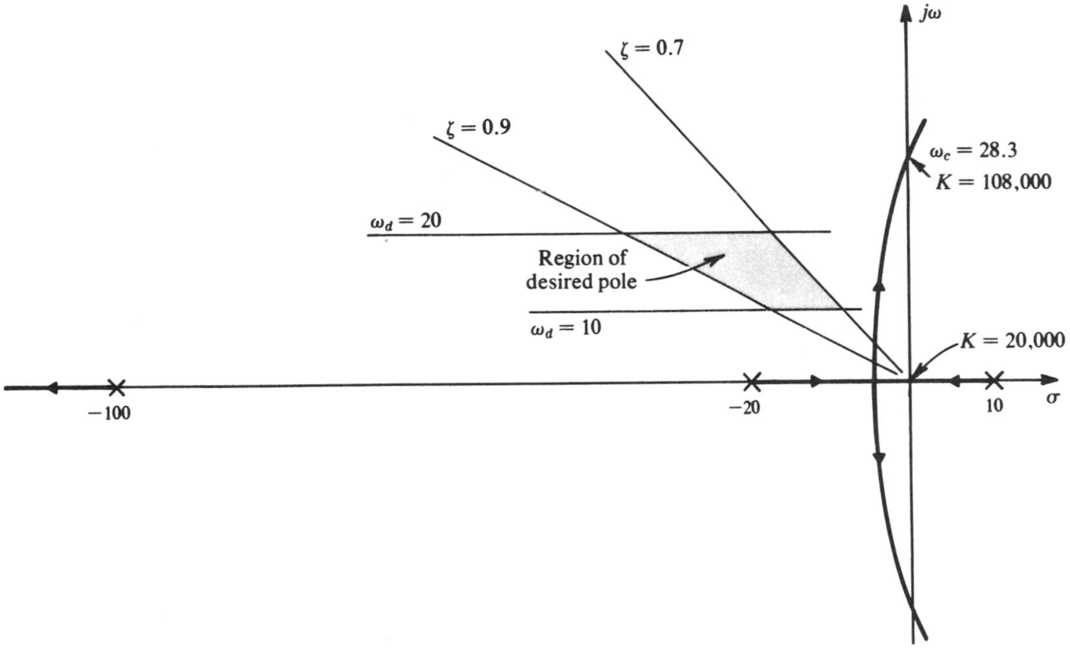

Figure 2.6 Approximate sketch of uncompensated root locus.

must be reshaped in order to meet the specifications. The angle criterion at points inside the desired region indicates that an additional 30–70° of phase lead is needed if the locus is to pass through this region. By placing a zero at $s = -20$ and a pole at $s = -45$, adequate phase lead is obtained. However, the lead network transfer function $G_{C1}(s) = (s + 20)/(s + 45)$ introduces a decrease in the error constant by a factor of $20/45 = 0.445$. This decrease can be made up, and the additional increase gained by using a lag filter, such as

$$G_{C2} = \frac{(s + 0.1)}{(s + 0.01)}$$

This pole-zero pair near the origin will have only a small effect on the locus in the region of interest since their angle contributions almost cancel each other. Using the compensator transfer function $G_C(s) = (s + 0.1)(s + 20)/[(s + 0.01)(s + 45)]$, the compensated root locus of Fig. 2.7 is obtained. Applying equation (*2.3*) at the point ⊡ indicates that the required

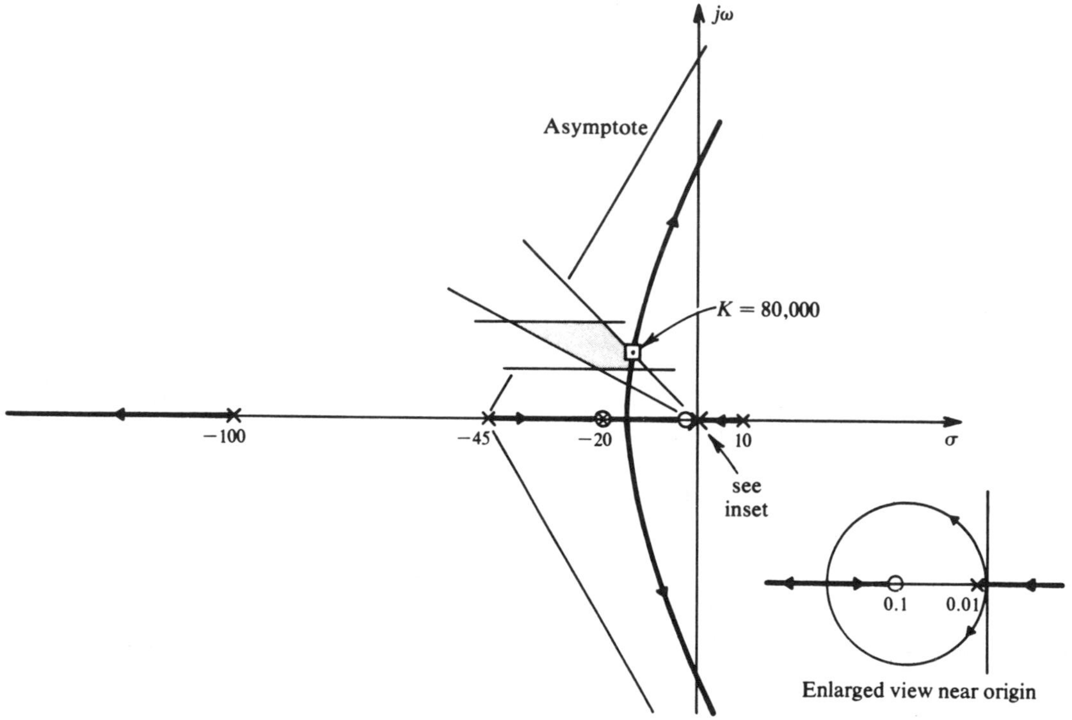

Figure 2.7

gain is $K = 80{,}000$. Using this result,

$$E(t)|_{ss} = \frac{1}{1 + \dfrac{K(0.1)}{(-10)(45)(100)(0.01)}} = -0.06 = 6\% \text{ error}$$

■

Example 2.6

Consider the error-sampled system of Fig. 2.8, which represents a linearized model of a position control system using an armature controlled dc motor with a time constant $\tau = 2$ seconds.

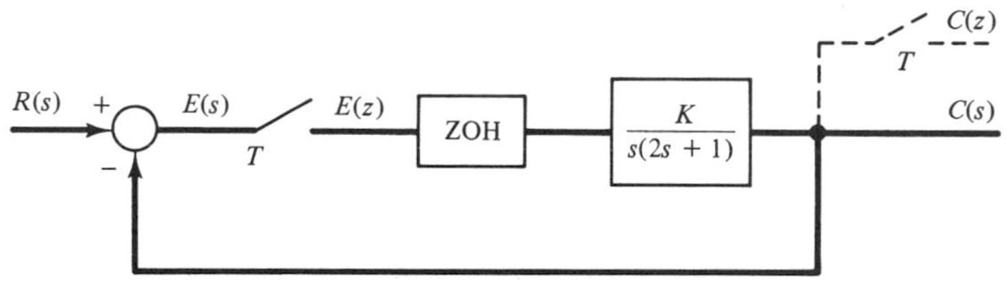

Figure 2.8

1. Find expressions for $C(s)$, $C(z)$, and $E(z)$.
2. Show that the steady-state error is zero for a step input and approaches a constant for ramp inputs, with the constant decreasing inversely with K, so long as the system remains stable.
3. Show that the maximum allowable gain for stability is inversely related to the sampling period T.
4. With $T = 2$ sec, find the maximum gain for stability.
5. Set the gain to $K = 0.5$ and find the steady-state error for a ramp input $R(t) = t$. The sampling period is $T = 2$.

1. *Using the two-step procedure of Sec. 2.3,*

$$E(s) = R(s) - E(z)(1 - z^{-1})K/[s^2(2s + 1)]$$

$$E(z) = R(z) - E(z)(1 - z^{-1})KZ\{1/[s^2(2s + 1)]\}$$

Define $Z\{.5K/[s^2(s + 0.5)]\} = G'(z)$. Using Z-transform tables,

$$G'(z) = \frac{K\{z[T - 2(1 - e^{-.5T})] + [2(1 - e^{-.5T}) - Te^{-.5T}]\}}{(z - 1)(z - e^{-.5T})} \tag{2.6}$$

Therefore

$$E(z) = \frac{R(z)}{1 + G'(z)}$$

$$C(s) = E(z)(1 - z^{-1})K/[s^2(2s + 1)]$$

and

$$C(z) = E(z)G'(z)$$

$$= \frac{G'(z)R(z)}{1 + G'(z)} \tag{2.7}$$

Note that the sampled signal $C(t_k)$ does not exist at any point in this system, so a fictitious sampler is added at the output to facilitate finding it. This sampler does not affect actual system operation.

2. *If $R(t)$ is a unit step, then $R(z) = z/(z-1)$ and the final value theorem gives*

$$\lim_{R\to\infty} E(t_k) = 1/[1 + \lim_{z\to 1} G'(z)] = 0$$

since $G'(z) \longrightarrow \infty$ as $z \longrightarrow 1$.

If $R(t) = t$, then $R(z) = TZ/(z-1)^2$ and the final value theorem now gives

$$\lim_{R\to\infty} E(t_k) = \lim_{z\to 1} Tz/[(z-1)G[(z)' = T/[KT] = 1/K$$

This result holds as long as all limits exist, which requires that the system be stable.

3. *Stability requires that the roots of the characteristic equation*

$$1 + G'(z) = 0$$

be inside the unit circle. The characteristic equation can be reduced to

$$F(z) = z^2 + \alpha z + \beta = 0 \tag{2.8}$$

where the α and β coefficients are

$$\alpha = K[T - 2(1 - e^{-.5T})] - (1 + e^{-.5T}) \tag{2.9}$$

$$\beta = e^{-.5T}(1 - KT) + 2K(1 - e^{-.5T}) \tag{2.10}$$

Rather than factor this quadratic directly, use the bilinear transformation $z = (w+1)/(w-1)$ to find a quadratic in w:

$$[1 + \alpha + \beta]w^2 + [2(1 - \beta)] + [1 + \beta - \alpha] = 0 \tag{2.11}$$

Applying Routh's criterion to the w quadratic shows that for stability

$$1 + \alpha + \beta > 0, 2(1 - \beta) > 0, \quad \text{and} \quad 1 + \beta - \alpha > 0$$

are required. In terms of the original z quadratic, these requirements are

$$F(1) > 0, \qquad F(0) < 1, \qquad \text{and} \qquad F(-1) > 0$$

These requirements are true for any second-order characteristic equation $F(z) = 0$, and are an example of the Schur-Cohn stability test [69]. For this problem $F(1) = KT(1 - e^{-.5T})$ is positive for all positive T and K.

The second condition states that $\beta < 1$, or

$$K[2 - 2e^{-.5T} - Te^{-.5T}] + e^{-.5T} < 1 \tag{2.12}$$

If $T = 0$, then $\beta = 1$, but T will never be zero. As $T \longrightarrow \infty$, $\beta \longrightarrow 2K$, so this condition must be checked in detail when specific values are given in part 4.

Finally, $F(-1) > 0$ leads to

$$K < \frac{2(1 + e^{-.5T})}{T(1 + e^{-.5T}) - 4(1 - e^{-.5T})} \tag{2.13}$$

For very small T this gives $K < 2/T$, and for very large T it gives $K < 2/(T - 4)$. This demonstrates the inverse relationship between K_{max} and T.

4. *With $T = 2$, the requirement of equation (2.12) gives $K_{max} = 1.196$, and equation (2.13) gives* $K < 13.19$. The most constraining result is the one which is operable.

5. *With $T = 2$*

$$G'(z) = 0.7357588K(z + 0.71828)/[(z - 1)(z - 0.36788)] \tag{2.14}$$

With $K = 0.5$, the steady-state error $\lim_{R \to 1} E(t_k) = 1/K = 2$. This motor control system will follow a commanded ramp in position, but the actual position will be offset by two units from the command. This may not be accurate enough. Even if the gain is increased to near its limit, the error only decreases to around one unit, and the transients will be very slow to die out with the system being that close to the stability limits. This system will need to have some form of compensation if the sampling time cannot be decreased. Note that if $T = 1$, then

$$G'(z) = 0.21306K(z + 0.84675)/[(z - 1)(z - 0.6065)] \tag{2.15}$$

and the maximum allowable stable gain increases to 2.18. The steady-state error due to a ramp input is still $1/K$ independent of T. ■

2.6 EXTENSION OF CLASSICAL TECHNIQUES TO MORE COMPLEX SYSTEMS

When multiple loops or multiple inputs and outputs must be considered, signal flow graph techniques can be used to reduce the problem to one of single-loop analysis. However, the resulting "open-loop" transfer function will usually not be in the convenient factored form. Even the simple techniques can become tedious in this case.

Linear multiple-input, multiple-output systems can be treated systematically using various transfer function matrix representations. Pursuing the subject in that direction quickly leads to the theory of polynomial matrices and the so-called matrix fraction description of systems. In this book, with very few exceptions, the alternate approach of using state space methods is pursued.

Simulation has long served as a supplement and extension to the classical analytical techniques, especially when dealing with complex and nonlinear systems. Analog computers were first historically, then came digital and hybrid methods. At present, the strong trend toward the digital computer continues, not only as an analysis and simulation aid, but also directly as a component in the control system.

ILLUSTRATIVE PROBLEMS

Properties of Feedback

2.1 Compare the open-loop and feedback control systems of Fig. 2.9 in terms of the sensitivity of the output C to variations in system parameters.

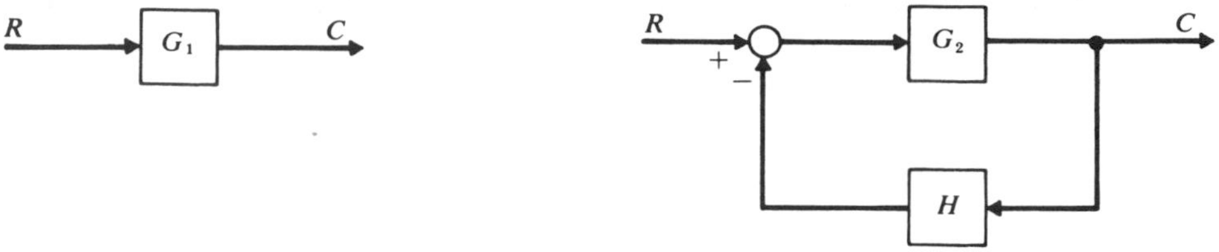

(*a*) Open loop

(*b*) Closed loop

Figure 2.9

In the open-loop case $C = G_1 R$ and $\partial C/\partial G_1 = R$ so that a change δG produces a change in the output $\delta C = R\delta G$. System sensitivity S is defined as percentage change in C/R divided by the percentage change in the process transfer function. For the open-loop system,

$$S = \frac{\delta G/G}{\delta G/G} = 1$$

For the closed-loop feedback system $C = G_2 R/(1 + G_2 H)$,

$$\delta G_2 \frac{\partial C}{\partial G_2} = \frac{R\delta G_2}{1 + G_2 H} - \frac{G_2 R H \delta G_2}{(1 + G_2 H)^2} = \frac{R\delta G_2}{(1 + G_2 H)^2}$$

The sensitivity is

$$S = \frac{\delta(C/R)}{C/R}\frac{G_2}{\delta G_2} = \frac{\delta G_2}{(1 + G_2 H)^2}\frac{(1 + G_2 H)}{G_2}\frac{G_2}{\delta G_2} = \frac{1}{1 + G_2 H}$$

As the loop gain, included in G_2, increases, $S \longrightarrow 0$. This demonstrates the decreased sensitivity of feedback systems to parameter variations.

2.2 Show how precise feedback coefficients can give precision feedback control even if gross errors exist in the forward loop system being controlled.

As the loop gain increases, the feedback transfer function becomes

$$\frac{C}{R} = \frac{KG}{1 + KGH} \longrightarrow \frac{1}{H}$$

Thus for very high loop gain the response is largely determined by H rather than G.

2.3 Show that the sensitivity of the output to errors in H approaches unity for high loop gain.

This could easily be shown by starting with the result of the last problem. Alternatively, define sensitivity to H as

$$S_H = \frac{\delta(C/R)/(C/R)}{\delta H/H} = -\frac{G^2\delta H}{(1+GH)^2}\frac{1+GH}{G}\frac{H}{\delta H} = -\frac{GH}{1+GH} \longrightarrow -1$$

as $|GH| \longrightarrow \infty$

This indicates the need for precision components in the feedback loop.

2.4 Compare performance of the systems shown in Fig. 2.10*a*,*b* as degraded by the unwanted disturbance D.

In the open-loop case, $C(s) = KG_1G_2R(s) + G_2D(s)$.

In the feedback case, $C(s) = \dfrac{KG_1G_2R(s)}{1+KG_1G_2H} + \dfrac{G_2D(s)}{1+KG_1G_2H}$.

In the second case the contribution of the disturbance to the output can be made small by increasing the gain K. Notice that feedback also introduces a loss of useful

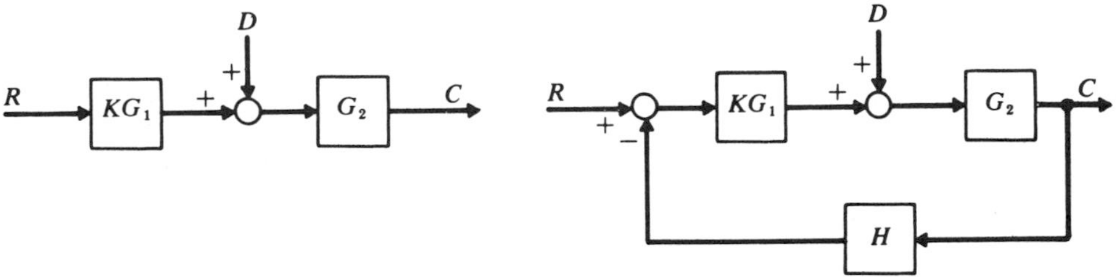

(*a*) Open loop (*b*) Feedback

Figure 2.10

gain between C and R. For example, if $H = 1$ and G_1 and G_2 are ideal amplifiers with constant gains, the open-loop system gain is KG_1G_2. The feedback system gain $KG_1G_2/(1 + KG_1G_2)$ would then be less than unity.

2.5 Use the dc motor of Problem 1.2, page 13, to demonstrate how feedback can favorably improve transient response. Neglect the inductance L.

When $L = 0$, the transfer function can be written as

$$\frac{\Omega(s)}{V(s)} = \frac{K/JR}{s + (bR + K^2)/JR}$$

This is considered as the open-loop system. If a step voltage $V(s) = V/s$ is applied, the open-loop speed response is, letting $K' = K/JR$ and $a = (bR + K^2)/JR$,

$$\Omega(s) = \frac{VK'}{s(s+a)} \quad \text{or} \quad \omega(t) = \frac{VK'}{a}(1 - e^{-at})$$

The speed of response is determined by a, which is determined by the load and motor characteristics. If response is too slow for a given load, a new motor with a larger value of K could be installed. Alternatively, consider the tachometer feedback system of Fig. 2.11. Here $\Omega(s) = [K'/(s + a + K'K_t)V(s)$. The system response time is now determined by $a' = a + K'K_t$ and can obviously be improved by proper choice of K_t.

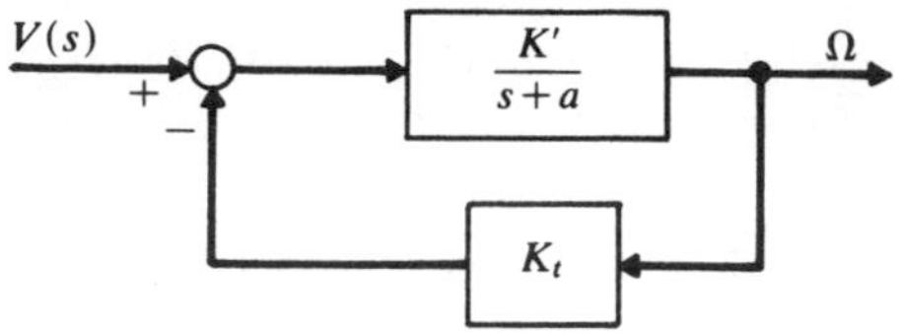

Figure 2.11

Routh's Criterion

2.6 Use Routh's criterion to determine the number of roots of

$$s^5 + 5s^4 - 2s^3 + 8s^2 + 10s + 3 = 0$$

which have positive real parts.

The Routhian array is given in Table 2.2.

TABLE 2.2

s^5	1	-2	10
s^4	5	8	3
s^3	$\dfrac{5(-2) - (1)(8)}{5} = -18/5 \sim -18$	$\dfrac{5(10) - 1(3)}{5} = \dfrac{47}{5} \sim 47$	—
s^2	$\dfrac{-18(8) - 5(47)}{-18} = \dfrac{379}{18}$	3	—
s	$\dfrac{(379/18)(47) - (-18)(3)}{379/18} = \dfrac{18785}{379}$	—	
s^0	3		

Note that any row can be normalized by multiplying or dividing by a positive constant, as in the s^3 row. For the terms in column one, the sign changes from row s^4 to row s^3 and again from row s^3 to row s^2. Two sign changes indicate there are two right-half-plane roots to this equation.

2.7 Does the following equation have any roots in the right-half plane?

$$s^4 + 2s^3 + 4s^2 + 8s + \alpha = 0$$

The Routhian array is shown in Table 2.3.

TABLE 2.3

s^4	1	4	α
s^3	2	8	
s^2	$\dfrac{2(4) - 8}{2} = 0^{\epsilon}$	α	
	$\dfrac{8\epsilon - 2\alpha}{\epsilon}$		
s^0	α		

Note that the leading term in the s^2 row is zero. Whenever this happens, the zero is replaced by a small number ϵ and the rest of the array is computed as usual. The limiting behavior as $\epsilon \longrightarrow 0$ is used to determine stability. Here the first column reduces to $\{1, 2, 0, \lim_{\epsilon \to 0} (-2\alpha/\epsilon), \alpha\}$. If $\alpha < 0$, there is just one sign change between the s and s^0 rows, and, therefore, just one right-half-plane root. If $\alpha > 0$, there are two sign changes and two right-half-plane roots.

Steady-State Error and Error Constants

2.8 Derive expressions for the steady-state value of $E(t)$ for the system of Fig. 2.1*a* when the input $R(s)$ is a step, ramp, and parabolic function, respectively.

The Laplace transform of the error is $E(s) = R(s)/[1 + KG(s)H(s)]$. For a unit step, $R(s) = 1/s$. Using the final value theorem and assuming a constant steady-state error exists, $E_{ss} \triangleq \lim_{t \to \infty} \{E(t)\} = 1/[1 + K \lim_{s \to 0} \{GH\}]$. Similarly, for a ramp input, $R(s) = 1/s^2$ and $E_{ss} = 1/K \lim_{s \to 0} \{sGH\}$. If the input is the parabola $t^2/2$, $R(s) = 1/s^3$ and $E_{ss} = 1/K \lim_{s \to 0} \{s^2 GH\}$.

To proceed, we must know the system type. The system type is the number of s terms that factor out of the denominator of $G(s)H(s)$. For a type 0 system there are no such factors, so $K \lim_{s \to 0} \{GH\} = K_b$, the Bode gain. Likewise, for type 0, $K \lim_{s \to 0} \{sGH\} = 0$ and $K \lim_{s \to 0} \{s^2 GH\} = 0$. For type 1 systems $K \lim_{s \to 0} \{GH\} = \infty$, $K \lim_{s \to 0} \{sGH\} = K_b$ and $K \lim_{s \to 0} \{s^2 GH\} = 0$. Similar results hold for type 2 and higher systems. The three limiting values for each system are called the position, velocity, and acceleration *error constants* K_p, K_v, and K_a. These are summarized in Table 2.4, along with the steady-state error values. Note that larger error constants give smaller steady-state error.

TABLE 2.4

	Error constants			Steady-state error		
System type	K_p	K_v	K_a	Step input	Ramp input	Parabolic input
0	K_b	0	0	$1/(1 + K_b)$	∞	∞
1	∞	K_b	0	0	$1/K_b$	∞
2	∞	∞	K_b	0	0	$1/K_b$

Miscellaneous Methods

2.9 Sketch the root locus for a system with

$$KG(s) = \frac{K}{s(s + 8)(s^2 + 8s + 32)}, \qquad H(s) = s + 4$$

The open-loop poles, plotted as $\times$, are located at $s = 0, -8, -4 + 4j$, and $-4 - 4j$. The open-loop zero $\odot$ is at $s = -4$. There are four branches of the loci, and Rule 3 gives the real axis portion, shown in Fig. 2.12. Three branches must approach

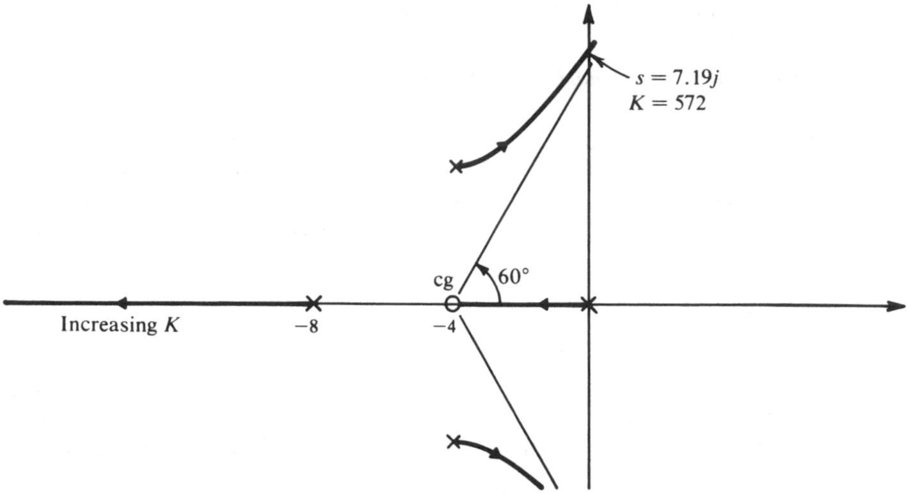

Figure 2.12

$s = \infty$ as $K \longrightarrow \infty$. One is on the negative real axis, and the others are at $\pm 60°$, since

$$\phi_i = \frac{(1 + 2k)180°}{3} = (1 + 2k)60°$$

$$= +60 \quad (k = 0), \qquad -60 \quad (k = -1), \qquad \text{and} \qquad 180° \quad (k = 1)$$

Other values of k give multiples of the same three asymptotic angles. Rule 6 gives

$$\text{cg} = \frac{1}{3}[(0 - 8 - 4 - 4) - (-4)] = -4$$

In general, the phase angle of the transfer function, equation (*2.2*), can be written as

$$\angle GH(s) = \phi_{z_1} + \phi_{z_2} + \cdots - \phi_{p_1} - \phi_{p_2} = \cdots = (1 + 2k)180°$$

where ϕ_{z_i} is the angle of the line segment from the zero z_i to a point s and ϕ_{p_i} is the angle from pole p_i to s. In order to determine the angle of departure of the locus from a complex pole, a test point s_1 is used which is infinitesimally close to the pole. The angles of the vectors from all zeros and poles except one are easily measured. The remaining angle, associated with that complex pole, can be computed. This is the angle of departure and in this case it is 0°. With this information, a few more test points allow an accurate sketch of the complete locus.

2.10 For the system shown in Fig. 2.13, select K so that the phase margin is greater than 30° and the gain margin is greater than 10 db.

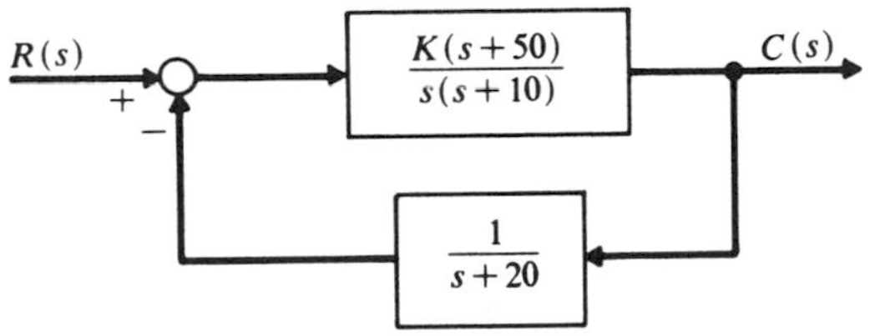

Figure 2.13

In Bode form,

$$KGH = \frac{50K}{200} \frac{(0.02s + 1)}{s(0.1s + 1)(0.05s + 1)}$$

The Bode plots are drawn in Fig. 2.14 with the Bode gain $K_b = K/4$ set to unity. At $\omega = 10$, the phase is $-150°$ and the gain is -24 db. This means that K_b could be increased from 0 db to $+24$ db, and the phase margin specification would just be satisfied. If this were done, then at $\omega = 24$ rad/sec, where the phase is $-180°$, the gain would increase from -39 db to -15 db. This gain margin of 15 db satisfies the specifications, so $K_b = 24$ db, which converts to a real gain of about 15. This means that $K = 4K_b = 60$ can be used to satisfy both specifications.

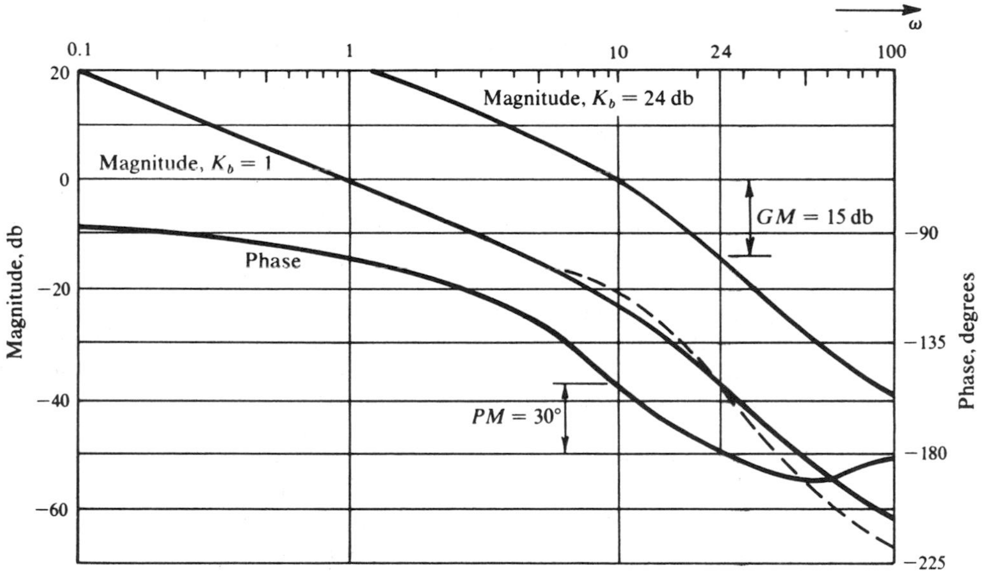

Figure 2.14

2.11 Use root locus to determine the closed-loop poles of Problem 2.10 when $K = 60$ is used.

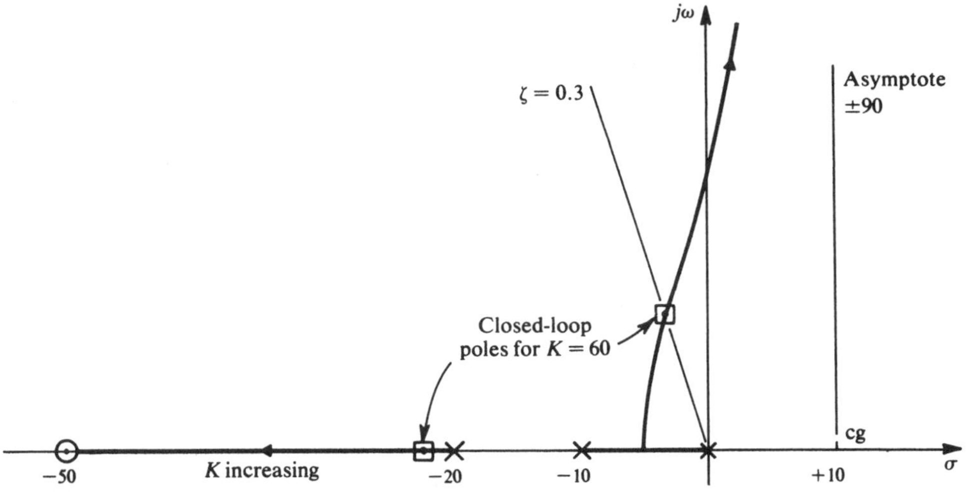

Figure 2.15

The upper half of the locus is sketched in Fig. 2.15 using the rules of Example 2.4 and a spirule to check equation (*2.2*) at a few additional points. The closed-loop poles are shown as ⊡. Notice that the complex poles lie almost exactly on the $\zeta = 0.3$ damping line. For this example, the rule of thumb $\zeta = 0.01\,PM$ is verified.

Polar Plots and Nyquist's Criterion

2.12 What information is readily available from a polar plot of $KG(j\omega)H(j\omega)$?

The number of closed-loop poles in the right-half plane can be determined in terms of the number of encirclements of -1, using Nyquist's criterion. The system type is indicated, assuming a minimum phase system, by the phase angle at $\omega = 0$. Type 0 systems have a finite magnitude and zero phase angle. Type 1 systems approach infinite magnitude at an angle of $-90°$, type 2 systems approach infinite magnitude at an angle of $-180°$, etc. The excess of open-loop poles compared to zeros is indicated by the behavior as $\omega \longrightarrow \infty$. If there is an equal number of poles and zeros, the magnitude approaches a finite constant. In all other cases the magnitude approaches zero, but if there is one more pole than zero, the approach is along the $-90°$ axis. For two more poles than zeros, it is along the $-180°$ axis, etc. Relative stability, in terms of gain and phase margins, is also readily apparent. For example, in the plot shown in Fig. 2.16 the system is type 1, and it has three more poles than zeros. Assuming no open-loop, right-half-plane poles, the system is stable since the plot does not encircle the point -1. The phase margin is 60° and the gain margin is 1.25, since the gain could be increased by that factor without causing the plot to encircle the -1 point.

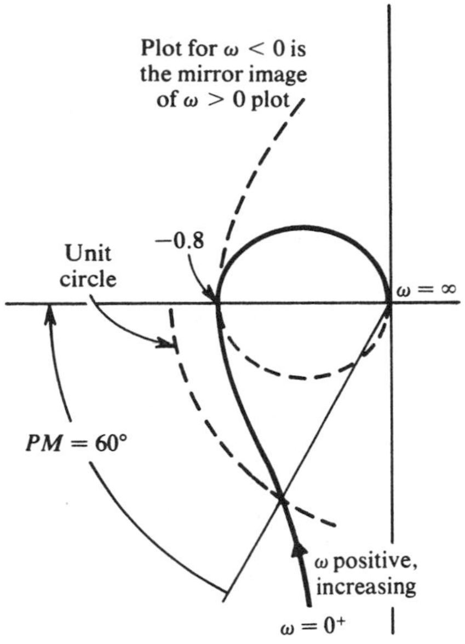

Figure 2.16

2.13 Draw the polar plot for

$$KGH = \frac{K(s + 0.5)(s + 10)}{(s + 1)(s + 2)(s^2 + 2s + 5)}$$

Determine the gain and phase margins when $K = 2$.

The polar plot for this type 0 system is given in Fig. 2.17. The -1 point is not

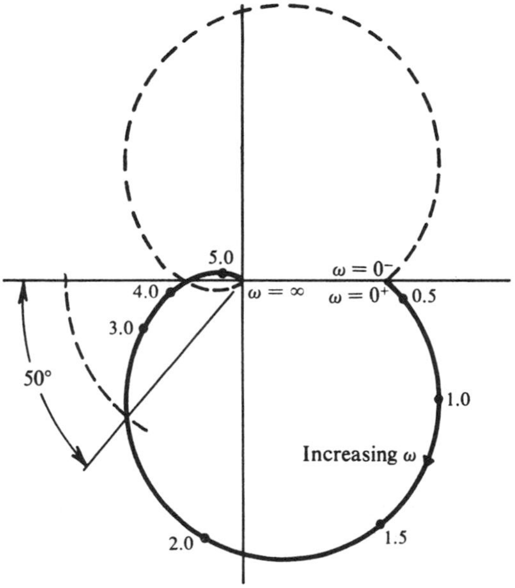

Figure 2.17

encircled, so $N = 0$. Since there are no open-loop, right-half-plane poles, $P_R = 0$. Therefore, Nyquist's criterion indicates that there are no unstable closed-loop poles. The phase margin is approximately 50°. The magnitude at 180° phase is 0.32, so the gain margin is $1/0.32 \cong 3.1$.

2.14 Analyze the polar plot for the nonminimum phase system with

$$KGH = \frac{K(s + 10)(s + 20)}{s(s - 10)(s + 40)}$$

Even though this is a type 1 system, the phase angle approaches −270° as $\omega \longrightarrow 0$ since the minus sign on the unstable pole contributes −180° phase. The general shape of the polar plot is shown in Fig. 2.18*a*.

The −1 point could be encircled by either the *a-d-c-b-a* circuit, counterclockwise, or by the infinite *a-g-f-e-a* circuit, clockwise. Which circumstance prevails depends upon K. If −1 is inside the small circuit, then $N = 1$. Since $P_R = 1$, there would be $Z_R = 1 - 1 = 0$ unstable closed-loop poles. If −1 is to the left of point *a*, then there is one clockwise encirclement so $N = -1$ and, therefore, $Z_R = 1 - (-1) = 2$, indicating two unstable closed-loop poles. An enlarged plot is also given in Fig. 2.18*b* for

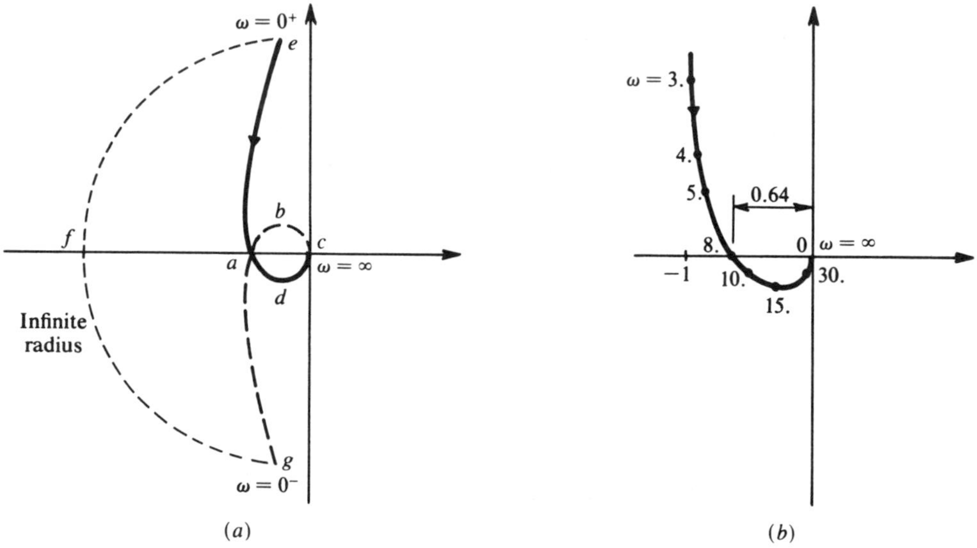

Figure 2.18

$K = 10$. The magnitude of 0.64 at −180° indicates that the system is unstable for values of $K < 10/0.64 \cong 15.6$, and stable if $K > 15.6$. This problem illustrates that the usual visualization of gain and phase margin is incorrect for nonminimum phase systems. This is why Bode plots should not be used with nonminimum phase transfer functions.

Compensation

2.15 Explain the essence of classical compensation using root locus techniques.

The basis for compensation is a knowledge of the correspondence between closed-loop pole locations and the type of transient time response terms they yield. Some typical closed-loop pole locations are indicated by □ in Fig. 2.19, and the corresponding time response terms are shown.

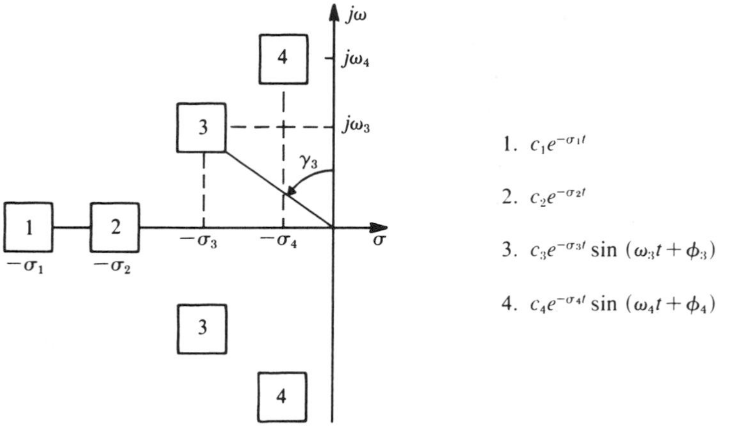

Figure 2.19

Poles farther to the left of the imaginary axis give terms which die out faster, i.e. faster response times. Complex poles give oscillating terms with a frequency equal to the distance from the real axis and decay time inversely proportional to the real part of the pole, σ. The damping ratio, related to overshoot, is defined in terms of the angle γ as $\zeta = \sin\gamma$. The undamped natural frequency ω_n is the radial distance from the origin, so that $\sigma_i = \zeta\omega_n$ and the damped frequency is $\omega_i = \omega_n\sqrt{1-\zeta^2}$.

Based on these relations, the desired locations of the most dominant closed-loop poles are selected. Checking the root locus angle criterion at that point indicates whether additional lag or lead is needed and how much. Compensating poles and zeros are then selected to provide this phase shift.

Suppose a damping ratio of $\zeta = 0.707$ and a frequency of 10 rad/sec are desired. Then a pair of complex closed-loop poles must be located as shown in Fig. 2.20. If the angle for the uncompensated KGH is $-160°$ at that point, then an additional $-20°$ phase must be provided by compensation. It is common practice to place the compensator zero directly below the desired closed-loop pole. Obviously, a single pole zero pair can provide up to 60–65° phase shift. If a greater shift is needed, more complicated compensation is required.

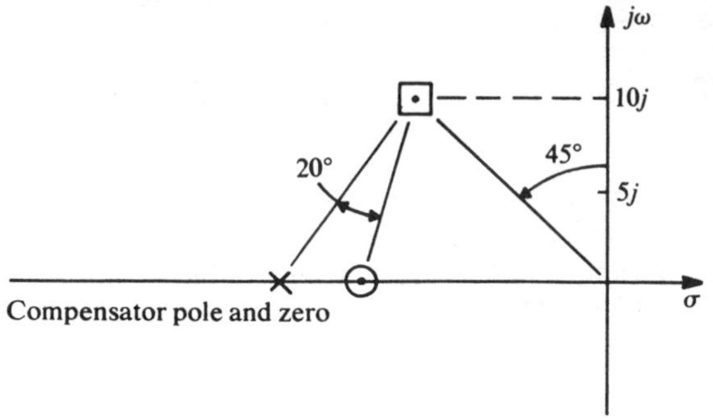

Figure 2.20

2.16 Discuss the compensation characteristics of the phase lag and phase lead circuits of Fig. 2.21 using Bode plots.

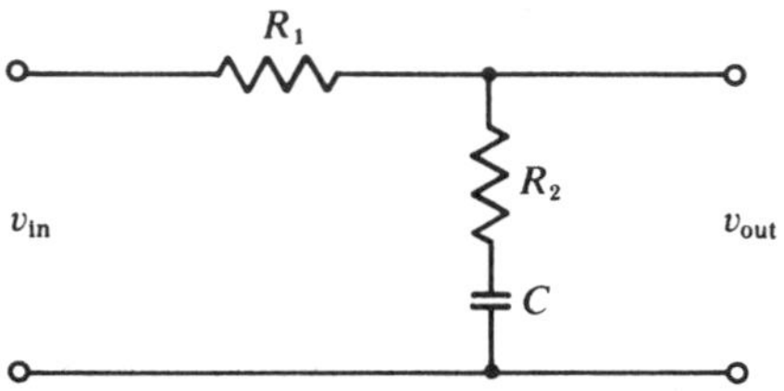

$$\frac{V_{\text{out}}}{V_{\text{in}}} = \frac{1+\tau_1 s}{1+\tau_2 s} \qquad \tau_2 > \tau_1$$

(*a*) Phase-lag circuit

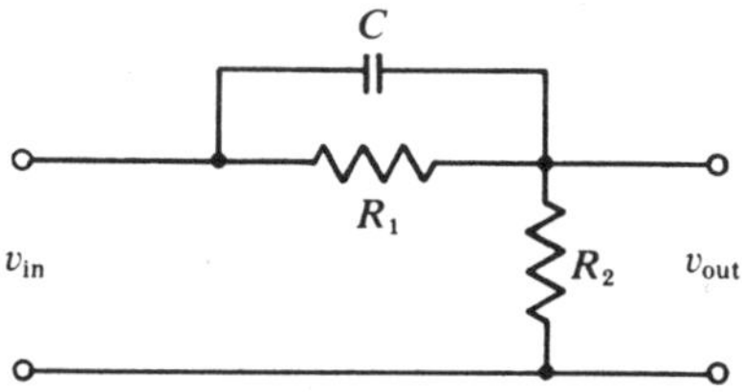

$$\frac{V_{\text{out}}}{V_{\text{in}}} = \left(\frac{\tau_2}{\tau_1}\right)\frac{1+\tau_1 s}{1+\tau_2 s} \qquad \tau_1 > \tau_2$$

(*b*) Phase-lead circuit

Figure 2.21

The lag circuit lowers the gain at high frequencies and leaves the phase unchanged for $\omega \gg 1/\tau_1$ (Fig. 2.22*a*). The corner frequencies $1/\tau_1$ and $1/\tau_2$ would be

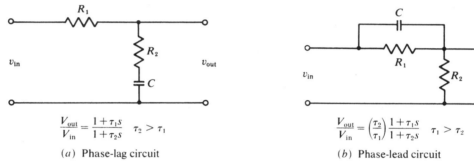

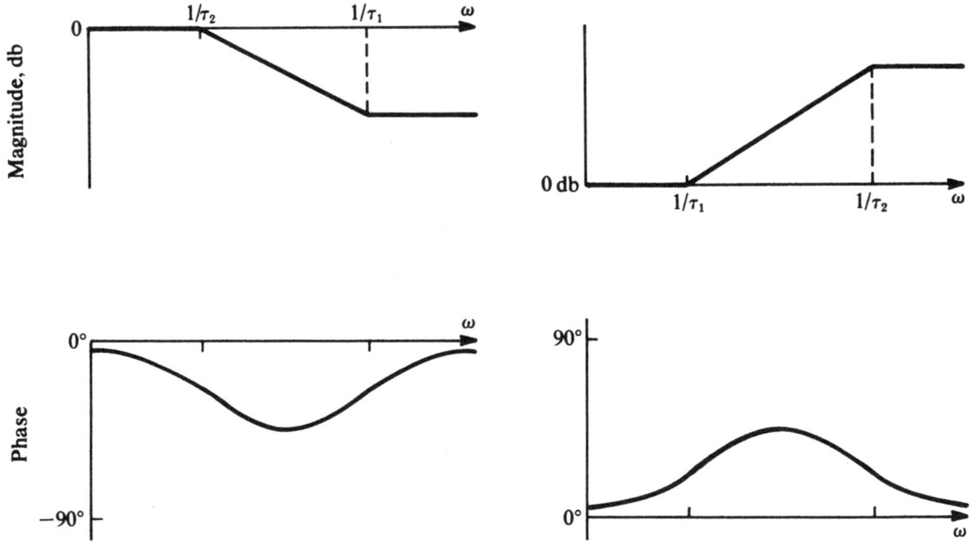

(*a*) Phase-lag circuit (*b*) Phase-lead circuit

Figure 2.22

chosen well below the critical cross-over frequency. This means the dc gain can be raised in order to improve steady-state accuracy. This increase in gain returns the high frequency gain to its uncompensated level and thus leaves the stability margins relatively unchanged.

The lead circuit Bode plot assumes an additional dc gain has been added to compensate for the τ_2/τ_1 reduction. By proper choice of $1/\tau_1$ and $1/\tau_2$ relative to the cross-over frequencies, the phase lead can be used to increase phase margin. This circuit also delays the 0 db cross-over frequency, thus increasing bandwidth and speed of response. It leaves the low frequency characteristics, such as steady-state error, unchanged.

Discrete-Time Problems

2.17 In the process of designing a digital control system, it was determined that the following compensator was desired:

$$G_c(z) = \frac{10(z + 0.4)(z - 0.5)}{(z + 0.5)(z + 1)}$$

Determine an algorithm which can be coded on the computer to implement $G_c(z)$.

There are several possible answers, two of which follow. Writing G_c in expanded polynomial form gives

$$G_c(z) = \frac{10(z^2 - 0.1z - 0.2)}{z^2 + 1.5z + 0.5} = \frac{10(1 - 0.1z^{-1} - 0.2z^{-2})}{1 + 1.5z^{-1} + 0.5z^{-2}}$$

If the input to G_c is $E(t_k)$ and the output is $y(t_k)$, then using z^{-1} as the delay operator, cross multiplication gives the so-called direct form realization

$$(1 + 1.5z^{-1} + 0.5z^{-2})Y(z) = (10 - z^{-1} - 2z^{-2})E(z)$$

or, in the time domain

$$y(t_k) = 10E(t_k) - E(t_{k-1}) - 2E(t_{k-2}) - 1.5y(t_{k-1}) - 0.5y(t_{k-2})$$

If $G_c(z)/z$ is expanded in partial fractions and the result is then multiplied by z, one obtains

$$G_c(z) = -4 - 4z/(z + 0.5) + 18z/(z + 1)$$

This represents three separate paths through the compensator, as shown in Fig. 2.23. This is the so-called parallel realization.

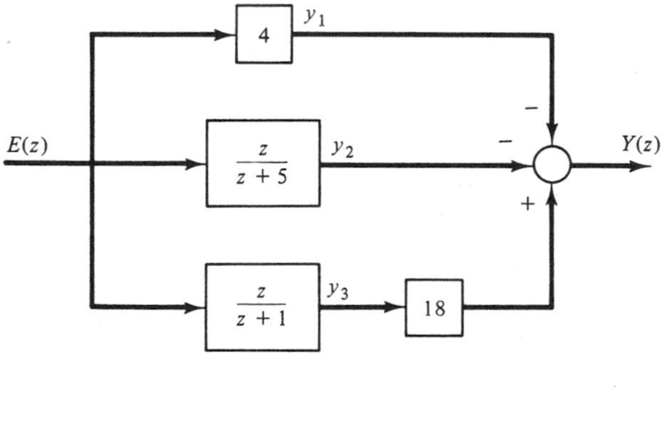

Figure 2.23

The algorithm thus consists of

$$y_1(t_k) = 4E(t_k)$$
$$y_2(t_k) = E(t_k) - 0.5y(t_{k-1})$$
$$y_3(t_k) = E(t_k) - y(t_{k-1})$$

and

$$y(t_k) = 18y_3(t_k) - y_1(t_k) - y_2(t_k)$$

Other forms are possible.

2.18 Derive the transfer function for the device which outputs the piecewise constant approximation to a continuous-time function $E(t)$ as shown in Fig. 2.24.

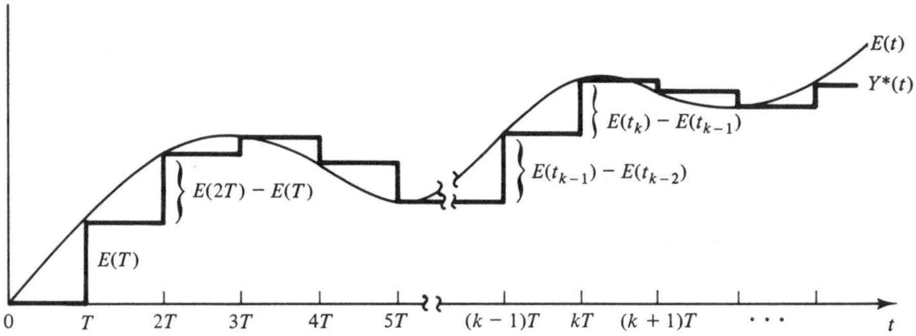

Figure 2.24

Let the unit step function starting at time $t = 0$ be $u(t)$. Then a shifted unit step function starting at time t_k is $u(t - t_k)$. The piecewise constant function can be written as

$$\begin{aligned} Y^*(t) &= E(0)u(t) + [E(1) - E(0)]u(t - t_1) + [E(2) - E(1)]u(t - t_2) + \cdots \\ &\quad + [E(t_k) - E(t_{k-1})]u(t - T_k) + \cdots \\ &= \sum_{k=0}^{\infty} [E(t_k) - E(t_{k-1})]u(t_k) \end{aligned}$$

Since $du(t - t_k)/dt = \delta(t - t_k)$, the derivative of the device output is

$$x(t) = dY^*(t)/dt = \sum_{k=0}^{\infty} [E(t_k) - E(t_{k-1})]\, \delta(t - t_k)$$

Using the definition of the Z-transform in Problem 1.15, the transform of $x(t)$ is $X(z) = E(z) - z^{-1}E(z) = (1 - z^{-1})E(z)$. The final relationship among these variables is shown in Fig. 2.25.

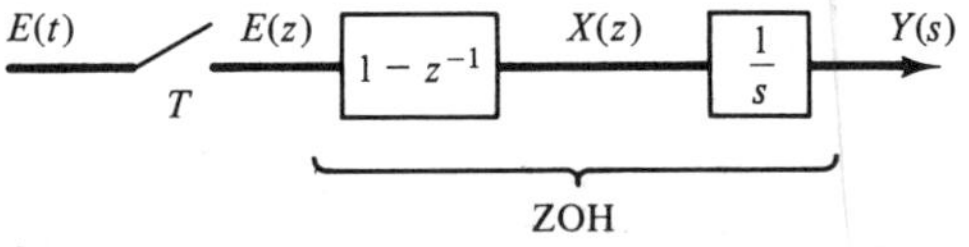

Figure 2.25

Since $y(t)$ is the integral of $x(t)$, $Y(s) = [(1 - z^{-1})/s]E(z)$. Clearly, then, the transfer function of the zero order hold is $G_0(s) = (1 - z^{-1})/s$.

2.19 Investigate methods of obtaining a z-transfer function $G'(z)$ which has approximately the same behavior as an s-transfer function $G(s)$. There are several approaches to this question.

a. One way is to determine the exact Z-transform that corresponds to the product of $G(s)$ and the ZOH transfer function $G_0(s)$, perhaps using transform tables. The ZOH should be included in most cases involving combined continuous and discrete systems, as shown in Fig. 2.1*b*. The reason is that Z-transforms are always applied to everything between two samplers (or the same sampler around a complete loop). For most physical systems of interest, a D/A will be included in this segment of the system.

b. An approximate conversion is obtained by treating s as a derivative operator and using a forward finite difference approximation on the sampled signal:

$$\dot{y}(t) \cong [y(t_{k+1}) - y(t_k)]/T$$

In the transform domain this means that $s \cong (z - 1)/T$, since z is the advance operator. If this approximation for s is used in $G(s)$, the approximate transfer function $G'_b(z)$ is obtained. Note that this result is also given directly from $z = e^{Ts} \cong 1 + Ts$, which is approximately true for small Ts.

c. A backward difference approximation is also possible. $\dot{y}(t) \cong [y(t_k) - y(t_{k-1})]/T$ leads to $s \cong (1 - z^{-1})/T$ and to $G'_c(z)$. This result is also obtainable from $z = e^{Ts} = 1/e^{-Ts} \cong 1/[1 - Ts]$.

d. If z is written as $z = e^{Ts/2}/e^{-Ts/2} \cong [1 + Ts]/[1 - Ts]$, then s can be solved for as $s \cong (2/T)[z - 1]/[z + 1]$. Using this gives $G'_d(z)$.

All of the above results can also be derived by approximating the integration operator instead of the derivative operator [39].

2.20 Apply the above techniques to determine discrete approximations for $G(s) = 1/(s + a)$.

a.1 $Z\{(1 - z^{-1})G(s)/s\} = [1 - e^{-aT}]/[a(z - e^{-aT})] = G'_a(z)$

Note that the pole is the exact s- to z-plane mapping of $s = -a$, so stability properties are preserved.

a.2 For comparison, the transform without the zero order hold is $Z\{G(s)\} = z/[z - e^{-aT}]$. Even though the denominators are the same, there is a one period delay difference and a gain difference. Unless one is working with true pulsed circuits, form a.1 is the appropriate one to use.

b. With $s \cong (z - 1)/T$, $G'_b(z) = T/[z - 1 + aT]$. The z-plane pole can be in the unstable region (outside the unit circle) even if the s-plane pole is stable.

c. With $s \cong (z - 1)/(Tz)$, $G'_c(z) = [T/(1 + aT)]z/\{z - [1/(1 + aT)]\}$. Here the z-plane pole is always stable (inside the unit circle) whenever the s-plane pole is stable (and sometimes even when the s-plane pole is unstable).

d. With $s \cong (2/T)[z-1]/[z+1]$, $G_d'(z) = [T/(aT+1)]z/[z-1/(1+aT)]$. In this case the z-plane pole is inside the unit circle if $a < 0$, is on the unit circle if $a = 0$, and is outside if $a > 0$. Thus $G_d'(z)$ inherits the exact stability properties of $G(s)$. It can be shown that this is true for all transfer functions formed using approximation (*d*). That is,

$$z = [1 + Ts/2]/[1 - Ts/2]$$

exactly maps the left-hand s-plane into the interior of the unit circle. Approximations (*a*) and (*d*) both preserve stability properties, while (*b*) and (*c*) do not. However, the behavior in (*c*) is preferable to that of (*b*) in this regard.

2.21 By using the mapping $z = e^{Ts}$, determine how the pole locations discussed in Problem 2.15 map into the Z-plane.

For poles [1] and [2] or any other on the negative real axis, $z = e^{-\sigma T}$ is a positive number between 0 and 1. Therefore Z-plane poles on the positive real axis correspond to exponential time functions. These are stable (decaying) exponentials for poles inside the unit circle. Poles on the positive real axis in the s-plane also give positive real axis Z-plane poles that are outside the unit circle. These correspond to unstable (growing) exponentials.

Poles like [3] and [4] give $z = e^{-\sigma T}e^{j\omega T} = e^{-\sigma T}[\cos(\omega T) \pm j\sin(\omega T)]$. These will always occur in conjugate pairs, and are on the unit circle if $\sigma = 0$ and give persistent oscillations. If $\sigma > 0$, they are inside the unit circle and give damped oscillations. For $\sigma < 0$, the poles are outside the unit circle and correspond to growing oscillations. The decay or growth rate depends directly on the radial distance of the poles from the unit circle. The frequency of oscillation is directly related to their angular position ωT. Note that if $\omega T = \pi$, the poles are on the negative real axis. No s-plane pole with $\omega > \pi/T$ will occur if the sampling rate satisfies the Nyquist sampling theorem. If such higher frequency poles do occur, they map into the Z-plane at points which also correspond to lower frequency poles. This ambiguity is called aliasing, and can be avoided by sampling at least twice per period of the highest frequency pole in the system.

As any s-plane pole's real part approaches $-\infty$, the Z-plane pole approaches the origin. Poles at $s = 0$ map into $z = 1$.

It is informative to consider several familiar contours in the s-plane and see how they map into the Z-plane. In Fig. 2.26*a*, a closed contour is considered, with arrows indicating the traverse direction and numbered points showing the correspondence at eight key points. In Fig. 2.26*b*, lines of constant frequency, constant settling time, and lines of constant damping ratio are shown for the s- and Z-planes.

2.22 The dc motor controller of Example 2.6 is to be designed using root locus. The sampling period is $T = 1$ second. Closed-loop Z-plane poles at $z = 0.19877 \pm 0.30956j$ are desired. These are the images of $s = -1 \pm j$. They are selected because of the desirable properties they give to continuous-time systems, namely, a settling time of about 4 seconds and the damping ratio of 0.7, which gives about 5% overshoot. The open-loop transfer function is given in equation (*2.15*).

The uncompensated root locus is sketched in Fig. 2.27*a*. To achieve the desired root locations, compensation is required to reshape the locus. A forward-path cascade compensator of the form $G_c(z) = K_c(z - 0.60653)/(z + a)$ is selected. The value of a will be selected so that the locus will be shifted over, as shown in Fig. 2.27*b*. This is a

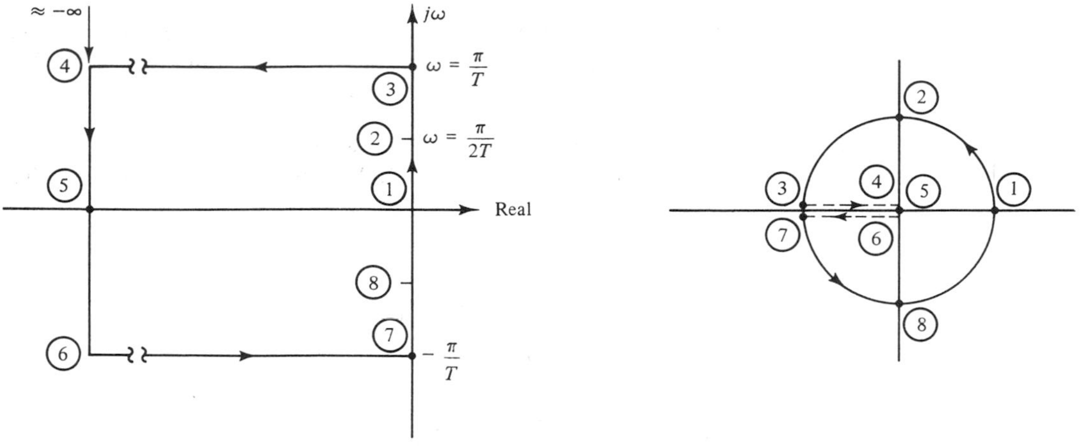

(a)

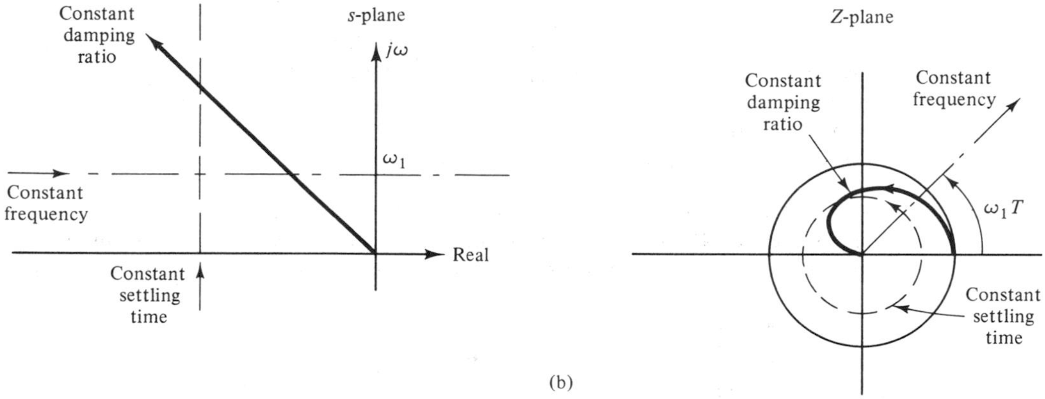

(b)

Figure 2.26

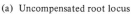

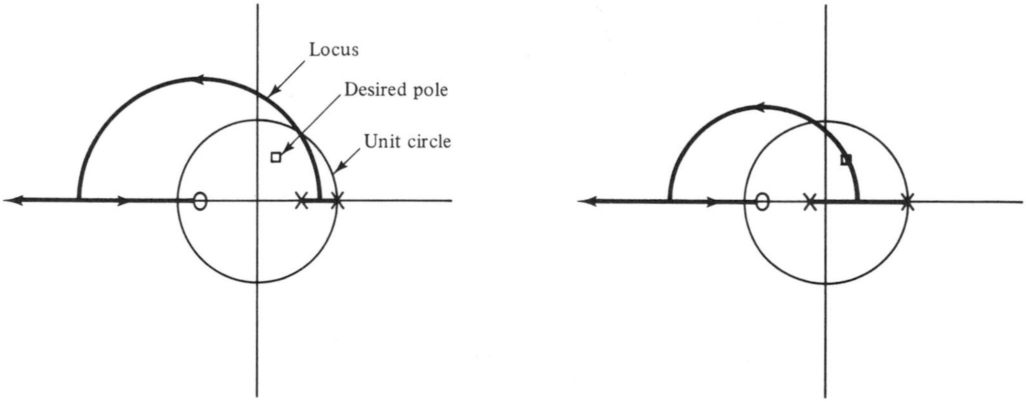

(a) Uncompensated root locus

(b) Compensated root locus

Figure 2.27

lead-type compensator. As shown in Fig. 2.28, if the desired point is to be on the locus, it must be true that $\phi_3 - \phi_1 - \phi_2 = -180$. But $\phi_1 = 180 - \tan^{-1}[0.30956/(1 - 0.19877)] = 158.88°$. Likewise, $\phi_3 = \tan^{-1}[0.30956/(0.84675 + 0.19877)] = 16.49°$. Therefore, ϕ_2 must be 37.61°. This means that the compensator pole must be at $z = -0.203$. The required root locus gain is computed as the product of the vector lengths from the poles, divided by the product of the vector lengths from the zeros to the desired

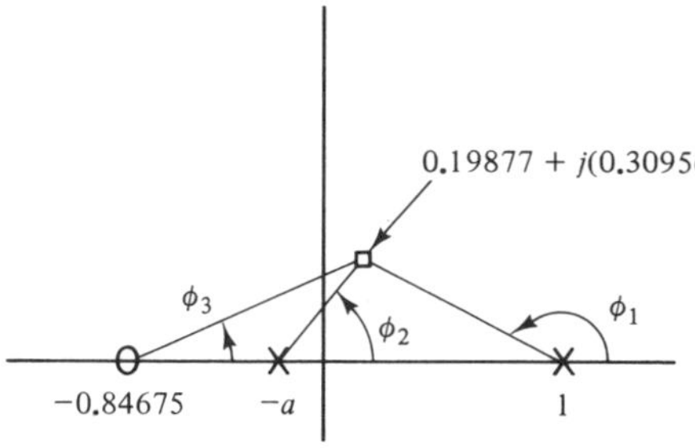

Figure 2.28

root location ⊡. This gives $K_{RL} = (0.7378)(0.2572)/1.1889 = 0.3995$. The total root locus gain for this system is $K_{RL} = K_c(0.21306K)$, so if, for example, the gain K in the open-loop transfer function has a value of 0.5, then $K_c = 3.7501$, giving the final compensator

$$G_c(z) = 3.7501(z - 0.6065)/(z + 0.2030).$$

The compensated system's response to a step input is shown in Fig. 2.29. It shows about 4% overshoot, just what might have been expected from the s-plane roots. This result is not necessarily typical. Quite often the discrete system will have

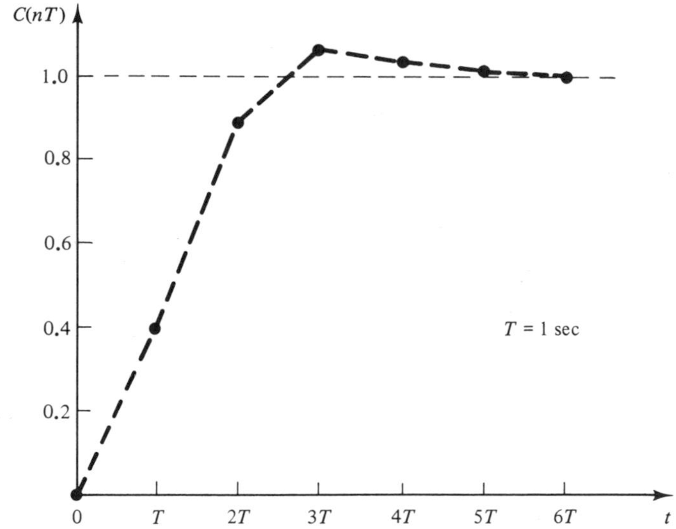

Figure 2.29 Step response.

far more overshoot than expected from the s-plane pole positions. This is due to the fact that the sampled system is essentially running open-loop for the intersample periods T. This allows larger overshoot to build up before corrective feedback action can occur. Also, the location of the zero is a major factor in determining the response [39]. The final value theorem shows that this system will have a steady-state error of 3.26 when the input is a ramp.

2.23 Design a cascade compensator for the system of Example 2.6 so that the steady-state value of the error $E(t_k)$ is zero for a ramp input, and such that E goes to zero in the minimum number of sampling periods. This is referred to as the deadbeat response controller.

The transfer function for the error is

$$E(z) = W(z)R(z)$$

where $W(z) = 1/[1 + G_c(z)G'(z)]$.

If $E(r_k)$ is to go to and remain at zero in some finite time, then $W(z)R(z)$ must be a *finite* polynomial in z^{-1}. Since steps, ramps, and other typically used inputs contain a denominator factor of $(1 - z^{-1})^{\alpha}$, $W(z)$ must contain as a factor $(1 - z^{-1})^{\alpha}$. Otherwise, an infinite series in z^{-1} would result for $E(z)$. In this particular case of a ramp input, $\alpha = 2$ and $W(z)$ must have the factor $(1 - z^{-1})^2$. It is generally necessary that $W(z)$ contain another factor $F(z^{-1})$ as well. If $W(z)$ is given, then it is easy to show that $G_c(z) = [1 - W(z)]/[W(z)G'(z)]$. The reason why the extra factor F may be required is that the resulting $G_c(z)$ must be *forced* to be physically realizable if it doesn't come out that way initially. The general rules for doing this are:

1. If $G'(z)$ has no poles or zeros on or outside the unit circle (a single pole at $z = 1$ is acceptable), then $F(z^{-1}) = 1$ is all that is required. A leading 1 is always assumed for the polynomial F, and if any unnecessary extra powers of z^{-1} are included, they only delay the time at which E reaches zero. Otherwise, the following three additional dictums must be met:
2. $F(z^{-1})$ must contain as zeros all the unstable poles of $G'(z)$.
3. $1 - W(z)$ must contain as zeros all the zeros of $G'(z)$ on or outside the unit circle.
4. $1 - W(z)$ must contain z^{-1} as a factor.

In the present case $G'(z)$ has no poles or zeros outside the unit circle so it is permissible to use $F = 1$. Then $W(z) = (1 - z^{-1})^2$ and the compensator is

$$G_c(z) = (2z - 1)(z - 0.6065)/[0.21306K(z + 0.84675)(z - 1)]$$

When this compensator is used, the closed-loop transfer function is $M(z) = 2z^{-1} - z^{-2}$. The responses to step and ramp inputs are as shown in Figs. 2.30 and 2.31, respectively. Note that the design objectives are met for the ramp input, but the step response may not be acceptable since it has 100% overshoot. This illustrates one drawback of deadbeat response controllers: they are tuned to specific inputs. The other potential drawback is that the intersample error is not necessarily zero just because $E(t_k)$ is zero. There are other methods of suppressing intersample ripple [69].

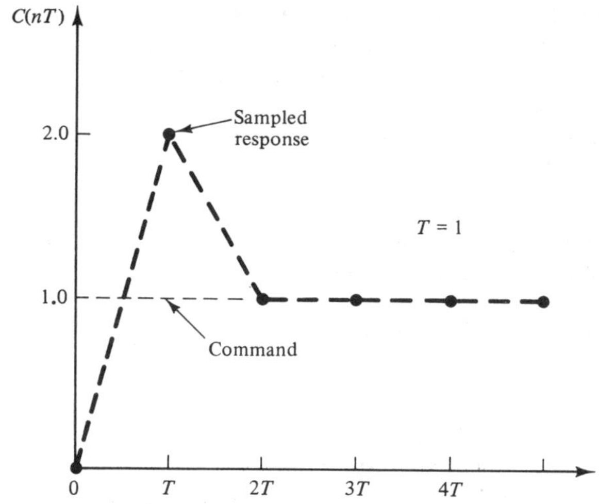

Figure 2.30 Step response.

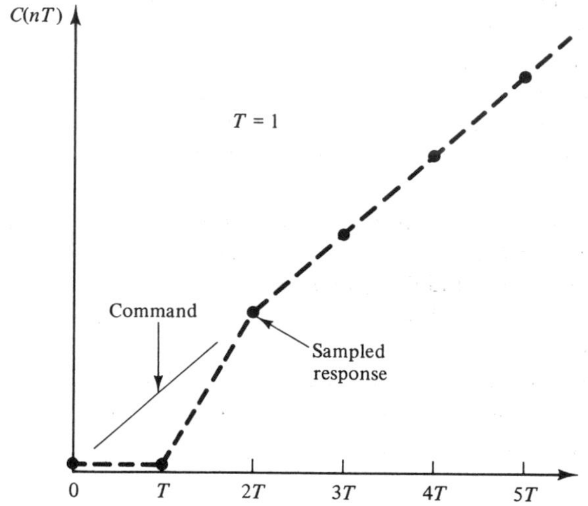

Figure 2.31 Ramp response.

2.24 Consider the same system as in the previous two problems. This time a cascade compensator is sought which meets the following specs: (1) closed-loop poles at $z = 0.19877 \pm 0.30956j$, (2) zero steady-state error for a step input, and (3) a velocity error constant K_v of 5 to insure an ability to follow ramp inputs with acceptably small error. This is an example of a direct synthesis method [69], sometimes called the method of Raggazini [39].

A summary of the rules associated with this design method is given before applying them to this specific problem. Let the final closed-loop transfer function be called $M(z)$.

1. In order to insure that the final compensator is physically realizable, $M(z)$ must have at least as many more poles than zeros as $G'(z)$ does.
2. In order to insure that $G_c(z)$ is stable, $M(z)$ must contain as zeros all the zeros of $G'(z)$ that are outside or on the unit circle. A consideration of the poles and zeros of equation (*2.5*), with $H = 1$, shows why this is necessary.
3. Of necessity, if $M(z)$ is to be stable, $1 - M(z)$ must have as zeros all the unstable poles of $G'(z)$. (Those on or outside the unit circle. A single pole at $z = 1$ is not considered unstable.) The reason why this is necessary is clear from a consideration of equation (*2.5*), rearranged as $M = G_c G'(1 - M)$.
4. In order to have zero steady-state error, the final value theorem indicates that $\lim_{z\to 1} \{(z-1)R(z)[1 - M(z)]\} = 0$. The implication of this depends on what $R(z)$ is, but for a step input, since $R(z) = z/(z-1)$, it means that $\lim_{z\to 1} M(z) = 1$.
5. The velocity error coefficient is defined as

$$K_v = (1/T)\lim_{z\to 1}\{(z-1)G_c(z)G'(z)\} = (1/T)\lim_{z\to 1}(z-1)M(z)/[1 - M(z)]$$

For the case where a step input was used in (4), $M(1) = 1$, so the expression for the error constant is indeterminant of the form 0/0. Use of L'Hospital's rule gives $K_v = -1/\{Td[M(z)]/dz\}|_{z=1}$. Solving gives a requirement on $M(z)$ in terms of a specified K_v:

$$\left.\frac{dM(z)}{dz}\right|_{z=1} = \frac{-1}{K_v T}$$

Now for the specific problem here, $M(z)$ must have at least one more pole than zero since $G'(z)$ does. The open-loop system $G'(z)$ has no poles or zeros outside the unit circle. In view of this and the desired closed-loop poles, a tentative $M(z)$ is selected which satisfies (1), (2), and (3) above.

$$M(z) = \frac{A(z+a)}{z^2 - 0.39754z + 0.13534}$$

There are two free design parameters A and a that can be used to satisfy (4) and (5). From (4), $A(1 + a)/0.737797 = 1$ and from (5)

$$A/0.737797 - A(1 + a)(2 - 0.39754)/(0.737797)^2 = -1/5$$

From these two equations the two unknowns are found to be $A = 1.4549$ and $a = -0.492888$. The compensated closed-loop transfer function is

$$M(z) = 1.4549(z - 0.492888)/(z^2 - 0.39754z + 0.135337)$$

Using this, equation (*2.5*) gives the following compensator, after algebraic simplification:

$$G_c(z) = \frac{6.82859(z - 0.60653)(z - 0.492888)}{K(z - 0.85244)(z + 0.84675)}$$

The transient response of the closed-loop system to step and ramp inputs are shown in Figs. 2.32 and 2.33, respectively. Note that the maximum percent overshoot (at the sampling times) due to a step input is 45.5% and the settling time is about $4T$, i.e., 4 seconds. A different set of design specifications may lead to improved performance.

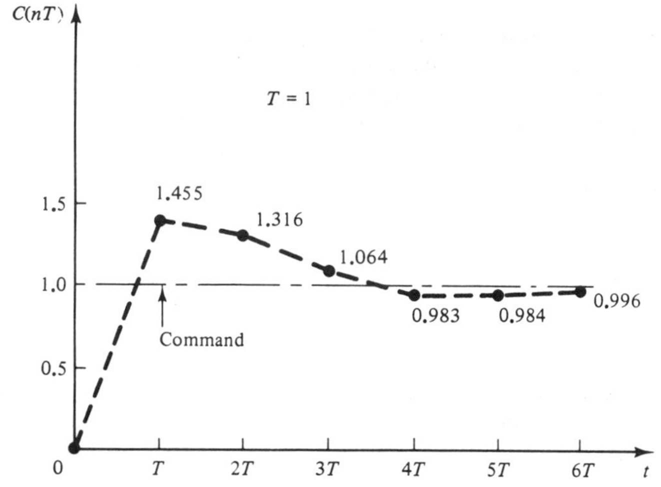

Figure 2.32 Step response.

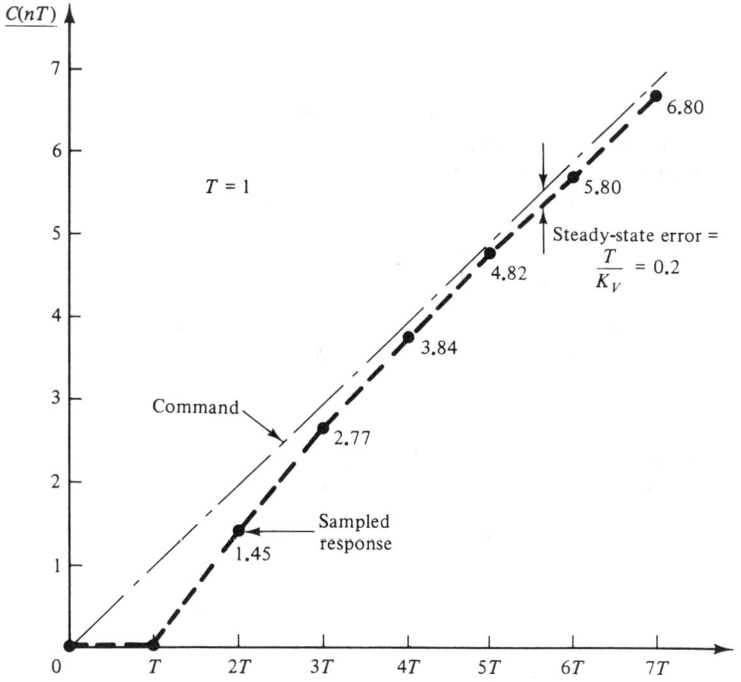

Figure 2.33 Ramp response.

PROBLEMS

2.25 A single-input, single-output system is described by $\dddot{y} + \ddot{y} + 6\dot{y} + (K - 3)y = u(t)$. What is the range of values of K for stability?

2.26 Find the gain K and the frequency ω at which the system of Fig. 2.34 becomes unstable. Consider only positive gains.

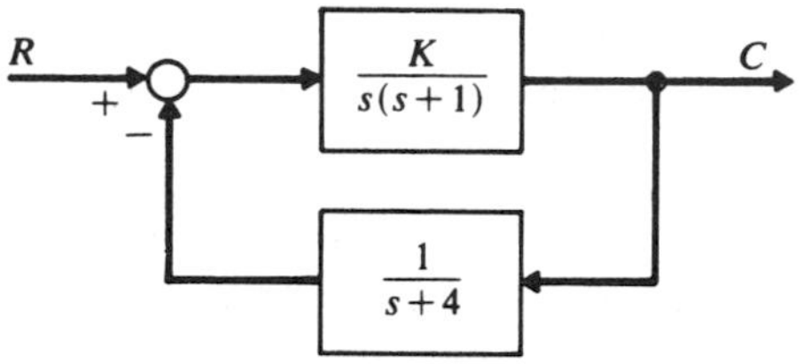

Figure 2.34

2.27 Use Routh's criterion to accurately compute the gain and frequency at the imaginary axis cross-over for the system of Problem 2.9.

2.28 The asymptotic gain portions of three Bode plots are shown in Fig. 2.35. Identify the system types and their error coefficients.

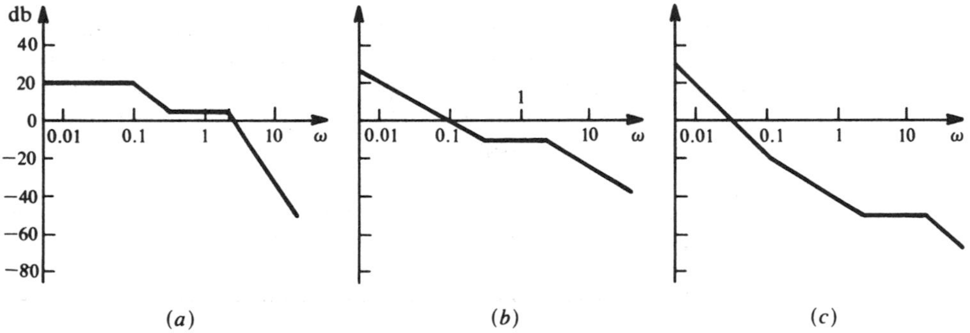

Figure 2.35

2.29 Use Bode's method to show that the system with $KGH = K/[s^2(s + 1)(s^2 + 2s + 225)]$ is unstable for all positive K.

2.30 Sketch the polar plot and use Nyquist's criterion to investigate the stability of a system with $KGH = K(s + 10)/s^2$.

2.31 Sketch the polar plot and use Nyquist's criterion to investigate the stability of a system with $KGH = K(s + 10)(s + 30)/s^3$.

2.32 Use the polar plot to determine the gain-phase margins for the system described by

$$KGH = \frac{5000(s + 2)}{s^2(s + 10)(s + 30)}$$

2.33 A feedback control system has an open-loop transfer function

$$KGH = \frac{65{,}000K}{s(s+25)(s^2+100s+2600)}$$

Find the value of K such that the exponential envelope of the dominant terms decays to 0.15% of its maximum value in one second. Also find the frequency of this damped oscillation.

2.34 Why should Bode plots not be used to infer stability margins for the system in Example 2.2, page 32?

2.35 Give three different algorithms for realizing a digital compensator

$$G_c(z) = (z - 0.5)/[z(z+0.5)]$$

whose input and output are E and Y, respectively.

2.36 Determine an expression for the output sequence $C(nT)$, valid for any nonnegative n, if $C(z) = 10z/[(z-1)(z-0.5)(z+0.5)]$.

2.37 For $C(z) = (z-a)(z-b)/[(z-\alpha)(z-\beta)]$, with a, b, α, and β all distinct:

a. Find $C(0)$, $C(T)$, and $C(2T)$ using long division.

b. Find an expression for $C(nT)$ valid for all $n > 0$.

c. Apply the final value theorem (assume $|\alpha|$ and $|\beta|$ are less than 1), and from its results verify the limiting value of your part (b) answer as $n \longrightarrow \infty$.

2.38 The system of Fig. 2.36 is open-loop unstable.

a. Determine the closed-loop transfer function $C(z)/R(z)$, assuming that $G_c(z) = 1$.

b. Sketch the root locus for the uncompensated system. From this sketch, show that the closed-loop system will be stable for some narrow range of K values, but that the resulting system's transient response will not be very desirable. Compensation will probably be required.

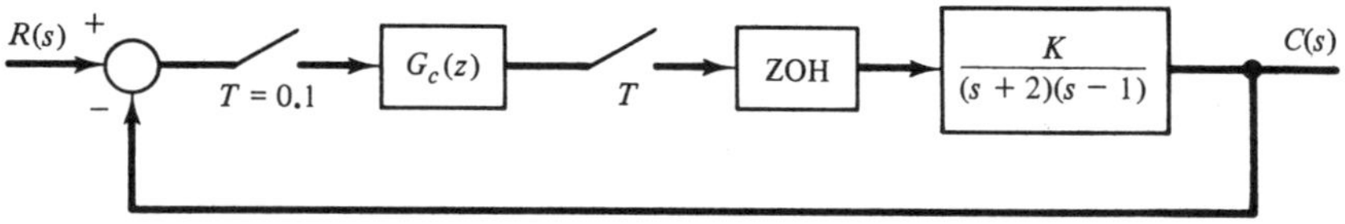

Figure 2.36

2.39 Determine $E(z)$, $C(s)$, and $C(z)$ for the system of Fig. 2.37.

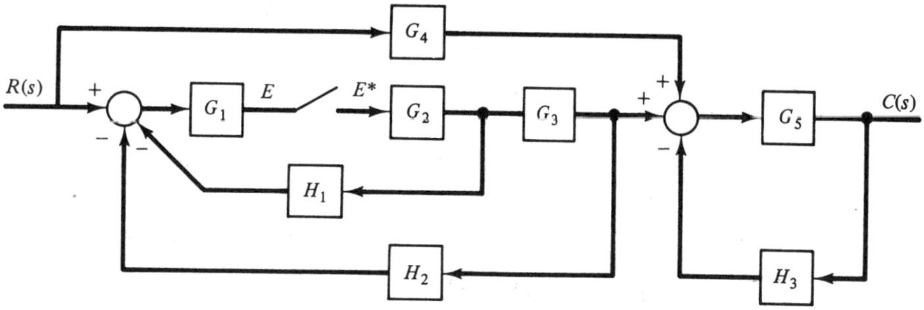

Figure 2.37

2.40 Consider the widely used PID controller (proportional, integral, and derivative control action) shown in Fig. 2.38. Use the stable forward difference approximation for s and

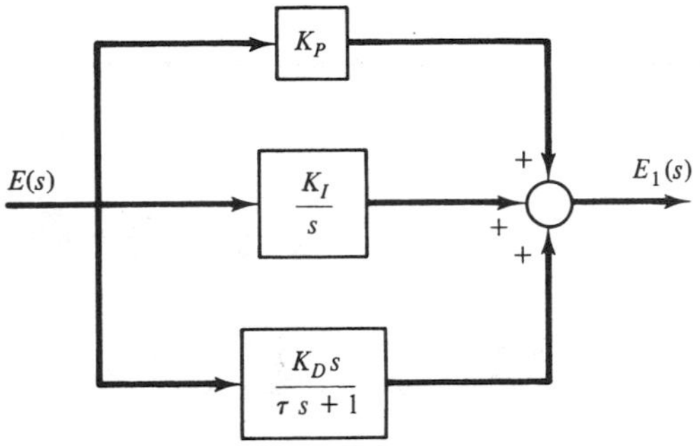

Figure 2.38

determine an equivalent digital controller transfer function. Note that a pure derivative does not have a physically realizable transfer function, so a small time constant term τ is included.

2.41 Using the sampler and zero order hold models suggested for A/D and D/A converters, show that the cascade connection of a D/A followed by an A/D has a Z-transfer function of 1, meaning no alteration of a digital signal sequence that passes through it.

2.42 A simple one-dimensional model of a digital tracking loop is shown in Fig. 2.39. The purpose of the control loop is to keep the angular rate of the antenna ω_a approximately equal to the angular rate ω_0 that the line of sight to the tracked object is making. The integrated angular rate difference is a pointing angle error E. This error angle is sampled, because of a time-shared control computer, and then used to command an antenna rate proportional to the error. The dynamics of the antenna drive system are so rapid that they are neglected, meaning that the actual ω_a is equal to its commanded value. Show that this loop is stable for $0 < KT < 2$.

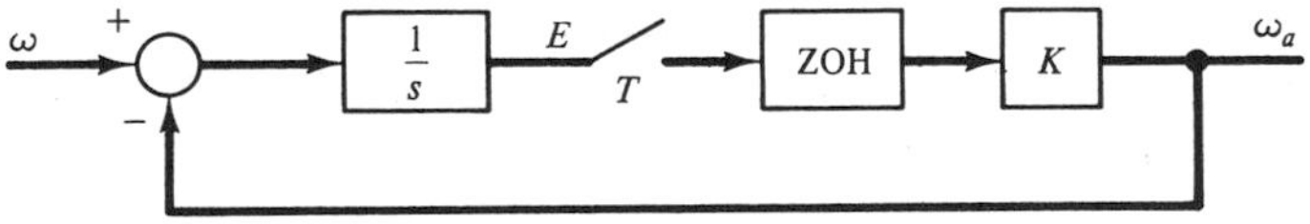

Figure 2.39

Chapter

3

FUNDAMENTALS OF MATRIX ALGEBRA

3.1 INTRODUCTION

This chapter introduces matrices and elementary matrix algebra. Matrices provide a convenient method of dealing with systems of many variables. Additional applications of matrix theory are presented in later chapters.

3.2 NOTATION

Matrices are rectangular arrays of elements. The elements of a matrix are referred to as *scalars*, and will be denoted by lowercase letters, a, b, α, β, etc. In order to define algebraic operations with matrices, it is necessary to restrict these scalar elements to be members of a field. A field $\mathfrak{F}$ is any set of two or more elements for which the operations of addition, multiplication, and division are defined, and for which the following axioms hold:

1. If $a \in \mathfrak{F}$ and $b \in \mathfrak{F}$, then $(a + b) = (b + a) \in \mathfrak{F}$.
2. $(ab) = (ba) \in \mathfrak{F}$.
3. There exists a unique null element $0 \in \mathfrak{F}$ such that $a + 0 = a$ and $0(a) = 0$.
4. If $b \neq 0$, then $(a/b) \in \mathfrak{F}$.
5. There exists a unique identity element $1 \in \mathfrak{F}$ such that $1(a) = (a)1 = (a/1) = a$.
6. For every $a \in \mathfrak{F}$ there is a unique negative element $-a \in \mathfrak{F}$ such that $a + (-a) = 0$.
7. The associative, commutative, and distributive laws of algebra are satisfied.

Note that the set of integers does not form a field because axiom 4 is not necessarily true. Some examples of fields are the set of all rational numbers, the set of all real numbers, and the set of all complex numbers. Unless otherwise stated, all matrices in this text will be assumed to be defined over the complex number field.

Boldface uppercase letters will be used to represent matrices, such as

$$\mathbf{A} = \begin{bmatrix} 42 & 16 \\ 5 & 3 \\ 8 & 1 \end{bmatrix}$$

Horizontal sets of entries such as (42 16) and (5 3) are called rows, while vertical sets of entries such as (42 5 8) are called columns. It will often be convenient to refer to the element in the ith row and jth column of $\mathbf{A}$ as a_{ij}. Rather than explicitly displaying all elements of $\mathbf{A}$, the shorthand notation $\mathbf{A} = [a_{ij}]$ will sometimes be used. If $\mathbf{A}$ has m rows and n columns, it is said to be an $m \times n$ (or m by n) matrix. In that case, the indices i and j in the shorthand notation indicate collectively the range of values $i = 1, 2, \ldots, m$ and $j = 1, 2, \ldots, n$. In particular, when $m = n = 1$, the matrix has a single element and is just a scalar. The subscripts are then unnecessary. If $n = 1$, the matrix has a single column and is called a *column matrix*. The column index j is then superfluous and is sometimes omitted. Similarly, when $m = 1$, the matrix is called a *row matrix*. Whenever $m = n$, the matrix is called a *square matrix*. In general, m and n can take on any finite integer values.

3.3 ALGEBRAIC OPERATIONS WITH MATRICES

Matrix Equality

Matrices $\mathbf{A}$ and $\mathbf{B}$ are equal, written $\mathbf{A} = \mathbf{B}$, if and only if their corresponding elements are equal. That is, $a_{ij} = b_{ij}$ for $1 \le i \le m$ and $1 \le j \le n$. Of course, this means that equality can exist only between matrices of the same size, $m \times n$ in this case.

Matrix Addition and Subtraction

Matrix addition and subtraction are performed on an element-by-element basis. That is, if $\mathbf{A} = [a_{ij}]$ and $\mathbf{B} = [b_{ij}]$ are both $m \times n$ matrices, then $\mathbf{A} + \mathbf{B} = \mathbf{C}$ and $\mathbf{A} - \mathbf{B} = \mathbf{D}$ indicate that the matrices $\mathbf{C} = [c_{ij}]$ and $\mathbf{D} = [d_{ij}]$ are also $m \times n$ matrices whose elements are given by $c_{ij} = a_{ij} + b_{ij}$ and $d_{ij} = a_{ij} - b_{ij}$ for $i = 1, 2, \ldots, m$ and $j = 1, 2, \ldots, n$.

Matrix Multiplication

Two types of multiplication can be defined. Multiplication of a matrix $\mathbf{A} = [a_{ij}]$ by an arbitrary scalar $\alpha \in \mathfrak{F}$ amounts to multiplying every element in $\mathbf{A}$ by α. That is, $\alpha\mathbf{A} = \mathbf{A}\alpha = [\alpha a_{ij}]$.

Multiplication of an $m \times n$ matrix $\mathbf{A} = [a_{ij}]$ by a $p \times q$ matrix $\mathbf{B} = [b_{ij}]$ is now considered. In forming the product $\mathbf{AB} = \mathbf{C}$, it is said that **A** *premultiplies* **B** or equivalently, **B** *postmultiplies* **A**. This product is only defined when **A** has the same number of columns as **B** has rows. When this is true, that is, when $n = p$, **A** and **B** are said to be *conformable*. The elements of $\mathbf{C} = [c_{ij}]$ are then computed according to

$$c_{ij} = \sum_{k=1}^{n} a_{ik} b_{kj}$$

Clearly, the product **C** is an $m \times q$ matrix.

Example 3.1

Let $\mathbf{A} = \begin{bmatrix} 2 & 3 \\ 4 & 5 \end{bmatrix}$, $\mathbf{B} = \begin{bmatrix} 1 & 3 & 5 \\ 2 & 4 & 8 \end{bmatrix}$, and $\mathbf{C} = \begin{bmatrix} 4 \\ -5 \end{bmatrix}$. Then

$$\mathbf{AB} = \begin{bmatrix} 2(1)+3(2) & 2(3)+3(4) & 2(5)+3(8) \\ 4(1)+5(2) & 4(3)+5(4) & 4(5)+5(8) \end{bmatrix} = \begin{bmatrix} 8 & 18 & 34 \\ 14 & 32 & 60 \end{bmatrix}$$

$$\mathbf{AC} = \begin{bmatrix} 2(4)+3(-5) \\ 4(4)+5(-5) \end{bmatrix} = \begin{bmatrix} -7 \\ -9 \end{bmatrix}. \qquad 10\mathbf{A} = \begin{bmatrix} 20 & 30 \\ 40 & 50 \end{bmatrix}$$

The products **BA**, **CA**, and **CB** are not defined. ■

Division

Division by a matrix, per se, is not defined. An operation somewhat analogous to division, called *matrix inversion*, is discussed later.

The Null Matrix and the Unit Matrix

As a necessary part of scalar algebra, axioms 3 and 5 for number fields introduce a null element and an identity element. Correspondingly, the null matrix **0** is one which has all of its elements equal to zero. Then, $\mathbf{A} + \mathbf{0} = \mathbf{A}$ and $\mathbf{0A} = \mathbf{0}$. Note, however, that the null matrix is not unique because the numbers of rows and columns it possesses can be any finite positive integers. Whenever necessary, the dimensions of the null matrix will be indicated by two subscripts, $\mathbf{0}_{mn}$. Another difference exists between the scalar zero and the null matrix. In scalar algebra, $ab = 0$ implies that either a or b or both are zero. No similar inference can be drawn from the matrix product $\mathbf{AB} = \mathbf{0}$. For a simple verification, let $\mathbf{A} = \begin{bmatrix} 1 & -1 \\ -1 & 1 \end{bmatrix}$ and $\mathbf{B} = \begin{bmatrix} 3 & 4 \\ 3 & 4 \end{bmatrix}$ and form the product **AB**.

The *identity* or *unit matrix* **I** is a square matrix with all elements zero, except those on the main diagonal ($i = j$ positions) are ones. The unit matrix is not unique because of its dimensions. When necessary, an $n \times n$ unit matrix will be denoted by $\mathbf{I}_n$. The unit matrix has algebraic properties similar to the scalar identity element, namely if **A** is $m \times n$, then $\mathbf{I}_m\mathbf{A} = \mathbf{A}$ and $\mathbf{AI}_n = \mathbf{A}$.

3.4 THE ASSOCIATIVE, COMMUTATIVE, AND DISTRIBUTIVE LAWS OF MATRIX ALGEBRA

Many of the associative, commutative, and distributive laws of scalar algebra carry over to matrix algebra, as summarized below:

$$\mathbf{A} + \mathbf{B} = \mathbf{B} + \mathbf{A}, \qquad \mathbf{A} - \mathbf{B} = \mathbf{A} + (-\mathbf{B}) = -\mathbf{B} + \mathbf{A}$$

$$\mathbf{A} + (\mathbf{B} + \mathbf{C}) = (\mathbf{A} + \mathbf{B}) + \mathbf{C}, \qquad \alpha(\mathbf{A} + \mathbf{B}) = \alpha\mathbf{A} + \alpha\mathbf{B}$$

$$\alpha\mathbf{A} = \mathbf{A}\alpha, \qquad \mathbf{A}(\mathbf{BC}) = (\mathbf{AB})\mathbf{C}$$

$$\mathbf{A}(\mathbf{B} + \mathbf{C}) = \mathbf{AB} + \mathbf{AC}, \qquad (\mathbf{B} + \mathbf{C})\mathbf{A} = \mathbf{BA} + \mathbf{CA}$$

One major difference exists between scalar and matrix algebra. Scalar multiplication is commutative, i.e., $ab = ba$. However, matrix multiplication is not commutative, i.e., $\mathbf{AB} \neq \mathbf{BA}$. In many cases the reversed product is not even defined because the conformability conditions are not satisfied. Even when both $\mathbf{A}$ and $\mathbf{B}$ are square so that $\mathbf{AB}$ and $\mathbf{BA}$ are both defined, they need not be equal. It is for this reason that it is necessary to distinguish between premultiplication and postmultiplication.

Example 3.2

Let $\mathbf{A} = \begin{bmatrix} 2 & 3 \\ 1 & 8 \end{bmatrix}$, $\mathbf{B} = \begin{bmatrix} -1 & 1 \\ 0 & 4 \end{bmatrix}$. Then $\mathbf{AB} = \begin{bmatrix} -2 & 14 \\ -1 & 33 \end{bmatrix}$ and $\mathbf{BA} = \begin{bmatrix} -1 & 5 \\ 4 & 32 \end{bmatrix}$. ■

3.5 MATRIX TRANSPOSE, CONJUGATE, AND THE ASSOCIATE MATRIX

The operation of matrix transposition is the interchanging of each row with the column of the same index number. If $\mathbf{A} = [a_{ij}]$, then the *transpose* of $\mathbf{A}$ is $\mathbf{A}^T = [a_{ji}]$. The matrix $\mathbf{A}$ is said to be *symmetric* if $\mathbf{A} = \mathbf{A}^T$. If $\mathbf{A} = -\mathbf{A}^T$, then $\mathbf{A}$ is *skew-symmetric.* An important property of matrix transposition of products is illustrated by

$$(\mathbf{AB})^T = \mathbf{B}^T\mathbf{A}^T, \quad (\mathbf{ABC})^T = \mathbf{C}^T\mathbf{B}^T\mathbf{A}^T, \quad \ldots$$

The *conjugate* of $\mathbf{A}$, written $\bar{\mathbf{A}}$, is the matrix formed by replacing every element in $\mathbf{A}$ by its complex conjugate. Thus $\bar{\mathbf{A}} = [\bar{a}_{ij}]$. If all elements of $\mathbf{A}$ are real, then $\bar{\mathbf{A}} = \mathbf{A}$. If all elements are purely imaginary, then $\bar{\mathbf{A}} = -\mathbf{A}$.

The *associate matrix* of $\mathbf{A}$ is the conjugate transpose of $\mathbf{A}$. The order of these two operations is immaterial. Matrices satisfying $\mathbf{A} = \bar{\mathbf{A}}^T$ are called *Hermitian matrices. Skew-Hermitian* matrices satisfy $\mathbf{A} = -\bar{\mathbf{A}}^T$. For real matrices, symmetric and Hermitian mean the same thing.

3.6 DETERMINANTS, MINORS, AND COFACTORS

Determinants are defined for square matrices only. The determinant of the $n \times n$ matrix **A**, written $|\mathbf{A}|$, is a scalar-valued function of **A**. The familiar form of the determinants for $n = 1$, 2, and 3 are

$$n = 1 \qquad |\mathbf{A}| = a_{11}$$

$$n = 2 \qquad |\mathbf{A}| = a_{11}a_{22} - a_{12}a_{21}$$

$$n = 3 \qquad |\mathbf{A}| = a_{11}a_{22}a_{33} + a_{12}a_{23}a_{31} + a_{13}a_{21}a_{32} - a_{13}a_{22}a_{31} - a_{12}a_{24}a_{33} - a_{11}a_{32}a_{23}$$

There is a common pattern which can be generalized for any n [91]. Each determinant has $n!$ terms, with each term consisting of n elements of **A**, one from each row and from each column. However, the general pattern is inefficient for evaluating large determinants. Usually, a larger-order determinant is first reduced to an expression involving one or more smaller determinants. The methods of *Laplace expansion* and *pivotal condensation* can be used for this purpose. Also, the basic properties of determinants can be used to simplify the evaluation task. Some of these methods are discussed later.

Minors

An $n \times n$ matrix **A** contains n^2 elements a_{ij}. Each of these has associated with it a unique scalar, called a *minor* M_{ij}. The minor M_{pq} is the determinant of the $n - 1 \times n - 1$ matrix formed from **A** by crossing out the pth row and qth column.

Cofactors

Each element a_{pq} of **A** has a *cofactor* C_{pq}, which differs from M_{pq} at most by a sign change. Cofactors are sometimes called signed minors for this reason, and are given by $C_{pq} = (-1)^{p+q}M_{pq}$.

Determinants by Laplace Expansion

If **A** is an $n \times n$ matrix, any arbitrary row k can be selected and $|\mathbf{A}|$ is then given by $|\mathbf{A}| = \sum_{j=1}^{n} a_{kj}C_{kj}$. Similarly, Laplace expansion can be carried out with respect to any arbitrary column l, to obtain $|\mathbf{A}| = \sum_{i=1}^{n} a_{il}C_{il}$. Laplace expansion reduces the evaluation of an $n \times n$ determinant down to the evaluation of a string of $(n - 1) \times (n - 1)$ determinants, namely the cofactors.

Example 3.3

Given $\mathbf{A} = \begin{bmatrix} 2 & 4 & 1 \\ 3 & 0 & 2 \\ 2 & 0 & 3 \end{bmatrix}$. Three of its minors are

$$M_{12} = \begin{vmatrix} 3 & 2 \\ 2 & 3 \end{vmatrix} = 5, \quad M_{22} = \begin{vmatrix} 2 & 1 \\ 2 & 3 \end{vmatrix} = 4, \quad \text{and} \quad M_{32} = \begin{vmatrix} 2 & 1 \\ 3 & 2 \end{vmatrix} = 1$$

The associated cofactors are

$$C_{12} = (-1)^3 5 = -5, \quad C_{22} = (-1)^4 4 = 4, \quad C_{32} = (-1)^5 1 = -1$$

Using Laplace expansion with respect to column 2 gives $|\mathbf{A}| = 4C_{12} = -20$. ■

Pivotal Condensation

Pivotal condensation [28], also called the method of Chio [53, 94], reduces an $n \times n$ determinant to a single $n - 1 \times n - 1$ determinant, and thus avoids the long string of determinants encountered with Laplace expansion. Let a_{pq} be any nonzero element of $\mathbf{A}$. This is called the *pivot element*. An $n - 1 \times n - 1$ determinant is formed, with each of its elements obtained from a 2×2 determinant. Each 2×2 determinant contains a_{pq}, one other element from row p, one other element from column q, and the 4th element is from the 4th corner of the rectangle defined by the previous three elements. Let the $n - 1 \times n - 1$ determinant be called $|\Delta|$. Then $|\mathbf{A}| = [1/(a_{pq})^{n-2}]\,|\Delta|$. The method is best illustrated by an example.

Example 3.4

Let $\mathbf{A}$ be a 4×4 matrix and assume that $a_{23} \neq 0$. Then

$$|\mathbf{A}| = \frac{1}{(a_{23})^2} \begin{vmatrix} \begin{vmatrix} a_{11} & a_{13} \\ a_{21} & a_{23} \end{vmatrix} & \begin{vmatrix} a_{12} & a_{13} \\ a_{22} & a_{23} \end{vmatrix} & \begin{vmatrix} a_{13} & a_{14} \\ a_{23} & a_{24} \end{vmatrix} \\ \begin{vmatrix} a_{21} & a_{23} \\ a_{31} & a_{33} \end{vmatrix} & \begin{vmatrix} a_{22} & a_{23} \\ a_{32} & a_{33} \end{vmatrix} & \begin{vmatrix} a_{23} & a_{24} \\ a_{33} & a_{34} \end{vmatrix} \\ \begin{vmatrix} a_{21} & a_{23} \\ a_{41} & a_{43} \end{vmatrix} & \begin{vmatrix} a_{22} & a_{23} \\ a_{42} & a_{43} \end{vmatrix} & \begin{vmatrix} a_{23} & a_{24} \\ a_{43} & a_{44} \end{vmatrix} \end{vmatrix}$$

Note that the pivot element a_{23} is in the same location relative to the other elements within each 2×2 determinant as it is in the original $\mathbf{A}$ matrix. ■

Useful Properties of Determinants

1. If $\mathbf{A}$ and $\mathbf{B}$ are both $n \times n$, then $|\mathbf{AB}| = |\mathbf{A}|\,|\mathbf{B}|$.
2. $|\mathbf{A}| = |\mathbf{A}^T|$.
3. If all the elements in any row or in any column are zero, then $|\mathbf{A}| = 0$.
4. If any two rows of $\mathbf{A}$ are proportional, $|\mathbf{A}| = 0$. If a row is a linear combination of any number of other rows, then $|\mathbf{A}| = 0$. Similar statements hold for columns.

5. Interchanging any two rows (or any two columns) of a matrix changes the sign of its determinant.
6. Multiplying all elements of any one row (or column) of a matrix $\mathbf{A}$ by a scalar α yields a matrix whose determinant is $\alpha|\mathbf{A}|$.
7. Any multiple of a row (column) can be added to any other row (column) without changing the value of the determinant.

3.7 RANK AND TRACE OF A MATRIX

The *rank* of $\mathbf{A}$, designated as r_A or rank($\mathbf{A}$), is defined as the size of the largest nonzero determinant that can be formed from $\mathbf{A}$. The maximum possible rank of an $m \times n$ matrix is obviously the smaller of m and n. If $\mathbf{A}$ is $n \times n$ and has its maximal rank n, then the matrix is said to be *nonsingular*.

The rank of the product of two or more matrices is never more than the smallest rank of the matrices forming the product. For example, if r_A and r_B are the ranks of $\mathbf{A}$ and $\mathbf{B}$, then $\mathbf{C} = \mathbf{AB}$ has rank r_C satisfying $0 \leq r_C \leq \min\{r_A, r_B\}$.

Let $\mathbf{A}$ be an $n \times n$ matrix. Then the *trace* of $\mathbf{A}$, denoted Tr ($\mathbf{A}$), is the sum of the diagonal elements of $\mathbf{A}$, $\text{Tr}(\mathbf{A}) = \sum_{i=1}^{n} a_{ii}$. If $\mathbf{A}$ and $\mathbf{B}$ are conformable square matrices, then $\text{Tr}(\mathbf{A} + \mathbf{B}) = \text{Tr}(\mathbf{A}) + \text{Tr}(\mathbf{B})$ and $\text{Tr}(\mathbf{AB}) = \text{Tr}(\mathbf{BA})$. From the definition of the trace it is obvious that $\text{Tr}(\mathbf{A}^T) = \text{Tr}(\mathbf{A})$. From this it follows that $\text{Tr}(\mathbf{AB}) = \text{Tr}(\mathbf{B}^T\mathbf{A}^T)$.

Example 3.5

Let $\mathbf{A} = \begin{bmatrix} 1 & 5 & 8 \\ 3 & -1 & 2 \\ 4 & -4 & 6 \end{bmatrix}$. $\mathbf{B} = \begin{bmatrix} 1 & -1 & 8 \\ 3 & -3 & 2 \\ 4 & -4 & 6 \end{bmatrix}$. Then $|\mathbf{A}| = -112$ so that $r_A = 3$ and $\mathbf{A}$ is nonsingular. Also $\text{Tr}(\mathbf{A}) = 6$. The matrix $\mathbf{B}$ has $|\mathbf{B}| = 0$, so $r_B < 3$. Crossing out column 2 and row 3 of $\mathbf{B}$ gives a 2×2 determinant with a value -22, so $r_B = 2$. The trace of $\mathbf{B}$ is $\text{Tr}(\mathbf{B}) = 4$. Forming $\mathbf{AB}$ and $\mathbf{BA}$ shows that $r_{AB} = 2$ and $r_{BA} = 2$. Also $\text{Tr}(\mathbf{A} + \mathbf{B}) = 10 = \text{Tr}(\mathbf{A}) + \text{Tr}(\mathbf{B})$ and $\text{Tr}(\mathbf{AB}) = 100 = \text{Tr}(\mathbf{BA})$. Note that $\text{Tr}(\mathbf{AB}) \neq \text{Tr}(\mathbf{A})\,\text{Tr}(\mathbf{B})$. ■

3.8 MATRIX INVERSION

The inverse of the scalar element a is $1/a$, or a^{-1}. It satisfies $a(a^{-1}) = (a^{-1})a = 1$. If an arbitrary matrix $\mathbf{A}$ is to havc an analogous inverse $\mathbf{B} = \mathbf{A}^{-1}$, then the following must hold:

$$\mathbf{BA} = \mathbf{AB} = \mathbf{I}$$

Because of conformability requirements, this can never be true if $\mathbf{A}$ is not square. In addition, $\mathbf{A}$ must have a nonzero determinant, i.e., $\mathbf{A}$ must be nonsingular. When this is true, $\mathbf{A}$ has a unique inverse given by

$$\mathbf{A}^{-1} = \frac{\mathbf{C}^T}{|\mathbf{A}|}$$

where $\mathbf{C}$ is the matrix formed by the cofactors C_{ij}. The matrix $\mathbf{C}^T$ is called the *adjoint matrix*, Adj ($\mathbf{A}$). Thus the inverse of a nonsingular matrix is

$$\mathbf{A}^{-1} = \text{Adj}\,(\mathbf{A})/|\mathbf{A}|$$

Example 3.6

Let $\mathbf{A} = \begin{bmatrix} 1 & 1 \\ 2 & 2 \end{bmatrix}$, $\mathbf{B} = \begin{bmatrix} 1 & 2 \\ 3 & 4 \end{bmatrix}$, $\mathbf{D} = \begin{bmatrix} 4 & 2 & 1 \\ 2 & 6 & 3 \\ 1 & 3 & 5 \end{bmatrix}$. Then $\mathbf{A}^{-1}$ does not exist since $|\mathbf{A}| = 0$. Since $|\mathbf{B}| = -2$, $\mathbf{B}^{-1}$ exists and is given by $\mathbf{B}^{-1} = \begin{bmatrix} -2 & 1 \\ 3/2 & -1/2 \end{bmatrix}$. Similarly, $\mathbf{D}^{-1} = \frac{1}{70}\begin{bmatrix} 21 & -7 & 0 \\ -7 & 19 & -10 \\ 0 & -10 & 20 \end{bmatrix}$. ■

Inversion of large matrices by direct application of the above definition is tedious. Numerical techniques such as *Gaussian elimination* are often used. Matrix partitioning can also be employed to obtain a matrix inverse in terms of several smaller inverses. Another method, based on the Cayley-Hamilton theorem, is given in Chapter 8.

The Inverse of a Product

Let $\mathbf{A}, \mathbf{B}, \mathbf{C}, \ldots, \mathbf{W}$ be any number of conformable nonsingular matrices. Then

$$(\mathbf{ABC}\cdots\mathbf{W})^{-1} = \mathbf{W}^{-1}\cdots\mathbf{C}^{-1}\mathbf{B}^{-1}\mathbf{A}^{-1}$$

Some Matrices with Special Relationships to Their Inverses

If $\mathbf{A}^{-1} = \mathbf{A}$, $\mathbf{A}$ is said to be *involutory*.
If $\mathbf{A}^{-1} = \mathbf{A}^T$, $\mathbf{A}$ is said to be *orthogonal*.
If $\mathbf{A}^{-1} = \bar{\mathbf{A}}^T$, $\mathbf{A}$ is said to be *unitary*.

3.9 PARTITIONED MATRICES

Any matrix $\mathbf{A}$ can be subdivided or partitioned into a number of smaller submatrices. If conformable matrices are partitioned in a compatible fashion, the submatrices can be treated just as if they were scalar elements when performing the operations of addition and multiplication. Of course, the order of the products is not arbitrary, as it would be with scalars.

Example 3.7

$\mathbf{AB} = \mathbf{C}$ can be partitioned in various ways. A few of them are given below:

$$(a)\quad \begin{bmatrix} \mathbf{A}_1 \\ \hline \mathbf{A}_2 \end{bmatrix} [\mathbf{B}_1 \,\vdots\, \mathbf{B}_2] = \left[\begin{array}{c|c} \mathbf{A}_1\mathbf{B}_1 & \mathbf{A}_1\mathbf{B}_2 \\ \hline \mathbf{A}_2\mathbf{B}_1 & \mathbf{A}_2\mathbf{B}_2 \end{array}\right] = \left[\begin{array}{c|c} \mathbf{C}_1 & \mathbf{C}_2 \\ \hline \mathbf{C}_3 & \mathbf{C}_4 \end{array}\right]$$

$$(b)\quad \left[\begin{array}{c|c} \mathbf{A}_1 & \mathbf{A}_2 \\ \hline \mathbf{A}_3 & \mathbf{A}_4 \end{array}\right] \begin{bmatrix} \mathbf{B}_1 \\ \hline \mathbf{B}_2 \end{bmatrix} = \begin{bmatrix} \mathbf{A}_1\mathbf{B}_1 + \mathbf{A}_2\mathbf{B}_2 \\ \hline \mathbf{A}_3\mathbf{B}_1 + \mathbf{A}_4\mathbf{B}_2 \end{bmatrix} = \begin{bmatrix} \mathbf{C}_1 \\ \hline \mathbf{C}_2 \end{bmatrix}$$

$$(c)\quad \left[\begin{array}{c|c} \mathbf{A}_1 & \mathbf{A}_2 \\ \hline \mathbf{A}_3 & \mathbf{A}_4 \end{array}\right] \left[\begin{array}{c|c} \mathbf{B}_1 & \mathbf{B}_2 \\ \hline \mathbf{B}_3 & \mathbf{B}_4 \end{array}\right] = \left[\begin{array}{c|c} \mathbf{A}_1\mathbf{B}_1 + \mathbf{A}_2\mathbf{B}_3 & \mathbf{A}_1\mathbf{B}_2 + \mathbf{A}_2\mathbf{B}_4 \\ \hline \mathbf{A}_3\mathbf{B}_1 + \mathbf{A}_4\mathbf{B}_3 & \mathbf{A}_3\mathbf{B}_2 + \mathbf{A}_4\mathbf{B}_4 \end{array}\right] = \left[\begin{array}{c|c} \mathbf{C}_1 & \mathbf{C}_2 \\ \hline \mathbf{C}_3 & \mathbf{C}_4 \end{array}\right] \quad ■$$

Partitioned matrices can be used to find an expression for the inverse of a nonsingular matrix $\mathbf{A}$. If $\mathbf{A}$ is partitioned into four submatrices, then $\mathbf{A}^{-1} = \mathbf{B}$ will also have four submatrices:

$$\mathbf{AB} = \mathbf{I} \quad \text{or} \quad \left[\begin{array}{c|c} \mathbf{A}_1 & \mathbf{A}_2 \\ \hline \mathbf{A}_3 & \mathbf{A}_4 \end{array}\right] \left[\begin{array}{c|c} \mathbf{B}_1 & \mathbf{B}_2 \\ \hline \mathbf{B}_3 & \mathbf{B}_4 \end{array}\right] = \left[\begin{array}{c|c} \mathbf{I} & \mathbf{0} \\ \hline \mathbf{0} & \mathbf{I} \end{array}\right]$$

The partitioned form implies four separate matrix equations, two of which are $\mathbf{A}_1\mathbf{B}_1 + \mathbf{A}_2\mathbf{B}_3 = \mathbf{I}$ and $\mathbf{A}_3\mathbf{B}_1 + \mathbf{A}_4\mathbf{B}_3 = \mathbf{0}$. These can be solved simultaneously for $\mathbf{B}_1$ and $\mathbf{B}_3$. The remaining two equations give $\mathbf{B}_2$ and $\mathbf{B}_4$ and lead to the result

$$\mathbf{A}^{-1} = \left[\begin{array}{c|c} (\mathbf{A}_1 - \mathbf{A}_2\mathbf{A}_4^{-1}\mathbf{A}_3)^{-1} & -\mathbf{A}_1^{-1}\mathbf{A}_2(\mathbf{A}_4 - \mathbf{A}_3\mathbf{A}_1^{-1}\mathbf{A}_2)^{-1} \\ \hline -\mathbf{A}_4^{-1}\mathbf{A}_3(\mathbf{A}_1 - \mathbf{A}_2\mathbf{A}_4^{-1}\mathbf{A}_3)^{-1} & (\mathbf{A}_4 - \mathbf{A}_3\mathbf{A}_1^{-1}\mathbf{A}_2)^{-1} \end{array}\right]$$

Several matrix identities can be derived by starting with the reversed order, $\mathbf{BA} = \mathbf{I}$, repeating the above process, and then using the uniqueness of $\mathbf{B} = \mathbf{A}^{-1}$ to equate the various terms. One such identity, called the *matrix inversion lemma*, is particularly useful. A general form is

$$(\mathbf{A}_1 - \mathbf{A}_2\mathbf{A}_4^{-1}\mathbf{A}_3)^{-1} = \mathbf{A}_1^{-1} + \mathbf{A}_1^{-1}\mathbf{A}_2(\mathbf{A}_4 - \mathbf{A}_3\mathbf{A}_1^{-1}\mathbf{A}_2)^{-1}\mathbf{A}_3\mathbf{A}_1^{-1}$$

By letting $\mathbf{A}_1 = \mathbf{P}^{-1}$, $\mathbf{A}_2 = \mathbf{H}^T$, $\mathbf{A}_3 = \mathbf{H}$, and $\mathbf{A}_4 = -\mathbf{Q}^{-1}$, an extremely useful special form of the inversion lemma that will be encountered in recursive weighted least squares is obtained:

$$[\mathbf{P}^{-1} + \mathbf{H}^T\mathbf{Q}\mathbf{H}]^{-1} = \mathbf{P} - \mathbf{P}\mathbf{H}^T[\mathbf{H}\mathbf{P}\mathbf{H}^T + \mathbf{Q}^{-1}]^{-1}\mathbf{H}\mathbf{P}$$

Diagonal, Block Diagonal, and Triangular Matrices

If the only nonzero elements of a square matrix $\mathbf{A}$ are on the main diagonal, then $\mathbf{A}$ is called a *diagonal matrix*. This is often written as $\mathbf{A} = \operatorname{diag}[a_{11} \quad a_{22} \quad \cdots \quad a_{nn}]$. For this case, $|\mathbf{A}| = a_{11}a_{22}\cdots a_{nn}$ and $\mathbf{A}^{-1} = \operatorname{diag}[1/a_{11} \quad 1/a_{22} \quad \cdots \quad 1/a_{nn}]$. The unit matrix is a special case with all $a_{ii} = 1$.

A *block diagonal* or *quasidiagonal* matrix is a square matrix that can be partitioned so that the only nonzero elements are contained in square submatrices along the main diagonal,

$$\mathbf{A} = \begin{bmatrix} \mathbf{A}_1 & & & \\ & \mathbf{A}_2 & & \\ & & \ddots & \\ & & & \mathbf{A}_k \end{bmatrix} = \text{diag}\,[\mathbf{A}_1 \quad \mathbf{A}_2 \quad \cdots \quad \mathbf{A}_k]$$

For this case $|\mathbf{A}| = |\mathbf{A}_1||\mathbf{A}_2| \cdots |\mathbf{A}_k|$ and $\mathbf{A}^{-1} = \text{diag}\,[\mathbf{A}_1^{-1} \quad \mathbf{A}_2^{-1} \quad \cdots \quad \mathbf{A}_k^{-1}]$, provided that $\mathbf{A}^{-1}$ exists.

A square matrix which has all of its elements below (above) the main diagonal equal to zero is called an *upper triangular* (*lower triangular*) matrix. The determinant of any triangular matrix is the product of its diagonal elements.

3.10 ELEMENTARY OPERATIONS AND ELEMENTARY MATRICES

Three basic operations on a matrix, called *elementary operations*, are:

1. The interchange of two rows (or of two columns).
2. The multiplication of every element in a given row (or column) by a constant α.
3. The multiplication of the elements of a given row (or column) by a constant α, and adding the result to another row (column). The original row (column) is unaltered.

When these row operations are applied to the unit matrix, the resultant matrices are called elementary matrices, and are denoted as follows:

$\mathbf{E}_{p,q}$: pth and qth rows of $\mathbf{I}$ interchanged
$\mathbf{E}_p(\alpha)$: pth row of $\mathbf{I}$ multiplied by α
$\mathbf{E}_{p,q}(\alpha)$: pth row of $\mathbf{I}$ multiplied by α and added to qth row

The elementary matrices are all nonsingular. In fact,

$$|\mathbf{E}_{p,q}| = -1, \qquad |\mathbf{E}_p(\alpha)| = \alpha, \qquad |\mathbf{E}_{p,q}(\alpha)| = 1$$

Premultiplication (postmultiplication) of a matrix by one of the elementary matrices performs the corresponding elementary row (column) operation on that matrix.

By performing a sequence of elementary row and column operations, any matrix of rank r can be reduced to one of the following normal forms:

$$\mathbf{I}_r, \qquad [\mathbf{I}_r \mid \mathbf{0}], \qquad \begin{bmatrix} \mathbf{I}_r \\ \hline \mathbf{0} \end{bmatrix}, \qquad \left[\begin{array}{c|c} \mathbf{I}_r & \mathbf{0} \\ \hline \mathbf{0} & \mathbf{0} \end{array}\right]$$

These are special cases of, or analogous to, matrices in the "row-reduced echelon" form. They are also sometimes called "Hermite normal forms." These will be defined more formally in Chapter 5, where their value will be more fully appreciated. Thus elementary operations provide a practical means of computing the rank of a matrix, but they have many other uses as well.

3.11 DIFFERENTIATION AND INTEGRATION OF MATRICES

When a matrix $\mathbf{A}$ has elements which are functions of a scalar variable (such as time), differentiation and integration of the matrix are defined on an element-by-element basis. If $\mathbf{A}(t) = [a_{ij}(t)]$, then $d\mathbf{A}/dt = \dot{\mathbf{A}} = [\dot{a}_{ij}(t)]$ and $\int \mathbf{A}(\tau)\, d\tau = [\int a_{ij}(\tau)\, d\tau]$. Because of the integration rule, Laplace transforms and inverse Laplace transforms of matrices are also found on element-by-element basis. This is also true for Z-transforms of matrices.

Useful rules for differentiating a determinant are

$$\frac{\partial |\mathbf{A}|}{\partial a_{ij}} = C_{ij} \qquad \text{(follows immediately from the Laplace expansion)}$$

If the $n \times n$ matrix $\mathbf{A}$ is a function of t, then $|\mathbf{A}|$ is also a function of t. Then $d|\mathbf{A}|/dt$ is just the sum of n separate determinants. The first determinant has row (or column) one differentiated, the second has row (or column) two differentiated, and so on through all n rows (columns).

References 6 and 91 are recommended for additional reading on matrix theory. Reference 4 combines matrix methods with transform techniques for linear systems, and Reference 53 provides an introduction to some computational aspects.

ILLUSTRATIVE PROBLEMS

Introductory Manipulations

3.1 Let

$$\mathbf{A} = \begin{bmatrix} 1 & 4 \\ 2 & 5 \end{bmatrix}, \quad \mathbf{B} = \begin{bmatrix} 3 & 1 \\ 1 & 3 \end{bmatrix}, \quad \text{and} \quad \mathbf{C} = \begin{bmatrix} 42 & 16 \\ 5 & 3 \\ 8 & 1 \end{bmatrix}$$

Compute $\mathbf{A}+\mathbf{B}$, $\mathbf{A}-\mathbf{B}$, $\mathbf{AB}$, $\mathbf{BA}$, $\mathbf{CA}$, $\mathbf{CB}$, $\mathbf{AC}$, and $\mathbf{AC}^T$.

$$\mathbf{A}+\mathbf{B}=\begin{bmatrix}1+3 & 4+1\\ 2+1 & 5+3\end{bmatrix}=\begin{bmatrix}4 & 5\\ 3 & 8\end{bmatrix},\qquad \mathbf{A}-\mathbf{B}=\begin{bmatrix}-2 & 3\\ 1 & 2\end{bmatrix}$$

$$\mathbf{AB}=\begin{bmatrix}1(3)+4(1) & 1(1)+4(3)\\ 2(3)+5(1) & 2(1)+5(3)\end{bmatrix}=\begin{bmatrix}7 & 13\\ 11 & 17\end{bmatrix},\qquad \mathbf{BA}=\begin{bmatrix}5 & 17\\ 7 & 19\end{bmatrix}$$

$$\mathbf{CA}=\begin{bmatrix}42(1)+16(2) & 42(4)+16(5)\\ 5(1)+\ 3(2) & 5(4)+\ 3(5)\\ 8(1)+\ 1(2) & 8(4)+\ 1(5)\end{bmatrix}=\begin{bmatrix}74 & 248\\ 11 & 35\\ 10 & 37\end{bmatrix},\qquad \mathbf{CB}=\begin{bmatrix}142 & 90\\ 18 & 14\\ 25 & 11\end{bmatrix}$$

$\mathbf{AC}$ is not defined because of the conformability rule: $(2\times 2)(3\times 2)$.

$$\mathbf{AC}^T=\begin{bmatrix}1 & 4\\ 2 & 5\end{bmatrix}\begin{bmatrix}42 & 5 & 8\\ 16 & 3 & 1\end{bmatrix}=\begin{bmatrix}1(42)+4(16) & 1(5)+4(3) & 1(8)+4(1)\\ 2(42)+5(16) & 2(5)+5(3) & 2(8)+5(1)\end{bmatrix}$$

$$=\begin{bmatrix}106 & 17 & 12\\ 164 & 25 & 21\end{bmatrix}$$

Multiple Variable Systems and Transfer Matrices

3.2 Consider the multiple-input, multiple-output feedback system shown in Fig. 3.1, where $\mathbf{G}_1$, $\mathbf{G}_2$, $\mathbf{H}_1$, and $\mathbf{H}_2$ are transfer function matrices and $\mathbf{R}$, $\mathbf{E}_1$, $\mathbf{E}_2$, $\mathbf{V}$, $\mathbf{W}$, $\mathbf{C}$, $\mathbf{D}$, $\mathbf{F}_1$, and

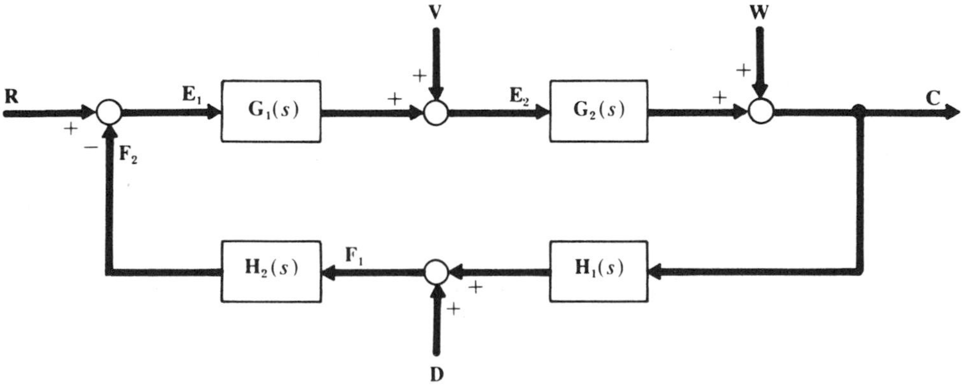

Figure 3.1

$\mathbf{F}_2$ are column matrices. If $\mathbf{R}$ has r components, $\mathbf{V}$ has m components, $\mathbf{C}$ has n components, and $\mathbf{D}$ has p components, determine the dimensions of all other matrices in the diagram.

In order for $\mathbf{R}-\mathbf{F}_2=\mathbf{E}_1$ to make sense, $\mathbf{F}_2$ and $\mathbf{E}_1$ must be $r\times 1$ matrices like $\mathbf{R}$. If $\mathbf{G}_1\mathbf{E}_1$ is to be conformable, $\mathbf{G}_1$ must have r columns. Since this product adds to $\mathbf{V}$, it must be an $m\times 1$ matrix. So $\mathbf{G}_1$ must be $m\times r$. Since $\mathbf{G}_1\mathbf{E}_1+\mathbf{V}=\mathbf{E}_2$, $\mathbf{E}_2$ must be an $m\times 1$ matrix. Conformability requires that $\mathbf{G}_2$ have m columns, and since $\mathbf{G}_2\mathbf{E}_2+\mathbf{W}=\mathbf{C}$ is an $n\times 1$ matrix, the transfer matrix $\mathbf{G}_2$ must be $n\times m$. Similar reasoning requires that $\mathbf{H}_1$ be a $p\times n$ matrix, $\mathbf{F}_1$ be a $p\times 1$ matrix, and $\mathbf{H}_2$ be an $r\times p$ matrix.

3.3 Referring to the system of Problem 3.2, derive the overall transfer matrix relating the input **R** to the output **C**.

Ignoring all inputs except **R**, this system is described by five matrix equations:

$$\mathbf{E}_1 = \mathbf{R} - \mathbf{F}_2, \quad \mathbf{E}_2 = \mathbf{G}_1\mathbf{E}_1, \quad \mathbf{C} = \mathbf{G}_2\mathbf{E}_2, \quad \mathbf{F}_1 = \mathbf{H}_1\mathbf{C}, \quad \text{and} \quad \mathbf{F}_2 = \mathbf{H}_2\mathbf{F}_1$$

Depending upon the sequence of algebraic manipulations used to eliminate all terms except **R** and **C**, different forms of the final result are obtained. Four different sequences are presented.

1. Eliminating $\mathbf{E}_2$ gives $\mathbf{C} = \mathbf{G}_2\mathbf{G}_1\mathbf{E}_1$ and eliminating $\mathbf{F}_1$ gives $\mathbf{F}_2 = \mathbf{H}_2\mathbf{H}_1\mathbf{C}$. Combining gives $\mathbf{C} = \mathbf{G}_2\mathbf{G}_1\mathbf{R} - \mathbf{G}_2\mathbf{G}_1\mathbf{H}_2\mathbf{H}_1\mathbf{C}$ or $[\mathbf{I}_n + \mathbf{G}_2\mathbf{G}_1\mathbf{H}_2\mathbf{H}_1]\mathbf{C} = \mathbf{G}_2\mathbf{G}_1\mathbf{R}$ so that

$$\mathbf{C} = [\mathbf{I}_n + \mathbf{G}_2\mathbf{G}_1\mathbf{H}_2\mathbf{H}_1]^{-1}\mathbf{G}_2\mathbf{G}_1\mathbf{R}$$

The similarity with the scalar closed-loop transfer function is apparent.

2. In this sequence $\mathbf{E}_2$ is first isolated and then, in turn, related to **C** as follows: $\mathbf{E}_2 = \mathbf{G}_1(\mathbf{R} - \mathbf{F}_2)$, but

$$\mathbf{F}_2 = \mathbf{H}_2\mathbf{H}_1\mathbf{G}_2\mathbf{E}_2 \quad \text{so that } [\mathbf{I}_m + \mathbf{G}_1\mathbf{H}_2\mathbf{H}_1\mathbf{G}_2]\mathbf{E}_2 = \mathbf{G}_1\mathbf{R}$$

or

$$\mathbf{E}_2 = [\mathbf{I}_m + \mathbf{G}_1\mathbf{H}_2\mathbf{H}_1\mathbf{G}_2]^{-1}\mathbf{G}_1\mathbf{R} \quad \text{so that } \mathbf{C} = \mathbf{G}_2\mathbf{E}_2$$
$$= \mathbf{G}_2[\mathbf{I}_m + \mathbf{G}_1\mathbf{H}_2\mathbf{H}_1\mathbf{G}_2]^{-1}\mathbf{G}_1\mathbf{R}$$

Note the different arrangement of terms and the size of the matrix to be inverted.

3. If $\mathbf{E}_1$ is first isolated in a similar way, we obtain

$$\mathbf{E}_1 = [\mathbf{I}_r + \mathbf{H}_2\mathbf{H}_1\mathbf{G}_2\mathbf{G}_1]^{-1}\mathbf{R}$$

from which

$$\mathbf{C} = \mathbf{G}_2\mathbf{G}_1[\mathbf{I}_r + \mathbf{H}_2\mathbf{H}_1\mathbf{G}_2\mathbf{G}_1]^{-1}\mathbf{R}$$

4. If $\mathbf{F}_1$ is first isolated, we obtain

$$\mathbf{F}_1 = [\mathbf{I}_p + \mathbf{H}_1\mathbf{G}_2\mathbf{G}_1\mathbf{H}_2]^{-1}\mathbf{H}_1\mathbf{G}_2\mathbf{G}_1\mathbf{R}$$

Premultiplying this by $\mathbf{H}_2$ gives $\mathbf{F}_2$. Subtracting $\mathbf{F}_2$ from **R** gives

$$\mathbf{E}_1 = \{\mathbf{I}_r - \mathbf{H}_2[\mathbf{I}_p + \mathbf{H}_1\mathbf{G}_2\mathbf{G}_1\mathbf{H}_2]^{-1}\mathbf{H}_1\mathbf{G}_2\mathbf{G}_1\}\mathbf{R}$$

Premultiplying this by $\mathbf{G}_2\mathbf{G}_1$ gives

$$\mathbf{C} = \{\mathbf{G}_2\mathbf{G}_1 - \mathbf{G}_2\mathbf{G}_1\mathbf{H}_2[\mathbf{I}_p + \mathbf{H}_1\mathbf{G}_2\mathbf{G}_1\mathbf{H}_2]^{-1}\mathbf{H}_1\mathbf{G}_2\mathbf{G}_1\}\mathbf{R}$$

3.4 Use the result of Problem 3.3 to establish some useful matrix identities.

In the previous problem four results of the form $\mathbf{C} = \mathbf{T}_i\mathbf{R}$ were found, with the differences contained in the four $\mathbf{T}_i$ matrices. Equating two forms for the output **C** gives

$$\mathbf{T}_i\mathbf{R} = \mathbf{T}_j\mathbf{R}$$

In general, this equation is not sufficient for concluding that $\mathbf{T}_i = \mathbf{T}_j$. However, in this case it must be true for *all* **R** that $(\mathbf{T}_i - \mathbf{T}_j)\mathbf{R} = \mathbf{0}$ so that $\mathbf{T}_i - \mathbf{T}_j$ must be the null matrix, or $\mathbf{T}_i = \mathbf{T}_j$. Therefore, the results of Problem 3.3 give the following matrix identities. They are true for any matrices which satisfy the conformability conditions, whenever the indicated inverses exist:

$$[\mathbf{I}_n + \mathbf{G}_2\mathbf{G}_1\mathbf{H}_2\mathbf{H}_1]^{-1}\mathbf{G}_2\mathbf{G}_1 = \mathbf{G}_2[\mathbf{I}_m + \mathbf{G}_1\mathbf{H}_2\mathbf{H}_1\mathbf{G}_2]^{-1}\mathbf{G}_1 = \mathbf{G}_2\mathbf{G}_1[\mathbf{I}_r + \mathbf{H}_2\mathbf{H}_1\mathbf{G}_2\mathbf{G}_1]^{-1}$$
$$= \mathbf{G}_2\mathbf{G}_1 - \mathbf{G}_2\mathbf{G}_1\mathbf{H}_2[\mathbf{I}_p + \mathbf{H}_1\mathbf{G}_2\mathbf{G}_1\mathbf{H}_2]^{-1}\mathbf{H}_1\mathbf{G}_2\mathbf{G}_1$$

3.5 Consider the system of Problem 3.2, and find the closed-loop transfer function matrices which relate the following input-output matrix pairs: **R** to $\mathbf{E}_1$, **V** to $\mathbf{E}_2$, **W** to **C**, and **D** to $\mathbf{F}_1$. Also determine the characteristic equations.

Starting with the basic relations in Problem 3.3 and using similar manipulations leads to

$$\mathbf{E}_1 = [\mathbf{I}_r + \mathbf{H}_2\mathbf{H}_1\mathbf{G}_2\mathbf{G}_1]^{-1}\mathbf{R} \qquad \mathbf{C} = [\mathbf{I}_n + \mathbf{G}_2\mathbf{G}_1\mathbf{H}_2\mathbf{H}_1]^{-1}\mathbf{W}$$
$$\mathbf{E}_2 = [\mathbf{I}_m + \mathbf{G}_1\mathbf{H}_2\mathbf{H}_1\mathbf{G}_2]^{-1}\mathbf{V} \qquad \mathbf{F}_1 = [\mathbf{I}_p + \mathbf{H}_1\mathbf{G}_2\mathbf{G}_1\mathbf{H}_2]^{-1}\mathbf{D}$$

Since, for example,

$$[\mathbf{I}_r + \mathbf{H}_2\mathbf{H}_1\mathbf{G}_2\mathbf{H}_1]^{-1} = \frac{\text{Adj}\,[\mathbf{I}_r + \mathbf{H}_2\mathbf{H}_1\mathbf{G}_2\mathbf{G}_1]}{|\mathbf{I}_r + \mathbf{H}_2\mathbf{H}_1\mathbf{G}_2\mathbf{G}_1|}$$

the characteristic equation is obtained by setting the determinant in the denominator to zero. The characteristic equation is a property of the system and not the particular inputs or outputs considered. It is reasonable to expect that the following determinant identities are true (this is proven in Chapter 7, Problem 7.18, by other means):

$$|\mathbf{I}_r + \mathbf{H}_2\mathbf{H}_1\mathbf{G}_2\mathbf{G}_1| = |\mathbf{I}_m + \mathbf{G}_1\mathbf{H}_2\mathbf{H}_1\mathbf{G}_2| = |\mathbf{I}_n + \mathbf{G}_2\mathbf{G}_1\mathbf{H}_2\mathbf{H}_1|$$
$$= |\mathbf{I}_p + \mathbf{H}_1\mathbf{G}_2\mathbf{G}_1\mathbf{H}_2|$$

The roots of any one of these determinants constitute the poles of the multivariable system, provided no cancellation with numerator terms has occurred.

3.6 A single-input, two-output feedback system has the form shown in Fig. 3.1, with

$$\mathbf{G}_2\mathbf{G}_1 = \begin{bmatrix} \dfrac{1}{s+1} \\ \dfrac{1}{s+2} \end{bmatrix} \qquad \mathbf{H}_2\mathbf{H}_1 = [s \quad 1]$$

Find the characteristic equation and the transfer function matrix relating **R** to **C**, using two different formulations.

Using $|\mathbf{I}_2 + \mathbf{G}_2\mathbf{G}_1\mathbf{H}_2\mathbf{H}_1| = \mathbf{0}$ gives

$$\left|\begin{bmatrix} 1 & 0 \\ 0 & 1 \end{bmatrix} + \begin{bmatrix} \dfrac{s}{s+1} & \dfrac{1}{s+1} \\ \dfrac{s}{s+2} & \dfrac{1}{s+2} \end{bmatrix}\right| = \left(1 + \frac{s}{s+1}\right)\left(1 + \frac{1}{s+2}\right) - \frac{s}{(s+1)(s+2)} = 0$$

Using $|\mathbf{I}_1 + \mathbf{H}_2\mathbf{H}_1\mathbf{G}_2\mathbf{G}_1| = 1 + s/(s+1) + 1/(s+2) = 0$ leads to the same characteristic equation with less effort.

Likewise, $\mathbf{C} = [\mathbf{I}_2 + \mathbf{G}_2\mathbf{G}_1\mathbf{H}_2\mathbf{H}_1]^{-1}\mathbf{G}_2\mathbf{G}_1\mathbf{R}$ gives

$$\mathbf{C} = \frac{\begin{bmatrix} 1 + \dfrac{1}{s+2} & -\dfrac{1}{s+1} \\ -\dfrac{s}{s+2} & 1 + \dfrac{s}{s+1} \end{bmatrix}\begin{bmatrix} \dfrac{1}{s+1} \\ \dfrac{1}{s+2} \end{bmatrix}\mathbf{R}}{1 + \dfrac{s}{s+1} + \dfrac{1}{s+2}}$$

Using $\mathbf{C} = \mathbf{G}_2\mathbf{G}_1[1 + \mathbf{H}_2\mathbf{H}_1\mathbf{G}_2\mathbf{G}_1]^{-1}\mathbf{R}$ gives

$$\mathbf{C} = \frac{\begin{bmatrix} \dfrac{1}{s+1} \\ \dfrac{1}{s+2} \end{bmatrix}\mathbf{R}}{1 + \dfrac{s}{s+1} + \dfrac{1}{s+2}}$$

These are identical, but the second form is obtained without the requirement of matrix inversion.

3.7 Consider the four forms of the closed-loop transfer matrix derived in Problem 3.3. What are the dimensions of the matrices that need to be inverted in each form if $\mathbf{G}_1$ is 10×1000, $\mathbf{G}_2$ is 50×10, $\mathbf{H}_1$ is 1×50, and $\mathbf{H}_2$ is 1000×1?

The first form requires inverting $\mathbf{I}_n + \mathbf{G}_2\mathbf{G}_1\mathbf{H}_2\mathbf{H}_1$, which is a 50×50 matrix. Form 2 requires inverting a 10×10 matrix since $m = 10$. The third form requires inversion of a 1000×1000 matrix since $r = 1000$. The fourth form requires only a scalar division since $p = 1$. The same possibilities exist for the size of the determinant to be used in finding the characteristic equation.

Determinants, Cramer's Rule, Rank

3.8 A 5×5 matrix decomposes into the unit matrix plus a product, as shown. Evaluate its determinant.

$$\mathbf{A} = \begin{bmatrix} 0 & -2 & -3 & -4 & -5 \\ -1 & -1 & -3 & -4 & -5 \\ 4 & 8 & 13 & 16 & 20 \\ 2 & 4 & 6 & 9 & 10 \\ 8 & 16 & 24 & 32 & 41 \end{bmatrix} = \mathbf{I}_5 + \begin{bmatrix} -1 \\ -1 \\ 4 \\ 2 \\ 8 \end{bmatrix}[1 \quad 2 \quad 3 \quad 4 \quad 5]$$

$$|\mathbf{A}| = |\mathbf{I}_5 + \mathbf{GH}| = 1 + \mathbf{HG} = 1 + [1 \quad 2 \quad 3 \quad 4 \quad 5]\begin{bmatrix} -1 \\ -1 \\ 4 \\ 2 \\ 8 \end{bmatrix} = 58$$

3.9 Does $\mathbf{A} = \begin{bmatrix} 16 & 0 & 4 & 7 \\ -3 & 8 & 8 & 2 \\ 1 & 0 & 5 & 2 \\ -7 & 6 & 5 & 4 \end{bmatrix}$ have an inverse?

$\mathbf{A}^{-1}$ exists if and only if $|\mathbf{A}| \neq 0$. To check the determinant, the method of Laplace expansion is used with respect to the second column, $|\mathbf{A}| = 8C_{22} + 6C_{42}$. The two cofactors are

$$C_{22} = (-1)^4 \begin{vmatrix} 16 & 4 & 7 \\ 1 & 5 & 2 \\ -7 & 5 & 4 \end{vmatrix} = 368 \qquad C_{42} = (-1)^6 \begin{vmatrix} 16 & 4 & 7 \\ -3 & 8 & 2 \\ 1 & 5 & 2 \end{vmatrix} = -33$$

Therefore, $|\mathbf{A}| = 8(368) - 6(33) = 2746 \neq 0$ and $\mathbf{A}^{-1}$ does exist.

3.10 Check the determinant in the previous problem using a_{31} as the pivot element.

$$|\mathbf{A}| = \frac{1}{1^2}\begin{vmatrix} \begin{vmatrix} 16 & 0 \\ 1 & 0 \end{vmatrix} & \begin{vmatrix} 16 & 4 \\ 1 & 5 \end{vmatrix} & \begin{vmatrix} 16 & 7 \\ 1 & 2 \end{vmatrix} \\ \begin{vmatrix} -3 & 8 \\ 1 & 0 \end{vmatrix} & \begin{vmatrix} -3 & 8 \\ 1 & 5 \end{vmatrix} & \begin{vmatrix} -3 & 2 \\ 1 & 2 \end{vmatrix} \\ \begin{vmatrix} 1 & 0 \\ -7 & 6 \end{vmatrix} & \begin{vmatrix} 1 & 5 \\ -7 & 5 \end{vmatrix} & \begin{vmatrix} 1 & 2 \\ -7 & 4 \end{vmatrix} \end{vmatrix} = \begin{vmatrix} 0 & 76 & 25 \\ -8 & -23 & -8 \\ 6 & 40 & 18 \end{vmatrix}$$

Using the new a_{31} as the pivot gives

$$|\mathbf{A}| = \frac{1}{6}\begin{vmatrix} \begin{vmatrix} 0 & 76 \\ 6 & 40 \end{vmatrix} & \begin{vmatrix} 0 & 25 \\ 6 & 18 \end{vmatrix} \\ \begin{vmatrix} -8 & -23 \\ 6 & 40 \end{vmatrix} & \begin{vmatrix} -8 & -8 \\ 6 & 18 \end{vmatrix} \end{vmatrix} = \frac{1}{6}\begin{vmatrix} -456 & -150 \\ -182 & -96 \end{vmatrix} = \begin{vmatrix} -76 & -25 \\ -182 & -96 \end{vmatrix} = 2746$$

3.11 $\mathbf{A}$, $\mathbf{B}$, $\mathbf{I}_p$, and $\mathbf{0}$ are submatrices. Show that

$$\left|\left[\begin{array}{c|c} \mathbf{A} & \mathbf{0} \\ \hline \mathbf{B} & \mathbf{I}_p \end{array}\right]\right| = |\mathbf{A}|$$

Using Laplace expansion p times with respect to the last p columns gives

$$\left|\begin{array}{c|c} \mathbf{A} & \mathbf{0} \\ \hline \mathbf{B} & \mathbf{I}_p \end{array}\right| = 1\left|\begin{array}{c|c} \mathbf{A} & \mathbf{0} \\ \hline \mathbf{B}' & \mathbf{I}_{p-1} \end{array}\right| = 1\cdot 1\left|\begin{array}{c|c} \mathbf{A} & \mathbf{0} \\ \hline \mathbf{B}'' & \mathbf{I}_{p-2} \end{array}\right| = \cdots = |\mathbf{A}|$$

3.12 Show that

$$\left|\begin{array}{c|c} \mathbf{A} & \mathbf{0} \\ \hline \mathbf{B} & \mathbf{C} \end{array}\right| = |\mathbf{A}|\cdot|\mathbf{C}|$$

and that

$$\left|\begin{array}{c|c} \mathbf{A} & \mathbf{B} \\ \hline \mathbf{C} & \mathbf{D} \end{array}\right| = |\mathbf{A}|\cdot|\mathbf{D} - \mathbf{C}\mathbf{A}^{-1}\mathbf{B}| \qquad \text{if } \mathbf{A}^{-1} \text{ exists}$$

$$= |\mathbf{D}|\cdot|\mathbf{A} - \mathbf{B}\mathbf{D}^{-1}\mathbf{C}| \qquad \text{if } \mathbf{D}^{-1} \text{ exists}$$

We have

$$\left|\left[\begin{array}{c|c} \mathbf{A} & \mathbf{0} \\ \hline \mathbf{B} & \mathbf{C} \end{array}\right]\right| = \left|\left[\begin{array}{c|c} \mathbf{A} & \mathbf{0} \\ \hline \mathbf{B} & \mathbf{I} \end{array}\right]\left[\begin{array}{c|c} \mathbf{I} & \mathbf{0} \\ \hline \mathbf{0} & \mathbf{C} \end{array}\right]\right| = \left|\begin{array}{c|c} \mathbf{A} & \mathbf{0} \\ \hline \mathbf{B} & \mathbf{I} \end{array}\right| \cdot \left|\begin{array}{c|c} \mathbf{I} & \mathbf{0} \\ \hline \mathbf{0} & \mathbf{C} \end{array}\right| = |\mathbf{A}|\cdot|\mathbf{C}|$$

using results of the previous problem.

If $\mathbf{A}^{-1}$ exists, the desired determinant can be multiplied by $\begin{vmatrix} \mathbf{I} & \mathbf{0} \\ -\mathbf{C}\mathbf{A}^{-1} & \mathbf{I} \end{vmatrix} = 1$:

$$\begin{vmatrix} \mathbf{A} & \mathbf{B} \\ \mathbf{C} & \mathbf{D} \end{vmatrix} = \begin{vmatrix} \mathbf{I} & \mathbf{0} \\ -\mathbf{C}\mathbf{A}^{-1} & \mathbf{I} \end{vmatrix}\begin{vmatrix} \mathbf{A} & \mathbf{B} \\ \mathbf{C} & \mathbf{D} \end{vmatrix} = \begin{vmatrix} \mathbf{A} & \mathbf{B} \\ \mathbf{0} & \mathbf{D} - \mathbf{C}\mathbf{A}^{-1}\mathbf{B} \end{vmatrix} = |\mathbf{A}|\cdot|\mathbf{D} - \mathbf{C}\mathbf{A}^{-1}\mathbf{B}|$$

If $\mathbf{D}^{-1}$ exists, use $\begin{vmatrix} \mathbf{A} & \mathbf{B} \\ \mathbf{C} & \mathbf{D} \end{vmatrix} = \begin{vmatrix} \mathbf{I} & -\mathbf{B}\mathbf{D}^{-1} \\ \mathbf{0} & \mathbf{I} \end{vmatrix}\begin{vmatrix} \mathbf{A} & \mathbf{B} \\ \mathbf{C} & \mathbf{D} \end{vmatrix}$ and repeat the above procedure.

3.13 Use Cramer's rule to find the solutions for x_1 and x_2, if $3x_1 + 2x_2 = 6$, and $x_1 - 5x_2 = 1$.

In matrix form,

$$\begin{bmatrix} 3 & 2 \\ 1 & -5 \end{bmatrix}\begin{bmatrix} x_1 \\ x_2 \end{bmatrix} = \begin{bmatrix} 6 \\ 1 \end{bmatrix} \qquad \text{or} \qquad \mathbf{AX} = \mathbf{Y}$$

If the coefficient matrix **A** is nonsingular, Cramer's rule gives the solution for the component x_i as

$$x_i = \frac{|\mathbf{B}_i|}{|\mathbf{A}|}$$

where $\mathbf{B}_i$ is formed from **A** by replacing column i by **Y**. In this example

$$x_1 = \frac{\begin{vmatrix} 6 & 2 \\ 1 & -5 \end{vmatrix}}{|\mathbf{A}|} = \frac{32}{17} \quad \text{and} \quad x_2 = \frac{\begin{vmatrix} 3 & 6 \\ 1 & 1 \end{vmatrix}}{|\mathbf{A}|} = \frac{3}{17}$$

3.14 Use Cramer's rule and partitioned matrices to prove

Theorem: Let **A** be an $n \times n$ matrix and let a_{ij} be any nonzero element. Define $\mathbf{B}_{ij}$ as the $n-1$ by $n-1$ matrix formed by deleting row i and column j of **A**. Let **R** and **C** be $1 \times n-1$ and $n-1 \times 1$ row and column matrices formed by deleting a_{ij} from the ith row and jth column of **A**. Then

$$|\mathbf{A}| = (-1)^{i+j} a_{ij} \left| \mathbf{B}_{ij} - \frac{1}{a_{ij}} \mathbf{CR} \right|$$

Proof: Consider the set of linear equations $\mathbf{AX} = \mathbf{Y}$, where **X** and **Y** are column matrices and **Y** is all zero except $y_i = 1$. Cramer's rule is used to solve for x_j:

$$\mathbf{x}_j = \frac{1}{|\mathbf{A}|} \begin{vmatrix} \mathbf{B}_1 & \begin{matrix} 0 \\ \vdots \end{matrix} & \mathbf{B}_2 \\ a_{i1} \cdots & 1 & \cdots a_{in} \\ \mathbf{B}_3 & \begin{matrix} \vdots \\ 0 \end{matrix} & \mathbf{B}_4 \end{vmatrix} \quad \text{where} \quad \mathbf{B}_{ij} = \begin{bmatrix} \mathbf{B}_1 & \mathbf{B}_2 \\ \mathbf{B}_3 & \mathbf{B}_4 \end{bmatrix}$$

Using the Laplace expansion method with respect to column j and then rearranging gives

$$|\mathbf{A}| = \frac{(-1)^{i+j}}{x_j} |\mathbf{B}_{ij}| \tag{1}$$

In order to determine $1/x_j$, the original equation $\mathbf{AX} = \mathbf{Y}$ can be written as

$$\mathbf{B}_{ij}\mathbf{X}_a + \mathbf{C}x_j = \mathbf{0}$$

$$\mathbf{RX}_a + a_{ij}x_j = 1$$

$\mathbf{X}_a$ is formed from **X** by deleting x_j. Thus

$$\mathbf{X}_a = -\mathbf{B}_{ij}^{-1}\mathbf{C}x_j$$

so that $(-\mathbf{RB}_{ij}^{-1}\mathbf{C} + a_{ij})x_j = 1$. Using this in equation (*1*) gives

$$|\mathbf{A}| = (-1)^{i+j}|\mathbf{B}_{ij}|\cdot(a_{ij} - \mathbf{RB}_{ij}^{-1}\mathbf{C}) = (-1)^{i+j}a_{ij}|\mathbf{B}_{ij}|\cdot\left|1 - \frac{1}{a_{ij}}\mathbf{RB}_{ij}^{-1}\mathbf{C}\right|$$

The determinant identities of Problem 3.5 give the final result:

$$|\mathbf{A}| = (-1)^{i+j} a_{ij} \left| \mathbf{B}_{ij} - \frac{1}{a_{ij}} \mathbf{CR} \right|$$

(A limiting process can be used to show the validity of this proof and the final result even if $|\mathbf{A}| = 0$ or $|\mathbf{B}_{ij}| = 0$.)

3.15 Use the previous theorem to evaluate $|\mathbf{A}|$, using a_{11} as the divisor.

$$|\mathbf{A}| = \begin{vmatrix} 4 & 8 & 1 & 3 \\ 2 & 5 & -1 & 3 \\ -1 & 6 & 7 & 9 \\ 1 & 1 & -3 & 3 \end{vmatrix} = 4\left|\begin{bmatrix} 5 & -1 & 3 \\ 6 & 7 & 9 \\ 1 & -3 & 3 \end{bmatrix} - \tfrac{1}{4}\begin{bmatrix} 2 \\ -1 \\ 1 \end{bmatrix}[8 \quad 1 \quad 3]\right|$$

$$= 4\begin{vmatrix} 1 & -3/2 & 3/2 \\ 8 & 29/4 & 39/4 \\ -1 & -13/4 & 9/4 \end{vmatrix} = 246$$

3.16 Find the rank and trace of

$$\mathbf{A} = \begin{bmatrix} 1 & 2 & 3 & 3 \\ 2 & -2 & 1 & 1 \\ 0 & 1 & 5 & 8 \\ 1 & -6 & 4 & -4 \end{bmatrix}$$

The sum of the diagonal terms gives $\text{Tr}(\mathbf{A}) = 1 - 2 + 5 - 4 = 0$. Using any one of several methods gives $|\mathbf{A}| = 338$. Since the determinant is nonzero, the rank of $\mathbf{A}$ is 4.

Matrix Inversion and Related Topics

3.17 Given

$$\begin{bmatrix} 2 & 0.5 & 2 \\ 3 & 3 & 0 \\ 1 & 0.5 & 2 \end{bmatrix}\begin{bmatrix} x_1 \\ x_2 \\ x_3 \end{bmatrix} = \begin{bmatrix} 3 \\ 1 \\ 5 \end{bmatrix} \quad \text{or} \quad \mathbf{AX} = \mathbf{Y}$$

find x_1, x_2, and x_3 using the definition of matrix inversion.

The solution is $\mathbf{X} = \mathbf{A}^{-1}\mathbf{Y}$, where $|\mathbf{A}| = 6$ and

$$\text{adj}\,\mathbf{A} = \begin{bmatrix} 6 & -6 & -1.5 \\ 0 & 2 & -0.5 \\ -6 & 6 & 4.5 \end{bmatrix}^T$$

so that

$$\mathbf{X} = \frac{1}{6}\begin{bmatrix} 6 & 0 & -6 \\ -6 & 2 & 6 \\ -1.5 & -0.5 & 4.5 \end{bmatrix}\begin{bmatrix} 3 \\ 1 \\ 5 \end{bmatrix} = \begin{bmatrix} -2 \\ 7/3 \\ 35/12 \end{bmatrix}$$

3.18 Explain the method of Gaussian elimination in terms of elementary matrices and partitioned matrices.

Gaussian elimination is used to solve $\mathbf{AX} = \mathbf{Y}$, where $\mathbf{A}$ is known, $n \times n$, and $\mathbf{Y}$ is known, $n \times p$. The $n \times p$ matrix $\mathbf{X}$ is unknown. Often $\mathbf{X}$ and $\mathbf{Y}$ are column vectors, with $p = 1$. Assuming that $\mathbf{A}^{-1}$ exists, $\mathbf{A}^{-1}\mathbf{AX} = \mathbf{A}^{-1}\mathbf{Y}$ or $\mathbf{IX} = \mathbf{A}^{-1}\mathbf{Y}$. The elementary row operations are equivalent to premultiplication by the elementary matrices. A sequence of these operations which reduces the coefficient of $\mathbf{X}$ from $\mathbf{A}$ to $\mathbf{I}$ will simultaneously change $\mathbf{Y}$ to $\mathbf{A}^{-1}\mathbf{Y}$, that is, $\mathbf{X}$. The operations are:

1. Form the $n \times n + p$ matrix $\mathbf{W}_0 = [\mathbf{A} \mid \mathbf{Y}]$.
2. Find the element in column one with maximum absolute value. Interchange that row with row one. This gives $\mathbf{W}_1 = \mathbf{E}_{1,q}\mathbf{W}_0 = [\mathbf{E}_{1,q}\mathbf{A} \mid \mathbf{E}_{1,q}\mathbf{Y}]$.
3. Divide the entire first row of $\mathbf{W}_1$ by w_{11}, assuming $w_{11} \neq 0$. This gives $W_2 =$

$\mathbf{E}_1(1/w_{11})\mathbf{W}_1$. If $w_{11} = 0$, then the entire first column is zero, indicating that $\mathbf{A}$ is singular. If this happens, skip to step 6.

4. Multiply the first row of $\mathbf{W}_2$ by $-w_{21}$ and add to the second row. $\mathbf{W}_3 = \mathbf{E}_{1,2}(-w_{21})\mathbf{W}_2$.
5. Repeat step 4 using $\mathbf{E}_{1,3}(-w_{31}), \ldots, \mathbf{E}_{1,n}(-w_{n1})$ in sequence. This reduces column one to a one followed by $n-1$ zeros.
6. Find the maximum absolute value element $w_{\alpha 2}$ from column 2, rows 2 through n. Interchange that row with row 2, $\mathbf{W}_{k+1} = \mathbf{E}_{2,\alpha}\mathbf{W}_k$.
7. Divide row 2 by the current value of w_{22}. This is analogous to step 3. Repeat steps 4 and 5 until w_{22} is 1 (possibly zero if $\mathbf{A}$ is singular) and all $w_{i2} = 0$ below w_{22}.
8. Repeat steps 6 and 7 until the first n columns form an $n \times n$ upper triangular matrix with $\delta_i = 1$ or 0 for the ith diagonal. From this, $|\mathbf{A}| = (-1)^\nu \mu_1\mu_2\mu_3 \cdots \mu_n$, where ν is the number of row interchanges used and μ_i is the divisor used for the ith row, steps 3 and 7. If a zero is encountered on the diagonal, then $|\mathbf{A}| = 0$. When this happens, the rank of $\mathbf{A}$ is still of interest. The triangular form is a convenient starting point for reducing to one of the normal forms to determine rank. If $|\mathbf{A}| \neq 0$, continue to step 9.
9. Multiply the current $\mathbf{W}$ by $\mathbf{E}_{2,n}(-w_{1n})$, then by $\mathbf{E}_{1,n}(-w_{2n}), \ldots$. This reduces the nth column of $\mathbf{W}$ to $n-1$ zeros but leaves the n, n element unity.
10. Repeat step 9 for other columns, until $\mathbf{W}$ has the unit matrix for its first n columns. The last n columns of this final $\mathbf{W}$ matrix contain $\mathbf{A}^{-1}\mathbf{Y}$.
    Note that $\mathbf{A}^{-1}$ can be found by using $\mathbf{Y} = \mathbf{I}$ when setting up $\mathbf{W}_0$.

3.19 Find $\begin{bmatrix} 0 & 3 \\ 4 & 2 \end{bmatrix}^{-1}$.

The sequence of matrices, interchanges, and divisors is

$$\mathbf{W}_0 = \left[\begin{array}{cc|cc} 0 & 3 & 1 & 0 \\ 4 & 2 & 0 & 1 \end{array}\right] \xrightarrow[\nu=1]{\text{interchange}} \left[\begin{array}{cc|cc} 4 & 2 & 0 & 1 \\ 0 & 3 & 1 & 0 \end{array}\right] \xrightarrow{\mu_1=4} \left[\begin{array}{cc|cc} 1 & 1/12 & 0 & 1/4 \\ 0 & 3 & 1 & 0 \end{array}\right]$$

$$\xrightarrow{\mu_2=3} \left[\begin{array}{cc|cc} 1 & 1/2 & 0 & 1/4 \\ 0 & 1 & 1/3 & 0 \end{array}\right] \longrightarrow \left[\begin{array}{cc|cc} 1 & 0 & -1/6 & 1/4 \\ 0 & 1 & 1/3 & 0 \end{array}\right]$$

This problem demonstrates why row interchanges are required to avoid dividing by zero. The results are

$$|\mathbf{A}| = (-1)^\nu \mu_1\mu_2 = -12, \qquad r_A = 2, \qquad \mathbf{A}^{-1} = \frac{1}{12}\begin{bmatrix} -2 & 3 \\ 4 & 0 \end{bmatrix}$$

3.20 Let $\mathbf{A} = \begin{bmatrix} 1 & 2 & 3 \\ 2 & 4 & 6 \\ 0 & 1 & 5 \end{bmatrix}$. Determine $|\mathbf{A}|$, r_A, and $\mathbf{A}^{-1}$ if it exists.

$$\mathbf{W}_0 = \left[\begin{array}{ccc|ccc} 1 & 2 & 3 & 1 & 0 & 0 \\ 2 & 4 & 6 & 0 & 1 & 0 \\ 0 & 1 & 5 & 0 & 0 & 1 \end{array}\right] \xrightarrow{\nu=1} \left[\begin{array}{ccc|ccc} 2 & 4 & 6 & 0 & 1 & 0 \\ 1 & 2 & 3 & 1 & 0 & 0 \\ 0 & 1 & 5 & 0 & 0 & 1 \end{array}\right]$$

$$\xrightarrow{\mu_1=2} \left[\begin{array}{ccc|ccc} 1 & 2 & 3 & 0 & 1/2 & 0 \\ 1 & 2 & 3 & 1 & 0 & 0 \\ 0 & 1 & 5 & 0 & 0 & 1 \end{array}\right] \longrightarrow \left[\begin{array}{ccc|ccc} 1 & 2 & 3 & 0 & 1/2 & 0 \\ 0 & 0 & 0 & 1 & -1/2 & 0 \\ 0 & 1 & 5 & 0 & 0 & 1 \end{array}\right]$$

$$\xrightarrow[\nu=2]{\text{interchange}} \left[\begin{array}{ccc|ccc} 1 & 2 & 3 & 0 & 1/2 & 0 \\ 0 & 1 & 5 & 0 & 0 & 1 \\ 0 & 0 & 0 & 1 & -1/2 & 0 \end{array}\right]$$

At this point we see that $\mathbf{A}$ is singular, that is, $|\mathbf{A}| = 0$ and $\mathbf{A}^{-1}$ does not exist. We can therefore drop the right half of $\mathbf{W}$ and use row or column operations to give

$$\begin{bmatrix} 1 & 2 & 3 \\ 0 & 1 & 5 \\ 0 & 0 & 0 \end{bmatrix} \longrightarrow \begin{bmatrix} 1 & 2 & -7 \\ 0 & 1 & 0 \\ 0 & 0 & 0 \end{bmatrix} \longrightarrow \begin{bmatrix} 1 & 2 & 0 \\ 0 & 1 & 0 \\ 0 & 0 & 0 \end{bmatrix} \longrightarrow \left[\begin{array}{cc|c} 1 & 0 & 0 \\ 0 & 1 & 0 \\ \hline 0 & 0 & 0 \end{array}\right] = \left[\begin{array}{c|c} \mathbf{I}_2 & \mathbf{0} \\ \hline \mathbf{0} & 0 \end{array}\right]$$

Therefore $r_A = 2$.

3.21 Use Gaussian elimination to solve for x_1 and x_2, if $\begin{bmatrix} 2 & -2 \\ 3 & 8 \end{bmatrix}\begin{bmatrix} x_1 \\ x_2 \end{bmatrix} = \begin{bmatrix} 10 \\ -5 \end{bmatrix}$.

$$\mathbf{W}_0 = \left[\begin{array}{cc|c} 2 & -2 & 10 \\ 3 & 8 & -5 \end{array}\right] \longrightarrow \left[\begin{array}{cc|c} 3 & 8 & -5 \\ 2 & -2 & 10 \end{array}\right] \longrightarrow \left[\begin{array}{cc|c} 1 & 8/3 & -5/3 \\ 2 & -2 & 10 \end{array}\right]$$

$$\longrightarrow \left[\begin{array}{cc|c} 1 & 8/3 & -5/3 \\ 0 & -22/3 & 40/3 \end{array}\right] \longrightarrow \left[\begin{array}{cc|c} 1 & 8/3 & -5/3 \\ 0 & 1 & -20/11 \end{array}\right] \longrightarrow \left[\begin{array}{cc|c} 1 & 0 & 35/11 \\ 0 & 1 & -20/11 \end{array}\right]$$

Therefore $x_1 = 35/11$ and $x_2 = -20/11$.

3.22 $\mathbf{G}$ is an $n \times n$ complex matrix. Find $\mathbf{G}^{-1}$ using only real matrix inversion routines.

Let $\mathbf{G} = \mathbf{A} + j\mathbf{B}$, with $\mathbf{A}$ and $\mathbf{B}$ real. Assuming that $\mathbf{G}^{-1}$ exists, it can also be expressed in terms of two real matrices $\mathbf{C}$ and $\mathbf{D}$, $\mathbf{G}^{-1} = \mathbf{C} + j\mathbf{D}$. The basic requirement of a matrix inverse is that

$$\mathbf{G}\mathbf{G}^{-1} = \mathbf{I} = (\mathbf{A} + j\mathbf{B})(\mathbf{C} + j\mathbf{D}) = (\mathbf{AC} - \mathbf{BD}) + j(\mathbf{AD} + \mathbf{BC})$$

Therefore, equating real parts to real parts, $\mathbf{AC} - \mathbf{BD} = \mathbf{I}$. Equating imaginary parts $\mathbf{AD} + \mathbf{BC} = \mathbf{0}$. If $\mathbf{A}^{-1}$ exists, the solution can be written as $\mathbf{C} = (\mathbf{A} + \mathbf{BA}^{-1}\mathbf{B})^{-1}$ and $\mathbf{D} = -\mathbf{CBA}^{-1}$. The rearrangement identities of Problem 3.4 are useful in proving this. If $\mathbf{B}^{-1}$ exists but $\mathbf{A}^{-1}$ doesn't, then $[j\mathbf{G}]^{-1}$ can be sought instead. This effectively reverses the roles of $\mathbf{A}$ and $\mathbf{B}$ so the above procedure can again be used. If both $\mathbf{A}$ and $\mathbf{B}$ are singular, but $\mathbf{G}$ is not, further modifications will be necessary [105, p. 355].

Cholesky Decomposition

3.23 If $\mathbf{A}$ is symmetric and positive definite (see Chapter 7), it can be uniquely (except for signs) factored into $\mathbf{A} = \mathbf{S}^T\mathbf{S}$, where $\mathbf{S}$ is an upper triangular matrix. $\mathbf{S}$ is called the square root matrix of $\mathbf{A}$. The procedure for factoring $\mathbf{A}$ is most commonly called Cholesky decomposition [114] (although it is sometimes called the method of Banachiewicz and Dwyer [68]). Deduce the algorithm for finding $\mathbf{S}$.

The algorithm for finding the s_{ij} entries in $\mathbf{S}$ is as follows:

$$s_{11} = [a_{11}]^{1/2};\ s_{1j} = a_{1j}/s_{11} \text{ for } j = 2, \ldots, n$$

$$s_{22} = [a_{22} - (s_{12})^2]^{1/2}$$

$$s_{2j} = [a_{2j} - s_{12}s_{1j}]/s_{22} \text{ for } j = 3, \ldots, n$$

$$\vdots$$

$$s_{ii} = [a_{ii} - \sum_{k=1}^{i-1} (s_{ki})^2]^{1/2} \text{ for } i = 2, \ldots, n$$

$$s_{ij} = [a_{ij} - \sum_{k=1}^{i-1} s_{ki}s_{kj}]/s_{ii} \text{ for } j = i + 1, \ldots, n$$

3.24 Show how Cholesky decomposition can be used in solving simultaneous equations of the form $\mathbf{Ax} = \mathbf{y}$. Assume y is known and that $\mathbf{A}$ is known, symmetric, and positive definite.

Assume $\mathbf{S}$ has been found such that $\mathbf{A} = \mathbf{S}^T\mathbf{S}$. Then $\mathbf{S}^T\mathbf{S}\mathbf{x} = \mathbf{y}$. Define $\mathbf{Sx} = \mathbf{v}$. Then $\mathbf{S}^T\mathbf{v} = \mathbf{y}$. Because $\mathbf{S}^T$ is lower triangular, the elements of $\mathbf{v}$ can easily be found one-by-one by back substitution,

$$v_1 = y_1/s_{11},\ v_2 = (y_2 - s_{12}v_1)/s_{22},\ v_3 = (y_3 - s_{13}v_1 - s_{23}v_2)/s_{33}, \ldots$$

Once $\mathbf{v}$ is determined, a similar procedure can be used to find the components of $\mathbf{x}$:

$$x_n = v_n/s_{nn},\ x_{n-1} = (v_{n-1} - s_{n-1,n}v_n)/s_{n-1,n-1},$$

$$x_{n-2} = (y_{n-2} - s_{n-2,n}v_n - s_{n-2,n-1}v_{n-1})/s_{n-2,n-2}, \ldots$$

PROBLEMS

3.25 Show that every real, square matrix $\mathbf{A}$ can be written as the sum of a symmetric matrix and a skew-symmetric matrix.

3.26 Let $\mathbf{E} = [e_1 \quad e_2 \quad \cdots \quad e_n]^T$ be a column of errors in a multivariable control system. Show that the sum of the squares of the errors can be written in several forms, $e_1^2 + e_2^2 + \cdots + e_n^2 = \mathbf{E}^T\mathbf{E} = \mathrm{Tr}\,(\mathbf{E}\mathbf{E}^T)$.

3.27 Consider the h-parameter model of a transistor, which is typical of many two-port devices (Fig. 3.2).

$$\begin{bmatrix} v_1 \\ i_2 \end{bmatrix} = \begin{bmatrix} h_{11} & h_{12} \\ h_{21} & h_{22} \end{bmatrix} \begin{bmatrix} i_1 \\ v_2 \end{bmatrix}$$

Add the third equation $v_2 = -R_L i_2$ and find i_1, i_2, and v_2 if the source is an ideal voltage source v_1.

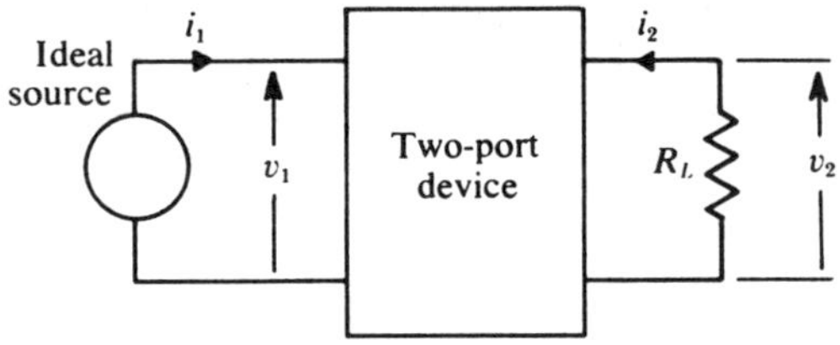

Figure 3.2

3.28 If the ideal source in the previous problem is a current source i_1, find v_1, i_2, and v_2.

3.29 Compute $|\mathbf{A}|$ using Laplace expansion, pivotal condensation, elementary operations, and the method of Problem 3.14. Draw conclusions about the effort required by each method.

$$\mathbf{A} = \begin{bmatrix} 1 & 3 & -1 & 4 \\ 2 & 0 & 1 & 5 \\ -1 & 6 & 10 & -8 \\ 0 & -2 & 7 & 1 \end{bmatrix}$$

3.30 Find the inverses of

$$\mathbf{A} = \begin{bmatrix} 4 & 1 & 1 \\ 2 & 0 & 3 \\ 1 & 1 & 5 \end{bmatrix}, \qquad \mathbf{B} = \begin{bmatrix} 0 & -2 & -3 & -4 & -5 \\ -1 & -1 & -3 & -4 & -5 \\ 4 & 8 & 13 & 16 & 20 \\ 2 & 4 & 6 & 9 & 10 \\ 8 & 16 & 24 & 32 & 41 \end{bmatrix},$$

$$\mathbf{C} = \left[\begin{array}{cc|cc|cc} 1 & 2 & 0 & 0 & 0 & 0 \\ -2 & 1 & 0 & 0 & 0 & 0 \\ \hline 0 & 0 & 5 & 7 & 0 & 0 \\ 0 & 0 & 1 & 0 & 0 & 0 \\ \hline 0 & 0 & 0 & 0 & 3 & 8 \\ 0 & 0 & 0 & 0 & 8 & 3 \end{array}\right]$$

(*Hint:* **B** is the matrix of Problem 3.8. Use an identity from Problem 3.4.)

3.31 Find the Laplace transforms of

$$\mathbf{A}(t) = \begin{bmatrix} 1 & t \\ e^{-at} & b \sin \beta t \\ t^2 & e^{-t} \cos \beta t \end{bmatrix}, \qquad \mathbf{B}(t) = \begin{bmatrix} \cosh \beta t & \sinh \beta t \\ te^{-at} & \cos \beta t \end{bmatrix}$$

3.32 Find the inverse Laplace transform of

$$\mathbf{A}(s) = \begin{bmatrix} \dfrac{24}{s^5} & \dfrac{s^2 - \beta^2}{(s^2 + \beta^2)^2} \end{bmatrix}, \qquad \mathbf{B}(s) = \begin{bmatrix} \dfrac{1}{(s+1)(s+2)} & \dfrac{s}{(s+1)(s+2)} \\ 0 & s \end{bmatrix}$$

3.33 Find the upper triangular square root matrix of

$$(a)\ \mathbf{A} = \begin{bmatrix} 4 & -1 & 2 \\ -1 & 8 & 4 \\ 2 & 4 & 9 \end{bmatrix} \qquad (b)\ \mathbf{A} = \begin{bmatrix} 4 & 6 & 1 \\ 6 & 1 & 2 \\ 1 & 2 & 2 \end{bmatrix}$$

$$(c)\ \mathbf{A} = \begin{bmatrix} 16 & 4 & 1 & -1 & 3 \\ 4 & 10 & 4 & 2 & -2 \\ 1 & 4 & 25 & 4 & 1 \\ -1 & 2 & 4 & 11 & 7 \\ 3 & -2 & 1 & 7 & 17 \end{bmatrix}$$

Chapter
4

VECTORS AND LINEAR VECTOR SPACES

4.1 INTRODUCTION

Every student of introductory physics is familiar with the concept of a vector as a quantity which possesses both a magnitude and a direction. This chapter begins with a brief account of vectors in this context and then develops the general case of n-dimensional vectors. State space, a name frequently used in modern control theory, is an example of an n-dimensional vector space. The response of a dynamical system can be represented by the motion of a time-dependent vector in this space, and this approach is presented in later chapters. To review vectors in the physical sense, see Chapter 15 of Reference 94. Abstract vector spaces are treated in References 23, 40, 47, 52, and 56, while developments related to state space can be found in 28 and 117.

4.2 PLANAR AND THREE-DIMENSIONAL REAL VECTOR SPACES

Many physical quantities, such as force and velocity, possess both a magnitude and a direction. Such entities are referred to as *vector* quantities. They are often represented by directed line segments or arrows. The length represents the magnitude, and the orientation indicates the direction.

If one point in the plane is defined as the origin, a unique vector can be associated with every point in the (2 dimensional) plane. The origin is defined as the zero vector, $\mathbf{0}$, and every other point can be associated with the directed line segment from the origin to the point. The same correspondence between points and directed line segments can obviously be made with points along the real line $\mathcal{R}$ (one dimension) and in three dimensions. Thus the terms "point" and "vector" can be used interchangeably.

If a coordinate system is defined in the plane, then each point can identified by a unique pair of ordered numbers. These coordinate numbers can be written as a column matrix. However, since many different coordinate systems could be selected, a given vector could be represented by many different column matrices. A vector is *not* a column matrix, but is a more basic entity which may be *represented* by a column matrix once a coordinate system is defined.

Vector Addition, Subtraction, and Multiplication by a Scalar

The coordinate-free description of vector addition is given by the parallelogram law. The sum of two vectors $\mathbf{v}_1$ and $\mathbf{v}_2$ is the diagonal of the parallelogram formed with $\mathbf{v}_1$ and $\mathbf{v}_2$ as sides. Since $-\mathbf{v}_2$ is a vector with the same magnitude and orientation, but the opposite direction of $\mathbf{v}_2$, the vector difference $\mathbf{v}_1 - \mathbf{v}_2$ is just the sum of $\mathbf{v}_1$ and $-\mathbf{v}_2$. Multiplication of a vector by a scalar alters the magnitude but not the orientation. In particular, any nonzero vector $\mathbf{v}$ can be used to form a *unit* vector $\hat{\mathbf{v}}$ with the same direction as $\mathbf{v}$ by multiplying $\mathbf{v}$ by the reciprocal of its magnitude.

Whenever vectors are represented as column matrices with respect to a common coordinate system, the usual rules apply for addition of matrices and multiplication of a matrix by a scalar.

Vector Products

Products such as $\mathbf{vw}$ are not defined because of matrix conformability requirements. Three types of vector products are defined.

The *inner product* (or scalar product or dot product) of $\mathbf{v}$ and $\mathbf{w}$ is defined as $\langle \mathbf{v}, \mathbf{w} \rangle = vw \cos \theta$, where v and w are the vector magnitudes and θ is the angle included between the two vectors. When real vectors are represented in orthogonal cartesian coordinates, the inner product may be computed in terms of the components as $\langle \mathbf{v}, \mathbf{w} \rangle = \mathbf{v}^T\mathbf{w} = \mathbf{w}^T\mathbf{v}$. If $\mathbf{v}$ and $\mathbf{w}$ are perpendicular, then $\theta = \pi/2$ so that $\langle \mathbf{v}, \mathbf{w} \rangle = 0$. Any two vectors which have a zero inner product are said to be perpendicular or orthogonal. The zero vector is considered to be orthogonal to every other vector. The magnitude of a vector $\mathbf{v}$ can be expressed as $v = \langle \mathbf{v}, \mathbf{v} \rangle^{1/2}$.

The *outer product* of two real vectors $\mathbf{v} = [v_1 \quad v_2 \quad v_3]^T$ and $\mathbf{w} = [w_1 \quad w_2 \quad w_3]^T$ is defined as

$$\mathbf{v}\rangle\langle\mathbf{w} = \mathbf{vw}^T = \begin{bmatrix} v_1w_1 & v_1w_2 & v_1w_3 \\ v_2w_1 & v_2w_2 & v_2w_3 \\ v_3w_1 & v_3w_2 & v_3w_3 \end{bmatrix}$$

Since matrix multiplication is not commutative, neither is the outer product, $\mathbf{vw}^T \neq \mathbf{wv}^T$.

The *cross product* $\mathbf{v} \times \mathbf{w}$ is defined only in three dimensions. This product yields another vector, perpendicular to the plane of $\mathbf{v}$ and $\mathbf{w}$. It points in the direction a right-hand screw would advance if $\mathbf{v}$ were rotated toward $\mathbf{w}$ through the smaller of the two angles θ between them. The magnitude is equal to the area of the parallelogram formed by $\mathbf{v}$ and $\mathbf{w}$, i.e., $vw \sin \theta$.

4.3 AXIOMATIC DEFINITION OF A LINEAR VECTOR SPACE

Concepts such as directed line segments, lengths, angles, and dimensions of the space are considered to be intuitively obvious in one, two, or three dimensions. In dimensions higher than three, visualization is no longer possible, and a more axiomatic approach is required. It will still be helpful to consider some of the general results for the particular cases of two or three dimensions. Geometrical descriptions of this nature will frequently be of use in gaining understanding of certain concepts.

Linear Vector Spaces

A linear vector space $\mathcal{X}$ is a set of elements, called vectors, defined over a scalar number field $\mathcal{F}$, which satisfies the following conditions for addition and multiplication by scalars.

1. For any two vectors $\mathbf{x} \in \mathcal{X}$ and $\mathbf{y} \in \mathcal{X}$, the sum $\mathbf{x} + \mathbf{y} = \mathbf{v}$ is also a vector belonging to $\mathcal{X}$.
2. Addition is commutative: $\mathbf{x} + \mathbf{y} = \mathbf{y} + \mathbf{x}$.
3. Vector addition is also associative: $(\mathbf{x} + \mathbf{y}) + \mathbf{z} = \mathbf{x} + (\mathbf{y} + \mathbf{z})$.
4. There is a zero vector, $\mathbf{0}$, contained in $\mathcal{X}$ which satisfies $\mathbf{x} + \mathbf{0} = \mathbf{0} + \mathbf{x} = \mathbf{x}$.
5. For every $\mathbf{x} \in \mathcal{X}$ there is a unique vector $\mathbf{y} \in \mathcal{X}$ such that $\mathbf{x} + \mathbf{y} = \mathbf{0}$. This vector $\mathbf{y}$ is $-\mathbf{x}$.
6. For every $\mathbf{x} \in \mathcal{X}$ and for any scalar $a \in \mathcal{F}$, the product $a\mathbf{x}$ gives another vector $\mathbf{y} \in \mathcal{X}$. In particular, if a is the unit scalar,

   $$1 \cdot \mathbf{x} = \mathbf{x} \cdot 1 = \mathbf{x}$$

7. For any scalars $a \in \mathcal{F}$ and $b \in \mathcal{F}$, and for any $\mathbf{x} \in \mathcal{X}$, $a(b\mathbf{x}) = (ab)\mathbf{x}$.
8. Multiplication by scalars is distributive,

   $$(a + b)\mathbf{x} = a\mathbf{x} + b\mathbf{x}$$

   $$a(\mathbf{x} + \mathbf{y}) = a\mathbf{x} + a\mathbf{y}$$

The sets of all real one-, two-, or three-dimensional vectors discussed in the previous section satisfy all of these conditions and, therefore, are linear vector spaces. Elements in these familiar spaces can be represented as ordered sets of real numbers $[\alpha_1]$, $[\alpha_1 \quad \alpha_2]^T$, and $[\alpha_1 \quad \alpha_2 \quad \alpha_3]^T$, respectively. A fairly obvious generalization is to consider spaces whose elements are ordered n-tuples of real numbers $[\alpha_1 \quad \alpha_2 \cdots \alpha_n]^T$, where n is a finite integer. This space is referred to as $\mathcal{R}^n$. If the scalars α_i are allowed to be complex, then the space is referred to as $\mathcal{C}^n$. Both of these possibilities will be simultaneously covered by referring to an ordered set of n-tuples $\alpha_i \in \mathcal{F}$ as belonging to the space $\mathcal{X}^n$.

Other vector spaces can be defined: for example, (1) the set of all $m \times n$ matrices with elements in $\mathfrak{F}$, (2) the set of all continuous functions $f(t)$ on the interval $a \leq t \leq b$, and (3) the set of all polynomials of degree less than or equal to n. The zero vectors in these spaces are the $m \times n$ null matrix, the function, and the polynomial which are identically zero respectively. It is not difficult to verify that these spaces, as well as many others, satisfy the eight requirements of a linear vector space. These three examples illustrate that the axiomatic definition of linear vector spaces does not require that the elements of the space be vectors in the usual sense. In general, they could be matrices, polynomials, or functions of various types. Whenever this degree of generality is intended, the space will be denoted by $\mathfrak{X}$. For the most part, the discussion here will deal with vectors in the sense of the previous section and their generalizations to $\mathfrak{X}^n$.

It is explicitly pointed out that the definition of a linear vector space makes no mention of products of two vector elements, such as the inner product. When the additional definition of inner product is introduced, the vector space is called an *inner product space*.

4.4 LINEAR DEPENDENCE AND INDEPENDENCE

Definition 4.1: Let a finite number of vectors belonging to a linear vector space $\mathfrak{X}$ be denoted by $\{\mathbf{x}_i\} = \{\mathbf{x}_1, \mathbf{x}_2, \ldots, \mathbf{x}_n\}$. If there exists a set of n scalars, a_i, at least one of which is not zero, which satisfies $a_1\mathbf{x}_1 + a_2\mathbf{x}_2 + \cdots + a_n\mathbf{x}_n = \mathbf{0}$, then the vectors $\{\mathbf{x}_i\}$ are said to be *linearly dependent*.

Definition 4.2: Any set of vectors $\{\mathbf{x}_i\}$ which is not linearly dependent is said to be *linearly independent*. That is, if $a_1\mathbf{x}_1 + a_2\mathbf{x}_2 + \cdots + a_n\mathbf{x}_n = \mathbf{0}$ implies that each $a_i = 0$, then $\{\mathbf{x}_i\}$ is a set of linearly independent vectors.

Example 4.1

Consider the set of n vectors $\mathbf{e}_i$, each of which has n components. All components of $\mathbf{e}_i$ are zero, except the ith component which is unity. Then

$$a_1\mathbf{e}_1 + a_2\mathbf{e}_2 + \cdots + a_n\mathbf{e}_n = a_1\begin{bmatrix}1\\0\\0\\ \cdot\\ \cdot\\ \cdot\\0\end{bmatrix} + a_2\begin{bmatrix}0\\1\\0\\ \cdot\\ \cdot\\ \cdot\\0\end{bmatrix} + \cdots + a_n\begin{bmatrix}0\\0\\ \cdot\\ \cdot\\ \cdot\\0\\1\end{bmatrix} = \begin{bmatrix}a_1\\a_2\\ \cdot\\ \cdot\\ \cdot\\ \\a_n\end{bmatrix}$$

The only way that this sum can give the $\mathbf{0}$ vector is if each and every $a_i = 0$. Thus the set $\{\mathbf{e}_i\}$ is linearly independent. This set of $\mathbf{e}_i$ vectors represents the natural extension of the cartesian coordinate directions often used in two- and three-dimensional spaces. They will be referred to as the natural cartesian coordinates. ■

Example 4.2

Let the components of three vectors with respect to the natural cartesian coordinates be

$$\mathbf{x}_1^T = [5 \quad 2 \quad 3], \qquad \mathbf{x}_2^T = [-1 \quad 7 \quad 4], \qquad \mathbf{x}_3^T = [14 \quad 50 \quad 36]$$

These vectors are linearly dependent because $2\mathbf{x}_1 + 3\mathbf{x}_2 - \frac{1}{2}\mathbf{x}_3 = 0$. ■

Lemma 4.1: Let $\mathcal{V} = \{\mathbf{x}_i, i = 1, n\}$ be a set of linearly dependent vectors. Then the set formed by adding any vector $\mathbf{x}_{n+1}$ to $\mathcal{V}$ is also linearly dependent.

Lemma 4.2: If a set of vectors $\{\mathbf{x}_i\}$ is linearly dependent, then one of the vectors can be written as a linear combination of the others.

Tests for Linear Dependence

Consider a set of n vectors $\{\mathbf{x}_i\}$, each having n components with respect to a given coordinate system. Let $\mathbf{A}$ be the $n \times n$ matrix which has the $\mathbf{x}_i$ vectors as columns. The set of vectors is linearly dependent if and only if $|\mathbf{A}| = 0$.

Example 4.3

Use the above test to show that the three vectors of Example 4.2 are linearly dependent.

We find $|\mathbf{A}| = \begin{vmatrix} 5 & -1 & 14 \\ 2 & 7 & 50 \\ 3 & 4 & 36 \end{vmatrix} = 0$. Therefore, the set is linearly dependent. ■

The previous test for linear independence is not applicable when considering a set of n vectors $\{\mathbf{x}_i\}$, each of which has m components, with $m \neq n$. The matrix $\mathbf{A}$ is $m \times n$ and $|\mathbf{A}|$ is not defined. Assume the set is linearly dependent so that $\sum_{i=1}^{n} a_i\mathbf{x}_i = \mathbf{0}$ with at least one nonzero a_i. Premultiplying by $\bar{\mathbf{x}}_1^T$ gives the scalar equation

$$a_1\bar{\mathbf{x}}_1^T\mathbf{x}_1 + a_2\bar{\mathbf{x}}_1^T\mathbf{x}_2 + \cdots + a_n\bar{\mathbf{x}}_1^T\mathbf{x}_n = 0$$

Repeated premultiplication by $\bar{\mathbf{x}}_2^T$, then $\bar{\mathbf{x}}_3^T$, and so on gives a set of n simultaneous equations which can be written in matrix form as

$$[\bar{\mathbf{x}}_i^T\mathbf{x}_j][\mathbf{a}] = \mathbf{0}$$

If the $n \times n$ matrix $\mathbf{G} \triangleq [\bar{\mathbf{x}}_i^T\mathbf{x}_j]$ has a nonzero determinant, then $\mathbf{G}^{-1}$ exists, and solving gives

$$\mathbf{a} = \mathbf{G}^{-1}\mathbf{0} = \mathbf{0}$$

This contradicts the assumption of at least one nonzero a_i. The matrix $\mathbf{G}$ is called the *Grammian matrix*. A necessary and sufficient condition for the set $\{\mathbf{x}_i\}$ to be linearly dependent is that $|\mathbf{G}| = 0$. An alternate means of determining linear independence is to reduce the matrix $\mathbf{A}$ to row-reduced echelon form, as mentioned in Sec. 3.10. This approach is convenient for computer applications and will be used in Chapter 5.

Geometrical Significances of Linear Dependence

Two vectors can normally be used to form sides of a parallelogram. If the vectors are linearly dependent, they have the same direction, so the parallelogram degenerates to a line. It is shown in Problem 4.14 that the 2×2 Grammian determinant is equal to the square of the area of the parallelogram formed by the vectors. Thus $|\mathbf{G}| = 0$ indicates that the parallelogram has degenerated to a single line. Three vectors can normally be used to define the sides of a parallelepiped. If there is one linear dependency relation (i.e., any two of the three vectors are linearly independent but the set of three is linearly dependent), then the parallelepiped has degenerated to a plane figure and hence has zero volume. $|\mathbf{G}| = 0$ indicates this. If there are two dependency relations, the parallelepiped degenerates to a single line. Similar significance can be attached in higher dimensional cases. The number of dependency relationships among a set of vectors (or the columns of a matrix) is called the *degeneracy*, q. For an $n \times n$ matrix $\mathbf{A}$, q, n, and the rank r_A are related by

$$n = r_A + q$$

Often it is easier to determine the rank first and then use that to determine $q = n - r_A$. The degeneracy q is the key to finding eigenvectors and generalized eigenvectors for a matrix with repeated eigenvalues. This is discussed in Chapter 7.

Sylvester's Law of Degeneracy

If $\mathbf{A}$ and $\mathbf{B}$ are square conformable matrices whose product is $\mathbf{AB} = \mathbf{C}$, Sylvester's law of degeneracy can be used to place bounds on the degeneracy of $\mathbf{C}$, q_C in terms of the degeneracy of $\mathbf{A}$, q_A, and of $\mathbf{B}$, q_B:

$$\max\{q_A, q_B\} \le q_C \le q_A + q_B$$

If the relations between n, rank, and degeneracy are used, the following limits on the rank of $\mathbf{C}$ can be obtained:

$$r_A + r_B - n \le r_C \le \min\{r_A, r_B\}$$

A similar result was presented in Chapter 3 for $\mathbf{A}$ and $\mathbf{B}$ not necessarily square, but the lower limit there was $0 \le r_C$.

4.5 VECTORS WHICH SPAN A VECTOR SPACE; BASIS VECTORS AND DIMENSIONALITY

The dimension of a vector space has been referred to several times. In two or three dimensions, the concept is obvious, but in higher dimensions a precise definition must be relied upon rather than intuition. It is first necessary to define what is meant by a set of vectors which *span* a vector space.

Let $\mathcal{X}$ be a linear vector space and let $\{\mathbf{u}_i, i = 1, m\}$ be a subset of vectors in $\mathcal{X}$. The set $\{\mathbf{u}_i\}$ is said to span the space $\mathcal{X}$ if for every vector $\mathbf{x} \in \mathcal{X}$ there is at least one set of scalars $a_i \in \mathcal{F}$ which permits $\mathbf{x}$ to be expressed as a linear combination of the $\mathbf{u}_i$,

$$\mathbf{x} = a_1\mathbf{u}_1 + a_2\mathbf{u}_2 + \cdots + a_m\mathbf{u}_m = \sum_{i=1}^{m} a_i\mathbf{u}_i$$

Example 4.4

Consider all vectors in the plane. Then any pair of noncollinear vectors such as $\{\mathbf{x}, \mathbf{y}\}$, $\{\mathbf{x}', \mathbf{y}'\}$, and $\{\mathbf{x}'', \mathbf{y}''\}$ spans the two-dimensional space, since every vector in the plane can be represented as a combination of any one of these pairs. Another set of vectors which spans this space is $\{\mathbf{x}, \mathbf{y}, \mathbf{y}''\}$. Two ways of expressing a vector $\mathbf{w}$ in terms of these three vectors are shown in Fig. 4.1. The coefficients in the linear expansion of a vector in terms of a set of spanning vectors need not be unique. There is an infinite number of possibilities in this example. ■

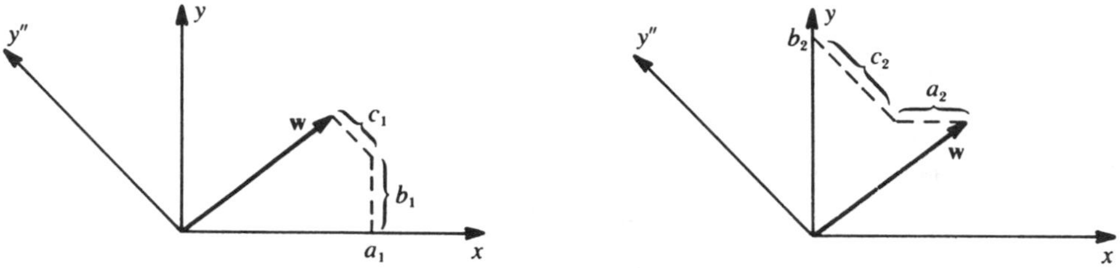

(a) $\mathbf{w} = a_1\mathbf{x} + b_1\mathbf{y} + c_1\mathbf{y}''$ (b) $\mathbf{w} = a_2\mathbf{x} + b_2\mathbf{y} - c_2\mathbf{y}''$

Figure 4.1

Basis Vectors

A set of basis vectors, $\mathcal{B} = \{\mathbf{v}_i\}$, for a space $\mathcal{X}$ is a subset of vectors in $\mathcal{X}$ which (*a*) spans the space $\mathcal{X}$ and (*b*) is a linearly independent set. Alternatively, a set of basis vectors is a set consisting of the minimum number of vectors required to span the space $\mathcal{X}$.

There are many different choices that could be made for basis vectors. Basis vectors for a given space are not unique, but every set of basis vectors will have the same number of elements. Once a basis is selected for $\mathcal{X}$, every vector $\mathbf{x} \in \mathcal{X}$ has a *unique* representation with respect to that basis. For this reason, basis vectors can be considered as the generalization of the notion of coordinate vectors discussed in two and three dimensions. Some sets of basis vectors are more convenient to use than others. One of the more convenient sets for $\mathcal{X}^n$ is the set $\{\mathbf{e}_i\}$ discussed in Example 4.1. Whenever a specific basis is not mentioned for $\mathcal{X}^n$, this natural cartesian basis set will be implied.

Dimension of a Vector Space

Definition 4.3: The dimension of a vector space $\mathcal{X}$, written dim ($\mathcal{X}$), is equal to the number of vectors in the basis set $\mathcal{B}$. Thus an n-dimensional linear vector space has n basis vectors.

Example 4.5

1. *Let $\mathcal{X}$ be the linear vector space consisting of all n component vectors*

$$\mathbf{x}^T = [x_1 \quad x_2 \quad x_3 \quad \cdots \quad x_n]$$

which satisfy $x_1 = x_2 = x_3 = \cdots = x_n$. Since the basis set for this space consists of the single vector $\mathbf{v}^T = [1 \quad 1 \quad 1 \quad \cdots \quad 1]$, this space is one dimensional.

2. *Let $\mathcal{X}$ be the linear space consisting of all polynomials of degree $n - 1$ or less*, $\{f(t) \mid f(t) = \alpha_1 + \alpha_2 t + \alpha_3 t^2 + \cdots + \alpha_n t^{n-1}, \alpha_i \in \mathcal{F}\}$. An obvious basis set is $\{1, t, t^2, \ldots, t^{n-1}\}$. Since the basis contains n elements, $\dim(\mathcal{X}) = n$, but $\mathcal{X}$ is not $\mathcal{X}^n$ as defined earlier.

It should be pointed out that a linear vector space can consist of a single element, the zero vector **0**. Such a space is said to be a zero-dimensional space. ■

4.6 SPECIAL OPERATIONS AND DEFINITIONS IN VECTOR SPACES

In order to generalize many of the useful concepts of familiar two- and three-dimensional spaces to n-dimensional spaces, some additional definitions are required.

Inner Product

Let $\mathcal{X}$ be an n-dimensional linear vector space defined over the scalar number field $\mathcal{F}$. If, to each pair of vectors **x** and **y** in $\mathcal{X}$, a unique scalar belonging to $\mathcal{F}$, called the inner product, is assigned, then $\mathcal{X}$ is said to be an inner product space. Various definitions for the inner product are possible. Any scalar valued function of **x** and **y** can be defined as the *inner product*, written $\langle \mathbf{x}, \mathbf{y} \rangle$, provided the following axioms are satisfied:

1. $\langle \mathbf{x}, \mathbf{y} \rangle = \overline{\langle \mathbf{y}, \mathbf{x} \rangle}$
2. $\langle \mathbf{x}, \alpha \mathbf{y}_1 + \beta \mathbf{y}_2 \rangle = \alpha \langle \mathbf{x}, \mathbf{y}_1 \rangle + \beta \langle \mathbf{x}, \mathbf{y}_2 \rangle$
3. $\langle \mathbf{x}, \mathbf{x} \rangle \geq 0$ for all **x** and $\langle \mathbf{x}, \mathbf{x} \rangle = 0$ if and only if $\mathbf{x} = \mathbf{0}$

If the scalar number field $\mathcal{F}$ is the set of complex numbers, then a commonly used definition of the complex inner product in $\mathcal{X}^n$ is

$$\langle \mathbf{x}, \mathbf{y} \rangle = \bar{\mathbf{x}}^T \mathbf{y}$$

If $\mathcal{F}$ is the set of reals, then the real inner product can be defined in the same way, but the complex conjugate on **x** is then superfluous. The inner product space defined on the real scalar field is called *Euclidean space*. Unless otherwise stated, the inner product will be assumed to be the complex inner product given above.

Combining axioms 1 and 2, it is easy to show that the inner product also satisfies

$$\langle \alpha \mathbf{x}_1 + \beta \mathbf{x}_2, \mathbf{y} \rangle = \bar{\alpha} \langle \mathbf{x}_1 \, \mathbf{y} \rangle + \bar{\beta} \langle \mathbf{x}_2, \mathbf{y} \rangle$$

The Grammian matrix introduced in Sec. 4.4 is generally defined in terms of the inner product as $\mathbf{G} = [\langle \mathbf{x}_i, \mathbf{x}_j \rangle]$. Some further definitions of inner products are given in the problems.

Vector Norm

Axioms 1 and 3 for inner products ensure that $\langle \mathbf{x}, \mathbf{x} \rangle$ is a nonnegative real number and is zero if and only if $\mathbf{x} = \mathbf{0}$. Because of these properties, the inner product can be used to define the *length* or *norm* of a vector as $\|\mathbf{x}\| = \langle \mathbf{x}, \mathbf{x} \rangle^{1/2}$. This norm will be used throughout this book unless an explicit statement to the contrary is made. It is called the quadratic norm, or in the case of real vector spaces, the Euclidean norm. In two or three dimensions, it is easy to see that this definition for the length of $\mathbf{x}$ satisfies the conditions of Euclidean geometry. It is a generalization to n dimensions of the theorem of Pythagoras.

Many other norms can be defined; the only requirements are that $\|\mathbf{x}\|$ be a nonnegative real scalar satisfying

1. $\|\mathbf{x}\| = 0$ if and only if $\mathbf{x} = 0$
2. $\|\alpha\mathbf{x}\| = |\alpha| \cdot \|\mathbf{x}\|$ for any scalar α
3. $\|\mathbf{x} + \mathbf{y}\| \leq \|\mathbf{x}\| + \|\mathbf{y}\|$

The last condition is called the triangle inequality for reasons which are obvious in two or three dimensions.

An important inequality, called the Cauchy-Schwarz inequality, can be expressed in terms of the norm

$$|\langle \mathbf{x}, \mathbf{y} \rangle| \leq \|\mathbf{x}\| \cdot \|\mathbf{y}\|$$

The equality holds if and only if $\mathbf{x}$ and $\mathbf{y}$ are linearly dependent.

Unit Vectors

A unit vector, $\hat{\mathbf{x}}$, is by definition a vector whose norm is unity, $\|\hat{\mathbf{x}}\| = 1$. Any nonzero vector $\mathbf{x}$ can be normalized to form a unit vector

$$\hat{\mathbf{x}} = \frac{\mathbf{x}}{\|\mathbf{x}\|}$$

Metric or Distance Measure

The concept of distance between two points (vectors are used synonymously with points) in a linear vector space can be introduced by using the norm. The distance between two points $\mathbf{x}$ and $\mathbf{y}$ is defined as the scalar function

$$\rho(\mathbf{x}, \mathbf{y}) = \|\mathbf{x} - \mathbf{y}\|$$

When the quadratic norm is used, this gives

$$\rho(\mathbf{x}, \mathbf{y}) = \langle \mathbf{x} - \mathbf{y}, \mathbf{x} - \mathbf{y} \rangle^{1/2}$$

Generalized Angles in n-Dimensional Spaces

The concept of angles between vectors can be generalized to real n-dimensional spaces by extending the notion of the dot product of two- or three-dimensional spaces,

$$\mathbf{x} \cdot \mathbf{y} = \langle \mathbf{x}, \mathbf{y} \rangle = \|\mathbf{x}\| \cdot \|\mathbf{y}\| \cos \theta$$

Thus, the cosine of the angle between $\mathbf{x} \in \mathcal{X}$ and $\mathbf{y} \in \mathcal{X}$ is

$$\cos \theta = \frac{1}{\|\mathbf{x}\| \cdot \|\mathbf{y}\|} \langle \mathbf{x}, \mathbf{y} \rangle = \langle \hat{\mathbf{x}}, \hat{\mathbf{y}} \rangle$$

It was mentioned earlier that various inner products can be defined. The particular choice of inner product dictates a specific meaning for the geometric concept of angle. Since $\langle \mathbf{x}, \mathbf{y} \rangle$ need not be real in spaces defined over the complex scalars, it is not particularly useful to try to place an interpretation upon angles in complex spaces.

Outer Product

The outer product (sometimes called the dyad product) of two vectors $\mathbf{x}$ and $\mathbf{y}$ belonging to $\mathcal{X}^n$ is

$$\mathbf{x}\rangle \langle \mathbf{y} = \mathbf{x}\bar{\mathbf{y}}^T$$

The brackets are motivated by a comparison with the usual definition for the inner product.

Multiplication of a Vector by an Arbitrary, Conformable Matrix

Since a vector can be represented as a column matrix with respect to a specific set of basis vectors (coordinate system), all the operations of matrix algebra must then apply. In particular, premultiplication by a conformable matrix yields another column matrix (vector). Column matrices are not the same thing as vectors, but just particular representations of vectors. Similar relations hold between matrices and linear transformations, and these are discussed in Chapter 6.

4.7 ORTHOGONAL VECTORS AND THEIR CONSTRUCTION

Any two vectors $\mathbf{x}$ and $\mathbf{y}$ which belong to a linear vector space $\mathcal{X}$ are said to be *orthogonal* if and only if

$$\langle \mathbf{x}, \mathbf{y} \rangle = 0$$

This is the natural generalization of the geometric concept of perpendicularity. Note that this definition of orthogonality indicates that the zero vector is orthogonal to every other vector. If each pair of vectors in a given set is mutually orthogonal, then the set is said to be an *orthogonal set.* If, in addition, each vector in this orthogonal set is a unit vector, then the set is said to be *orthonormal.* Each pair of orthonormal vectors $\hat{\mathbf{v}}_i$ and $\hat{\mathbf{v}}_j$ satisfies $\langle \hat{\mathbf{v}}_i, \hat{\mathbf{v}}_j \rangle = \delta_{ij}$, where δ_{ij} is the Kronecker delta and equals 1 if $i = j$ and 0 otherwise. Orthonormal vectors are convenient choices for basis vectors. The natural set of cartesian basis vectors $\mathbf{e}_i$ of Example 4.1 is the simplest example of an orthonormal set.

The Gram-Schmidt Process

Orthonormal vectors form a convenient basis set, so it is of interest to know how to construct an orthonormal set. Given any set of n linearly independent vectors $\{\mathbf{y}_i, i = 1, n\}$, an orthonormal set $\{\hat{\mathbf{v}}_i, i = 1, n\}$ can be constructed by using the Gram-Schmidt process. The process consists of two steps. First an orthogonal set $\{\mathbf{v}_i\}$ is constructed, and second, each vector in this set is normalized. Let $\mathbf{v}_1 = \mathbf{y}_1$ and select $\mathbf{v}_2$ as the vector formed from $\mathbf{y}_2$ by subtracting out the component in the direction of $\mathbf{v}_1$. This is equivalent to requiring that $\langle \mathbf{v}_1, \mathbf{v}_2 \rangle = 0$. Let $\mathbf{v}_2 = \mathbf{y}_2 - a\mathbf{v}_1$. Then in order to satisfy orthogonality,

$$a = \frac{\langle \mathbf{v}_1, \mathbf{y}_2 \rangle}{\langle \mathbf{v}_1, \mathbf{v}_1 \rangle}$$

so that

$$\mathbf{v}_2 = \mathbf{y}_2 - \frac{\langle \mathbf{v}_1, \mathbf{y}_2 \rangle}{\langle \mathbf{v}_1, \mathbf{v}_1 \rangle}\mathbf{v}_1$$

the next vector is chosen as

$$\mathbf{v}_3 = \mathbf{y}_3 - a_1\mathbf{v}_1 - a_2\mathbf{v}_2$$

and the two scalars a_i are chosen to satisfy

$$\langle \mathbf{v}_1, \mathbf{v}_3 \rangle = 0 \qquad \text{and} \qquad \langle \mathbf{v}_2, \mathbf{v}_3 \rangle = 0$$

This leads to

$$\mathbf{v}_3 = \mathbf{y}_3 - \frac{\langle \mathbf{v}_1, \mathbf{y}_3 \rangle}{\langle \mathbf{v}_1, \mathbf{v}_1 \rangle}\mathbf{v}_1 - \frac{\langle \mathbf{v}_2, \mathbf{y}_3 \rangle}{\langle \mathbf{v}_2, \mathbf{v}_2 \rangle}\mathbf{v}_2$$

Continuing in this manner leads to the general equation

$$\mathbf{v}_i = \mathbf{y}_i - \sum_{k=1}^{i-1} \frac{\langle \mathbf{v}_k, \mathbf{y}_i \rangle}{\langle \mathbf{v}_k, \mathbf{v}_k \rangle}\mathbf{v}_k$$

After all n of the vectors $\mathbf{v}_i$ are computed, the normalization

$$\hat{\mathbf{v}}_i = \frac{\mathbf{v}_i}{\|\mathbf{v}_i\|}, \qquad i = 1, \ldots, n$$

gives the desired orthonormal set.

Example 4.6

Construct a set of orthonormal vectors from

$$\mathbf{y}_1^T = [1 \quad 0 \quad 1], \qquad \mathbf{y}_2^T = [-1 \quad 2 \quad 1], \qquad \mathbf{y}_3^T = [0 \quad 1 \quad 2]$$

Since the Grammian gives $|\mathbf{G}| = 4$, these vectors are linearly independent.

Step 1. Let

$$\mathbf{v}_1 = \mathbf{y}_1 = [1 \quad 0 \quad 1]^T$$

$$\mathbf{v}_2 = \mathbf{y}_2 - \frac{\cancelto{0}{\langle \mathbf{v}_1, \mathbf{y}_2 \rangle}}{\langle \mathbf{v}_1, \mathbf{v}_1 \rangle}\mathbf{v}_1 = \mathbf{y}_2$$

In this case $\mathbf{y}_1$ and $\mathbf{y}_2$ are already orthogonal:

$$\mathbf{v}_3 = \mathbf{y}_3 - \frac{\langle \mathbf{v}_1, \mathbf{y}_3 \rangle}{\langle \mathbf{v}_1, \mathbf{v}_1 \rangle}\mathbf{v}_1 - \frac{\langle \mathbf{v}_2, \mathbf{y}_3 \rangle}{\langle \mathbf{v}_2, \mathbf{v}_2 \rangle}\mathbf{v}_2 = \begin{bmatrix} -\frac{1}{3} & -\frac{1}{3} & \frac{1}{3} \end{bmatrix}^T$$

Step 2. Normalize $\mathbf{v}_i$ to get $\hat{\mathbf{v}}_i$:

$$\hat{\mathbf{v}}_1 = \frac{\mathbf{v}_1}{\|\mathbf{v}\|} = \frac{1}{\sqrt{2}}\begin{bmatrix} 1 \\ 0 \\ 1 \end{bmatrix}, \qquad \hat{\mathbf{v}}_2 = \frac{\mathbf{v}_2}{\|\mathbf{v}\|} = \frac{1}{\sqrt{6}}\begin{bmatrix} -1 \\ 2 \\ 1 \end{bmatrix}, \qquad \hat{\mathbf{v}}_3 = \frac{\mathbf{v}_3}{\|\mathbf{v}_3\|} = \frac{1}{\sqrt{3}}\begin{bmatrix} -1 \\ -1 \\ 1 \end{bmatrix}$$

■

4.8 VECTOR EXPANSIONS AND THE RECIPROCAL BASIS VECTORS

Every vector $\mathbf{x} \in \mathcal{X}$ has a unique expansion

$$\mathbf{x} = \sum_{i=1}^{n} a_i \mathbf{v}_i$$

with respect to the basis set $\mathcal{B} = \{\mathbf{v}_i, i = 1, n\}$. Taking the inner product of $\mathbf{v}_j$ and $\mathbf{x}$ gives

$$\langle \mathbf{v}_j, \mathbf{x} \rangle = \left\langle \mathbf{v}_j, \sum_{i=1}^{n} a_i \mathbf{v}_i \right\rangle = \sum_{i=1}^{n} a_i \langle \mathbf{v}_j, \mathbf{v}_i \rangle$$

If the basis set is orthonormal so that $\langle \mathbf{v}_j, \mathbf{v}_i \rangle = \delta_{ij}$, then the jth expansion coefficient is $a_j = \langle \mathbf{v}_j, \mathbf{x} \rangle$.

Example 4.7

Use the set of orthonormal vectors generated in Example 4.6 as basis vectors and find the three coefficients a_i which allow $\mathbf{z} = [4 \quad -8 \quad 1]^T$ to be written as

$$\mathbf{z} = a_1 \hat{\mathbf{v}}_1 + a_2 \hat{\mathbf{v}}_2 + a_3 \hat{\mathbf{v}}_3$$

Because $\{\hat{\mathbf{v}}_i\}$ is an orthonormal set,

$$a_1 = \langle \hat{\mathbf{v}}_1, \mathbf{z} \rangle, \qquad a_2 = \langle \hat{\mathbf{v}}_2, \mathbf{z} \rangle \qquad \text{and} \qquad a_3 = \langle \hat{\mathbf{v}}_3, \mathbf{z} \rangle$$

so that

$$\mathbf{z} = \frac{5}{\sqrt{2}}\hat{\mathbf{v}}_1 - \frac{19}{\sqrt{6}}\hat{\mathbf{v}}_2 + \frac{5}{\sqrt{3}}\hat{\mathbf{v}}_3$$

■

Reciprocal Basis Vectors

When the basis set $\mathcal{B} = \{\mathbf{v}_i\}$ is not orthonormal, the preceding simple results no longer hold, but every vector $\mathbf{z} \in \mathcal{X}$ still has a unique expansion

$$\mathbf{z} = \sum_{i=1}^{n} a_i \mathbf{v}_i$$

Another set of n vectors, called the *reciprocal basis vectors* $\{\mathbf{r}_1, \mathbf{r}_2, \ldots, \mathbf{r}_n\}$, is introduced to facilitate finding the expansion coefficients. These reciprocal or dual basis vectors are defined by n^2 equations, each of the form

$$\langle \mathbf{r}_i, \mathbf{v}_j \rangle = \delta_{ij}$$

In matrix form this set of equations becomes

$$\mathbf{RB} = \mathbf{I}$$

where $\mathbf{B}$ is the $n \times n$ matrix whose columns are $\mathbf{v}_i$ and $\mathbf{R}$ is the $n \times n$ matrix whose *rows* are $\bar{\mathbf{r}}_i^T$. Thus $\mathbf{R} = \mathbf{B}^{-1}$, so the reciprocal basis vector $\mathbf{r}_i$ is the conjugate transpose of the ith row of $\mathbf{B}^{-1}$. With the reciprocal basis vectors available, it is apparent that the expansion coefficients are given by $a_i = \langle \mathbf{r}_i, \mathbf{z} \rangle$ so that

$$\mathbf{z} = \sum_{i=1}^{n} \langle \mathbf{r}_i, \mathbf{z} \rangle \mathbf{v}_i$$

The $\mathbf{B}$ matrix need not always be square. For example, the basis set for a three-dimensional subspace of a five-dimensional space would consist of three vectors, each with five components. The definition of the reciprocal basis set still requires that $\mathbf{RB} = \mathbf{I}$. Now, however, $\mathbf{R}$ is not the usual inverse of $\mathbf{B}$ but, rather, is what is called the left inverse. This is discussed more fully in Sec. 5.6.

Example 4.8

Let $\mathbf{v}_1 = [1 \quad 0]^T$ and $\mathbf{v}_2 = [-1 \quad 1]^T$. Express the vector $\mathbf{z} = [3 \quad 3]^T$ in terms of this basis set.

First, the reciprocal basis set is found from

$$\mathbf{R} = \begin{bmatrix} 1 & -1 \\ 0 & 1 \end{bmatrix}^{-1} = \begin{bmatrix} 1 & 1 \\ 0 & 1 \end{bmatrix}$$

so that $\mathbf{r}_1 = [1 \quad 1]^T$ and $\mathbf{r}_2 = [0 \quad 1]^T$. The coefficients are

$$a_1 = \langle \mathbf{r}_1, \mathbf{z} \rangle = 6 \qquad \text{and} \qquad a_2 = \langle \mathbf{r}_2, \mathbf{z} \rangle = 3$$

so that $\mathbf{z} = 6\mathbf{v}_1 + 3\mathbf{v}_2$. A sketch of the $\mathbf{r}_i$ and $\mathbf{v}_i$ vectors may be informative. ■

4.9 LINEAR MANIFOLDS, SUBSPACES, AND PROJECTIONS

Let $\mathcal{X}$ be a linear vector space defined over the number field $\mathcal{F}$. A nonempty subset, $\mathcal{M}$, of $\mathcal{X}$ is called a *linear manifold* if for each vector $\mathbf{x}$ and $\mathbf{y}$ in $\mathcal{M}$, the combination $\alpha\mathbf{x} + \beta\mathbf{y}$ is also in $\mathcal{M}$ for arbitrary $\alpha, \beta \in \mathcal{F}$. The zero vector, of necessity, is included in every linear manifold.

A closed linear manifold is called a *subspace*. In finite dimensional spaces, there is no distinction between linear manifolds and subspaces, because every finite dimensional manifold is closed.

A subspace of an n-dimensional linear vector space $\mathcal{X}^n$ is itself a linear vector space contained within $\mathcal{X}^n$, but with dimension $m \leq n$. A *proper* subspace has $m < n$.

Example 4.9

The space defined in Problem 4.11 is a three-dimensional subspace of $\mathcal{X}^5$, and the first space defined in Example 4.5 is a one-dimensional subspace of $\mathcal{X}^n$. In general, since $\mathcal{X}^n$ has n basis vectors, deleting any one of the basis vectors leaves a basis set for an $n - 1$ dimensional subspace, deleting two allows the definition of an $n - 2$ dimensional subspace, etc. Note that $\mathbf{0}$ must be an element of every subspace. If it is the only element, then that subspace is zero dimensional. ■

Given any two linear vector spaces $\mathcal{U}$ and $\mathcal{V}$ defined over the same number field $\mathcal{F}$, a new vector space $\mathcal{X}$ can be constructed from their sum:

$$\mathcal{X} = \mathcal{U} + \mathcal{V}$$

This means that every vector $\mathbf{x}$ in $\mathcal{X}$ can be written as

$$\mathbf{x} = \mathbf{u} + \mathbf{v}, \qquad \mathbf{u} \in \mathcal{U}, \qquad \mathbf{v} \in \mathcal{V}$$

If there is one and only one pair $\mathbf{u}, \mathbf{v}$ for each $\mathbf{x}$, then $\mathcal{X}$ is called the *direct sum* of $\mathcal{U}$ and $\mathcal{V}$, written

$$\mathcal{X} = \mathcal{U} \oplus \mathcal{V}$$

This implies that the only vector common to both $\mathcal{U}$ and $\mathcal{V}$ is $\mathbf{0}$. In this case,

$$\dim(\mathcal{X}) = \dim(\mathcal{U}) + \dim(\mathcal{V})$$

Conversely the space $\mathcal{X}$ can be decomposed into two (or more) subspaces. For the vector $\mathbf{x}$ defined above, $\mathbf{u}$ is called the projection of $\mathbf{x}$ on $\mathcal{U}$ and $\mathbf{v}$ is the projection of $\mathbf{x}$ on $\mathcal{V}$.

The Projection Theorem

Let $\mathcal{X}^n$ be an n-dimensional vector space, and let $\mathcal{U}$ be a subspace of dimension $m < n$. Then for every $\mathbf{x} \in \mathcal{X}^n$ there exists a vector $\mathbf{u} \in \mathcal{U}$, called the projection of $\mathbf{x}$ on $\mathcal{U}$, which satisfies

$$\langle \mathbf{x} - \mathbf{u}, \mathbf{y} \rangle = 0$$

for every vector $\mathbf{y} \in \mathcal{U}$. This says that $\mathbf{w} = \mathbf{x} - \mathbf{u}$ is orthogonal to $\mathbf{y}$. In other words, $\mathbf{u}$ is the orthogonal projection of $\mathbf{x}$ on $\mathcal{U}$, and $\mathbf{w}$ is orthogonal to the subspace $\mathcal{U}$.

For proofs of the projection theorem, see References 40 and 47.

For a given $\mathbf{x}$ there is a unique projection $\mathbf{u}$, but there are infinitely many $\mathbf{x}$ vectors which have the same projection. The set of all vectors in $\mathcal{X}^n$ which are orthogonal to $\mathcal{U}$ forms an $n - m$ dimensional subspace of $\mathcal{X}^n$, called the *orthogonal complement* of $\mathcal{U}$, written $\mathcal{U}^\perp$. Every $\mathbf{w} \in \mathcal{U}^\perp$ is orthogonal to every $\mathbf{y} \in \mathcal{U}$. The set of all vectors which are orthogonal to $\mathcal{U}^\perp$ is the subspace $\mathcal{U}$, that is, $(\mathcal{U}^\perp)^\perp = \mathcal{U}$. By using any subspace $\mathcal{U}$ and its orthogonal complement, an n-dimensional space can be expressed as the direct sum $\mathcal{X}^n = \mathcal{U} \oplus \mathcal{U}^\perp$. Each vector $\mathbf{x} \in \mathcal{X}^n$ can be written uniquely as $\mathbf{x} = \mathbf{u} + \mathbf{v}$. It is easy to show that $\|\mathbf{x}\|^2 = \|\mathbf{u}\|^2 + \|\mathbf{v}\|^2$ because of the orthogonal nature of this decomposition.

The projection theorem and related concepts can be used to develop the theory of least squares estimation and the theory of *generalized* or *pseudo-inverses* of non-square or singular matrices. Some of these applications appear in the next chapter.

4.10 PRODUCT SPACES

Let $\mathcal{X}$ and $\mathcal{Y}$ be arbitrary linear vector spaces defined over the field $\mathcal{F}$. Let $\mathbf{x} \in \mathcal{X}$ and $\mathbf{y} \in \mathcal{Y}$. Then the *product space* $\mathcal{X} \times \mathcal{Y}$ is defined as all ordered pairs of vectors $(\mathbf{x}, \mathbf{y})$. It can be verified that the product space satisfies the conditions of Sec. 4.3 and is therefore a linear vector space. Let $\mathbf{z}_1 = (\mathbf{x}_1, \mathbf{y}_1)$ and $\mathbf{z}_2 = (\mathbf{x}_2, \mathbf{y}_2)$ belong to $\mathcal{X} \times \mathcal{Y}$. Then addition and scalar multiplication are defined by $\mathbf{z}_1 + \mathbf{z}_2 = (\mathbf{x}_1 + \mathbf{x}_2, \mathbf{y}_1 + \mathbf{y}_2) = \mathbf{z}_2 + \mathbf{z}_1$ and $\alpha\mathbf{z}_1 = (\alpha\mathbf{x}_1, \alpha\mathbf{y}_1)$. The zero vector in $\mathcal{X} \times \mathcal{Y}$ is the ordered pair of zero elements $\mathbf{0} \in \mathcal{X}$ and $\mathbf{0} \in \mathcal{Y}$.

Product spaces can be formed as the product of any number of spaces. The familiar Euclidean three-dimensional space is a product space formed from products of the real line $\mathcal{R}^1$, $\mathcal{R}^3 = \mathcal{R}^1 \times \mathcal{R}^1 \times \mathcal{R}^1$. Another common product space is formed from n products of the space of square integrable functions, $\mathcal{L}_2[a, b] \times \mathcal{L}_2[a, b] \times \cdots \times \mathcal{L}_2[a, b]$. Each element in this space is of the form $(f_1(t), f_2(t), \ldots, f_n(t))$, where $f_i(t) \in \mathcal{L}_2[a, b]$. Elements in this product space are usually written more simply as n component vectors $\mathbf{f}(t)$.

The spaces used in forming a product space need not be the same types of spaces. If $\mathcal{R}^1$ is considered as a vector space with elements t and if $\mathbf{x} \in \mathcal{X}^m$, $\mathbf{y} \in \mathcal{Y}$, then elements of $\mathcal{R}^1 \times \mathcal{X}^m \times \mathcal{Y}$ are $\mathbf{z} = (t, \mathbf{x}, \mathbf{y})$.

ILLUSTRATIVE PROBLEMS

Vectors in Two and Three Dimensions

4.1 If two vectors $\mathbf{v}^T = [3 \quad -5 \quad 6]$ and $\mathbf{w}^T = [\alpha \quad 2 \quad 2]$ are known to be orthogonal, what is α?

Assuming that the given components are expressed with respect to a common coordinate system, orthogonality requires

$$\langle \mathbf{v}, \mathbf{w} \rangle = \mathbf{v}^T\mathbf{w} = 0 \qquad \text{or} \qquad 3\alpha - 10 + 12 = 0$$

so that $\alpha = -\frac{2}{3}$.

4.2 If $\mathbf{v}^T = [3 \quad -5 \quad 6]$ and $\mathbf{w}^T = [5 \quad 8]$, find $\langle \mathbf{v}, \mathbf{w} \rangle$ and the two outer products.

Since $\mathbf{v}$ and $\mathbf{w}$ have different numbers of components and hence belong to different dimensional spaces, their inner product is not defined. The outer products are, however,

$$\mathbf{v}\mathbf{w}^T = \begin{bmatrix} 3 \\ -5 \\ 6 \end{bmatrix} [5 \quad 8] = \begin{bmatrix} 15 & 24 \\ -25 & -40 \\ 30 & 48 \end{bmatrix} = (\mathbf{w}\mathbf{v}^T)^T$$

4.3 Consider the nonzero vector $\mathbf{v}$ with complex components $\mathbf{v} = \begin{bmatrix} j \\ 1 \end{bmatrix}$. Compute $\mathbf{v}^T\mathbf{v}$ and $\bar{\mathbf{v}}^T\mathbf{v}$.

Vector multiplication gives

$$\mathbf{v}^T\mathbf{v} = (j)(j) + 1 = 0$$

$$\bar{\mathbf{v}}^T\mathbf{v} = (-j)(j) + 1 = 2$$

Therefore, if the real form of the inner product $\langle \mathbf{v}, \mathbf{v} \rangle = \mathbf{v}^T\mathbf{v}$ is used to define length, nonzero vectors can have zero "length." When the complex form of the inner product $\langle \mathbf{v}, \mathbf{v} \rangle = \bar{\mathbf{v}}^T\mathbf{v}$ is used, this cannot happen.

4.4 Find the component of $\mathbf{v}^T = [2 \quad -3 \quad -4]$ in the direction of the vector $\mathbf{w}^T = [1 \quad 2 \quad 1]$.

First find the unit vector $\hat{\mathbf{w}}$ in the desired direction:

$$\|\mathbf{w}\| = \langle \mathbf{w}, \mathbf{w} \rangle^{1/2} = \sqrt{6}$$

$$\hat{\mathbf{w}} = \frac{1}{\sqrt{6}}\mathbf{w}$$

Then form the inner product:

$$\langle \mathbf{v}, \hat{\mathbf{w}} \rangle = \frac{1}{\sqrt{6}}(2 \quad -6 \quad -4) = \frac{-1}{\sqrt{6}}8$$

This result indicates that $\mathbf{v}$ has a component along the negative $\mathbf{w}$ direction and its magnitude is $8/\sqrt{6}$.

4.5 Find the cross products $\mathbf{v} \times \mathbf{w}$ and $\mathbf{w} \times \mathbf{v}$ if $\mathbf{v}^T = [v_1 \quad v_2 \quad v_3]$ and $\mathbf{w}^T = [w_1 \quad w_2 \quad w_3]$ are real vectors.

If we let $\mathbf{e}_1$, $\mathbf{e}_2$, and $\mathbf{e}_3$ be unit vectors along the three mutually orthogonal coordinate axes, the cross product can be computed using the following determinant:

$$\mathbf{v} \times \mathbf{w} = \left| \begin{bmatrix} \mathbf{e}_1 & \mathbf{e}_2 & \mathbf{e}_3 \\ v_1 & v_2 & v_3 \\ w_1 & w_2 & w_3 \end{bmatrix} \right|$$

Using Laplace expansion with respect to row one gives

$$\mathbf{v} \times \mathbf{w} = \mathbf{e}_1(v_2w_3 - v_3w_2) - \mathbf{e}_2(v_1w_3 - v_3w_1) + \mathbf{e}_3(v_1w_2 - v_2w_1)$$

Since $\mathbf{e}_1 = \begin{bmatrix} 1 \\ 0 \\ 0 \end{bmatrix}$, $\mathbf{e}_2 = \begin{bmatrix} 0 \\ 1 \\ 0 \end{bmatrix}$, $\mathbf{e}_3 = \begin{bmatrix} 0 \\ 0 \\ 1 \end{bmatrix}$, the column matrix representations are

$$\mathbf{v} \times \mathbf{w} = \begin{bmatrix} v_2 w_3 - v_3 w_2 \\ v_3 w_1 - v_1 w_3 \\ v_1 w_2 - v_2 w_1 \end{bmatrix}$$

$$\mathbf{w} \times \mathbf{v} = \begin{bmatrix} v_3 w_2 - v_2 w_3 \\ v_1 w_3 - v_3 w_1 \\ v_2 w_1 - v_1 w_2 \end{bmatrix}$$

4.6 Show that $\mathbf{v} \times \mathbf{w}$ of the previous problem can be written as the product of a skew-symmetric matrix and $\mathbf{w}$, or another skew-symmetric matrix and $\mathbf{v}$.

Direct matrix multiplication verifies that

$$\mathbf{v} \times \mathbf{w} = \begin{bmatrix} 0 & -v_3 & v_2 \\ v_3 & 0 & -v_1 \\ -v_2 & v_1 & 0 \end{bmatrix} \begin{bmatrix} w_1 \\ w_2 \\ w_3 \end{bmatrix} = \begin{bmatrix} 0 & w_3 & -w_2 \\ -w_3 & 0 & w_1 \\ w_2 & -w_1 & 0 \end{bmatrix} \begin{bmatrix} v_1 \\ v_2 \\ v_3 \end{bmatrix}$$

4.7 From the results of the preceding problem, what can be said about the product $\mathbf{x}^T\mathbf{A}\mathbf{x}$ if $\mathbf{x}$ is a real three-component vector and $\mathbf{A}$ is skew-symmetric?

Any 3×3 skew-symmetric matrix could be used to define a 3×1 vector, and then $\mathbf{Ax}$ would represent a cross product. Therefore, $\mathbf{Ax}$ is a vector perpendicular to $\mathbf{x}$. Therefore, $\langle \mathbf{x}, \mathbf{Ax} \rangle = \mathbf{x}^T\mathbf{A}\mathbf{x} = \mathbf{0}$ for every real $\mathbf{x}$. In fact, if $\mathbf{A}$ is skew-symmetric of arbitrary dimension, the result $\mathbf{x}^T\mathbf{A}\mathbf{x} = \mathbf{0}$ is true for any real, conformable vector $\mathbf{x}$.

Defining a Vector Space

4.8 Does the set of all vectors in the first and fourth quadrants of the plane form a vector space?

No. The first four conditions of Sec. 4.3, page 94, hold. However, if $\mathbf{x}$ is in the first or fourth quadrant, $-\mathbf{x}$ is in the second or third quadrant, so condition 5 is not satisfied and neither is 6 when negative scalars are considered.

4.9 Does the set of all three-dimensional vectors inside a sphere of finite radius constitute a linear vector space?

No. If $\mathbf{x}$ and $\mathbf{y}$ are inside the sphere, $\mathbf{x} + \mathbf{y}$ need not be. If a is sufficiently large, $a\mathbf{x}$ will also extend outside the sphere. Conditions 1 and 6 are not satisfied.

4.10 Consider all vectors defined by points in a plane passing through a three-dimensional space. Is this set a linear vector space?

This is a linear vector space if and only if the plane passes through the origin. If it does not, condition 4 is not satisfied.

4.11 What is the dimension of the space $\mathcal{X}$ defined as the set of all linear combinations of

$$\mathbf{x}_1^T = [1 \quad 2 \quad 3 \quad 4 \quad 5]$$

$$\mathbf{x}_2^T = [1 \quad 0 \quad 0 \quad 0 \quad 1]$$

$$\mathbf{x}_3^T = [0 \quad 1 \quad 1 \quad 0 \quad 0]$$

The manner in which $\mathcal{X}$ is defined guarantees that $\mathbf{x}_1$, $\mathbf{x}_2$, and $\mathbf{x}_3$ span this space. Since the Grammian gives $|\mathbf{G}| = 98$, the set is linearly independent and thus constitutes a basis set. Since there are three vectors in the basis set, the dimension of $\mathscr{X}$, written dim ($\mathscr{X}$), is three even though every vector in $\mathscr{X}$ has five components.

Linear Dependence, Independence, and Degeneracy

4.12 Prove that the addition of the zero vector $\mathbf{0}$ to any set of linearly independent vectors yields a set of linearly dependent vectors.

Let $\mathcal{V} = \{\mathbf{x}_i, i = 1, n\}$ be a set of linearly independent vectors. This means that $a_1\mathbf{x}_1 + a_2\mathbf{x}_2 + \cdots + a_n\mathbf{x}_n = \mathbf{0}$ requires $a_1 = 0$, $i = 1, n$. Select $a_{n+1} \neq 0$. Then

$$a_1\mathbf{x}_1 + a_2\mathbf{x}_2 + \cdots + a_n\mathbf{x}_n + a_{n+1}\mathbf{0} = \mathbf{0}$$

so the set of vectors $\{\mathbf{0}, \mathbf{x}_i, i = 1, n\}$ is a linearly dependent set.

4.13 Use the Grammian to test the following vectors for linear dependence:

$$\mathbf{x}_1^T = [1 \quad 1 \quad 0 \quad 0], \qquad \mathbf{x}_2^T = [1 \quad 1 \quad 1 \quad 1], \qquad \mathbf{x}_3^T = [0 \quad 0 \quad 1 \quad 1]$$

The Grammian determinant is

$$|\mathbf{G}| = \left\| \begin{bmatrix} 2 & 2 & 0 \\ 2 & 4 & 2 \\ 0 & 2 & 2 \end{bmatrix} \right\| = 0$$

Hence the three vectors are linearly dependent. Note that the Grammian is symmetric for vectors with real components. In general, it is a Hermitian matrix.

4.14 Consider two real vectors $\mathbf{x}_1$ and $\mathbf{x}_2$ expressed as 2×1 column vectors in terms of the natural coordinate vectors. Show that $|\mathbf{G}|$ is the square of the area of the parallelogram which has $\mathbf{x}_1$ and $\mathbf{x}_2$ as sides.

We have

$$|\mathbf{G}| = \begin{vmatrix} \mathbf{x}_1^T\mathbf{x}_1 & \mathbf{x}_1^T\mathbf{x}_2 \\ \mathbf{x}_2^T\mathbf{x}_1 & \mathbf{x}_2^T\mathbf{x}_2 \end{vmatrix} = \begin{vmatrix} x_1^2 & x_1x_2\cos\theta \\ x_1x_2\cos\theta & x_2^2 \end{vmatrix}$$

where x_1 and x_2 are the magnitudes of $\mathbf{x}_1$ and $\mathbf{x}_2$ and θ is the included angle. The definition of the inner product of Sec. 4.2 has been used in arriving at this result. Expanding gives

$$|\mathbf{G}| = x_1^2x_2^2(1 - \cos^2\theta) = x_1^2x_2^2\sin^2\theta$$

Considering x_1 as the base of the parallelogram, the height is $x_2 \sin\theta$, so the result is proven.

4.15 What are the rank and degeneracy of the following matrices?

a. $\mathbf{A} = \begin{bmatrix} 6 & 2 & 4 \\ 2 & 0 & 2 \\ 1 & -1 & 2 \end{bmatrix}$ b. $\mathbf{B} = \begin{bmatrix} 4 & 3 & 7 & 1 \\ 2 & 6 & 2 & 10 \\ 8 & 6 & 14 & 2 \\ 1 & 3 & 1 & 5 \end{bmatrix}$

c. $\mathbf{C} = \begin{bmatrix} 6 & -4 & -4 & -9 \\ 24 & 3 & 0 & -9 \\ -14 & 3 & 4 & 12 \\ 48 & 25 & 16 & 9 \end{bmatrix}$

a. $|\mathbf{A}| = \mathbf{0}$ so that $r_A < 3$. Picking the submatrix $\mathbf{A}_1 = \begin{bmatrix} 6 & 2 \\ 2 & 0 \end{bmatrix}$ gives $|\mathbf{A}_1| = -4 \neq 0$, so $r_A = 2$. The degeneracy is $q_A = n - r_A$ or $q_A = 1$. The one linear dependency relation between columns can be written as $\mathbf{x}_2 = \mathbf{x}_1 - \mathbf{x}_3$.

b. $|\mathbf{B}| = \mathbf{0}$ (row 2 is twice row 4). Therefore $r_B < 4$. Any 3×3 matrix $\mathbf{B}_1$ formed by crossing out a row and column also has $|\mathbf{B}_1| = 0$ since row 3 is twice row 1. Therefore $r_B < 3$. It is easy to find a nonzero 2×2 determinant, so $r_B = 2$ and $q_B = 4 - 2 = 2$.

c. $|\mathbf{C}| = 0$ as does the determinant of every 3×3 submatrix. In fact, there are just two linearly independent column vectors

$$\mathbf{x}_1 = [1 \quad 3 \quad -2 \quad 5]^T \quad \text{and} \quad \mathbf{x}_2 = [-1 \quad 0 \quad 1 \quad 4]^T$$

which can be used to generate $\mathbf{C}$. Then column 1 of $\mathbf{C}$ is $\mathbf{c}_1 = 8\mathbf{x}_1 + 2\mathbf{x}_2$. Likewise, $\mathbf{c}_2 = 1\mathbf{x}_1 + 5\mathbf{x}_2$, $\mathbf{c}_3 = 4\mathbf{x}_2$, and $\mathbf{c}_4 = -3\mathbf{x}_1 + 6\mathbf{x}_2$. In this case $r_C = 2$ and $q_C = 2$.

Gram-Schmidt Process

4.16 Use the Gram-Schmidt process to construct a set of orthonormal vectors from

$$\mathbf{x}_1 = \begin{bmatrix} 1+j \\ 1-j \\ j \end{bmatrix}, \quad \mathbf{x}_2 = \begin{bmatrix} 2j \\ 1-2j \\ 1+2j \end{bmatrix}, \quad \mathbf{x}_3 = \begin{bmatrix} 1 \\ j \\ 5j \end{bmatrix}$$

The Grammian is first used to verify linear independence of the set $\{\mathbf{x}_i\}$. The complex form of the inner product must be used:

$$\mathbf{G} = [\langle \mathbf{x}_i, \mathbf{x}_j \rangle] = [\bar{\mathbf{x}}_i^T \mathbf{x}_j] = \begin{bmatrix} 5 & 7 & 5 \\ 7 & 14 & 8+4j \\ 5 & 8-4j & 27 \end{bmatrix}$$

The determinant is $|\mathbf{G}| = 377 \neq 0$; therefore, the $\mathbf{x}_i$ are linearly independent.
Step 1. Construct an orthogonal set of $\mathbf{v}_i$:

$$\mathbf{v}_1 = \mathbf{x}_1$$

$$\mathbf{v}_2 = \begin{bmatrix} 2j \\ 1-2j \\ 1+2j \end{bmatrix} - \frac{7}{5}\begin{bmatrix} 1+j \\ 1-j \\ j \end{bmatrix} = \frac{1}{5}\begin{bmatrix} -7+3j \\ -2-3j \\ 5+3j \end{bmatrix}$$

Anticipating their need in advance, the products $\langle \mathbf{v}_2, \mathbf{v}_2 \rangle = 21/5$ and $\langle \mathbf{v}_2, \mathbf{x}_3 \rangle = 1 + 4j$ are computed.

$$\mathbf{v}_3 = \begin{bmatrix} 1 \\ j \\ 5j \end{bmatrix} - \frac{5}{5}\begin{bmatrix} 1+j \\ 1-j \\ j \end{bmatrix} - \frac{5(1+4j)}{21(5)}\begin{bmatrix} -7+3j \\ -2-3j \\ 5+3j \end{bmatrix} = \frac{1}{21}\begin{bmatrix} 19+4j \\ -31+53j \\ 7+61j \end{bmatrix}$$

Step 2. Normalize to obtain

$$\hat{\mathbf{v}}_1 = \frac{1}{\sqrt{5}}\begin{bmatrix} 1+j \\ 1-j \\ j \end{bmatrix}, \quad \hat{\mathbf{v}}_2 = \frac{1}{\sqrt{105}}\begin{bmatrix} -7+3j \\ -2-3j \\ 5+3j \end{bmatrix}$$

Using $\langle \mathbf{v}_3, \mathbf{v}_3 \rangle = \dfrac{7917}{441}$ gives

$$\hat{\mathbf{v}}_3 = \sqrt{\frac{441}{7917}}\,\mathbf{v}_3 = \frac{1}{\sqrt{7917}}\begin{bmatrix} 19 + 4j \\ -31 + 53j \\ 7 + 61j \end{bmatrix}$$

It is a good exercise in the use of the complex inner product to verify that these results satisfy $\langle \hat{\mathbf{v}}_i, \hat{\mathbf{v}}_j \rangle = \delta_{ij}$.

Geometry in n-Dimensional Spaces

4.17 The equation of a plane in n-dimensions is $\langle \mathbf{c}, \mathbf{x} \rangle = a$, where $\mathbf{c}$ is the normal to the plane and a is a scalar constant. Find the point on the plane nearest the origin and find the distance to this point.

Any $\mathbf{x}$ can be decomposed into a component $\mathbf{x}_n$ normal to the plane plus $\mathbf{x}_p$ parallel to the plane. Then $\langle \mathbf{c}, \mathbf{x} \rangle = \langle \mathbf{c}, \mathbf{x}_n \rangle + \langle \mathbf{c}, \mathbf{x}_p \rangle = a$ for every $\mathbf{x}$ terminating on the plane. Since $\mathbf{c}$ and $\mathbf{x}_p$ are orthogonal, $\langle \mathbf{c}, \mathbf{x}_p \rangle = 0$. Since $\mathbf{c}$ and $\mathbf{x}_n$ are parallel, $\mathbf{x}_n = \pm\|\mathbf{x}_n\|\mathbf{c}/\|\mathbf{c}\|$. Then $\langle \mathbf{c}, \mathbf{x}_n \rangle = \pm\langle \mathbf{c}, \mathbf{c} \rangle\|\mathbf{x}_n\|/\|\mathbf{c}\| = a$. The $+$ or $-$ sign must be selected to agree with the sign of a. Solving gives the minimum distance as $\|\mathbf{x}_n\| = |a|/\|\mathbf{c}\|$ and the closest point is $\mathbf{x}_n = a\mathbf{c}/\langle \mathbf{c}, \mathbf{c} \rangle$.

4.18 Find the minimum distance from the origin to a point (x_1, x_2) on the line $6x_1 + 2x_2 = 4$, and find the coordinates of that point.

The normal to the line is $\mathbf{c} = [6 \quad 2]^T$, and so $\|\mathbf{c}\| = \sqrt{40}$. The results of Problem 4.17 apply, and the minimum distance is $\|\mathbf{x}_n\| = 4/\sqrt{40} = 2/\sqrt{10}$. The point nearest the origin is

$$\mathbf{x}_n = \frac{1}{5}\begin{bmatrix} 3 \\ 1 \end{bmatrix}$$

4.19 Generalize the concepts of lines, planes, spheres, cones, and convex sets to n-dimensional Euclidean spaces.

The generalization of a line is the set of all vectors satisfying

$$\mathbf{x} = a\mathbf{v} + \mathbf{k} \tag{1}$$

where $\mathbf{v}$ and $\mathbf{k}$ are constant vectors and a is a scalar.

The generalization of a plane is called a *hyperplane*. An $n - 1$ dimensional hyperplane consists of the set of n-dimensional vectors $\mathbf{x}$ satisfying

$$\langle \mathbf{c}, \mathbf{x} \rangle = a \tag{2}$$

where $\mathbf{c}$ and a are a constant vector and scalar, respectively. Since a subspace always contains the $\mathbf{0}$ vector, equation (*1*) represents a subspace only if $\mathbf{k}$ is zero. Equation (*2*) represents a subspace only if a is zero.

Points on or inside a hypersphere of radius R are defined by the set of all $\mathbf{x}$ satisfying

$$\langle \mathbf{x}, \mathbf{x} \rangle \le R^2$$

A right circular cone of semi-vertex angle θ, with its axis in the direction of a unit vector $\mathbf{n}$, and with vertex at the origin, consists of the set of all $\mathbf{x}$ satisfying

$$\langle \mathbf{n}, \mathbf{x} \rangle / \|\mathbf{x}\| = \cos\theta$$

If $\theta = \pi/2$, the cone degenerates to a hyperplane containing the origin.

If the line segments connecting every two points in a set contain only points in the set, the set is *convex*. A convex set of vectors in n-dimensional space is a set for which the vector

$$\mathbf{z} = a\mathbf{x}_1 + (1 - a)\mathbf{x}_2$$

belongs to the set for every $\mathbf{x}_1, \mathbf{x}_2$ in the set and for every real scalar satisfying $0 \leq a \leq 1$.

Some Generalizations

4.20 Let $\mathfrak{L}_2[a, b]$ be the linear space consisting of all real square integrable functions of t, that is, all functions $f(t)$ satisfying $\int_a^b f(\tau)^2\,d\tau < \infty$. Define a suitable inner product, norm, and metric.

The three requirements for an inner product can be shown to be satisfied by

$$\langle f, g\rangle = \int_a^b f(\tau)g(\tau)\,d\tau \qquad \text{where } f, g \in \mathfrak{L}_2[a, b]$$

As in other cases, a norm can always be defined as

$$\|f\| = \langle f, f\rangle^{1/2}$$

and the metric or distance measure between two functions can be defined in terms of the norm,

$$\rho(f, g) = \|f - g\|$$

The inner product space defined by this set of functions and the inner product definition is a complete infinite dimensional linear vector space. A space $\mathfrak{X}$ is *complete* if every convergent (Cauchy) sequence of elements in $\mathfrak{X}$ converges to a limit which is also in $\mathfrak{X}$. Any infinite dimensional inner product space which is complete is called a *Hilbert space*.

4.21 Use the results of the previous problem to prove that the mean value of a real function is always less than or equal to its root-mean-square value (rms).

We are to prove that

$$\frac{1}{T}\int_0^T f(\tau)\,d\tau \leq \left[\frac{1}{T}\int_0^T f^2(\tau)\,d\tau\right]^{1/2}$$

For any functions f and $g \in \mathfrak{L}_2[0, T]$,

$$\|f - g\| \geq 0$$

or

$$\langle f - g, f - g\rangle \geq 0 \qquad \text{or} \qquad \langle f, f\rangle \geq 2\langle f, g\rangle \quad \langle g, g\rangle$$

Choosing the particular function $g = \text{constant} = \dfrac{1}{T}\displaystyle\int_0^T f(\tau)\,d\tau$ gives

$$\int_0^T f^2\,dt \geq \frac{1}{T}\left[\int_0^T f\,dt\right]^2$$

Dividing both sides by T and taking the square root gives the desired result.

4.22 Is it always necessary to define the norm in terms of the inner product?

No. Linear spaces can be defined with a norm, but without any mention of an inner product. Such spaces are called *normed linear spaces*. If they are infinite dimensional spaces, and are complete, then they are usually referred to as *Banach spaces*.

Two examples of other valid norms for finite dimensional spaces whose elements $\mathbf{x}$ are ordered n-tuples of scalars belonging to $\mathfrak{F}$ are

$$\|\mathbf{x}\| = \max_i \{|x_1|, |x_2|, \ldots, |x_n|\}$$

and

$$\|\mathbf{x}\|_p = [|x_1|^p + |x_2|^p + \cdots + |x_n|^p]^{1/p}$$

where p is real, $1 \leq p \leq \infty$. The quadratic norm is a special case with $p = 2$. All the required axioms for a norm are satisfied for these examples.

4.23 Let $\mathfrak{X}$ be a linear space consisting of all $n \times n$ matrices defined over $\mathfrak{F}$. (Let $n = 2$ for simplicity.) Show that

a. $\langle \mathbf{A}, \mathbf{B} \rangle = \operatorname{Tr}(\bar{\mathbf{A}}^T\mathbf{B})$ is a valid inner product; and

b. an orthonormal basis for this space is the set of matrices

$$\left\{\begin{bmatrix} 1 & 0 \\ 0 & 0 \end{bmatrix}, \begin{bmatrix} 0 & 1 \\ 0 & 0 \end{bmatrix}, \begin{bmatrix} 0 & 0 \\ 1 & 0 \end{bmatrix}, \begin{bmatrix} 0 & 0 \\ 0 & 1 \end{bmatrix}\right\}$$

a. The inner product of two elements must yield a scalar. Obviously the trace gives a scalar. In addition, the three axioms must be satisfied:

i. $\langle \mathbf{A}, \mathbf{B} \rangle = \overline{\langle \mathbf{B}, \mathbf{A} \rangle}$ but

$$\langle \mathbf{A}, \mathbf{B} \rangle = \operatorname{Tr}(\bar{\mathbf{A}}^T\mathbf{B})$$

and

$$\overline{\langle \mathbf{B}, \mathbf{A} \rangle} = \overline{\operatorname{Tr}(\bar{\mathbf{B}}^T\mathbf{A})} = \operatorname{Tr}(\mathbf{B}^T\bar{\mathbf{A}}) = \operatorname{Tr}(\mathbf{B}^T\bar{\mathbf{A}})^T = \operatorname{Tr}(\bar{\mathbf{A}}^T\mathbf{B})$$

ii. $\langle \mathbf{A}, \alpha\mathbf{B}_1 + \beta\mathbf{B}_2 \rangle = \alpha\langle \mathbf{A}, \mathbf{B}_1 \rangle + \beta\langle \mathbf{A}, \mathbf{B}_2 \rangle$ but

$$\begin{aligned} \langle \mathbf{A}, \alpha\mathbf{B}_1 + \beta\mathbf{B}_2 \rangle &= \operatorname{Tr}[\bar{\mathbf{A}}^T(\alpha\mathbf{B}_1 + \beta\mathbf{B}_2)] = \alpha \operatorname{Tr}(\bar{\mathbf{A}}^T\mathbf{B}_1) + \beta \operatorname{Tr}(\bar{\mathbf{A}}^T\mathbf{B}_2) \\ &= \alpha\langle \mathbf{A}, \mathbf{B}_1 \rangle + \beta\langle \mathbf{A}, \mathbf{B}_2 \rangle \end{aligned}$$

iii. $\langle \mathbf{A}, \mathbf{A} \rangle \geq 0$ for all $\mathbf{A}$ and equals zero if and only if $\mathbf{A} = \mathbf{0}$. Let $\mathbf{A} = [a_{ij}]$. Then

$$\begin{aligned} \langle \mathbf{A}, \mathbf{A} \rangle &= \operatorname{Tr}(\mathbf{A}^T\mathbf{A}) = \bar{a}_{11}a_{11} + \bar{a}_{21}a_{21} + \bar{a}_{12}a_{12} + \bar{a}_{22}a_{22} \\ &= |a_{11}|^2 + |a_{21}|^2 + |a_{12}|^2 + |a_{22}|^2 \end{aligned}$$

This is obviously nonnegative, and can vanish only if $a_{ij} = 0$ for all i and j.

b. First show that the set is orthonormal. $\mathbf{G} = [\langle \mathbf{V}_i, \mathbf{V}_j \rangle]$, where $\mathbf{V}_i$ are the four indicated matrices. Simple calculation shows that $\mathbf{G}$ is the 4×4 unit matrix, thus the set is orthonormal. They span the space since every 2×2 matrix $[a_{ji}]$ can be written as

$$\mathbf{A} = a_{11}\mathbf{V}_1 + a_{12}\mathbf{V}_2 + a_{21}\mathbf{V}_3 + a_{22}\mathbf{V}_4$$

An orthonormal set which spans the space is an orthonormal basis set.

Miscellaneous Applications

4.24 Find the projection of $\mathbf{y} = [1 \quad -3 \quad 4 \quad 2 \quad 8]^T$ on the subspace spanned by

$$\mathbf{x}_1 = [1 \quad 2 \quad -3 \quad 1 \quad 0]^T \quad \text{and} \quad \mathbf{x}_2 = [0 \quad 1 \quad 3 \quad 3 \quad 1]^T$$

The dimension of the subspace is two, since $|\mathbf{G}| = 284 \neq 0$. An orthonormal basis is constructed:

$$\hat{\mathbf{v}}_1 = \frac{\mathbf{x}_1}{\|\mathbf{x}_1\|} = \frac{1}{\sqrt{15}}[1 \quad 2 \quad -3 \quad 1 \quad 0]^T$$

$$\hat{\mathbf{v}}_2 = \frac{\mathbf{x}_2 - \langle \mathbf{x}_2, \hat{\mathbf{v}}_1 \rangle \hat{\mathbf{v}}_1}{\|\mathbf{x}_2 - \langle \mathbf{x}_2, \hat{\mathbf{v}}_1 \rangle \hat{\mathbf{v}}_1\|} = \frac{1}{\sqrt{4260}} \begin{bmatrix} 4 \\ 23 \\ 33 \\ 49 \\ 15 \end{bmatrix}$$

The projection of $\mathbf{y}$ on this subspace is

$$\mathbf{y}_p = \langle \hat{\mathbf{v}}_1, \mathbf{y} \rangle \hat{\mathbf{v}}_1 + \langle \hat{\mathbf{v}}_2, \mathbf{y} \rangle \hat{\mathbf{v}}_2$$

$$= -\sqrt{15}\,\hat{\mathbf{v}}_1 + \frac{285}{\sqrt{4260}}\hat{\mathbf{v}}_2 = \begin{bmatrix} -1 \\ -2 \\ 3 \\ -1 \\ 0 \end{bmatrix} + \frac{285}{4260} \begin{bmatrix} 4 \\ 23 \\ 33 \\ 49 \\ 15 \end{bmatrix} = \frac{1}{284} \begin{bmatrix} -208 \\ -131 \\ 1479 \\ 647 \\ 285 \end{bmatrix}$$

The projection $\mathbf{y}_p$ is the closest vector in the subspace to $\mathbf{y}$, in the sense that

$$\|\mathbf{y} - \mathbf{y}_p\|^2 \leq \|\mathbf{y} - \mathbf{z}\|^2$$

for every $\mathbf{z}$ in the subspace. These results are directly related to the problem of least squares approximations, considered in the next chapter.

4.25 Consider the dc motor of Problem 2.5, with transfer function

$$\frac{\Omega(s)}{V(s)} = \frac{K'}{s + a}$$

If the motor is initially at rest, $\omega(0) = 0$, find the input $v(t)$ which gives an angular velocity $\omega(T) = 100$ at a fixed time T, while minimizing a measure of the input energy,

$$J = \int_0^T v^2(\tau)\, d\tau$$

The input-output relationship is written in terms of the *system weighting function.* Since $W(t, 0) = \mathcal{L}^{-1}\{K'/(s + a)\} = K'e^{-at}$, the weighting function is $W(t, \tau) = K'e^{-a(t-\tau)}$. Then

$$\omega(T) = \int_0^T W(T, \tau) v(\tau)\, d\tau$$

This is in the form of an inner product, $100 = \langle W(T, \tau), v(\tau) \rangle$, so the Cauchy-Schwarz inequality can be used to give

$$100 = |\langle W(T, \tau), v(\tau) \rangle| \leq \|W(T, \tau)\| \cdot \|v(\tau)\|$$

The minimum value of $\|v(\tau)\|$ is obtained when the equality holds. Therefore,

$$\|v(\tau)\| = \frac{100}{\|W(T, \tau)\|} = \frac{100}{\left\{\int_0^T [K'e^{-a(T-\tau)}]^2\, d\tau\right\}^{1/2}} = \frac{100\sqrt{2a}}{K'[1 - e^{-2aT}]^{1/2}}$$

The equality holds if and only if $v(t)$ and $W(T, t)$ are linearly dependent. This means

$v(t) = kW(T, t)$ for some scalar k. Comparing gives $\|v\| = |k|\|W\| = 100/\|W\|$ and so

$$|k| = \frac{100}{\|W\|^2} \quad \text{and} \quad v_{\text{optimal}}(t) = \frac{100}{\|W\|^2} W(T, t) = \frac{200ae^{-a(T-t)}}{K'[1-e^{-2aT}]}$$

PROBLEMS

Linear Independence, Orthonormal Basis Vectors, and Reciprocal Basis Vectors

4.26 Consider $\mathbf{x}_1 = [1 \quad 2 \quad 3]^T$, $\mathbf{x}_2 = [1 \quad -2 \quad 3]^T$, $\mathbf{x}_3 = [0 \quad 1 \quad 1]^T$.
a. Show that this set is linearly independent.
b. Generate an orthonormal set using the Gram-Schmidt procedure.

4.27 Considering $\mathbf{x}_1$, $\mathbf{x}_2$, and $\mathbf{x}_3$ of Problem 4.26 as a basis set, find the reciprocal basis set.

4.28 Express the vector $\mathbf{z} = [6 \quad 4 \quad -3]^T$ in terms of the orthonormal basis set $\{\hat{\mathbf{v}}_i\}$ of Problem 4.26.

4.29 Express the vector $\mathbf{z} = [6 \quad 4 \quad -3]^T$ in terms of the original basis set $\{\mathbf{x}_i\}$ of Problem 4.26 by using the reciprocal basis vectors $\{\mathbf{r}_i\}$ found in Problem 4.27.

4.30 Find the reciprocal basis set if the basis vectors are $\mathbf{x}_1 = [4 \quad 2 \quad 1]^T$, $\mathbf{x}_2 = [2 \quad 6 \quad 3]^T$, $\mathbf{x}_3 = [1 \quad 3 \quad 5]^T$.

4.31 Given $\mathbf{x}_1 = [1 \quad 1 \quad 1]^T$, $\mathbf{x}_2 = [1 \quad -1 \quad 1]^T$, $\mathbf{x}_3 = [1 \quad 0 \quad 0]^T$. Use these as basis vectors and find reciprocal basis vectors. Also, express $\mathbf{z} = [6 \quad 3 \quad 1]^T$ in terms of the basis vectors.

4.32 Show that an orthonormal basis set and the corresponding set of reciprocal basis vectors are the same.

4.33 Verify that the following four $\mathbf{y}_i$ vectors are linearly independent by computing their Grammian. Then use them to construct an orthonormal set using the Gram-Schmidt process. Then use the orthonormal vectors to expand the vector $\mathbf{x} = [12.3 \quad 9.8 \quad -4.03 \quad 33.33]^T$.

The given **y** *vectors*

$$\mathbf{y}_1 = \begin{bmatrix} 4.4400002\text{E}+01 \\ 1.2800000\text{E}+01 \\ 1.5000000\text{E}+00 \\ -2.1000000\text{E}+01 \end{bmatrix}, \quad \mathbf{y}_2 = \begin{bmatrix} 7.7700000\text{E}+00 \\ 2.1500000\text{E}+01 \\ 1.0000000\text{E}+01 \\ 0.0000000\text{E}+00 \end{bmatrix},$$

$$\mathbf{y}_3 = \begin{bmatrix} -3.3329999\text{E}+00 \\ 4.1250000\text{E}+00 \\ 6.6670001\text{E}-01 \\ 1.0000000\text{E}+00 \end{bmatrix}, \quad \mathbf{y}_4 = \begin{bmatrix} 9.1250000\text{E}+00 \\ 2.1222000\text{E}+00 \\ -3.0500000\text{E}+00 \\ 4.4400001\text{E}+00 \end{bmatrix}$$

4.34 Compute the Grammian for the following vectors and draw conclusions about their linear independence.

The given **y** *vectors*

$$\mathbf{y}_1 = \begin{bmatrix} 1.0000000E+00 \\ 1.0000000E+00 \\ 1.0000000E+00 \\ 1.0000000E+00 \end{bmatrix}, \quad \mathbf{y}_2 = \begin{bmatrix} 1.0001000E+00 \\ 9.9989998E-01 \\ 1.0000000E+00 \\ 1.0000000E+00 \end{bmatrix},$$

$$\mathbf{y}_3 = \begin{bmatrix} -2.0000000E+00 \\ -1.9999000E+00 \\ -2.0000000E+00 \\ -2.0000000E+00 \end{bmatrix}$$

4.35 Find the orthogonal projection of the vectors

$$\mathbf{x}_1 = \begin{bmatrix} 1 \\ 1 \\ 1 \\ 1 \\ 1 \end{bmatrix} \quad \text{and} \quad \mathbf{x}_2 = \begin{bmatrix} -4 \\ 2 \\ -8 \\ 3 \\ 9 \end{bmatrix}$$

on the subspace spanned by the following three vectors.

The given **y** *vectors*

$$\mathbf{y}_1 = \begin{bmatrix} 1.0000000E+01 \\ 5.0000000E+00 \\ -5.0000000E+00 \\ 1.0000000E+00 \\ 6.0000000E+00 \end{bmatrix}, \quad \mathbf{y}_2 = \begin{bmatrix} 1.1000000E+01 \\ 4.0000000E+00 \\ -1.1000000E+01 \\ -2.0000000E+00 \\ 6.0000000E+00 \end{bmatrix},$$

$$\mathbf{y}_3 = \begin{bmatrix} 1.0000000E+00 \\ 0.0000000E+00 \\ -1.0000000E+00 \\ -1.0000000E+00 \\ -1.0000000E+00 \end{bmatrix}$$

Miscellaneous

4.36 Determine the dimension of the vector space spanned by $\mathbf{x}_1 = [1 \quad 2 \quad 2 \quad 1]^T$, $\mathbf{x}_2 = [1 \quad 0 \quad 0 \quad 1]^T$, $\mathbf{x}_3 = [3 \quad 4 \quad 4 \quad 3]^T$.

4.37 Find the minimum distance from the origin to the plane $2x_1 + 3x_2 - x_3 = -5$ and find coordinates of the point on the plane nearest the origin.

4.38 Let $\mathbf{c} = [1 \quad 2 \quad -1]^T$ and $\mathbf{y} = [2 \quad 5 \quad 3]^T$. Find the projection of $\mathbf{y}$ that is parallel to the family of planes defined by $\langle \mathbf{c}, \mathbf{x} \rangle = \text{constant}$.

4.39 Show that the various Fourier series expansion formulas are special cases of the general expansion formula in an infinite dimensional linear inner product space:

$$\mathbf{x} = \sum_{i=1}^{\infty} \langle \mathbf{r}_i, \mathbf{x} \rangle \mathbf{v}_i$$

• **4.40** Under what conditions does $\langle \mathbf{x}, \mathbf{y} \rangle = \mathbf{x}^T \mathbf{A} \mathbf{y}$ define a valid inner product for an n-dimension vector space defined over the real number field?

4.41 If $\mathcal{X}^m$ and $\mathcal{X}^n$ are m- and n-dimensional linear vector spaces, respectively, then the product space $\mathcal{X}^m \times \mathcal{X}^n$ is itself a linear vector space consisting of all ordered pairs of $\mathbf{x} \in \mathcal{X}^m$, $\mathbf{y} \in \mathcal{X}^n$. That is,

$$\mathcal{X}^m \times \mathcal{X}^n = \{(\mathbf{x}, \mathbf{y}); \mathbf{x} \in \mathcal{X}^m, \mathbf{y} \in \mathcal{X}^n\}$$

If $\mathcal{X}^m$ has an inner product $\langle \mathbf{x}_1, \mathbf{x}_2 \rangle_m$ and $\mathcal{X}^n$ has an inner product $\langle \mathbf{y}_1, \mathbf{y}_2 \rangle_n$, show that the appropriate inner product for $\mathcal{X}^m \times \mathcal{X}^n$ is

$$\left\langle \begin{bmatrix} \mathbf{x}_1 \\ \mathbf{y}_1 \end{bmatrix}, \begin{bmatrix} \mathbf{x}_2 \\ \mathbf{y}_2 \end{bmatrix} \right\rangle = \langle \mathbf{x}_1, \mathbf{x}_2 \rangle_m + \langle \mathbf{y}_1\ \mathbf{y}_2 \rangle_n$$

4.42 Let $\mathcal{L}_2^n$ be the linear space consisting of all complex valued square integrable n component vector functions of a scalar variable $t \in [a, b]$, $\mathbf{f}(t) = [f_i(t)]$, where $i = 1, n$. Define an appropriate inner product and norm for this space.

4.43 Prove the Cauchy-Schwarz inequality given on page 100.

4.44 Let $\mathcal{R}^n$ be an n-dimensional Euclidean space with an orthonormal basis $\mathcal{B} = \{\mathbf{v}_i, i = 1, n\}$. Prove that for any $\mathbf{x} \in \mathcal{R}^n$,

$$\|\mathbf{x}\|^2 \geq \sum_{i=1}^{m} |\langle \mathbf{v}_i, \mathbf{x} \rangle|^2$$

where the summation is over any subset of m basis vectors. This is called Bessel's inequality. If $m = n$, the equality holds.

4.45 Let $\mathbf{A}$ be a real matrix whose columns are linearly independent. Is it possible to construct an orthogonal matrix $\mathbf{T}$ such that $\mathbf{TA}$ is upper triangular?

4.46 Let $\mathbf{y}_i$ be a set of linearly independent real vectors. Show that an orthonormal set $\hat{\mathbf{v}}_i$ can be generated by the following matrix version of the Gram-Schmidt procedure.

$$\mathbf{T}_1 = \mathbf{I}; \qquad \mathbf{v}_i = \mathbf{T}_i \mathbf{y}_i; \qquad \hat{\mathbf{v}}_i = \mathbf{v}_i / [\mathbf{v}_i^T \mathbf{v}_i]$$

$$\mathbf{T}_{i+1} = \mathbf{T}_i - \hat{\mathbf{v}}_i \hat{\mathbf{v}}_i^T$$

The sequence of matrices $\mathbf{T}_i$ are projection operators. See Sec. 6.7.

Chapter

5

SIMULTANEOUS LINEAR ALGEBRAIC EQUATIONS

5.1 INTRODUCTION

The task of solving a set of simultaneous linear algebraic equations is frequently encountered by engineers and scientists in all fields. In modern control theory many problems of system identification, estimation, pole placement, and optimization can be cast in this framework.

5.2 STATEMENT OF THE PROBLEM AND CONDITIONS FOR SOLUTIONS

Consider the set of simultaneous linear algebraic equations

$$\begin{aligned}
&a_{11}x_1 + a_{12}x_2 + \cdots + a_{1n}x_n = y_1\\
&a_{21}x_1 + a_{22}x_2 + \cdots + a_{2n}x_n = y_2\\
&\vdots\\
&a_{m1}x_1 + a_{m2}x_2 + \cdots + a_{mn}x_n = y_m
\end{aligned}$$

In matrix notation this is simply $\mathbf{Ax} = \mathbf{y}$, where the elements a_{ij} of the $m \times n$ matrix $\mathbf{A}$ are known, as are the scalar components y_i of the $m \times 1$ vector $\mathbf{y}$. The $n \times 1$ vector $\mathbf{x}$ contains the unknowns which are to be determined if possible. Any vector, say $\mathbf{x}_1$, which satisfies all m of these equations is called a *solution*. Not every set of simultaneous equations has a solution. The *augmented matrix*, defined by $\mathbf{W} = [\mathbf{A} \mid \mathbf{y}]$, indicates whether or not solutions exist. In fact,

1. If $r_W \neq r_A$, no solution exists. The equations are *inconsistent*.
2. If $r_W = r_A$, at least one solution exists.
   (a) If $r_W = r_A = n$, there is a *unique* solution for **x**.
   (b) If $r_W = r_A < n$, then there is an infinite set of solution vectors.

It is clearly impossible for r_A to exceed n, so the only possibilities are that there are no solutions, or exactly one solution, or an infinity of solutions.

5.3 THE ROW-REDUCED ECHELON FORM OF A MATRIX

The rank of certain matrices plays a vital role in the above discussion and in many other contexts in modern control theory. An efficient method of determining the rank of a matrix is to put the matrix into *row-reduced echelon* form [91]. This form is obtained by using elementary row operations, just as described in Problem 3.18, page 87, for Gaussian elimination. Elementary row operations are performed until the first nonzero term in each row is unity and all terms below these ones are zero. This form of the matrix is the *echelon form*, but is not yet in row-reduced echelon form. Even in this intermediate form, the rank of the matrix is obvious by inspection. It is the number of nonzero rows in the matrix. Recall from Chapter 3 that the elementary matrices are nonsingular, and therefore multiplication of a matrix by them does not change the rank of that matrix.

Additional row operations are carried out until the leading ones in each row are the only nonzero terms in their respective columns. That is, any nonzero terms *above* the leading ones of the echelon form are now removed. The result is the desired row-reduced echelon form. Every matrix has a *unique* row-reduced echelon form. Some texts refer to this as the Hermite normal form of the matrix. The rank of a matrix is certainly obvious from its row-reduced echelon form, but its usefulness goes far beyond that. Any nonsingular **A** matrix just reduces to the unit matrix **I**. Therefore, aside from confirming the rank, it seems that all the information in **A** is "lost" by this reduction. The usefulness of the technique usually comes about by applying the reduction to some matrix other than just the coefficient matrix **A** in a set of simultaneous equations. If it is applied to the composite matrix $\mathbf{W} = [\mathbf{A} \mid \mathbf{y}]$ defined above, and if the resultant row-reduced echelon form is called $\mathbf{W}' = [\mathbf{A}' \mid \mathbf{y}']$, then

$$\text{rank } \mathbf{W} = \text{rank } \mathbf{W}' \qquad \text{and} \qquad \text{rank } \mathbf{A} = \text{rank } \mathbf{A}'$$

so that inspection of $\mathbf{W}'$ reveals instantly which of the previous categories applies. Further, when solutions do exist they are obtained directly from $\mathbf{W}'$ with little or no additional effort.

Example 5.1

In the following nine situations a set of simultaneous equations of the form $\mathbf{Ax} = \mathbf{y}$ is being considered. In each case the **W** matrix containing **A** and **y** is displayed, followed by the row-reduced echelon form $\mathbf{W}'$. These are then used to draw conclusions about the original set of equations. Note that in cases 1, 2, and 3 the number of equations, m, is equal to the number of unknowns in $\mathbf{x}$, n. In cases 4 and 5 m is less than n and in cases 6 through 9 m is greater than n.

1.

$$\mathbf{W} = \left[\begin{array}{ccc|c} 1 & 0 & 1 & 3 \\ 0 & 1 & 1 & -1 \\ 1 & 0 & 1 & 5 \end{array}\right] \qquad \mathbf{W}' = \left[\begin{array}{ccc|c} 1 & 0 & 0 & 0 \\ 0 & 1 & 1 & 0 \\ 0 & 0 & 0 & 1 \end{array}\right]$$

From the reduced form, $r_A = 2$ and $r_W = 3$, so no solutions exist.

2.

$$\mathbf{W} = \left[\begin{array}{ccc|c} 1 & 0 & 1 & 3 \\ 0 & 1 & 1 & -1 \\ 1 & 0 & 1 & 3 \end{array}\right] \qquad \mathbf{W}' = \left[\begin{array}{ccc|c} 1 & 0 & 1 & 3 \\ 0 & 1 & 1 & -1 \\ 0 & 0 & 0 & 0 \end{array}\right]$$

From this, $r_A = 2 = r_W < n = 3$. Therefore, an infinite number of solutions exist, and $\mathbf{W}'$ tells us that $\mathbf{x}_1 + \mathbf{x}_3 = 3$ and $\mathbf{x}_2 + \mathbf{x}_3 = -1$.

3.

$$\mathbf{W} = \left[\begin{array}{ccc|c} 1 & 2 & 3 & 4 \\ 2 & 1 & 2 & 7 \\ 3 & 2 & 1 & 1 \end{array}\right] \qquad \mathbf{W}' = \left[\begin{array}{ccc|c} 1 & 0 & 0 & 2.125 \\ 0 & 1 & 0 & -4.5 \\ 0 & 0 & 1 & 3.625 \end{array}\right]$$

Thus $r_A = r_W = n = 3$, so a unique solution exists and it is just $\mathbf{y}'$.

4.

$$\mathbf{W} = \left[\begin{array}{ccc|c} 1 & -1 & 2 & 8 \\ -1 & 2 & 0 & 2 \end{array}\right] \qquad \mathbf{W}' = \left[\begin{array}{ccc|c} 1 & 0 & 4 & 18 \\ 0 & 1 & 2 & 10 \end{array}\right]$$

Here $r_A = r_W = 2 < n$, so solutior exist but are not unique. They all must satisfy $\mathbf{x}_1 + 4\mathbf{x}_3 = 18$ and $\mathbf{x}_2 + 2\mathbf{x}_3 = 10$.

5.

$$\mathbf{W} = \left[\begin{array}{cccc|c} 1 & 2 & 3 & 1 & 1 \\ -4 & 5 & 1 & 9 & 2 \\ -2 & 8 & 6 & 10 & 3 \end{array}\right] \qquad \mathbf{W}' = \left[\begin{array}{cccc|c} 1 & 0 & 1 & -1 & 0 \\ 0 & 1 & 1 & 1 & 0 \\ 0 & 0 & 0 & 0 & 1 \end{array}\right]$$

This case has $r_A = 2$, $r_W = 3$, so no solutions exist.

6.

$$\mathbf{W} = \left[\begin{array}{cc|c} 1 & 2 & 2 \\ 3 & 4 & 3 \\ 5 & 6 & 4 \end{array}\right] \qquad \mathbf{W}' = \left[\begin{array}{cc|c} 1 & 0 & -1 \\ 0 & 1 & 1.5 \\ 0 & 0 & 0 \end{array}\right]$$

Since $r_A = r_W = 2 = n$, there is a unique solution given by the first two components of $\mathbf{y}'$, namely $\mathbf{x} = [-1 \quad 1.5]^T$.

7.

$$\mathbf{W} = \left[\begin{array}{cc|c} 1 & 2 & 2 \\ 3 & 4 & 3 \\ 5 & 6 & -4 \end{array}\right] \qquad \mathbf{W}' = \left[\begin{array}{cc|c} 1 & 0 & 0 \\ 0 & 1 & 0 \\ 0 & 0 & 1 \end{array}\right]$$

Note that this is the same as case 6 except for the sign of one component of $\mathbf{y}$. The results are quite different. Since $r_A = 2$ and $r_W = 3$, there is no solution.

8.

$$\mathbf{W} = \left[\begin{array}{rrr|l} 1 & 3 & 5 & 3 \\ 1 & 4 & 6 & 3.5 \\ -1 & 5 & 3 & 1 \\ -1 & 4 & 2 & 0.5 \\ 1 & 3 & 5 & 3 \end{array}\right] \qquad \mathbf{W}' = \left[\begin{array}{rrr|l} 1 & 0 & 2 & 1.5 \\ 0 & 1 & 1 & 0.5 \\ 0 & 0 & 0 & 0 \\ 0 & 0 & 0 & 0 \\ 0 & 0 & 0 & 0 \end{array}\right]$$

Here, even though there are more equations than unknowns, there are still solutions—in fact, an infinite number—all of which satisfy $\mathbf{x}_1 + 2\mathbf{x}_3 = 1.5$ and $\mathbf{x}_2 + \mathbf{x}_3 = 0.5$.

9. Changing only the $\mathbf{y}$ vector of the previous case to $[3 \quad 3 \quad 1 \quad 1 \quad 3]^T$ leads to

$$\mathbf{W}' = \left[\begin{array}{rrr|l} 1 & 0 & 2 & 0 \\ 0 & 1 & 1 & 0 \\ 0 & 0 & 0 & 1 \\ 0 & 0 & 0 & 0 \\ 0 & 0 & 0 & 0 \end{array}\right]$$

Since $r_A = 2$ and $r_W = 3$, there is no solution. In addition to demonstrating the various categories of simultaneous equations and illustrating row-reduced echelon forms, this example is intended to show that characterizations strictly in terms of numbers of equations and numbers of unknowns are clearly inadequate. ■

5.4 SOLUTION BY PARTITIONING

It is assumed in this section that $r_A = r_W$ so that one or more solutions exist. By definition, the $m \times n$ coefficient matrix $\mathbf{A}$ contains a nonsingular $r_A \times r_A$ matrix. The original equations $\mathbf{Ax} = \mathbf{y}$ can always be rearranged and partitioned into

$$\left[\begin{array}{c|c} \mathbf{A}_1 & \mathbf{A}_2 \\ \hline \mathbf{A}_3 & \mathbf{A}_4 \end{array}\right] \left[\begin{array}{c} \mathbf{x}_1 \\ \hline \mathbf{x}_2 \end{array}\right] = \left[\begin{array}{c} \mathbf{y}_1 \\ \hline \mathbf{y}_2 \end{array}\right]$$

where $\mathbf{A}_1$ is $r_A \times r_A$ and nonsingular. Depending on the relations between m, n, and r_A, some of the terms in the partitioned equation will not be required. For example, if $m = n = r_A$, then $\mathbf{A}_1 = \mathbf{A}$, $\mathbf{x}_1 = \mathbf{x}$, and $\mathbf{y}_1 = \mathbf{y}$. The general case is treated here, and then

$$\mathbf{A}_1\mathbf{x}_1 + \mathbf{A}_2\mathbf{x}_2 = \mathbf{y}_1 \qquad \text{or} \qquad \mathbf{x}_1 = \mathbf{A}_1^{-1}[\mathbf{y}_1 - \mathbf{A}_2\mathbf{x}_2]$$

The *degeneracy* of $\mathbf{A}$ is $q_A = n - r_A$. The values of the q_A components of $\mathbf{x}_2$ are completely arbitrary, and generate the q_A parameter family of solutions for $\mathbf{x}$ mentioned above in case 2(*b*). If $r_A = n$, as in case 2(*a*), then $\mathbf{A}_2$, $\mathbf{A}_4$, and $\mathbf{x}_2$ will not be present in the partitioned equation. In that case the unique solution is $\mathbf{x} = \mathbf{A}_1^{-1}\mathbf{y}_1$. If in addition $m = n$, then $\mathbf{A}_3$ and $\mathbf{y}_2$ will not be present and $\mathbf{x} = \mathbf{A}^{-1}\mathbf{y}$. This is the simple case mentioned in Chapter 3, and $\mathbf{x}$ could be computed by using Cramer's rule or various matrix inversion techniques. However, Gaussian elimination or similar reduction techniques are more efficient for large values of n [53].

Example 5.2

Consider once more the situation of case 2 in Example 5.1, and find all solutions $\mathbf{x}$ for

$$\begin{bmatrix} 1 & 0 & 1 \\ 0 & 1 & 1 \\ 1 & 0 & 1 \end{bmatrix} \begin{bmatrix} x_1 \\ x_2 \\ x_3 \end{bmatrix} = \begin{bmatrix} 3 \\ -1 \\ 3 \end{bmatrix}$$

Here $r_A = 2$ (rows 1 and 3 are identical) and $r_W = 2$ also. An infinite set of solutions exists. Let

$$\mathbf{A}_1 = \begin{bmatrix} 1 & 0 \\ 0 & 1 \end{bmatrix} \qquad \mathbf{x}_1 = \begin{bmatrix} x_1 \\ x_2 \end{bmatrix} \qquad \mathbf{y}_1 = \begin{bmatrix} 3 \\ -1 \end{bmatrix}$$

$$\mathbf{A}_2 = \begin{bmatrix} 1 \\ 1 \end{bmatrix} \qquad \mathbf{x}_2 = x_3 \qquad \mathbf{y}_2 = 3$$

Then

$$\mathbf{x}_1 = \begin{bmatrix} 1 & 0 \\ 0 & 1 \end{bmatrix}^{-1} \left\{ \begin{bmatrix} 3 \\ -1 \end{bmatrix} - \begin{bmatrix} 1 \\ 1 \end{bmatrix} x_3 \right\} = \begin{bmatrix} 3 - x_3 \\ -1 - x_3 \end{bmatrix}$$

The one parameter family of solutions is $\mathbf{x} = [3 - x_3 \quad -1 - x_3 \quad x_3]^T$ with x_3 arbitrary. ■

5.5 A GRAM-SCHMIDT EXPANSION METHOD OF SOLUTION

The set of m simultaneous equations in n unknowns $\mathbf{x}$ is again considered.

$$\mathbf{A}\mathbf{x} = \mathbf{y} \tag{5.1}$$

No special assumptions are made at the outset about r_A and r_W relative to each other or to m and n. By definition, there are r_A independent a_j columns in the matrix $\mathbf{A}$. These vectors can be used as a basis set for a linear vector space $L(a_j)$ called the *column space* of $\mathbf{A}$. The number of vectors in this basis set could be n or any smaller positive integer in a given case. In addition to these r_A vectors a_j, the $\mathbf{y}$ vector is considered, giving a set of $r_A + 1$ vectors. The Gram-Schmidt procedure is used on this set to form an orthonormal basis set $\{\hat{\mathbf{v}}_j\}$. The only possible exception is the last vector $\hat{\mathbf{v}}_{r_A+1}$. Since $\mathbf{y}$ may be linearly dependent on the columns $\mathbf{a}_j$, it might not be possible to form a nonzero vector from $\mathbf{y}$ which is orthogonal to all the $\mathbf{a}_j$ vectors. If $\mathbf{y}$ is linearly dependent on the $\mathbf{a}_j$ vectors, then the unnormalized vector $\hat{\mathbf{v}}_{r_A+1}$ will automatically come out zero during the Gram-Schmidt construction. Since the zero vector is orthogonal to every other vector, an orthogonal set of $\{\hat{\mathbf{v}}_j\}$ can thus be constructed in all cases. Each vector in the set is a unit vector with the possible exception of a zero vector as the last entry. Form the $m \times (r_A + 1)$ matrix $\mathbf{V}$ from this set. Premultiplying equation (*5.1*) by V^T is equivalent to premultiplying the previously defined $\mathbf{W}$ matrix. The result is

$$
\mathbf{V}^T\mathbf{W} = \left[\begin{array}{cccccc:ccc:c}
\langle \hat{\mathbf{v}}_1, \mathbf{a}_1\rangle & \langle \hat{\mathbf{v}}_1, \mathbf{a}_2\rangle & & \cdots & & \langle \hat{\mathbf{v}}_1, \mathbf{a}_r\rangle & \langle \hat{\mathbf{v}}_1, \mathbf{a}_{r+1}\rangle & \cdots & \langle \hat{\mathbf{v}}_1, \mathbf{a}_n\rangle & \langle \hat{\mathbf{v}}_1, \mathbf{y}\rangle \\
0 & \langle \hat{\mathbf{v}}_2, \mathbf{a}_2\rangle & & \cdots & & \langle \hat{\mathbf{v}}_2, \mathbf{a}_r\rangle & \langle \hat{\mathbf{v}}_2, \mathbf{a}_{r+1}\rangle & \cdots & \langle \hat{\mathbf{v}}_2, \mathbf{a}_n\rangle & \langle \hat{\mathbf{v}}_2, \mathbf{y}\rangle \\
0 & 0 & \langle \hat{\mathbf{v}}_3, \mathbf{a}_3\rangle & \cdots & & \langle \hat{\mathbf{v}}_3, \mathbf{a}_r\rangle & \langle \hat{\mathbf{v}}_3, \mathbf{a}_{r+1}\rangle & \cdots & \langle \hat{\mathbf{v}}_3, \mathbf{a}_n\rangle & \langle \hat{\mathbf{v}}_3, \mathbf{y}\rangle \\
\vdots & \vdots & \vdots & & & & & & & \\
0 & 0 & 0 & \cdots & & \langle \hat{\mathbf{v}}_r, \mathbf{a}_r\rangle & \langle \hat{\mathbf{v}}_r, \mathbf{a}_{r+1}\rangle & \cdots & \langle \hat{\mathbf{v}}_r, \mathbf{a}_n\rangle & \langle \hat{\mathbf{v}}_r, \mathbf{y}\rangle \\
\hdashline
0 & 0 & 0 & \cdots & & 0 & \cdots & & 0 & \langle \hat{\mathbf{v}}_{r+1}, \mathbf{y}\rangle
\end{array}\right]
\begin{array}{l} \left.\begin{array}{c}\\ \\ \\ \\ \\ \end{array}\right\} r \text{ rows} \\ \\ \end{array}
\left.\begin{array}{c}\\ \\ \\ \\ \\ \\ \end{array}\right\} r+1 \text{ rows}
$$

$$\underbrace{\hphantom{xxxxxxxxxx}}_{r \text{ columns}}\quad\underbrace{\hphantom{xxxxxxxx}}_{n-r \text{ columns}}$$

$$\underbrace{\hphantom{xxxxxxxxxxxxxxxxxxxxxx}}_{n+1 \text{ columns}}$$

In writing this semitriangular form it is assumed that the first r columns of $\mathbf{A}$ are the r_A independent ones used in the Gram-Schmidt process. The entire last row of the above matrix will be zero with the possible exception of the very last term $\langle \hat{\mathbf{v}}_{r+1}, \mathbf{y}\rangle$. If $\mathbf{y}$ is dependent on the columns of $\mathbf{A}$, then this term will be zero, since then $\hat{\mathbf{v}}_{r+1}$ is exactly zero. This is the case for which there are solutions, since then $r_A = r_W$. When this last inner product is not zero, $r_W > r_A$, so no solutions exist. The last inner product is the component of $\mathbf{y}$ normal to the column space of $\mathbf{A}$. There is no $\mathbf{x}$ vector which will cause $\mathbf{Ax}$ to equal this part of $\mathbf{y}$. The vector $\mathbf{y}$ can be decomposed into a component $\mathbf{y}_p$ parallel to $L(\mathbf{a}_j)$ and a component $\mathbf{y}_e$ normal to $L(\mathbf{a}_j)$, $\mathbf{y} = \mathbf{y}_p + \mathbf{y}_e$. See Fig. 5.1.

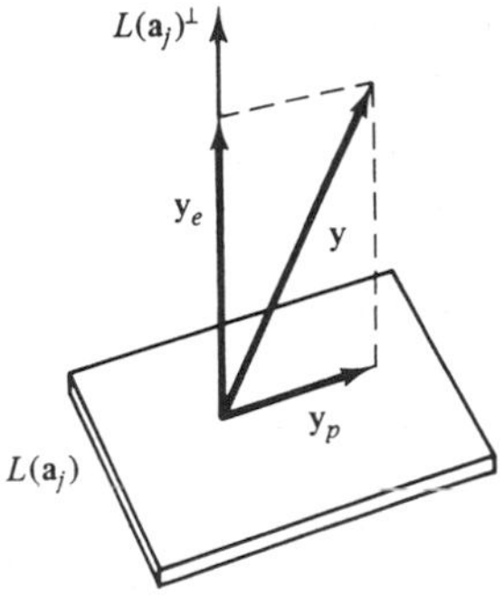

Figure 5.1

The best that can be done by choice of $\mathbf{x}$ is to force $\mathbf{Ax} = \mathbf{y}_p$. The unavoidable error commited in doing this is the residual $\mathbf{Ax} - \mathbf{y} = \mathbf{Ax} - \mathbf{y}_p - \mathbf{y}_e = \mathbf{y}_e$. The length or norm of this residual error is the lower corner element in $\mathbf{W}'$, namely $\|\mathbf{y}_e\| = \langle \hat{\mathbf{v}}_{r+1}, \mathbf{y}\rangle$. Thus, a glance at $\mathbf{W}'$ tells whether or not solutions exist, and if they do not, then the magnitude of the smallest possible error is also given.

The solution which satisfies $\mathbf{y}$, or $\mathbf{y}_p$ if necessary, is found from $\mathbf{W}'$ as in Sec. 5.3, except the final row is ignored. This represents r_A equations in n unknowns. At least one solution always exists, and it will be unique if and only if $r_A = n$. When more than one solution exists, some additional criterion may be used to select one particular solution. These underdetermined problems are discussed further in Sec. 5.7. In cases where $\mathbf{y}$ has a nonzero component normal to $L(\mathbf{a}_j)$, the procedure given here leads to the least-squares solution (or solutions). This topic is pursued further in Sec. 5.8.

Example 5.3

Analyze the following set of equations using the Gram-Schmidt expansion method (GSE).

$$\begin{bmatrix} 1 & 3 & 2 \\ 2 & 5 & 3 \\ 3 & 7 & 4 \\ 4 & 9 & 5 \end{bmatrix} \begin{bmatrix} x_1 \\ x_2 \\ x_3 \end{bmatrix} = \begin{bmatrix} 1 \\ 0 \\ -1 \\ 1 \end{bmatrix}$$

In this example it is easy to see that column 3 of **A** is the difference between columns 2 and 1, so **A** has rank 2. Columns 1 and 2 of **A** are used, along with **y**, to form the following orthonormal set:

THE ORTHONORMAL BASIS SET

$$\mathbf{V} = \begin{bmatrix} 1.8257418\text{E}-01 & 8.1649667\text{E}-01 & 3.6514840\text{E}-01 \\ 3.6514837\text{E}-01 & 4.0824848\text{E}-01 & -1.8257420\text{E}-01 \\ 5.4772252\text{E}-01 & 5.8400383\text{E}-07 & -7.3029685\text{E}-01 \\ 7.3029673\text{E}-01 & -4.0824789\text{E}-01 & 5.4772240\text{E}-01 \end{bmatrix}$$

Then $\mathbf{V}^T\mathbf{W} = \mathbf{W}'$

THE EXPANSION COEFFICIENT VECTOR(S)

$$\mathbf{W}' = \begin{bmatrix} 5.4772258\text{E}+00 & 1.2780193\text{E}+01 & 7.3029675\text{E}+00 & 3.6514840\text{E}-01 \\ 3.9339066\text{E}-06 & 8.1650591\text{E}-01 & 8.1650186\text{E}-01 & 4.0824819\text{E}-01 \\ -9.5367432\text{E}-07 & -2.3841858\text{E}-06 & -1.1920929\text{E}-06 & 1.6431677\text{E}+00 \end{bmatrix}$$

Since the first three entries in the last row and the first entry in row 2 are theoretically zero, a "machine zero" can be defined. Here any number of magnitude less than 4×10^{-6} is set to zero.

From this it is seen that

1. *There is no solution to the given set of equations, since* $r_A = 2$ *and* $r_W = 3$. The equations are inconsistent.
2. *The best that can be done is to satisfy the column space portion of the equations.* Let this be called a projected solution. If this is done the residual error will have the norm

$$\|\mathbf{Ax} - \mathbf{y}\| = 1.6431677$$

3. *There are an infinite number of projected solutions, all of which will give exactly this same residual error norm.* They all must satisfy (rounded off)

$$0.8165(x_2 + x_3) = 0.40825$$

$$5.4772x_1 + 12.7802x_2 + 7.3030x_3 = 0.36515$$

If the row-reduced-echelon (RRE) method of Sec. 5.3 is applied to this problem instead, the resulting **W**′ matrix is

$$\mathbf{W}' = \begin{bmatrix} 1 & 0 & -1 & 0 \\ 0 & 1 & 1 & 0 \\ 0 & 0 & 0 & 1 \\ 0 & 0 & 0 & 0 \end{bmatrix}$$

This also indicates that the equations are inconsistent, but gives no clue about how closely a solution can be approached. It also indicates, apparently, that if one row of inconsistent equations could be ignored, an infinite set of solutions would exist, and they would all satisfy $x_1 - x_3 = 0$ and $x_2 + x_3 = 0$. Although the RRE method has yielded somewhat less information than the GSE method, it has yielded the one-dimensional null space spanned by $[1 \quad -1 \quad 1]^T$. It can be shown that any constant times this vector can be added to any projected solution found from the GSE method (or any other method) and the result will still be a projected solution. ■

5.6 HOMOGENEOUS LINEAR EQUATIONS

The set of homogeneous equations $\mathbf{Ax} = \mathbf{0}$ always has at least one solution, $\mathbf{x} = \mathbf{0}$. This is true because $\mathbf{A}$ and $\mathbf{W}$ always have the same rank. However, $\mathbf{x} = \mathbf{0}$ is called the *trivial solution*. In order for *nontrivial solutions* to exist, it must be true that $r_A < n$. Of course, if one such nontrivial solution exists, there will be an infinite set of solutions with $n - r_A$ free parameters. The methods of the previous sections apply to the homogeneous case without modification.

It is pointed out that any set of nonhomogeneous equations $\mathbf{Ax} = \mathbf{y}$ can always be written as an equivalent set of homogeneous equations:

$$[\mathbf{A} \mid \mathbf{y}]\begin{bmatrix} \mathbf{x} \\ \hline -1 \end{bmatrix} = \mathbf{0} \qquad \text{or} \qquad \mathbf{W}\begin{bmatrix} \mathbf{x} \\ \hline -1 \end{bmatrix} = \mathbf{0}$$

Example 5.4

Find all nontrivial solutions to the equations $\mathbf{Ax} = \mathbf{0}$, if $\mathbf{A} = \begin{bmatrix} 0 & 2 & 1 \\ 0 & 2 & 1 \\ 0 & -4 & -2 \end{bmatrix}$. The $\mathbf{W}$ matrix and its RRE form $\mathbf{W}'$ are

THE W MATRIX

$$\begin{bmatrix} 0.0000000E+00 & 2.0000000E+00 & 1.0000000E+00 & 0.0000000E+00 \\ 0.0000000E+00 & 2.0000000E+00 & 1.0000000E+00 & 0.0000000E+00 \\ 0.0000000E+00 & -4.0000000E+00 & -2.0000000E+00 & 0.0000000E+00 \end{bmatrix}$$

RANK OF W IS 1: THE HERMITE FORM W′ FOLLOWS

$$\begin{bmatrix} 0.0000000E+00 & 1.0000000E+00 & 5.0000000E-01 & 0.0000000E+00 \\ 0.0000000E+00 & 0.0000000E+00 & 0.0000000E+00 & 0.0000000E+00 \\ 0.0000000E+00 & 0.0000000E+00 & 0.0000000E+00 & 0.0000000E+00 \end{bmatrix}$$

Therefore, all nontrivial solutions are linear combinations of the following two vectors, which constitute a basis set for the null space of $\mathbf{A}$.

DEGENERACY OF A IS 2; NULL SPACE BASIS IS

$$\begin{bmatrix} -1.0000000E+00 \\ 0.0000000E+00 \\ 0.0000000E+00 \end{bmatrix} \begin{bmatrix} 0.0000000E+00 \\ 5.0000000E-01 \\ -1.0000000E+00 \end{bmatrix}$$

■

5.7 THE UNDERDETERMINED CASE

When the matrix **A** has $m < n$, there is no possibility of a unique solution **x**. The underdetermined case which has an infinite number of solutions ($r_A = r_W$) is discussed here. The methods of Sec. 5.3, 5.4, or 5.5 can be used to find the family of solutions. This section presents methods for singling out two particular types of solutions:

1. The solution with the minimum norm, $\|\mathbf{x}\|$.
2. The solution which maximizes a specified linear combination of the x_i components. This is the *linear programming* problem.

The Minimum Norm Solution

Consider the equation $\mathbf{Ax} = \mathbf{y}$, where **A** is $m \times n$ with $m < n$ and with $r_A = r_W$. If $r_A < m$, some rows of **W** are linearly dependent. This means that some of the original equations are redundant and can be deleted without losing information. Assume that these deletions have been made and as a result $r_A = r_W = m$. The conjugate transpose of the m rows of any **A** can be used to define the n-component vectors $\{\mathbf{c}_i, i = 1, \ldots, m\}$. These vectors belong to $\mathfrak{X}^n$. The space spanned by the set of $\mathbf{c}_i$ vectors is called the *row space* [91] $L(\mathbf{c}_i)$ of **A**. In general, $L(\mathbf{c}_i)$ will be a subspace of $\mathfrak{X}^n$, and here $r_A = m$ means that it is an m-dimensional subspace with the $\mathbf{c}_i$ vectors forming a basis. The space $\mathfrak{X}^n$, which contains all possible **x** vectors, can be written as the direct sum

$$\mathfrak{X}^n = L(\mathbf{c}_i) \oplus L(\mathbf{c}_i)^\perp$$

Every vector in $\mathfrak{X}^n$ can be written

$$\mathbf{x} = \mathbf{x}_1 + \mathbf{x}_2 \qquad \text{where } \mathbf{x}_1 \in L(\mathbf{c}_i),\ \mathbf{x}_2 \in L(\mathbf{c}_i)^\perp$$

Figure 5.2 illustrates this decomposition for $n = 3$ and $m = 2$.

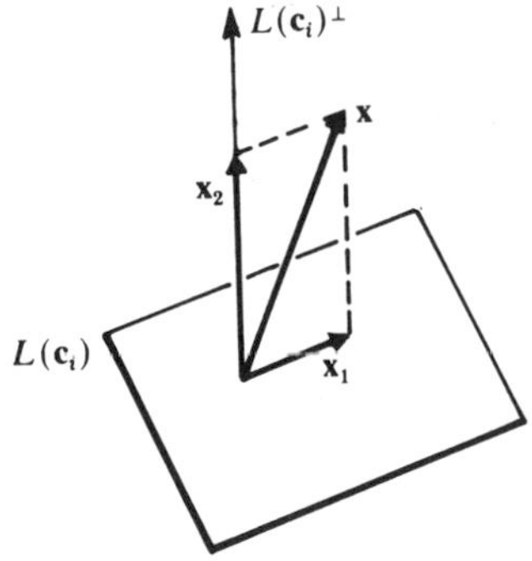

Figure 5.2

The norm of **x** satisfies

$$\|\mathbf{x}\|^2 = \|\mathbf{x}_1\|^2 + \|\mathbf{x}_2\|^2$$

Since $\mathbf{x}_2 \in L(\mathbf{c}_i)^\perp$, $\langle \mathbf{c}_i, \mathbf{x}_2 \rangle = 0$ for each $\mathbf{c}_i$, so $\mathbf{A}\mathbf{x}_2 = \mathbf{0}$. Thus

$$\mathbf{A}\mathbf{x} = \mathbf{A}(\mathbf{x}_1 + \mathbf{x}_2) = \mathbf{A}\mathbf{x}_1 = \mathbf{y}$$

For every $\mathbf{x}_1 \in L(\mathbf{c}_i)$, $\mathbf{x}_1 = \sum_{i=1}^{m} \alpha_i \mathbf{c}_i = \bar{\mathbf{A}}^T \boldsymbol{\alpha}$. Using $\mathbf{A}\mathbf{x}_1 = \mathbf{y}$ gives

$$\mathbf{A}\bar{\mathbf{A}}^T \boldsymbol{\alpha} = \mathbf{y}$$

But $\mathbf{A}\bar{\mathbf{A}}^T$ is an $m \times m$ matrix with rank m and is therefore nonsingular. Solving for $\boldsymbol{\alpha}$ gives $\boldsymbol{\alpha} = (\mathbf{A}\bar{\mathbf{A}}^T)^{-1}\mathbf{y}$ and therefore $\mathbf{x}_1 = \bar{\mathbf{A}}^T\boldsymbol{\alpha} = \bar{\mathbf{A}}^T(\mathbf{A}\bar{\mathbf{A}}^T)^{-1}\mathbf{y}$. This $\mathbf{x}_1$ is the unique $\mathbf{x} \in L(\mathbf{c}_i)$ which satisfies $\mathbf{A}\mathbf{x} = \mathbf{y}$. From the norm relations, it is clear that $\mathbf{x}_1$ is the minimum norm solution, since any other solution must have a component in $L(\mathbf{c}_i)^\perp$ and this would increase the norm. The minimum norm solution then is

$$\mathbf{x} = \bar{\mathbf{A}}^T(\mathbf{A}\bar{\mathbf{A}}^T)^{-1}\mathbf{y}$$

This result can also be derived by using Lagrange multipliers [111] and straightforward minimization of $\|\mathbf{x}\|^2$ subject to the constraint $\mathbf{A}\mathbf{x} - \mathbf{y} = \mathbf{0}$.

Example 5.5

The minimum norm solution of $\begin{bmatrix} 1 & 1 & 0 \\ 0 & 0 & 1 \end{bmatrix} \begin{bmatrix} x_1 \\ x_2 \\ x_3 \end{bmatrix} = \begin{bmatrix} 1 \\ 4 \end{bmatrix}$ is

$$\mathbf{x}_1 = \begin{bmatrix} 1 & 0 \\ 1 & 0 \\ 0 & 1 \end{bmatrix} \begin{bmatrix} 2 & 0 \\ 0 & 1 \end{bmatrix}^{-1} \begin{bmatrix} 1 \\ 4 \end{bmatrix} = \frac{1}{2} \begin{bmatrix} 1 \\ 1 \\ 8 \end{bmatrix}$$

■

Example 5.6

Find the minimum norm solution to the projected problem derived in Example 5.3. From the previous analysis the projected problem is

$$\begin{bmatrix} 5.4772 & 12.7802 & 7.3030 \\ 0 & 0.8165 & 0.8165 \end{bmatrix} \begin{bmatrix} x_1 \\ x_2 \\ x_3 \end{bmatrix} = \begin{bmatrix} 0.3651 \\ 0.4082 \end{bmatrix}$$

A direct calculation of $\mathbf{x} = \mathbf{A}^T[\mathbf{A}\mathbf{A}^T]^{-1}\mathbf{y}$ can be made in this simple case. Alternately one can form

$$\mathbf{A}\mathbf{A}^T = \begin{bmatrix} 246.6670 & 16.3979 \\ 16.3979 & 1.3333 \end{bmatrix}$$

and then solve $\mathbf{A}\mathbf{A}^T\mathbf{x}' = \mathbf{y}$ for $\mathbf{x}'$. Then, finally, $\mathbf{x} = \mathbf{A}^T\mathbf{x}'$. The advantage of doing this is that a matrix inversion is not directly needed, and a routine such as the RRE package can instead be used to solve for x'. This is the approach used here.

$$\mathbf{W} = \left[\begin{array}{cc:c} 246.6670 & 16.3979 & 0.3651 \\ 16.3979 & 1.3333 & 0.4082 \end{array}\right] \qquad \mathbf{W}' = \left[\begin{array}{cc:c} 1 & 0 & -0.10346 \\ 0 & 1 & 1.57862 \end{array}\right]$$

Therefore, $\mathbf{x}' = \begin{bmatrix} -0.10346 \\ 1.57862 \end{bmatrix}$ and $\mathbf{x} = \begin{bmatrix} -0.5667 \\ -0.0334 \\ 0.5334 \end{bmatrix}$.

■

Linear Programming Problems

A solution, $\mathbf{x}$, is sought for the underdetermined set of simultaneous linear equations

$$\mathbf{A}\mathbf{x} = \mathbf{y} \tag{5.1}$$

where $\mathbf{A}$ is $m \times n$ and has rank m, and $\mathbf{y}$ is a known $m \times 1$ vector. It is assumed that $r_A = r_W = m$, so that an infinite number of solutions exist. The *optimal solution* is defined as the one which maximizes

$$Z = p_1x_1 + p_2x_2 + \cdots + p_nx_n = \mathbf{p}^T\mathbf{x} \tag{5.2}$$

and has all components nonnegative

$$x_i \geq 0 \qquad i = 1, 2, \ldots, n \tag{5.3}$$

These three equations constitute a linear programming problem and the stated form will be referred to as *the standard form*. (See Problem 5.13 for other forms and conversion to standard form.)

A common interpretation for problems of this type arises in the field of economics. Each component x_i could represent the amount of one ingredient, the quantity of one product, or the level of an activity. Then p_i is the profit accrued per unit of x_i and Z is the total profit. The restrictions placed on $\mathbf{x}$ by equation (*5.1*) indicate that limits exist on available money, manpower, time, or machine capacity, or that restrictions exist on the product mix, etc. The restrictions expressed by equation (*5.3*) reflect the fact that negative amounts of production or other problem ingredients are not meaningful.

Definitions

A *feasible solution* is any vector $\mathbf{x}$ which satisfies equations (*5.1*) and (*5.3*).

A *basic solution* is any vector $\mathbf{x}$ which is obtained by deleting any $n - m$ columns of $\mathbf{A}$, setting the corresponding x_j components equal to zero, and solving for the remaining m components x_j.

A *basic feasible solution* is any vector $\mathbf{x}$ which is simultaneously basic and feasible.

A *degenerate basic feasible solution* is one which has less than m nonzero components.

Nondegenerate basic feasible solutions have exactly m nonzero components.

Only nondegenerate problems are considered here.

A fundamental theorem of linear programming states that if a finite optimal (maximum) value of Z exists, then there will exist a basic feasible solution that yields that maximum value [41]. This solution is called an *optimal solution*. Two or more basic feasible solutions could give the same maximum value to Z.

Each row of equation (*5.1*) defines a hyperplane in the n-dimensional space $\mathcal{X}^n$; the set of m such hyperplanes along with the nonnegative requirements of equation

(*5.3*) defines a convex polytope. The basic feasible solutions are those **x** vectors at the corners of this polytope. Since there are

$$\binom{n}{m} = \frac{n!}{m!\,(n-m)!}$$

ways of choosing m nonzero components at a time from the n components of $\mathbf{x}$, this value represents the number of basic solutions. The number of basic feasible solutions will be some subset of these, and is certainly finite. Linear programming is a systematic examination of a sequence of basic feasible solutions, and therefore will usually lead to the optimal solution, when it exists, in a finite number of steps. The exceptional case of cycling through a sequence of solutions repetitively can occur in pathological cases, but this possibility is not considered here.

Linear programming routines consist of three major tasks:

(a) *The starting procedure*. Any basic feasible solution may be used as an initial starting point. Methods of obtaining a starting point can be found in the literature [41, 110].

(b) *The introduction and replacement rules*. These rules indicate which basic feasible solution should be examined next. The criterion is that Z never be allowed to decrease; instead, it should increase at the maximum possible rate.

(c) *The termination rules*. These rules indicate when an optimum solution has been found, or that no finite optimum solution exists.

Vector-Matrix Approach to Linear Programming Problems

The columns of $\mathbf{A}$ are referred to as $\mathbf{a}_j$. These vectors, along with $\mathbf{y}$, belong to $\mathfrak{X}^m$, which is sometimes referred to as the *requirements* or *commodity space*. The vectors $\mathbf{x}$ and $\mathbf{p}$ belong to $\mathfrak{X}^n$, called the *solutions* or *activity space*. Since equation (*5.1*) can be expanded as

$$x_1\mathbf{a}_1 + x_2\mathbf{a}_2 + \cdots + x_n\mathbf{a}_n = \mathbf{y}$$

a basic solution can always be found by selecting any m columns $\mathbf{a}_j$ as the basis set. Let $\mathbf{A}_B$ be the nonsingular $m \times m$ matrix formed from m columns, and let $\mathbf{A}_R$ be the $m \times n - m$ matrix of the remaining columns. For bookkeeping convenience, an index set of ordered integers is defined as

$$\mathcal{J} = \{j_1, j_2, \ldots, j_m\}$$

where the values j_1 through j_m refer to column numbers of the original $\mathbf{A}$ matrix that are contained in $\mathbf{A}_B$. The same components of $\mathbf{x}$ and $\mathbf{p}$ are used to define the $m \times 1$ vectors $\mathbf{x}_B$ and $\mathbf{p}_B$. Then

$$[\mathbf{A}_B \mid \mathbf{A}_R]\begin{bmatrix}\mathbf{x}_B \\ \hline \mathbf{x}_R\end{bmatrix} = \mathbf{y} \qquad \text{or} \qquad \mathbf{A}_B\mathbf{x}_B + \mathbf{A}_R\mathbf{x}_R = \mathbf{y}$$

A basic solution is obtained by setting $\mathbf{x}_R = \mathbf{0}$, giving

$$\mathbf{x}_B = \mathbf{A}_B^{-1}\mathbf{y}$$

The corresponding profit or pay-off is

$$Z = \mathbf{p}_B^T\mathbf{x}_B = \mathbf{p}_B^T\mathbf{A}_B^{-1}\mathbf{y}$$

The above manipulations do not guarantee that $\mathbf{x}_B$ is feasible. However, if the initial $\mathbf{x}_B$ is feasible, the following development will ensure that successive $\mathbf{x}_B$ vectors remain feasible.

The linear programming procedure consists of a systematic replacement of one column in $\mathbf{A}_B$ by another in $\mathbf{A}_R$. Let k be a column number not currently in $\mathcal{J}$. Since the columns of $\mathbf{A}_B$ form a basis, $\mathbf{a}_k$ can be written as

$$\mathbf{a}_k = d_1\mathbf{a}_{j_1} + d_2\mathbf{a}_{j_2} + \cdots + d_m\mathbf{a}_{j_m} \qquad \text{or} \qquad \mathbf{a}_k = \mathbf{A}_B\mathbf{d}_k$$

Therefore $\mathbf{d}_k = \mathbf{A}_B^{-1}\mathbf{a}_k$.

If a component $x_k = \alpha$ along $\mathbf{a}_k$ is to be added to the solution while still satisfying equation (*5.1*), then $\mathbf{x}_B$ must change. Denote the modification of $\mathbf{x}_B$, which depends upon the value α, as $\mathbf{x}_\alpha$. Then

$$\mathbf{A}_B\mathbf{x}_\alpha + \alpha\mathbf{a}_k = \mathbf{y}$$

so that

$$\mathbf{x}_\alpha = \mathbf{A}_B^{-1}\mathbf{y} - \alpha\mathbf{A}_B^{-1}\mathbf{a}_k = \mathbf{x}_B - \alpha\mathbf{d}_k$$

The modified pay-off is

$$Z_\alpha = \mathbf{p}_B^T\mathbf{x}_\alpha + p_k\alpha = \mathbf{p}_B^T\mathbf{x}_B + \alpha(p_k - \mathbf{p}_B^T\mathbf{d}_k) = Z + \alpha(p_k - \mathbf{p}_B^T\mathbf{d}_k)$$

Since a feasible solution requires $\alpha > 0$, the introduction of $\mathbf{a}_k$ will cause Z_α to be larger than Z if

$$\beta_k \triangleq p_k - \mathbf{p}_B^T\mathbf{d}_k > 0$$

Since $(Z_\alpha - Z)/\alpha = \boldsymbol{\beta}_k$, this represents the rate of change of Z per unit α. The particular column $\mathbf{a}_k$ that should be introduced into the basis set is the one for which this rate is maximum:

$$\beta_k = \max_{1 \le i \le n} \{\beta_i\}$$

The entire set of β_i components can be expressed as the vector

$$\boldsymbol{\beta}^T = \mathbf{p}^T - \mathbf{p}_B^T\mathbf{A}_B^{-1}A$$

Those components β_i for which $i \in \mathcal{J}$ are automatically zero.

The increase in Z caused by adding $\mathbf{a}_k$ to the basis set depends on the rate β_k and also upon the value of α. The larger α is, the more Z increases. However, α is limited by the requirement that the new solution must be feasible. If α increases sufficiently, some components of $\mathbf{x}_\alpha$ could become negative. The value of α is chosen as large as possible while maintaining all components of $\mathbf{x}_\alpha$ nonnegative. The first component that goes to zero is the column which is to be replaced by $\mathbf{a}_k$. Since the ith component of $\mathbf{x}_\alpha$ is $x_{\alpha i} = x_{Bi} - \alpha d_{ki}$, this means that $\alpha = x_{Bi}/d_{ki}$ will cause $x_{\alpha i}$ to equal zero, provided $d_{ki} > 0$. Therefore, column $\mathbf{a}_l$ is the one which is replaced by $\mathbf{a}_k$, where l is given by

$$\frac{x_{Bl}}{d_{kl}} = \min_i \left\{\frac{x_{Bi}}{d_{ki}}\right\} \qquad \text{(only those } i \text{ for which } d_i > 0 \text{ are considered)}$$

A new $\mathbf{p}_B$ vector is also introduced by replacing p_l by p_k. The index set $\mathcal{J}$ is also modified accordingly. This completes one step of the procedure. The stopping rules are

(a) If all components of β_k are negative or zero, the value of Z cannot be increased further, and the optimum has been found. If one or more components are positive, then either more steps are required or no finite solution exists. Which possibility applies is determined by the next result.

(b) If all $d_i \leq 0$, then α can be made arbitrarily large while maintaining a feasible solution. There is no finite optimal solution in this case.

The above method is essentially the simplex method, which is normally discussed in terms of the *Simplex Tableau*. Details on this and on the many generalizations and extensions may be found in the references [41, 110].

5.8 THE OVERDETERMINED CASE

When there are more equations than unknowns, the $m \times n$ coefficient matrix $\mathbf{A}$ has $m > n$. If the equations are inconsistent, no solution exists. This situation often arises because of inaccuracies in measuring the components of the $\mathbf{y}$ vector, or because the relationship assumed to exist between $\mathbf{x}$ and $\mathbf{y}$, as expressed by $\mathbf{A}$, is oversimplified or wrong. Approximate solution vectors $\mathbf{x}$ are desired in this case. Three approaches are presented. The first method ignores some equations and places total reliance on those remaining. The second method (least squares) places equal reliance on all equations with the hope that the errors will average out. The third method (weighted least squares) uses all of the equations but weights some more heavily than others. An alternative computational procedure (recursive weighted least squares) is also given for obtaining the latter two approximations.

A considerable amount of information which is useful for the overdetermined case has already been given. The GSE method of Sec. 5.5 applies to this case, as already demonstrated. When the problem is overdetermined, $\langle \hat{\mathbf{v}}_{r+1}, \mathbf{y} \rangle \neq 0$. The so-called projected solution is then sought. If the projected solution is nonunique, the minimum norm solution is often singled out, as was done in Example 5.3 and continued in Example 5.6. This combination of GSE plus minimum norm solution always gives a solution to equation (*5.1*). It is true for the underdetermined, overdetermined, or uniquely determined cases. The only difficulty that might remain on a machine solution is the ability to recognize the difference between 0 and a very small number, or the difference between vectors that are linearly dependent or nearly linearly dependent. The notion of machine zero was introduced, and an example of how it can be determined on a given problem has been given. The GSE-mininum norm solution combination has many things in common with the method of *singular value decomposition* [38], but there are also some unique differences.

In this section a more traditional approach to the overdetermined problem is presented. It is assumed that $\mathbf{A}$ is of full rank n, and that $m > n$.

Ignore Some Equations

If a subset of n equations is selected, and the remaining $m - n$ are ignored, an approximate solution can be obtained. The basis for ignoring certain equations is a subjective matter. Perhaps certain equations are more reliable for one reason or another. Perhaps other results are obviously "wild points" and can be discarded. If $\mathbf{A}_1$ is a nonsingular $n \times n$ matrix formed by deleting rows from $\mathbf{A}$ and if $\mathbf{y}_1$ is the $n \times 1$ vector obtained from $\mathbf{y}$ by deleting the corresponding elements, then a result which satisfies n of the original equations is

$$\mathbf{x} = \mathbf{A}_1^{-1}\mathbf{y}_1$$

Least-Squares Approximate Solution

If all of the equations are used correctly, errors may tend to average out and a good approximation for $\mathbf{x}$ results. Since no one $\mathbf{x}$ can satisfy all of the simultaneous equations, it is inappropriate to write the equality $\mathbf{Ax} = \mathbf{y}$. Rather, an $n \times 1$ error vector $\mathbf{e}$ is introduced:

$$\mathbf{e} = \mathbf{y} - \mathbf{Ax}$$

The least-squares approach yields that one $\mathbf{x}$ which minimizes the sum of the squares of the e_i components. That is, $\mathbf{x}$ is chosen to minimize

$$\|\mathbf{e}\|^2 = \mathbf{e}^T\mathbf{e} = (\mathbf{y} - \mathbf{Ax})^T(\mathbf{y} - \mathbf{Ax})$$

The vectors $\mathbf{e}$, $\mathbf{y}$, and $\mathbf{Ax}$ all belong to $\mathcal{X}^m$. But $\mathbf{Ax}$ belongs to the *column space* [91] of $\mathbf{A}$. This is the space spanned by the columns $\mathbf{a}_j$ of $\mathbf{A}$ and is denoted by $L(\mathbf{a}_j)$. $\mathcal{X}^m$ can be written as the direct sum

$$\mathcal{X}^m = L(\mathbf{a}_j) \oplus L(\mathbf{a}_j)^\perp$$

The error has a unique decomposition:

$$\mathbf{e} = \mathbf{e}_1 + \mathbf{e}_2, \qquad \mathbf{e}_1 \in L(\mathbf{a}_j), \qquad \mathbf{e}_2 \in L(\mathbf{a}_j)^\perp$$

(The vector $\mathbf{e}_2$ is the $\mathbf{y}_e$ vector of Sec. 5.5). The norm of $\mathbf{e}$ satisfies $\|\mathbf{e}\|^2 = \|\mathbf{e}_1\|^2 + \|\mathbf{e}_2\|^2$.

Since $\mathbf{y}$ is given and since $\mathbf{Ax} \in L(\mathbf{a}_j)$, the choice of $\mathbf{x}$ cannot affect $\mathbf{e}_2$. The least-squares solution vector $\mathbf{x}$ is the one for which $\|\mathbf{e}_1\|^2 = 0$, so $\mathbf{e}_1 = 0$. This means that the projection of $\mathbf{y}$ on $L(\mathbf{a}_j)$, call it $\mathbf{y}_1$, must equal $\mathbf{Ax} = \sum_{j=1}^{n} x_j \mathbf{a}_j$. Since $r_A = n$, the columns $\mathbf{a}_j$ form a *basis* for $L(\mathbf{a}_j)$ so that $\mathbf{y}_1 = \sum_{j=1}^{n} \alpha_j \mathbf{a}_j$. Because of the uniqueness of this expansion, $\alpha_j = x_j$, that is, $\mathbf{x} = \boldsymbol{\alpha}$. The set of n reciprocal basis vectors $\mathbf{r}_i$ is defined by $\langle \mathbf{r}_i, \mathbf{a}_j \rangle = \delta_{ij}$, or in matrix form

$$\underset{(n\times m)}{\mathbf{R}} \cdot \underset{(m\times n)}{\mathbf{A}} = \mathbf{I}$$

Since $\mathbf{A}$ is not square, it cannot be inverted to find $\mathbf{R}$ as in Chapter 4. It is still true that $\alpha_j = \langle \mathbf{r}_j, \mathbf{y} \rangle = x_j$ so that

$$\boldsymbol{\alpha} = \mathbf{Ry} = \mathbf{x} \tag{5.4}$$

Therefore $\mathbf{Ax} = \mathbf{ARy} = \mathbf{y}_1$. Using $\mathbf{e}_2 = \mathbf{y} - \mathbf{y}_1$ and the fact that $\langle \mathbf{a}_j, \mathbf{e}_2 \rangle = 0$ gives $\mathbf{A}^T[\mathbf{y} - \mathbf{ARy}] = \mathbf{0}$, or

$$\mathbf{Ry} = (\mathbf{A}^T\mathbf{A})^{-1}\mathbf{A}^T\mathbf{y} \tag{5.5}$$

Combining equations (*5.4*) and (*5.5*) gives the least-squares solution

$$\mathbf{x} = (\mathbf{A}^T\mathbf{A})^{-1}\mathbf{A}^T\mathbf{y}$$

The amount of error in this approximate solution is indicated by

$$\|\mathbf{e}\|^2 = \|\mathbf{e}_2\|^2 = \mathbf{y}^T[\mathbf{I} - \mathbf{A}(\mathbf{A}^T\mathbf{A})^{-1}\mathbf{A}^T]\mathbf{y} = \|\mathbf{y} - \mathbf{Ax}\|^2$$

Recall that the square root of this quantity was given directly in the GSE method.

The matrix $\mathbf{R} = (\mathbf{A}^T\mathbf{A})^{-1}\mathbf{A}^T$ is a particular example of the *generalized* or *pseudo-inverse* [89] of $\mathbf{A}$, written $\mathbf{A}^\dagger$. If $\mathbf{A}^{-1}$ exists, then $\mathbf{A}^\dagger = \mathbf{A}^{-1}$ and $\|\mathbf{e}\|^2 = 0$. The minimum norm solution of Sec. 5.7 provides another example of the pseudo-inverse which was appropriate to those circumstances, namely $\mathbf{A}^\dagger = \mathbf{A}^T(\mathbf{A}\bar{\mathbf{A}}^T)^{-1}$. The general solutions to the $n \times n$ nonsingular case, the underdetermined minimum norm case, and the overdetermined least-squares case can all be expressed in terms of the pseudo-inverse as $\mathbf{x} = \mathbf{A}^\dagger\mathbf{y}$.

Weighted Least-Squares Approximation to the Solution

Ignoring some equations or placing equal reliance on all equations represents two extremes. If some equations are more reliable than others, but all equations are to be retained, a weighted least-squares approximation can be used. That is, $\mathbf{x}$ should minimize $\mathbf{e}^T\mathbf{R}^{-1}\mathbf{e} = (\mathbf{y} - \mathbf{Ax})^T\mathbf{R}^{-1}(\mathbf{y} - \mathbf{Ax})$. $\mathbf{R}^{-1}$ is symmetric, $m \times m$, nonsingular, and often is diagonal. Those familiar with random processes should know that $\mathbf{R}$ is

*The matrix R in this section is unrelated to the matrix of reciprocal basis vectors of previous sections. This choice of symbols is made for consistency with Problems 11.10 and 11.11.

generally selected as the covariance matrix for the noise on the vector $\mathbf{y}$. Smaller values of r_{ii} will cause e_i^2 to be smaller and the ith equation is more nearly satisfied. If a norm $\|\mathbf{e}\|_{\mathbf{R}^{-1}}^2 = \mathbf{e}^T\mathbf{R}^{-1}\mathbf{e}$ is defined, the method of orthogonal projections immediately leads to

$$\mathbf{A}^T\mathbf{R}^{-1}\mathbf{A}\mathbf{x} = \mathbf{A}^T\mathbf{R}^{-1}\mathbf{y}$$

If $r_A = n$, as assumed here, the $n \times n$ matrix $\mathbf{A}^T\mathbf{R}^{-1}\mathbf{A}$ is nonsingular and the weighted least-squares solution is

$$\mathbf{x} = (\mathbf{A}^T\mathbf{R}^{-1}\mathbf{A})^{-1}\mathbf{A}^T\mathbf{R}^{-1}\mathbf{y}$$

Notice that if $\mathbf{A}$ is not full rank, the required inverse will not exist, signaling that the least-squares solution is not unique.

The least-squares and weighted least-squares formulas can also be derived simply by setting $\partial\|\mathbf{e}\|^2/\partial x_i = 0$.

Recursive Weighted Least-Squares Solutions

The preceding sections dealt with what is commonly called "batch least squares," because all data equations are treated in one batch. A recursive method of using each new set of data as it is received is now presented.

Assume that a set of m equations

$$\mathbf{y}_k = \mathbf{A}\mathbf{x} + \mathbf{e}$$

has been used to obtain a weighted least-squares estimate for $\mathbf{x}$, denoted by $\mathbf{x}_k$:

$$\mathbf{x}_k = (\mathbf{A}^T\mathbf{R}^{-1}\mathbf{A})^{-1}\mathbf{A}^T\mathbf{R}^{-1}\mathbf{y}_k$$

As is often the case, assume that an additional set of relations

$$\mathbf{y}_{k+1} = \mathbf{H}_{k+1}\mathbf{x} + \mathbf{e}_{k+1}$$

then becomes available. It is desired to obtain a new estimate for $\mathbf{x}$, denoted as $\mathbf{x}_{k+1}$, which combines both sets of data and minimizes

$$J = [\mathbf{e}^T \mid \mathbf{e}_{k+1}^T]\left[\begin{array}{c|c} \mathbf{R}^{-1} & \mathbf{0} \\ \hline \mathbf{0} & \mathbf{R}_{k+1}^{-1} \end{array}\right]\left[\begin{array}{c} \mathbf{e} \\ \hline \mathbf{e}_{k+1} \end{array}\right]$$

$\mathbf{R}_{k+1}^{-1}$ is the weighting matrix, analogous to $\mathbf{R}^{-1}$ but applied to the new data $\mathbf{y}_{k+1}$. It is not necessary to reprocess the whole set of equations involving $[\mathbf{y}_k \mid \mathbf{y}_{k+1}]$ in order to determine $\mathbf{x}_{k+1}$. It is shown in Problem 5.18, using partitioned matrices and a matrix inversion identity, that

$$\mathbf{x}_{k+1} = \mathbf{x}_k + \mathbf{K}_k[\mathbf{y}_{k+1} - \mathbf{H}_{k+1}\mathbf{x}_k]$$

where $\mathbf{K}_k = \mathbf{P}_k\mathbf{H}_{k+1}^T[\mathbf{H}_{k+1}\mathbf{P}_k\mathbf{H}_{k+1}^T + \mathbf{R}_{k+1}]^{-1}$ and $\mathbf{P}_k \triangleq (\mathbf{A}^T\mathbf{R}^{-1}\mathbf{A})^{-1}$, which is available from the computation of $\mathbf{x}_k$. If still other sets of equations are to be incorporated, the above relations can be used recursively. A new matrix, $\mathbf{P}_{k+1}$, is then needed and is given by

$$\mathbf{P}_{k+1} = [\mathbf{P}_k^{-1} + \mathbf{H}_{k+1}^T\mathbf{R}_{k+1}^{-1}\mathbf{H}_{k+1}]^{-1} \tag{5.6}$$

Using the matrix inversion lemma, page 78, this can also be written as

$$\mathbf{P}_{k+1} = \mathbf{P}_k - \mathbf{P}_k\mathbf{H}_{k+1}^T[\mathbf{H}_{k+1}\mathbf{P}_k\mathbf{H}_{k+1}^T + \mathbf{R}_{k+1}]^{-1}\mathbf{H}_{k+1}\mathbf{P}_k \tag{5.7}$$

The latter form is often more convenient. For example, if $\mathbf{y}_{k+1}$ is a scalar, then matrix inversion is not required, just a scalar division.

Data Deweighting

A common occurrence is that data are received sequentially over time. As each new group of data is received, it is used to improve the estimate of $\mathbf{x}$. If this process is carried out over a sufficient number of steps, one will find that the $\mathbf{P}_{k+1}$ matrix has decreased to very small values due to the repeated addition of a nonnegative term to its inverse in equation (*5.6*). This in turn will cause the value of $\mathbf{K}_{k+1}$ to become small. This means that the corrections made to $\mathbf{x}_k$ in order to determine $\mathbf{x}_{k+1}$ get small, independent of what new or surprising information may be contained in the latest measurements. In order to prevent the recursive estimator from failing to respond adequately to new data (called going to sleep), some form of data deweighting is often used. Two types will be presented here, additive deweighting and multiplicative deweighting. In both cases the $\mathbf{P}$ matrix is prevented from getting too small. The easiest way to change the former algorithm is to introduce another matrix $\mathbf{M}_k$, given by

$$\mathbf{M}_k = \mathbf{P}_k/\beta \qquad \text{with } \beta < 1\text{; this is multiplicative deweighting}$$

or

$$\mathbf{M}_k = \mathbf{P}_k + \mathbf{Q} \qquad \text{with } \mathbf{Q} \text{ a positive definite matrix; this is additive deweighting}$$

The gain is now computed as

$$\mathbf{K}_k = \mathbf{M}_k\mathbf{H}_{k+1}^T[\mathbf{H}_{k+1}\mathbf{M}_k\mathbf{H}_{k+1}^T + \mathbf{R}]^{-1}$$

and the new $\mathbf{P}_{k+1}$ is given by either equation (*5.6*) or (*5.7*), but with $\mathbf{P}_k$ on the right-hand side replaced everywhere by $\mathbf{M}_k$. The formula for updating the estimate of $\mathbf{x}$ remains the same. If $\beta = 1$ or $\mathbf{Q} = 0$, both of these deweighting schemes revert to the original algorithm. Values used for the so-called forgetting factor β depend on how much deweighting is desired. A concept called asymptotic sample length, ASL, is a measure of how much past data are having a significant effect on the current estimate of $\mathbf{x}$. A relation between ASL and β is

$$\text{ASL} = 1/(1 - \beta)$$

Therefore, the commonly used values of β between 0.999 and 0.95 correspond to asymptotic sample lengths of 1000 past measurements down to 20. Although **Q** is often selected as a diagonal matrix with small diagonal elements, an idea of the appropriate magnitudes can be obtained by assuming that $\mathbf{Q} = \alpha\mathbf{P}$, with α a scalar. Then comparison of the two forms of deweighting shows that $\alpha = (1/\beta) - 1$). Therefore, to get an ASL of 1000, α would be 0.001 or **Q** should be about 0.1% of **P**. Of course, since **P** is changing, this kind of comparison is not perfect. It does indicate roughly that a small **Q** matrix can be an effective deweighting scheme. Figure 5.3 shows the relative weight

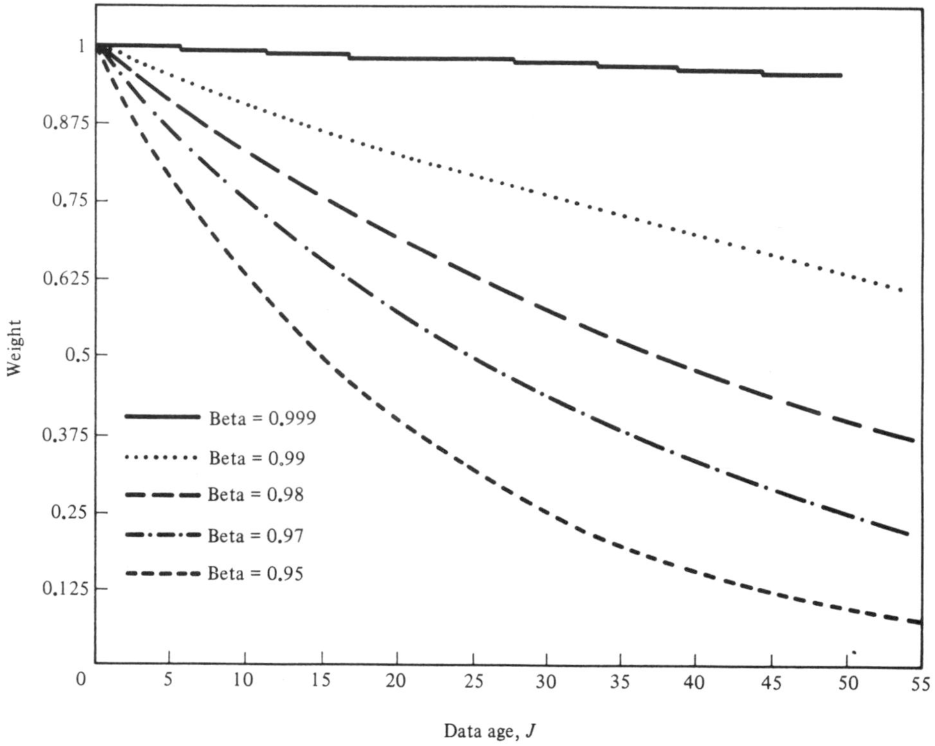

Figure 5.3 Deweighting for common forgetting factors

applied to past measurements when forming the current estimate of **x**. The curve is normalized so that the current measurement weight is one. This is computed by using the recursive estimation algorithm backwards in time for a scalar **x**. Qualitatively similar deweighting occurs in the vector case, but it is harder to normalize and display.

If the matrices **P**, **Q**, and **R** are given the appropriate statistical interpretation as covariance matrices of certain signals, then the recursive equations constitute a simple example of the discrete *Kalman filtering* equations. The Kalman filter is used extensively in modern control theory in order to estimate the internal (state) variables of a linear system based on noisy measurements of the output variables [57, 80]. The least-squares approach finds many other applications in modern control theory as well.

ILLUSTRATIVE PROBLEMS

General Considerations

5.1 Use arguments in $\mathfrak{X}^m$ to draw conclusions about solutions to $\mathbf{Ax} = \mathbf{y}$.

Let the columns of $\mathbf{A}$ be m component vectors, $\mathbf{a}_j$. Then

$$x_1\mathbf{a}_1 + x_2\mathbf{a}_2 + \cdots + x_n\mathbf{a}_n = \mathbf{y}$$

At least one set of scalars $\{x_i\}$ satisfying this equation will exist if $\mathbf{y}$ belongs to the column space $L(\mathbf{a}_j)$. This is equivalent to saying that $\mathbf{y}$ is linearly dependent on the set $\{\mathbf{a}_j\}$ so that $\mathbf{A}$ and $\mathbf{W} = [\mathbf{A} \mid \mathbf{y}]$ have the same rank. If $r_A \neq r_W$, no solution exists.

If the set $\{\mathbf{a}_j, j = 1, \ldots, n\}$ forms a basis for $L(\mathbf{a}_j)$, then there is a unique set of expansion coefficients x_i for every $\mathbf{y} \in L(\mathbf{a}_j)$. But this requires that $m = n$, and $r_A = r_W = n$. The difference between "solutions" and "one unique solution" hinges upon whether the set of $\mathbf{a}_j$ vectors merely spans the column space or forms a basis for it.

5.2 Use arguments in $\mathfrak{X}^n$ to draw conclusions about solutions to $\mathbf{Ax} = \mathbf{y}$.

Let the rows of $\mathbf{A}$ be considered as the conjugate transpose of n component vectors $\mathbf{c}_i$. Then the set of simultaneous equations is equivalent to m scalar equations of the form

$$\langle \mathbf{c}_i, \mathbf{x} \rangle = y_i \qquad \text{where } i = 1, \ldots, m$$

Each of these equations defines an $n - 1$ dimensional hyperplane in $\mathfrak{X}^n$, with a normal $\mathbf{c}_i$.

The existence of a solution means that there exists a vector $\mathbf{x}$ which simultaneously terminates in all m of the hyperplanes. If the set $\{\mathbf{c}_i\}$ is linearly independent, so that $r_A = m$, the intersection of m hyperplanes of dimension $n - 1$ defines an $n - m$ dimensional hyperplane. Every vector $\mathbf{x}$ terminating in this hyperplane is a solution. If $m - n$, thc hyperplane is of zero dimension, i.e. a single point, and defines a unique solution. Obviously, whenever $r_A = m$, $r_W = m$ also.

If $r_A < m$, two or more of the $n - 1$ dimensional hyperplanes are either parallel or they coincide. If parallel but distinct, they never intersect and the equations are inconsistent. It can be shown that these geometrical conditions are equivalent to the algebraic conditions given in terms of the $\mathbf{W}$ matrix.

Homogeneous Equations

5.3 Let $\mathbf{A}$ be an $n \times n$ matrix with $r_A = n - 1$. Show that a nontrivial solution to $\mathbf{Ax} = \mathbf{0}$ can be selected as any nonzero column of the matrix Adj $\mathbf{A}$.

This is shown using a limiting process. Since $\mathbf{A}$ has rank $n - 1$, at least one cofactor $C_{ij} \neq 0$. Assume $C_{pq} \neq 0$ and construct a new matrix $\tilde{\mathbf{A}}$ by adding ϵ to the a_{pq} element. Then $|\tilde{\mathbf{A}}| \neq 0$ since $|\tilde{\mathbf{A}}| = \sum_{i=1}^{n} \tilde{a}_{iq}C_{iq} = |\mathbf{A}| + \epsilon C_{pq}$, using Laplace expansion. Consider a slightly different problem $\tilde{\mathbf{A}}\mathbf{z} = \boldsymbol{\epsilon}_j$, where $\boldsymbol{\epsilon}_j$ is an $n \times 1$ vector, all elements of which are zero except the jth one, which is ϵ. As $\epsilon \longrightarrow 0$, this problem reduces to the original one, with $\mathbf{z} \longrightarrow \mathbf{x}$. But

$$\mathbf{z} = \tilde{\mathbf{A}}^{-1}\boldsymbol{\epsilon}_j = \frac{(\text{Adj }\tilde{\mathbf{A}})}{|\tilde{\mathbf{A}}|}\boldsymbol{\epsilon}_j = \frac{(\text{Adj }\tilde{\mathbf{A}})}{C_{pq}}\begin{bmatrix} 0 \\ 0 \\ \vdots \\ 1 \\ 0 \\ \vdots \\ 0 \end{bmatrix} = \frac{1}{C_{pq}}[\text{Adj }\tilde{\mathbf{A}}]_{j\text{th column}}$$

As $\epsilon \longrightarrow 0$, $\mathbf{z} \longrightarrow \frac{1}{C_{pq}}[j\text{th column of Adj }\mathbf{A}] = \mathbf{x}$. This $\mathbf{x}$ vector is not a trivial solution, provided the selected jth column is not all zeros.

5.4 Use the results of Problem 5.3 to find a nontrivial solution for

$$\begin{bmatrix} 1 & 0 & 1 \\ 0 & 1 & 0 \\ 2 & 2 & 2 \end{bmatrix}\begin{bmatrix} x_1 \\ x_2 \\ x_3 \end{bmatrix} = \mathbf{0}$$

Since $r_A = 2$, the degeneracy is $n - r_A = 1$, so a one-parameter family of nontrivial solutions exists. It is found by computing $\text{Adj }\mathbf{A} = \begin{bmatrix} 2 & 2 & -1 \\ 0 & 0 & 0 \\ -2 & -2 & 1 \end{bmatrix}$. Thus

$\mathbf{x} = k\begin{bmatrix} 1 \\ 0 \\ -1 \end{bmatrix}$, $k \neq 0$, generates the set of all nontrivial solutions.

5.5 If an $n \times n$ matrix has rank $r_A < n$, it can be shown that $\mathbf{Ax} = \mathbf{0}$ has $q = n - r_A$ linearly independent solutions. They may be chosen as linearly independent columns of

$$\frac{d^{q-1}}{d\epsilon^{q-1}}\{\text{Adj }[\mathbf{A} - \mathbf{I}\epsilon]\}\Big|_{\epsilon=0}$$

Find nontrivial solutions for this problem when $\mathbf{A} = \begin{bmatrix} 1 & 1 & 1 \\ 2 & 2 & 2 \\ -2 & -2 & -2 \end{bmatrix}$.

The rank of $\mathbf{A}$ is 1, so $q = 2$ and

$$\lim_{\epsilon\to 0}\frac{d}{d\epsilon}\{\text{Adj }[\mathbf{A} - \mathbf{I}\epsilon]\} = \begin{bmatrix} 0 & 1 & 1 \\ 2 & 1 & 2 \\ -2 & -2 & -3 \end{bmatrix}$$

There are just two linearly independent columns, and nontrivial solutions are $\mathbf{x}_1 = [0 \quad 2 \quad -2]^T$, $\mathbf{x}_2 = [1 \quad 1 \quad -2]^T$ or any linear combination of these two. $\mathbf{x}_1$ and $\mathbf{x}_2$ form a basis for the two-dimensional subspace defined by $\langle \mathbf{c}, \mathbf{x} \rangle = 0$, where $\mathbf{c}$ is the transpose of any row of $\mathbf{A}$.

5.6 Find all nontrivial solutions to

$$\begin{bmatrix} 1 & 3 & 5 \\ 1 & 4 & 6 \\ -1 & 5 & 3 \\ -1 & 4 & 2 \\ 1 & 3 & 5 \end{bmatrix}\mathbf{x} = \begin{bmatrix} 0 \\ 0 \\ 0 \\ 0 \\ 0 \end{bmatrix}$$

This is the same **A** matrix as appeared in Example 5.1(8). The RRE form for **W** is the same as given in that example, except that the last column is all zeros. Therefore, all nontrivial solutions must satisfy $x_1 + 2x_3 = 0$ and $x_2 + x_3 = 0$. Thus $\mathbf{x} = a[-2 \quad -1 \quad -]^T$, for any scalar a, constitutes the one parameter family of nontrivial solutions. From $\mathbf{W}'$ it can be seen that the rank of **A** is 2 and the number of unknowns is $n = 3$, so the degeneracy or nullity is $q = 3 - 2 = 1$. This is the dimension of the null space of **A**.

5.7 Find all nontrivial solutions to

$$\begin{bmatrix} 4 & -2 & 3 \\ 1 & 3 & 1 \\ 1 & 3 & 1 \end{bmatrix} \mathbf{x} = \begin{bmatrix} 0 \\ 0 \\ 0 \end{bmatrix}$$

For this problem the RRE form of **W** is

$$\mathbf{W}' = \left[\begin{array}{ccc|c} 1 & 0 & 0.78571427 & 0 \\ 0 & 1 & 0.071428575 & 0 \\ 0 & 0 & 0 & 0 \end{array}\right]$$

Therefore $x_1 + 0.78571427x_3 = 0$ and $x_2 + 0.07142857x_3 = 0$. All nontrivial solutions must be proportional to

$$\mathbf{x} = [11 \quad 1 \quad -14]^T$$

5.8 Do nontrivial solutions exist for the following?

$$\begin{bmatrix} 2 & -2 & 3 \\ 1 & 1 & 1 \\ 1 & 3 & -1 \end{bmatrix} \mathbf{x} = \begin{bmatrix} 0 \\ 0 \\ 0 \end{bmatrix}$$

Using the RRE method, or just computing its determinant, shows that the rank of **A** is 3. Its degeneracy is zero, it is nonsingular, and so the only solution to this problem is the trivial solution $\mathbf{x} = \mathbf{0}$.

Minimum Norm Solutions

5.9 Find the minimum norm solution for $[1 \quad 2]\begin{bmatrix} x_1 \\ x_2 \end{bmatrix} = 1$.

Identifying $\mathbf{A} = [1 \quad 2]$, the minimum norm solution is

$$\mathbf{x} = \begin{bmatrix} 1 \\ 2 \end{bmatrix} \left\{ [1 \quad 2] \begin{bmatrix} 1 \\ 2 \end{bmatrix} \right\}^{-1} (1) = \frac{1}{5} \begin{bmatrix} 1 \\ 2 \end{bmatrix}$$

5.10 A specified amount of constant current, i, must be delivered to the ground point of Fig. 5.4. Specify v_1, v_2, and v_3 so that the total energy dissipated in the resistors is minimized.

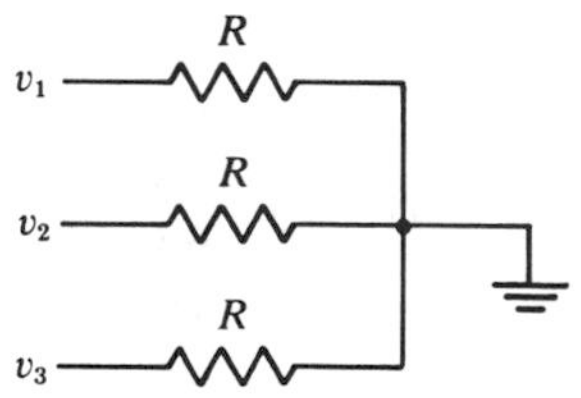

Figure 5.4

It is required that

$$v_1/R + v_2/R + v_3/R = i \quad \text{or} \quad \frac{1}{R}[1 \quad 1 \quad 1]\mathbf{v} = i$$

The total energy dissipated per unit time is

$$\dot{\mathcal{E}} = \frac{1}{R}[v_1^2 + v_2^2 + v_3^2] = \frac{1}{R}\|\mathbf{v}\|^2$$

The desired solution is thus the minimum norm solution.

$$\mathbf{v} = \frac{1}{R}\begin{bmatrix}1\\1\\1\end{bmatrix}\left\{\frac{1}{R^2}[1 \quad 1 \quad 1]\begin{bmatrix}1\\1\\1\end{bmatrix}\right\}^{-1} i = \frac{R}{3}\begin{bmatrix}1\\1\\1\end{bmatrix} i$$

Frequently, the norm can be given a physical interpretation of energy or power. This is one reason why minimum norm solutions are often sought for underdetermined problems.

5.11 A small microcomputer has five terminals connected to it. The fraction of the time devoted to each terminal is x_i, so

$$x_1 + x_2 + x_3 + x_4 + x_5 = 1$$

The programmer at terminal 2 types four times as fast as the programmer at 1. They are both typing in the same code and must finish at the same time so $x_1 = 4x_2$. Both terminals 3 and 4 are sending mail files to 5, so $x_3 + x_4 = x_5$. Find the minimum norm solution for allocating CPU time.

In matrix form the constraints are

$$\begin{bmatrix}1 & 1 & 1 & 1 & 1\\1 & -4 & 0 & 0 & 0\\0 & 0 & 1 & 1 & -1\end{bmatrix}\mathbf{x} = \begin{bmatrix}1\\0\\0\end{bmatrix}$$

The minimum norm solution is

$$\mathbf{x} = \mathbf{A}^T[\mathbf{A}\mathbf{A}^T]^{-1}\mathbf{y} = \begin{bmatrix}1 & 1 & 0\\1 & -4 & 0\\1 & 0 & 1\\1 & 0 & 1\\1 & 0 & -1\end{bmatrix}\begin{bmatrix}5 & -3 & 1\\-3 & 17 & 0\\1 & 0 & 3\end{bmatrix}^{-1}\begin{bmatrix}1\\0\\0\end{bmatrix} = \begin{bmatrix}0.28436\\0.07109\\0.15602\\0.15602\\0.32739\end{bmatrix}$$

5.12 Find the shortest four-dimensional vector from the origin to the four-dimensional hyperplane described by

$$5x_1 - 2x_2 + x_3 + 7x_4 = 12$$

This is the same as asking for the minimum norm solution to

$$[5 \quad -2 \quad 1 \quad 7]\mathbf{x} = 12$$

Therefore

$$\mathbf{x} = \begin{bmatrix}5\\-2\\1\\7\end{bmatrix}[79]^{-1}(12) = \begin{bmatrix}0.7595\\-0.3038\\0.1519\\1.0633\end{bmatrix}$$

Linear Programming

5.13 Show how a variety of linear programming problems can be reduced to the standard form [equations (*5.1*), (*5.2*), and (*5.3*)].

If it is desired to minimize, rather than maximize, a linear function of x_i, a new set of coefficients $p'_i = -p_i$ is introduced. Then minimizing $Z' = \mathbf{p}^T\mathbf{x}$ is equivalent to maximizing $Z = \mathbf{p}'^T\mathbf{x}$.

The constraints which arise in linear programming take one of the following three forms:

$$h_{i1}x_1 + h_{i2}x_2 + \cdots + h_{in}x_n = y_i$$

or

$$h_{j1}x_1 + h_{j2}x_2 + \cdots + h_{jn}x_n \leq y_j$$

or

$$h_{k1}x_1 + h_{k2}x_2 + \cdots + h_{kn}x_n \geq y_k$$

The first equation is already in the form of one row of the standard form. The third equation can be written in the form of the second equation by multiplying by -1. The minus signs can be absorbed into the definitions for a new set of coefficients $h'_{kj} = -h_{kj}$ and $y'_k = -y_k$. A typical inequality constraint of the second type can be converted to an equality constraint by adding a nonnegative quantity x_s, called a slack variable:

$$h_{j1}x_1 + h_{j2}x_2 + \cdots + h_{jn}x_n + x_s = y_j$$

In vector-matrix form the complete set of inequality constraints can be written as

$$\mathbf{Hx} \leq \mathbf{y}$$

Letting $\mathbf{x}_s$ be a vector of slack variables, this can be modified to

$$\mathbf{Hx} + \mathbf{x}_s = \mathbf{y} \qquad \text{or} \qquad [\mathbf{H} \mid \mathbf{I}]\begin{bmatrix} \mathbf{x} \\ \hline \mathbf{x}_s \end{bmatrix} = \mathbf{y}$$

By defining a new $\mathbf{A}$ and $\mathbf{x}$ in the obvious way, this is in the standard form of equation (*5.1*).

Since the slack variables contribute nothing to the pay-off function Z, a new $n \times 1$ $\mathbf{p}$ vector, corresponding to the enlarged $\mathbf{x}$ vector, can be defined by adding additional zero components to the original $\mathbf{p}$ vector.

5.14 During each step of linear programming solutions, an inversion of an $m \times m$ matrix is required. Each step requires the introduction of one new column (basis vector) and the deletion of one old column. Derive the inverse of the $m \times m$ matrix $\mathbf{A}_1$ which differs from another $m \times m$ matrix $\mathbf{A}_B$ only in the lth column. Make use of the fact that $\mathbf{A}_B^{-1}$ is known.

Let

$$\mathbf{A}_B = [\mathbf{a}_1 \quad \mathbf{a}_2 \quad \cdots \quad \mathbf{a}_l \quad \cdots \quad \mathbf{a}_m] \qquad \text{and} \qquad \mathbf{A}_1 = [\mathbf{a}_1 \quad \mathbf{a}_2 \quad \cdots \quad \mathbf{a}_k \quad \cdots \quad \mathbf{a}_m]$$

Then

$$\mathbf{A}_1 = \mathbf{A}_B - [\mathbf{0} \mid \Delta\mathbf{a} \mid \mathbf{0}]$$

where $\Delta\mathbf{a} = \mathbf{a}_l - \mathbf{a}_k$ is in the lth column. Define the *row* vector

$$\mathbf{r} = [0 \quad 0 \quad \cdots \quad 0 \quad \underset{\uparrow\, l\text{th element}}{1} \quad 0 \quad \cdots \quad 0]$$

Then

$$\mathbf{A}_1 = \mathbf{A}_B - \Delta\mathbf{a}\,\mathbf{r}$$

The following identity can be derived using the methods suggested in Sec. 3.9 or it can be obtained from results presented in Problem 3.4 by making the identifications $\mathbf{G}_2\mathbf{G}_1 = \mathbf{A}_B^{-1}$, $\mathbf{H}_2 = -\Delta\mathbf{a}$, $\mathbf{H}_1 = \mathbf{r}$:

$$\mathbf{A}_1^{-1} = [\mathbf{A}_B - \Delta\mathbf{a}\,\mathbf{r}]^{-1} = \mathbf{A}_B^{-1} + \mathbf{A}_B^{-1}\,\Delta\mathbf{a}[1 - \mathbf{r}\,\mathbf{A}_B^{-1}\,\Delta\mathbf{a}]^{-1}\,\mathbf{r}\,\mathbf{A}_B^{-1}$$

For convenience the lth row of $\mathbf{A}_B^{-1}$ is defined as the *row* vector $\mathbf{v} = \mathbf{r}\mathbf{A}_B^{-1}$. Then

$$\mathbf{A}_1^{-1} = \mathbf{A}_B^{-1}\left\{\mathbf{I}_m + \frac{\Delta\mathbf{a}\,\mathbf{v}}{[1 - \mathbf{v}\,\Delta\mathbf{a}]}\right\}$$

Since $\mathbf{A}_B^{-1}$ is already known, $\mathbf{A}_1^{-1}$ can be obtained by using only scalar division and some matrix manipulations.

5.15 Let $\mathbf{A}_B = \mathbf{I}_3$. Find the inverse of $\mathbf{A}_1 = \begin{bmatrix} 3 & 0 & 0 \\ 2 & 1 & 0 \\ 1 & 0 & 1 \end{bmatrix}$.

Since $\mathbf{A}_1$ differs from $\mathbf{A}_B$ only in the first column, the results of the previous problem apply, with $\mathbf{A}_B^{-1} = \mathbf{I}_3$, $\mathbf{r} = [1 \quad 0 \quad 0] = \mathbf{v}$ and $\Delta\mathbf{a} = [-2 \quad -2 \quad -1]^T$. Thus

$$\mathbf{A}_1^{-1} = \mathbf{I}_3 - \frac{1}{3}\begin{bmatrix} 2 & 0 & 0 \\ 2 & 0 & 0 \\ 1 & 0 & 0 \end{bmatrix} = \begin{bmatrix} \frac{1}{3} & 0 & 0 \\ -\frac{2}{3} & 1 & 0 \\ -\frac{1}{3} & 0 & 1 \end{bmatrix}$$

5.16 A company makes two versions of a product, a deluxe and a standard version. The numbers of each to be made in a month are x_1 and x_2. The profit per unit on these items is \$3.00 and \$2.00, respectively. Because of limited facilities, no more than 2000 total units can be handled each month. A total of 3000 man-hours are available; and each deluxe unit requires three man-hours, while one man-hour is required for the standard version. Because of tool wear, only one deluxe unit can be machined with a given tool bit. However, the bit can be used to produce ten more lower-quality standard units before it must be discarded. An older machine can also be used to produce up to 1000 standard units per month. Find the quantities x_1 and x_2 which maximize profit.

Problem Set-up:

$$
\begin{aligned}
Z &= 3x_1 + 2x_2 && \text{(profit)} \\
x_1 + x_2 &\le 2000 && \text{(facilities limitation)} \\
3x_1 + x_2 &\le 3000 && \text{(man-hours limitation)} \\
-10x_1 + x_2 &\le 1000 && \text{(machining limitation)} \\
x_1 \ge 0,\ x_2 &\ge 0 && \text{(nonnegative production)}
\end{aligned}
$$

Adding slack variables x_3, x_4, and x_5 allows the constraints to be written as

$$\left[\begin{array}{cc|ccc} 1 & 1 & 1 & 0 & 0 \\ 3 & 1 & 0 & 1 & 0 \\ -10 & 1 & 0 & 0 & 1 \end{array}\right]\begin{bmatrix} x_1 \\ x_2 \\ x_3 \\ x_4 \\ x_5 \end{bmatrix} = \begin{bmatrix} 2000 \\ 3000 \\ 1000 \end{bmatrix}$$

The profit vector $\mathbf{p}$ is also expanded to five components by adding three zeros. Then

$$Z = \mathbf{p}^T\mathbf{x}$$

Starting Solution: An obvious basic feasible solution is

$$\mathbf{x}_B^T = [x_3 \quad x_4 \quad x_5] = [2000 \quad 3000 \quad 1000]$$

so that

$$\mathcal{J} = \{3, \ 4, \ 5\}, \qquad \mathbf{A}_B = \mathbf{I}_3, \qquad \mathbf{p}_B^T = [0 \quad 0 \quad 0]$$

The associated profit is $Z = 0$.

First Replacement: To find which new column should be brought into $\mathbf{A}_B$, compute

$$\boldsymbol{\beta}^T = \mathbf{p}^T - \mathbf{p}_B^T\mathbf{A}_B^{-1}\mathbf{A} = [3 \quad 2 \quad 0 \quad 0 \quad 0]$$

Since β_1 is the largest component, $\mathbf{a}_1$ will be added to the basis set. To determine which column should be removed from $\mathbf{A}_B$, compute

$$\mathbf{d}_1 = \mathbf{I}\mathbf{a}_1 = \begin{bmatrix} 1 \\ 3 \\ -10 \end{bmatrix}$$

Since the third component of $\mathbf{d}_1$ is negative, it need not be considered. $\alpha_1 = 2000$, $\alpha_2 = 3000/3 = 1000$. Since α_2 is the smallest, the second column of $\mathbf{A}_B$ is deleted, giving

$$\mathcal{J} = \{3, \ 1, \ 5\}, \qquad \mathbf{p}_B^T = [0 \quad 3 \quad 0],$$

$$\mathbf{A}_B = \begin{bmatrix} 1 & 1 & 0 \\ 0 & 3 & 0 \\ 0 & -10 & 1 \end{bmatrix}, \qquad \mathbf{x}_B = \begin{bmatrix} x_3 \\ x_1 \\ x_5 \end{bmatrix} = \begin{bmatrix} 1{,}000 \\ 1{,}000 \\ 11{,}000 \end{bmatrix}, \qquad Z = 3000$$

Second Replacement: The new vector $\boldsymbol{\beta}$ is

$$\boldsymbol{\beta} = \begin{bmatrix} 3 \\ 2 \\ 0 \\ 0 \\ 0 \end{bmatrix} - \frac{1}{3}\begin{bmatrix} 1 & 3 & -10 \\ 1 & 1 & 1 \\ 1 & 0 & 0 \\ 0 & 1 & 0 \\ 0 & 0 & 1 \end{bmatrix}\begin{bmatrix} 3 & 0 & 0 \\ -1 & 1 & 10 \\ 0 & 0 & 3 \end{bmatrix}\begin{bmatrix} 0 \\ 3 \\ 0 \end{bmatrix} = \begin{bmatrix} 0 \\ 1 \\ 0 \\ -1 \\ 0 \end{bmatrix}$$

Since β_2 is the largest positive component, $\mathbf{a}_2$ should be added to the basis set. It is found that $\mathbf{d}_2 = \frac{1}{3}[2 \quad 1 \quad 13]^T$ and

$$\frac{x_{B_1}}{d_{2_1}} = 1500, \qquad \frac{x_{B_2}}{d_{2_2}} = 3000, \qquad \frac{x_{B_3}}{d_{2_3}} = \frac{33{,}000}{13}$$

Since the first term is smallest, $\mathbf{a}_2$ will occupy column one of the new $\mathbf{A}_B$. Then $\mathcal{J} = \{2, 1, 5\}$, and

$$\mathbf{A}_B = \begin{bmatrix} 1 & 1 & 0 \\ 1 & 3 & 0 \\ 1 & -10 & 1 \end{bmatrix}, \qquad \mathbf{p}_B = \begin{bmatrix} 2 \\ 3 \\ 0 \end{bmatrix}, \qquad \mathbf{x}_B = \begin{bmatrix} x_2 \\ x_1 \\ x_5 \end{bmatrix} = \begin{bmatrix} 1500 \\ 500 \\ 4500 \end{bmatrix}$$

The corresponding profit is $Z = 4500$.

Termination: Calculating a new $\boldsymbol{\beta}$ will show that all components are negative or zero. Z cannot be increased further.

Final Results: The optimal solution consists of

$$x_1 = 500, \qquad x_2 = 1500, \qquad x_3 = x_4 = 0$$

$$x_5 = 4500, \qquad Z_{\max} = 4500$$

5.17 Consider a system described by the pair of difference equations [113]

$$\begin{bmatrix} y_1(t_{k+1}) \\ y_2(t_{k+1}) \end{bmatrix} = \begin{bmatrix} 1/3 & 0 \\ 0 & 1/2 \end{bmatrix} \begin{bmatrix} y_1(t_k) \\ y_2(t_k) \end{bmatrix} + \begin{bmatrix} 1 \\ 1 \end{bmatrix} u(t_k)$$

where t_k represents discrete time points, 0, 1, 2, ... Initial conditions are $\begin{bmatrix} y_1(0) \\ y_2(0) \end{bmatrix} = \begin{bmatrix} 10 \\ 2 \end{bmatrix}$. It is required that $\mathbf{y}(3) = \mathbf{0}$. Use linear programming to find the sequence of inputs $u(0)$, $u(1)$, and $u(2)$ which satisfies the terminal constraints and which also minimizes

$$J = |u(0)| + |u(1)| + |u(2)|$$

There is no restriction on the signs of $u(t_k)$.

This is a simple example of a minimum fuel type problem. To put it into standard form, several things are required. By successive substitution the form of $\mathbf{y}(3)$ is

$$\mathbf{y}(3) = \begin{bmatrix} 1/27 & 0 \\ 0 & 1/8 \end{bmatrix} \mathbf{y}(0) + \begin{bmatrix} 1/9 \\ 1/4 \end{bmatrix} u(0) + \begin{bmatrix} 1/3 \\ 1/2 \end{bmatrix} u(1) + \begin{bmatrix} 1 \\ 1 \end{bmatrix} u(2)$$

Setting this to $\mathbf{0}$ and rearranging gives

$$\begin{bmatrix} 3 \\ 2 \end{bmatrix} u(0) + \begin{bmatrix} 9 \\ 4 \end{bmatrix} u(1) + \begin{bmatrix} 27 \\ 8 \end{bmatrix} u(2) = -\mathbf{y}(0) = \begin{bmatrix} -10 \\ -2 \end{bmatrix}$$

This represents two simultaneous constraints in the three unknowns $u(k)$. Before they can be put into the standard form, the substitutions $u(k) = u(k)^+ - u(k)^-$ are made, where all $u(k)^+$ and $u(k)^-$ are required to be nonnegative. Instead of minimizing J, we maximize $Z = -J$, which can be written as

$$Z = -u(0)^+ - u(0)^- - u(1)^+ - u(1)^- - u(2)^+ - u(2)^-$$

This uses the fact that $u(t_k)^+$ and $u(t_k)^-$ will never simultaneously be nonzero, so that $-|u(t_k)| = -u(t_k)^+ - u(t_k)^-$. Then the constraints are $u(k)^+ \geq 0$, $u(k)^- \geq 0$, and

$$\begin{bmatrix} 3 & -3 & 9 & -9 & 27 & -27 \\ 2 & -2 & 4 & -4 & 8 & -8 \end{bmatrix} \begin{bmatrix} u(0)^+ \\ u(0)^- \\ u(1)^+ \\ u(1)^- \\ u(2)^+ \\ u(2)^- \end{bmatrix} = \begin{bmatrix} -10 \\ -2 \end{bmatrix}$$

An initial basis set can be formed from $\mathbf{a}_1$ and $\mathbf{a}_4$, so that

$$\mathcal{J} = \{1, \; 4\}, \qquad \mathbf{p}_B^T = [-1 \quad -1],$$

$$\mathbf{A}_B = \begin{bmatrix} 3 & -9 \\ 2 & -4 \end{bmatrix}, \quad \mathbf{A}_B^{-1} = \frac{1}{6}\begin{bmatrix} -4 & 9 \\ -2 & 3 \end{bmatrix}, \qquad \mathbf{x}_B = \begin{bmatrix} u(0)^+ \\ u(1)^- \end{bmatrix} = \frac{1}{3}\begin{bmatrix} 11 \\ 7 \end{bmatrix}$$

The "profit," which is the negative of the fuel used, is $Z = -6$. The vector $\boldsymbol{\beta}^T = \mathbf{p}^T - \mathbf{p}_B^T\mathbf{A}_B^{-1}\mathbf{A}$ is determined to be $\boldsymbol{\beta}^T = [0 \quad -2 \quad -2 \quad 0 \quad -12 \quad 10]$. Since β_6 is the largest, $\mathbf{a}_6$ will become a basis vector. To determine which vector it replaces, we compute $\mathbf{d}_6 = \mathbf{A}_B^{-1}\mathbf{a}_6 = \begin{bmatrix} 6 \\ 5 \end{bmatrix}$ and $\alpha_1 = 11/18$, $\alpha_2 = 7/15$. Since α_2 is the smallest, $\mathcal{J} = \{1, 6\}$ and

$$\mathbf{A}_B = \begin{bmatrix} 3 & -27 \\ 2 & -8 \end{bmatrix}, \qquad \mathbf{x}_B = \begin{bmatrix} u(0)^+ \\ u(2)^- \end{bmatrix} = \frac{1}{15}\begin{bmatrix} 13 \\ 7 \end{bmatrix}$$

with $Z = -4/3$. An additional calculation shows that all $\beta_k \le 0$, so the optimal solution has been found. It is

$$u(0) = \frac{13}{15}, \qquad u(1) = 0, \qquad u(2) = -\frac{7}{15}$$

and since $Z_{\max} = -4/3$, we have $J_{\min} = 4/3$. When this sequence of optimal controls is used, the corresponding values of $y(t_k)$ are

$$\mathbf{y}(0) = \begin{bmatrix} 10 \\ 2 \end{bmatrix}, \qquad \mathbf{y}(1) = \begin{bmatrix} 21/5 \\ 28/15 \end{bmatrix}, \qquad \mathbf{y}(2) = \begin{bmatrix} 7/5 \\ 14/15 \end{bmatrix}, \qquad \mathbf{y}(3) = \begin{bmatrix} 0 \\ 0 \end{bmatrix}$$

Least Squares, Weighted Least Squares, and Recursive Least Squares

5.18 Consider the set of simultaneous linear equations

$$\begin{bmatrix} \mathbf{y}_k \\ \hline \mathbf{y}_{k+1} \end{bmatrix} = \begin{bmatrix} \mathbf{A} \\ \hline \mathbf{H}_{k+1} \end{bmatrix} \mathbf{x} + \begin{bmatrix} \mathbf{e} \\ \hline \mathbf{e}_{k+1} \end{bmatrix}$$

Find the vector $\mathbf{x}$ which minimizes

$$J = [\mathbf{e}^T \mid \mathbf{e}_{k+1}^T] \begin{bmatrix} \mathbf{R}^{-1} & \mathbf{0} \\ \mathbf{0} & \mathbf{R}_{k+1}^{-1} \end{bmatrix} \begin{bmatrix} \mathbf{e} \\ \hline \mathbf{e}_{k+1} \end{bmatrix}$$

The weighted least-squares estimate is

$$\mathbf{x}_{k+1} = \left\{ [\mathbf{A}^T \mid \mathbf{H}_{k+1}^T] \begin{bmatrix} \mathbf{R}^{-1} & \mathbf{0} \\ \mathbf{0} & \mathbf{R}_{k+1}^{-1} \end{bmatrix} \begin{bmatrix} \mathbf{A} \\ \hline \mathbf{H}_{k+1} \end{bmatrix} \right\}^{-1} [\mathbf{A}^T \mid \mathbf{H}_{k+1}^T] \begin{bmatrix} \mathbf{R}^{-1} & \mathbf{0} \\ \mathbf{0} & \mathbf{R}_{k+1}^{-1} \end{bmatrix} \begin{bmatrix} \mathbf{y}_k \\ \hline \mathbf{y}_{k+1} \end{bmatrix}$$

$$= [\mathbf{A}^T\mathbf{R}^{-1}\mathbf{A} + \mathbf{H}_{k+1}^T\mathbf{R}_{k+1}^{-1}\mathbf{H}_{k+1}]^{-1}[\mathbf{A}^T\mathbf{R}^{-1}\mathbf{y}_k + \mathbf{H}_{k+1}^T\mathbf{R}_{k+1}^{-1}\mathbf{y}_{k+1}]$$

Defining $\mathbf{A}^T\mathbf{R}^{-1}\mathbf{A} = \mathbf{P}_k^{-1}$ and using the matrix inversion identity of Sec. 3.9 gives

$$\mathbf{x}_{k+1} = \{\mathbf{P}_k - \mathbf{P}_k\mathbf{H}_{k+1}^T[\mathbf{H}_{k+1}\mathbf{P}_k\mathbf{H}_{k+1}^T + \mathbf{R}_{k+1}]^{-1}\mathbf{H}_{k+1}\mathbf{P}_k\}\{\mathbf{A}^T\mathbf{R}^{-1}\mathbf{y}_k + \mathbf{H}_{k+1}^T\mathbf{R}_{k+1}^{-1}\mathbf{y}_{k+1}\}$$

Note that $\mathbf{P}_k\mathbf{A}^T\mathbf{R}^{-1}\mathbf{y}_k = \mathbf{x}_k$ is the weighted least-squares solution when only the first group of equations is used. Therefore

$$\mathbf{x}_{k+1} = \mathbf{x}_k - \mathbf{P}_k\mathbf{H}_{k+1}^T[\mathbf{H}_{k+1}\mathbf{P}_k\mathbf{H}_{k+1}^T + \mathbf{R}_{k+1}]^{-1}\mathbf{H}_{k+1}\mathbf{x}_k$$
$$+ \mathbf{P}_k\mathbf{H}_{k+1}^T\{\mathbf{I} - [\mathbf{H}_{k+1}\mathbf{P}_k\mathbf{H}_{k+1}^T + \mathbf{R}_{k+1}]^{-1}\mathbf{H}_{k+1}\mathbf{P}_k\mathbf{H}_{k+1}^T\}\mathbf{R}_{k+1}^{-1}\mathbf{y}_{k+1}$$

The unit matrix in the last equation is written as

$$\mathbf{I} = [\mathbf{H}_{k+1}\mathbf{P}_k\mathbf{H}_{k+1}^T + \mathbf{R}_{k+1}]^{-1}\,[\mathbf{H}_{k+1}\mathbf{P}_k\mathbf{H}_{k+1}^T + \mathbf{R}_{k+1}]$$

This step is analogous to finding the common denominator in scalar algebra and leads to

$$\mathbf{x}_{k+1} = \mathbf{x}_k + \mathbf{P}_k\mathbf{H}_{k+1}^T[\mathbf{H}_{k+1}\mathbf{P}_k\mathbf{H}_{k+1}^T + \mathbf{R}_{k+1}]^{-1}\{\mathbf{y}_{k+1} - \mathbf{H}_{k+1}\mathbf{x}_k\}$$

If this recursive process is to be continued, then an expression for $\mathbf{P}_{k+1}$ is needed. If $\mathbf{A}$ is replaced by $\begin{bmatrix} \mathbf{A} \\ \hline \mathbf{H}_{k+1} \end{bmatrix}$ and if $\mathbf{R}^{-1}$ is replaced by $\begin{bmatrix} \mathbf{R}^{-1} & \mathbf{0} \\ \mathbf{0} & \mathbf{R}_{k+1}^{-1} \end{bmatrix}$, then the definition for $\mathbf{P}_k$ is modified to read

$$\mathbf{P}_{k+1} = \left\{ [\mathbf{A}^T \mid \mathbf{H}_{k+1}^T] \begin{bmatrix} \mathbf{R}^{-1} & \mathbf{0} \\ \mathbf{0} & \mathbf{R}_{k+1}^{-1} \end{bmatrix} \begin{bmatrix} \mathbf{A} \\ \hline \mathbf{H}_{k+1} \end{bmatrix} \right\}^{-1}$$

$$= [\mathbf{A}^T\mathbf{R}^{-1}\mathbf{A} + \mathbf{H}_{k+1}^T\mathbf{R}_{k+1}^{-1}\mathbf{H}_{k+1}]^{-1} = [\mathbf{P}_k^{-1} + \mathbf{H}_{k+1}^T\mathbf{R}_{k+1}^{-1}\mathbf{H}_{k+1}]^{-1}$$

5.19 A tracking station measures $\dot{r}$, the time derivative of the range to a satellite, every second. The measurements are noisy. Find the least-squares fit to a straight line.

The measurements are assumed to fit the equation $\dot{r}(t) = a + bt + e(t)$, where a and b are to be determined. Measurement times are $t = 1, 2, \ldots, k$:

$$\begin{bmatrix} \dot{r}_1 \\ \dot{r}_2 \\ \cdot \\ \cdot \\ \cdot \\ \dot{r}_k \end{bmatrix} = \begin{bmatrix} 1 & 1 \\ 1 & 2 \\ \cdot & \cdot \\ \cdot & \cdot \\ \cdot & \cdot \\ 1 & k \end{bmatrix} \begin{bmatrix} a \\ b \end{bmatrix} + \begin{bmatrix} e_1 \\ e_2 \\ \cdot \\ \cdot \\ \cdot \\ e_k \end{bmatrix}$$

The least-squares solution gives

$$\begin{bmatrix} a \\ b \end{bmatrix} = \begin{bmatrix} k & \sum_{j=1}^{k} j \\ \sum_{j=1}^{k} j & \sum_{j=1}^{k} j^2 \end{bmatrix}^{-1} \begin{bmatrix} \sum_{j=1}^{k} \dot{r}_i \\ \sum_{j=1}^{k} j\dot{r}_j \end{bmatrix}$$

Using the identities $\sum_{j=1}^{k} j = \frac{k(k+1)}{2}$, $\sum_{j=1}^{k} j^2 = \frac{k(k+1)(2k+1)}{6}$, and carrying out the matrix inversion gives

$$a = \frac{2(2k+1)\sum_{j=1}^{k} \dot{r}_j - 6\sum_{j=1}^{k} j\dot{r}_j}{k(k-1)}, \qquad b = \frac{-6(k+1)\sum_{j=1}^{k} \dot{r}_j + 12\sum_{j=1}^{k} j\dot{r}_j}{k(k^2-1)}$$

If $k = 1$, the results are indeterminate, indicating that an infinite number of lines can be passed through a single point.

5.20 Investigate the following equations:

$$\begin{bmatrix} 1 & 2 \\ 3 & 4 \\ 5 & 6 \end{bmatrix} \mathbf{x} = \begin{bmatrix} 2 \\ 3 \\ 14 \end{bmatrix}$$

The RRE form of $\mathbf{W}$ is $\mathbf{W}' = \left[\begin{array}{cc:c} 1 & 0 & 0 \\ 0 & 1 & 0 \\ 0 & 0 & 1 \end{array}\right]$.

Since the rank of $\mathbf{A}$ is 2 and the rank of $\mathbf{W}$ is 3, the equations are inconsistent. Least-squares solutions are investigated by several methods.

a. The normal equation $\mathbf{A}^T\mathbf{A}\mathbf{x} = \mathbf{A}^T\mathbf{y}$ is in this case

$$\begin{bmatrix} 35 & 44 \\ 44 & 56 \end{bmatrix} \begin{bmatrix} \mathbf{x}_1 \\ \mathbf{x}_2 \end{bmatrix} = \begin{bmatrix} 81 \\ 100 \end{bmatrix}$$

Straightforward matrix inversion could be used, or the RRE method as follows:

$$\mathbf{W} = \left[\begin{array}{cc:c} 35 & 44 & 81 \\ 44 & 56 & 100 \end{array}\right] \longrightarrow \left[\begin{array}{cc:c} 1 & 0 & 5.6666 \\ 0 & 1 & -2.6666 \end{array}\right] \quad \text{so} \quad \begin{bmatrix} x_1 \\ x_2 \end{bmatrix} = \begin{bmatrix} 5.6666 \\ -2.6666 \end{bmatrix}$$

b. Cholesky decomposition gives

$$\mathbf{A}^T\mathbf{A} = \begin{bmatrix} 5.91608 & 0 \\ 7.43736 & 0.82808 \end{bmatrix} \begin{bmatrix} 5.91608 & 7.43736 \\ 0 & 0.82808 \end{bmatrix} = \mathbf{S}^T\mathbf{S}$$

Solving $\mathbf{S}^T\mathbf{v} = \begin{bmatrix} 81 \\ 100 \end{bmatrix}$ gives $\mathbf{v} = \begin{bmatrix} 13.69150 \\ -2.20818 \end{bmatrix}$

Then solving $\mathbf{S}\mathbf{x} = \mathbf{v}$ gives the same answer for $\mathbf{x}$ as in part (a).

c. Using the GSE method, the two columns of the original **A** matrix, plus the vector **y**, are used to generate an orthonormal basis set which is shown as columns of the matrix **V**.

$$\mathbf{V} = \begin{bmatrix} 1.6903085E{-}01 & 8.9708525E{-}01 & 4.0824756E{-}01 \\ 5.0709254E{-}01 & 2.7602646E{-}01 & -8.1649655E{-}01 \\ 8.4515423E{-}01 & -3.4503257E{-}01 & 4.0824878E{-}01 \end{bmatrix}$$

Then $\mathbf{V}^T\mathbf{A}\mathbf{x} = \mathbf{V}^T\mathbf{y}$ can be compactly written as $\mathbf{V}^T\mathbf{W} = \mathbf{W}'$. Rounding terms of order 10^{-6} to zero gives

$$\mathbf{W}' = \left[\begin{array}{cc|c} 5.91608 & 7.43736 & 13.69150 \\ 0 & 0.82808 & -2.20821 \\ 0 & 0 & 4.08249 \end{array}\right].$$

Compare this with the Cholesky results.

The last row indicates that the norm of the residual error is $\|\mathbf{y}_e\| = 4.08256$. Ignoring the last row and solving the first two obviously again give the same answer.

5.21 After seven semesters of college a student surmises that his cumulative grade point average (GPA) is a cubic function of the number of semesters completed. His record to date is

Semester, s	1	2	3	4	5	6	7	8
Cum. GPA	2.5	3.1	2.9	2.8	2.8	3.0	3.1	??

Find the coefficients for the least-squares fit to a cubic. Also determine the residual error norm. Then use the cubic to predict his GPA on graduation day after semester 8.

It is assumed that GPA $= a_0 + a_1 s + a_2 s^2 + a_3 s^3$. In matrix form this student's data and postulated model are

$$\begin{bmatrix} 1 & 1 & 1 & 1 \\ 1 & 2 & 4 & 8 \\ 1 & 3 & 9 & 27 \\ 1 & 4 & 16 & 64 \\ 1 & 5 & 25 & 125 \\ 1 & 6 & 36 & 216 \\ 1 & 7 & 49 & 343 \end{bmatrix} \begin{bmatrix} a_0 \\ a_1 \\ a_2 \\ a_3 \end{bmatrix} = \begin{bmatrix} 2.5 \\ 3.1 \\ 2.9 \\ 2.8 \\ 2.8 \\ 3.0 \\ 3.1 \end{bmatrix}$$

Using the GSE method leads to the following $\mathbf{W}'$ matrix, rounded off.

$$\mathbf{W}' = \left[\begin{array}{cccc|c} 2.646 & 10.583 & 52.915 & 296.324 & 7.635 \\ 0 & 5.292 & 42.322 & 291.033 & 0.283 \\ 0 & 0 & 9.165 & 109.982 & -0.033 \\ 0 & 0 & 0 & 14.700 & 0.327 \\ \hline 0 & 0 & 0 & 0 & 0.284 \end{array}\right]$$

From this the coefficients are determined as

$$\mathbf{a} = [1.828 \quad 0.993 \quad -0.270 \quad 0.022]^T \qquad \text{and} \qquad \|\mathbf{y}_e\| = 0.284$$

Using these coefficients the estimated GPA after eight semesters is

$1.828 + 8(0.993) - 64(0.27) + 512(0.022) = 3.756$. The optimist!

5.22 Use the recursive least-squares algorithm to estimate $\mathbf{x}$ from Problem 5.20. Start with an initial estimate of $\mathbf{x} = \mathbf{0}$ and set $\mathbf{P}_0 = \text{diag}\,[10000, 10000]$. Use $\mathbf{R} = 10$ and $\mathbf{Q} = 0$

(no deweighting). Also add the following additional equations to be processed:

$$0 = 3x_1 + 6x_2,\ 10 = 2x_1 + x_2$$

The recursive calculations are rounded off and tabulated in the order performed.

k	$\mathbf{M}_k$	$\mathbf{H}_k$	$\mathbf{y}_k$	$\mathbf{K}_k$	$\mathbf{y}_k - \mathbf{H}_k\mathbf{x}_{k-1}$	$\mathbf{x}_k$	$\mathbf{P}_k$
0	—	—	—	$\begin{pmatrix} 0 \\ 0 \end{pmatrix}$	—	—	$\begin{pmatrix} 10000 & 0 \\ 0 & 10000 \end{pmatrix}$
1	$\begin{pmatrix} 10000 & 0 \\ 0 & 10000 \end{pmatrix}$	$(1 \quad 2)$	2	$\begin{pmatrix} 0.2 \\ 0.4 \end{pmatrix}$	2	$\begin{pmatrix} 0.4 \\ 0.8 \end{pmatrix}$	$\begin{pmatrix} 8000 & -4000 \\ -4000 & 2000 \end{pmatrix}$
2	$\begin{pmatrix} 8000 & -4000 \\ -4000 & 2000 \end{pmatrix}$	$(3 \quad 4)$	3	$\begin{pmatrix} 0.993 \\ -0.495 \end{pmatrix}$	-1.4	$\begin{pmatrix} -0.990 \\ 1.493 \end{pmatrix}$	$\begin{pmatrix} 49.628 & -34.474 \\ -34.474 & 24.815 \end{pmatrix}$
3	$\begin{pmatrix} 49.628 & -34.474 \\ -34.474 & 24.815 \end{pmatrix}$	$(5 \quad 6)$	14	$\begin{pmatrix} 0.664 \\ -0.415 \end{pmatrix}$	9.992	$\begin{pmatrix} 5.648 \\ -2.652 \end{pmatrix}$	$\begin{pmatrix} 23.245 & -18.264 \\ -18.264 & 14.529 \end{pmatrix}$
4	$\begin{pmatrix} 23.245 & -18.264 \\ -18.264 & 14.529 \end{pmatrix}$	$(3 \quad 6)$	0	$\begin{pmatrix} -0.470 \\ 0.382 \end{pmatrix}$	-1.033	$\begin{pmatrix} 6.134 \\ -3.046 \end{pmatrix}$	$\begin{pmatrix} 4.507 & -3.037 \\ -3.037 & 2.155 \end{pmatrix}$
5	$\begin{pmatrix} 4.507 & -3.037 \\ -3.037 & 2.155 \end{pmatrix}$	$(2 \quad 1)$	10	$\begin{pmatrix} 0.331 \\ -0.217 \end{pmatrix}$	0.779	$\begin{pmatrix} 6.392 \\ -3.216 \end{pmatrix}$	$\begin{pmatrix} 2.526 & -1.739 \\ -1.739 & 1.304 \end{pmatrix}$

Notice that the estimate of $\mathbf{x}$ after three measurements is not exactly the same as was found in Problem 5.20. The difference is due to the initial estimates used for $\mathbf{x}$ and $\mathbf{P}$.

PROBLEMS

Miscellaneous

5.23 Solve for $\mathbf{x}$ if $\begin{bmatrix} 8 & 2 & 1 \\ 1 & 1 & 3 \\ 2 & 5 & 4 \end{bmatrix} \mathbf{x} = \begin{bmatrix} 10 \\ 5 \\ 1 \end{bmatrix}$.

5.24 Assume that a dynamic system can be described by a vector of time-varying parameters $\mathbf{x}(t)$, with initial conditions $\mathbf{x}(0)$. The relation between $\mathbf{x}(t)$ and $\mathbf{x}(0)$ is $\mathbf{x}(t) = \mathbf{\Phi}(t)\mathbf{x}(0)$, where $\mathbf{\Phi}(t)$ is an $n \times n$ matrix. Let $\mathbf{x}$ be partitioned into $\begin{bmatrix} \mathbf{x}_1(t) \\ \hline \mathbf{x}_2(t) \end{bmatrix}$. If $\mathbf{x}_1(0)$ and $\mathbf{x}_2(T)$ are known, find $\mathbf{x}_2(0)$.

Homogeneous Equations

5.25 If $\mathbf{Ax} = \mathbf{0}$ has q linearly independent solutions $\mathbf{x}_i$ and $\mathbf{Ax} = \mathbf{y}$ has $\mathbf{x}_0$ as a solution, show that

a. $\mathbf{x}_c = \sum_{i=1}^{q} \alpha_i \mathbf{x}_i$ is also a solution of $\mathbf{Ax} = \mathbf{0}$,

b. $\mathbf{x} = \mathbf{x}_0 + \sum_{i=1}^{q} \alpha_i \mathbf{x}_i$ is a solution of $\mathbf{Ax} = \mathbf{y}$.

5.26 Find the nontrivial solutions for $\begin{bmatrix} 1 & 0 & 4 \\ 2 & 3 & 8 \end{bmatrix}\begin{bmatrix} x_1 \\ x_2 \\ x_3 \end{bmatrix} = \mathbf{0}$.

5.27 Find all nontrivial solutions for $\begin{bmatrix} 1 & 1 & 1 \\ 2 & 2 & 2 \end{bmatrix}\begin{bmatrix} x_1 \\ x_2 \\ x_3 \end{bmatrix} = \mathbf{0}$.

5.28 Determine whether nontrivial solutions exist for $\begin{bmatrix} 1 & 2 \\ 2 & 4 \\ 0 & 0 \\ -1 & -2 \end{bmatrix}\begin{bmatrix} x_1 \\ x_2 \end{bmatrix} = \mathbf{0}$.

5.29 Find all nontrivial solutions of $\mathbf{Ax} = \mathbf{0}$, i.e., the null space, of

$$\mathbf{A} = \begin{bmatrix} 26 & 17 & 8 & 39 & 35 \\ 17 & 13 & 9 & 29 & 28 \\ 8 & 9 & 10 & 19 & 21 \\ 39 & 29 & 19 & 65 & 62 \\ 35 & 28 & 21 & 62 & 61 \end{bmatrix}.$$

Linear Programming

5.30 Find nonnegative values of $x_1, x_2, \ldots, x_6$ which satisfy

$$\begin{bmatrix} 3 & 0 & 0 & 14 & 1 & 1 \\ 0 & 1 & 0 & 16 & \frac{1}{2} & -2 \\ 0 & 0 & 1 & 3 & 0 & 0 \end{bmatrix}\mathbf{x} = \begin{bmatrix} 7 \\ 5 \\ 0 \end{bmatrix}$$

while maximizing $Z = 28x_4 + x_5 + 2x_6$.

5.31 Find nonnegative values of x_1 through x_8 which maximize $14x_1 + 10x_2 + x_3$ subject to

$$\begin{bmatrix} 100 & -7 & 1 & 1 & 0 & 0 & 0 & 0 \\ 1 & 101 & 3 & 0 & 1 & 0 & 0 & 0 \\ 0 & 4 & -9 & 0 & 0 & 1 & 0 & 0 \\ -5 & 8 & 16 & 0 & 0 & 0 & 1 & 0 \\ 1 & 3 & 5 & 0 & 0 & 0 & 0 & 1 \end{bmatrix}\begin{bmatrix} x_1 \\ x_2 \\ \vdots \\ x_8 \end{bmatrix} = \begin{bmatrix} 16 \\ 7 \\ 178 \\ 4 \\ 1 \end{bmatrix}$$

Least Squares

5.32 Solve for x_1 and x_2 if

a. $2x_1 - x_2 = 5$, $x_1 + 2x_2 = 3$,

b. in addition to the equations in a, a third equation is $-x_1 + x_2 = -1$. Use least squares.

5.33 Given that $\begin{bmatrix} y_1 \\ y_2 \end{bmatrix} = \begin{bmatrix} 2 \\ 1 \end{bmatrix}x + \begin{bmatrix} e_1 \\ e_1 \end{bmatrix}$. Measurements give $[y_1 \quad y_2] = [3 \quad 4]$. Find the least-squares estimate for x. Use a sketch in the y_1, y_2 plane to indicate the geometrical interpretation.

5.34 Verify the result of Problem 4.24, page 113 by determining the least-squares solution for $\mathbf{x}$:

$$\begin{bmatrix} 1 & 0 \\ 2 & 1 \\ -3 & 3 \\ 1 & 3 \\ 0 & 1 \end{bmatrix} \begin{bmatrix} x_1 \\ x_2 \end{bmatrix} = \begin{bmatrix} 1 \\ -3 \\ 4 \\ 2 \\ 8 \end{bmatrix}$$

and then use the fact that the orthogonal projection of $\mathbf{y}$ on the column space of $\mathbf{A}$ is $\mathbf{y}_p = \mathbf{A}\mathbf{x}$.

5.35 A physical device is shown in Fig. 5.5. It is believed that the output y is linearly related to the input u. That is, $y = au + b$. What are the values of a and b if the following data are taken?

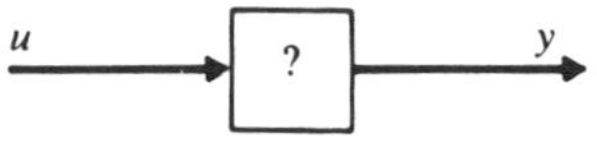

Figure 5.5

u	2	-2
y	5	1

5.36 The same device as in Problem 5.35 is considered. One more set of readings is taken as

$$u = 5, \qquad y = 7$$

Find a least-squares estimate of a and b. Also find the minimum mean-squared error in this straight line fit to the three points.

5.37 Consider the data of Problem 5.36, but assume that the first two equations are much more reliable. Use

$$\mathbf{R}^{-1} = \begin{bmatrix} 10 & 0 & 0 \\ 0 & 10 & 0 \\ 0 & 0 & 1 \end{bmatrix}$$

and show that the resultant weighted least-squares estimate is much closer to the values obtained in Problem 5.35.

5.38 Estimate the initial current $i(0)$ in the circuit of Fig. 5.6, if $R = 10$ ohms, $L = 3.56$ henrys, and the following voltmeter readings are taken:

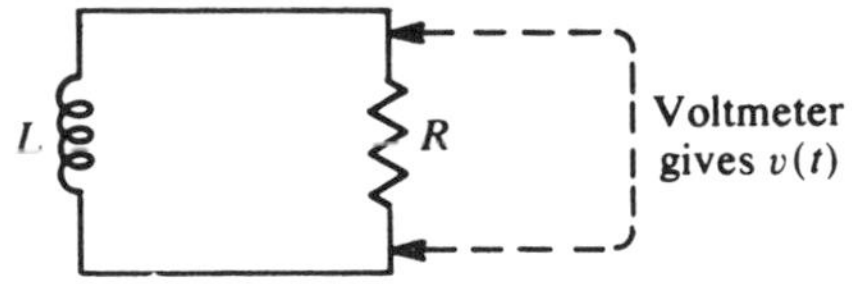

Figure 5.6

t	0	1	2	3
$v(t)$	167.9	95.5	88.8	55.3

5.39 Least-squares fit a quadratic function to the data of Problem 5.21. Determine the coefficients and the norm of the residual error. Then predict the GPA after semester eight.

5.40 An empirical theory used by many distance runners states that the time T_i required to race a distance D_i can be expressed as $T_i = C(D_i)^\alpha$, where C and α are constants for a given person, determined by lung capacity, body build, etc. Obtain a least-squares fit to the following data for one middle-aged jogger. (Convert to a linear equation in the unknowns C and α by taking the logarithm of the above expression.) Predict the time for one mile.

Time	185 min	79.6 min	60 min	37.9 min	11.5 min
Distance	26.2 mi	12.4 mi	9.5 mi	6.2 mi	2 mi

5.41 Apply the recursive least-squares algorithm to the data of Examples 5.3 and 5.6. Try different starting assumptions to determine their effect on the estimate. Recall that **A** is not full rank and thus a unique least-squares solution does not exist. Add a fifth equation,

$$4x_1 - x_2 + 6x_3 = 2$$

so that the enlarged **A** matrix is of full rank. How do your results compare with Example 5.6?

Chapter

6

TRANSFORMATIONS

6.1 INTRODUCTION

In previous chapters the product of a matrix and a vector, represented as a column matrix, was frequently used. Such a product yields another column matrix. It has been pointed out that column matrices may be used to represent a vector with respect to a given basis set. However, a vector may have many different representations depending on the choice of basis. The vector itself is a more abstract quantity which belongs to a linear vector space, since it exists independently of any particular basis for the space. This chapter develops the notion of transforming one vector into another. The transformation operator is an abstract entity, analogous to the vector. An important class of transformations can be represented as matrices once basis vectors are specified. The words *mapping* or *function* are synonymous with the word *transformation*, and sometimes the word *operator* is used.

In modern control theory the behavior of a system is usually described in terms of a vector in a linear vector space, the state space. The relationship between input functions and the system state, the change of the state with time, and the relationship between the state and the system output can be conveniently expressed in terms of transformations. Important properties such as stability, controllability, and observability are better understood after gaining an appreciation of transformations.

6.2 SIMPLE FUNCTIONS AND GENERALIZATIONS

A function is a rule by which elements in one set are associated with elements in another set. A function consists of three things: two specified sets of elements $\mathcal{X} = \{x_i\}$ and $\mathcal{Y} = \{y_i\}$ and a rule relating elements $x_i \in \mathcal{X}$ to elements $y_i \in \mathcal{Y}$. The rule must be unambiguous. That is, for every $x \in \mathcal{X}$, there is associated a single element $y \in \mathcal{Y}$. The rule is often written as

$$y = f(x)$$

A common notation which indicates all three aspects of a function is

$$f: \mathcal{X} \longrightarrow \mathcal{Y}$$

This states that there is a rule, f, by which every element in $\mathcal{X}$ is mapped into some element in $\mathcal{Y}$. The set $\mathcal{X}$ is called the *domain* of the function, and the set $\mathcal{Y}$ is called the *codomain*. For a particular x, $y = f(x)$ is called the *image* of x, or conversely x is the *pre-image* of y. When the function is applied to every element in $\mathcal{X}$, a set of image points in $\mathcal{Y}$ is generated. This set of images is called the *range* of the function, and is sometimes expressed as $f(\mathcal{X})$.

The words *into* and *onto* are frequently used in conjunction with functions or mappings. The set $f(\mathcal{X})$ is always contained within or possibly equal to $\mathcal{Y}$, written $f(\mathcal{X}) \subseteq \mathcal{Y}$. Thus f is said to map $\mathcal{X}$ *into* $\mathcal{Y}$. If every element in $\mathcal{Y}$ is the image of at least one $x \in \mathcal{X}$, then $f(\mathcal{X}) = \mathcal{Y}$, and the function is said to map $\mathcal{X}$ *onto* $\mathcal{Y}$.

In order that the function be unambiguously defined, there is always just one y associated with each x. However, it is possible that two or more distinct elements $x \in \mathcal{X}$ have the same image point $y \in \mathcal{Y}$. The special case for which distinct elements $x \in \mathcal{X}$ map into distinct elements $y \in \mathcal{Y}$ is called a *one-to-one* mapping. That is, if f is one-to-one, then $x_1 \neq x_2$ implies that $f(x_1) \neq f(x_2)$. Implications such as this are more concisely written as

$$x_1 \neq x_2 \Longrightarrow f(x_1) \neq f(x_2)$$

As in all logical arguments, negating both propositions reverses the implication, so that f is one-to-one if

$$f(x_1) = f(x_2) \Longrightarrow x_1 = x_2$$

If a function is both one-to-one and onto, then for each $y \in \mathcal{Y}$ there is a *unique* pre-image $x \in \mathcal{X}$. The unique relation between y and x defines the inverse function $g = f^{-1}$ with

$$g: \mathcal{Y} \longrightarrow \mathcal{X}, \qquad \text{where} \quad g[f(x)] = x$$

Example 6.1

Consider the graphs shown in Fig. 6.1.

In (*a*) it is impossible to write $y = f(x)$, because for a given value of x there are multiple values of y. It would be possible to write the function $x = g(y)$, since for each y there is a single value of x. In fact, $x = \sin y$.

In (*b*) the functional relation $y = f(x)$ can be written. In fact, $y = x^2$. This function is not one-to-one since $x = 2$ and $x = -2$ both have the same image, $y = 4$. If the sets $\mathcal{X}$ and $\mathcal{Y}$ are taken to be the real line $(-\infty, \infty)$, then this function is not onto since $y = -3$ is not the image of any real x. If $\mathcal{Y}$ is taken as the half line $[0, \infty)$, then the function is onto, but still not one-to-one.

In (*c*) $\mathcal{X}$ and $\mathcal{Y}$ are the set of all real numbers, and the function shown is $y = x^3$. This function is both one-to-one and onto. ■

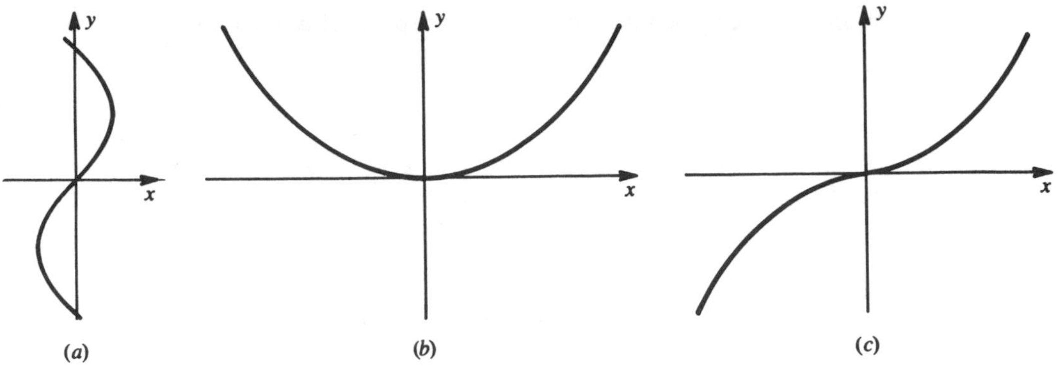

Figure 6.1

Example 6.2

Let $\mathcal{C}$ be the set of all complex numbers, z, and let $\mathcal{R}$ be the set of all real numbers. Then the absolute value function $f(z) = |z|$ maps $\mathcal{C}$ into $\mathcal{R}$,

$$f\colon \mathcal{C} \longrightarrow \mathcal{R}$$

The function is not onto since the negative reals are not images of any $z \in \mathcal{C}$. It is not one-to-one since z, $-z$, and $\bar{z}$ all have the same absolute value. ■

The previous concepts are now generalized to abstract vector spaces. Let $\mathfrak{X}$ and $\mathfrak{Y}$ be linear vector spaces (not necessarily distinct) which are defined over the same scalar number field $\mathfrak{F}$. If for each vector $\mathbf{x} \in \mathfrak{X}$ there is associated, according to some rule, a vector $\mathbf{y} \in \mathfrak{Y}$, then that "rule" defines a mapping of $\mathbf{x}$ into $\mathbf{y}$. This mapping rule is referred to as a transformation (or an operator or a function). This relationship is expressed by

$$\mathcal{A}\colon \mathfrak{X} \longrightarrow \mathfrak{Y}$$

The transformation is $\mathcal{A}$ and the mapping rule is $\mathcal{A}(\mathbf{x}) = \mathbf{y}$. The spaces $\mathfrak{X}$ and $\mathfrak{Y}$ are called the domain and codomain of $\mathcal{A}$, respectively. The domain is often written as $\mathfrak{D}(\mathcal{A})$, and the range of $\mathcal{A}$ is $\mathcal{A}(\mathfrak{X})$ or $\mathfrak{R}(\mathcal{A})$. Obviously $\mathfrak{R}(\mathcal{A})$ is contained within or equal to $\mathfrak{Y}$. This is written as $\mathfrak{R}(\mathcal{A}) \subseteq \mathfrak{Y}$. In general, $\mathcal{A}$ maps $\mathfrak{X}$ *into* $\mathfrak{Y}$, but if the equality holds, it maps $\mathfrak{X}$ *onto* $\mathfrak{Y}$. Again, if $\mathcal{A}(\mathbf{x}) = \mathbf{y}$, $\mathbf{y}$ is called the image of $\mathbf{x}$ or $\mathbf{x}$ is the pre-image of $\mathbf{y}$. The transformation $\mathcal{A}$ is said to be one-to-one if

$$\mathbf{x}_1 \neq \mathbf{x}_2 \Longrightarrow \mathcal{A}(\mathbf{x}_1) \neq \mathcal{A}(\mathbf{x}_2)$$

or equivalently, if

$$\mathcal{A}(\mathbf{x}_1) = \mathcal{A}(\mathbf{x}_2) \Longrightarrow \mathbf{x}_1 = \mathbf{x}_2$$

If $\mathcal{A}$ is both one-to-one and onto, then for each $\mathbf{y} \in \mathfrak{Y}$ there is a unique pre-image $\mathbf{x} \in \mathfrak{X}$, and an inverse transformation $\mathcal{A}^{-1}$ maps $\mathbf{y}$ into $\mathbf{x}$. In this case, $\mathcal{A}(\mathbf{x}) = \mathbf{y}$ and

$\mathcal{A}^{-1}(\mathbf{y}) = \mathbf{x}$, so $\mathcal{A}^{-1}(\mathcal{A}(\mathbf{x})) = \mathbf{x}$. Thus

$$\mathcal{A}^{-1}\mathcal{A} = \mathcal{I}$$

is the identity transformation which maps each vector in its domain into itself.

The *null space* $\mathfrak{N}(\mathcal{A})$ of the transformation $\mathcal{A}$ is the set of all vectors $\mathbf{x} \in \mathcal{X}$ which are mapped into the zero vector in $\mathcal{Y}$:

$$\mathfrak{N}(\mathcal{A}) \triangleq \{\mathbf{x} \in \mathcal{X} \mid \mathcal{A}(\mathbf{x}) = \mathbf{0}\}$$

6.3 LINEAR TRANSFORMATIONS

A transformation $\mathcal{A}: \mathcal{X} \longrightarrow \mathcal{Y}$ is said to be *linear* if the following two conditions are satisfied:

1. For any $\mathbf{x}_1$ and $\mathbf{x}_2 \in \mathcal{X}$, $\mathcal{A}(\mathbf{x}_1 + \mathbf{x}_2) = \mathcal{A}(\mathbf{x}_1) + \mathcal{A}(\mathbf{x}_2)$.
2. For any $\mathbf{x} \in \mathcal{X}$ and any scalar $\alpha \in \mathcal{F}$, $\mathcal{A}(\alpha\mathbf{x}) = \alpha\mathcal{A}(\mathbf{x})$.

Although nonlinear functions or transformations arise in connection with nonlinear control systems, major emphasis in this book is on linear systems or linear approximations to nonlinear ones. For this reason the rest of this chapter is devoted to linear transformations.

By far the most useful linear transformation for the purposes of this book is one whose domain and codomain are finite dimensional vector spaces. Every linear transformation of this type can be represented as a matrix, once suitable bases are selected.

Consider the linear transformation $\mathcal{A}: \mathcal{X}^n \longrightarrow \mathcal{X}^m$ such that for $\mathbf{x} \in \mathcal{X}^n$ and $\mathbf{y} \in \mathcal{X}^m$, $\mathbf{y} = \mathcal{A}(\mathbf{x})$. Let $\{\mathbf{v}_i, i = 1, \ldots, n\}$ and $\{\mathbf{u}_i, i = 1, \ldots, m\}$ be basis sets for $\mathcal{X}^n$ and $\mathcal{X}^m$, respectively. Then $\mathbf{x} = \sum_{i=1}^{n} \alpha_i \mathbf{v}_i$ and the linearity properties of $\mathcal{A}$ give

$$\mathbf{y} = \sum_{i=1}^{n} \alpha_i \mathcal{A}(\mathbf{v}_i) = [\mathcal{A}(\mathbf{v}_1) \mid \mathcal{A}(\mathbf{v}_2) \mid \cdots \mid \mathcal{A}\mathbf{v}_n)] \begin{bmatrix} \alpha_1 \\ \alpha_2 \\ \cdot \\ \cdot \\ \cdot \\ \alpha_n \end{bmatrix} \tag{6.1}$$

The vectors $\mathcal{A}(\mathbf{v}_i)$ are images of the basis vectors $\mathbf{v}_i$ under the transformation $\mathcal{A}$. Since $\mathbf{y}$ and each $\mathcal{A}(\mathbf{v}_i)$ belong to $\mathcal{X}^m$, they have unique expansions with respect to the basis set $\{\mathbf{u}_i\}$,

$$\mathbf{y} = \sum_{j=1}^{m} \beta_j \mathbf{u}_j \quad \text{and} \quad \mathcal{A}(\mathbf{v}_i) = \sum_{j=1}^{m} a_{ji} \mathbf{u}_j \tag{6.2}$$

Combining equations (*6.1*) and (*6.2*) gives

$$\mathbf{y} = \sum_{j=1}^{m} \beta_j \mathbf{u}_j = \sum_{i=1}^{n} \alpha_i \left[\sum_{j=1}^{m} a_{ji} \mathbf{u}_j \right]$$

Interchanging the order of summation and using the fact that the expansion coefficients β_j are unique lead to

$$\beta_j = \sum_{i=1}^{n} a_{ji}\alpha_i, \qquad j = 1, 2, \dots, m \tag{6.3}$$

Let $[\mathbf{x}]_{\mathrm{v}} \triangleq [\alpha_1 \quad \alpha_2 \quad \cdots \quad \alpha_n]^T$ and $[\mathbf{y}]_{\mathrm{u}} \triangleq [\beta_1 \quad \beta_2 \quad \cdots \quad \beta_m]^T$ be the coordinate representations of the vectors $\mathbf{x}$ and $\mathbf{y}$ with respect to the basis sets $\{\mathbf{v}_i\}$ and $\{\mathbf{u}_i\}$, respectively. (When the natural cartesian basis of Example 4.1 is used, this cumbersome notation is not necessary since then $\mathbf{x} = [x_1 \quad x_2 \quad \cdots \quad x_n]^T$ and $\mathbf{y} = [y_1 \quad y_2 \quad \cdots \quad y_m]^T$.) Regardless of which basis sets are selected, the transformation $\mathbf{y} = \mathcal{A}(\mathbf{x})$, or equivalently the set of equations (*6.3*), can be represented by the matrix equation $[\mathbf{y}]_{\mathrm{u}} = \mathbf{A}[\mathbf{x}]_{\mathrm{v}}$. The matrix $\mathbf{A}$ is $m \times n$, and a typical element a_{ji} is seen to be the jth component (with respect to the basis $\{\mathbf{u}_i\}$) of the image of $\mathbf{v}_i$. The particular matrix representation $\mathbf{A}$ for $\mathcal{A}$ obviously depends on the choice of basis in both $\mathcal{X}^n$ and $\mathcal{X}^m$. Changing either basis set changes the resultant representation $\mathbf{A}$. However, many properties of $\mathcal{A}$ are independent of the particular representation $\mathbf{A}$. For example, the rank of $\mathcal{A}$ equals the rank of $\mathbf{A}$ regardless of which representation $\mathbf{A}$ is used. This is also the dimension of the range space of $\mathcal{A}$:

$$\operatorname{rank}(\mathcal{A}) = r_A = \dim(\mathfrak{R}(\mathcal{A}))$$

The range of $\mathcal{A}$ is the same space referred to as the column space of $\mathbf{A}$ in Chapter 5.

6.4 CHANGE OF BASIS

Consider the n-dimensional linear vector space $\mathcal{X}^n$. Let $\{\mathbf{v}_i, i = 1, \dots, n\}$ and $\{\mathbf{v}'_i, i = 1, \dots, n\}$ be two basis sets. Each vector $\mathbf{x} \in \mathcal{X}^n$ can be expressed with respect to either basis; for example,

$$\mathbf{x} = \sum_{j=1}^{n} x_j \mathbf{v}_j = \sum_{i=1}^{n} x'_i \mathbf{v}'_i$$

where x_i and x'_i are scalar components. Since the basis vectors themselves belong to $\mathcal{X}^n$, one set can be expressed in terms of the other. For example,

$$\mathbf{v}_j = \sum_{i=1}^{n} b_{ij}\mathbf{v}'_i$$

Using this result to eliminate $\mathbf{v}_j$ in the expression for $\mathbf{x}$ gives

$$\sum_{j=1}^{n} x_j \sum_{i=1}^{n} b_{ij}\mathbf{v}'_i = \sum_{i=1}^{n} x'_i\mathbf{v}'_i \qquad \text{or} \qquad \sum_{i=1}^{n}\left(\sum_{j=1}^{n} b_{ij}x_j - x'_i\right)\mathbf{v}'_i = \mathbf{0}$$

Linear independence of the set $\{\mathbf{v}'_i\}$ requires that

$$\sum_{j=1}^{n} b_{ij}x_j = x'_i$$

The component vectors $[\mathbf{x}]_{\mathrm{v}}$ and $[\mathbf{x}]_{\mathrm{v}'}$ are thus related by a matrix multiplication:

$$[\mathbf{x}]_{\mathrm{v}'} = [\mathbf{B}]\,[\mathbf{x}]_{\mathrm{v}}$$

A change of basis is seen to be equivalent to a matrix multiplication. The effect of a change of basis on the representation of a linear transformation is now considered. Let $\mathcal{A}$ map vectors in $\mathfrak{X}^n$ into other vectors also in $\mathfrak{X}^n$: $\mathcal{A}: \mathfrak{X}^n \longrightarrow \mathfrak{X}^n$, where $\mathcal{A}(\mathbf{x}) = \mathbf{y}$. Let $\mathbf{A}$ be the representation of $\mathcal{A}$ when the basis $\{\mathbf{v}_i\}$ is used, and let $\mathbf{A}'$ be the representation when $\{\mathbf{v}_i'\}$ is used. The relation between $\mathbf{A}$ and $\mathbf{A}'$ is to be found. When using the unprimed basis set, the transformation is represented as

$$\mathbf{A}[\mathbf{x}]_{\mathrm{v}} = [\mathbf{y}]_{\mathrm{v}}$$

When using the primed basis set,

$$\mathbf{A}'[\mathbf{x}]_{\mathrm{v}'} = [\mathbf{y}]_{\mathrm{v}'}$$

But it was shown earlier that coordinate representations of any vector with respect to two sets of basis vectors are related by

$$[\mathbf{x}]_{\mathrm{v}'} = [\mathbf{B}]\,[\mathbf{x}]_{\mathrm{v}} \qquad [\mathbf{y}]_{\mathrm{v}'} = [\mathbf{B}]\,[\mathbf{y}]_{\mathrm{v}}$$

Thus

$$\mathbf{A}'[\mathbf{B}]\,[\mathbf{x}]_{\mathrm{v}} = [\mathbf{B}]\,[\mathbf{y}]_{\mathrm{v}}$$

The matrix $\mathbf{B}$ which represents the change of basis always has an inverse (see Problem 6.4), so

$$[\mathbf{B}^{-1}]\mathbf{A}'[\mathbf{B}]\,[\mathbf{x}]_{\mathrm{v}} = [\mathbf{y}]_{\mathrm{v}}$$

This is the representation of the transformation in the unprimed system. The two representations for $\mathcal{A}$ are related by

$$\mathbf{B}^{-1}\mathbf{A}'\mathbf{B} = \mathbf{A}$$

This relationship between $\mathbf{A}'$ and $\mathbf{A}$ is called a *similarity transformation*. Any two matrices which are related by a similarity transformation are said to be *similar matrices*. In the present context similar matrices are representations of a linear transformation with respect to different basis vectors.

If both basis sets $\{\mathbf{v}_i\}$ and $\{\mathbf{v}_i'\}$ are orthonormal, then it can be shown (see Problem 6.5) that the matrix $\mathbf{B}$ is an orthogonal matrix. That is,

$$\mathbf{B}^{-1} = \mathbf{B}^T$$

In this case the two representations of $\mathcal{A}$ are related by an *orthogonal transformation*,

$$\mathbf{B}^T\mathbf{A}'\mathbf{B} = \mathbf{A}$$

6.5 OPERATIONS WITH LINEAR TRANSFORMATIONS

Every linear transformation on finite dimensional spaces can be represented as a matrix. It is natural to expect that algebraic operations with linear transformations are governed by rules much like those of matrix algebra. Let $\mathcal{X}^n$, $\mathcal{X}^m$, and $\mathcal{X}^p$ be linear vector spaces defined over the same scalar number field $\mathcal{F}$. Then if

$$\mathcal{A}_1: \mathcal{X}^n \longrightarrow \mathcal{X}^m, \qquad \mathcal{A}_2: \mathcal{X}^n \longrightarrow \mathcal{X}^m \qquad \text{and} \qquad \mathcal{A}_1(\mathbf{x}) = \mathbf{y}_1, \qquad \mathcal{A}_2(\mathbf{x}) = \mathbf{y}_2$$

then

$$(\mathcal{A}_1 + \mathcal{A}_2)(\mathbf{x}) = \mathcal{A}_1(\mathbf{x}) + \mathcal{A}_2(\mathbf{x}) = \mathbf{y}_1 + \mathbf{y}_2$$

If $\mathcal{A}_1: \mathcal{X}^n \longrightarrow \mathcal{X}^m$ and $\mathcal{A}_2: \mathcal{X}^m \longrightarrow \mathcal{X}^p$, then $\mathcal{A}_2\mathcal{A}_1: \mathcal{X}^n \longrightarrow \mathcal{X}^p$ and, in general, $\mathcal{A}_1\mathcal{A}_2$ is not defined. Thus linear transformations are distributive but not commutative.

A norm can be defined for linear transformations, as follows. If

$$\mathcal{A}(\mathbf{x}) = \mathbf{y}$$

then $\|\mathbf{y}\| = \|\mathcal{A}(\mathbf{x})\|$. If there exists a finite number K such that $\|\mathcal{A}(\mathbf{x})\| \leq K\|\mathbf{x}\|$ for all $\mathbf{x}$, the linear transformation is said to be *bounded.* (Every linear transformation on finite dimensional spaces is bounded. See Problem 6.15.) Assuming that $\mathcal{A}$ is bounded, the norm of $\mathcal{A}$, written $\|\mathcal{A}\|$, is the smallest value of K which provides such a bound. Alternatively, $\|\mathcal{A}\|$ is the least upper bound (supremum or sup) of $\|\mathcal{A}(\mathbf{x})\|/\|\mathbf{x}\|$ for nonzero $\mathbf{x}$. Two equivalent formulas are

$$\|\mathcal{A}\| = \sup_{\mathbf{x}\neq \mathbf{0}} \frac{\|\mathcal{A}(\mathbf{x})\|}{\|\mathbf{x}\|} \qquad \text{or} \qquad \|\mathcal{A}\| = \sup_{\|\mathbf{x}\|=1} \|\mathcal{A}(\mathbf{x})\| \tag{6.4}$$

There are various possible choices for the norm of the vector $\mathcal{A}(\mathbf{x})$. Each particular choice induces a different form for $\|\mathcal{A}\|$. If $\mathbf{A}$ is the matrix representation for a finite dimensional transformation, $\mathcal{A}$, and if the quadratic vector norm is used, then

$$\|\mathcal{A}\|^2 = \max_{\|\mathbf{x}\|=1} \{\bar{\mathbf{x}}^T\bar{\mathbf{A}}^T\mathbf{A}\mathbf{x}\}$$

Some properties satisfied by the norm of a linear transformation are

$$\begin{aligned} \|\mathcal{A}(\mathbf{x})\| &\leq \|\mathcal{A}\| \cdot \|\mathbf{x}\| \qquad \text{for all } \mathbf{x} \\ \|\mathcal{A}_1 + \mathcal{A}_2\| &\leq \|\mathcal{A}_1\| + \|\mathcal{A}_2\| \\ \|\mathcal{A}_1\mathcal{A}_2\| &\leq \|\mathcal{A}_1\| \cdot \|\mathcal{A}_2\| \\ \|\alpha\mathcal{A}\| &= |\alpha| \cdot \|\mathcal{A}\| \end{aligned}$$

As defined above, the norm of every linear transformation is a nonnegative number, and is zero only for a null transformation, i.e., a transformation which maps every vector into the zero vector.

A particular class of linear transformations is that which maps vectors into the one-dimensional vector space formed by the scalar number field. These transformations are called *linear functionals*. The set of all linear functionals, whose domain is a particular vector space $\mathcal{X}$, can be used to define another vector space, called the *conjugate space*. This topic will not be pursued here [96, 112].

6.6 ADJOINT TRANSFORMATIONS

Let $\mathcal{A}: \mathcal{X}_1 \longrightarrow \mathcal{X}_2$ be a linear transformation, where $\mathcal{X}_1$ and $\mathcal{X}_2$ are inner product spaces, with inner products $\langle\ ,\ \rangle_1$ and $\langle\ ,\ \rangle_2$, respectively. For each $\mathbf{x} \in \mathcal{X}_1$, $\mathcal{A}(\mathbf{x}) = \mathbf{y} \in \mathcal{X}_2$. If $\mathbf{z}$ is an arbitrary vector in $\mathcal{X}_2$, then the inner product $\langle \mathbf{z}, \mathbf{y}\rangle_2 = \langle \mathbf{z}, \mathcal{A}(\mathbf{x})\rangle_2$ is well defined and can be used to define the *adjoint transformation* $\mathcal{A}^*: \mathcal{X}_2 \longrightarrow \mathcal{X}_1$, according to $\langle \mathbf{z}, \mathcal{A}(\mathbf{x})\rangle_2 = \langle \mathcal{A}^*(\mathbf{z}), \mathbf{x}\rangle_1$. It can be shown for finite dimensional spaces that $\mathcal{A}^*$ is also a linear transformation, i.e., if $\mathcal{A}^*(\mathbf{z}_1) = \mathbf{w}_1$ and $\mathcal{A}^*(\mathbf{z}_2) = \mathbf{w}_2$, then $\mathcal{A}^*(\alpha_1\mathbf{z}_1 + \alpha_2\mathbf{z}_2) = \alpha_1\mathbf{w}_1 + \alpha_2\mathbf{w}_2$ for arbitrary scalars α_1 and $\alpha_2 \in \mathcal{F}$. This follows from the linearity properties of the inner product.

Example 6.3

Let $\mathcal{A}$ be a transformation from an n-dimensional vector space to an m-dimensional space, with the usual definition of the complex inner products,

$$\langle \mathbf{x}_1, \mathbf{x}_2\rangle \triangleq \bar{\mathbf{x}}_1^T\mathbf{x}_2, \qquad \langle \mathbf{y}_1, \mathbf{y}_2\rangle \triangleq \bar{\mathbf{y}}_1^T\mathbf{y}_2$$

The operator $\mathcal{A}$ can be represented by an $m \times n$ matrix $\mathbf{A}$, so that

$$\langle \mathbf{z}, \mathcal{A}(\mathbf{x})\rangle = \bar{\mathbf{z}}^T(\mathbf{A}\mathbf{x}) = (\overline{\bar{\mathbf{A}}^T\mathbf{z}})^T\mathbf{x} = \langle \bar{\mathbf{A}}^T\mathbf{z}, \mathbf{x}\rangle$$

For this example $\mathcal{A}^*$ is represented by the matrix $\bar{\mathbf{A}}^T$. ■

The adjoint transformation defined by the inner product should not be confused with the adjoint matrix defined and used in Chapter 3. Adjoint transformations appear in several roles in modern control theory, and some of these will be developed later. Only a few properties of adjoint transformations are presented here.

If $\mathcal{A}: \mathcal{X}_1 \longrightarrow \mathcal{X}_2$, then $\mathcal{A}^*: \mathcal{X}_2 \longrightarrow \mathcal{X}_1$. Also, $\mathcal{A}^*\mathcal{A}: \mathcal{X}_1 \longrightarrow \mathcal{X}_1$ and $\mathcal{A}\mathcal{A}^*: \mathcal{X}_2 \longrightarrow \mathcal{X}_2$. If $\mathcal{X}_1 = \mathcal{X}_2$ and $\mathcal{A} = \mathcal{A}^*$, then $\mathcal{A}$ is said to be *self-adjoint*. In all cases, it can be shown that $\|\mathcal{A}\| = \|\mathcal{A}^*\|$, and that $(\mathcal{A}^*)^* = \mathcal{A}$. It is clear that $\mathcal{A}^*\mathcal{A}$ is generally not equal to $\mathcal{A}\mathcal{A}^*$. Those particular transformations for which $\mathcal{A}^*\mathcal{A} = \mathcal{A}\mathcal{A}^*$ are said to be *normal transformations*.

Let $\mathcal{A}: \mathcal{X}_1 \longrightarrow \mathcal{X}_2$ be an arbitrary linear transformation. Then the linear vector spaces $\mathcal{X}_1$ and $\mathcal{X}_2$ can be written as direct sums

$$\begin{aligned} \mathcal{X}_1 &= \mathfrak{N}(\mathcal{A}) \oplus \bar{\mathfrak{R}}(\mathcal{A}^*) \\ \mathcal{X}_2 &= \mathfrak{N}(\mathcal{A}^*) \oplus \bar{\mathfrak{R}}(\mathcal{A}) \end{aligned} \tag{6.5}$$

where $\mathfrak{N}(\cdot)$ and $\mathfrak{R}(\cdot)$ are the null space and range of the indicated transformations. $\bar{\mathfrak{R}}(\cdot)$ denotes the closure of the range $\mathfrak{R}(\cdot)$, that is, $\mathfrak{R}(\cdot)$ plus the limit of all convergent sequences of elements in $\mathfrak{R}(\cdot)$. In finite dimensional spaces every subspace is closed, so that $\bar{\mathfrak{R}}(\cdot) = \mathfrak{R}(\cdot)$. Equation (*6.5*) constitutes an *orthogonal* decomposition of $\mathfrak{X}_1$ into two linear subspaces. That is, for any vector $\mathbf{x} \in \mathfrak{N}(\mathcal{A})$ and any vector $\mathbf{y} \in \mathfrak{R}(\mathcal{A}^*)$, $\langle \mathbf{x}, \mathbf{y} \rangle = 0$. Equation (*6.5*) also provides an orthogonal decomposition for $\mathfrak{X}_2$.

6.7 SOME FINITE DIMENSIONAL TRANSFORMATIONS

Every linear transformation from one finite dimensional space to another finite dimensional space can be represented as a matrix. Within this general category, a few special transformations are now discussed.

Rotations

A particular transformation that frequently arises in control applications is a pure rotation. This can often be viewed in two ways. The result can be considered as a new vector obtained by rotating the original vector, or it can be considered as the same vector expressed in terms of a new coordinate system which is rotated with respect to the original coordinate system. The latter point of view is adopted for the time being, and the treatment is restricted to real, three-dimensional space, $\mathcal{R}^3$. Let $\{\mathbf{x}_1, \mathbf{x}_2, \mathbf{x}_3\}$ be an orthonormal basis set, and more specifically, let it define a right-handed cartesian coordinate system. Let $\{\mathbf{y}_1, \mathbf{y}_2, \mathbf{y}_3\}$ be another right-handed cartesian coordinate system.

The set $\{\mathbf{x}_i\}$ might represent an orthogonal triad fixed to an aerospace vehicle or a tracking antenna. The set $\{\mathbf{y}_i\}$ might represent an inertially fixed coordinate system. These two sets can be brought into coincidence by a sequence of rotations. The most familiar set of angles of rotation are the Euler angles [44], although the present discussion applies to any sequence of finite rotations such as those of Fig. 6.2. A rotation θ about the $\mathbf{x}_2$ axis rotates $\mathbf{x}_1$ and $\mathbf{x}_3$ into $\mathbf{x}'_1$ and $\mathbf{x}'_3$, and leaves $\mathbf{x}'_2 = \mathbf{x}_2$. A rotation ψ about $\mathbf{x}'_1$ gives $\mathbf{x}''_1 = \mathbf{x}'_1$ and $\mathbf{x}''_2$, $\mathbf{x}''_3$. The final rotation ϕ about $\mathbf{x}''_3$ gives $\mathbf{y}_1, \mathbf{y}_2$, and $\mathbf{y}_3 = \mathbf{x}''_3$.

Let $\mathbf{z}$ be an arbitrary vector and let $[\mathbf{z}]$, $[\mathbf{z}]'$, $[\mathbf{z}]''$, and $[\mathbf{z}]'''$ be its coordinate representations in the $\{\mathbf{x}_i\}$, $\{\mathbf{x}'_i\}$, $\{\mathbf{x}''_i\}$, and $\{\mathbf{y}_i\}$ coordinate systems, respectively. It is easily verified that

$$[\mathbf{z}]' = \begin{bmatrix} \cos\theta & 0 & -\sin\theta \\ 0 & 1 & 0 \\ \sin\theta & 0 & \cos\theta \end{bmatrix}[\mathbf{z}], \qquad [\mathbf{z}]'' = \begin{bmatrix} 1 & 0 & 0 \\ 0 & \cos\psi & \sin\psi \\ 0 & -\sin\psi & \cos\psi \end{bmatrix}[\mathbf{z}]'$$
$$[\mathbf{z}]''' = \begin{bmatrix} \cos\phi & \sin\phi & 0 \\ -\sin\phi & \cos\phi & 0 \\ 0 & 0 & 1 \end{bmatrix}[\mathbf{z}]'' \tag{6.6}$$

The overall transformation from $[\mathbf{z}]$ to $[\mathbf{z}]'''$ is given by the product of the three transformation matrices.

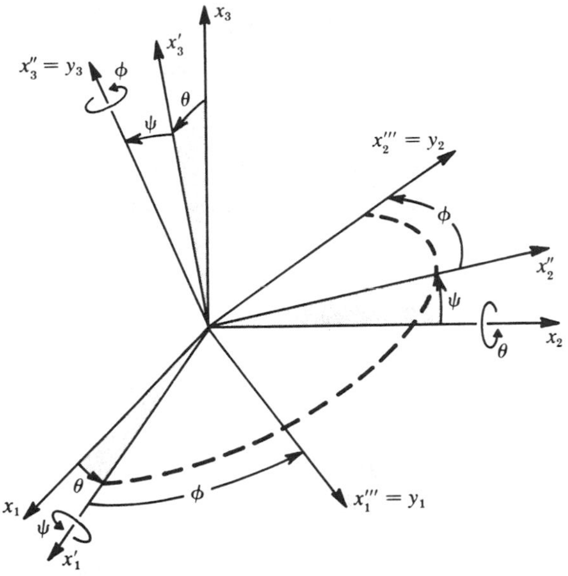

Figure 6.2

A symbolic method of representing coordinate rotations has been developed [92, 93]. These resolver-like diagrams, called Piograms, make it possible to write vector components in the new coordinate system without resorting to successive matrix multiplications (see Problem 6.23).

Reflections

The relation between a vector $\mathbf{x}$ and its reflection $\mathbf{x}_r$ at a plane surface defined by a unit normal $\mathbf{n}$ is $\mathbf{x}_r = \mathbf{x} - 2\langle\mathbf{n}, \mathbf{x}\rangle\mathbf{n}$. That is, $\mathbf{x}$ and $\mathbf{x}_r$ are equal except for a sign change in the component along $\mathbf{n}$. This can be rewritten as

$$\mathbf{x}_r = \mathbf{x} - 2\mathbf{n}\rangle\langle\mathbf{n}\mathbf{x} = [\mathbf{I} - 2\mathbf{n}\rangle\langle\mathbf{n}]\mathbf{x}$$

The matrix $\mathbf{A}_r = [\mathbf{I} - 2\mathbf{n}\rangle\langle\mathbf{n}]$ is the general representation of a reflection transformation. It is characterized by the fact that $|\mathbf{A}_r| = -1$, as verified by using the results of Problem 3.5.

Projections

A simple example of a transformation $\mathcal{A}(\mathbf{x})$ which maps $\mathbf{x}$ into its orthogonal projection on a hyperplane with a unit normal vector $\mathbf{n}$ is

$$\mathbf{A}_p = [\mathbf{I} - \mathbf{n}\rangle\langle\mathbf{n}]$$

This is fairly obvious since $\mathbf{A}_p\mathbf{x} = \mathbf{x} - \langle \mathbf{n}, \mathbf{x} \rangle \mathbf{n}$ has the effect of subtracting out the component of $\mathbf{x}$ along $\mathbf{n}$. It is easily verified that $\mathbf{A}_p\mathbf{A}_p = \mathbf{A}_p^2 = \mathbf{A}_p$. In general, any linear transformation which satisfies

$$\mathcal{A}^2 = \mathcal{A}$$

is a projection, although it need not be an orthogonal projection as in the above case. It is always possible to express a linear vector space as a direct sum $\mathfrak{X} = \mathfrak{U} \oplus \mathfrak{V}$, where $\mathfrak{U}$ and $\mathfrak{V}$ are nonvoid subspaces of $\mathfrak{X}$. This means that for each $\mathbf{x} \in \mathfrak{X}$ there is one and only one way of writing

$$\mathbf{x} = \mathbf{u} + \mathbf{v}, \qquad \text{where} \quad \mathbf{u} \in \mathfrak{U}, \qquad \mathbf{v} \in \mathfrak{V}$$

A transformation $\mathcal{P}$ satisfying $\mathcal{P}(\mathbf{x}) = \mathbf{u}$ is said to be the projection on $\mathfrak{U}$ along $\mathfrak{V}$.

6.8 SOME TRANSFORMATIONS ON INFINITE DIMENSIONAL SPACES

Most of the analysis of lumped-parameter systems in modern control theory can be considered in terms of a finite dimensional linear space, the state space. Consequently, the major emphasis is on finite dimensional transformations. However, transformations on infinite dimensional spaces do arise, and two of the more important ones are mentioned here.

It is recalled that the dimension of a space is equal to the number of elements in its basis set. The set of all periodic functions with period π is an example of an infinite dimensional space and its basis could be selected as the functions $\{\sin nt, n = 0, 1, \ldots\}$. The expansion with respect to this basis is the Fourier series. The set of all continuous functions, or of integrable functions, or of all square integrable functions are other examples of infinite dimensional spaces. A space is not necessarily infinite dimensional just because its elements are functions of time. For example, the space of all polynomials of degree 3 or less, i.e. $\{f(t) \mid f(t) = \alpha_0 + \alpha_1 t + \alpha_2 t^2 + \alpha_3 t^3, \alpha_i \in \mathfrak{F}\}$, is a four-dimensional space.

Integral Relations

The integral form of a system's input-output equation was mentioned in Sec. 1.5 and a particular example was used in Problem 4.25. For a linear system the input and output can be related by

$$\mathbf{y}(T) = \int_{-\infty}^{T} \mathbf{W}(T, \tau)\mathbf{u}(\tau)\, d\tau$$

where $\mathbf{y}(T)$ is the $m \times 1$ output vector at time T, $\mathbf{u}(t)$ is an $r \times 1$ input vector for each value of t, and $\mathbf{W}(t, \tau)$ is the $m \times r$ weighting matrix. At each time T, $\mathbf{y}(T)$ is a vector in an m-dimensional space. A particular input function, $u(t)$, $t \in (-\infty, T]$ can be considered as an element of the (infinite dimensional) input function space $\mathcal{U}$. The input-output integral represents a transformation $\mathcal{A}: \mathcal{U} \longrightarrow \mathcal{Y}^m$, where $\mathcal{A}(u) = \mathbf{y}(T)$.

The equation for the Laplace transform

$$\mathbf{y}(s) = \int_0^\infty e^{-st} y(t)\, dt$$

provides another example of a linear transformation on infinite dimensional spaces. The domain of these transformations must be suitably restricted so that the indicated operations "make sense." In other words, nonintegrable functions cannot be integrated, and functions which cannot be bounded by some exponential function do not have Laplace transforms.

Differential Relations

A linear differential equation can be considered to be a transformation, but again infinite dimensional spaces (function spaces) are involved. The simplest case

$$\frac{dx}{dt} = u$$

maps a function $x(t)$ into another function $u(t)$. Of course, the domain of this transformation must be restricted to the class of functions which are differentiable. Other restrictions may be necessary as well. Perhaps only those functions for which $x(0) = 0$ are considered. This constitutes an initial condition. The relation

$$\left[\mathbf{I}\frac{d}{dt} - \mathbf{A}\right]\mathbf{x}(t) = \mathbf{Bu}(t)$$

is another example of a differential transformation which maps $\mathbf{x}(t)$ into $\mathbf{Bu}(t)$. Although this equation appears repeatedly in the modern formulation of control problems, it will not be necessary to consider it as an abstract mapping on function spaces. Rather, this brief section dealing with transformations on function spaces is intended only to hint at a direction for an abstract treatment of all linear transformations. If the function spaces are Hilbert spaces (i.e., complete inner product spaces), then the results parallel the finite dimensional results to a large degree [47]. In general, however, there will be some major differences. Every finite dimensional linear vector space is complete, and every transformation on finite dimensional spaces is bounded. These are not generally true in the infinite dimensional case. For example, the space of square integrable functions $\mathcal{L}_2[a, b]$ of Problem 4.20 is complete. But the space of all continuous functions $C[a, b]$ is not complete because a sequence of continuous functions may converge to a discontinuous function [40]. Differential operators are examples of *unbounded* linear transformations. Some of the other differences arise because of the greater variety of definitions that can be given for norms and distance measures in function spaces. A complete treatment of these topics can be found in texts on functional analysis [40, 66, 96, 112].

ILLUSTRATIVE PROBLEMS

Introductory Considerations

6.1 Let $y = \sin(x)$ be a mapping from the real line to itself, $f: \mathcal{R}^1 \longrightarrow \mathcal{R}^1$. Is this a valid transformation? Is it one-to-one? Is it onto? What is the null space?

Since for each $x \in \mathcal{R}^1$ there is associated a single value $y = \sin(x)$, this relation defines a valid function or transformation. It is not one-to-one since $x_1 = \pi/2$ and $x_2 = 5\pi/2$ satisfy $x_1 \neq x_2$, but still $\sin(x_1) = \sin(x_2) = 1$. This function is not onto since $y = 2$ (or any real number with absolute value greater than 1) is not the image of any x. The null space is the set of values $\{x_n = \pm n\pi, n = 0, 1, 2, \dots\}$.

6.2 Does the function $f: \mathcal{R}^1 \longrightarrow \mathcal{R}^1$ with $y = f(t)$ defined by $y = e^{at}$ have an inverse?

A graph of this function for any real, nonzero value of a shows that it is one-to-one. That is, on a t-y plot each value of t intersects the graph exactly once. (This is necessary if it is a function.) Each positive value of y intersects the graph exactly once and thus defines a unique t. However, the function does not map $\mathcal{R}^1$ onto $\mathcal{R}^1$ since negative values of y are not the image of any t. In order to have an inverse, a function must be both one-to-one and onto, so no inverse exists. However, if we use the same rule to define a function $f: \mathcal{R}^1 \longrightarrow \mathcal{R}^+$, where $\mathcal{R}^+$ is the set of positive real numbers, then an inverse exists for $a \neq 0$, and is given by $f^{-1}(y) = t = (1/a)\ln y$.

6.3 Let $\mathcal{X}$ be an n-dimensional linear vector space with $\mathbf{x} \in \mathcal{X}$. Discuss the transformation $\mathcal{A}(\mathbf{x}) = \|\mathbf{x}\|$.

This function maps $\mathcal{X}$ into the real line, $\mathcal{A}: \mathcal{X} \longrightarrow \mathcal{R}^1$. In this case the range of $\mathcal{A}$ is the nonnegative half line, so $\mathfrak{R}(\mathcal{A}) \neq \mathcal{R}^1$ and the mapping is not onto. It is not one-to-one either, because many different vectors can have the same norm. Thus, given a value for $\|\mathbf{x}\|$, it is not possible to determine which vector $\mathbf{x}$ is the pre-image, i.e., $\mathcal{A}^{-1}$ does not exist. This transformation is not a linear transformation since, in general,

$$\mathcal{A}(\mathbf{x}_1 + \mathbf{x}_2) = \|\mathbf{x}_1 + \mathbf{x}_2\| \neq \|\mathbf{x}_1\| + \|\mathbf{x}_2\| = \mathcal{A}(\mathbf{x}_1) + \mathcal{A}(\mathbf{x}_2)$$

The null space of this transformation consists of the single vector $\mathbf{x} = \mathbf{0}$, by the properties of the norm.

Change of Basis

6.4 A vector $\mathbf{x}$ is represented by the $n \times 1$ column matrix $[\mathbf{x}]_{\mathbf{v}}$ with respect to the basis $\{\mathbf{v}_i\}$ and by the $n \times 1$ column matrix $[\mathbf{x}]_{\mathbf{v}'}$ with respect to another basis $\{\mathbf{v}'_i\}$. Show that the matrix $\mathbf{B}$ satisfying $[\mathbf{x}]_{\mathbf{v}'} = \mathbf{B}[\mathbf{x}]_{\mathbf{v}}$ always has an inverse.

Expand a typical member of the first basis in terms of the second basis, $\mathbf{v}_j = \sum_{i=1}^n b_{ij}\mathbf{v}'_j$. Likewise, a typical $\mathbf{v}'_j$ can be expanded as $\mathbf{v}'_j = \sum_{i=1}^n c_{ij}\mathbf{v}_i$. Thus $\mathbf{v}'_i = \sum_{k=1}^n c_{ki}\mathbf{v}_k$. Eliminating $\mathbf{v}'_i$ from the first equation gives

$$\mathbf{v}_j = \sum_{i=1}^n b_{ij} \sum_{k=1}^n c_{ki}\mathbf{v}_k = \sum_{i=1}^n \sum_{k=1}^n b_{ij}c_{ki}\mathbf{v}_k.$$

Because the vectors $\mathbf{v}_j$ are linearly independent, this requires

$$\sum_{i=1}^n b_{ij}c_{ki} = \begin{cases} 1 & \text{if } k = j \\ 0 & \text{if } k \neq j \end{cases}$$

Letting $\mathbf{B} = [b_{ij}]$ and $\mathbf{C} = [c_{ij}]$, this can be written as $\mathbf{CB} = \mathbf{I}$. If the above procedure is modified slightly, by eliminating $\mathbf{v}_i$ from the second equation, we obtain $\mathbf{BC} = \mathbf{I}$. Taken together, the last two results imply that $\mathbf{C} = \mathbf{B}^{-1}$.

6.5 Show that if $\{\mathbf{v}_i\}$ and $\{\mathbf{v}'_i\}$ are both real, orthonormal basis sets, then $\mathbf{B}^{-1} = \mathbf{B}^T$.

Results of the previous problem give $\mathbf{B}^{-1} = \mathbf{C}$. But $c_{ij} = \langle \mathbf{r}_i, \mathbf{v}'_j \rangle$, where $\{\mathbf{r}_i\}$ are the reciprocal bases associated with $\{\mathbf{v}_i\}$. Since $\{\mathbf{v}_i\}$ is an orthonormal set, $\mathbf{r}_j = \mathbf{v}_i$ and $c_{ij} = \langle \mathbf{v}_i, \mathbf{v}'_j \rangle$. Similarly, $b_{ij} = \langle \mathbf{r}'_i, \mathbf{v}_j \rangle$ in the general case, where $\{\mathbf{r}'_i\}$ is the set of reciprocal bases for $\{\mathbf{v}'_i\}$. For the orthonormal case, $b_{ij} = \langle \mathbf{v}'_i, \mathbf{v}_j \rangle$. Interchanging subscripts gives $b_{ji} = \langle \mathbf{v}'_j, \mathbf{v}_i \rangle$. For the case of real vectors this shows that $b_{ji} = c_{ij}$, or $\mathbf{B}^T = \mathbf{C} = \mathbf{B}^{-1}$.

Properties of Linear Transformations

6.6 Prove that if $\mathcal{A}$ is any linear transformation there are only two possibilities regarding the null space: either *(1)* $\mathfrak{N}(\mathcal{A})$ consists of only the zero vector or *(2)* $\mathfrak{N}(\mathcal{A})$ contains an infinite number of vectors.

Consider $\mathcal{A}(\mathbf{x}) = \mathbf{y}$. Since $\mathcal{A}$ is linear, $\mathcal{A}(\mathbf{x} - \mathbf{x}) = \mathcal{A}(\mathbf{x}) - \mathcal{A}(\mathbf{x}) = \mathbf{y} - \mathbf{y}$ so $\mathbf{x} = \mathbf{0}$ is *always* a solution to $\mathcal{A}(\mathbf{x}) = \mathbf{0}$. Thus the null space always contains the zero vector. If $\mathbf{x}_1$ also belongs to the null space, $\mathcal{A}(\mathbf{x}_1) = \mathbf{0}$. Assume $\mathbf{x}_1 \neq \mathbf{0}$. Then the infinite set of vectors defined by $\mathbf{x} = \alpha\mathbf{x}_1$, with α any scalar, satisfies $\mathcal{A}(\mathbf{x}) = \mathcal{A}(\alpha\mathbf{x}_1) = \alpha\mathcal{A}(\mathbf{x}_1) = \mathbf{0}$. Thus if one nonzero vector belongs to the null space, then so does the infinite set of scalar multiples of it. These need not be the only vectors in $\mathfrak{N}(\mathcal{A})$.

6.7 Consider the linear transformation $\mathcal{A}: \mathfrak{X} \longrightarrow \mathcal{Y}$. Prove that the null space of $\mathcal{A}$ is a subspace of $\mathfrak{X}$.

The solution to this problem consists of combining the results of the previous problem with the definition of a subspace. A set of vectors is a subspace if for each $\mathbf{x}_1, \mathbf{x}_2$ in the set, $\alpha\mathbf{x}_1 + \beta\mathbf{x}_2$ is also in the set, for arbitrary scalars $\alpha, \beta \in \mathfrak{F}$. Let $\mathbf{x}_1$ and $\mathbf{x}_2 \in \mathfrak{N}(\mathcal{A})$. That is, $\mathcal{A}(\mathbf{x}_1) = \mathbf{0}$ and $\mathcal{A}(\mathbf{x}_2) = \mathbf{0}$. Then $\mathcal{A}(\alpha\mathbf{x}_1 + \beta\mathbf{x}_2) = \alpha\mathcal{A}(\mathbf{x}_1) + \beta\mathcal{A}(\mathbf{x}_2) = \mathbf{0}$. Thus $\alpha\mathbf{x}_1 + \beta\mathbf{x}_2 \in \mathfrak{N}(\mathcal{A})$ and so the null space is a subspace. It is a zero dimensional subspace if its only element is the zero vector.

6.8 Let $\mathcal{A}: \mathfrak{X} \longrightarrow \mathcal{Y}$ be a linear transformation, with $\mathfrak{X}$ and $\mathcal{Y}$ finite dimensional. Prove that the space $\mathfrak{X}$ can be written as a direct sum $\mathfrak{X} = \mathfrak{N}(\mathcal{A}) \oplus \mathfrak{R}(\mathcal{A}^*)$.

It was shown in Problem 6.7 that $\mathfrak{N}(\mathcal{A})$ is a subspace of $\mathfrak{X}$. Let $\mathfrak{N}(\mathcal{A})^\perp$ be the orthogonal complement of $\mathfrak{N}(\mathcal{A})$. Then from the results of Sec. 4.9, $\mathfrak{X} = \mathfrak{N}(\mathcal{A}) \oplus \mathfrak{N}(\mathcal{A})^\perp$.

It remains to be shown that $\mathfrak{R}(\mathcal{A}^*) = \mathfrak{N}(\mathcal{A})^\perp$. Let $\mathbf{x}$ be an arbitrary vector in $\mathfrak{N}(\mathcal{A})$ and let $\mathbf{z}$ be an arbitrary vector in $\mathcal{Y}$. Then $\mathcal{A}(\mathbf{x}) = \mathbf{0}$ so that $\langle \mathbf{z}, \mathcal{A}(\mathbf{x}) \rangle = \langle \mathcal{A}^*(\mathbf{z}), \mathbf{x} \rangle = \mathbf{0}$. From this we see that $\mathbf{x}$ is orthogonal to $\mathcal{A}^*(\mathbf{z})$, which shows that $\mathfrak{N}(\mathcal{A}) = \mathfrak{R}(\mathcal{A}^*)^\perp$. The orthogonal complements are also equal, $\mathfrak{N}(\mathcal{A})^\perp = (\mathfrak{R}(\mathcal{A}^*)^\perp)^\perp$, but $(\mathfrak{R}(\mathcal{A}^*)^\perp)^\perp = \mathfrak{R}(\mathcal{A}^*)$. This completes the proof for $\mathfrak{X}$ finite dimensional.

For the infinite dimensional case, the decomposition is valid if the closure of $\mathfrak{R}(\mathcal{A}^*)$ is used in place of $\mathfrak{R}(\mathcal{A}^*)$. Every finite dimensional space is closed.

6.9 Consider the linear transformation $\mathcal{A}(\mathbf{x}) = \mathbf{y}$, where $\mathcal{A}: \mathfrak{X} \longrightarrow \mathcal{Y}$. Show that there is no unique solution for $\mathbf{x}$ if $\mathfrak{N}(\mathcal{A})$ contains vectors other than the zero vector.

Suppose $\mathbf{x}_1 \in \mathfrak{N}(\mathcal{A})$ and $\mathbf{x}_1 \neq \mathbf{0}$. If $\mathbf{x}$ satisfies $\mathcal{A}(\mathbf{x}) = \mathbf{y}$, then $\mathbf{x} + \mathbf{x}_1$ is also a solution since $\mathcal{A}(\mathbf{x} + \mathbf{x}_1) = \mathcal{A}(\mathbf{x}) + \mathcal{A}(\mathbf{x}_1) = \mathbf{y} + \mathbf{0} = \mathbf{y}$.

6.10 Prove that the linear transformation $\mathcal{A}: \mathcal{X} \longrightarrow \mathcal{Y}$, with $\mathcal{A}(\mathbf{x}) = \mathbf{y}$, has a solution for every $\mathbf{y} \in \mathcal{Y}$ if and only if $\mathfrak{N}(\mathcal{A}^*) = \{\mathbf{0}\}$. Assume that $\mathcal{Y}$ is finite dimensional.

The linear transformation $\mathcal{A}^*$ and its finite dimensional domain $\mathcal{Y}$ can be used in place of $\mathcal{A}$ and $\mathcal{X}$ in the result of Problem 6.8. That is, $\mathcal{Y} = \mathfrak{N}(\mathcal{A}^*) \oplus \mathfrak{R}((\mathcal{A}^*)^*) = \mathfrak{N}(\mathcal{A}^*) \oplus \mathfrak{R}(\mathcal{A})$. If $\mathfrak{N}(\mathcal{A}^*) = \{\mathbf{0}\}$, then $\mathcal{Y} = \mathfrak{R}(\mathcal{A})$, so that every $\mathbf{y} \in \mathcal{Y}$ is the image of at least one $\mathbf{x} \in \mathcal{X}$. Note that $\mathbf{y} \in \mathfrak{R}(\mathcal{A})$ is equivalent to the requirement for existence of solutions given in Chapter 5: rank $[\mathbf{A}]$ = rank $[\mathbf{A} \mid \mathbf{y}]$. If $\mathfrak{N}(\mathcal{A}^*) \neq \{\mathbf{0}\}$, then $\mathcal{Y} \neq \mathfrak{R}(\mathcal{A})$; so there exists some $\mathbf{y} \in \mathcal{Y}$ but $\mathbf{y} \notin \mathfrak{R}(\mathcal{A})$. Such a $\mathbf{y}$ is not the image of any $\mathbf{x} \in \mathcal{X}$.

6.11 Prove: $\mathcal{A}(\mathbf{x}) = \mathbf{y}$ has a *unique* solution for every $\mathbf{y} \in \mathcal{Y}$ if $\mathfrak{N}(\mathcal{A}^*) = \{\mathbf{0}\}$ and $\mathfrak{N}(\mathcal{A}) = \{\mathbf{0}\}$.

The results of Problem 6.10 guarantee that at least one solution exists. Assume $\mathbf{x}_1$ and $\mathbf{x}_2$ are two solutions, that is, $\mathcal{A}(\mathbf{x}_1) = \mathbf{y}$ and $\mathcal{A}(\mathbf{x}_2) = \mathbf{y}$. Then $\mathcal{A}(\mathbf{x}_1) - \mathcal{A}(\mathbf{x}_2) = \mathbf{0}$ or $\mathcal{A}(\mathbf{x}_1 - \mathbf{x}_2) = \mathbf{0}$. But since $\mathfrak{N}(\mathcal{A}) = \{\mathbf{0}\}$, this requires that $\mathbf{x}_1 - \mathbf{x}_2 = \mathbf{0}$ or $\mathbf{x}_1 = \mathbf{x}_2$ is the unique solution. Note that if the domain $\mathcal{X}$ is n-dimensional, $\mathfrak{N}(\mathcal{A}) = \{\mathbf{0}\}$ implies that rank $(\mathcal{A}^*) = \dim \mathfrak{R}(\mathcal{A}^*) = \dim \mathcal{X} = n$. Also $\mathfrak{N}(\mathcal{A}^*) = \{\mathbf{0}\}$ implies that rank $(\mathcal{A}) = \dim \mathfrak{R}(\mathcal{A}) = \dim \mathcal{Y}$. But rank $(\mathcal{A})$ = rank $(\mathcal{A}^*)$, so a unique solution requires rank $(\mathcal{A}) = n$ as shown in Chapter 5, using a matrix representation $\mathbf{A}$ for $\mathcal{A}$.

6.12 If the domain and codomain for $\mathcal{A}$ are restricted so that $\mathcal{A}: \mathfrak{R}(\mathcal{A}^*) \longrightarrow \mathfrak{R}(\mathcal{A})$, show that $\mathcal{A}$ is one-to-one and onto, and thus possesses an inverse.

Since, in general, $\mathcal{X} = \mathfrak{N}(\mathcal{A}) \oplus \mathfrak{R}(\mathcal{A}^*)$ and $\mathcal{Y} = \mathfrak{N}(\mathcal{A}^*) \oplus \mathfrak{R}(\mathcal{A})$, restricting $\mathcal{A}$ as stated guarantees that the conditions for a unique solution, as stated in Problem 6.11, are satisfied, so the restriction of $\mathcal{A}$ is one-to-one and onto.

6.13 Let $\mathcal{A}: \mathcal{X} \longrightarrow \mathcal{Y}$ be a linear transformation for which $\mathfrak{N}(\mathcal{A}) \neq \{\mathbf{0}\}$, but $\mathfrak{N}(\mathcal{A}^*) = \{\mathbf{0}\}$.

a. Show that the most general solution to $\mathcal{A}(\mathbf{x}) = \mathbf{y}$ is given by $\mathbf{x} = \mathbf{x}_0 + \mathbf{x}_1$, where $\mathbf{x}_1 \in \mathfrak{R}(\mathcal{A}^*)$, $\mathbf{x}_0 \in \mathfrak{N}(\mathcal{A})$.

b. Show that $\mathbf{x}_1$ is the minimum norm solution given by $\mathbf{x}_1 = \mathcal{A}^*(\mathcal{A}\mathcal{A}^*)^{-1}\mathbf{y}$.

(1) For every $\mathbf{x} \in \mathcal{X}$, a unique decomposition is possible, $\mathbf{x} = \mathbf{x}_0 + \mathbf{x}_1$ with $\mathbf{x}_1 \in \mathfrak{R}(\mathcal{A}^*)$, $\mathbf{x}_0 \in \mathfrak{N}(\mathcal{A})$.

(2) Using this decomposition gives $\mathcal{A}(\mathbf{x}) = \mathcal{A}(\mathbf{x}_0 + \mathbf{x}_1) = \mathcal{A}(\mathbf{x}_1) = \mathbf{y}$. Since $\mathbf{x}_1 \in \mathfrak{R}(\mathcal{A}^*)$, there exists a $\mathbf{y}_1 \in \mathcal{Y}$ such that $\mathcal{A}^*(\mathbf{y}_1) = \mathbf{x}_1$, from which $\mathcal{A}(\mathbf{x}_1) = \mathbf{y} = \mathcal{A}\mathcal{A}^*(\mathbf{y}_1)$. Since $\mathcal{A}\mathcal{A}^*$ is a one-to-one linear transformation from $\mathfrak{R}(\mathcal{A}) = \mathcal{Y}$ onto itself, and thus possesses an inverse, $\mathbf{y}_1 = (\mathcal{A}\mathcal{A}^*)^{-1}y$. Using $\mathbf{x}_1 = \mathcal{A}^*(\mathbf{y}_1)$ gives the minimum norm solution $\mathbf{x}_1 = \mathcal{A}^*(\mathcal{A}\mathcal{A}^*)^{-1}\mathbf{y}$. Any vector $\mathbf{x}_0 \in \mathfrak{N}(\mathcal{A})$ can be added to $\mathbf{x}_1$ and the result is still a solution, but with a larger norm.

6.14 If $\mathfrak{N}(\mathcal{A}^*) \neq \{\mathbf{0}\}$ and $\mathbf{y} \notin \mathfrak{R}(\mathcal{A})$, no solution to $\mathcal{A}(\mathbf{x}) = \mathbf{y}$ exists. Give a geometrical interpretation of the least-squares approximate solution, $\mathbf{x} = (\mathcal{A}^*\mathcal{A})^{-1}\mathcal{A}^*\mathbf{y}$.

Figure 6.3 illustates the decomposition of $\mathcal{X}$ and $\mathcal{Y}$. For any $\mathbf{y} \in \mathcal{Y}$, $\mathbf{y} = \mathbf{z} + \mathbf{w}$ with $\mathbf{z} \in \mathfrak{R}(\mathcal{A})$ and $\mathbf{w} \in \mathfrak{N}(\mathcal{A}^*)$. Therefore

$$\mathcal{A}^*(\mathbf{y}) = \mathcal{A}^*(\mathbf{z} + \mathbf{w}) = \mathcal{A}^*(\mathbf{z})$$

Since $\mathbf{z} \in \mathfrak{R}(\mathcal{A})$, there exists some $\mathbf{x} \in \mathcal{X}$ such that $\mathcal{A}(\mathbf{x}) = \mathbf{z}$. Combining these results gives

$$\mathcal{A}^*(\mathbf{z}) = \mathcal{A}^*\mathcal{A}(\mathbf{x}) = \mathcal{A}^*(\mathbf{y})$$

If $\mathfrak{N}(\mathcal{A}) \neq \{\mathbf{0}\}$ there will be many vectors $\mathbf{x}$ satisfying $\mathcal{A}(\mathbf{x}) = \mathbf{z}$. The one such vector with minimum norm is selected, i.e. the one belonging to $\mathfrak{R}(\mathcal{A}^*)$. With this restriction, $\mathcal{A}^*\mathcal{A}$ is a one-to-one transformation from $\mathfrak{R}(\mathcal{A}^*)$ onto itself and thus possesses an inverse. The least-squares solution is $\mathbf{x} = (\mathcal{A}^*\mathcal{A})^{-1}\mathcal{A}^*(\mathbf{y})$. This solution is the pre-image of the projection of $\mathbf{y}$ on $\mathfrak{R}(\mathcal{A})$.

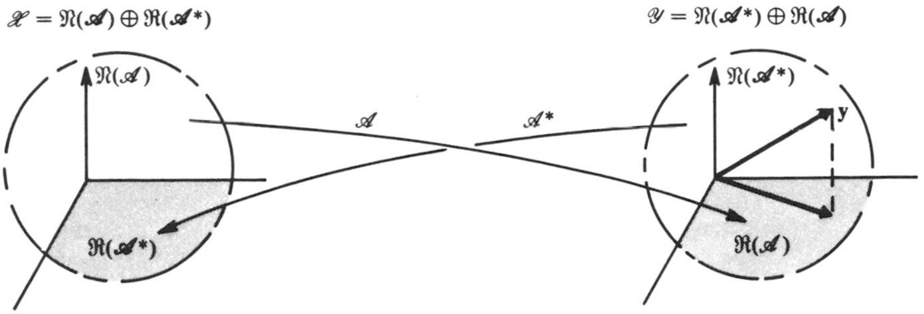

Figure 6.3

6.15 Prove that every finite dimensional linear transformation is bounded.

Let $\mathcal{A}: \mathfrak{X}^n \longrightarrow \mathfrak{X}^m$, with $\mathbf{x} \in \mathfrak{X}^n$. Let $\{\mathbf{v}_i, i = 1, 2, \ldots, n\}$ be an orthonormal basis for $\mathfrak{X}^n$. Then $\mathbf{x} = \sum_{i=1}^{n} \alpha_i \mathbf{v}_i$ and $\mathcal{A}(\mathbf{x}) = \sum_{i=1}^{n} \alpha_i \mathcal{A}(\mathbf{v}_i)$. Therefore

$$\|\mathcal{A}(\mathbf{x})\| = \left\| \sum_{i=1}^{n} \alpha_i \mathcal{A}(\mathbf{v}_i) \right\| \leq \sum_{i=1}^{n} |\alpha_i| \, \|\mathcal{A}(\mathbf{v}_i)\|$$

Since the vectors $\mathbf{y}_i \triangleq \mathcal{A}(\mathbf{v}_i)$ belong to $\mathfrak{X}^m$, they have a finite norm. Define $K = \max_{i=1,\ldots,n} \|\mathbf{y}_i\|$. Then

$$\|\mathcal{A}(\mathbf{x})\| \leq \sum_{i=1}^{n} |\alpha_i| \, \|\mathbf{y}_i\| \leq K \sum_{i=1}^{n} |\alpha_i|$$

But $\alpha_i = \langle \mathbf{v}_i, \mathbf{x} \rangle$ so that the Cauchy-Schwarz inequality gives $|\alpha_i| \leq \|\mathbf{v}_i\| \cdot \|\mathbf{x}\| = \|\mathbf{x}\|$ since $\|\mathbf{v}_i\| = 1$. Using this gives $\|\mathcal{A}(\mathbf{x})\| \leq Kn\|\mathbf{x}\|$, so that $\mathcal{A}$ is clearly bounded [equation (*6.4*)]. Notice the dependence on the finite dimension n.

6.16 Let $\mathcal{A}: \mathfrak{X} \longrightarrow \mathcal{Y}$ be a linear transformation such that $(\mathcal{A}^*\mathcal{A})^{-1}$ exists. Prove that $\mathcal{A}(\mathcal{A}^*\mathcal{A})^{-1}\mathcal{A}^*$ is the orthogonal projection of $\mathcal{Y}$ onto $\mathfrak{R}(\mathcal{A})$.

The indicated transformation is a projection if and only if

$$[\mathcal{A}(\mathcal{A}^*\mathcal{A})^{-1}\mathcal{A}^*][\mathcal{A}(\mathcal{A}^*\mathcal{A})^{-1}\mathcal{A}^*] = \mathcal{A}(\mathcal{A}^*\mathcal{A})^{-1}\mathcal{A}^*$$

This condition is obviously satisfied. Decompose $\mathcal{Y} = \mathfrak{N}(\mathcal{A}^*) \oplus \mathfrak{R}(\mathcal{A})$ and let $\mathbf{y} = \mathbf{y}_0 + \mathbf{y}_1$, where $\mathbf{y}_0 \in \mathfrak{N}(\mathcal{A}^*)$, $\mathbf{y}_1 \in \mathfrak{R}(\mathcal{A})$. The orthogonal projection of $\mathbf{y}$ onto $\mathfrak{R}(\mathcal{A})$ is $\mathbf{y}_1$ by definition.

We must show that $\mathcal{A}(\mathcal{A}^*\mathcal{A})^{-1}\mathcal{A}^*\mathbf{y} = \mathbf{y}_1$. First note that

$$\mathcal{A}^*(\mathbf{y}) = \overrightarrow{\mathcal{A}^*(\mathbf{y}_0)}^{\mathbf{0}} + \mathcal{A}^*(\mathbf{y}_1)$$

Both $\mathbf{y}_1$ and $\mathcal{A}(\mathcal{A}^*\mathcal{A})^{-1}\mathcal{A}^*\mathbf{y}_1 \in \mathfrak{R}(\mathcal{A}) = \mathfrak{N}(\mathcal{A}^*)^\perp$ so their difference, $\mathbf{e}$, does also. Applying the adjoint transformation $\mathcal{A}^*$ to their difference gives

$$\mathcal{A}^*[\mathcal{A}(\mathcal{A}^*\mathcal{A})^{-1}\mathcal{A}^*\mathbf{y}_1 - \mathbf{y}_1] = \mathbf{0}$$

Since $\mathbf{e} \in \mathfrak{N}(\mathcal{A}^*)^\perp$ and $\mathcal{A}^*(\mathbf{e}) = \mathbf{0}$, the conclusion is that $\mathbf{e} = \mathbf{0}$. This leads to $\mathbf{y}_1 = \mathcal{A}(\mathcal{A}^*\mathcal{A})\mathcal{A}^{-1}\mathcal{A}^*\mathbf{y}$.

Adjoint Transformations

6.17 Consider the linear transformation $\mathcal{A}: \mathfrak{X} \longrightarrow \mathfrak{Y}$, with the adjoint transformation $\mathcal{A}^*$. Prove that $\|\mathcal{A}\| = \|\mathcal{A}^*\|$.

The definition of the adjoint requires that $\langle \mathbf{y}, \mathcal{A}(\mathbf{x})\rangle = \langle \mathcal{A}^*(\mathbf{y}), \mathbf{x}\rangle$ for all $\mathbf{x} \in \mathfrak{X}$, $\mathbf{y} \in \mathfrak{Y}$. The Cauchy-Schwarz inequality and the definition of $\|\mathcal{A}^*\|$ gives

$$|\langle \mathbf{y}, \mathcal{A}(\mathbf{x})\rangle| = |\langle \mathcal{A}^*(\mathbf{y}), \mathbf{x}\rangle| \leq \|\mathcal{A}^*(\mathbf{y})\| \cdot \|\mathbf{x}\| \leq \|\mathcal{A}^*\| \cdot \|\mathbf{y}\| \cdot \|\mathbf{x}\|$$

This must be true for all $\mathbf{y}, \mathbf{x}$ including the particular pair related by $\mathcal{A}(\mathbf{x}) = \mathbf{y}$. Using this gives

$$\langle \mathcal{A}(x), \mathcal{A}(x)\rangle \leq \|\mathcal{A}^*\| \cdot \|\mathcal{A}(\mathbf{x})\| \cdot \|\mathbf{x}\| \quad \text{or} \quad \frac{\|\mathcal{A}(\mathbf{x})\|}{\|\mathbf{x}\|} \leq \|\mathcal{A}^*\|$$

for all $\mathbf{x} \neq \mathbf{0}$. Therefore

$$\|\mathcal{A}^*\| \geq \|\mathcal{A}\| \tag{1}$$

Similarly, if particular $\mathbf{x}, \mathbf{y}$ pairs are chosen such that $\mathbf{x} = \mathcal{A}^*(\mathbf{y})$, then

$$|\langle \mathcal{A}^*(\mathbf{y}), \mathcal{A}^*(\mathbf{y})\rangle| \leq \|\mathbf{y}\| \cdot \|\mathcal{A}(x)\| \leq \|\mathbf{y}\| \cdot \|\mathbf{x}\| \cdot \|\mathcal{A}\|$$

or

$$\|\mathcal{A}^*(y)\|^2 \leq \|\mathbf{y}\| \cdot \|\mathcal{A}^*(y)\| \cdot \|\mathcal{A}\| \quad \text{or} \quad \frac{\|\mathcal{A}^*(\mathbf{y})\|}{\|\mathbf{y}\|} \leq \|\mathcal{A}\| \quad \text{for all } \mathbf{y} \neq \mathbf{0}$$

Therefore $\|\mathcal{A}\| \geq \|\mathcal{A}^*\|$. This, together with equation (*1*), gives $\|\mathcal{A}\| = \|\mathcal{A}^*\|$.

6.18 Let $\mathfrak{X}$ be the set of all n-component real functions defined over $[t_0, t_f]$ which have continuous first derivatives and let $\mathfrak{Y}$ be the set of all continuous functions defined over the same interval. Find the adjoint of $\mathcal{A}: \mathfrak{X} \longrightarrow \mathfrak{Y}$ if $\mathcal{A}(\mathbf{x}) = (d/dt)\mathbf{x} - \mathbf{A}\mathbf{x}$, where $\mathbf{A}$ is an $n \times n$ matrix.

The inner product for $\mathfrak{X}$ and $\mathfrak{Y}$ takes the form of equation (*1*). The adjoint is defined by equation (*2*):

$$\langle \mathbf{x}_1, \mathbf{x}_2\rangle = \int_{t_0}^{t_f} \mathbf{x}_1^T(\tau)\mathbf{x}_2(\tau)\, d\tau \tag{1}$$

$$\langle \mathbf{y}, \mathcal{A}(\mathbf{x})\rangle = \langle \mathcal{A}^*(\mathbf{y}), \mathbf{x}\rangle \tag{2}$$

Integrating the left-hand side of equation (*2*) gives

$$\int_{t_0}^{t_f} \mathbf{y}^T(\tau)\left[\frac{d\mathbf{x}}{dt} - \mathbf{A}\mathbf{x}(\tau)\right] d\tau = \mathbf{y}^T(\tau)\,\mathbf{x}(\tau)\Big|_{t_0}^{t_f} - \int_{t_0}^{t_f} \left[\frac{d\mathbf{y}^T}{dt} + \mathbf{y}(\tau)^T\mathbf{A}\right] \mathbf{x}(\tau)\, d\tau$$

The term involving the limits of integration, $\mathbf{y}^T(t_f)\mathbf{x}(t_f) - \mathbf{y}^T(t_0)\mathbf{x}(t_0)$, could be made to vanish by specifying appropriate boundary conditions. Ignoring this term, the remaining term is in the form of the right-hand side of equation (*2*). Therefore $\mathcal{A}^*(\mathbf{y}) = -(d\mathbf{y}/dt) - \mathbf{A}^T\mathbf{y}(t)$. We conclude that the formal adjoint of $\dot{\mathbf{x}} = \mathbf{A}\mathbf{x}$ is $\dot{\mathbf{y}} = -\mathbf{A}^T\mathbf{y}$. This adjoint differential equation arises frequently in optimal control theory and in other applications.

6.19 Consider the equation $\dot{\mathbf{x}} = \mathbf{A}\mathbf{x}$ with $\mathbf{x}(t_0) = \mathbf{x}_0$. Show that $\mathbf{y}^T(t)\mathbf{x}(t)$ is a constant for all time, where $\mathbf{y}$ satisfies the adjoint equation $\dot{\mathbf{y}} = -\mathbf{A}^T\mathbf{y}$.

A necessary and sufficient condition that $\mathbf{y}^T\mathbf{x}$ be constant is that $(d/dt)(\mathbf{y}^T\mathbf{x}) = 0$. Using the chain rule for differentiating gives $(d/dt)(\mathbf{y}^T\mathbf{x}) = \dot{\mathbf{y}}^T\mathbf{x} + \mathbf{y}^T\dot{\mathbf{x}}$. Substituting in for $\dot{\mathbf{y}}$ and $\dot{\mathbf{x}}$ gives

$$(d/dt)(\mathbf{y}^T\mathbf{x}) = -\mathbf{y}^T\mathbf{A}\mathbf{x} + \mathbf{y}^T\mathbf{A}\mathbf{x} = 0$$

6.20 The input-output equation $\mathbf{y}(t) = \int_{t_0}^{\infty} \mathbf{W}(t, \tau)\mathbf{u}(\tau)\, d\tau$ defines a linear transformation from the space of r-component square integrable functions $\mathbf{u}(t) \in \mathcal{U}$ to the set of m-component continuous functions $\mathbf{y}(t) \in \mathcal{Y}$. The functions are defined over (t_0, ∞). Consider $\mathcal{A}: \mathcal{U} \longrightarrow \mathcal{Y}$ and find $\mathcal{A}^*$, if the inner product for each space has the form

$$\langle \mathbf{x}_1, \mathbf{x}_2 \rangle = \int_{t_0}^{\infty} \mathbf{x}_1^T(t)\mathbf{x}_2(t)\, dt$$

The transformation $\mathcal{A}$ is $\mathcal{A}(\mathbf{u}) = \int_{t_0}^{\infty} \mathbf{W}(t, \tau)\mathbf{u}(\tau)\, d\tau$ and $\langle \mathbf{z}, \mathcal{A}(\mathbf{u}) \rangle = \langle \mathcal{A}^*(\mathbf{z}), \mathbf{u} \rangle$. Using the inner product definition gives

$$\langle \mathbf{z}, \mathcal{A}(\mathbf{u}) \rangle = \int_{t_0}^{\infty} \mathbf{z}^T(t) \int_{t_0}^{\infty} \mathbf{W}(t, \tau)\mathbf{u}(\tau)\, d\tau\, dt = \int_{t_0}^{\infty} \int_{t_0}^{\infty} \mathbf{z}^T(t)\mathbf{W}(t, \tau)\, dt\, \mathbf{u}(\tau)\, d\tau$$

Therefore $\mathcal{A}^*(\mathbf{z}) = \int_{t_0}^{\infty} \mathbf{W}^T(t, \tau)\mathbf{z}(t)\, dt$. The weighting matrix for the adjoint equation is the transpose of $\mathbf{W}$ and the integration variable is t rather than τ as in the original transformation. This operator is self-adjoint if $\mathbf{W}(t, \tau) = \mathbf{W}^T(\tau, t)$.

Matrix Norms

6.21 Consider a linear transformation which maps vectors from one finite dimensional space to another. Let $\mathbf{A}$ be its matrix representation. If $\|\mathbf{x}\| = \max_i |x_i|$, find $\|\mathbf{A}\|$.

With this definition for the vector norm,

$$\|\mathbf{A}\mathbf{x}\| = \max_i \left| \sum_j a_{ij}x_j \right| \le \left\{ \max_i \left| \sum_j a_{ij} \right| \right\} \left\{ \max_j |x_j| \right\} \le \left\{ \max_i \sum_j |a_{ij}| \right\} \|\mathbf{x}\|$$

The norm of $\mathbf{A}$ must satisfy $\|\mathbf{A}\mathbf{x}\| \le \|\mathbf{A}\| \cdot \|\mathbf{x}\|$. We see that $\max_i \sum_j |a_{ij}|$ qualifies as a bound for $\|\mathbf{A}\mathbf{x}\|/\|\mathbf{x}\|$, with $\|\mathbf{x}\| \neq 0$. To show that it is the *least* upper bound, and hence is equal to the norm, we must show that no smaller bound is possible. This can be done by demonstrating that the bound is actually attained for some $\mathbf{x}$.

Let i^* be the row i which maximizes the above sum. Let

$$x_j = \begin{cases} \mathbf{0} & \text{if } a_{i^*j} = \mathbf{0} \\ \dfrac{\bar{a}_{i^*j}}{a_{i^*j}} & \text{otherwise} \end{cases}$$

For this choice, $\|\mathbf{x}\| = 1$ and

$$\max_i \left| \sum_j a_{ij}x_j \right| = \left| \sum_j \frac{a_{i^*j}\bar{a}_{i^*j}}{|a_{i^*j}|} \right| = \sum_j |a_{i^*j}|$$

Therefore $\|\mathbf{A}\| = \max_i \sum_j |a_{ij}|$.

6.22 Let the vector norm be defined by $\|\mathbf{x}\| = \sum_i |x_i|$ and find a bound for $\|\mathbf{A}\|$.

Beginning with the given definition for the norm, a series of manipulations gives

$$\|\mathbf{A}\mathbf{x}\| = \sum_i \left| \sum_j a_{ij}x_j \right| \le \sum_i \left| \sum_j a_{ij} \right| \cdot \left| \sum_j x_j \right| \le \sum_i \sum_j |a_{ij}| \sum_j |x_j| = \sum_i \sum_j |a_{ij}| \cdot \|\mathbf{x}\|$$

Since $||\mathbf{Ax}|| \leq ||\mathbf{x}|| \sum_i \sum_j |a_{ij}|$, the double summation term is an upper bound for $||\mathbf{Ax}||/||\mathbf{x}||$, $||\mathbf{x}|| \neq 0$. Since $||\mathbf{A}||$ is the least such bound, $||\mathbf{A}|| \leq \sum_i \sum_j |a_{ij}|$.

Orthogonal Transformations in $\mathcal{R}^3$

6.23 A satellite position vector has components (X, Y, Z) with respect to an inertially fixed coordinate system with origin at the earth's center. Determine the components of this position vector as measured by a tracking station at longitude L degrees east and latitude λ degrees north, if the angle between the X_I inertial axis and the 0° longitudinal meridian is ϕ degrees and the earth's radius is R_e. See Fig. 6.4.

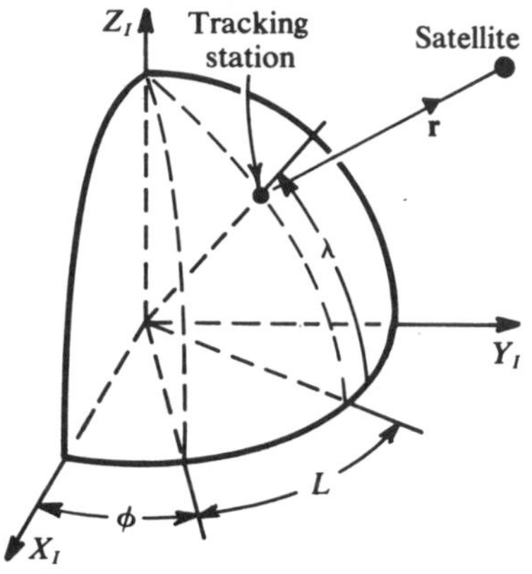

Figure 6.4

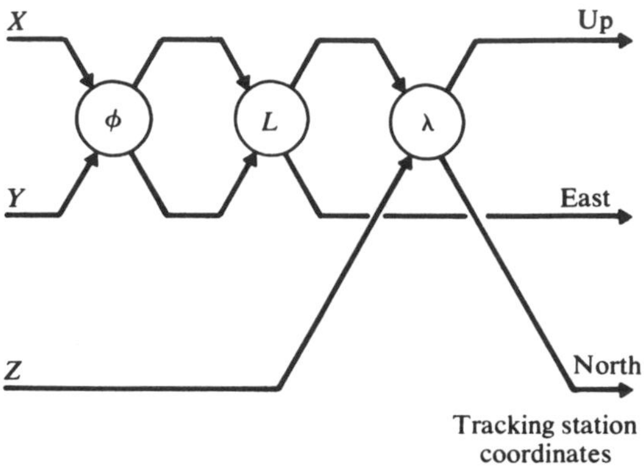

Figure 6.5

The coordinate transformations can be represented by the Piogram of Fig. 6.5. The measured range vector has the following components in the Up-East-North coordinate system. (These components can be read directly from the Piogram by using a few standard conventions [92, 93].) The origin is shifted by subtracting R_e from the Up component:

$$\mathbf{r} = \begin{bmatrix} X[\cos(\phi + L)\cos\lambda] + Y\sin(\phi + L)\cos\lambda + Z\sin\lambda - R_e \\ -X\sin(\phi + L) + Y\cos(\phi + L) \\ -X\cos(\phi + L)\sin\lambda - Y\sin(\phi + L)\sin\lambda + Z\cos\lambda \end{bmatrix}$$

6.24 Demonstrate how the spatial attitude orientation of a vehicle (aircraft, satellite, etc.) can be determined by sighting two known stars.

The attitude of the vehicle will be characterized by the 3 × 3 transformation matrix $\mathbf{T}_{BE}$ which relates a set of orthonormal body fixed axes $\{x, y, z\}$ to a fixed orthonormal inertial coordinate system $\{X, Y, Z\}$. This means a vector **v** expressed in the two coordinate systems satisfies

$$\begin{bmatrix} v_x \\ v_y \\ v_z \end{bmatrix} = \mathbf{T}_{BE} \begin{bmatrix} v_X \\ v_Y \\ v_Z \end{bmatrix}$$

$\mathbf{T}_{BE}$ is to be found. The unit vectors which point towards the two stars are assumed available from star catalogs, in inertial components. That is,

$$\hat{\mathbf{u}} = [u_X \quad u_Y \quad u_Z]$$

$$\hat{\mathbf{v}} = [v_X \quad v_Y \quad v_Z]$$

The pointing direction of a telescope mounted in the vehicle is described by two gimbal angles α and β as shown in Fig. 6.6.

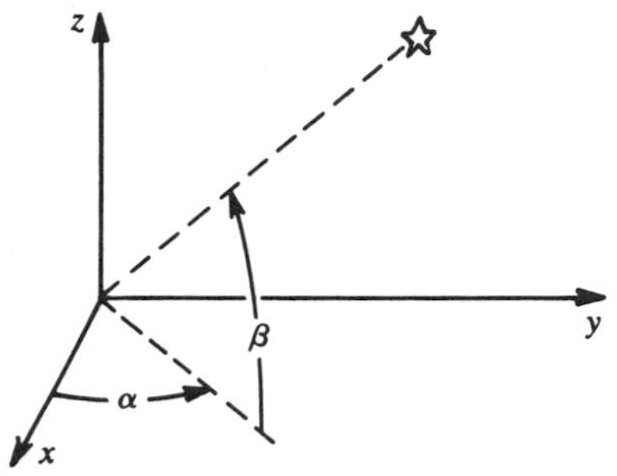

Figure 6.6

The unit vectors to the given stars are expressed in vehicle fixed coordinates as

$$\hat{\mathbf{u}}_B = \begin{bmatrix} \cos\alpha_1 \cos\beta_1 \\ \sin\alpha_1 \cos\beta_1 \\ \sin\beta_1 \end{bmatrix}, \qquad \hat{\mathbf{v}}_B = \begin{bmatrix} \cos\alpha_2 \cos\beta_2 \\ \sin\alpha_2 \cos\beta_2 \\ \sin\beta_2 \end{bmatrix}$$

where α_1, β_1 and α_2, β_2 are the pointing angles for the two stars. $\hat{\mathbf{u}}$ and $\hat{\mathbf{v}}$ are known, and $\hat{\mathbf{u}}_B$ and $\hat{\mathbf{v}}_B$ are available from measurements.

In order to determine $\mathbf{T}_{BE}$, one more relation is necessary. A vector normal to the plane defined by $\hat{\mathbf{u}}$ and $\hat{\mathbf{v}}$ is constructed using the cross product. Thus

$$\mathbf{w} \triangleq \hat{\mathbf{u}} \times \hat{\mathbf{v}} \qquad \text{and} \qquad \mathbf{w}_B \triangleq \hat{\mathbf{u}}_B \times \hat{\mathbf{v}}_B$$

Now $\hat{\mathbf{u}}_B = \mathbf{T}_{BE}\hat{\mathbf{u}}$, $\hat{\mathbf{v}}_B = \mathbf{T}_{BE}\hat{\mathbf{v}}$, and $\mathbf{w}_B = \mathbf{T}_{BE}\mathbf{w}$. These can be combined into

$$[\hat{\mathbf{u}}_B \quad \hat{\mathbf{v}}_B \quad \mathbf{w}_B] = \mathbf{T}_{BE}[\hat{\mathbf{u}} \quad \hat{\mathbf{v}} \quad \mathbf{w}]$$

If the vectors to the two stars are linearly independent, then the 3×3 matrix $[\hat{\mathbf{u}} \quad \hat{\mathbf{v}} \quad \mathbf{w}]$ is nonsingular and

$$\mathbf{T}_{BE} = [\hat{\mathbf{u}}_B \quad \hat{\mathbf{v}}_B \quad \mathbf{w}_B][\hat{\mathbf{u}} \quad \hat{\mathbf{v}} \quad \mathbf{w}]^{-1}$$

PROBLEMS

• **6.25** Let $\mathcal{X}$ be a real n-dimensional linear vector space with $\mathbf{x} \in \mathcal{X}$ and with $\mathbf{b}$ a fixed, nonzero element of $\mathcal{X}$.

a. Show that the transformation $\mathcal{A}(\mathbf{x}) = \langle \mathbf{b}, \mathbf{x} \rangle$ is linear.

b. What are the domain, codomain, range, and null space of $\mathcal{A}$?

6.26 Consider the linear transformation $\mathcal{A}: \mathcal{R}^3 \longrightarrow \mathcal{R}^3$. The basis vectors are selected as

$$\mathbf{v}_1 = [1 \quad 0 \quad 1]^T, \qquad \mathbf{v}_2 = [1 \quad 1 \quad 0]^T, \qquad \mathbf{v}_3 = [1 \quad 1 \quad 1]^T$$

The images of the basis vectors under the transformation $\mathcal{A}$ are

$$\mathbf{u}_1 = [2 \quad -1 \quad 3]^T, \qquad \mathbf{u}_2 = [-1 \quad -1 \quad 2]^T, \qquad \mathbf{u}_3 = [1 \quad 1 \quad 5]^T$$

with respect to the same basis. What is the matrix representation for $\mathcal{A}$?

6.27 With respect to a particular basis, the linear transformation $\mathcal{A}$ has the matrix representation $\mathbf{A} = \dfrac{1}{70}\begin{bmatrix} 56 & 18 & 20 \\ -42 & 124 & 60 \\ -21 & -43 & 240 \end{bmatrix}$. If $\mathbf{B} = \begin{bmatrix} 4 & 2 & 1 \\ 2 & 6 & 3 \\ 1 & 3 & 5 \end{bmatrix}$ represents a change of basis, what is the representation for $\mathcal{A}$ with respect to the new basis?

6.28 The coordinate representations of a real vector $\mathbf{x}$ with respect to two different sets of orthonormal basis vectors are related by

$$[\mathbf{x}]' = \begin{bmatrix} \cos\alpha & \sin\alpha & 0 \\ -\sin\alpha & \cos\alpha & 0 \\ 0 & 0 & 1 \end{bmatrix}[\mathbf{x}]$$

Verify that $\sum_{i=1}^{3} x_i^2 = \sum_{i=1}^{3} (x_i')^2$. Would this be true for nonorthonormal bases?

6.29 If $\hat{\mathbf{u}}$ and $\hat{\mathbf{v}}$ are real unit vectors in three-dimensional space with an included angle α, and if $\mathbf{w} = \hat{\mathbf{u}} \times \hat{\mathbf{v}}$, find an expression for $\mathbf{A}^{-1} = [\hat{\mathbf{u}} \quad \hat{\mathbf{v}} \quad \mathbf{w}]^{-1}$.

6.30 Derive a matrix differential equation for the time rate of change $\dot{\mathbf{T}}_{BE}$ for the transformation of Problem 6.24.

6.31 A mirror lies in the plane defined by $-2x_1 + 3x_2 + x_3 = 0$. Find the reflected image of $\mathbf{y} = [4 \quad -2 \quad 3]^T$ and also find the orthogonal projection of $\mathbf{y}$ on the plane of the mirror.

6.32 Show that $\mathcal{A}^*(\mathcal{A}\mathcal{A}^*)^{-1}\mathcal{A}$ gives the orthogonal projection of $\mathcal{X}$ onto $\mathcal{R}(\mathcal{A}^*)$.

6.33 If $\mathbf{A}$ is the matrix representation of an operator which acts on $\mathbf{x} \in \mathcal{X}$, and if $\|\mathbf{x}\|^2 = \bar{\mathbf{x}}^T\mathbf{x}$, show that $\|\mathbf{A}\| \leq [\sum_i \sum_j |a_{ij}|^2]^{1/2}$.

6.34 The vector norm is defined by $\|\mathbf{x}\|^2 = \bar{\mathbf{x}}^T\mathbf{x}$ and $\mathbf{Ax} = \mathbf{y}$.

a. Show that $\|\mathbf{A}\|^2 = \max\{\gamma_i\}$, where $\{\gamma_i\}$ is the set of eigenvalues for $\bar{\mathbf{A}}^T\mathbf{A}$.

b. Show that if $\mathbf{A}$ is normal, $\|\mathbf{A}\| = \max_i |\lambda_i|$, where $\{\lambda_i\}$ is the set of eigenvalues for $\mathbf{A}$.

6.35 Let $\mathcal{X}$ and $\mathcal{Y}$ be linear vector spaces whose elements are summable n-component vector sequences, $\{\mathbf{x}(k),\ k = 0, 1, 2, \ldots\}$ and $\{\mathbf{y}(k),\ k = 0, 1, 2, \ldots\}$. A linear transformation $\mathcal{A}: \mathcal{X} \longrightarrow \mathcal{Y}$ is defined by the first-order difference equation $\mathcal{A}(\mathbf{x}) = \mathbf{x}(k+1) - \mathbf{A}_k\mathbf{x}(k)$.

Find $\mathcal{A}^*$ if the inner product is defined as $\langle \mathbf{x}_1, \mathbf{x}_2 \rangle = \sum_{k=0}^{\infty} \mathbf{x}_1(k)^T\mathbf{x}_2(k)$.

6.36 How should the results for the minimum norm and least-squares solutions of Problems 6.13 and 6.14 be modified if a weighted norm or weighted least-squares solution is desired?

Chapter 7

EIGENVALUES AND EIGENVECTORS

7.1 INTRODUCTION

This chapter defines eigenvalues and eigenvectors (also called proper or characteristic values and vectors). Methods of determining eigenvalues and eigenvectors are presented, as well as some of their important properties and uses. They are extremely important in modern control theory. For linear systems the eigenvalues indicate whether a system is stable or not. Eigenvectors (and in some cases, generalized eigenvectors) represent a most convenient set of basis vectors for state space. When this basis set is used, the system representation takes on its simplest possible form, and greater insight into the system's structure is thus attained.

7.2 DEFINITION OF THE EIGENVALUE-EIGENVECTOR PROBLEM

Let $\mathcal{A}$ be any linear transformation with domain $\mathfrak{D}(\mathcal{A})$ and range $\mathfrak{R}(\mathcal{A})$, both contained within the same linear vector space $\mathfrak{X}$. Let elements of $\mathfrak{X}$ be denoted as $\mathbf{x}_i$. Those particular elements $\mathbf{x}_i \neq \mathbf{0}$ and the particular scalars $\lambda_i \in \mathfrak{F}$ which satisfy

$$\mathcal{A}(\mathbf{x}_i) = \lambda_i \mathbf{x}_i \tag{7.1}$$

are called *eigenvectors* and *eigenvalues*, respectively. Note that the trivial case $\mathbf{x} = \mathbf{0}$ is explicitly excluded. Thus λ_i is an eigenvalue if and only if the transformation $\mathcal{A} - \mathcal{g}\lambda_i$ has no inverse. The set of all scalars λ for which this is true is called the *spectrum of* $\mathcal{A}$.

With the exception of the material on singular value decomposition, the transformations considered in this chapter map elements in $\mathfrak{X}^n$ into other elements in $\mathfrak{X}^n$, so $\mathcal{A}$ can be represented by an $n \times n$ matrix $\mathbf{A}$. The identity transformation is represented by the unit matrix **I**. The matrix representation of equation (*7.1*) is

$$(\mathbf{A} - \mathbf{I}\lambda_i)\mathbf{x}_i = \mathbf{0} \tag{7.2}$$

and the determination of eigenvectors is a matter of finding nontrivial solutions to a set of n homogeneous equations. This can be accomplished with any of the techniques of Secs. 5.3 through 5.6 or by the adjoint matrix technique of Problems 5.3 to 5.5. The method most adaptable to machine computation for large matrices is probably the row-reduced echelon method of Sec. 5.3 or the GSE method of Sec. 5.5.

7.3 EIGENVALUES

It was shown in Chapter 5 that a necessary condition for the existence of nontrivial solutions to the set of n homogeneous equations (*7.2*) is that rank$(\mathbf{A} - \mathbf{I}\lambda_i) < n$. This is equivalent to requiring $|\mathbf{A} - \mathbf{I}\lambda_i| = 0$. When the determinant is expanded, it yields an nth degree polynomial in the scalar λ_i. That is

$$|\mathbf{A} - \mathbf{I}\lambda| = (-\lambda)^n + c_{n-1}\lambda^{n-1} + c_{n-2}\,\lambda^{n-2} + \cdots + c_1\lambda + c_0 = \Delta(\lambda) \tag{7.3}$$

The roots of this algebraic equation are the eigenvalues λ_i. A fundamental result in algebra states that an nth degree polynomial has exactly n roots, so every $n \times n$ matrix **A** has exactly n eigenvalues. The nth degree polynomial in λ is called the *characteristic polynomial* and the characteristic equation is $\Delta(\lambda) = 0$. In factored form,

$$\Delta(\lambda) = (-1)^n(\lambda - \lambda_1)(\lambda - \lambda_2) \cdots (\lambda - \lambda_n) = 0$$

and the roots are $\lambda_1, \lambda_2, \ldots, \lambda_n$. In general, some of these roots may be equal. If there are $p < n$ distinct roots, $\Delta(\lambda)$ takes the form

$$\Delta(\lambda) = (-1)^n(\lambda - \lambda_1)^{m_1}(\lambda - \lambda_2)^{m_2} \cdots (\lambda - \lambda_p)^{m_p}$$

This indicates that $\lambda = \lambda_1$ is an m_1 order root, $\lambda = \lambda_2$ is an m_2 order root, etc. The integer m_i is called the *algebraic multiplicity* of the eigenvalue λ_i. Of course, $m_1 + m_2 + \cdots + m_p = n$.

The problem of determining eigenvalues for $\mathcal{A}$ amounts to factoring an nth degree polynomial. For large n this is not an easy computational problem, although it is conceptually simple and will not be discussed here. Problem 7.33 presents a direct iterative method of determining eigenvalues and eigenvectors which, when applicable, avoids factoring the characteristic polynomial.

Many relationships exist between **A**, c_i, and λ_i, three of which are:

1. If the scalars c_i in equation (*7.3*) are real (the usual case), then if λ_i is a complex eigenvalue, so is $\bar{\lambda}_i$.
2. $\text{Tr}(\mathbf{A}) = \lambda_1 + \lambda_2 + \cdots + \lambda_n = (-1)^{n\mp 1}c_{n-1}$
3. $|\mathbf{A}| = \lambda_1\lambda_2 \cdots \lambda_n = c_0$.

The characteristic polynomial is often defined by $|\mathbf{I}\lambda_i - \mathbf{A}|$ rather than $|\mathbf{A} - \mathbf{I}\lambda_i|$. This leaves the roots unaltered, but changes equation (*7.3*) by a factor $(-1)^n$ (see Problem 7.43).

7.4 DETERMINATION OF EIGENVECTORS

The procedure for determining eigenvectors can be divided into two possible cases, depending on the results of the eigenvalue calculations.

Case I: All of the eigenvalues are distinct.

Case II: Some eigenvalues are multiple roots of the characteristic equation.

Case I: When each of the n eigenvalues has an algebraic multiplicity of one (i.e., they are all simple, distinct roots), then $\text{rank}(\mathbf{A} - \mathbf{I}\lambda_i) = n - 1$. As was shown in Problem 5.3, a nontrivial solution to the homogeneous equation

$$(\mathbf{A} - \mathbf{I}\lambda_i)\mathbf{x}_i = \mathbf{0}$$

is given by any nonzero column of the matrix

$$\text{Adj}\,(\mathbf{A} - \mathbf{I}\lambda_i)$$

Substitution of each of the n eigenvalues into the adjoint matrix gives a full set of n linearly independent eigenvectors (see Problem 7.13). Alternatively, $\mathbf{A} - \mathbf{I}\lambda_i$ can be reduced to purely numeric form by substituting in values for λ_i. The homogeneous equations are easily solved using the row-reduced echelon form.

If $\mathbf{x}_i$ is an eigenvector, then so is $\alpha\mathbf{x}_i$, for any nonzero scalar α. This fact is sometimes employed to normalize the eigenvectors so that $\|\mathbf{x}_i\| = 1$.

The full set of n eigenvalues and eigenvectors satisfies $\mathbf{A}\mathbf{x}_1 = \lambda_1\mathbf{x}_1$, $\mathbf{A}\mathbf{x}_2 = \lambda_2\mathbf{x}_2$, $\ldots$, $\mathbf{A}\mathbf{x}_n = \lambda_n\mathbf{x}_n$. If an $n \times n$ matrix, called the *modal matrix*, is defined by $\mathbf{M} = [\mathbf{x}_1 \mid \mathbf{x}_2 \mid \cdots \mid \mathbf{x}_n]$, then the n separate equations can be written as one matrix equation $\mathbf{AM} = \mathbf{M}\mathbf{\Lambda}$, where $\mathbf{\Lambda} = \text{diag}\,[\lambda_1, \lambda_2, \lambda_3, \ldots, \lambda_n]$. Since the eigenvectors are linearly independent, $\mathbf{M}$ has rank n, so its inverse exists and gives $\mathbf{\Lambda} = \mathbf{M}^{-1}\mathbf{AM}$. This is referred to as a *similarity transformation*. A square matrix $\mathbf{A}$ with distinct eigenvalues can always be diagonalized by a similarity transformation. In some cases (see Problems 7.21 and 7.23) the eigenvectors are orthogonal, and thus can be normalized to form an orthonormal set. In this case $\mathbf{M}^{-1} = \mathbf{M}^T$, so $\mathbf{A}$ can be diagonalized by the *orthogonal transformation* $\mathbf{\Lambda} = \mathbf{M}^T\mathbf{AM}$.

Case II: When one or more eigenvalues is a repeated root of the characteristic equation, a full set of n linearly independent eigenvectors may or may not exist. The number of linearly independent eigenvectors associated with a given eigenvalue λ_i of algebraic multiplicity m_i is equal to the degeneracy q_i of $\mathbf{A} - \mathbf{I}\lambda_i$:

$$q_i \triangleq n - \text{rank}(\mathbf{A} - \mathbf{I}\lambda_i) \qquad \text{(see Problem 7.14)}$$

Note that $1 \leq q_i \leq m_i$. The degeneracy q_i just defined is often referred to as the *geometric multiplicity* of λ_i. The distinction is necessary because, for example, $\lambda_1 = 5$ might be a triple root of $\Delta(\lambda) = 0$ ($m_1 = 3$ = algebraic multiplicity) and yet there may be only two independent eigenvectors ($q_1 = 2$ = dimension of space spanned by the eigenvectors = geometric multiplicity). It is convenient to consider three subsclassifications for Case II.

Case II_1: Full Degeneracy, $q_i = m_i$. The fully degenerate case has a full set of m_i eigenvectors associated with λ_i. They can be found from the nonzero columns of

$$\frac{1}{(m_i - 1)!}\left\{\frac{d^{m_i-1}}{d\lambda^{m_i-1}}[\text{Adj}\,(\mathbf{A} - \mathbf{I}\lambda)]\right\}_{\lambda=\lambda_i}$$

or by applying the row-reduced echelon form to $(\mathbf{A} - \mathbf{I}\lambda_i)\mathbf{x}_i = \mathbf{0}$. Note that simple eigenvalues provide a special case of this category, with $m_i = 1$. If $\mathcal{A}$ is normal, special cases of which are matrices which are real and symmetric or Hermitian, then if $\Delta(\lambda)$ has repeated roots, Case II_1 will always apply (see Problem 7.24).

Case II_2: Simple Degeneracy, $q_i = 1$. For this case there is only one eigenvector associated with λ_i regardless of the multiplicity m_i. It may be found as in Case I, or by any of the methods of solving homogeneous equations, discussed in Sec. 5.6. Let the single eigenvector be $\mathbf{x}_1$. An additional set of $m_i - 1$ vectors, called *generalized eigenvectors*, can be found for λ_i. If this is done for each eigenvalue, a set of n linearly independent vectors which are either eigenvectors or generalized eigenvectors, will be determined for $\mathbf{A}$. This set constitutes a convenient basis in many situations. A generalized eigenvector of rank k is defined as a nonzero vector satisfying $(\mathbf{A} - \lambda_i\mathbf{I})^k\mathbf{x}_k = \mathbf{0}$ and $(\mathbf{A} - \lambda_i\mathbf{I})^{k-1}\mathbf{x}_k \neq \mathbf{0}$. If the generalized eigenvector $\mathbf{x}_{m_i}$ of rank m_i is found, then the entire set of generalized eigenvectors is generated by the rule $\mathbf{x}_{m_i-1} = (\mathbf{A} - \mathbf{I}\lambda_i)\mathbf{x}_{m_i}$, $\mathbf{x}_{m_i-2} = (\mathbf{A} - \mathbf{I}\lambda_i)\mathbf{x}_{m_i-1}, \ldots, \mathbf{x}_{j-1} = (\mathbf{A} - \mathbf{I}\lambda_i)\mathbf{x}_j, \ldots, \mathbf{x}_1 = (\mathbf{A} - \mathbf{I}\lambda_i)\mathbf{x}_2$. The eigenvector $\mathbf{x}_1$ is the final vector found in this sequence and satisfies $(\mathbf{A} - \mathbf{I}\lambda_i)\mathbf{x}_1 = \mathbf{0}$. Whenever Case II_2 applies, it is not possible to diagonalize $\mathbf{A}$ by a similarity transformation. If the modal matrix $\mathbf{M}$ is formed as before, but the generalized eigenvectors are included where required, then

$$\mathbf{AM} = \mathbf{MJ} \quad \text{or} \quad \mathbf{M}^{-1}\mathbf{AM} = \mathbf{J}$$

where $\mathbf{J}$ is an $n \times n$ block diagonal matrix, called the *Jordan form*:

$$\mathbf{J} = \text{diag}\,[\mathbf{J}_1, \mathbf{J}_2, \ldots, \mathbf{J}_p]$$

Each submatrix $\mathbf{J}_i$ has the same eigenvalue for all of its main diagonal elements and ones for all elements in the diagonal above the main diagonal. All other elements are zero:

$$\mathbf{J}_i = \begin{bmatrix} \lambda_i & 1 & & & \\ & \lambda_i & 1 & & \mathbf{0} \\ & & \ddots & \ddots & \\ & \mathbf{0} & & \ddots & 1 \\ & & & & \lambda_i \end{bmatrix}$$

There will be one $m_i \times m_i$ submatrix of this form, called a *Jordan block*, for each eigenvalue of multiplicity m_i satisfying the conditions of Case II_2. Eigenvalues satisfying Case II_1, and having multiplicity m_i, will yield m_i separate 1×1 Jordan blocks, $\mathbf{J}_i = [\lambda_j]$.

Case II_3: The general case for an eigenvalue of algebraic multiplicity m_i and degeneracy q_i satisfying $1 \le q_i \le m_i$ still has q_i eigenvectors associated with λ_i. There will be one Jordan block for each eigenvector; that is, λ_i will have q_i blocks associated with it. Case II_3 is really just a combination of the previous two cases, but knowledge of m_i and q_i still leaves some ambiguity. Assume λ_1 is a fourth-order root of the characteristic equation and assume $q_1 = 2$. Then it is known that there are two eigenvectors and two generalized eigenvectors. The eigenvectors satisfy $\mathbf{A}\mathbf{x}_a = \lambda_1\mathbf{x}_a$ and $\mathbf{A}\mathbf{x}_b = \lambda_1\mathbf{x}_b$, but it is still uncertain whether the generalized eigenvectors are both associated with $\mathbf{x}_a$ or both with $\mathbf{x}_b$ or one with each. That is, the two Jordan blocks could take one of the following forms:

$$\mathbf{J}_1 = \begin{bmatrix} \lambda_1 & 1 & 0 \\ 0 & \lambda_1 & 1 \\ 0 & 0 & \lambda_1 \end{bmatrix}, \quad \mathbf{J}_2 = [\lambda_1] \quad \text{or} \quad \mathbf{J}_1 = \begin{bmatrix} \lambda_1 & 1 \\ 0 & \lambda_1 \end{bmatrix}, \quad \mathbf{J}_2 = \begin{bmatrix} \lambda_1 & 1 \\ 0 & \lambda_1 \end{bmatrix}$$

The first pair corresponds to the equations

$$\mathbf{A}\mathbf{x}_1 = \lambda_1\mathbf{x}_1, \quad \mathbf{A}\mathbf{x}_2 = \lambda_1\mathbf{x}_2 + \mathbf{x}_1, \quad \mathbf{A}\mathbf{x}_3 = \lambda_1\mathbf{x}_3 + \mathbf{x}_2, \quad \mathbf{A}\mathbf{x}_4 = \lambda_1\mathbf{x}_4$$

The second pair corresponds to

$$\mathbf{A}\mathbf{x}_1 = \lambda_1\mathbf{x}_1, \quad \mathbf{A}\mathbf{x}_2 = \lambda_1\mathbf{x}_2 + \mathbf{x}_1, \quad \mathbf{A}\mathbf{x}_3 = \lambda_1\mathbf{x}_3, \quad \mathbf{A}\mathbf{x}_4 = \lambda_1\mathbf{x}_4 + \mathbf{x}_3$$

Ambiguities such as this can be resolved by a trial and error process [28] or they can be avoided by using a systematic method from Sec. 7.5. Combinations of the preceding cases may be required for a given $n \times n$ matrix $\mathbf{A}$. The applicable case can be different for each eigenvalue. For example, if the eigenvalues for some 9×9 matrix were $\{2, 3, 3, 5, 5, 5, 6, 6, 6\}$, $\lambda = 2$ is of necessity an example of Case I. $\lambda = 3$ might have two eigenvectors, i.e., case II_1; $\lambda = 5$ might have only one eigenvector, i.e., Case II_2; and $\lambda = 6$ might have two eigenvectors and one generalized eigenvector, i.e., Case II_3. Finding a total of n vectors, m_i for each eigenvalue with multiplicity m_i, allows the nonsingular modal matrix $\mathbf{M}$ to be formed. The similarity transformation $\mathbf{M}^{-1}\mathbf{A}\mathbf{M}$ again gives the Jordan form $\mathbf{J}$. The diagonal matrix $\mathbf{\Lambda}$ of Case I is considered a special case of the Jordan form with all of its Jordan blocks being 1×1.

Summary: Every $n \times n$ matrix has n eigenvalues and n linearly independent vectors, either eigenvectors or generalized eigenvectors. The eigenvalues are roots of an nth degree polynomial. For each repeated eigenvalue the degeneracy q_i should be found. There will be q_i eigenvectors and Jordan blocks associated with λ_i. If $q_i < m_i$, then generalized eigenvectors will be required. This type of analysis makes it clear how many eigenvectors, generalized eigenvectors, and Jordan blocks there are, as demonstrated in Example 7.1. The actual determination of the generalized eigenvectors is discussed in more detail in Section 7.5.

Example 7.1

Let **A** be an 8 × 8 matrix and assume that the eigenvalues have been found as $\lambda_1 = \lambda_2 = 2$, $\lambda_3 = \lambda_4 = \lambda_5 = -3$, $\lambda_6 = 8$, $\lambda_7 = \lambda_8 = 4$. If rank $[\mathbf{A} - 2\mathbf{I}] = 7$, rank $[\mathbf{A} + 3\mathbf{I}] = 6$, and rank $[\mathbf{A} - 4\mathbf{I}] = 6$, find the degeneracies and determine how many eigenvectors and generalized eigenvectors there are. Also, write down the Jordan form.

For $\lambda_1 = 2$, $q_1 = 8 - 7 = 1$. This is the simple degeneracy Case II_2, so $\mathbf{x}_1$ is one eigenvector and $\mathbf{x}_2$ must be a generalized eigenvector. For $\lambda_3 = -3$, $q_3 = 8 - 6 = 2$. This falls into Case II_3 and there are two eigenvectors (and Jordan blocks) and one generalized eigenvector must be associated with this triple root. For $\lambda_6 = 8$, there is one eigenvector (Case I). For $\lambda_7 = 4$, $q_7 = 8 - 6 = 2$. This is Case II_1, since $q_7 = m_7 = 2$. There are two eigenvectors and no generalized eigenvectors associated with this eigenvalue. There are a total of six eigenvectors (and Jordan blocks), and two generalized eigenvectors. The Jordan form is

$$\mathbf{J} = \begin{bmatrix} 2 & 1 & & & & & & \\ 0 & 2 & & & & & & \\ & & -3 & 1 & & & \mathbf{0} & \\ & & 0 & -3 & & & & \\ & & & & -3 & & & \\ & & \mathbf{0} & & & 8 & & \\ & & & & & & 4 & \\ & & & & & & & 4 \end{bmatrix}$$

■

7.5 DETERMINATION OF GENERALIZED EIGENVECTORS

It is assumed throughout this section that one or more multiple eigenvalues exist for a matrix **A** and that a need for generalized eigenvectors has already been established. Three alternative methods are presented for finding generalized eigenvectors.

1. *The first method is a "bottom-up" method in that the eigenvectors are found first and then a chain of one or more generalized eigenvectors is built up from these.* That is, first find all solutions of the homogeneous equation

$$(\mathbf{A} - \mathbf{I}\lambda_i)\mathbf{x}_i = \mathbf{0}$$

for the repeated eigenvalue λ_i. For each $\mathbf{x}_i$ thus determined, try to construct a generalized eigenvector using

$$(\mathbf{A} - \mathbf{I}\lambda_i)\mathbf{x}_{i+1} = \mathbf{x}_i$$

If the resultant vector $\mathbf{x}_{i+1}$ is linearly independent of all vectors already found, it is a valid generalized eigenvector. If still more generalized eigenvectors are needed for λ_i, then solve

$$(\mathbf{A} - \mathbf{I}\lambda_i)\mathbf{x}_{i+2} = \mathbf{x}_{i+1}$$

and so on until all needed vectors are found.

Example 7.2

Consider the matrix $\mathbf{A} = \begin{bmatrix} 1 & 2 & 3 \\ 0 & 1 & 4 \\ 0 & 0 & 1 \end{bmatrix}$. Clearly, $\lambda_i = 1$ has an algebraic multiplicity $m = 3$. The rank of $\mathbf{A} - \mathbf{I}\lambda_i$ is 2, so $q = 1$. There is only one eigenvector, so two generalized eigenvectors are needed. One way of finding the eigenvector is to compute

$$\operatorname{Adj}(\mathbf{A} - \mathbf{I}\lambda) = \begin{bmatrix} (1-\lambda)^2 & -2(1-\lambda) & 8-3(1-\lambda) \\ 0 & (1-\lambda)^2 & -4(1-\lambda) \\ 0 & 0 & (1-\lambda)^2 \end{bmatrix}$$

Using $\lambda = 1$, it is clear that the only possible eigenvector is

$$\mathbf{x}_1 = \alpha\begin{bmatrix} 1 \\ 0 \\ 0 \end{bmatrix}, \text{ with } \alpha \text{ arbitrary.}$$

The first generalized eigenvector $\mathbf{x}_2$ is a solution of $(\mathbf{A} - \lambda\mathbf{I})\mathbf{x}_2 = \mathbf{x}_1$. Letting the unknown components of $\mathbf{x}_2$ be $[a \quad b \quad c]^T$ leads to $2b + 3c = \alpha$ and $4c = 0$. Therefore, $c = 0$, $b = \alpha/2$, and a is arbitrary. Normally specific values for α and a would be selected before proceeding, but to illustrate fully the choices, use $\mathbf{x}_2 = [a \quad \alpha/2 \quad 0]^T$ in finding the components d, e, and f of $\mathbf{x}_3$.

$$\begin{bmatrix} 0 & 2 & 3 \\ 0 & 0 & 4 \\ 0 & 0 & 0 \end{bmatrix}\begin{bmatrix} d \\ e \\ f \end{bmatrix} = \begin{bmatrix} a \\ \alpha/2 \\ 0 \end{bmatrix}. \text{ This gives } 2e + f = a,\ 4f = \alpha/2, \text{ and } d \text{ is arbitrary.}$$

Clearly, the number of possibilities multiplies at each step. Two valid choices are presented here. $\alpha = 8$, $a = 1$, $d = 1$ gives

$$\mathbf{x}_1 = \begin{bmatrix} 8 \\ 0 \\ 0 \end{bmatrix}, \quad \mathbf{x}_2 = \begin{bmatrix} 1 \\ 4 \\ 0 \end{bmatrix}, \quad \mathbf{x}_3 = \begin{bmatrix} 1 \\ 5/2 \\ 1 \end{bmatrix}.$$

If, instead, $\alpha = 1$ had been picked, along with $a = 1$ and $d = 1$, then the three vectors would be $[1 \quad 0 \quad 0]^T$, $[1 \quad 1/2 \quad 0]^T$, and $[1 \quad 5/16 \quad 1/8]^T$. ■

2. *A second method of proceeding is to use the adjoint matrix and its various derivatives.* This is also a bottom-up method, since the eigenvectors are found first. This is done by selecting all linearly independent columns of $\operatorname{Adj}(\mathbf{A} - \mathbf{I}\lambda)$. A given column of successively differentiated adjoint matrices is an eigenvector if that column is nonzero for the first time. It is a generalized eigenvector if that same column was selected on the previous differentiation step as either an eigenvector or a generalized eigenvector. The factorial divisor shown for Case II_1 is required to maintain correct scaling. The vectors in each step cannot be

arbitrarily scaled since they are all tied together in a chain. Also note that the same vector can show up in different columns on different steps.

Example 7.3

The same $\mathbf{A}$ matrix as the previous example is used. Column 3 of the adjoint matrix is used, without any rescaling, to give $\mathbf{x}_1 = [8 \quad 0 \quad 0]^T$. Then

$$\frac{d}{d\lambda}\{\text{Adj}(\mathbf{A} - \mathbf{I}\lambda)\} = \begin{bmatrix} -2(1-\lambda) & 2 & 3 \\ 0 & -2(1-\lambda) & 4 \\ 0 & 0 & -2(1-\lambda) \end{bmatrix}$$

With $\lambda = 1$, this reduces to $\begin{bmatrix} 0 & 2 & 3 \\ 0 & 0 & 4 \\ 0 & 0 & 0 \end{bmatrix}$. Column 3 is used again to give $\mathbf{x}_2$. Note that column 2 has given $\mathbf{x}_1$ again, except for a different scale factor. Of course, this vector cannot be used twice since a linearly independent set must ultimately be selected. One more differentiation, including the factorial divisor, gives $\frac{1}{2}\frac{d}{d\lambda}\{\text{Adj}(\mathbf{A} - \mathbf{I}\lambda)\} = \frac{1}{2}\begin{bmatrix} 2 & 0 & 0 \\ 0 & 2 & 0 \\ 0 & 0 & 2 \end{bmatrix}$. Column 3 must be selected again, so $\mathbf{x}_3 = [0 \quad 0 \quad 1]^T$. Column 1 is nonzero for the first time and therefore is an eigenvector. However it is one that has already been selected. Column 2 at this step would be a generalized eigenvector if column 2 had been picked at the previous step as the eigenvector, i.e., if $\mathbf{x}_1$ had been scaled differently. The final modal matrix obtained from this method is $\mathbf{M} = \begin{bmatrix} 8 & 3 & 0 \\ 0 & 4 & 0 \\ 0 & 0 & 1 \end{bmatrix}$. The previous method also gives this result when the choices made are $\alpha = 8$, $a = 3$, $d = 0$. ■

3. *The third possible method of finding generalized eigenvectors is a "top-down" method.* First the index k of the repeated eigenvalue is found. This requires finding k such that $\text{rank}[\mathbf{A} - \mathbf{I}\lambda]^k = n - m$. Then all linearly independent solutions of $[\mathbf{A} - \mathbf{I}\lambda]^k\mathbf{x}_i = \mathbf{0}$ are found. These $\mathbf{x}_i$ are either eigenvectors or generalized eigenvectors. Which possibility applies will become apparent. Each of the found vectors $\mathbf{x}_i$ is then subjected to a simple matrix multiplication $[\mathbf{A} - \mathbf{I}\lambda]\mathbf{x}_i$; call the result $\mathbf{x}_j$. If $\mathbf{x}_j$ turns out to be the zero vector then $\mathbf{x}_i$ is an eigenvector. If $\mathbf{x}_j$ is nonzero, then $\mathbf{x}_i$ is a generalized eigenvector and $\mathbf{x}_j$ may be either another generalized eigenvector or an eigenvector, as can be determined by another matrix product.

Example 7. 4

The previous example is reexamined using the top-down method. Since $n = 3$ and $m = 3$, the index k must give $\text{rank}\,[\mathbf{A} - \mathbf{I}\lambda]^k = 0$. It easily follows that $k = 3$ for this example, and since $[\mathbf{A} - \mathbf{I}]^3 = [0]$, *any* vector is a solution to $[\mathbf{A} - \mathbf{I}\lambda]\mathbf{x}_i = 0$. Three independent choices can be made, and one possible set is $\mathbf{x}_a = [1 \quad 0 \quad 0]^T$, $\mathbf{x}_b = [0 \quad 1 \quad 0]^T$, $\mathbf{x}_c = [0 \quad 0 \quad 1]^T$. It must now be determined which of these are eigenvectors and which are generalized eigenvectors. Since $[\mathbf{A} - \mathbf{I}\lambda]\mathbf{x}_a = \mathbf{0}$, $\mathbf{x}_a$ is an eigenvector. It is not going to end up in our final set, however. Continuing the testing shows that $[\mathbf{A} - \mathbf{I}\lambda]\mathbf{x}_b = [2 \quad 0 \quad 0]^T \neq \mathbf{0}$, so $\mathbf{x}_b$ is a generalized eigenvector associated with the eigenvector $[2 \quad 0 \quad 0]^T$ (as we already knew from Example 7.3, column 2). Finally, $[\mathbf{A} - \mathbf{I}\lambda]\mathbf{x}_c = [3 \quad 4 \quad 0]^T$. Since this is not zero, $\mathbf{x}_c$ is a generalized eigenvector associated with a vector $[3 \quad 4 \quad 0]^T$. In general, the right-hand side vectors formed by the above matrix products can be eigenvectors or generalized eigenvectors and they too need to be tested to determine which. For example $[\mathbf{A} - \mathbf{I}\lambda]\begin{bmatrix} 3 \\ 4 \\ 0 \end{bmatrix} = \begin{bmatrix} 8 \\ 0 \\ 0 \end{bmatrix}$. Since the result is nonzero, $[3 \quad 4 \quad 0]^T$ is not an eigenvector, but rather a generalized one. Then the nonzero result $[8 \quad 0 \quad 0]^T$ would be tested, and so on until the chain of vectors terminates with an eigenvector, as indicated by a zero matrix product. Here the result is already known from the preceding examples and the same modal matrix is obtained as in Example 7.3. ■

7.6 SPECTRAL DECOMPOSITION AND INVARIANCE PROPERTIES

Definition 7.1: Let $\mathcal{X}$ be a linear vector space defined over the complex number field. Let $\mathcal{A}: \mathcal{X} \longrightarrow \mathcal{X}$ be a linear transformation and let $\mathcal{X}_1$ be a subspace of $\mathcal{X}$. Then $\mathcal{X}_1$ is said to be *invariant* under the transformation $\mathcal{A}$ if for every $\mathbf{x} \in \mathcal{X}_1$, $\mathcal{A}(\mathbf{x})$ also belongs to $\mathcal{X}_1$.

Definition 7.2: The set of all vectors $\mathbf{x}_i$ satisfying

$$\mathcal{A}(\mathbf{x}_i) = \lambda_i \mathbf{x}_i$$

for a particular λ_i is called the *eigenspace* of λ_i.

It consists of all the eigenvectors associated with that particular eigenvalue λ_i, plus the zero vector. The eigenspace of λ_i is a subspace of $\mathcal{X}$, and may alternatively be characterized as the null space of the transformation $(\mathcal{A} - \mathcal{I}\lambda_i)$, denoted as $\mathfrak{N}_i$ for brevity.

Theorem 7.1: $\mathfrak{N}_i$ is a q_i dimensional subspace of $\mathcal{X}$ which is invariant under $\mathcal{A}$, where q_i is the degeneracy, $q_i = n - \text{rank}(\mathcal{A} - \mathcal{I}\lambda_i)$.

Definition 7.3: If the linear transformation $\mathcal{A}$ has a complete set of n linearly independent eigenvectors (i.e., Case I or Case II_1), then $\mathcal{A}$ is said to be a *simple* linear transformation.

Theorem 7.2: If $\mathcal{A}$ is simple, then

$$\mathfrak{X} = \mathfrak{N}_1 \oplus \mathfrak{N}_2 \oplus \cdots \oplus \mathfrak{N}_p$$

where the direct sum is taken over the p distinct eigenvalues. (Note that $p < n$ for Case II_1.)

Theorem 7.3: If $\mathcal{A}$ is normal, $\mathfrak{N}_i$ and $\mathfrak{N}_j$ are orthogonal to each other, for all $i \neq j$. Note that normal transformations are a subset of simple transformations.

Let $\mathcal{A}: \mathfrak{X} \longrightarrow \mathfrak{X}$ be a simple linear transformation, with the matrix representation $\mathbf{A}$. Then the eigenvectors of $\mathbf{A}$, $\{\mathbf{x}_i\}$, form a basis for $\mathfrak{X}$. Let $\{\mathbf{r}_i\}$ be the reciprocal basis vectors. Then, for every $\mathbf{z} \in \mathfrak{X}$,

$$\mathbf{z} = \sum_{i=1}^{n} \langle \mathbf{r}_i, \mathbf{z} \rangle \mathbf{x}_i \qquad \text{and} \qquad \mathbf{A}\mathbf{z} = \sum_{i=1}^{n} \langle \mathbf{r}_i, \mathbf{z} \rangle \mathbf{A}\mathbf{x}_i = \sum_{i=1}^{n} \lambda_i \langle \mathbf{r}_i, \mathbf{z} \rangle \mathbf{x}_i$$

This allows $\mathbf{A}$ to be written as

$$\mathbf{A} = \sum_{i=1}^{n} \lambda_i \mathbf{x}_i \rangle \langle \mathbf{r}_i \tag{7.4}$$

This is called the *spectral representation* of $\mathbf{A}$. If $\mathcal{A}$ is normal, then its eigenvectors are mutually orthogonal (see Problem 7.23) so that the reciprocal basis vector $\mathbf{r}_i$ can be made equal to $\mathbf{x}_i$ by normalizing the eigenvectors. Then

$$\mathbf{A} = \sum_{i=1}^{n} \lambda_i \mathbf{x}_i \rangle \langle \mathbf{x}_i \tag{7.5}$$

When a linear transformation is not simple, its eigenvectors do not form a basis for $\mathfrak{X}$ (Cases II_2 and II_3). It is still possible to construct a basis by adding generalized eigenvectors, as discussed in Sec. 7.5. These vectors, along with the eigenvectors, belong to $\mathfrak{N}_i^{k_i}$, the null space of $(\mathcal{A} - \mathcal{I}\lambda_i)^{k_i}$ (see Problem 7.31). The power k_i is called the *index* of the eigenvalue λ_i and is one for simple transformations. Using this generalization, it is again possible to write $\mathfrak{X}$ as a direct sum of invariant subspaces of $\mathcal{A}$.

$$\mathfrak{X} = \mathfrak{N}_1^{k_1} \oplus \mathfrak{N}_2^{k_2} \oplus \cdots \oplus \mathfrak{N}_p^{k_p} \tag{7.6}$$

Alternatively, the space $\mathfrak{X}$ can be decomposed into

$$\mathfrak{X} = \mathfrak{N}_i^{k_i} \oplus \mathfrak{N}_i^{k_i \perp}$$

Since all m_i eigenvectors and generalized eigenvectors associated with λ_i belong to $\mathfrak{N}_i^{k_i}$, $\dim(\mathfrak{N}_i^{k}) = m_i$. Thus, rank $(\mathbf{A} - \mathbf{I}\lambda_i)^{k_i} = n - m_i$ and the index k_i is the smallest integer for which is true. The ambiguity regarding Jordan blocks mentioned in Sec. 7.4

can be avoided by first finding all generalized eigenvectors satisfying $(\mathbf{A} - \mathbf{I}\lambda_i)^{k_i}\mathbf{x} = \mathbf{0}$ *and* $(\mathbf{A} - \mathbf{I}\lambda_i)^{k_i-1}\mathbf{x} \neq \mathbf{0}$. The remaining vectors are found by simple multiplication. The largest Jordan block associated with λ_i will be $k_i \times k_i$ (see method 3, page 180).

Example 7.5

Let

$$\mathbf{A} = \begin{bmatrix} 0 & 0 & 1 & 0 \\ 0 & 0 & 0 & 1 \\ 0 & 0 & 0 & 0 \\ 0 & 0 & 0 & 0 \end{bmatrix}$$

The characteristic equation is $\lambda^4 = 0$, so $\lambda_1 = 0$ with $m_1 = 4$. Since rank$(\mathbf{A} - \mathbf{I}\lambda_1) = 2$, there are $q = 2$ eigenvectors and also 2 generalized eigenvectors. Since $n - m = 0$, and since rank $[\mathbf{A} - \mathbf{I}\lambda_1]^2 = 0$, $k_1 = 2$. The largest Jordan block is 2×2. Since there are just two blocks, they both must be 2×2. To find the eigenvectors and generalized eigenvectors, consider $(\mathbf{A} - \mathbf{I}\lambda_1)^2\mathbf{x} = \mathbf{0}$. Any vector satisfies this equation, but there are at most four linearly independent solutions. Select $\mathbf{x}_a = [1 \quad 0 \quad 0 \quad 0]^T$. This is *not* a generalized eigenvector since $(\mathbf{A} - \mathbf{I}\lambda_1)\mathbf{x}_a = \mathbf{0}$. Similarly $\mathbf{x}_b = [0 \quad 1 \quad 0 \quad 0]^T$ is not a generalized eigenvector. Select $\mathbf{x}_c = [0 \quad 0 \quad 1 \quad 0]^T$. Then since $[\mathbf{A} - \mathbf{I}\lambda_1]\mathbf{x}_c = [1 \quad 0 \quad 0 \quad 0]^T \neq \mathbf{0}$, $\mathbf{x}_c$ is a generalized eigenvector associated with the eigenvector $[1 \quad 0 \quad 0 \quad 0]^T$. Finally $\mathbf{x}_d = [0 \quad 0 \quad 0 \quad 1]^T$ is a generalized eigenvector and $[\mathbf{A} - \mathbf{I}\lambda_1]\mathbf{x}_d = [0 \quad 1 \quad 0 \quad 0]^T$ is the associated eigenvector. Thus the modal matrix and the Jordan form are

$$\mathbf{M} = \begin{bmatrix} 1 & 0 & 0 & 0 \\ 0 & 0 & 1 & 0 \\ 0 & 1 & 0 & 0 \\ 0 & 0 & 0 & 1 \end{bmatrix} \quad \text{and} \quad \mathbf{J} = \left[\begin{array}{cc:cc} 0 & 1 & 0 & 0 \\ 0 & 0 & 0 & 0 \\ \hdashline 0 & 0 & 0 & 1 \\ 0 & 0 & 0 & 0 \end{array}\right]$$

See page 259 of Reference 28 for a trial and error solution to the same problem. ■

Theorem 7.4: If $\mathcal{A}: \mathcal{X} \longrightarrow \mathcal{X}$, with dim $(\mathcal{X}) = n$, and if $\mathcal{X}$ can be expressed as the direct sum of p invariant subspaces, as in equation (7.6), then $\mathcal{A}$ can be represented by a block diagonal matrix, with p blocks, each of dimension k_i. provided a suitable basis is selected.

The block diagonal representation for $\mathcal{A}$ is the Jordan form, and "the suitable basis" consists of eigenvectors, and if necessary, generalized eigenvectors. This provides the simplest possible representation for a linear transformation, and will be most useful in analyzing systems in later chapters.

7.7 BILINEAR AND QUADRATIC FORMS

The expression $\langle \mathbf{y}, \mathbf{Ax} \rangle$ is called a *bilinear form* since if $\mathbf{y}$ is held fixed, it is linear in $\mathbf{x}$: and if $\mathbf{x}$ is held fixed, it is linear in $\mathbf{y}$. When $\mathbf{x} = \mathbf{y}$, the result is the *quadratic form*, $Q(\mathbf{x}) = \langle \mathbf{x}, \mathbf{Ax} \rangle$. Every matrix $\mathbf{A}$ can be written as the sum of a Hermitian matrix and a skew-Hermitian matrix. It will be assumed that all quadratic forms are defined in terms of a Hermitian matrix **A**. For real quadratic forms, there is no loss of generality since $\langle \mathbf{x}, \mathbf{Ax} \rangle = 0$ for all $\mathbf{x}$ if **A** is skew-symmetric.

Quadratic forms arise in connection with performance criteria in optimal control problems, in consideration of system stability, and in other applications. Here the several types of quadratic forms are defined and means for establishing the type of a given quadratic form are summarized.

Definitions:

1. Q (or the defining matrix **A**) is said to be *positive definite* if and only if $\langle \mathbf{x}, \mathbf{Ax} \rangle > 0$ for all $\mathbf{x} \neq 0$.
2. Q is *positive semidefinite* if $\langle \mathbf{x}, \mathbf{Ax} \rangle \geq 0$ for all $\mathbf{x}$. That is, $Q = 0$ is possible for some $\mathbf{x} \neq \mathbf{0}$.
3. Q is *negative definite* if and only if $\langle \mathbf{x}, \mathbf{Ax} \rangle < 0$ for all $\mathbf{x} \neq \mathbf{0}$.
4. Q is *negative semidefinite* if $\langle \mathbf{x}, \mathbf{Ax} \rangle \leq 0$ for all $\mathbf{x}$.
5. Q is said to be *indefinite* if $\langle \mathbf{x}, \mathbf{Ax} \rangle > 0$ for same $\mathbf{x}$ and $\langle \mathbf{x}, \mathbf{Ax} \rangle < 0$ for other $\mathbf{x}$.

Tests for Definiteness: Let **A** be an $n \times n$ real symmetric matrix with eigenvalues λ_i. Define

$$\Delta_1 = a_{11}, \qquad \Delta_2 = \begin{vmatrix} a_{11} & a_{12} \\ a_{21} & a_{22} \end{vmatrix}, \qquad \Delta_3 = \begin{vmatrix} a_{11} & a_{12} & a_{13} \\ a_{21} & a_{22} & a_{23} \\ a_{31} & a_{32} & a_{33} \end{vmatrix}, \quad \ldots, \quad \Delta_n = |\mathbf{A}|$$

The Δ_i are called the *prinicipal minors* of **A**. Two alternative methods of determining the definiteness of a Hermitian matrix **A** are given in Table 7.1.

TABLE 7.1

Class	Tests Using:	
	Eigenvalues of **A**	Principal Minors of **A**
1. Positive definite	all $\lambda_i > 0$	$\Delta_1 > 0, \Delta_2 > 0, \ldots, \Delta_n > 0$
2. Positive semidefinite	all $\lambda_i \geq 0$	$\Delta_1 \geq 0, \Delta_2 \geq 0, \ldots, \Delta_n \geq 0$
3. Negative definite	all $\lambda_i < 0$	$\Delta_1 < 0, \Delta_2 > 0, \Delta_3 < 0, \ldots$ (note alternating signs)
4. Negative semidefinite	all $\lambda_i \leq 0$	$\Delta_1 \leq 0, \Delta_2 \geq 0, \Delta_3 \leq 0, \ldots$
5. Indefinite	some $\lambda_i > 0$, some $\lambda_j < 0$	None of the above

7.8 MISCELLANEOUS USES OF EIGENVALUES AND EIGENVECTORS

Eigenvalues and eigenvectors are useful in many contexts. Four of the more important uses in modern control theory are mentioned.

1. *Existence of solutions for sets of linear equations:* In Chapter 5, the existence of nontrivial solutions for homogeneous equations and of a unique solution for

nonhomogeneous equations was seen to depend upon whether or not the coefficient matrix **A** had a zero determinant. In Chapter 6, the existence of unique solutions depended upon whether or not the null space of a linear transformation contained nonzero vectors. Both of these conditions are related to the question of whether or not zero is an eigenvalue. Even when the transformation maps vectors from a space of one dimension to a space of another dimension (and thus cannot define an eigenvalue problem), conditions can be expressed in terms of the eigenvalues of transformations $\mathcal{A}\mathcal{A}^*$ and/or $\mathcal{A}^*\mathcal{A}$. This will be done in Chapter 12 when discussing controllability and observability.

2. *Stability of linear differential and difference equations:* It is not difficult to show that the characteristic equation of the companion matrix (Problem 7.32) is the same as the denominator of the input-output transfer function for the nth order differential equation. Thus the system poles are the same as the eigenvalues of the matrix. The influence of pole locations on system stability has been discussed in Chapter 2. It will be seen in Chapter 14 that the eigenvalues of a system matrix, not necessarily the companion matrix, determine the stability of linear systems described either by differential or difference equations.

3. *Eigenvectors are convenient basis vectors:* When eigenvectors and, if needed, generalized eigenvectors are used as basis vectors, a linear transformation assumes its simplest possible form. In this simple form independent modes of system behavior become apparent. As a simple example, consider the equation

$$\mathbf{y} = \mathbf{A}\mathbf{x}$$

where **A** is an $n \times n$ matrix, and **y** might be a set of time derivatives or any other $n \times 1$ vector. If a change of basis is used, $\mathbf{x} = \mathbf{M}\mathbf{z}$, $\mathbf{y} = \mathbf{M}\mathbf{w}$, where **M** is the modal matrix, then

$$\mathbf{M}\mathbf{w} = \mathbf{A}\mathbf{M}\mathbf{z} \qquad \text{or} \qquad \mathbf{w} = \mathbf{M}^{-1}\mathbf{A}\mathbf{M}\mathbf{z} = \mathbf{J}\mathbf{z}$$

J is the Jordan form in general, but will simply be the diagonal matrix $\mathbf{\Lambda}$ in many cases. The simultaneous equations are now as nearly uncoupled as is possible.

When considering the real quadratic form $Q = \langle \mathbf{x}, \mathbf{A}\mathbf{x} \rangle$, with **A** symmetric, all second-order products of the components of **x** are usually present. If a change of basis $\mathbf{x} = \mathbf{M}\mathbf{z}$ is used, then $Q = \langle \mathbf{M}\mathbf{z}, \mathbf{A}\mathbf{M}\mathbf{z} \rangle = \langle \mathbf{z}, \mathbf{M}^T\mathbf{A}\mathbf{M}\mathbf{z} \rangle$. Since **A** is real and symmetric, it can always be diagonalized by an orthogonal transformation, so $Q = \langle \mathbf{z}, \mathbf{\Lambda}\mathbf{z} \rangle$ reduces to the sum of the squares of the z_i components, weighted by the eigenvalues λ_i. This makes the relationships between eigenvalues and the various kinds of definiteness rather transparent.

4. *Sufficient conditions for relative maximum or minimum:* When considering a smooth function of a single variable on an open interval, the necessary condition for a relative maximum or minimum is that the first derivative vanish. To determine whether a maximum, minimum, or saddle point exists, the sign of the second derivative must be determined. In multidimensional cases it is again necessary that the first derivatives (all or them) vanish at a point of relative maximum or minimum. The test of the sign of the second derivative in the scalar

case is replaced by a test for positive or negative definiteness of a matrix of second derivative terms. Eigenvalue–eigenvector theory plays an important part in the investigation of these and many other questions.

Additional material related to this chapter may be found in References 23, 28, 52, 53, 99, and 117.

ILLUSTRATIVE PROBLEMS

Determination of Eigenvalues, Eigenvectors, and the Jordan Form

7.1 Find the eigenvalues and eigenvectors and then use a similarity transformation to diagonalize $\mathbf{A} = \begin{bmatrix} 0 & 1 \\ -3 & -4 \end{bmatrix}$.

The characteristic equation is $|\mathbf{A} - \mathbf{I}\lambda| = \lambda^2 + 4\lambda + 3 = 0$. Therefore $\lambda_1 = -1$, $\lambda_2 = -3$. For simple roots (Case I) compute

$$\text{Adj}\,[\mathbf{A} - \mathbf{I}\lambda] = \begin{bmatrix} -4-\lambda & -1 \\ 3 & -\lambda \end{bmatrix}$$

Substituting in $\lambda = -1$ gives $\mathbf{x}_1 = [-1 \quad 1]^T$ or any vector proportional to this. Using $\lambda = -3$ gives $\mathbf{x}_2 = [-1 \quad 3]^T$:

$$\Lambda = \begin{bmatrix} -1 & -1 \\ 1 & 3 \end{bmatrix}^{-1} \mathbf{A} \begin{bmatrix} -1 & -1 \\ 1 & 3 \end{bmatrix} = \begin{bmatrix} -1 & 0 \\ 0 & -3 \end{bmatrix}$$

7.2 Consider the eigenvalue–eigenvector problem for $\mathbf{A} = \begin{bmatrix} 1 & 2 \\ -2 & -3 \end{bmatrix}$.

The characteristic equation is $|\mathbf{A} - \mathbf{I}\lambda| = \lambda^2 + 2\lambda + 1 = (\lambda + 1)^2$. Therefore, $\lambda = -1$ with algebraic multiplicity 2. Rank $[\mathbf{A} - \mathbf{I}\lambda]|_{\lambda=-1} = 1$ and degeneracy $q = 2 - 1 = 1$. This is an example of simple degeneracy, so there is one eigenvector $\mathbf{x}_1$ and one generalized eigenvector $\mathbf{x}_2$:

$$\text{Adj}\,[\mathbf{A} - \mathbf{I}\lambda]\Big|_{\lambda=-1} = \begin{bmatrix} -2 & -2 \\ 2 & 2 \end{bmatrix}$$

Select $\mathbf{x}_1 = \begin{bmatrix} -1 \\ 1 \end{bmatrix}$.

Generalized eigenvector, method 1:

Set $\mathbf{A}\mathbf{x}_2 = -\mathbf{x}_2 + \mathbf{x}_1$ and let $\mathbf{x}_2 = [a \quad b]^T$. Then $a + 2b = -a - 1$ or $2a + 2b = -1$. We could set $a = 1$, then $b = -3/2$ and $\mathbf{x}_2 = [1 \quad -3/2]^T$, or if $a = -1$, $b = 1/2$. If $\mathbf{x}_1$ was selected as $\mathbf{x}_1 = [2 \quad -2]^T$, then $\mathbf{x}_2 = [1 \quad 0]^T$. Either of these choices is valid and each gives $\mathbf{J} = \mathbf{M}^{-1}\mathbf{A}\mathbf{M} = \begin{bmatrix} -1 & 1 \\ 0 & -1 \end{bmatrix}$.

Generalized eigenvector, method 2:

If $\mathbf{x}_1$ is selected as column 1 of Adj $[\mathbf{A} + \mathbf{I}]$, i.e., $\mathbf{x}_1 = [-2 \quad 2]^T$, then $\mathbf{x}_2$ is column 1 of the differentiated adjoint matrix

$$\mathbf{x}_2 = \frac{d}{d\lambda}\begin{bmatrix} -3-\lambda \\ 2 \end{bmatrix}\bigg|_{\lambda=-1} = \begin{bmatrix} -1 \\ 0 \end{bmatrix}$$

If $\mathbf{x}_1 = [-2 \quad 2]^T$ (column 2 of Adj $[\mathbf{A} + \mathbf{I}]$), then $\mathbf{x}_2$ is column 2 of the differentiated adjoint matrix

$$\mathbf{x}_2 = \frac{d}{d\lambda}\begin{bmatrix} -2 \\ 1-\lambda \end{bmatrix}\bigg|_{\lambda=-1} = \begin{bmatrix} 0 \\ -1 \end{bmatrix}$$

Eigenvector and generalized eigenvector, method 3:

Since $n = m = 2$, we seek the smallest k such that rank $[\mathbf{A} - \mathbf{I}\lambda]^k|_{\lambda_1=-1} = 0$. The index k is 2. Then $(\mathbf{A} - \mathbf{I}\lambda_1)^2\mathbf{x} = \mathbf{0}$ has two independent solutions, one of which is $\mathbf{x}_2 = [1 \quad 0]^T$. Since $[\mathbf{A} - \mathbf{I}\lambda_1]\mathbf{x}_2 = [2 \quad -2]^T \neq \mathbf{0}$, $\mathbf{x}_2$ is a generalized eigenvector and $\mathbf{x}_1$ is the eigenvector. Alternatively, one could select $\mathbf{x}_2 = [0 \quad 1]^T$ and this also leads to $\mathbf{x}_1 = [2 \quad -2]^T$. All of these methods give the same Jordan form.

7.3 Find the eigenvalues–eigenvectors and the Jordan form for $\mathbf{A} = \begin{bmatrix} 1 & 1 \\ 1 & 1 \end{bmatrix}$.

The characteristic equation is $|\mathbf{A} - \mathbf{I}\lambda| = \lambda^2 - 2\lambda = \lambda(\lambda - 2)$. Therefore $\lambda_1 = 0$, $\lambda_2 = 2$. Since $\lambda_1 \neq \lambda_2$, Case I applies:

$$\text{Adj}\,[\mathbf{A} - \mathbf{I}\lambda] = \begin{bmatrix} 1-\lambda & -1 \\ -1 & 1-\lambda \end{bmatrix}$$

Using the first column with $\lambda = 0$ gives $\mathbf{x}_1 = [1 \quad -1]^T$. Using the first column with $\lambda = 2$ gives $\mathbf{x}_2 = [-1 \quad -1]^T$ or $[1 \quad 1]^T$:

$$\mathbf{M} = \begin{bmatrix} 1 & 1 \\ -1 & 1 \end{bmatrix}, \qquad \mathbf{M}^{-1} = \frac{1}{2}\begin{bmatrix} 1 & -1 \\ 1 & 1 \end{bmatrix}, \qquad \mathbf{J} = \mathbf{M}^{-1}\mathbf{A}\mathbf{M} = \begin{bmatrix} 0 & 0 \\ 0 & 2 \end{bmatrix}$$

7.4 Find the eigenvalues, eigenvectors, and Jordan form for

$$\mathbf{A} = \begin{bmatrix} 1 & 0 & 0 & -3 \\ 0 & 1 & -3 & 0 \\ -0.5 & -3 & 1 & 0.5 \\ -3 & 0 & 0 & 1 \end{bmatrix}$$

The characteristic equation is $\lambda^4 - 4\lambda^3 - 12\lambda^2 + 32\lambda + 64 = 0$. The roots are found to be $\lambda_i = -2, -2, 4, 4$. With $\lambda = -2$

$$\mathbf{A} - \lambda\mathbf{I} = \mathbf{A} + 2\mathbf{I} = \begin{bmatrix} 3 & 0 & 0 & -3 \\ 0 & 3 & -3 & 0 \\ -0.5 & -3 & 3 & 0.5 \\ -3 & 0 & 0 & 3 \end{bmatrix}$$

This matrix has rank 2, so $q = n - r = 2$. This shows that there are two eigenvectors associated with $\lambda = -2$, and since the multiplicity of that root is also 2, no generalized eigenvectors are needed for this eigenvalue. Two linearly independent solutions of $[\mathbf{A} + 2\mathbf{I}]\mathbf{x}_i = \mathbf{0}$ are $\mathbf{x}_1 = [0 \quad 1 \quad 1 \quad 0]^T$ and $\mathbf{x}_2 = [1 \quad 0 \quad 0 \quad 1]^T$. With $\lambda = 4$, $\mathbf{A} - \lambda\mathbf{I}$

$$= \begin{bmatrix} -3 & 0 & 0 & -3 \\ 0 & -3 & -3 & 0 \\ -0.5 & -3 & -3 & 0.5 \\ -3 & 0 & 0 & -3 \end{bmatrix}.$$

The rank is 3 and $q = 1$. There is only one eigenvector,

and since the algebraic multiplicity $m = 2$, a generalized eigenvector is needed. Solving $[\mathbf{A} - 4\mathbf{I}]\mathbf{x}_i = \mathbf{0}$ gives only one independent solution, $\mathbf{x}_3 = [0 \quad 1 \quad -1 \quad 0]^T$. Therefore, a generalized eigenvector is needed and it can be found by solving

$$[\mathbf{A} - 4\mathbf{I}]\mathbf{x}_4 = \mathbf{x}_3.$$

The result is $\mathbf{x}_4 = [2 \quad -1/3 \quad 0 \quad -2]^T$. The modal matrix is thus

$$\mathbf{M} = [\mathbf{x}_1 \quad \mathbf{x}_2 \quad \mathbf{x}_3 \quad \mathbf{x}_4].$$

and the Jordan form is

$$\mathbf{J} = \begin{bmatrix} -2 & 0 & 0 & 0 \\ 0 & -2 & 0 & 0 \\ 0 & 0 & 4 & 1 \\ 0 & 0 & 0 & 4 \end{bmatrix}$$

7.5 A is a 5×5 matrix for which the following information has been found:

$$\lambda_1 = \lambda_2 = 2, \qquad \text{Rank}\,[\mathbf{A} - 2\mathbf{I}] = 4$$

$$\lambda_3 = \lambda_4 = \lambda_5 = -2, \qquad \text{Rank}\,[\mathbf{A} + 2\mathbf{I}] = 3$$

Determine the Jordan form for **A**.

For $\lambda = 2$, the degeneracy is $q_1 = 1$, so there is a single Jordan block $\mathbf{J}_1 = \begin{bmatrix} 2 & 1 \\ 0 & 2 \end{bmatrix}$. For $\lambda = -2$, the degeneracy is $q_3 = 2$. Thus there are two Jordan blocks $\mathbf{J}_2 = \begin{bmatrix} -2 & 1 \\ 0 & -2 \end{bmatrix}$ and $\mathbf{J}_3 = [-2]$. The arrangement of these blocks within the Jordan form depends upon the ordering of the eigenvectors and generalized eigenvectors within **M**. Assuming that $\mathbf{x}_1$ and $\mathbf{x}_2$ are an eigenvector and generalized eigenvector for $\lambda = 2$, $\mathbf{x}_3$ and $\mathbf{x}_4$ are an eigenvector and generalized eigenvector for $\lambda = -2$, and $\mathbf{x}_5$ is the second eigenvector associated with $\lambda = -2$, then

$$\text{diag}\,[\mathbf{J}_1, \mathbf{J}_2, \mathbf{J}_3] = \begin{bmatrix} 2 & 1 & & & \\ 0 & 2 & & \mathbf{0} & \\ & & -2 & 1 & \\ & \mathbf{0} & 0 & -2 & \\ & & & & -2 \end{bmatrix} = \mathbf{J}$$

7.6 Let

$$\mathbf{A} = \begin{bmatrix} 3 & 1 & 0 & 0 & 0 & 0 & 0 \\ 0 & 3 & 0 & 0 & 0 & 0 & 0 \\ 0 & 0 & 3 & 0 & 0 & 0 & 0 \\ 0 & 0 & 0 & 4 & 1 & 0 & 0 \\ 0 & 0 & 0 & 0 & 4 & 0 & 0 \\ 0 & 0 & 0 & 0 & 0 & 4 & 1 \\ 0 & 0 & 0 & 0 & 0 & 0 & 4 \end{bmatrix}$$

a. What are the eigenvalues? b. How many linearly independent eigenvectors does **A** have? c. How many generalized eigenvectors?

a. **A** is upper triangular and so is $\mathbf{A} - \mathbf{I}\lambda$. Thus the eigenvalues are the diagonal elements of **A**, $\lambda = 3, 3, 3, 4, 4, 4, 4$. b. The matrix **A** is already in Jordan form with four Jordan blocks. There are four eigenvectors. c. There are three generalized eigenvectors. The number of "ones" above the main diagonal is always equal to the number of generalized eigenvectors.

7.7 Find the eigenvalues, eigenvectors, and, if needed, the generalized eigenvectors for $\mathbf{A} = \begin{bmatrix} 4 & 2 & 1 \\ 0 & 6 & 1 \\ 0 & -4 & 2 \end{bmatrix}$. Also find the Jordan form.

The characteristic equation is $|\mathbf{A} - \mathbf{I}\lambda| = (4 - \lambda)^3$. Then $\lambda_1 = \lambda_2 = \lambda_3 = 4$ with algebraic multiplicity 3.

$[\mathbf{A} - 4\mathbf{I}] = \begin{bmatrix} 0 & 2 & 1 \\ 0 & 2 & 1 \\ 0 & -4 & -2 \end{bmatrix}$ has rank $r = 1$. The degeneracy is $q = 2$. This is an example of the general Case II_3 with two eigenvectors and one generalized eigenvector.

Method 1:

$$\text{Adj}\,[\mathbf{A} - \mathbf{I}\lambda] = \begin{bmatrix} (6-\lambda)(2-\lambda)+4 & 2\lambda - 8 & \lambda - 4 \\ 0 & (4-\lambda)(2-\lambda) & \lambda - 4 \\ 0 & 16 - 4\lambda & (4-\lambda)(6-\lambda) \end{bmatrix}$$

With $\lambda = 4$, this reduces to the null matrix.

$$\frac{d}{d\lambda}\{\text{Adj}\,[\mathbf{A} - \mathbf{I}\lambda]\}\Big|_{\lambda=4} = \begin{bmatrix} 2\lambda - 8 & 2 & 1 \\ 0 & 2\lambda - 6 & 1 \\ 0 & -4 & 2\lambda - 10 \end{bmatrix}\Bigg|_{\lambda=4} = \begin{bmatrix} 0 & 2 & 1 \\ 0 & 2 & 1 \\ 0 & -4 & -2 \end{bmatrix}$$

One eigenvector can be selected as $\mathbf{x}_2 = [1 \quad 1 \quad -2]^T$.

$$\frac{1}{2}\frac{d^2}{d\lambda^2}\{\text{Adj}\,[\mathbf{A} - \mathbf{I}\lambda]\}\Big|_{\lambda=4} = \frac{1}{2}\begin{bmatrix} 2 & 0 & 0 \\ 0 & 2 & 0 \\ 0 & 0 & 2 \end{bmatrix}$$

The first column can be selected as another eigenvector, call it $\mathbf{x}_1 = [1 \quad 0 \quad 0]^T$. A generalized eigenvector is given by the third column $\mathbf{x}_3 = [0 \quad 0 \quad 1]^T$. Note that $\mathbf{x}_1$ and $\mathbf{x}_2$ are selected as columns which are nonzero for the first time as the adjoint matrix is repeatedly differentiated. Note also that $\mathbf{x}_3$ is selected from the same column that gave $\mathbf{x}_2$, but with one more differentiation.

Method 2:

We require two independent solutions of $[\mathbf{A} - 4\mathbf{I}]\mathbf{x} = \mathbf{0}$. Let $\mathbf{x} = [a \quad b \quad c]^T$. Then $2b + c = 0$ is the only restriction placed on a, b and c. Since a is arbitrary, set $a = 1$ and $b = c = 0$, or $\mathbf{x}_1 = [1 \quad 0 \quad 0]^T$. Another solution is $a = 1$, $b = 1$, $c = -2$, or $\mathbf{x}_2 = [1 \quad 1 \quad -2]^T$.

A generalized eigenvector is needed, and it must satisfy $(\mathbf{A} - 4\mathbf{I})\mathbf{x}_3 = \mathbf{x}_2$ or $(\mathbf{A} - 4\mathbf{I})^2\mathbf{x}_3 = (\mathbf{A} - 4\mathbf{I})\mathbf{x}_2 = \mathbf{0}$. This reduces to $[\mathbf{0}]\mathbf{x}_3 = \mathbf{0}$, so $\mathbf{x}_3$ is arbitrary, except it must be nonzero and linearly independent of $\mathbf{x}_1$ and $\mathbf{x}_2$. $\mathbf{x}_3 = [0 \quad 0 \quad 1]^T$ is one such vector.

Setting $\mathbf{M} = \begin{bmatrix} 1 & 1 & 0 \\ 0 & 1 & 0 \\ 0 & -2 & 1 \end{bmatrix}$ gives $\mathbf{M}^{-1} = \begin{bmatrix} 1 & -1 & 0 \\ 0 & 1 & 0 \\ 0 & 2 & 1 \end{bmatrix}$ so that $\mathbf{J} = \mathbf{M}^{-1}\mathbf{A}\mathbf{M} = \left[\begin{array}{c|cc} 4 & 0 & 0 \\ \hline 0 & 4 & 1 \\ 0 & 0 & 4 \end{array}\right]$.

Alternatively, the index of $\lambda = 4$ is $k = 2$ since rank $[\mathbf{A} - 4\mathbf{I}]^2 = n - m = 0$. A generalized eigenvector satisfying $[\mathbf{A} - 4\mathbf{I}]^2\mathbf{x}_3 = \mathbf{0}$, $[\mathbf{A} - 4\mathbf{I}]\mathbf{x}_3 \neq \mathbf{0}$ is $\mathbf{x}_3 = [0 \quad 0 \quad 1]^T$. Then $\mathbf{x}_2 = [\mathbf{A} - 4\mathbf{I}]\mathbf{x}_3 = [1 \quad 1 \quad -2]^T$. This is one eigenvector, with the other one obviously being $\mathbf{x}_1 = [1 \quad 0 \quad 0]^T$.

Similar Matrices

7.8 Prove that two similar matrices have the same eigenvalues.

A and B are similar matrices if they are related by $\mathbf{A} = \mathbf{Q}^{-1}\mathbf{B}\mathbf{Q}$ for some nonsingular matrix $\mathbf{Q}$. The characteristic equation for $\mathbf{A}$ is

$$|\mathbf{A} - \mathbf{I}\lambda| = |\mathbf{Q}^{-1}\mathbf{B}\mathbf{Q} - \mathbf{Q}^{-1}\mathbf{Q}\lambda| = 0$$

or

$$|\mathbf{Q}^{-1}[\mathbf{B} - \mathbf{I}\lambda]\mathbf{Q}| = |\mathbf{Q}^{-1}|\cdot|\mathbf{Q}|\,|\mathbf{B} - \mathbf{I}\lambda| = |\mathbf{B} - \mathbf{I}\lambda| = 0$$

The characteristic equations for $\mathbf{A}$ and $\mathbf{B}$ are the same so they have the same eigenvalues.

7.9 Are the following matrices similar?

$$\begin{bmatrix} 2 & 0 & 0 & 0 \\ 0 & 2 & 0 & 0 \\ 0 & 0 & 2 & 0 \\ 0 & 0 & 0 & 2 \end{bmatrix}, \quad \begin{bmatrix} 2 & 1 & 0 & 0 \\ 0 & 2 & 0 & 0 \\ 0 & 0 & 2 & 1 \\ 0 & 0 & 0 & 2 \end{bmatrix}, \quad \begin{bmatrix} 2 & 0 & 0 & 0 \\ 0 & 2 & 1 & 0 \\ 0 & 0 & 2 & 1 \\ 0 & 0 & 0 & 2 \end{bmatrix}, \quad \begin{bmatrix} 2 & 1 & 0 & 0 \\ 0 & 2 & 1 & 0 \\ 0 & 0 & 2 & 1 \\ 0 & 0 & 0 & 2 \end{bmatrix}$$

All four matrices have the same characteristic equation $(2 - \lambda)^4 = 0$, so $\lambda = 2$ is the eigenvalue with algebraic multiplicity $m = 4$. The given matrices are expressed in Jordan form. Since similar matrices must have the same Jordan form, the answer is no. Note that the index of the eigenvalue is 1, 2, 3, and 4, respectively, for these matrices.

Miscellaneous Properties

7.10 Prove that $|\mathbf{A}| = \lambda_1\lambda_2\cdots\lambda_n$.

Any $n \times n$ matrix $\mathbf{A}$ can be reduced to the Jordan form $\mathbf{J} = \mathbf{M}^{-1}\mathbf{A}\mathbf{M}$; so $\mathbf{A} = \mathbf{M}\mathbf{J}\mathbf{M}^{-1}$. From this $|\mathbf{A}| = |\mathbf{M}\mathbf{J}\mathbf{M}^{-1}| = |\mathbf{M}||\mathbf{J}||\mathbf{M}^{-1}| = |\mathbf{J}|$. Since $\mathbf{J}$ is upper triangular, with the eigenvalues on the main diagonal, $|\mathbf{A}| = |\mathbf{J}| = \lambda_1\lambda_2 \cdots \lambda_n$. Thus $|\mathbf{A}| = 0 \Leftrightarrow$ at least one $\lambda_i = 0$.

7.11 Prove $\mathrm{Tr}(\mathbf{A}) = \lambda_1 + \lambda_2 + \cdots + \lambda_n$.

Since $\mathbf{A} = \mathbf{M}\mathbf{J}\mathbf{M}^{-1}$, $\mathrm{Tr}(\mathbf{A}) = \mathrm{Tr}(\mathbf{M}\mathbf{J}\mathbf{M}^{-1})$.

But $\mathrm{Tr}(\mathbf{AB}) = \mathrm{Tr}(\mathbf{BA})$, so $\mathrm{Tr}(\mathbf{A}) = \mathrm{Tr}(\mathbf{J}\mathbf{M}^{-1}\mathbf{M})$ or $\mathrm{Tr}(\mathbf{A}) = \mathrm{Tr}(\mathbf{J}) = \lambda_1 + \lambda_2 + \cdots + \lambda_n$.

7.12 Prove that if $\mathbf{A}$ is nonsingular with eigenvalues λ_i, then $1/\lambda_i$ are the eigenvalues of $\mathbf{A}^{-1}$.

The n roots λ_i are defined by $|\mathbf{A} - \mathbf{I}\lambda| = 0$. But $|\mathbf{A} - \mathbf{I}\lambda| = |\mathbf{A}[\mathbf{I} - \mathbf{A}^{-1}\lambda]| = |\mathbf{A}|\cdot|\mathbf{I} - \mathbf{A}^{-1}\lambda| = 0$. Since $\mathbf{A}$ is nonsingular, $|\mathbf{A}|$ can be divided out leaving $|\mathbf{I} - \mathbf{A}^{-1}\lambda| = |\mathbf{I}(1/\lambda) - \mathbf{A}^{-1}|\lambda^n = 0$. Since $|\mathbf{A}|$, and hence λ, are not zero, the characteristic equation for $\mathbf{A}$ leads to $|\mathbf{A}^{-1} - \mathbf{I}(1/\lambda)| = 0$. Thus if λ_i is an eigenvalue of $\mathbf{A}$, then $1/\lambda_i$ is an eigenvalue of $\mathbf{A}^{-1}$.

7.13 Let $\mathbf{A}$ be an $n \times n$ matrix with n distinct eigenvalues. Prove that the set of n eigenvectors $\mathbf{x}_i$ are linearly independent.

Let

$$a_1\mathbf{x}_1 + a_2\mathbf{x}_2 + \cdots + a_n\mathbf{x}_n = \mathbf{0} \tag{1}$$

If it can be shown that this implies that $a_1 = a_2 = \cdots = a_n = 0$, then the set $\{\mathbf{x}_i\}$ is linearly independent. Define $\mathbf{T}_i = \mathbf{A} - \mathbf{I}\lambda_i$ and note that $\mathbf{T}_i\mathbf{x}_i = \mathbf{0}$. $\mathbf{T}_i\mathbf{x}_j = (\lambda_j - \lambda_i)\mathbf{x}_j$ if $i \neq j$. Multiplying equation *(1)* by $\mathbf{T}_1$ gives

$$a_2(\lambda_2 - \lambda_1)\mathbf{x}_2 + a_3(\lambda_3 - \lambda_1)\mathbf{x}_3 + \cdots + a_n(\lambda_n - \lambda_1)\mathbf{x}_n = \mathbf{0}$$

Multiplying this in turn by $\mathbf{T}_2$, then $\mathbf{T}_3, \ldots, \mathbf{T}_{n-1}$ gives

$$a_3(\lambda_3 - \lambda_1)(\lambda_3 - \lambda_2)\mathbf{x}_3 + \cdots + a_n(\lambda_n - \lambda_1)(\lambda_n - \lambda_2)\mathbf{x}_n = \mathbf{0}$$

$$\vdots$$

$$a_{n-1}(\lambda_{n-1} - \lambda_1)(\lambda_{n-1} - \lambda_2) \cdots (\lambda_{n-1} - \lambda_{n-2})\mathbf{x}_{n-1} + a_n(\lambda_n - \lambda_1)(\lambda_n - \lambda_2) \cdots (\lambda_n - \lambda_{n-2})\mathbf{x}_n = \mathbf{0} \tag{2}$$

$$a_n(\lambda_n - \lambda_1)(\lambda_n - \lambda_2) \cdots (\lambda_n - \lambda_{n-2})(\lambda_n - \lambda_{n-1})\mathbf{x}_n = \mathbf{0} \tag{3}$$

Since $\mathbf{x}_n \neq \mathbf{0}$, and $\lambda_n \neq \lambda_i$ for $i \neq n$, equation *(3)* requires that $a_n = 0$. This plus equation *(2)* requires that $a_{n-1} = 0$. Continuing this reasoning shows that equation *(1)* requires $a_i = 0$ for $i = 1, 2, \ldots, n$, so the eigenvectors are linearly independent.

7.14 Let $\mathbf{T}_i\colon \mathcal{X} \longrightarrow \mathcal{X}$ be defined by $\mathbf{T}_i = \mathbf{A} - \mathbf{I}\lambda_i$, where $\mathcal{X}$ is an n-dimensional space. Prove that there are always $q_i = n - \text{rank}(\mathbf{T}_i)$ linearly independent eigenvectors associated with eigenvalue λ_i.

The space $\mathcal{X}$ can be written as the direct sum $\mathcal{X} = \mathfrak{R}(\mathbf{T}_i^*) \oplus \mathfrak{N}(\mathbf{T}_i)$ and $\dim(\mathcal{X}) = \dim(\mathfrak{R}(\mathbf{T}_i^*)) + \dim(\mathfrak{N}(\mathbf{T}_i))$ or $n = \text{rank}(\mathbf{T}_i^*) + \dim(\mathfrak{N}(\mathbf{T}_i))$. But since $\text{rank}(\mathbf{T}_i) = \text{rank}(\mathbf{T}_i^*)$, $n - \text{rank}(\mathbf{T}_i) = q_i = \dim(\mathfrak{N}(\mathbf{T}_i))$. The null space of $\mathbf{T}_i$ is of dimension q_i and, therefore, it contains q_i linearly independent vectors, all of which are eigenvectors.

7.15 Let $\mathbf{A}$ be an arbitrary $n \times r$ matrix and let $\mathbf{B}$ be an arbitrary $r \times n$ matrix, so that $\mathbf{AB}$ and $\mathbf{BA}$ are $n \times n$ and $r \times r$ matrices respectively. Assume that $n \geq r$ and prove:

a. The scalar λ is a nonzero eigenvalue of $\mathbf{AB}$ if and only if it is a nonzero eigenvalue of $\mathbf{BA}$.

b. If $\mathbf{x}_i$ is an eigenvector (or generalized eigenvector) of $\mathbf{AB}$ associated with a nonzero eigenvalue, then $\zeta_i \triangleq \mathbf{B}\mathbf{x}_i$ is an eigenvector (or generalized eigenvector) or $\mathbf{BA}$.

c. $\mathbf{AB}$ has at least $n - r$ zero eigenvalues.

Assume $\mathbf{AB}\mathbf{x}_i = \lambda\mathbf{x}_i$ with $\lambda \neq 0$, $\mathbf{x}_i \neq \mathbf{0}$. Then multiplying by $\mathbf{B}$ gives $\mathbf{BA}(\mathbf{B}\mathbf{x}_i) = \lambda\mathbf{B}\mathbf{x}_i$ or $\mathbf{BA}\zeta_i = \lambda\zeta_i$. Thus λ and ζ_i are an eigenvalue and eigenvector of $\mathbf{BA}$ provided $\zeta_i \neq \mathbf{0}$. But since $\lambda\mathbf{x}_i \neq \mathbf{0}$, $\mathbf{B}\mathbf{x}_i \neq \mathbf{0}$; otherwise $\mathbf{AB}\mathbf{x}_i = \mathbf{0}$. Therefore, λ and ζ_i are an eigenvalue and eigenvector of $\mathbf{BA}$, provided $\lambda \neq 0$ and $\mathbf{x}_i$ are an eigenvalue and eigenvector of $\mathbf{AB}$. Now assume $\mathbf{BA}\zeta_i = \lambda\zeta_i$, $\lambda \neq 0$, and $\zeta_i \neq \mathbf{0}$. Using the same kind of arguments as above show that λ and $\mathbf{x}_i = \mathbf{A}\zeta_i$ are an eigenvalue and eigenvector of $\mathbf{AB}$. These results can be generalized for the case of generalized eigenvectors. This proves a and b. To prove c, it is only necessary to note that $\mathbf{AB}$ has n eigenvalues and each nonzero eigenvalue is simultaneously an eigenvalue of $\mathbf{BA}$. Since $\mathbf{BA}$ has r eigenvalues, $\mathbf{AB}$ has at most r nonzero eigenvalues and, therefore, at least $n - r$ zero eigenvalues.

7.16 Let $\mathbf{A}$ and $\mathbf{B}$ be defined as in Problem 7.15. Define $\mathbf{N} = \mathbf{I}_n + \mathbf{AB}$ and $\mathbf{R} = \mathbf{I}_r + \mathbf{BA}$. Prove:

a. $\mathbf{x}_i$ is an eigenvector of $\mathbf{AB}$ if and only if it is an eigenvector of $\mathbf{N}$ and λ is an eigenvalue of $\mathbf{AB}$ if and only if $1 + \lambda$ is an eigenvalue of $\mathbf{N}$.
b. ζ_i is an eigenvector of $\mathbf{BA}$ if and only if it is an eigenvector of $\mathbf{R}$, and λ is an eigenvalue of $\mathbf{BA}$ if and only if $1 + \lambda$ is an eigenvalue of $\mathbf{R}$.

a. The proof requires showing that

$$(\mathbf{AB} - \lambda_i \mathbf{I}_n)\mathbf{x}_i = \mathbf{0} \Leftrightarrow (\mathbf{N} - (\lambda_i + 1)\mathbf{I}_n)\mathbf{x}_i = \mathbf{0}$$

But $\mathbf{N} - (\lambda_i + 1)\mathbf{I}_n = \mathbf{AB} + \mathbf{I}_n - \lambda_i \mathbf{I}_n - \mathbf{I}_n = \mathbf{AB} - \lambda_i \mathbf{I}_n$.

b. The proof is a simple matter of applying the definitions of the two eigenvalue problems in question, just as in part a.

7.17 Show that the r eigenvalues of $\mathbf{R}$ defined in Problem 7.16 are also eigenvalues of $\mathbf{N}$. The remaining $n - r$ eigenvalues of $\mathbf{N}$ are all equal to one.

If λ is an eigenvalue of $\mathbf{BA}$, then it is also an eigenvalue of $\mathbf{AB}$. Since the eigenvalues of $\mathbf{N}$ and $\mathbf{R}$ are shifted by one from these eigenvalues, the result is proven for the r eigenvalues of $\mathbf{BA}$. It was shown in Problem 7.15 that the remaining $n - r$ eigenvalues of $\mathbf{AB}$ must be zero, so the corresponding eigenvalues of $\mathbf{N}$ are one.

7.18 Let $\mathbf{A}$ and $\mathbf{B}$ be as defined in Problem 7.15. Prove the determinant identity of Problem 3.5, page 83, i.e. prove $|\mathbf{I}_n \pm \mathbf{AB}| = |\mathbf{I}_r \pm \mathbf{BA}|$.

Since the determinant is equal to the product of the eigenvalues (see Problem 7.10) and since $\mathbf{N} = \mathbf{I}_n + \mathbf{AB}$ and $\mathbf{R} = \mathbf{I}_r + \mathbf{BA}$ have been shown to have r eigenvalues in common with the rest of the $n - r$ eigenvalues equal to one, the identity is proven for the plus sign. The identity also holds for the minus sign, since we can always define $\mathbf{B}_1 = -\mathbf{B}$ or $\mathbf{A}_1 = -\mathbf{A}$. The previous results placed no restrictions on $\mathbf{A}$ and $\mathbf{B}$ other than their dimensions. Another form of the same identity is

$$|\mathbf{A}_1\mathbf{B} \pm \lambda \mathbf{I}_n| = (\pm\lambda)^{n-r}\,|\mathbf{BA}_1 \pm \lambda \mathbf{I}_r|$$

which is established by defining $\lambda\mathbf{A} = \mathbf{A}_1$ and using the rule for multiplying a determinant by a scalar.

7.19 Let $\mathbf{P}$ be a nonsingular $n \times n$ matrix whose determinant and inverse are known. Let $\mathbf{C}$ and $\mathbf{D}$ be arbitrary $n \times r$ and $r \times n$ matrices, respectively. Show that $|\mathbf{P} + \mathbf{CD}| = |\mathbf{P}|\cdot|\mathbf{I}_r + \mathbf{DP}^{-1}\mathbf{C}|$.

Simple manipulations show $|\mathbf{P} + \mathbf{CD}| = |\mathbf{P}[\mathbf{I}_n + \mathbf{P}^{-1}\mathbf{CD}]| = |\mathbf{P}|\cdot|\mathbf{I}_n + \mathbf{P}^{-1}\mathbf{CD}|$. Using the result of Problem 7.18 allows the interchange of factors $(\mathbf{P}^{-1}\mathbf{C})$ and $(\mathbf{D})$ and the corresponding change from an $n \times n$ determinatnt to an $r \times r$ determinant.

7.20 Let $\mathbf{N}$ be an $n \times n$ matrix given by $\mathbf{N} = \mathbf{I}_n + \mathbf{AB}$, where $\mathbf{A}$ and $\mathbf{B}$ are $n \times 1$ and $1 \times n$ matrices. Show that the $n \times n$ determinant can be expressed in terms of the easily evaluated trace: $|\mathbf{N}| = \mathrm{Tr}\,(\mathbf{N}) + 1 - n$.

Since $|\mathbf{N}| = |\mathbf{I}_n + \mathbf{AB}| = |\mathbf{I}_r + \mathbf{BA}|$ and since in this case the dimension r of $\mathbf{BA}$ is one, $|\mathbf{N}| = 1 + \mathbf{BA} = 1 + \mathrm{Tr}\,(\mathbf{BA})$. But $\mathrm{Tr}\,(\mathbf{BA}) = \mathrm{Tr}\,(\mathbf{AB})$ and $\mathrm{Tr}\,(\mathbf{N}) = \mathrm{Tr}\,(\mathbf{AB}) + \mathrm{Tr}\,(\mathbf{I}_n)$ or $\mathrm{Tr}\,(\mathbf{AB}) = \mathrm{Tr}\,(\mathbf{N}) - n$, so $|\mathbf{N}| = 1 + \mathrm{Tr}\,(\mathbf{N}) - n$.

Self-Adjoint Transformation

7.21 If $\mathcal{A}$ is a self-adjoint transformation (see Sec. 6.6), show that all of its eigenvalues are real and that the eigenvectors associated with two different eigenvalues are orthogonal.

Consider $\mathcal{A}(\mathbf{x}_i) = \lambda_i \mathbf{x}_i$ and form the inner product $\langle \mathbf{x}_i, \mathcal{A}(\mathbf{x}_i)\rangle = \langle \mathbf{x}_i, \lambda_i \mathbf{x}_i\rangle =$

$\lambda_i\langle \mathbf{x}_i, \mathbf{x}_i\rangle$. The definition of $\mathcal{A}^*$ ensures that $\langle \mathbf{x}_i, \mathcal{A}(\mathbf{x}_i)\rangle = \langle \mathcal{A}^*(\mathbf{x}_i), \mathbf{x}_i\rangle$ and if $\mathcal{A} = \mathcal{A}^*$, this gives $\langle \lambda_i\mathbf{x}_i, \mathbf{x}_i\rangle = \bar{\lambda}_i\langle \mathbf{x}_i, \mathbf{x}_i\rangle$. Subtracting gives $0 = (\bar{\lambda}_i - \lambda_i)\langle \mathbf{x}_i, \mathbf{x}_i\rangle$. Since $\mathbf{x}_i$ is an eigenvector $\|\mathbf{x}_i\|^2 \neq 0$, so $\bar{\lambda}_i = \lambda_i$ and all eigenvalues are real.

Now consider $\mathcal{A}(\mathbf{x}_i) = \lambda_i\mathbf{x}_i$, $\mathcal{A}(\mathbf{x}_j) = \lambda_j\mathbf{x}_j$ with $\lambda_i \neq \lambda_j$. Then $\langle \mathbf{x}_j, \mathcal{A}(\mathbf{x}_i)\rangle = \lambda_i\langle \mathbf{x}_j, \mathbf{x}_i\rangle$. Also $\langle \mathbf{x}_j, \mathcal{A}(\mathbf{x}_i)\rangle = \langle \mathcal{A}^*(\mathbf{x}_j), \mathbf{x}_i\rangle = \langle \mathcal{A}(\mathbf{x}_j), \mathbf{x}_i\rangle = \bar{\lambda}_j\langle \mathbf{x}_j, \mathbf{x}_i\rangle$. But $\bar{\lambda}_j = \lambda_j$, so subtracting gives $0 = (\lambda_i - \lambda_j)\langle \mathbf{x}_j, \mathbf{x}_i\rangle$. Since $\lambda_i \neq \lambda_j$, we have $\langle \mathbf{x}_j, \mathbf{x}_i\rangle = 0$ and $\mathbf{x}_j$ is orthogonal to $\mathbf{x}_i$.

Normal Transformation

7.22 Let $\mathcal{A}$ be a normal transformation (see Sec. 6.6). Prove that $\mathcal{A}(\mathbf{x}_i) = \lambda_i\mathbf{x}_i$ if and only if $\mathcal{A}^*(\mathbf{x}_i) = \bar{\lambda}_i\mathbf{x}_i$.

This is equivalent to showing $(\mathcal{A} - \mathcal{I}\lambda_i)\mathbf{x}_i = \mathbf{0} \Leftrightarrow (\mathcal{A}^* - \mathcal{I}\bar{\lambda}_i)\mathbf{x}_i = \mathbf{0}$.

$$\begin{aligned}\langle(\mathcal{A} - \mathcal{I}\lambda_i)\mathbf{x}_i, (\mathcal{A} - \mathcal{I}\lambda_i)\mathbf{x}_i\rangle &= \langle \mathcal{A}(\mathbf{x}_i), \mathcal{A}(\mathbf{x}_i)\rangle - \langle \lambda_i\mathbf{x}_i, \mathcal{A}(\mathbf{x}_i)\rangle \\ &\quad - \langle \mathcal{A}(\mathbf{x}_i), \lambda_i\mathbf{x}_i\rangle + \langle \lambda_i\mathbf{x}_i, \lambda_i\mathbf{x}_i\rangle \\ &= \langle \mathcal{A}^*\mathcal{A}(\mathbf{x}_i), \mathbf{x}_i\rangle - \bar{\lambda}_i\langle \mathcal{A}^*(\mathbf{x}_i), \mathbf{x}_i\rangle \\ &\quad - \lambda_i\langle \mathbf{x}_i, \mathcal{A}^*(\mathbf{x}_i)\rangle + \bar{\lambda}_i\lambda_i\langle \mathbf{x}_i, \mathbf{x}_i\rangle\end{aligned}$$

Using $\mathcal{A}^*\mathcal{A} = \mathcal{A}\mathcal{A}^*$ allows this to be rewritten as

$$\begin{aligned}&\langle \mathcal{A}^*(\mathbf{x}_i), \mathcal{A}^*(\mathbf{x}_i)\rangle - \langle \mathcal{A}^*(\mathbf{x}_i). \bar{\lambda}_i\mathbf{x}_i\rangle - \langle \bar{\lambda}_i\mathbf{x}_i, \mathcal{A}^*(\mathbf{x}_i)\rangle + \langle \bar{\lambda}_i\mathbf{x}_i, \bar{\lambda}_i\mathbf{x}_i\rangle \\ &\qquad = \langle(\mathcal{A}^* - \mathcal{I}\bar{\lambda}_i)\mathbf{x}_i, (\mathcal{A}^* - \mathcal{I}\bar{\lambda}_i)\mathbf{x}_i\rangle\end{aligned}$$

or

$$\|(\mathcal{A} - \mathcal{I}\lambda_i)\mathbf{x}_i\|^2 = \|(\mathcal{A}^* - \mathcal{I}\bar{\lambda}_i)\mathbf{x}_i\|^2$$

The desired result follows.

7.23 Prove that the eigenvectors $\mathbf{x}_i$ and $\mathbf{x}_j$ associated with eigenvalues λ_i and λ_j are orthogonal for any normal transformation, provided $\lambda_i \neq \lambda_j$.

A normal transformation satisfies $\mathcal{A}^*\mathcal{A} = \mathcal{A}\mathcal{A}^*$, and therefore the class of normal transformations includes self-adjoint transformations as a subclass. Consider $\mathcal{A}(\mathbf{x}_i) = \lambda_i\mathbf{x}_i$ and $\mathcal{A}(\mathbf{x}_j) = \lambda_j\mathbf{x}_j$ with $\lambda_i \neq \lambda_j$. Then

$$\langle \mathbf{x}_j, \mathcal{A}(\mathbf{x}_i)\rangle = \lambda_i\langle \mathbf{x}_j, \mathbf{x}_i\rangle = \langle \mathcal{A}^*(\mathbf{x}_j), \mathbf{x}_i\rangle \tag{1}$$

From the previous problem $\mathcal{A}^*(\mathbf{x}_j) = \bar{\lambda}_j\mathbf{x}_j$, so

$$\langle \mathcal{A}^*(\mathbf{x}_j), \mathbf{x}_i\rangle = \langle \bar{\lambda}_j\mathbf{x}_j, \mathbf{x}_i\rangle = \lambda_j\langle \mathbf{x}_j, \mathbf{x}_i\rangle \tag{2}$$

Subtracting equation (*2*) from equation (*1*) gives $0 = (\lambda_i - \lambda_j)\langle \mathbf{x}_j, \mathbf{x}_i\rangle$. Since $\lambda_i \neq \lambda_j$, it follows that $\langle \mathbf{x}_j, \mathbf{x}_i\rangle = 0$ and, therefore, $\mathbf{x}_i$ and $\mathbf{x}_j$ are orthogonal.

7.24 Prove that if $\mathbf{A}$ is the $n \times n$ matrix representation of a normal transformation, then $\mathbf{A}$ has a full set of n linearly independent eigenvectors, regardless of the multiplicity of the eigenvalues.

Let $\mathbf{T}_i \triangleq \mathbf{A} - \mathbf{I}\lambda_i$. The eigenvectors satisfy $\mathbf{T}_i\mathbf{x}_i = \mathbf{0}$, and a generalized eigenvector $\mathbf{x}_{i+1}$ must satisfy $\mathbf{T}_i\mathbf{x}_{i+1} = \mathbf{x}_i$. The required proof consists of showing that if $\mathbf{A}$ is normal, the condition on $\mathbf{x}_{i+1}$ leads to a contradiction and thus cannot be satisfied. For $\mathbf{A}$ normal, $\mathbf{A}^*\mathbf{x}_i = \bar{\lambda}_i\mathbf{x}_i$ for each eigenvector $\mathbf{x}_i$; that is, $\mathbf{T}_i^*\mathbf{x}_i = \mathbf{0}$. Then $\mathbf{T}_i^*\mathbf{T}_i\mathbf{x}_{i+1} = \mathbf{T}_i^*\mathbf{x}_i = \mathbf{0}$. This means that

$$\langle \mathbf{x}_{i+1}, \mathbf{T}_i^*\mathbf{T}_i\mathbf{x}_{i+1}\rangle = \langle \mathbf{T}_i\mathbf{x}_{i+1}, \mathbf{T}_i\mathbf{x}_{i+1}\rangle = 0 \qquad \text{or} \qquad \|\mathbf{T}_i\mathbf{x}_{i+1}\|^2 = 0$$

This requires that $\mathbf{T}_i\mathbf{x}_{i+1} = \mathbf{0}$, but this contradicts the original assumption, since $\mathbf{x}_i \neq \mathbf{0}$. When $\mathbf{A}$ is normal, it cannot have generalized eigenvectors, and, therefore, must have a full set of n linearly independent eigenvectors.

Singular Value Decomposition SVD [38]

7.25 Consider an $m \times n$ matrix $\mathbf{A}$ with $m \geq n$, and with rank ($\mathbf{A}$) $= r$. Show that $\mathbf{A}$ can be written as $\mathbf{A} = \mathbf{U\Sigma V}^T$, where $\mathbf{U}$ and $\mathbf{V}$ are $m \times m$ and $n \times n$ orthogonal matrices respectively and where $\mathbf{\Sigma}$ is $m \times n$ and "diagonal." (A nonsquare matrix is diagonal if all i, j entries are zero for $i \neq j$.)

Since $\mathbf{A}$ is not square, it cannot be used directly in an eigenvalue problem. However, two related problems are pertinent. Consider $\mathbf{AA}^T\boldsymbol{\xi}_i = \sigma_i^2\boldsymbol{\xi}_i$ and $\mathbf{A}^T\mathbf{A}\boldsymbol{\eta}_i = \lambda_i^2\boldsymbol{\eta}_i$. $\mathbf{AA}^T$ is $m \times m$, symmetric, and hence normal. It is also positive semidefinite, and thus the σ_i^2 notation for the eigenvalue is justified. From Problem 7.24 there is a full set of m eigenvectors. From Problem 7.21 or 7.23 these eigenvectors are mutually orthogonal, at least for two different eigenvalues. They still can be selected as orthogonal even if there are repeated eigenvalues. For a multiplicity k we are assured there are k independent eigenvectors, and Gram-Schmidt can be used to construct k orthogonal vectors from them. The new vectors are still eigenvectors. By proper normalization, all $\boldsymbol{\xi}_i$ are also unit vectors, i.e., they are orthonormal. These vectors form the columns of an $m \times m$ orthogonal matrix $\mathbf{U}$.

$\mathbf{A}^T\mathbf{A}$ is $n \times n$ symmetric and at least positive semidefinite. Thus it also has a full set of n orthonormal eigenvectors $\boldsymbol{\eta}_i$ and nonnegative eigenvalues λ_i^2. Use the $\boldsymbol{\eta}_i$ vectors to form columns of an $n \times n$ orthogonal matrix $\mathbf{V}$. From Problem 7.15 it is known that $\mathbf{A}^T\boldsymbol{\xi}_i \triangleq \boldsymbol{\zeta}_i$ will be an eigenvector of $\mathbf{A}^T\mathbf{A}$, at least in the case where $\sigma_i \neq 0$. This is still true even for $\sigma_i = 0$, as will be seen when the length of $\boldsymbol{\zeta}_i$ is computed below. From that same problem the nonzero values of σ_i and λ_i are the same. Thus $\mathbf{AA}^T\boldsymbol{\xi}_i = \sigma_i^2\boldsymbol{\xi}_i$ becomes $\mathbf{A}\boldsymbol{\zeta}_i = \sigma_i^2\boldsymbol{\xi}_i$. But $\boldsymbol{\zeta}_i$ is generally not a unit vector. In fact its length is found from $\|\boldsymbol{\zeta}_i\|^2 = \langle\boldsymbol{\zeta}_i, \boldsymbol{\zeta}_i\rangle = \langle\mathbf{A}^T\boldsymbol{\xi}_i, \mathbf{A}^T\boldsymbol{\xi}_i\rangle = \langle\mathbf{AA}^T\boldsymbol{\xi}_i, \boldsymbol{\xi}_i\rangle = \sigma_i^2\langle\boldsymbol{\xi}_i, \boldsymbol{\xi}_i\rangle = \sigma_i^2$. Therefore, dividing the earlier equation by σ_i gives $\mathbf{A}\boldsymbol{\eta}_i = \sigma_i\boldsymbol{\xi}_i$. The entire set of such equations is

$$\mathbf{A}[\boldsymbol{\eta}_1\boldsymbol{\eta}_2 \ldots \boldsymbol{\eta}_n] = [\sigma_1\boldsymbol{\xi}_1 \ldots \sigma_m\boldsymbol{\xi}_m] = [\boldsymbol{\xi}_1\boldsymbol{\xi}_2 \ldots \boldsymbol{\xi}_m]\begin{bmatrix} \sigma_1 & & & \\ & \sigma_2 & & 0 \\ & & \ddots & \\ 0 & & & \sigma_n \\ \hline & & 0 & \end{bmatrix}\begin{matrix} \left.\begin{matrix} \\ \\ \\ \\ \end{matrix}\right\} n \times n \text{ diagonal, rank } r \\ \}\,(m-n) \times n \text{ zero} \end{matrix}$$

or $\mathbf{AV} = \mathbf{U\Sigma}$. Using the orthogonality of $\mathbf{V}$ gives $\mathbf{V}^{-1} = \mathbf{V}^T$. The final result is $\mathbf{A} = \mathbf{U\Sigma V}^T$. For exposition purposes it was assumed that $m > n$. This was not at all essential. For example, if $m < n$, define $\mathbf{B} = \mathbf{A}^T$ and then all the above applies to $\mathbf{B}$.

The positive square roots of the eigenvalues of $\mathbf{A}^T\mathbf{A}$, namely the $\sigma_i = \lambda_i$, are called the singular values of $\mathbf{A}$. The eigenvectors of $\mathbf{AA}^T$, namely the $\boldsymbol{\xi}_i$, are called the left singular vectors of $\mathbf{A}$ and the $\boldsymbol{\eta}_i$ are called the right singular vectors of $\mathbf{A}$ [38].

7.26 Show that any $m \times n$ matrix $\mathbf{A}$ of rank r can be written as

$$\mathbf{A} = \mathbf{U}'\mathbf{\Sigma}'\mathbf{V}'^T$$

where $\mathbf{U}'$ and $\mathbf{V}'$ are $m \times r$ and $r \times n$ matrices, respectively, with orthonormal columns, and where $\boldsymbol{\Sigma}'$ is an $r \times r$ full rank diagonal matrix.

Starting with the previous problem results, all the zero columns of $\boldsymbol{\Sigma}$ can be deleted as long as the corresponding rows of $\mathbf{V}^T$ are also deleted to maintain conformability. The values in the resulting matrix product are unchanged. Likewise, all the zero rows of $\boldsymbol{\Sigma}$ can be deleted without changing the answer, so long as the corresponding columns of $\mathbf{U}$ are deleted to maintain a conformable product. Actually this last set of deletions can be done in every case, whether $\mathbf{A}$ is full rank n or not. The first set of deletions only applies when $\mathbf{A}$ is of less than full rank, say r, because in the full rank case there are no zero columns in $\boldsymbol{\Sigma}$. The row-deleted version of $\mathbf{V}^T$ is $\mathbf{V}'^T$ and the column-deleted version of $\mathbf{U}$ is $\mathbf{U}'$. Likewise for $\boldsymbol{\Sigma}$ and $\boldsymbol{\Sigma}'$. The primed matrices form what has been called the economy-sized version of singular value decomposition [38]. It can save a lot of computer storage. Even though $\mathbf{U}'$ and $\mathbf{V}'$ are no longer square, it is still true that $\mathbf{U}'^T\mathbf{U}' = \mathbf{I}$ and $\mathbf{V}'^T\mathbf{V}' = \mathbf{I}$. In both this form of the singular value decomposition and the previous full-sized form, the rank of $\mathbf{A}$ is the number of nonzero singular values in $\boldsymbol{\Sigma}$ or $\boldsymbol{\Sigma}'$.

7.27 Show how SVD can be used to solve simultaneous linear equations $\mathbf{Ax} = \mathbf{y}$.

Using the SVD form for $\mathbf{A}$ gives $\mathbf{U\Sigma V}^T\mathbf{x} = \mathbf{y}$. Using the orthogonality of $\mathbf{U}$ and defining $\mathbf{U}^T\mathbf{y} \triangleq \mathbf{w}$ and $\mathbf{V}^T\mathbf{x} \triangleq \mathbf{v}$ gives $\boldsymbol{\Sigma}\mathbf{v} = \mathbf{w}$. Because of the diagonal nature of $\boldsymbol{\Sigma}$, these are easily solved for $\mathbf{v}$ in most cases. Then a simple matrix product gives $\mathbf{x} = \mathbf{Vv}$. An expanded form of the crucial equation is

$$\boldsymbol{\Sigma}\mathbf{v} = \mathbf{w} \Longrightarrow \left[\begin{array}{cccc|c} \sigma_1 & & & & \\ & \cdot & & & \\ & & \cdot & & 0 \\ & & & \cdot & \\ & & & \sigma_r & \\ \hline & & 0 & & \\ \hline & & 0 & & \end{array}\right] \left[\begin{array}{c} v_1 \\ \cdot \\ \cdot \\ \cdot \\ v_r \\ \hline v_{r+1} \\ \cdot \\ \cdot \\ \cdot \\ v_n \end{array}\right] = \left[\begin{array}{c} w_1 \\ \cdot \\ \cdot \\ \cdot \\ w_r \\ \hline w_{r+1} \\ \cdot \\ \cdot \\ \cdot \\ w_m \end{array}\right]$$

From this it can be seen that the original equations are inconsistent and have no solution if $\mathbf{A}$ and $\boldsymbol{\Sigma}$ have rank $r < n$ unless $\mathbf{w}$ also has these last $m - r$ rows zero. A least-squares solution is still possible. The solution (or least-squares solution) for $\mathbf{v}$ will have some arbitrary components whenever there are zero columns in $\boldsymbol{\Sigma}$. Setting these components of $\mathbf{v}$ to zero will give the minimum norm solution (or least-squares solution if required) for $\mathbf{v}$. Since $\mathbf{x}$ and $\mathbf{v}$ are related by an orthogonal matrix, they have the same norm, so $\mathbf{x}$ is also minimum norm in that case.

7.28 Show that the SVD provides the means for extending the spectral representation of equation (*7.4*) to nonsquare matrices.

Starting with $\mathbf{A} = \mathbf{U\Sigma V}^T$ as defined in Problem 7.25

$$\mathbf{A} = [\boldsymbol{\xi}_1\boldsymbol{\xi}_2 \ldots \boldsymbol{\xi}_m][\boldsymbol{\Sigma}]\begin{bmatrix} \boldsymbol{\eta}_1^T \\ \cdot \\ \cdot \\ \cdot \\ \boldsymbol{\eta}_n^T \end{bmatrix} = [\sigma_1\boldsymbol{\xi}_1 \quad \sigma_2\boldsymbol{\xi}_2 \quad \cdots \quad \sigma_n\boldsymbol{\xi}_n]\begin{bmatrix} \boldsymbol{\eta}_1^T \\ \cdot \\ \cdot \\ \cdot \\ \boldsymbol{\eta}_n^T \end{bmatrix}$$

$$= \sum_{i=1}^{n} \sigma_i\boldsymbol{\xi}_i\boldsymbol{\eta}_i^T = \sum_{i=1}^{n} \sigma_i\boldsymbol{\xi}_i\rangle\langle\boldsymbol{\eta}_i$$

This is of the form of equation (*7.4*). It has similar uses. For example, in approximation theory this series can be truncated prior to including all n terms if some values of σ_i are considered to be sufficiently small to be neglected.

7.29 Find the singular value decomposition for the matrix **A** of Problem 5.20.

$$\mathbf{A} = \begin{bmatrix} 1 & 2 \\ 3 & 4 \\ 5 & 6 \end{bmatrix} \quad \text{so} \quad \mathbf{A}^T\mathbf{A} = \begin{bmatrix} 35 & 44 \\ 44 & 56 \end{bmatrix} \quad \text{and} \quad \mathbf{A}\mathbf{A}^T = \begin{bmatrix} 5 & 11 & 17 \\ 11 & 25 & 39 \\ 17 & 39 & 61 \end{bmatrix}$$

The nonzero eigenvalues are approximately $\sigma_1^2 = 90.7355$ and $\sigma_2^2 = 0.2645$. The square roots of these form the diagonal terms in $\mathbf{\Sigma}$ below, and the two sets of normalized eigenvectors are shown as columns of **U** and rows of $\mathbf{V}^T$ below.

$$\mathbf{A} = \overbrace{\begin{bmatrix} 0.2298 & -0.8835 & 0.4082 \\ 0.5247 & -0.2408 & -0.8165 \\ 0.8196 & 0.4019 & 0.4082 \end{bmatrix}}^{\mathbf{U}} \overbrace{\begin{bmatrix} 9.5255 & 0 \\ 0 & 0.5143 \\ \hdashline 0 & 0 \end{bmatrix}}^{\mathbf{\Sigma}} \overbrace{\begin{bmatrix} 0.6196 & 0.7849 \\ 0.7849 & -0.6196 \end{bmatrix}}^{\mathbf{V}^T}$$

The efficient computational determination of the SVD form is crucial if it is to be useful. The indicated eigenvectors can be determined by the means presented in this chapter. This easily leads to the SVD form in simple cases like this one. However, Reference 38 should be consulted for a superior algorithm for use in more realistic cases. The real value of the discussion of this and the four previous problems is in understanding the concepts of the method, not in developing a general-purpose algorithm.

7.30 Resolve the equations of Problem 5.20 using the SVD results of Problem 7.29 and the method of Problem 7.27.

The simultaneous equations are

$$\begin{bmatrix} 1 & 2 \\ 3 & 4 \\ 5 & 6 \end{bmatrix} \mathbf{x} = \begin{bmatrix} 2 \\ 3 \\ 14 \end{bmatrix}$$

Using **U** from Problem 7.29 gives $\mathbf{w} \triangleq \mathbf{U}^T\mathbf{y} = \begin{bmatrix} 13.5095 \\ 3.1372 \end{bmatrix}$

Then $\mathbf{v}_1 = 13.5095/\sigma_1 = 1.4182$ and $\mathbf{v}_2 = 3.1372/\sigma_2 = 6.0999$. A matrix product then gives $\mathbf{x} = \mathbf{V}\mathbf{v} = \begin{bmatrix} 5.666 \\ -2.666 \end{bmatrix}$.

Independence of Generalized Eigenvectors

7.31 If $\mathbf{x}_1$ is an eigenvector and $\mathbf{x}_2, \mathbf{x}_3, \ldots, \mathbf{x}_k$ are generalized eigenvectors, all associated with the same eigenvalue λ_1, show that:

a. All of these vectors belong to the null space of $(\mathbf{A} - \mathbf{I}\lambda_1)^k$.

b. This set of vectors is linearly independent.

a. The defining equations for the set of vectors are

$$\mathbf{A}\mathbf{x}_1 = \lambda_1\mathbf{x}_1, \quad \mathbf{x}_1 \neq \mathbf{0}, \quad \text{or} \quad (\mathbf{A} - \mathbf{I}\lambda_1)\mathbf{x}_1 = \mathbf{0} \qquad (1)$$

$$\mathbf{A}\mathbf{x}_2 = \lambda_1\mathbf{x}_2 + \mathbf{x}_1 \qquad (\mathbf{A} - \mathbf{I}\lambda_1)\mathbf{x}_2 = \mathbf{x}_1 \qquad (2)$$

$$\mathbf{A}\mathbf{x}_3 = \lambda_1\mathbf{x}_3 + \mathbf{x}_2 \qquad (\mathbf{A} - \mathbf{I}\lambda_1)\mathbf{x}_3 = \mathbf{x}_2 \qquad (3)$$

$$\vdots \qquad\qquad \vdots$$

$$\mathbf{A}\mathbf{x}_k = \lambda_1\mathbf{x}_k + \mathbf{x}_{k-1} \qquad (\mathbf{A} - \mathbf{I}\lambda_1)\mathbf{x}_k = \mathbf{x}_{k-1}$$

From equations (*1*) and (*2*), $(\mathbf{A} - \mathbf{I}\lambda_1)^2\mathbf{x}_2 = (\mathbf{A} - \mathbf{I}\lambda_1)\mathbf{x}_1 = \mathbf{0}$. Multiplying equation (*3*) by $(\mathbf{A} - \mathbf{I}\lambda_1)^2$ gives $(\mathbf{A} - \mathbf{I}\lambda_1)^3\mathbf{x}_3 = (\mathbf{A} - \mathbf{I}\lambda_1)^2\mathbf{x}_2 = \mathbf{0}$. In general, it can be seen that $(\mathbf{A} - \mathbf{I}\lambda_1)^p\mathbf{x}_p = \mathbf{0}$, and $(\mathbf{A} - \mathbf{I}\lambda_1)^{p-1}\mathbf{x}_p = \mathbf{x}_1$. Since $(\mathbf{A} - \mathbf{I}\lambda_1)^k = (\mathbf{A} - \mathbf{I}\lambda_1)^{k-p}(\mathbf{A} - \mathbf{I}\lambda_1)^p$, $(\mathbf{A} - \mathbf{I}\lambda_1)^k\mathbf{x}_p = \mathbf{0}$ for $p = 1, 2, \ldots, k$ and part a is proven.

b. Let

$$a_1\mathbf{x}_1 + a_2\mathbf{x}_2 + \cdots + a_k\mathbf{x}_k = \mathbf{0} \tag{4}$$

and show that this implies that each $a_i = 0$. Multiplying equation (*4*) by $(\mathbf{A} - \mathbf{I}\lambda_1)^{k-1}$ gives $a_k(\mathbf{A} - \mathbf{I}\lambda_1)^{k-1}\mathbf{x}_k = \mathbf{0}$. Since $(\mathbf{A} - \mathbf{I}\lambda_1)^{k-1}\mathbf{x}_k = \mathbf{x}_1 \neq \mathbf{0}$, a_k must be zero. Using this fact and then multiplying equation (*4*) by $(\mathbf{A} - \mathbf{I}\lambda_1)^{k-2}$ shows that $a_{k-1} = 0$. Continuing this process shows that if $\sum_{i=1}^{k} a_i\mathbf{x}_i = \mathbf{0}$, then $a_i = 0$ for $i = 1, 2, \ldots, k$. This means the set $\{\mathbf{x}_i\}$ is linearly independent.

Companion Matrix

7.32 When considering *n*th order linear differential equations of the type

$$\frac{d^n x}{dt^n} + a_{n-1}\frac{d^{n-1}x}{dt^{n-1}} + a_{n-2}\frac{d^{n-2}x}{dt^{n-2}} + \cdots + a_1\frac{dx}{dt} + a_0 x = u(t)$$

the $n \times n$ matrix $\mathbf{A} = \begin{bmatrix} 0 & 1 & 0 & 0 & \cdots & 0 \\ 0 & 0 & 1 & 0 & & 0 \\ 0 & 0 & 0 & 1 & & 0 \\ \vdots & & & & & \vdots \\ -a_0 & -a_1 & -a_2 & -a_3 & \cdots & -a_{n-1} \end{bmatrix}$

will often arise. $\mathbf{A}$ is called the companion matrix. Show that the companion matrix always has just one eigenvector for each eigenvalue, regardless of its algebraic multiplicity.

The matrix $\mathbf{A} - \mathbf{I}\lambda$ always has rank r which satisfies $r \geq n - 1$. To see this, delete the first column and the last row, leaving a lower triangular $n - 1 \times n - 1$ matrix with ones on the diagonal. If λ is an eigenvalue, rank $(\mathbf{A} - \mathbf{I}\lambda) < n$. Together, these results imply rank $(\mathbf{A} - \mathbf{I}\lambda_i) = n - 1$, so the degeneracy is $q = n - r = 1$. The case of simple degeneracy always applies, so there is exactly one eigenvector for each eigenvalue.

Iterative Approximation

7.33 Because of the difficulty of factoring high-degree polynomials to obtain eigenvalues, computer solutions are often required. This being the case, develop an iterative computational scheme which gives the eigenvalues and eigenvectors directly. Assume $\mathbf{A}$ is real and symmetric.

$\mathbf{A}$ being real and symmetric guarantees that all λ_i are real and $\{\mathbf{x}_i\}$ is an orthogonal set. Even if there are repeated roots, since $\mathbf{A}$ is normal, there is a full set of eigenvectors. The method described here assumes distinct eigenvalues; in fact, it is assumed that the magnitudes are distinct, $|\lambda_i| \neq |\lambda_j|$ if $i \neq j$. Any vector $\mathbf{z}_0$ can be written as

$$\mathbf{z}_0 = \alpha_1\mathbf{x}_1 + \alpha_2\mathbf{x}_2 + \cdots + \alpha_n\mathbf{x}_n$$

Therefore

$$\mathbf{A}\mathbf{z}_0 = \alpha_1\lambda_1\mathbf{x}_1 + \alpha_2\lambda_2\mathbf{x}_2 + \cdots + \alpha_n\lambda_n\mathbf{x}_n \triangleq \mathbf{z}_1$$
$$\mathbf{A}\mathbf{z}_1 = \alpha_1\lambda_1^2\mathbf{x}_1 + \alpha_2\lambda_2^2\mathbf{x}_2 + \cdots + \alpha_n\lambda_n^2\mathbf{x}_n \triangleq \mathbf{z}_2$$
$$\vdots$$
$$\mathbf{A}\mathbf{z}_k = \alpha_1\lambda_1^{k+1}\mathbf{x}_1 + \alpha_2\lambda_2^{k+1}\mathbf{x}_2 + \cdots + \alpha_n\lambda_n^{k+1}\mathbf{x}_n \triangleq \mathbf{z}_{k+1}$$

If λ_1 is the eigenvalue with the largest absolute value, then for k sufficiently large,

$$\mathbf{z}_{k+1} \cong \alpha_1\lambda_1^{k+1}\mathbf{x}_1 = \beta\mathbf{x}_1 \qquad \text{and} \qquad \mathbf{A}\mathbf{z}_{k+1} \cong \lambda_1\mathbf{z}_{k+1} = \lambda_1\beta\mathbf{x}_1$$

Hence starting with an arbitrary vector $\mathbf{z}_0$ and repeatedly calculating $\mathbf{z}_{\text{new}} = \mathbf{A}\mathbf{z}_{\text{old}}$ until $\mathbf{z}_{\text{new}}$ is proportional to $\mathbf{z}_{\text{old}}$ leads to the maximum magnitude eigenvalue (the constant of proportionality) and the corresponding eigenvector. At each step of the iterative calculations, the vectors $\mathbf{z}$ can be normalized in any number of ways.

The next largest eigenvalue and its eigenvector can be computed in a similar way, except at each step the vector $\mathbf{z}$ must be forced to be orthogonal to the previously found $\mathbf{x}_1$. Setting $\langle \mathbf{x}_1, \mathbf{z} \rangle = 0$ gives

$$z_1 = -\frac{1}{x_1}[x_2z_2 + x_3z_3 + \cdots + x_nz_n]$$

or

$$\mathbf{z}_{\text{constrained}} = \left[\begin{array}{c|cccc} 0 & -x_2/x_1 & -x_3/x_1 & \cdots & -x_n/x_1 \\ \hline \mathbf{0} & & \mathbf{I}_{n-1} & & \end{array}\right]\mathbf{z}_{\text{free}} = \mathbf{S}\mathbf{z}_{\text{free}}$$

The matrix $\mathbf{S}$, called the sweep matrix, can be combined with $\mathbf{A}$ to give the inerative procedure $\mathbf{A}\mathbf{S}[\mathbf{z}_k] = \mathbf{z}_{k+1}$, where $\mathbf{z}_0$ can be freely selected. This can be extended to find all eigenvalues and eigenvectors. The sweep matrix must be modified to force each new eigenvector to be orthogonal to all those already found.

Quadratic Form

7.34 Show that the eigenvalue problem for a real, symmetric matrix can be characterized as one of maximizing or minimizing a quadratic form subject to the constraint that $\mathbf{x}$ be a unit vector.

Consider $Q = \langle \mathbf{x}, \mathbf{A}\mathbf{x} \rangle$. If the change of basis $\mathbf{x} = \mathbf{M}\mathbf{z}$ is used, $Q = \sum_{i=1}^{n} \lambda_i z_i^2$. Because of the orthogonal transformation, $\mathbf{z}$ is also a unit vector. Since all $z_i^2 \leq 1$, this suggests that the maximum value of Q will be attained if all $z_i = 0$ except $\mathbf{z}_k^2 = 1$, where $\lambda_k = \lambda_{\max}$. Then $Q = \lambda_{\max}$. Alternatively, adjoining the constraint $\|\mathbf{x}\|^2 = 1$ to Q by means of the Lagrange multiplier λ shows this directly. That is, maximizing $\mathbf{x}^T\mathbf{A}\mathbf{x} - \lambda(\mathbf{x}^T\mathbf{x} - 1)$ requires that the derivative with respect to each component x_i must vanish. This gives the set $\mathbf{A}\mathbf{x} - \lambda\mathbf{x} = \mathbf{0}$ which is the eigenvalue-eigenvector equation. If $\mathbf{x}$ satisfies this condition, then $Q = \mathbf{x}^T\mathbf{A}\mathbf{x} = \mathbf{x}^T\mathbf{x}\lambda = \lambda$, so $Q_{\max} = \lambda_{\max}$. Also $Q_{\min} = \lambda_{\min}$.

If $\mathbf{x}_1$ is the eigenvector associated wigh $\lambda_{\max}$, then selecting $\mathbf{x}$ to maximize Q subject to $\langle \mathbf{x}_1, \mathbf{x} \rangle = 0$ and $\|\mathbf{x}\| = 1$ will lead to the second largest eigenvalue and its associated eigenvector. The remaining eigenvalues-eigenvectors are found in a similar way by requiring orthogonality with all previously found eigenvectors.

7.35 Reduce the quadratic form $Q = \frac{1}{3}[16y_1^2 + 10y_2^2 + 16y_3^2 - 4y_1y_2 + 16y_1y_3 + 4y_2y_3]$ to a sum of squared terms only by selecting a suitable change of coordinates.

This quadratic form can be expressed in matrix form as $Q = \mathbf{y}^T\mathbf{A}\mathbf{y}$, where $\mathbf{y} = [y_1 \quad y_2 \quad y_3]^T$ and $\mathbf{A} = \frac{1}{3}\begin{bmatrix} 16 & -2 & 8 \\ -2 & 10 & 2 \\ 8 & 2 & 16 \end{bmatrix}$. The eigenvalues of $\mathbf{A}$ are $\lambda_1 = 8$, $\lambda_2 = 4$, $\lambda_3 = 2$, and since $\mathbf{A}$ is real and symmetric, a set of orthonormal eigenvectors can be found. They are used as columns of the modal matrix

$$\mathbf{M} = \begin{bmatrix} 1/\sqrt{2} & -1/\sqrt{6} & -1/\sqrt{3} \\ 0 & 2/\sqrt{6} & -1/\sqrt{3} \\ 1/\sqrt{2} & 1/\sqrt{6} & 1/\sqrt{3} \end{bmatrix}$$

Since $\mathbf{M}$ is orthogonal, $\mathbf{M}^{-1} = \mathbf{M}^T$ and $\mathbf{A}$ is diagonalized by the orthogonal transformation $\mathbf{M}^T\mathbf{A}\mathbf{M} = \text{diag}\,[8, 4, 2]$. If the change of variables $\mathbf{y} = \mathbf{M}\mathbf{z}$ is used, then

$$Q = \mathbf{z}^T\mathbf{M}^T\mathbf{A}\mathbf{M}\mathbf{z} = 8z_1^2 + 4z_2^2 + 2z_3^2$$

PROBLEMS

7.36 Find the eigenvalues, eigenvectors, and Jordan form for $\mathbf{A} = \begin{bmatrix} 2 & -2 & 3 \\ 1 & 1 & 1 \\ 1 & 3 & -1 \end{bmatrix}$.

7.37 Find the eigenvectors of $\mathbf{A} = \begin{bmatrix} 2 & 0 \\ 0 & 2 \end{bmatrix}$.

7.38 Analyze the eigenvalue-eigenvector problem for $\mathbf{A} = \begin{bmatrix} 2 & 0 & 1 & 0 \\ 0 & 0 & 0 & 1 \\ 0 & 0 & 0 & 0 \\ 0 & 0 & 0 & 0 \end{bmatrix}$.

7.39 Compute the eigenvalues, eigenvectors, and Jordan form for $\mathbf{A} = \begin{bmatrix} 4 & -2 & 0 \\ 1 & 2 & 0 \\ 0 & 0 & 6 \end{bmatrix}$.

7.40 Are $\mathbf{A} = \frac{1}{2}\begin{bmatrix} 3 & 1 \\ -1 & 5 \end{bmatrix}$ and $\mathbf{B} = \begin{bmatrix} 2 & 0 \\ 0 & 2 \end{bmatrix}$ similar matrices?

7.41 Use the iterative technique of Problem 7.33 to find approximate eigenvalues and eigenvectors for $\mathbf{A} = \begin{bmatrix} 8 & 2 & -5 \\ 2 & 11 & -2 \\ -5 & -2 & 8 \end{bmatrix}$.

7.42 Find an approximate set of eigenvalues and eigenvectors for $\mathbf{A} = \begin{bmatrix} 3 & 2 & 1 \\ 2 & 2 & 1 \\ 1 & 1 & 1 \end{bmatrix}$ using iteration.

7.43 If $\Delta(\lambda)$ is defined as in equation (*7.3*) and if

$$\Delta'(\lambda) \triangleq |\mathbf{I}\lambda - \mathbf{A}| = \lambda^n + c'_{n-1}\lambda^{n-1} + c'_{n-2}\lambda^{n-2} + \cdots + c'_1\lambda + c'_0$$

show that

a. $c_0 = |\mathbf{A}|;\ c'_0 = |-\mathbf{A}| = (-1)^n|\mathbf{A}|$.
b. $(-1)^{n-1}c_{n-1} = \operatorname{Tr}(\mathbf{A});\ c'_{n-1} = -\operatorname{Tr}(\mathbf{A})$.

7.44 Draw conclusions about the sign definiteness of

a. $\mathbf{A} = \begin{bmatrix} -6 & 2 \\ 2 & -1 \end{bmatrix}$,

b. $\mathbf{A} = \begin{bmatrix} 13 & 4 & -13 \\ 4 & 22 & -4 \\ -13 & -4 & 13 \end{bmatrix}$,

c. $\mathbf{A} = \begin{bmatrix} -1 & 3 & 0 & 0 \\ 3 & -9 & 0 & 0 \\ 0 & 0 & -6 & 2 \\ 0 & 0 & 2 & -1 \end{bmatrix}$,

d. $\mathbf{A} = \begin{bmatrix} 8 & 2 & -5 \\ 2 & 11 & -2 \\ -5 & -2 & 8 \end{bmatrix}$,

e. $\mathbf{A} = \begin{bmatrix} -3 & 2 & 0 & 1 & 7 \\ 2 & 1 & -2 & 1 & 0 \\ 0 & -2 & 6 & 3 & 8 \\ 1 & 1 & 3 & 2 & 4 \\ 7 & 0 & 8 & 4 & 5 \end{bmatrix}$.

7.45 Let $\mathbf{A} = \begin{bmatrix} 3 & -1 \\ -1 & 3 \end{bmatrix}$ and let $\mathbf{x} = [x_1 \quad x_2]^T$ be any *unit* vector. Consider $Q = \langle \mathbf{x}, \mathbf{Ax} \rangle$ as a scalar function of $\mathbf{x}$. From among all unit vectors find the one which gives Q its maximum value. Also determine $Q_{\max}$.

7.46 Show that any simple linear transformation can be represented as $\mathbf{A} = \sum_{i=1}^{n} \lambda_i \mathbf{E}_i$, where $\mathbf{E}_i$ is a projection onto $\mathfrak{N}(\mathbf{A} - \mathbf{I}\lambda_i)$.

7.47 Show that any normal linear transformation can be written as a sum of orthogonal projection transformations.

7.48 Analyze the simultaneous equations of Example 5.3 using the SVD method.

Chapter

8

FUNCTIONS OF SQUARE MATRICES AND THE CAYLEY-HAMILTON THEOREM

8.1 INTRODUCTION

Functions of square matrices arise in connection with the solution of vector-matrix differential and difference equations. Some scalar-valued functions of matrices have already been considered, namely, $|\mathbf{A}|$, $\mathrm{Tr}(\mathbf{A})$, $\|\mathbf{A}\|$, etc. In this chapter matrix-valued functions of square matrices are considered. This material will be found useful in later chapters during the analysis of multivariable control systems.

8.2 POWERS OF A MATRIX AND MATRIX POLYNOMIALS

A matrix is conformable with itself only if it is square. Only square, $n \times n$ matrices are considered in this chapter. The product $\mathbf{AA}$ will be referred to as $\mathbf{A}^2$ for obvious reasons. The product of k such factors, $\mathbf{AA} \cdots \mathbf{A}$, is defined as $\mathbf{A}^k$. By definition, $\mathbf{A}^0 = \mathbf{I}$. With these definitions all the usual rules of exponents apply. That is,

$$\mathbf{A}^m\mathbf{A}^n = \underbrace{(\mathbf{AA}\cdots\mathbf{A})}_{m \text{ factors}}\underbrace{(\mathbf{AA}\cdots\mathbf{A})}_{n \text{ factors}} = A^{m+n}$$

$$(\mathbf{A}^m)^n = \underbrace{(\mathbf{AA}\cdots\mathbf{A})^n}_{m \text{ factors}} = \underbrace{(\mathbf{A}^n\mathbf{A}^n\cdots\mathbf{A}^n)}_{m \text{ factors}}$$

$$= \underbrace{(\mathbf{AA}\cdots\mathbf{A})}_{n \text{ factors}}\underbrace{(\mathbf{AA}\cdots\mathbf{A})}_{n \text{ factors}}\cdots\underbrace{(\mathbf{AA}\cdots\mathbf{A})}_{n \text{ factors}} = \mathbf{A}^{mn}$$

If $\mathbf{A}$ is nonsingular, then $(\mathbf{A}^{-1})^n = \mathbf{A}^{-n}$ and $\mathbf{A}^n\mathbf{A}^{-n} = \mathbf{A}^n(\mathbf{A}^{-1})^n = (\mathbf{A}\mathbf{A}^{-1})^n = \mathbf{I}^n = \mathbf{I}$ or $\mathbf{A}^{n-n} = \mathbf{A}^0 = \mathbf{I}$ as agreed upon earlier.

The notion of matrix powers can be used to define matrix polynomials in a natural way. For example, if $P(x) = c_m x^m + c_{m-1}x^{m-1} + \cdots + c_1 x + c_0$ is an mth degree polynomial in the scalar variable x, then a corresponding matrix polynomial can be defined as

$$P(\mathbf{A}) = c_m\mathbf{A}^m + c_{m-1}\mathbf{A}^{m-1} + \cdots + c_1\mathbf{A} + c_0\mathbf{I}$$

If the scalar polynomial can be written in factored form as

$$P(x) = c(x - a_1)(x - a_2)\cdots(x - a_m)$$

then this is also true for the matrix polynomial

$$P(\mathbf{A}) = c(\mathbf{A} - \mathbf{I}a_1)(\mathbf{A} - \mathbf{I}a_2)\cdots(\mathbf{A} - \mathbf{I}a_m)$$

Thus a matrix polynomial of an $n \times n$ matrix $\mathbf{A}$ is just another $n \times n$ matrix whose elements depend on $\mathbf{A}$ as well as on the coefficients of the polynomial.

8.3 INFINITE SERIES AND ANALYTIC FUNCTIONS OF MATRICES

Let $\mathbf{A}$ be an $n \times n$ matrix with eigenvalues $\lambda_1, \lambda_2, \ldots, \lambda_n$. Consider the infinite series in a scalar variable x,

$$\sigma(x) = a_0 + a_1 x + a_2 x^2 + \cdots + a_k x^k + \cdots$$

It is well known that a given infinite series may converge or diverge depending on the value of x. For example, the geometric series

$$1 + x + x^2 + x^3 + \cdots + x^k + \cdots$$

converges for $|x| < 1$ and diverges otherwise. Some infinite series are convergent for all values of x, such as

$$1 + x + \frac{x^2}{2!} + \frac{x^3}{3!} + \cdots + \frac{x^k}{k!} + \cdots$$

The various test for convergence of series will not be considered here. The following theorem is of major importance.

Theorem 8.1: Let $\mathbf{A}$ be an $n \times n$ matrix with eigenvalues λ_i. If the infinite series $\sigma(x) = a_0 + a_1 x + a_2 x^2 + \cdots$ is convergent for each of the n values $x = \lambda_i$, then the corresponding matrix infinite series

$$\sigma(\mathbf{A}) = a_0\mathbf{I} + a_1\mathbf{A} + a_2\mathbf{A}^2 + \cdots + a_k\mathbf{A}^k + \cdots = \sum_{k=0}^{\infty} a_k\mathbf{A}^k$$

converges.

Definition 8.1 [21]: A single-valued function $f(z)$, with z a complex scalar, is said to be *analytic* at a point z_0 if and only if its derivative exists at every point in some neighborhood of z_0. Points at which the function is not analytic are called *singular points*. For example, $f(z) = 1/z$ has $z = 0$ as its only singluar point and is analytic at every other point.

The result which makes Theorem 8.1 useful for the purposes of this book is Theorem 8.2.

Theorem 8.2: If a function $f(z)$ is analytic (contains no singularities) at every point in some circle Ω in the complex plane, then $f(z)$ can be represented as a convergent power series (the Taylor series) at every point z inside Ω [21].

Taken together, Theorems 8.1 and 8.2 give Theorem 8.3.

Theorem 8.3: If $f(z)$ is any function which is analytic within a circle in the complex plane which contains all eigenvalues λ_i of $\mathbf{A}$, then a corresponding matrix function $f(\mathbf{A})$ can be defined by a convergent power series.

Example 8.1

The function $e^{\alpha x}$ is analytic for all values of x. Thus it has a convergent series representation

$$e^{\alpha x} = 1 + \alpha x + \frac{\alpha^2 x^2}{2!} + \frac{\alpha^3 x^3}{3!} + \cdots + \frac{\alpha^k x^k}{k!} + \cdots$$

The corresponding matrix function is defined as

$$e^{\alpha \mathbf{A}} = \mathbf{I} + \alpha \mathbf{A} + \frac{\alpha^2 \mathbf{A}^2}{2!} + \frac{\alpha^3 \mathbf{A}^3}{3!} + \cdots + \frac{\alpha^k \mathbf{A}^k}{k!} + \cdots$$ ■

At any point within the circles of convergence, the power series representations for two functions $f(z)$ and $g(z)$ can be added, differentiated, or integrated term by term. Also, the product of the two series gives another series which converges to $f(z)g(z)$ [21]. Because of Theorems 8.1 and 8.2 the same properties are valid for analytic functions of the matrix $\mathbf{A}$. The usual cautions with matrix algebra must be observed, however.

Example 8.2

$e^{\mathbf{A}t}e^{\mathbf{B}t} = e^{(\mathbf{A}+\mathbf{B})t}$ if and only if $\mathbf{AB} = \mathbf{BA}$. This is easily verified by multiplying out the first few terms for $e^{\mathbf{A}t}$ and $e^{\mathbf{B}t}$. ■

Example 8.3

Find $\frac{d}{dt}[e^{\mathbf{A}t}]$. The series representation is

$$e^{\mathbf{A}t} = \mathbf{I} + \mathbf{A}t + \frac{\mathbf{A}^2 t^2}{2!} + \frac{\mathbf{A}^3 t^3}{3!} + \cdots$$

Term-by-term differentiation gives

$$\frac{de^{\mathbf{A}t}}{dt} = \mathbf{A} + \frac{2\mathbf{A}^2 t}{2!} + \frac{3\mathbf{A}^3 t^2}{3!} + \cdots$$

Since **A** can be factored out on either the left or the right,

$$\frac{de^{\mathbf{A}t}}{dt} = \mathbf{A}\left[\mathbf{I} + \mathbf{A}t + \frac{\mathbf{A}^2 t^2}{2!} + \cdots\right] = \left[\mathbf{I} + \mathbf{A}t + \frac{\mathbf{A}^2 t^2}{2!} + \cdots\right]\mathbf{A} = \mathbf{A}e^{\mathbf{A}t} = e^{\mathbf{A}t}\mathbf{A}$$ ■

Example 8.4

Compute $\int_0^t e^{\mathbf{A}\tau}\, d\tau$. Using the series representation, term-by-term integration gives

$$\int_0^t e^{\mathbf{A}\tau}\, d\tau = \int_0^t \mathbf{I}\, d\tau + \mathbf{A}\int_0^t \tau\, d\tau + \frac{\mathbf{A}^2}{2!}\int_0^t \tau^2\, d\tau + \cdots = \mathbf{I}t + \frac{\mathbf{A}t^2}{2} + \frac{\mathbf{A}^2 t^3}{3!} + \cdots$$

Therefore $\mathbf{A}\int_0^t e^{\mathbf{A}\tau}\, d\tau + \mathbf{I} = e^{\mathbf{A}t}$ or, if $\mathbf{A}^{-1}$ exists,

$$\int_0^t e^{\mathbf{A}\tau}\, d\tau = \mathbf{A}^{-1}[e^{\mathbf{A}t} - \mathbf{I}] = [e^{\mathbf{A}t} - \mathbf{I}]\mathbf{A}^{-1}$$ ■

Since the exponential function is analytic for all finite arguments, it is possible to define

$$e^{-\alpha\mathbf{A}} = \mathbf{I} - \alpha\mathbf{A} + \frac{\alpha^2\mathbf{A}^2}{2} - \frac{\alpha^3\mathbf{A}^3}{3!} + \cdots - \cdots$$

Since **A** commutes with itself,

$$e^{-\alpha\mathbf{A}}e^{\alpha\mathbf{A}} = e^{\alpha(\mathbf{A}-\mathbf{A})} = e^{\alpha[\mathbf{0}]} = \mathbf{I}$$

which is analogous to the scalar result.

Although the exponential function of a matrix is the one which will be most useful in this text, many other functions can be defined, for example:

$$\sin\mathbf{A} = \mathbf{A} - \frac{\mathbf{A}^3}{3!} + \frac{\mathbf{A}^5}{5!} - \cdots \qquad \sinh\mathbf{A} = \mathbf{A} + \frac{\mathbf{A}^3}{3!} + \frac{\mathbf{A}^5}{5!} + \cdots$$

$$\cos\mathbf{A} = \mathbf{I} - \frac{\mathbf{A}^2}{2!} + \frac{\mathbf{A}^4}{4!} - \cdots \qquad \cosh\mathbf{A} = \mathbf{I} + \frac{\mathbf{A}^2}{2!} + \frac{\mathbf{A}^4}{4!} + \cdots$$

Using these definitions, it can be verified that relations analogous to the scalar results hold. For example,

$$\sin^2\mathbf{A} + \cos^2\mathbf{A} = \mathbf{I}$$

$$\sin\mathbf{A} = \frac{e^{j\mathbf{A}} - e^{-j\mathbf{A}}}{2j}, \quad \cos\mathbf{A} = \frac{e^{j\mathbf{A}} + e^{-j\mathbf{A}}}{2}$$

$$\cosh^2\mathbf{A} - \sinh^2\mathbf{A} = \mathbf{I}$$

$$\sinh\mathbf{A} = \frac{e^{\mathbf{A}} - e^{-\mathbf{A}}}{2}, \qquad \cosh\mathbf{A} = \frac{e^{\mathbf{A}} + e^{-\mathbf{A}}}{2}$$

Although all of the preceding matrix functions are defined in terms of their power series representations, the series converge to $n \times n$ matrices.

It is useful to know how to evaluate these closed form expressions for analytic functions of square matrices. This is developed in Sec. 8.5.

8.4 THE CHARACTERISTIC POLYNOMIAL AND CAYLEY-HAMILTON THEOREM

Although arbitrary matrix polynomials have been discussed, one very special polynomial is the characteristic polynomial. If the characteristic polynomial for the matrix **A** is written as

$$|\mathbf{A} - \mathbf{I}\lambda| = (-\lambda)^n + c_{n-1}\lambda^{n-1} + c_{n-2}\lambda^{n-2} + \cdots + c_1\lambda + c_0 = \Delta(\lambda)$$

then the corresponding matrix polynomial is

$$\Delta(\mathbf{A}) = (-1)^n\mathbf{A}^n + c_{n-1}\mathbf{A}^{n-1} + c_{n-2}\mathbf{A}^{n-2} + \cdots + c_1\mathbf{A} + c_0\mathbf{I}$$

Cayley-Hamilton Theorem: Every matrix satisfies its own characteristic equation; that is, $\Delta(\mathbf{A}) = [\mathbf{0}]$.

Proof: (Valid when **A** is similar to a diagonal matrix. For the general case, see Problems 8.5 and 8.6.) A similarity transformation reduces **A** to the diagonal matrix **Λ**, so

$$\mathbf{A} = \mathbf{M\Lambda M}^{-1}, \qquad \mathbf{A}^2 = \mathbf{M\Lambda}^2\mathbf{M}^{-1}, \qquad \ldots, \qquad \mathbf{A}^k = \mathbf{M\Lambda}^k\mathbf{M}^{-1}$$

Therefore

$$\Delta(\mathbf{A}) = \mathbf{M}[(-1)^n\mathbf{\Lambda}^n + c_{n-1}\mathbf{\Lambda}^{n-1} + c_{n-2}\mathbf{\Lambda}^{n-2} + \cdots + c_1\mathbf{\Lambda} + c_0\mathbf{I}]\mathbf{M}^{-1}$$

Each term inside the brackets is a diagonal matrix. The sum of a typical i, i element is

$$(-\lambda_i)^n + c_{n-1}\lambda_i^{n-1} + \cdots + c_1\lambda_i + c_0$$

which is zero because λ_i is a root of the characteristic equation. Therefore

$$\Delta(\mathbf{A}) = \mathbf{M}[\mathbf{0}]\mathbf{M}^{-1} = [\mathbf{0}]$$

Example 8.5

Let $\mathbf{A} = \begin{bmatrix} 3 & 1 \\ 1 & 2 \end{bmatrix}$, $|\mathbf{A} - \mathbf{I}\lambda| = (3-\lambda)(2-\lambda) - 1 = \lambda^2 - 5\lambda + 5$. Then

$$\Delta(\mathbf{A}) = \mathbf{A}^2 - 5\mathbf{A} + 5\mathbf{I} = \begin{bmatrix} 10 & 5 \\ 5 & 5 \end{bmatrix} - 5\begin{bmatrix} 3 & 1 \\ 1 & 2 \end{bmatrix} + 5\begin{bmatrix} 1 & 0 \\ 0 & 1 \end{bmatrix} = \begin{bmatrix} 0 & 0 \\ 0 & 0 \end{bmatrix}$$ ■

Definition 8.2 [117]: The minimum polynomial of the square matrix **A** is the lowest degree polynomial, with the coefficient of the highest power normalized to one, such that

$$m(\mathbf{A}) = [\mathbf{0}]$$

It would be slightly more efficient to use the minimum polynomial rather than the characteristic polynomial in some cases (in Sec. 8.5, for example). The minimum polynomial $m(\mathbf{A})$ and the characteristic polynomial $\Delta'(\mathbf{A})$ are often the same (if all λ_i are distinct, or in the simple degeneracy case). In factored form the only possible differences between $m(\mathbf{A})$ and $\Delta'(\mathbf{A})$ are the powers of the terms involving repeated roots. Since for the definitions given in Chapter 7 $\Delta(\lambda) = (-1)^n\Delta'(\lambda)$, then

$$\Delta(\lambda) = (-1)^n(\lambda - \lambda_1)^{m_1}(\lambda - \lambda_2)^{m_2} \cdots (\lambda - \lambda_p)^{m_p}$$
$$m(\lambda) = (\lambda - \lambda_1)^{k_1}(\lambda - \lambda_2)^{k_2} \cdots (\lambda - \lambda_p)^{k_p}$$

In all cases $k_i \leq m_i$. In this text the characteristic polynomial will be used in most cases rather than the minimum polynomial because it is more familiar, and only occasionally requires a small amount of extra calculations (see Problem 8.18).

8.5 SOME USES OF THE CAYLEY-HAMILTON THEOREM

Matrix Inversion

Let $\mathbf{A}$ be an $n \times n$ matrix with the characteristic equation $\Delta(\lambda) = (-\lambda)^n + c_{n-1}\lambda^{n-1} + \cdots + c_1\lambda + c_0 = 0$. Recall that the constant $c_0 = \lambda_1\lambda_2 \cdots \lambda_n = |\mathbf{A}|$ and is zero if and only if $\mathbf{A}$ is singular. Using $\Delta(\mathbf{A}) = (-1)^n\mathbf{A}^n + c_{n-1}\mathbf{A}^{n-1} + \cdots + c_1\mathbf{A} + c_0\mathbf{I} = \mathbf{0}$, and assuming $\mathbf{A}^{-1}$ exists, multiplication by $\mathbf{A}^{-1}$ gives

$$(-1)^n\mathbf{A}^{n-1} + c_{n-1}\mathbf{A}^{n-2} + \cdots + c_1\mathbf{I} + c_0\mathbf{A}^{-1} = \mathbf{0}$$

or

$$\mathbf{A}^{-1} = \frac{-1}{c_0}[(-1)^n\mathbf{A}^{n-1} + c_{n-1}\mathbf{A}^{n-2} + \cdots + c_1\mathbf{I}]$$

Example 8.6

Let $\mathbf{A} = \begin{bmatrix} 3 & 1 \\ 1 & 2 \end{bmatrix}$, $\Delta(\lambda) = \lambda^2 - 5\lambda + 5$. Then

$$\Delta(\mathbf{A}) = \mathbf{A}^2 - 5\mathbf{A} + 5\mathbf{I} = \mathbf{0}, \qquad \mathbf{A}^{-1} = -\tfrac{1}{5}[\mathbf{A} - 5\mathbf{I}] = -\frac{1}{5}\begin{bmatrix} -2 & 1 \\ 1 & -3 \end{bmatrix}$$ ■

Reduction of a Polynomial in A to One of Degree n − 1 or Less

Let $P(x)$ be a scalar polynomial of degree m. Let $P_1(x)$ be another polynomial of degree n, where $n < m$. Then $P(x)$ can always be written $P(x) = Q(x)P_1(x) + R(x)$, where $Q(x)$ is a polynomial of degree $m - n$ and $R(x)$ is a remainder polynomial of degree $n - 1$ or less. For this scalar case, $Q(x)$ and $R(x)$ could be found by formally dividing $P(x)$ by $P_1(x)$, since this gives $P(x)/P_1(x) = Q(x) + R(x)/P_1(x)$.

Example 8.7

Let $P(x) = 3x^4 + 2x^2 + x + 1$ and $P_1(x) = x^2 - 3$. Then it is easily verified that

$$P(x) = (3x^2 + 11)(x^2 - 3) + (x + 34)$$

so that $Q(x) = 3x^2 + 11$ and $R(x) = x + 34$. ■

Similarly, the matrix polynomial $P(\mathbf{A})$ can be written

$$P(\mathbf{A}) = Q(\mathbf{A})P_1(\mathbf{A}) + R(\mathbf{A})$$

since it is always defined in the same manner as its scalar counterpart. If the arbitrary polynomial P_1 used above is selected as the characteristic polynomial of $\mathbf{A}$, then the scalar version of P is

$$P(x) = Q(x)\Delta(x) + R(x)$$

Note that $\Delta(x) \neq 0$ except for those specific values $x = \lambda_i$, the eigenvalues. The matrix version of P is

$$P(\mathbf{A}) = Q(\mathbf{A})\Delta(\mathbf{A}) + R(\mathbf{A})$$

By the Cayley-Hamilton theorem, $\Delta(\mathbf{A}) = \mathbf{0}$, so $P(\mathbf{A}) = R(\mathbf{A})$. The coefficients of the matrix remainder polynomial R can be found by long division of the corresponding scalar polynomials. Alternatively, the Cayley-Hamilton theorem can be used to reduce each individual term in $P(\mathbf{A})$ to one of degree $n - 1$ or less.

Example 8.8

Let $\mathbf{A} = \begin{bmatrix} 3 & 1 \\ 1 & 2 \end{bmatrix}$, $\Delta(\lambda) = \lambda^2 - 5\lambda + 5$. Compute $P(\mathbf{A}) = \mathbf{A}^4 + 3\mathbf{A}^3 + 2\mathbf{A}^2 + \mathbf{A} + \mathbf{I}$.

Method 1: By long division,

$$\frac{P(x)}{\Delta(x)} = x^2 + 8x + 37 + \frac{146x - 184}{x^2 - 5x + 5} \quad \text{or} \quad P(x) = (x^2 + 8x + 37)\Delta(x) + (146x - 184)$$

Therefore $R(x) = 146x - 184$ and the Cayley-Hamilton theorem guarantees that $P(\mathbf{A}) = R(\mathbf{A}) = 146\mathbf{A} - 184\mathbf{I}$.

Method 2: From the Cayley-Hamilton theorem,

$$\mathbf{A}^2 - 5\mathbf{A} + 5\mathbf{I} = [\mathbf{0}] \quad \text{or} \quad \mathbf{A}^2 = 5(\mathbf{A} - \mathbf{I})$$

Hence

$$\mathbf{A}^4 = \mathbf{A}^2\mathbf{A}^2 = 25(\mathbf{A} - \mathbf{I})(\mathbf{A} - \mathbf{I}) = 25(\mathbf{A}^2 - 2\mathbf{A} + \mathbf{I}) = 25[5(\mathbf{A} - \mathbf{I}) - 2\mathbf{A} + \mathbf{I}] = 25[3\mathbf{A} - 4\mathbf{I}]$$

$$\mathbf{A}^3 = \mathbf{A}(\mathbf{A}^2) = 5(\mathbf{A}^2 - \mathbf{A}) = 5[5(\mathbf{A} - \mathbf{I}) - \mathbf{A}] = 5[4\mathbf{A} - 5\mathbf{I}]$$

Thus

$$P(\mathbf{A}) = 25[3\mathbf{A} - 4\mathbf{I}] + 15[4\mathbf{A} - 5\mathbf{I}] + 10(\mathbf{A} - \mathbf{I}) + \mathbf{A} + \mathbf{I} = 146\mathbf{A} - 184\mathbf{I}$$ ■

Closed Form Solution for Analytic Functions of Matrices

Let $f(x)$ be a function which is analytic in a region Ω of the complex plane and let $\mathbf{A}$ be an $n \times n$ matrix whose eigenvalues $\lambda_i \in \Omega$. Then $f(x)$ has a power series representation

$$f(x) = \sum_{k=0}^{\infty} \alpha_k x^k$$

It is possible to regroup the infinite series for $f(x)$ so that

$$f(x) = \Delta(x) \sum_{k=0}^{\infty} \beta_k x^k + R(x)$$

The remainder R will have degree less than or equal to $n - 1$. The analytic function of the square matrix $\mathbf{A}$ is defined by the same series as its scalar counterpart, but with $\mathbf{A}$ replacing x. Therefore $f(\mathbf{A}) = R(\mathbf{A})$, since $\Delta(\mathbf{A})$ is always the null matrix.

Although the form of $R(x)$ is known to be

$$R(x) = \alpha_0 + \alpha_1 x + \alpha_2 x^2 + \cdots + \alpha_{n-1} x^{n-1}$$

it is clearly impossible to find the coefficients α_i by long division as in Example 8.8. However, if the n eigenvalues λ_i are distinct, n equations for determining the n α_i terms are available. Since $\Delta(\lambda_i) = 0$, setting $x = \lambda_i$ gives $f(\lambda_i) = R(\lambda_i)$, $i = 1, 2, \ldots, n$.

Example 8.9

Find the closed form expression for $\sin \mathbf{A}$ if $\mathbf{A} = \begin{bmatrix} -3 & 1 \\ 0 & -2 \end{bmatrix}$.

$\Delta(\lambda) = (-3 - \lambda)(-2 - \lambda)$, so $\lambda_1 = -3$, $\lambda_2 = -2$. Since $\mathbf{A}$ is a 2×2 matrix, it is known that R is of degree one (or less):

$$R(x) = \alpha_0 + \alpha_1 x$$

Also, using $x = \lambda_1$ and $x = \lambda_2$ gives

$$\sin \lambda_1 = R(\lambda_1) = \alpha_0 + \alpha_1 \lambda_1$$

$$\sin \lambda_2 = R(\lambda_2) = \alpha_0 + \alpha_1 \lambda_2$$

Solving gives

$$\alpha_0 = \frac{\lambda_1 \sin \lambda_2 - \lambda_2 \sin \lambda_1}{\lambda_1 - \lambda_2}, \qquad \alpha_1 = \frac{\sin \lambda_1 - \sin \lambda_2}{\lambda_1 - \lambda_2}$$

or $\alpha_0 = 3 \sin(-2) - 2 \sin(-3)$, $\alpha_1 = \sin(-2) - \sin(-3)$. Using these in $f(\mathbf{A}) = R(\mathbf{A})$ gives

$$\sin \mathbf{A} = \alpha_0 \mathbf{I} + \alpha_1 \mathbf{A} = \begin{bmatrix} \alpha_0 - 3\alpha_1 & \alpha_1 \\ 0 & \alpha_0 - 2\alpha_1 \end{bmatrix} = \begin{bmatrix} \sin(-3) & \sin(-2) - \sin(-3) \\ 0 & \sin(-2) \end{bmatrix}$$ ■

When λ_i is a repeated root, the above procedure must be modified. Some of the equations $f(\lambda_i) = R(\lambda_i)$ will be repeated, so they do not form a set of n linearly independent equations. However, for λ_i a repeated root, $\left.\dfrac{d\Delta(\lambda)}{d\lambda}\right|_{\lambda=\lambda_i} = 0$ also, and so

$$\left.\frac{df(\lambda)}{d\lambda}\right|_{\lambda=\lambda_i} = \left.\frac{d\Delta}{d\lambda}\sum_{k=0}^{\infty}\beta_k\lambda_i^k + \Delta(\lambda_i)\frac{d}{d\lambda}[\sum \beta_k\lambda^k]\right|_{\lambda=\lambda_i} + \left.\frac{dR}{d\lambda}\right|_{\lambda=\lambda_i} = \left.\frac{dR}{d\lambda}\right|_{\lambda=\lambda_i}$$

For an eigenvalue with algebraic multiplicity m_i, the first $m_i - 1$ derivatives of Δ all vanish and thus

$$f(\lambda_i) = R(\lambda_i), \quad \left.\frac{df}{d\lambda}\right|_{\lambda_i} = \left.\frac{dR}{d\lambda}\right|_{\lambda_i}, \quad \left.\frac{d^2f}{d\lambda^2}\right|_{\lambda_i} = \left.\frac{d^2R}{d\lambda^2}\right|_{\lambda_i}, \ldots, \left.\frac{d^{m_i-1}f}{d\lambda^{m_i-1}}\right|_{\lambda_i} = \left.\frac{d^{m_i-1}R}{d\lambda^{m_i-1}}\right|_{\lambda_i}$$

form a set of m_i linearly independent equations. Thus a full set of n equations is always available for finding the α_i coefficients of the remainder term R.

Example 8.10

Find the closed form expression for $e^{\mathbf{A}t}$ if $\mathbf{A} = \begin{bmatrix} 0 & 1 & 0 \\ 0 & 0 & 1 \\ 27 & -27 & 9 \end{bmatrix}$.

We have

$$|\mathbf{A} - \mathbf{I}\lambda| = \Delta(\lambda) = -\lambda^3 + 9\lambda^2 - 27\lambda + 27 = (3 - \lambda)^3$$

Therefore $\lambda_1 = \lambda_2 = \lambda_3 = 3$, and

$$e^{\mathbf{A}t} = R(\mathbf{A}) = \alpha_0\mathbf{I} + \alpha_1\mathbf{A} + \alpha_2\mathbf{A}^2$$

where

$$e^{3t} = \alpha_0 + 3\alpha_1 + 9\alpha_2$$

$$\left.\frac{de^{\lambda t}}{d\lambda}\right|_{\lambda=3} = \left.\frac{d}{d\lambda}[\alpha_0 + \lambda\alpha_1 + \lambda^2\alpha_2]\right|_{\lambda=3} \quad \text{or} \quad te^{3t} = \alpha_1 + 6\alpha_2$$

$$\left.\frac{d^2e^{\lambda t}}{d\lambda^2}\right|_{\lambda=3} = \left.\frac{d^2}{d\lambda^2}[\alpha_0 + \lambda\alpha_1 + \lambda^2\alpha_2]\right|_{\lambda=3} \quad \text{or} \quad t^2e^{3t} = 2\alpha_2$$

Solving for α_0, α_1, and α_2 gives

$$\alpha_2 = \tfrac{1}{2}t^2e^{3t}, \qquad \alpha_1 = te^{3t} - 6\alpha_2 = (t - 3t^2)e^{3t},$$

$$\alpha_0 = e^{3t} - 3\alpha_1 - 9\alpha_2 = (1 - 3t + \tfrac{9}{2}t^2)e^{3t}$$

Using these coefficients in $R(\mathbf{A})$ gives

$$e^{\mathbf{A}t} = \begin{bmatrix} 1 - 3t + \frac{9}{2}t^2 & t - 3t^2 & \frac{1}{2}t^2 \\ \frac{27}{2}t^2 & 1 - 3t - 9t^2 & t + \frac{3}{2}t^2 \\ 27t + \frac{81}{2}t^2 & -27t - 27t^2 & 1 + 6t + \frac{9}{2}t^2 \end{bmatrix} e^{3t}$$

■

When some eigenvalues are repeated and others are simple roots, a full set of n independent equations are still available for computing the α_i coefficients.

The method presented above for computing functions of a matrix requires a knowledge of the eigenvalues. If the eigenvectors are also known, an alternative method can be used (see Problem 8.21). For the particular function $f(\mathbf{A}) = e^{\mathbf{A}t}$, a third alternative is available (see Problems 8.19 and 8.20). Finally, an approximation can be obtained by truncating the infinite series after a finite number of terms.

The reason for emphasizing the function $e^{\mathbf{A}t}$ is that it plays an important part in the solution of linear constant coefficient differential equations. The matrix $\mathbf{A}^k$ plays the corresponding role in solutions of linear constant coefficient difference equations [28, 86, 117].

ILLUSTRATIVE PROBLEMS

Inversion of Matrices and Reduction of Polynomials

8.1 If $\mathbf{A} = \begin{bmatrix} 1 & -1 \\ 1 & 1 \end{bmatrix}$, use the Cayley-Hamilton theorem to compute a. $\mathbf{A}^{-1}$ and b. $P(\mathbf{A}) = \mathbf{A}^5 + 16\mathbf{A}^4 + 32\mathbf{A}^3 + 16\mathbf{A}^2 + 4\mathbf{A} + \mathbf{I}$.

a. $\Delta(\lambda) = |\mathbf{A} - \mathbf{I}\lambda| = \lambda^2 - 2\lambda + 2$. Therefore

$$\mathbf{A}^2 - 2\mathbf{A} + 2\mathbf{I} = [\mathbf{0}] \qquad \text{or} \qquad \mathbf{A}^{-1} = -\tfrac{1}{2}[\mathbf{A} - 2\mathbf{I}] = \frac{1}{2}\begin{bmatrix} 1 & 1 \\ -1 & 1 \end{bmatrix}$$

b. $P(\mathbf{A}) = R(\mathbf{A})$, where R is the remainder term in

$$\frac{P(x)}{\Delta(x)} = x^3 + 18x^2 + 66x + 112 + \frac{96x - 223}{\Delta(x)}$$

Therefore $P(\mathbf{A}) = 96\mathbf{A} - 223\mathbf{I} = \begin{bmatrix} -127 & -96 \\ 96 & -127 \end{bmatrix}$.

8.2 $\mathbf{A} = \begin{bmatrix} 1 & 2 \\ 3 & 4 \end{bmatrix}$. Find a. $\mathbf{A}^{-1}$ and b. $P(\mathbf{A}) = \mathbf{A}^5 + \mathbf{A}^3 + \mathbf{A} + \mathbf{I}$.

a. $\Delta(\lambda) = \lambda^2 - 5\lambda - 2$. Therefore $\Delta(\mathbf{A}) = \mathbf{A}^2 - 5\mathbf{A} - 2\mathbf{I} = [\mathbf{0}]$ by the Cayley-Hamilton theorem and $\mathbf{A} - 5\mathbf{I} - 2\mathbf{A}^{-1} = [\mathbf{0}]$ or $\mathbf{A}^{-1} = \tfrac{1}{2}[\mathbf{A} - 5\mathbf{I}] = \begin{bmatrix} -2 & 1 \\ 3/2 & -1/2 \end{bmatrix}$.

b. $P(x)/\Delta(x) = x^3 + 5x^2 + 28x + 150 + [(807x + 301)/\Delta(x)]$. Therefore

$$P(x) = \Delta(x)[x^3 + 5x^2 + 28x + 150] + \underbrace{807x + 301}_{R(x)}$$

and $P(\mathbf{A}) = R(\mathbf{A}) = 807\mathbf{A} + 301\mathbf{I} = \begin{bmatrix} 1108 & 1614 \\ 2421 & 3529 \end{bmatrix}$.

Functions with Singularities

8.3 Comment on the following functions in view of Theorems 8.1, 8.2, and 8.3: a. $f_a(x) = 1/(1 - x)$, and b. $f_b(x) = \tan x$.

a. $f_a(x)$ has a singularity (pole) at $x = 1$, but is analytic elsewhere. In particular, it is analytic at all points x in the complex plane inside the circle $|x| < 1$, and for these points

$$f_a(x) = 1 + x + x^2 + x^3 + \cdots$$

If all the eigenvalues of $\mathbf{A}$ satisfy $|\lambda_i| < 1$, then $f_a(\mathbf{A}) = (\mathbf{I} - \mathbf{A})^{-1}$ exists and can be written as a convergent series

$$f_a(\mathbf{A}) = \mathbf{I} + \mathbf{A} + \mathbf{A}^2 + \mathbf{A}^3 + \cdots$$

Note that $(\mathbf{I} - \mathbf{A})f_a(\mathbf{A}) = \mathbf{I} + (\mathbf{A} - \mathbf{A}) + (\mathbf{A}^2 - \mathbf{A}^2) + \cdots = \mathbf{I}$ as required. When $\mathbf{A}$ has $\lambda = 1$ as an eigenvalue, then $(\mathbf{I} - \mathbf{A})$ is singular and the inverse does not exist. If $\lambda = 1$ is not an eigenvalue, but at least one eigenvalue has a magnitude larger than unity, then $(\mathbf{I} - \mathbf{A})^{-1}$ exists, but cannot be represented by the above infinite series.

b. $f_b(x) = \sin x/\cos x$ has a singularity (pole) at each zero of $\cos x$, that is, at $x = \pm\pi/2, \pm 3\pi/2, \ldots$, but is analytic elsewhere. If all eigenvalues of $\mathbf{A}$ satisfy $|\lambda| < \pi/2$, then we could defined $\tan \mathbf{A} = \mathbf{A} + \mathbf{A}^3/3 + 2\mathbf{A}^5/15 + \cdots$ and be assured that this series is convergent (see Problem 8.11).

Powers of a Jordan Block

8.4 Let $\mathbf{J}_1$ be an $m \times m$ Jordan block. Show that for any integer $k > 0$,

$$\mathbf{J}_1^k = \begin{bmatrix} \lambda^k & k\lambda^{k-1} & \frac{1}{2!}k(k-1)\lambda^{k-2} & \frac{1}{3!}k(k-1)(k-2)\lambda^{k-3} & \cdots & \frac{k!\lambda^{k-m+1}}{(k-m+1)!(m-1)!} \\ 0 & \lambda^k & k\lambda^{k-1} & \frac{1}{2!}k(k-1)\lambda^{k-2} & & \\ 0 & 0 & \lambda^k & k\lambda^{k-1} & & \vdots \\ 0 & 0 & 0 & \lambda^k & & \\ 0 & 0 & 0 & & \ddots & \\ \vdots & & & & & k\lambda^{k-1} \\ 0 & 0 & 0 & & & \lambda^k \end{bmatrix}$$

With $k = 1$, $\mathbf{J}_1$ satisfies the above equation. To see this, it must be recalled that $(k - m + 1)! = \infty$ if $k - m + 1 < 0$ [94]. Multiplying gives $\mathbf{J}_1$ squared:

$$\mathbf{J}_1 = \begin{bmatrix} \lambda & 1 & 0 & \cdots & 0 \\ 0 & \lambda & 1 & & \\ 0 & 0 & \lambda & & \vdots \\ 0 & 0 & 0 & & \\ \vdots & & & & 1 \\ 0 & 0 & 0 & & \lambda \end{bmatrix}, \quad \mathbf{J}_1^2 = \begin{bmatrix} \lambda^2 & 2\lambda & 1 & 0 & \cdots & 0 \\ 0 & \lambda^2 & 2\lambda & 1 & & \\ 0 & 0 & \lambda^2 & 2\lambda & \ddots & \vdots \\ \vdots & & & & \ddots & 1 \\ & & & & & 2\lambda \\ 0 & 0 & 0 & 0 & & \lambda^2 \end{bmatrix}$$

Use induction, assuming the stated form holds for k, and show that it holds for $k + 1$ by computing

$$\begin{bmatrix} \lambda & 1 & 0 & \cdots & 0 \\ 0 & \lambda & 1 & & \\ 0 & 0 & \lambda & & \vdots \\ 0 & 0 & 0 & & \\ \vdots & & & & \\ 0 & 0 & 0 & & \lambda \end{bmatrix} \begin{bmatrix} \lambda^k & k\lambda^{k-1} & \frac{1}{2!}k(k-1)\lambda^{k-2} & \cdots & \\ 0 & \lambda^k & k\lambda^{k-1} & & \\ 0 & 0 & k^k & & \vdots \\ 0 & 0 & 0 & \ddots & \\ \vdots & \vdots & & & k\lambda^{k-1} \\ 0 & 0 & & & \lambda^k \end{bmatrix} =$$

$$\begin{bmatrix} \lambda^{k+1} & (k+1)\lambda^k & \frac{1}{2!}k(k-1)\lambda^{k-1} + k\lambda^{k-1} & \frac{1}{3!}k(k-1)(k-2)\lambda^{k-2} + \frac{1}{2!}k(k-1)\lambda^{k-2} & \cdots \\ 0 & \lambda^{k+1} & (k+1)\lambda^k & & \\ 0 & 0 & \lambda^{k+1} & & \vdots \\ 0 & 0 & 0 & \vdots & \\ \vdots & & & & \\ 0 & 0 & 0 & & \lambda^{k+1} \end{bmatrix}$$

But

$$\left[\frac{1}{2!}k(k-1) + k\right]\lambda^{k-1} = \frac{k}{2!}(k - 1 + 2)\lambda^{k-1} = \frac{(k+1)k}{2!}\lambda^{k-1}$$

$$\left[\frac{1}{3!}k(k-1)(k-2) + \frac{1}{2!}k(k-1)\right]\lambda^{k-2} = \frac{1}{3!}k(k-1)(k+1)^{k-2}$$

etc.

so the stated result holds.

Proof of Cayley-Hamilton Theorem

8.5 Prove the Cayley-Hamilton theorem when $\mathbf{A}$ is not diagonalizable.

Any square $\mathbf{A}$ can be reduced to Jordan canonical form $\mathbf{J} = \mathbf{M}^{-1}\mathbf{A}\mathbf{M}$, so that $\mathbf{A} = \mathbf{M}\mathbf{J}\mathbf{M}^{-1}$, $\mathbf{A}^2 = \mathbf{M}\mathbf{J}\mathbf{M}^{-1}\mathbf{M}\mathbf{J}\mathbf{M}^{-1} = \mathbf{M}\mathbf{J}^2\mathbf{M}^{-1}$, $\mathbf{A}^k = \mathbf{M}\mathbf{J}^k\mathbf{M}^{-1}$. If the characteristic polynomial is $\Delta(\lambda) = c_n\lambda^n + c_{n-1}\lambda^{n-1} + \cdots + c_1\lambda + c_0$, then

$$\begin{aligned} \Delta(\mathbf{A}) &= c_n\mathbf{A}^n + c_{n-1}\mathbf{A}^{n-1} + \cdots + c_1\mathbf{A} + c_0\mathbf{I} \\ &= \mathbf{M}[c_n\mathbf{J}^k + c_{n-1}\mathbf{J}^{n-1} + \cdots + c_1\mathbf{J} + c_0\mathbf{I}]\mathbf{M}^{-1} \end{aligned}$$

It is to be shown that the matrix polynomial inside the brackets sums to the zero matrix. The Jordan form is $\mathbf{J} = \text{diag}\,[\mathbf{J}_1, \mathbf{J}_2, \ldots, \mathbf{J}_p]$, where $\mathbf{J}_i$ are Jordan blocks. Also, $\mathbf{J}^k = \text{diag}\,[\mathbf{J}_1^k, \mathbf{J}_2^k, \ldots, \mathbf{J}_p^k]$. Therefore, to prove the theorem it is only necessary to show that $c_n\mathbf{J}_i^n + c_{n-1}\mathbf{J}_i^{n-1} + \cdots + c_1\mathbf{J}_i + c_0\mathbf{I} = [\mathbf{0}]$ for a typical block $\mathbf{J}_i$. Using the result of the previous problem, all terms below the main diagonal are zero. The sum of the terms in a typical main diagonal position is $c_n\lambda_i^n + c_{n-1}\lambda_i^{n-1} + \cdots + c_1\lambda_i + c_0$ and equals zero because this is just the characteristic polynomial evaluated with a root λ_i. A typical term of the sum in the diagonal just above the main diagonal is $nc_n\lambda_i^{n-1} + (n-1)c_{n-1}\lambda_i^{n-2} + \cdots + c_2\lambda_i + c_1$. This sum is zero since it equals $\left.\dfrac{d\Delta(\lambda)}{d\lambda}\right|_{\lambda=\lambda_i}$. The root λ_i is a multiple root, so

$$\Delta(\lambda) = c(\lambda - \lambda_i)^m(\lambda - \lambda_j)(\lambda - \lambda_k)\ldots, \qquad \text{and} \qquad \left.\frac{d\Delta}{d\lambda}\right|_{\lambda=\lambda_i} = 0$$

If $\mathbf{J}_i$ is an $r \times r$ block, λ_i is at least an rth order root, so

$$\left.\frac{1}{2}\frac{d^2\Delta(\lambda)}{d\lambda^2}\right|_{\lambda=\lambda_i} = 0, \qquad \left.\frac{1}{3!}\frac{d^3\Delta(\lambda)}{d\lambda^3}\right|_{\lambda=\lambda_i} = 0, \qquad \ldots, \qquad \left.\frac{1}{(r-1)!}\frac{d^{r-1}\Delta(\lambda)}{d\lambda^{r-1}}\right|_{\lambda=\lambda_i} = 0$$

The successive diagonals above the main diagonal give terms which sum to these derivatives of the characteristic equation. Hence

$$c_n\mathbf{J}_i^n + \cdots + c_1\mathbf{J}_i + c_0\mathbf{I} = [\mathbf{0}] \qquad \text{and so}$$

$$c_n\mathbf{A}^n + c_{n-1}\mathbf{A}^{n-1} + \cdots + c_1\mathbf{A} + c_0\mathbf{I} = [\mathbf{0}]$$

8.6 Give a general proof of the Cayley-Hamilton theorem without using the Jordan form.

$\mathbf{A}$ is an $n \times n$ matrix with n eigenvalues λ_i, some of which may be equal. The characteristic polynomial is

$$\Delta(\lambda) = |\mathbf{A} - \mathbf{I}\lambda| = (-\lambda)^n + c_{n-1}\lambda^{n-1} + c_{n-2}\lambda^{n-2} + \cdots + c_1\lambda + c_0 \tag{1}$$

We are to show that

$$\Delta(\mathbf{A}) = (-1)^n\mathbf{A}^n + c_{n-1}\mathbf{A}^{n-1} + c_{n-2}\mathbf{A}^{n-2} + \cdots + c_1\mathbf{A} + c_0\mathbf{I} = [\mathbf{0}] \tag{2}$$

Consider Adj $[\mathbf{A} - \mathbf{I}\lambda]$. Its elements are formed from $n - 1 \times n - 1$ determinants obtained by deleting a row and a column of $\mathbf{A} - \mathbf{I}\lambda$. Therefore, the highest power of λ that can be in any element of Adj $[\mathbf{A} - \mathbf{I}\lambda]$ is λ^{n-1}. This means it is possible to write

$$\text{Adj}\,[\mathbf{A} - \mathbf{I}\lambda] = \mathbf{B}_{n-1}\lambda^{n-1} + \mathbf{B}_{n-2}\lambda^{n-2} + \cdots + \mathbf{B}_1\lambda + \mathbf{B}_0 \tag{3}$$

where the $\mathbf{B}_i$ terms are $n \times n$ matrices not containing λ, but are otherwise unknown. We use the known result

$$[\mathbf{A} - \mathbf{I}\lambda]\,\text{Adj}\,[\mathbf{A} - \mathbf{I}\lambda] = |\mathbf{A} - \mathbf{I}\lambda|\mathbf{I} \tag{4}$$

Substituting equation (*3*) into the left side of equation (*4*) gives

$$[\mathbf{A} - \mathbf{I}\lambda]\,\text{Adj}\,[\mathbf{A} - \mathbf{I}\lambda] = -\mathbf{B}_{n-1}\lambda^n + (\mathbf{AB}_{n-1} - \mathbf{B}_{n-2})\lambda^{n-1} + (\mathbf{AB}_{n-2} - \mathbf{B}_{n-3})\lambda^{n-2}$$
$$+ \cdots + (\mathbf{AB}_2 - \mathbf{B}_1)\lambda^2 + (\mathbf{AB}_1 - \mathbf{B}_0)\lambda + \mathbf{AB}_0$$

Using equation (*1*) on the right side of equation (*4*) gives

$$\Delta(\lambda)\mathbf{I} = (-\lambda)^n\mathbf{I} + c_{n-1}\lambda^{n-1}\mathbf{I} + c_{n-2}\lambda^{n-2}\mathbf{I} + \cdots + c_1\lambda\mathbf{I} + c_0\mathbf{I}$$

The left side equals the right side, and the coefficients of like powers of λ on the two sides must be equal. This leads to the following set of equations:

$$\begin{aligned}
-\mathbf{B}_{n-1} &= (-1)^n\mathbf{I} \\
\mathbf{AB}_{n-1} - \mathbf{B}_{n-2} &= c_{n-1}\mathbf{I} \\
\mathbf{AB}_{n-2} - \mathbf{B}_{n-3} &= c_{n-2}\mathbf{I} \\
&\vdots \\
\mathbf{AB}_2 - \mathbf{B}_1 &= c_2\mathbf{I} \\
\mathbf{AB}_1 - \mathbf{B}_0 &= c_1\mathbf{I} \\
\mathbf{AB}_0 &= c_0\mathbf{I}
\end{aligned}$$

If the first of these equations is premultiplied by $\mathbf{A}^n$, the second by $\mathbf{A}^{n-1}$, etc., the sum of the right-side terms is $\Delta(\mathbf{A})$. The left-side sum must also equal $\Delta(\mathbf{A})$, and takes the form

$$(-\mathbf{A}^n\mathbf{B}_{n-1} + \mathbf{A}^n\mathbf{B}_{n-1}) + (-\mathbf{A}^{n-1}\mathbf{B}_{n-2} + \mathbf{A}^{n-1}\mathbf{B}_{n-2}) + (-\mathbf{A}^{n-2}\mathbf{B}_{n-3} + \mathbf{A}^{n-2}\mathbf{B}_{n-3})$$
$$+ \cdots + (-\mathbf{A}^2\mathbf{B}_1 + \mathbf{A}^2\mathbf{B}_1) + (-\mathbf{A}\mathbf{B}_0 + \mathbf{A}\mathbf{B}_0) \equiv [\mathbf{0}]$$

Therefore $\Delta(\mathbf{A}) = [\mathbf{0}]$.

Functions of a Jordan Block

8.7 If $\mathbf{J}_1$ is an $m \times m$ Jordan block with eigenvalue λ_1, find a general expression for the coefficients α_i for $e^{\mathbf{J}_1 t}$.

The characteristic equation is $\Delta(\lambda) = (\lambda - \lambda_1)^m$ and

$$e^{\mathbf{J}_1 t} = \alpha_0 \mathbf{I} + \alpha_1 \mathbf{J}_1 + \alpha_2 \mathbf{J}_1^2 + \cdots + \alpha_{m-1}\mathbf{J}_1^{m-1}$$

where

$$e^{\lambda_1 t} = \alpha_0 + \alpha_1\lambda_1 + \alpha_2\lambda_1^2 + \cdots + \alpha_{m-1}\lambda_1^{m-1}$$

$$\left.\frac{de^{\lambda t}}{d\lambda}\right|_{\lambda_1} = te^{\lambda_1 t} = \alpha_1 + 2\alpha_2\lambda_1 + 3\alpha_3\lambda_1^2 + \cdots + (m-1)\alpha_{m-1}\lambda_1^{m-2}$$

More symmetry is achieved if the kth derivative term is divided by $1/k!$:

$$\tfrac{1}{2}t^2 e^{\lambda_1 t} = \alpha_2 + 3\alpha_3\lambda_1 + \cdots + \tfrac{1}{2}(m-1)(m-2)\alpha_{m-1}\lambda_1^{m-3}$$

$$\frac{1}{3!}t^3 e^{\lambda_1 t} = \alpha_3 + 4\lambda_1 + \cdots + \frac{1}{3!}(m-1)(m-2)(m-3)\alpha_{m-1}\lambda_1^{m-4}$$

$$\vdots$$

$$\frac{1}{(m-1)!}t^{m-1}e^{\lambda_1 t} = \alpha_{m-1}$$

or

$$\begin{bmatrix} 1 \\ t \\ \tfrac{1}{2}t^2 \\ \frac{1}{3!}t^3 \\ \vdots \\ \frac{1}{(m-1)!}t^{m-1} \end{bmatrix} e^{\lambda_1 t} = \begin{bmatrix} 1 & \lambda_1 & \lambda_1^2 & & \cdots & \lambda_1^{m-1} \\ 0 & 1 & 2\lambda_1 & 3\lambda_1^2 & & (m-1)\lambda_1^{m-2} \\ 0 & 0 & 1 & 3\lambda_1 & & \tfrac{1}{2}(m-1)(m-2)\lambda_1^{m-3} \\ \vdots & & & & & \vdots \\ 0 & 0 & 0 & 0 & & 1 \end{bmatrix} \begin{bmatrix} \alpha_0 \\ \alpha_1 \\ \alpha_2 \\ \vdots \\ \alpha_{m-1} \end{bmatrix} \triangleq \mathbf{F}\boldsymbol{\alpha}$$

$$\boldsymbol{\alpha} = [\mathbf{F}]^{-1} \begin{bmatrix} 1 \\ t \\ \tfrac{1}{2}t^2 \\ \frac{1}{3!}t^3 \\ \vdots \\ \frac{1}{(m-1)!}t^{m-1} \end{bmatrix} e^{\lambda_1 t}$$

But using Gaussian elimination, for example, gives

$$\mathbf{F}^{-1} = \begin{bmatrix} 1 & -\lambda_1 & \lambda_1^2 & -\lambda_1^3 & \lambda_1^4 & \cdots & (-\lambda_1)^{m-1} \\ 0 & 1 & -2\lambda_1 & 3\lambda_1^2 & -4\lambda_1^3 & & (m-1)(-\lambda_1)^{m-2} \\ 0 & 0 & 1 & -3\lambda_1 & 6\lambda_1^2 & & \frac{1}{2}(m-1)(m-2)(-\lambda_1)^{m-3} \\ 0 & 0 & 0 & 1 & -4\lambda_1 & & \frac{1}{3!}(m-1)(m-2)(m-3)(-\lambda_1)^{m-4} \\ \vdots & & & & \ddots & & \vdots \\ & & & & & \cdot & -(m-1)\lambda_1 \\ 0 & 0 & 0 & 0 & 0 & \cdots & 1 \end{bmatrix}$$

so

$$\alpha_0 = e^{\lambda_1 t}\left[1 - \lambda_1 t + \frac{\lambda_1^2 t^2}{2!} - \frac{\lambda_1^3 t^3}{3!} + \frac{\lambda_1^4 t^4}{4!} - \cdots + \frac{(-\lambda_1)^{m-1} t^{m-1}}{(m-1)!}\right]$$

$$\alpha_1 = e^{\lambda_1 t}\left[t - \lambda_1 t^2 + \frac{\lambda_1^2 t^3}{2} - \frac{\lambda_1^3 t^4}{3!} + \cdots + \frac{t^{m-1}(-\lambda_1)^{m-2}}{(m-2)!}\right]$$

$$\alpha_2 = e^{\lambda_1 t}\left[\frac{1}{2}t^2 - \frac{\lambda_1 t^3}{2} + \frac{\lambda_1^2 t^4}{4} + \cdots + \frac{t^{m-1}(-\lambda)^{m-3}}{2(m-3)!}\right]$$

$$\vdots$$

$$\alpha_{m-2} = e^{\lambda_1 t}\left[\frac{1}{(m-2)!}t^{m-2} - \frac{1}{(m-2)!}\lambda_1 t^{m-1}\right]$$

$$\alpha_{m-1} = e^{\lambda_1 t}\frac{t^{m-1}}{(m-1)!}$$

8.8 Use the results of the previous problem to find $e^{\mathbf{J}_1 t}$ if a. $\mathbf{J}_1$ is a 3×3 Jordan block, and b. $\mathbf{J}_1$ is a 4×4 Jordan block.

a. $$\mathbf{J}_1 = \begin{bmatrix} \lambda_1 & 1 & 0 \\ 0 & \lambda_1 & 1 \\ 0 & 0 & \lambda_1 \end{bmatrix}, \qquad \mathbf{J}_1^2 = \begin{bmatrix} \lambda_1^2 & 2\lambda_1 & 1 \\ 0 & \lambda_1^2 & 2\lambda_1 \\ 0 & 0 & \lambda_1^2 \end{bmatrix}$$

$$e^{\mathbf{J}_1 t} = \alpha_0\mathbf{I} + \alpha_1\mathbf{J}_1 + \alpha_2\mathbf{J}_1^2$$

$$= \begin{bmatrix} \alpha_0 + \alpha_1\lambda_1 + \alpha_2\lambda_1^2 & \alpha_1 + 2\lambda_1\alpha_2 & \alpha_2 \\ 0 & \alpha_0 + \alpha_1\lambda_1 + \alpha_2\lambda_1^2 & \alpha_1 + 2\lambda_1\alpha_2 \\ 0 & 0 & \alpha_0 + \alpha_1\lambda_1 + \alpha_2\lambda_1^2 \end{bmatrix}$$

$$= e^{\lambda_1 t}\begin{bmatrix} 1 & t & \frac{1}{2}t^2 \\ 0 & 1 & t \\ 0 & 0 & 1 \end{bmatrix}$$

b. $$\mathbf{J}_1 = \begin{bmatrix} \lambda_1 & 1 & 0 & 0 \\ 0 & \lambda_1 & 1 & 0 \\ 0 & 0 & \lambda_1 & 1 \\ 0 & 0 & 0 & \lambda_1 \end{bmatrix}, \qquad \mathbf{J}_1^2 = \begin{bmatrix} \lambda_1^2 & 2\lambda_1 & 1 & 0 \\ 0 & \lambda_1^2 & 2\lambda_1 & 1 \\ 0 & 0 & \lambda_1^2 & 2\lambda_1 \\ 0 & 0 & 0 & \lambda_1^2 \end{bmatrix},$$

$$\mathbf{J}_1^3 = \begin{bmatrix} \lambda_1^3 & 3\lambda_1^2 & 3\lambda_1 & 1 \\ 0 & \lambda_1^3 & 3\lambda_1^2 & 3\lambda_1 \\ 0 & 0 & \lambda_1^3 & 3\lambda_1^2 \\ 0 & 0 & 0 & \lambda_1^3 \end{bmatrix}$$

so

$$e^{\mathbf{J}_1 t} = \alpha_0 \mathbf{I} + \alpha_1 \mathbf{J}_1 + \alpha_2 \mathbf{J}_1^2 + \alpha_3 \mathbf{J}_1^3$$

$$= e^{\lambda_1 t}\begin{bmatrix} 1 & t & \frac{1}{2}t^2 & \frac{1}{3!}t^3 \\ 0 & 1 & t & \frac{1}{2}t^2 \\ 0 & 0 & 1 & t \\ 0 & 0 & 0 & 1 \end{bmatrix}$$

The pattern illustrated by these two cases continues for any $m \times m$ Jordan block.

Some Matrix Identities

8.9 a. If $\mathbf{A} = \begin{bmatrix} 1 & 2 \\ -2 & -3 \end{bmatrix}$, compute the 2×2 matrices $\sin \mathbf{A}$ and $\cos \mathbf{A}$.

b. Verify that $\sin^2 \mathbf{A} + \cos^2 \mathbf{A} = \mathbf{I}$.

a. From Problem 7.2, $\Delta(\lambda) = \lambda^2 + 2\lambda + 1$ and $\lambda_1 = \lambda_2 = -1$. Now $\sin \mathbf{A} = \alpha_0 \mathbf{I} + \alpha_1 \mathbf{A}$ and $\cos \mathbf{A} = \alpha_2 \mathbf{I} + \alpha_3 \mathbf{A}$, where

$$\left.\begin{aligned} \sin \lambda_1 &= \alpha_0 + \alpha_1 \lambda_1 \\ \frac{d}{d\lambda}(\sin \lambda)\Big|_{\lambda=\lambda_1} &= \cos \lambda_1 = \alpha_1 \end{aligned}\right\} \Rightarrow \begin{cases} \alpha_1 = \cos(-1) = \cos(1) \\ \alpha_0 = \sin \lambda_1 + \alpha_1 = \cos(1) - \sin(1) \end{cases}$$

and

$$\left.\begin{aligned} \cos \lambda_1 &= \alpha_2 + \lambda_1 \alpha_3 \\ \frac{d}{d\lambda}(\cos \lambda)\Big|_{\lambda=\lambda_1} &= -\sin \lambda_1 = \alpha_3 \end{aligned}\right\} \Rightarrow \begin{cases} \alpha_3 = \sin(1) \\ \alpha_2 = \cos(1) + \alpha_3 = \cos(1) + \sin(1) \end{cases}$$

Therefore

$$\sin \mathbf{A} = \begin{bmatrix} 2\cos(1) - \sin(1) & 2\cos(1) \\ -2\cos(1) & -2\cos(1) - \sin(1) \end{bmatrix}$$

and

$$\cos \mathbf{A} = \begin{bmatrix} \cos(1) + 2\sin(1) & 2\sin(1) \\ -2\sin(1) & \cos(1) - 2\sin(1) \end{bmatrix}$$

b. Multiplication gives

$$\sin^2 \mathbf{A} = \begin{bmatrix} \sin^2(1) - 4\cos(1)\sin(1) & -4\cos(1)\sin(1) \\ 4\cos(1)\sin(1) & \sin^2(1) + 4\cos(1)\sin(1) \end{bmatrix}$$

and

$$\cos^2 \mathbf{A} = \begin{bmatrix} \cos^2(1) + 4\cos(1)\sin(1) & 4\cos(1)\sin(1) \\ -4\cos(1)\sin(1) & \cos^2(1) - 4\cos(1)\sin(1) \end{bmatrix}$$

so $\sin^2 \mathbf{A} + \cos^2 \mathbf{A} = \begin{bmatrix} \sin^2(1) + \cos^2(1) & 0 \\ 0 & \sin^2(1) + \cos^2(1) \end{bmatrix} = \mathbf{I}.$

8.10 a. Compute $e^{\mathbf{A}t}$, $e^{-\mathbf{A}t}$, $\sinh \mathbf{A}t$, and $\cosh \mathbf{A}t$ for $\mathbf{A} = \begin{bmatrix} 1 & 1 \\ 1 & 1 \end{bmatrix}$.

b. Verify that $\sinh \mathbf{A}t = (e^{\mathbf{A}t} - e^{-\mathbf{A}t})/2$, $\cosh \mathbf{A}t = (e^{\mathbf{A}t} + e^{-\mathbf{A}t})/2$, and $\cosh^2 \mathbf{A}t - \sinh^2 \mathbf{A}t = \mathbf{I}$.

a. The eigenvalues are $\lambda_1 = 0$ and $\lambda_2 = 2$, so $f(\mathbf{A}) = \alpha_0\mathbf{I} + \alpha_1\mathbf{A} = \begin{bmatrix} \alpha_0 + \alpha_1 & \alpha_1 \\ \alpha_1 & \alpha_0 + \alpha_1 \end{bmatrix}$. The coefficients α_0 and α_1 are found from $f(\lambda_1) = \alpha_0$ and $f(\lambda_2) = \alpha_0 + 2\alpha_1$. Therefore $\alpha_1 = [f(\lambda_2) - f(\lambda_1)]/2$ and $\alpha_0 + \alpha_1 = [f(\lambda_2) + f(\lambda_1)]/2$ for any analytic $f(\lambda)$. Letting $f(\lambda)$ be each of the four required functions gives, in turn,

$$e^{\mathbf{A}t} = \begin{bmatrix} \dfrac{e^{2t}+1}{2} & \dfrac{e^{2t}-1}{2} \\ \dfrac{e^{2t}-1}{2} & \dfrac{e^{2t}+1}{2} \end{bmatrix}, \qquad e^{-\mathbf{A}t} = \begin{bmatrix} \dfrac{e^{-2t}+1}{2} & \dfrac{e^{-2t}-1}{2} \\ \dfrac{e^{-2t}-1}{2} & \dfrac{e^{-2t}+1}{2} \end{bmatrix}$$

$$\sinh \mathbf{A}t = \begin{bmatrix} \dfrac{\sinh 2t}{2} & \dfrac{\sinh 2t}{2} \\ \dfrac{\sinh 2t}{2} & \dfrac{\sinh 2t}{2} \end{bmatrix}, \qquad \cosh \mathbf{A}t = \begin{bmatrix} \dfrac{\cosh 2t+1}{2} & \dfrac{\cosh 2t-1}{2} \\ \dfrac{\cosh 2t-1}{2} & \dfrac{\cosh 2t+1}{2} \end{bmatrix}$$

b. Since $(e^{2t} - e^{-2t})/2 = \sinh 2t$ and $(e^{2t} + e^{-2t})/2 = \cosh 2t$, it is clear that $(e^{\mathbf{A}t} + e^{-\mathbf{A}t})/2 = \cosh \mathbf{A}t$ and $(e^{\mathbf{A}t} - e^{-\mathbf{A}t})/2 = \sinh \mathbf{A}t$. Computing

$$\cosh^2 \mathbf{A}t = \begin{bmatrix} \dfrac{\cosh^2 2t+1}{2} & \dfrac{\cosh^2 2t-1}{2} \\ \dfrac{\cosh^2 2t-1}{2} & \dfrac{\cosh^2 2t+1}{2} \end{bmatrix}$$

$$\sinh^2 \mathbf{A}t = \begin{bmatrix} \dfrac{\sinh^2 2t}{2} & \dfrac{\sinh^2 2t}{2} \\ \dfrac{\sinh^2 2t}{2} & \dfrac{\sinh^2 2t}{2} \end{bmatrix}$$

and using $\cosh^2 2t - \sinh^2 2t = 1$ gives $\cosh^2 \mathbf{A}t - \sinh^2 \mathbf{A}t = \mathbf{I}$.

8.11 Let $\mathbf{A} = \begin{bmatrix} -1 & 1 \\ 1 & 1 \end{bmatrix}$.

a. Find sin **A**, cos **A**, and tan **A**.

b. Show that $\tan \mathbf{A} = (\sin \mathbf{A})(\cos \mathbf{A})^{-1}$.

a. The eigenvalues are $\lambda_1 = \sqrt{2}$ and $\lambda_2 = -\sqrt{2}$. Since both eigenvalues satisfy $|\lambda_i| < \pi/2$, use of the results for tan **A** of Problem 8.3 is justified. For all three functions, $f(\mathbf{A}) = \alpha_0\mathbf{I} + \alpha_1\mathbf{A}$, where $\alpha_0 = (1/2)[f(\sqrt{2}) + f(-\sqrt{2})]$ and $\alpha_1 = (1/\sqrt{2})[\alpha_0 - f(-\sqrt{2})]$, so

$$\sin \mathbf{A} = \frac{\sin\sqrt{2}}{\sqrt{2}}\begin{bmatrix} -1 & 1 \\ 1 & 1 \end{bmatrix}, \qquad \cos \mathbf{A} = (\cos\sqrt{2})\begin{bmatrix} 1 & 0 \\ 0 & 1 \end{bmatrix}, \qquad \tan \mathbf{A} = \frac{\tan\sqrt{2}}{\sqrt{2}}\begin{bmatrix} -1 & 1 \\ 1 & 1 \end{bmatrix}$$

b. Using $(\cos \mathbf{A})^{-1} = (1/\cos\sqrt{2})\mathbf{I}$ shows that $\tan \mathbf{A} = (\sin \mathbf{A})(\cos \mathbf{A})^{-1}$.

Closed Form for Functions of a Matrix

8.12 If $\mathbf{A} = \begin{bmatrix} -2 & 2 \\ 1 & -3 \end{bmatrix}$, what is $\sin \mathbf{A}t$?

Since $\Delta(\lambda) = \lambda^2 + 5\lambda + 4 = (\lambda + 4)(\lambda + 1)$, the eigenvalues are $\lambda_1 = -1$ and $\lambda_2 = -4$. The general form of the solution is

$$\sin \mathbf{A}t = \alpha_0\mathbf{I} + \alpha_1\mathbf{A}$$

where

$$\left.\begin{array}{l}\sin(-4t) = \alpha_0 - 4\alpha_1\\ \sin(-t) = \alpha_0 - \alpha_1\end{array}\right\} \Rightarrow \begin{cases}\alpha_1 = -\frac{1}{3}[\sin(-4t) - \sin(-t)]\\ \alpha_0 = -\frac{1}{3}[\sin(-4t) - 4\sin(-t)]\end{cases}$$

Then $\sin \mathbf{A}t = \dfrac{1}{3}\begin{bmatrix}\sin(-4t) + 2\sin(-t) & -2\sin(-4t) + 2\sin(-t)\\ -\sin(-4t) + \sin(-t) & 2\sin(-4t) + \sin(-t)\end{bmatrix}$.

8.13 If $\mathbf{A} = \begin{bmatrix}2 & 0 & 0\\ 0 & -2 & 2\\ 0 & 1 & -3\end{bmatrix}$, find $e^{\mathbf{A}t}$.

Note that $\mathbf{A}$ is block diagonal and the lower block is the same matrix as in the previous problem. Since the algebra involved in finding the α's is the same when finding any analytic function, the answer can be written down by replacing $\sin \lambda_i t$ by $e^{\lambda_i t}$, so

$$e^{\mathbf{A}t} = \left[\begin{array}{c|cc} e^{2t} & 0 & 0\\ \hline 0 & \dfrac{(e^{-4t} + 2e^{-t})}{3} & \dfrac{2(-e^{-4t} + e^{-t})}{3}\\ 0 & \dfrac{(-e^{-4t} + e^{-t})}{3} & \dfrac{(2e^{-4t} + e^{-t})}{3}\end{array}\right]$$

A useful partial check is that $e^{\mathbf{A}t}$ must always equal $\mathbf{I}$ when $t = 0$.

8.14 Compute $\begin{bmatrix}1 & -1 & 1\\ 0 & 1 & 1\\ 0 & 0 & 1\end{bmatrix}^k$ for an arbitrary integer k.

Computing $\Delta(\lambda) = (1 - \lambda)^3$ gives $\lambda_1 = \lambda_2 = \lambda_3 = 1$. Since $\mathbf{A}$ is 3×3, $\mathbf{A}^k$ can always be written as a polynomial of degree 2 (or less):

$$\mathbf{A}^k = \alpha_0\mathbf{I} + \alpha_1\mathbf{A} + \alpha_2\mathbf{A}^2 = \begin{bmatrix}\alpha_0 + \alpha_1 + \alpha_2 & -\alpha_1 - 2\alpha_2 & \alpha_1 + \alpha_2\\ 0 & \alpha_0 + \alpha_1 + \alpha_2 & \alpha_1 + 2\alpha_2\\ 0 & 0 & \alpha_0 + \alpha_1 + \alpha_2\end{bmatrix}$$

where

$$(\lambda_1)^k = 1 = \alpha_0 + \alpha_1 + \alpha_2$$

$$\left.\frac{d(\lambda)^k}{d\lambda}\right|_{\lambda=\lambda_1} = k\lambda_1^{k-1} = k = \alpha_1 + 2\alpha_2$$

$$\left.\frac{d^2(\lambda)^k}{d\lambda^2}\right|_{\lambda=\lambda_1} = k(k-1)\lambda_1^{k-2} = k(k-1) = 2\alpha_2$$

Note that in this case all the individual α's need not be found. Combining the last two equations gives $\alpha_1 + \alpha_2 = k - k(k-1)/2 = k(3-k)/2$. All the combinations of α_i that are required in $\mathbf{A}^k$ are now available, and

$$\mathbf{A}^k = \begin{bmatrix}1 & -k & k(3-k)/2\\ 0 & 1 & k\\ 0 & 0 & 1\end{bmatrix}$$

8.15 Compute $e^{\mathbf{A}t}$ if $\mathbf{A} = \begin{bmatrix}-1 & 1\\ 1 & 1\end{bmatrix}$.

We know that $e^{\mathbf{A}t} = \alpha_0\mathbf{I} + \alpha_1\mathbf{A} = \begin{bmatrix}\alpha_0 - \alpha_1 & \alpha_1\\ \alpha_1 & \alpha_0 + \alpha_1\end{bmatrix}$. Since this is the same

A matrix as in Problem 8.11, the coefficients are

$$\alpha_0 = \tfrac{1}{2}[e^{\sqrt{2}t} + e^{-\sqrt{2}t}] = \cosh\sqrt{2}\,t, \qquad \alpha_1 = \frac{e^{\sqrt{2}t} - e^{\sqrt{2}t}}{2\sqrt{2}} = \frac{1}{\sqrt{2}}\sinh\sqrt{2}\,t$$

$$\text{so } e^{\mathbf{A}t} = \begin{bmatrix} \cosh\sqrt{2}\,t - \frac{1}{\sqrt{2}}\sinh\sqrt{2}\,t & \frac{1}{\sqrt{2}}\sinh\sqrt{2}\,t \\ \frac{1}{\sqrt{2}}\sinh\sqrt{2}\,t & \cosh\sqrt{2}\,t + \frac{1}{\sqrt{2}}\sinh\sqrt{2}\,t \end{bmatrix}.$$

8.16 Given $\mathbf{A} = \begin{bmatrix} 0 & -3 & 0 \\ 3 & 0 & 0 \\ 0 & 0 & -1 \end{bmatrix}$. a. Find $\mathbf{A}^{-1}$ using the Cayley-Hamilton theorem. b. Compute $e^{\mathbf{A}t}$.

a. The characteristic polynomial is $\Delta(\lambda) = (1 + \lambda)(\lambda^2 + 9) = \lambda^3 + \lambda^2 + 9\lambda + 9$. Therefore $\Delta(\mathbf{A}) = [\mathbf{0}] = \mathbf{A}^3 + \mathbf{A}^2 + 9\mathbf{A} + 9\mathbf{I}$ so

$$\mathbf{A}^{-1} = -\frac{1}{9}[\mathbf{A}^2 + \mathbf{A} + 9\mathbf{I}] = \begin{bmatrix} 0 & 1/3 & 0 \\ -1/3 & 0 & 0 \\ 0 & 0 & -1 \end{bmatrix}$$

b. The eigenvalues are roots of $\Delta(\lambda) = 0$, so $\lambda_1 = -1$, $\lambda_2 = 3j$, $\lambda_3 = -3j$. Since $\mathbf{A}$ is block diagonal,

$$e^{\mathbf{A}t} = \left[\begin{array}{cc|c} \multicolumn{2}{c|}{e^{\left[\begin{smallmatrix} 0 & -3 \\ 3 & 0 \end{smallmatrix}\right]t}} & \begin{matrix} 0 \\ 0 \end{matrix} \\ \hline 0 & 0 & e^{-t} \end{array}\right]$$

and

$$e^{\left[\begin{smallmatrix} 0 & -3 \\ 3 & 0 \end{smallmatrix}\right]t} = \alpha_0\mathbf{I} + \alpha_1\begin{bmatrix} 0 & -3 \\ 3 & 0 \end{bmatrix} = \begin{bmatrix} \alpha_0 & -3\alpha_1 \\ 3\alpha_1 & \alpha_0 \end{bmatrix}$$

where $e^{3jt} = \alpha_0 + \alpha_1 3j$ and $e^{-3jt} = \alpha_0 - \alpha_1 3j$. From this,

$$\alpha_0 = \tfrac{1}{2}[e^{3jt} + e^{-3jt}] = \cos 3t$$

$$\alpha_1 = \frac{1}{3}\left[\frac{e^{3jt} - e^{-3jt}}{2j}\right] = \tfrac{1}{3}\sin 3t$$

$$\text{and so } e^{\mathbf{A}t} = \begin{bmatrix} \cos 3t & -\sin 3t & 0 \\ \sin 3t & \cos 3t & 0 \\ 0 & 0 & e^{-t} \end{bmatrix}.$$

8.17 Find $e^{\mathbf{A}t}$ if $\mathbf{A} = \left[\begin{array}{cc|cc} 0 & -\Omega & a & 0 \\ \Omega & 0 & 0 & a \\ \hline 0 & 0 & 0 & -\Omega \\ 0 & 0 & \Omega & 0 \end{array}\right]$.

The algebra can be simplified in this case if it is noted that

$$\mathbf{A} = \left[\begin{array}{c|c} \mathbf{B}_1 & \mathbf{0} \\ \hline \mathbf{0} & \mathbf{B}_1 \end{array}\right] + \left[\begin{array}{c|c} \mathbf{0} & \mathbf{B}_2 \\ \hline \mathbf{0} & \mathbf{0} \end{array}\right] \triangleq \mathbf{A}_1 + \mathbf{A}_2 \qquad \text{and} \qquad \mathbf{A}_1\mathbf{A}_2 = \mathbf{A}_2\mathbf{A}_1 = \left[\begin{array}{c|c} \mathbf{0} & a\mathbf{B}_1 \\ \hline \mathbf{0} & \mathbf{0} \end{array}\right]$$

Since $\mathbf{A}_1$ and $\mathbf{A}_2$ commute, $e^{(\mathbf{A}_1 + \mathbf{A}_2)t} = e^{\mathbf{A}_1 t}e^{\mathbf{A}_2 t}$. But $\mathbf{A}_1$ is block diagonal, so

$$e^{\mathbf{A}_1 t} = \left[\begin{array}{c|c} e^{\mathbf{B}_1 t} & \mathbf{0} \\ \hline \mathbf{0} & e^{\mathbf{B}_1 t} \end{array}\right]$$

$e^{\mathbf{B}_1 t} = \alpha_0 \mathbf{I} + \alpha_1 \mathbf{B}_1$ and eigenvalues of $\mathbf{B}_1$ are $\pm j\Omega$:

$$\left.\begin{aligned} e^{j\Omega t} &= \alpha_0 + j\Omega\alpha_1 \\ e^{-j\Omega t} &= \alpha_0 - j\Omega\alpha_1 \end{aligned}\right\} \Longrightarrow e^{\mathbf{B}_1 t} = \begin{bmatrix} \cos\Omega t & -\sin\Omega t \\ \sin\Omega t & \cos\Omega t \end{bmatrix}$$

Consider $\mathbf{A}_2 : |\mathbf{A}_2 - \mathbf{I}\lambda| = \lambda^4 = 0$. $\mathbf{A}_2$ has $\lambda_i = 0$ with algebraic multiplicity of four:

$$e^{\mathbf{A}_2 t} = \beta_0 \mathbf{I} + \beta_1 \mathbf{A}_2 + \beta_2 \mathbf{A}_2^2 + \beta_3 \mathbf{A}_2^3$$

Since $\mathbf{A}_2^2 = \mathbf{A}_2^3 = [\mathbf{0}]$, coefficients β_2 and β_3 are not needed. The remaining coefficients are found to be $e^{0t} = 1 = \beta_0$ and $\left.\dfrac{de^{\lambda t}}{d\lambda}\right|_{\lambda=0} = t = \beta_1$. Thus

$$e^{\mathbf{A}_2 t} = \left[\begin{array}{cc|cc} 1 & 0 & at & 0 \\ 0 & 1 & 0 & at \\ \hline 0 & 0 & 1 & 0 \\ 0 & 0 & 0 & 1 \end{array}\right]$$

Combining gives $e^{\mathbf{A}t} = e^{\mathbf{A}_1 t} e^{\mathbf{A}_2 t} = \left[\begin{array}{cc|cc} \cos\Omega t & -\sin\Omega t & at\cos\Omega t & -at\sin\Omega t \\ \sin\Omega t & \cos\Omega t & at\sin\Omega t & at\cos\Omega t \\ \hline 0 & 0 & \cos\Omega t & -\sin\Omega t \\ 0 & 0 & \sin\Omega t & \cos\Omega t \end{array}\right]$.

Minimum Polynomial

8.18 In computing $e^{\mathbf{A}_2 t}$ in the previous problem, it was found that only first-order terms in $\mathbf{A}_2$ were needed even though $\mathbf{A}_2$ was a 4×4 matrix. This is a case where the minimum polynomial is of lower order than the characteristic polynomial. Find the minimal polynomial of $\mathbf{A}_2$.

The minimal polynomial will have the same factors as the characteristic polynomial, but perhaps raised to smaller powers. (There could also be a sign difference, depending on how the characteristic polynomial is defined.)

Here $\Delta(\lambda) = \lambda^4$, so $m(\lambda) = \lambda^k$, where k is the smallest integer for which $m(\mathbf{A}) = [\mathbf{0}]$. From the results of the previous problem, $k = 2$ and therefore $m(\lambda) = \lambda^2$. If this had been known in advance, then it would have been known that $e^{\mathbf{A}_2 t} = \beta_0 \mathbf{I} + \beta_1 \mathbf{A}_2$ and a slight amount of matrix algebra would be avoided. This savings is usually offset by the effort required to find $m(\lambda)$, and in this text the minimum polynomial is seldom used.

Alternative Methods

8.19 It was shown in the text that $de^{\mathbf{A}t}/dt = \mathbf{A}e^{\mathbf{A}t}$ if $\mathbf{A}$ is a constant matrix. Use this result, plus the fact that $e^{\mathbf{A}\cdot 0} = \mathbf{I}$, to derive an alternative method of computing $e^{\mathbf{A}t}$.

Let $e^{\mathbf{A}t} = \mathbf{F}(t)$. Then $\mathbf{F}$ satisfies the matrix differential equation $\dot{\mathbf{F}} = \mathbf{AF}$ with initial conditions $\mathbf{F}(0) = \mathbf{I}$. Laplace transforms can be used to solve this equation:

$$\mathcal{L}\{\dot{\mathbf{F}}(t)\} = s\mathbf{F}(s) - \mathbf{F}(0) = s\mathbf{F}(s) - \mathbf{I}$$

$$\mathcal{L}\{\mathbf{AF}(t)\} = \mathbf{A}\mathcal{L}\{\mathbf{F}(t)\} = \mathbf{AF}(s)$$

Therefore $[s\mathbf{I} - \mathbf{A}]\mathbf{F}(s) = \mathbf{I}$ or $\mathbf{F}(s) = [s\mathbf{I} - \mathbf{A}]^{-1}$, and

$$e^{\mathbf{A}t} = \mathbf{F}(t) = \mathcal{L}^{-1}\{[s\mathbf{I} - \mathbf{A}]^{-1}\}$$

8.20 Use the method of the previous problem to compute $e^{\mathbf{A}t}$ for $\mathbf{A} = \begin{bmatrix} -2 & -2 & 0 \\ 0 & 0 & 1 \\ 0 & -3 & -4 \end{bmatrix}$.

Form $s\mathbf{I} - \mathbf{A} = \begin{bmatrix} s+2 & 2 & 0 \\ 0 & s & -1 \\ 3 & 3 & s+4 \end{bmatrix}$ and $|s\mathbf{I} - \mathbf{A}| = (s+2)(s^2 + 4s + 3) = (s+1)(s+2)(s+3)$. Then

$$\mathbf{F}(s) = [s\mathbf{I} - \mathbf{A}]^{-1}$$

$$= \frac{1}{(s+1)(s+2)(s+3)} \begin{bmatrix} (s+1)(s+3) & -2(s+4) & -2 \\ 0 & (s+2)(s+4) & s+2 \\ 0 & -3(s+2) & s(s+2) \end{bmatrix}$$

The inverse Laplace transforms are computed term by term:

$$F_{11}(t) = \mathcal{L}^{-1}\left\{\frac{1}{s+2}\right\} = e^{-2t}; \qquad F_{21} = F_{31} = 0$$

$$F_{12}(t) = \mathcal{L}^{-1}\left\{\frac{-2(s+4)}{(s+1)(s+2)(s+3)}\right\} = -3e^{-t} + 4e^{-2t} - e^{-3t}$$

$$F_{13}(t) = \mathcal{L}^{-1}\left\{\frac{-2}{(s+1)(s+2)(s+3)}\right\} = -e^{-t} + 2e^{-2t} - e^{-3t}$$

$$F_{22}(t) = \mathcal{L}^{-1}\left\{\frac{s+4}{(s+1)(s+3)}\right\} = \tfrac{3}{2}e^{-t} - e^{-3t}$$

$$F_{32}(t) = \mathcal{L}^{-1}\left\{\frac{-3}{(s+1)(s+3)}\right\} = -\tfrac{3}{2}e^{-t} + \tfrac{3}{2}e^{-3t}$$

$$F_{23}(t) = \mathcal{L}^{-1}\left\{\frac{1}{(s+1)(s+3)}\right\} = \tfrac{1}{2}e^{-t} - \tfrac{1}{2}e^{-3t}$$

$$F_{33}(t) = \mathcal{L}^{-1}\left\{\frac{s}{(s+1)(s+3)}\right\} = -\tfrac{1}{2}e^{-t} + \tfrac{3}{2}e^{-3t}$$

8.21 Find the closed-form expression for $e^{\mathbf{A}t}$, where $\mathbf{A} = \begin{bmatrix} 0 & 1 & 0 & 0 \\ 0 & 0 & 1 & 0 \\ 0 & 0 & 0 & 1 \\ -27 & 54 & -36 & 10 \end{bmatrix}$.

Writing $|\mathbf{A} - \mathbf{I}\lambda| = \Delta(\lambda) = \lambda^4 - 10\lambda^3 + 36\lambda^2 - 54\lambda + 27 = (\lambda - 1)(\lambda - 3)^3$ shows that $\lambda_1 = 1, \lambda_2 = \lambda_3 = \lambda_4 = 3$. For the repeated eigenvalue, rank $[\mathbf{A} - \mathbf{I}\lambda_2] = 3$ and $q_2 = 1$. Thus the Jordan form is $\mathbf{J} = \left[\begin{array}{c|ccc} 1 & 0 & 0 & 0 \\ \hline 0 & 3 & 1 & 0 \\ 0 & 0 & 3 & 1 \\ 0 & 0 & 0 & 3 \end{array}\right]$. Using the results of Problem 8.8,

$$e^{\mathbf{J}t} = \left[\begin{array}{c|ccc} e^t & 0 & 0 & 0 \\ \hline 0 & & & \\ 0 & & e^{\mathbf{J}_2 t} & \\ 0 & & & \end{array}\right] = \begin{bmatrix} e^t & 0 & 0 & 0 \\ 0 & e^{3t} & te^{3t} & \frac{1}{2}t^2e^{3t} \\ 0 & 0 & e^{3t} & te^{3t} \\ 0 & 0 & 0 & e^{3t} \end{bmatrix}$$

and $e^{\mathbf{A}t} = \mathbf{M}e^{\mathbf{J}t}\mathbf{M}^{-1}$, where $\mathbf{M}$ is the modal matrix consisting of two eigenvectors and two generalized eigenvectors. The first column of Adj $[\mathbf{A} - \mathbf{I}\lambda]$ is

$$\begin{bmatrix} \lambda^2(10 - \lambda) - 36\lambda + 54 \\ 27 \\ 27\lambda \\ 27\lambda^2 \end{bmatrix}, \quad \text{so with } \lambda = 1,\ \mathbf{x} = \begin{bmatrix} 1 \\ 1 \\ 1 \\ 1 \end{bmatrix}$$

and with $\lambda = 3$, $\mathbf{x}_2 = [1 \quad 3 \quad 9 \quad 27]^T$ is an eigenvector. Generalized eigenvectors are solutions of $\mathbf{A}\mathbf{x}_3 = 3\mathbf{x}_3 + \mathbf{x}_2$ and $\mathbf{A}\mathbf{x}_4 = 3\mathbf{x}_4 + \mathbf{x}_3$. Solutions are $\mathbf{x}_3 = [1 \quad 4 \quad 15 \quad 54]^T$ and $\mathbf{x}_4 = [0 \quad 1 \quad 7 \quad 36]^T$, so

$$\mathbf{M} = \begin{bmatrix} 1 & 1 & 1 & 0 \\ 1 & 3 & 4 & 1 \\ 1 & 9 & 15 & 7 \\ 1 & 27 & 54 & 36 \end{bmatrix} \quad \text{and} \quad \mathbf{M}^{-1} = \frac{1}{8}\begin{bmatrix} 27 & -27 & 9 & -1 \\ -85 & 133 & -55 & 7 \\ 66 & -106 & 46 & -6 \\ -36 & 60 & -28 & 4 \end{bmatrix}$$

Thus

$$e^{\mathbf{A}t} = \mathbf{M}e^{\mathbf{J}t}\mathbf{M}^{-1} =$$

$$\frac{1}{8}\begin{bmatrix} 27e^t+(-19+30t-18t^2)e^{3t} & -27e^t+(27-46t+30t^2)e^{3t} & 9e^t+(-9+18t-14t^2)e^{3t} & -e^t+(1-2t+2t^2)e^{3t} \\ 27e^t+(-27+54t-54t^2)e^{3t} & -27e^t+(35-78t+90t^2)e^{3t} & 9e^t+(-9+26t-42t^2)e^{3t} & -e^t+(1-2t+6t^2)e^{3t} \\ 27e^t+(-27+54t-162t^2)e^{3t} & -27e^t+(27-54t+270t^2)e^{3t} & 9e^t+(-1-6t-126t^2)e^{3t} & -e^t+(1+6t+18t^2)e^{3t} \\ 27e^t+(-27-162t-486t^2)e^{3t} & -27e^t+(27+378t+810t^2)e^{3t} & 9e^t+(-9-270t-378t^2)e^{3t} & -e^t+(9+54t+54t^2)e^{3t} \end{bmatrix}$$

8.22 Another method of computing an analytic function $f(\mathbf{A})$ of an $n \times n$ matrix $\mathbf{A}$ with distinct eigenvalues is given by

$$f(\mathbf{A}) = \sum_{i=1}^{n} f(\lambda_i)\mathbf{Z}_i(\lambda)$$

where λ_i are the eigenvalues of $\mathbf{A}$ and the $n \times n$ $\mathbf{Z}_i$ matrices are given by

$$\mathbf{Z}_i(\lambda) = \frac{\prod_{\substack{j=1 \\ j\neq i}}^{n} (\mathbf{A} - \lambda_j\mathbf{I})}{\prod_{\substack{j=1 \\ j\neq i}}^{n} (\lambda_i - \lambda_j)}$$

This method is sometimes called Sylvester's expansion [27] and sometimes it is referred to as Lagrange interpolation [117]. This is a restricted form, but it can be extended to the case of repeated eigenvalues [117, Appendix D]. Use it to compute $e^{\mathbf{A}t}$ for

$$\mathbf{A} = \begin{bmatrix} 2 & -2 & 3 \\ 1 & 1 & 1 \\ 1 & 3 & -1 \end{bmatrix}$$

The eigenvalues are $\lambda_i = 1, -2, 3$ (see Problem 7.36).

$$\mathbf{Z}_1 = \frac{(\mathbf{A}+2\mathbf{I})(\mathbf{A}-3\mathbf{I})}{(1+2)(1-3)} = -\frac{1}{6}\begin{bmatrix} 4 & -2 & 3 \\ 1 & 3 & 1 \\ 1 & 3 & 1 \end{bmatrix}\begin{bmatrix} -1 & -2 & 3 \\ 1 & -2 & 1 \\ 1 & 3 & -4 \end{bmatrix} = -\frac{1}{6}\begin{bmatrix} -3 & 5 & -2 \\ 3 & -5 & 2 \\ 3 & -5 & 2 \end{bmatrix}$$

$$\mathbf{Z}_2 = \frac{(\mathbf{A}-\mathbf{I})(\mathbf{A}-3\mathbf{I})}{(-2-1)(-2-3)} = \frac{1}{15}\begin{bmatrix} 0 & 11 & -11 \\ 0 & 1 & -1 \\ 0 & -14 & 14 \end{bmatrix}$$

$$\mathbf{Z}_3 = \frac{(\mathbf{A}-\mathbf{I})(\mathbf{A}+2\mathbf{I})}{(3-1)(3+2)} = \frac{1}{10}\begin{bmatrix} 5 & 1 & 4 \\ 5 & 1 & 4 \\ 5 & 1 & 4 \end{bmatrix}$$

Therefore

$$e^{\mathbf{A}t} = \frac{1}{6}e^{t}\begin{bmatrix} 3 & -5 & 2 \\ -3 & 5 & -2 \\ -3 & 5 & -2 \end{bmatrix} + \frac{1}{15}e^{-2t}\begin{bmatrix} 0 & 11 & -11 \\ 0 & 1 & -1 \\ 0 & -14 & 14 \end{bmatrix} + \frac{1}{10}e^{3t}\begin{bmatrix} 5 & 1 & 4 \\ 5 & 1 & 4 \\ 5 & 1 & 4 \end{bmatrix}$$

8.23 Find $\mathbf{A}^k$ for the $\mathbf{A}$ matrix of the previous problem.

The expansion matrices $\mathbf{Z}_i$ depend only on $\mathbf{A}$, not on the particular function of $\mathbf{A}$ that is being computed. They are the same as in the previous problem, so with $f(\mathbf{A}) = \mathbf{A}^k$, the result is

$$\mathbf{A}^k = \frac{1}{6}\begin{bmatrix} 3 & -5 & 2 \\ -3 & 5 & -2 \\ -3 & 5 & -2 \end{bmatrix} + \frac{(-2)^k}{15}\begin{bmatrix} 0 & 11 & -11 \\ 0 & 1 & -1 \\ 0 & -14 & 14 \end{bmatrix} + \frac{(3)^k}{10}\begin{bmatrix} 5 & 1 & 4 \\ 5 & 1 & 4 \\ 5 & 1 & 4 \end{bmatrix}$$

PROBLEMS

8.24 Find the inverse of $\mathbf{A} = \begin{bmatrix} -1 & 2 & 0 \\ 1 & 1 & 0 \\ 2 & -1 & 2 \end{bmatrix}$ using the Cayley-Hamilton theorem.

8.25 Find $e^{\mathbf{A}t}$ if $\mathbf{A} = \begin{bmatrix} -1 & \frac{1}{2} \\ 0 & 1 \end{bmatrix}$.

8.26 Find $e^{\mathbf{A}t}$ with $\mathbf{A} = \begin{bmatrix} -3 & 2 \\ 0 & -3 \end{bmatrix}$.

8.27 Show that $e^{\left[\begin{smallmatrix} 0 & 1 \\ -3 & -4 \end{smallmatrix}\right]t} = \begin{bmatrix} \frac{1}{2}(3e^{-t} - e^{-3t}) & \frac{1}{2}(e^{-t} - e^{-3t}) \\ -\frac{3}{2}(e^{-t} - e^{-3t}) & \frac{1}{2}(3e^{-3t} - e^{-t}) \end{bmatrix}$.

8.28 Use $\mathbf{A} = \begin{bmatrix} 3 & 5 \\ 16/5 & 3 \end{bmatrix}$ to compute $e^{\mathbf{A}t}$.

8.29 Compute $e^{\mathbf{A}t}$ for $\mathbf{A} = \begin{bmatrix} -3 & 1 \\ 2 & -2 \end{bmatrix}$.

8.30 Find $\begin{bmatrix} \frac{1}{2} & -\frac{1}{2} & 1 \\ 0 & \frac{1}{2} & 2 \\ 0 & 0 & \frac{1}{2} \end{bmatrix}^k$.

8.31 Compute $e^{\mathbf{A}t}$ for $\mathbf{A} = \begin{bmatrix} -5 & -6 & 0 \\ 2 & 2 & 0 \\ 0 & 0 & -3 \end{bmatrix}$.

8.32 Find $e^{\mathbf{A}t}$ with $\mathbf{A} = \begin{bmatrix} 0 & 1 & 0 \\ 0 & 0 & 1 \\ 1 & -3 & 3 \end{bmatrix}$.

8.33 Find $e^{\mathbf{A}t}$ for $\mathbf{A} = \begin{bmatrix} -10 & 0 & -10 & 0 \\ 0 & -0.7 & 9 & 0 \\ 0 & -1 & -0.7 & 0 \\ 1 & 0 & 0 & 0 \end{bmatrix}$.

8.34 Find a closed-form expression for

$$\mathbf{A}^k = \begin{bmatrix} 0 & 1 & 0 \\ 0 & 0 & 1 \\ 0 & -0.5 & 1.5 \end{bmatrix}^k.$$

Chapter

9

STATE VARIABLES AND THE STATE SPACE DESCRIPTION OF DYNAMIC SYSTEMS

9.1 INTRODUCTION

The concepts of state variables, state vectors, and state space are introduced in this chapter. The state variable approach is then used to treat multiple-input, multiple-output control systems. The model of such a physical system can be described in a variety of forms, including simultaneous differential or difference equations, transfer functions, and linear graphs. Methods are presented for obtaining the state space description from each of these.

9.2 THE CONCEPT OF STATE

The concept of state occupies a central position in modern control theory. However, it appears in many other technical and nontechnical contexts as well. In thermodynamics the equations of *state* are prominently used. Binary sequential networks are normally analyzed in terms of their *states*. In everyday life, monthly financial *state*ments are commonplace. The President's *State* of the Union message is another familiar example.

In all of these examples the concept of state is essentially the same. It is a complete summary of the status of the system at a particular point in time. Knowledge of the state at some initial time t_0, plus knowledge of the system inputs after t_0, allows the determination of the state at a later time t_1. As far as the state at t_1 is concerned, it makes no difference how the initial state was attained. Thus the state at t_0 constitutes a complete history of the system behavior prior to t_0, insofar as that history affects future behavior. Knowledge of the present state allows a sharp separation between the past and the future.

At any fixed time the state of a system can be described by the values of a set of variables x_i, called *state variables* [28, 117]. One of the state variables of a thermodynamic system is temperature and its value can range over the continuum of real numbers $\mathcal{R}$. In a binary network state variables can take on only two discrete values, 0 or 1. Note that the state of your checking account at the end of the month can be represented by a single number, the balance. The state of the Union can be represented by such things as gross national product, percent unemployment, the balance of trade deficit, etc. For the systems considered in this book the state variables may take on any scalar value, real or complex. That is, $x_i \in \mathcal{F}$. Although some systems require an infinite number of state variables, only systems which can be described by a finite number n of state variables will be considered here. Then the state can be represented by an n component *state vector* $\mathbf{x} = [x_1 \quad x_2 \quad \cdots \quad x_n]^T$.

The state at a given time belongs to an n-dimensional vector space defined over the field $\mathcal{F}$. In previous chapters such a space was denoted by $\mathcal{X}^n$. However, because of its great importance, the *state space* will be referred to as Σ from now on.

The systems of interest in this book are dynamic systems. Although a more precise definition is given in the next section, the usual connotation of the word "dynamic" refers to something active or changing with time. *Continuous-time* systems have their state defined for all times in some interval, for example, a continually varying temperature or voltage. For *discrete-time* systems the state is defined only at discrete times, as with the monthly financial statement or the annual State of the Union message. Continuous-time and discrete-time systems can be discussed simultaneously by defining the times of interest as $\mathscr{T}$. For continuous-time systems $\mathscr{T}$ consists of the set of all real numbers $t \in [t_0, t_f]$. For discrete-time systems $\mathscr{T}$ consists of a discrete set of times $\{t_0, t_1, t_2, \ldots, t_k, \ldots, t_N\}$. In either case the initial time could be $-\infty$ and the final time could be ∞ in some circumstances.

The state vector $\mathbf{x}(t)$ is defined only for those $t \in \mathscr{T}$. At any given t, it is simply an ordered set of n numbers. However, the character of a system could change with time, causing the *number* of required state variables (and not just the values) to change. If the dimension of the state space varies with time, the notation Σ_t could be used. It is assumed here that Σ is the same n-dimensional state space at all $t \in \mathscr{T}$.

9.3 STATE SPACE REPRESENTATION OF DYNAMIC SYSTEMS

A general class of multivariable control systems is considered. There are r *real*-valued inputs or control variables $u_i(t)$, referred to collectively as the $r \times 1$ vector $\mathbf{u}(t)$. For a fixed time $t \in \mathscr{T}$, $\mathbf{u}(t)$ belongs to the real r-dimensional space $\mathcal{U}^r$.

There are m real-valued ouptuts $y_i(t)$, referred to collectively as the $m \times 1$ vector $\mathbf{y}(t)$ For any $t \in \mathscr{T}$, $\mathbf{y}(t)$ belongs to the real m-dimensional space $\mathcal{Y}^m$. Let the subset of times $t \in \mathscr{T}$ which satisfy $t_a \leq t \leq t_b$ be denoted as $[t_a, t_b]_{\mathscr{T}}$. For continuous-time systems $[t_a, t_b]_{\mathscr{T}} = [t_a, t_b]$, and for discrete-time systems $[t_a, t_b]_{\mathscr{T}}$ is the intersection of $[t_a, t_b]$ and $\mathscr{T}$.

It is necessary to distinguish between an input function (the graph of $\mathbf{u}(t)$ versus $t \in \mathscr{T}$) and the value of $\mathbf{u}(t)$ at a particular time. The notation $\mathbf{u}_{[t_a, t_b]}$ will be used to indicate a *segment* of an input function over the set of times in $[t_a, t_b]_{\mathscr{T}}$. Similarly, a segment of an output function will be indicated by $\mathbf{y}_{[t_a, t_b]}$. The admissible input functions (actually sequences in the discrete-time case) are elements of the input function space $\mathcal{U}$, that is, $\mathbf{u}: \mathscr{T} \longrightarrow \mathcal{U}$. The output functions (or sequences) are elements of the output function space $\mathcal{Y}$, $\mathbf{y}: \mathscr{T} \longrightarrow \mathcal{Y}$. Heuristically, $\mathcal{U}^r$ and $\mathcal{Y}^m$ can be thought of as "cross-sections" through $\mathcal{U}$ and $\mathcal{Y}$ at a particular time t. Similarly, the graph of $\mathbf{x}(t)$ versus t, the so-called *state trajectory*, could be considered as an element of a function space $\mathcal{S}$. The state space Σ can be visualized as a cross-section through $\mathcal{S}$ at a particular $t \in \mathscr{T}$.

The primary interest in a control system may be in the relationship between inputs, which can be manipulated, and outputs, which determine whether or not system goals are met. This relationship may be thought of as a mapping or transformation $\mathcal{W}: \mathcal{U} \longrightarrow \mathcal{Y}$, that is, $\mathbf{y}(t) = \mathcal{W}(\mathbf{u}(t))$. Examples of this type of relationship have been mentioned in Secs. 1.5 and 6.8 and used in Problems 4.25 and 6.20. It is easy to show that a given $\mathbf{u}_{[t_a, t_b]}$ need *not* define a unique ouptut function $\mathbf{y}_{[t_a, t_b]}$.

Example 9.1

The input-output differential equation for the circuit of Fig. 9.1 is $dy/dt + (1/RC)y = u(t)/RC$. The solution is

$$y(t) = e^{-(t-t_0)/RC}y(t_0) + (1/RC)\int_{t_0}^{t} e^{-(t-\tau)/RC}u(\tau)\,d\tau$$

A given input function $u(t) \in \mathcal{U}$ defines a *family* of output functions $y(t) \in \mathcal{Y}$. In order to relate one unique output with each input, additional information, such as the value of $y(t_0)$,

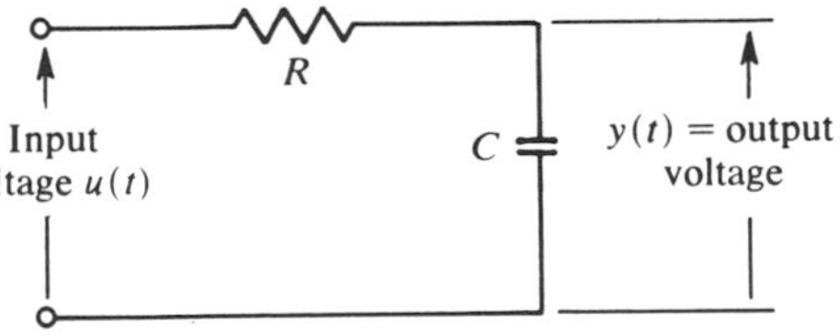

Figure 9.1

must be specified. Unless this is done, the nonuniqueness prevents $\mathcal{W}$ from being a transformation in the sense defined in Chapter 6. ■

State variables are important because they resolve the nonuniqueness problem illustrated in Example 9.1 and at the same time completely summarize the *internal* status of the system.

Definition 9.1: The state variables of a system consist of a minimum set of parameters which completely summarize the system's status in the following sense. If at any time $t_0 \in \mathscr{T}$, the values of the state variables $x_i(t_0)$ are known, then the output $\mathbf{y}(t_1)$ and the values $x_i(t_1)$ can be *uniquely* determined for any time $t_1 \in \mathscr{T}$, $t_1 > t_0$, provided $\mathbf{u}_{[t_0,t_1]}$ is known.

Definition 9.2: The state at any time t_0 is a set of the minimum number of parameters $x_i(t_0)$ which allows a *unique* output segment $\mathbf{y}_{[t_0,t]}$ to be associated with each input segment $\mathbf{u}_{[t_0,t]}$ for every $t_0 \in \mathscr{T}$ and for all $t > t_0$, $t \in \mathscr{T}$.

The implications of Definitions 9.1 and 9.2 can be stated in terms of transformations on the input, state, and output spaces. Given (1) a pair of times, t_0 and t_1 in $\mathscr{T}$, (2) $\mathbf{x}(t_0) \in \Sigma$, and (3) a segment $\mathbf{u}_{[t_0,t_1]}$ of an input in $\mathcal{U}$, both the state and the output must be uniquely determinable. This requires that there be a transformation $\mathbf{g}$ which maps the elements $(t_0, t_1, \mathbf{x}(t_0), \mathbf{u}_{[t_0,t_1]})$, which can be treated collectively as a single element of the product space (Sec. 4.10) $\mathscr{T} \times \mathscr{T} \times \Sigma \times \mathcal{U}$, into a unique element in Σ; that is, $\mathbf{g}: \mathfrak{I} \times \mathfrak{I} \times \Sigma \times \mathcal{U} \longrightarrow \Sigma$, where

$$\mathbf{x}(t_1) = \mathbf{g}(t_0, t_1, \mathbf{x}(t_0), \mathbf{u}_{[t_0,t_1]}) \tag{9.1}$$

Furthermore, since $\mathbf{y}(t_1)$ is uniquely determined, a second transformation exists, $\mathbf{h}: \mathscr{T} \times \Sigma \times \mathcal{U}^r \longrightarrow \mathcal{Y}^m$ with

$$\mathbf{y}(t_1) = \mathbf{h}(t_1, \mathbf{x}(t_1), \mathbf{u}(t_1)) \tag{9.2}$$

The transformation $\mathbf{h}$ has no memory, rather $\mathbf{y}(t_1)$ depends only on the instantaneous values of $\mathbf{x}(t_1)$, $\mathbf{u}(t_1)$, and t_1. The transformation $\mathbf{g}$ is *nonanticipative* (also called causal). This means that the state, and hence the output at t_1, do not depend on inputs occurring after t_1.

Definition 9.3: The model of a physical system is called a *dynamical system* [29, 59] if a set of times $\mathscr{T}$, spaces $\mathcal{U}$, Σ, and $\mathcal{Y}$, and transformations $\mathbf{g}$ and $\mathbf{h}$ can be associated with it. The transformations are those of equations (*9.1*) and (*9.2*) and must have the following properties:

$$\mathbf{x}(t_0) = \mathbf{g}(t_0, t_0, \mathbf{x}(t_0), \mathbf{u}_{[t_0,t_1]}) \qquad \text{for any } t_0, t_1 \in \mathscr{T} \tag{9.3}$$

If $\mathbf{u} \in \mathcal{U}$ and $\mathbf{v} \in \mathcal{U}$ with $\mathbf{u} = \mathbf{v}$ over some segment $[t_0, t_1]_{\mathscr{T}}$, then

$$\mathbf{g}(t_0, t_1, \mathbf{x}(t_0), \mathbf{u}_{[t_0,t_1]}) = \mathbf{g}(t_0, t_1, \mathbf{x}(t_0), \mathbf{v}_{[t_0,t_1]}) \tag{9.4}$$

If $t_0, t_1, t_2 \in \mathscr{T}$ and $t_0 < t_1 < t_2$, then

$$\begin{aligned}\mathbf{x}(t_2) &= \mathbf{g}(t_0, t_2, \mathbf{x}(t_0), \mathbf{u}_{[t_0,t_2]}) \\ &= \mathbf{g}(t_1, t_2, \mathbf{x}(t_1), \mathbf{u}_{[t_1,t_2]}) \\ &= \mathbf{g}(t_1, t_2, \mathbf{g}(t_0, t_1, \mathbf{u}_{[t_0,t_1]}), \mathbf{u}_{[t_1,t_2]})\end{aligned} \tag{9.5}$$

Equation (*9.3*) indicates that $\mathbf{g}$ is the identity transformation whenever its two time arguments are the same. Equation (*9.4*), called the *state transition property*, indicates that $\mathbf{x}(t_1)$ does not depend on inputs prior to t_0 except insofar as the past is summarized by $\mathbf{x}(t_0)$. It also indicates that $\mathbf{x}(t_1)$ does not depend on inputs after t_1. Equation (*9.5.*) called the *semigroup property*, states that it is immaterial whether $\mathbf{x}(t_2)$ is computed directly from $\mathbf{x}(t_0)$ and $\mathbf{u}_{[t_0,t_2]}$ or if $\mathbf{x}(t_1)$ is first obtained from $\mathbf{x}(t_0)$ and $\mathbf{u}_{[t_0,t_1]}$, and then this state is used, along with $\mathbf{u}_{[t_1,t_2]}$.

Lumped-parameter continuous-time dynamical systems can be represented in state space notation by a set of first-order differential equations (*9.6*) and a set of single-valued algebraic output equations (*9.7*):

$$\dot{\mathbf{x}} = \mathbf{f}(\mathbf{x}, \mathbf{u}, t) \tag{9.6}$$

$$\mathbf{y}(t) = \mathbf{h}(\mathbf{x}, \mathbf{u}, t) \tag{9.7}$$

In order that $\mathbf{x}$ be a valid state vector, equation (*9.6*) must have a unique solution. In essence, this means that the components of $\mathbf{f}$ are restricted so that equation (*9.8*) defines a unique $\mathbf{x}(t)$ for $t \geq t_0$:

$$\mathbf{x}(t) = \mathbf{x}(t_0) + \int_{t_0}^{t} \mathbf{f}(\mathbf{x}(\tau), \mathbf{u}(\tau), \tau)\, d\tau \tag{9.8}$$

The theory of differential equations [22] indicates that it is sufficient if $\mathbf{f}$ satisfies a Lipschitz condition with respect to $\mathbf{x}$, is continuous with respect to $\mathbf{u}$, and is piecewise continuous with respect to t. Then there will be a unique $\mathbf{x}(t)$ for any t_0, $\mathbf{x}(t_0)$ provided $\mathbf{u}(t)$ is piecewise continuous. The unique solution of equation (9.6) defines the transformation $\mathbf{g}$ of equation (9.1).

Example 9.2

Consider a point mass falling in a vacuum. The input is the constant gravitational attraction a and the output is the altitude $y(t)$ shown in Fig. 9.2. The input-output relation

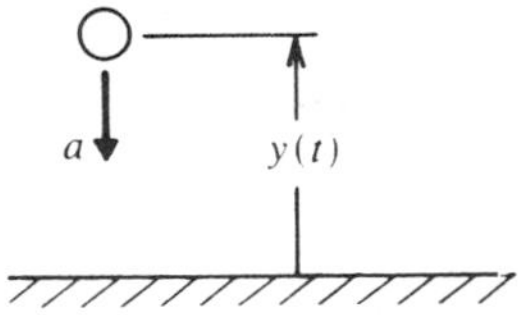

..gure 9.2

is $\ddot{y} = -a$. Integrating gives

$$\dot{y}(t) = \dot{y}(t_0) - a(t - t_0)$$

Integrating again gives

$$y(t) = y(t_0) + \dot{y}(t_0)(t - t_0) - a(t - t_0)^2/2$$

The variable y does not constitute the state since its initial value is not sufficient for uniquely determining $y(t)$. The set of variables $y, \dot{y}$, and $\ddot{y}$ does not form the state since this is not the *minimum* set of variables required to uniquely determine $y(t)$. The parameter $\ddot{y}$ is not needed. A valid state vector is $\mathbf{x}(t) = [y(t) \;\; \dot{y}(t)]^T$. It is easily seen that the form of equations (*9.1*) and (*9.2*) for this example is

$$\mathbf{x}(t) = \begin{bmatrix} 1 & t - t_0 \\ 0 & 1 \end{bmatrix} \mathbf{x}(t_0) - \begin{bmatrix} (t - t_0)^2/2 \\ t - t_0 \end{bmatrix} a, \qquad y(t) = [1 \quad 0]\mathbf{x}(t)$$ ■

Lumped-parameter discrete-time dynamical systems can be described in an analogous way by a set of first-order difference equations (*9.9*) and a set of algebraic output equations (*9.10*):

$$\mathbf{x}(t_{k+1}) = \mathbf{f}(\mathbf{x}(t_k), \mathbf{u}(t_k), t_k) \tag{9.9}$$

$$\mathbf{y}(t_k) = \mathbf{h}(\mathbf{x}(t_k), \mathbf{u}(t_k), t_k) \tag{9.10}$$

In general, the functions $\mathbf{f}$ and $\mathbf{h}$ of equations (*9.6*), (*9.7*), (*9.9*), and (*9.10*) can be nonlinear. However, the linear case is of major importance and also lends itself to more detailed analysis. The most general state space representation of a *linear* continuous-time dynamical system is given by equations (*9.11*) and (*9.12*):

$$\dot{\mathbf{x}} = \mathbf{A}(t)\mathbf{x}(t) + \mathbf{B}(t)\mathbf{u}(t) \tag{9.11}$$

$$\mathbf{y}(t) = \mathbf{C}(t)\mathbf{x}(t) + \mathbf{D}(t)\mathbf{u}(t) \tag{9.12}$$

A, **B**, **C**, and **D** are matrices of dimension $n \times n, n \times r, m \times n$, and $m \times r$ respectively. Equations (*9.11*) and (*9.12*) are shown in block diagram form in Fig. 9.3. The heavier

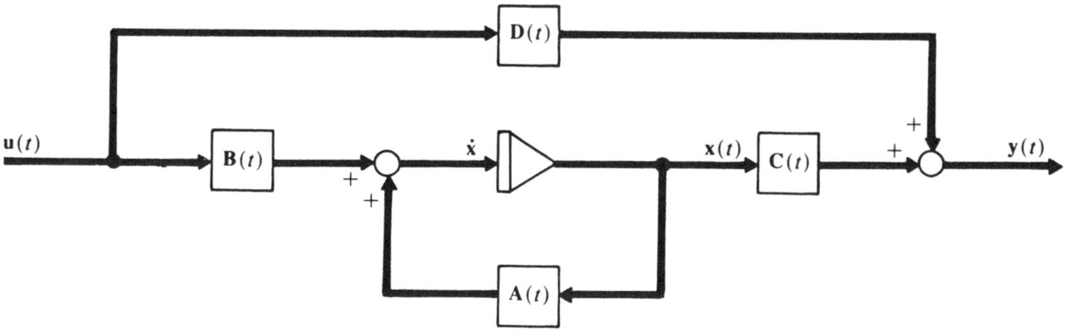

Figure 9.3 State space representation of continuous-time linear system.

lines indicate that the signals are vectors, and the integrator symbol really indicates n scalar integrators.

The state space representation of a discrete-time *linear* system is given by equations (*9.13*) and (*9.14*). The simplified notation k refers to a general time $t_k \in \mathscr{T}$:

$$\mathbf{x}(k+1) = \mathbf{A}(k)\mathbf{x}(k) + \mathbf{B}(k)\mathbf{u}(k) \tag{9.13}$$

$$\mathbf{y}(k) = \mathbf{C}(k)\mathbf{x}(k) + \mathbf{D}(k)\mathbf{u}(k) \tag{9.14}$$

The matrices **A**, **B**, **C**, and **D** have the same dimensions as in the continuous-time case, but their meanings are different. The block diagram representation of equations (*9.13*) and (*9.14*) is given in Fig. 9.4. The delay symbol is analogous to the integrator in Fig. 9.3 and really symbolizes n scalar delays.

Linear continuous-time systems will be analyzed in Chapter 10. The remainder of this chapter develops methods of obtaining state space representations such as given by equations (*9.6*) and (*9.7*) or equations (*9.11*) and (*9.12*). Chapter 11 provides a similar development and analysis for discrete-time systems.

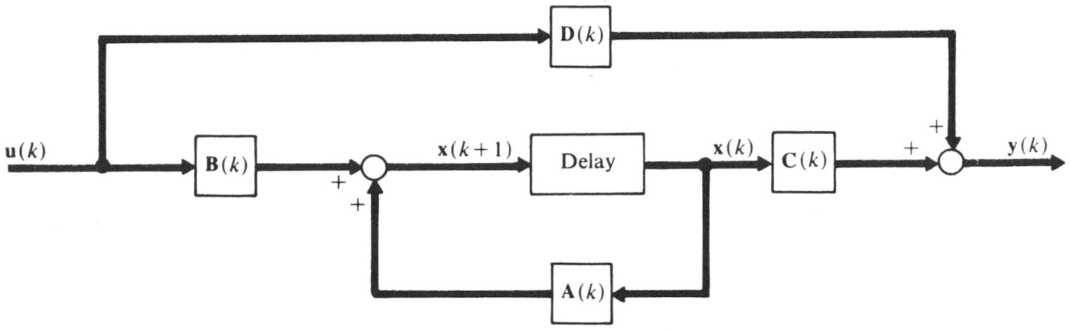

Figure 9.4 State space representation of discrete-time linear system.

9.4 OBTAINING THE STATE EQUATIONS

From Input-Output Differential Equations

A class of single-input, single-output systems can be described by an nth order linear ordinary differential equation:

$$\frac{d^n y}{dt^n} + a_{n-1}\frac{d^{n-1} y}{dt^{n-1}} + \cdots + a_2\frac{d^2 y}{dt^2} + a_1\frac{dy}{dt} + a_0 y = u(t)$$

This class of systems can be reduced to the form of n first-order state equations as follows. Define the state variables as

$$x_1 = y, \qquad x_2 = \frac{dy}{dt}, \qquad x_3 = \frac{d^2 y}{dt^2}, \qquad \ldots, \qquad x_n = \frac{d^{n-1} y}{dt^{n-1}}$$

As a direct result of this definition, $n - 1$ first-order differential equations are $\dot{x}_1 = x_2$, $\dot{x}_2 = x_3, \ldots, \dot{x}_{n-1} = x_n$. The nth equation is $\dot{x}_n = d^n y/dt^n$. Using the original differential equation and the preceding definitions gives

$$\dot{\mathbf{x}}_n = -a_0 x_1 - a_1 x_2 - \cdots - a_{n-1} x_n + u(t)$$

so that

$$\dot{\mathbf{x}} = \begin{bmatrix} 0 & 1 & 0 & 0 & \cdots & 0 \\ 0 & 0 & 1 & 0 & \cdots & 0 \\ 0 & 0 & 0 & 1 & \cdots & 0 \\ \vdots & & & & & \vdots \\ 0 & 0 & 0 & 0 & \cdots & 1 \\ -a_0 & -a_1 & -a_2 & -a_3 & \cdots & -a_{n-1} \end{bmatrix} \mathbf{x} + \begin{bmatrix} 0 \\ 0 \\ 0 \\ \vdots \\ 0 \\ 1 \end{bmatrix} u(t) = \mathbf{A}\mathbf{x} + \mathbf{B}u(t)$$

The output is $y(t) = x_1(t) = [1 \quad 0 \quad 0 \quad \cdots \quad 0]\mathbf{x}(t) = \mathbf{C}\mathbf{x}(t)$. In this case the coefficient matrix $\mathbf{A}$ is the companion matrix mentioned in Sec. 7.8 and Problem 7.32.

The same method of defining the state variables can be applied to multiple-input, multiple-output systems described by several coupled differential equations if the inputs are not differentiated.

Example 9.3

A system has three inputs u_1, u_2, u_3 and three outputs y_1, y_2, y_3. The input-output equations are

$$\dddot{y}_1 + a_1 \ddot{y}_1 + a_2(\dot{y}_1 + \dot{y}_2) + a_3(y_1 - y_3) = u_1(t)$$

$$\ddot{y}_2 + a_4(\dot{y}_2 - \dot{y}_1 + 2\dot{y}_3) + a_5(y_2 - y_1) = u_2(t)$$

$$\dot{y}_3 + a_6(y_3 - y_1) = u_3(t)$$

Notice that in the second equation $\dot{y}_3$ can be eliminated by using the third equation. State variables are selected as the outputs and their derivatives up to the $(n - 1)$th, where n is the order of the highest derivative of a given output.

Select $x_1 = y_1$, $x_2 = \dot{y}_1$, $x_3 = \ddot{y}_1$, $x_4 = y_2$, $x_5 = \dot{y}_2$, $x_6 = y_3$. Then

$$\dot{x}_1 = x_2, \qquad \dot{x}_2 = x_3, \qquad \dot{x}_4 = x_5$$

$$\dot{x}_3 = -a_1 x_3 - a_2(x_2 + x_5) - a_3(x_1 - x_6) + u_1$$

$$\dot{x}_5 = -a_4(x_5 - x_2 + 2\dot{x}_6) - a_5(x_4 - x_1) + u_2$$

$$\dot{x}_6 = -a_6(x_6 - x_1) + u_3$$

Eliminating $\dot{x}_6$ from the $\dot{x}_5$ equation leads to

$$\begin{bmatrix} \dot{x}_1 \\ \dot{x}_2 \\ \dot{x}_3 \\ \dot{x}_4 \\ \dot{x}_5 \\ \dot{x}_6 \end{bmatrix} = \begin{bmatrix} 0 & 1 & 0 & 0 & 0 & 0 \\ 0 & 0 & 1 & 0 & 0 & 0 \\ -a_3 & -a_2 & -a_1 & 0 & -a_2 & a_3 \\ 0 & 0 & 0 & 0 & 1 & 0 \\ a_5 - 2a_4a_6 & a_4 & 0 & -a_5 & -a_4 & 2a_4a_6 \\ a_6 & 0 & 0 & 0 & 0 & -a_6 \end{bmatrix} \begin{bmatrix} x_1 \\ x_2 \\ x_3 \\ x_4 \\ x_5 \\ x_6 \end{bmatrix} + \begin{bmatrix} 0 & 0 & 0 \\ 0 & 0 & 0 \\ 1 & 0 & 0 \\ 0 & 0 & 0 \\ 0 & 1 & -2a_4 \\ 0 & 0 & 1 \end{bmatrix} \begin{bmatrix} u_1 \\ u_2 \\ u_3 \end{bmatrix}$$

The output equation is

$$\begin{bmatrix} y_1 \\ y_2 \\ y_3 \end{bmatrix} = \begin{bmatrix} 1 & 0 & 0 & 0 & 0 & 0 \\ 0 & 0 & 0 & 1 & 0 & 0 \\ 0 & 0 & 0 & 0 & 0 & 1 \end{bmatrix} \mathbf{x}$$

■

When derivatives of the input appear in the system differential equation, the system is sometimes described as having numerator dynamics. This terminology is due to the fact that the transfer function then has zeros. The technique of state variable selection requires modification. The previous method would lead to a set of first-order equations, but the input *derivatives* would still be present. The state equation must express $\dot{\mathbf{x}}$ as a function of $\mathbf{x}$, $\mathbf{u}$, and t (and not $\dot{\mathbf{u}}$). Obviously, the derivatives of u_i must somehow be absorbed into the definition for the state variables. As an example, consider $\ddot{y} + a\dot{y} + by = u + c\dot{u}$. Rearranging gives $\ddot{y} - c\dot{u} = -a\dot{y} - by + u$. One could use $x_1 = y$, $x_2 = \dot{y} - cu$ so that, assuming c is constant, $\dot{x}_1 = \dot{y} = x_2 + cu$ and $\dot{x}_2 = -a[x_2 + cu] - bx_1 + u$; or

$$\begin{bmatrix} \dot{x}_1 \\ \dot{x}_2 \end{bmatrix} = \begin{bmatrix} 0 & 1 \\ -b & -a \end{bmatrix} \begin{bmatrix} x_1 \\ x_2 \end{bmatrix} + \begin{bmatrix} c \\ 1 - ac \end{bmatrix} u$$

$$y = [1 \quad 0]\mathbf{x}$$

Other choices are possible, since the *selection* of state variables is *not* a unique process. For more complex systems simulation diagrams are a useful aid in selecting state variables.

Using Simulation Diagrams

Equation (*9.8*) indicates that state variables for continuous-time systems are always determined by integrating a function of state variables and inputs. The simulation diagram approach makes use of this fact. Six ideal elements are used as building blocks in the simulation diagrams. They are described in Table 9.1.

Note that a differentiating element $\longrightarrow \boxed{d/dt} \longrightarrow$ is *not* included. If the equations for a continuous-time system can be simulated using any combination of these elements except the delay, and if no *unnecessary* integrators are used, then the output of each integrator can be selected as a state variable. For discrete-time systems the ideal delay is used as the only memory device instead of the integrator. Outputs of delays can be selected as discrete-time state variables. Discrete systems are considered in Chapter 11.

TABLE 9.1

Element	Symbol	Input-output relation
1. Integrator	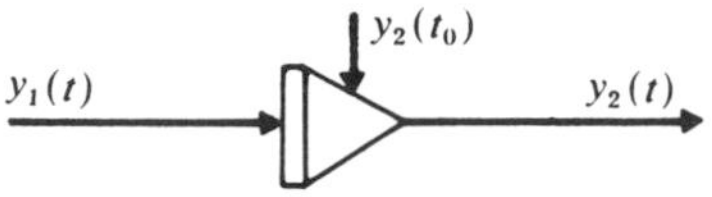	$y_2(t) = y_2(t_0) + \int_{t_0}^{t} y_1(\tau)\, d\tau$
2. Delay	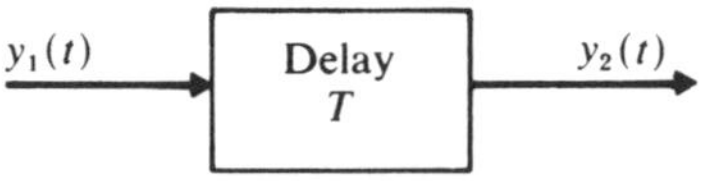	$y_2(t) = y_1(t - T)$
3. Summing junction	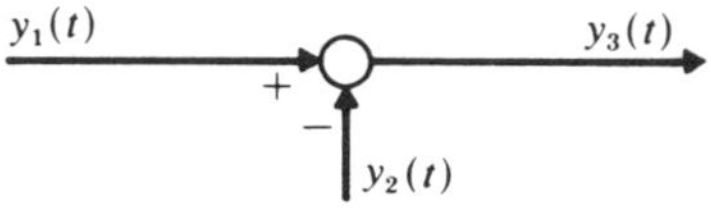	$y_3(t) = y_1(t) - y_2(t)$
4. Gain change	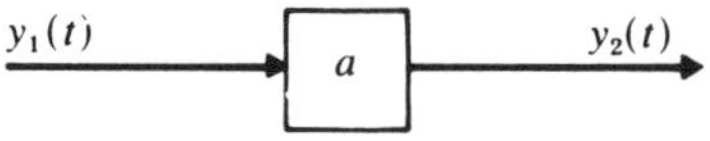	$y_2(t) = a y_1(t)$
5. Multiplier	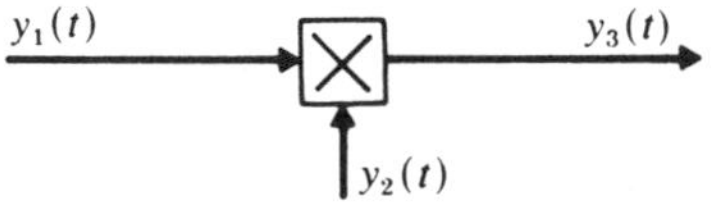	$y_3(t) = y_1(t) y_2(t)$
6. Single-valued function generator	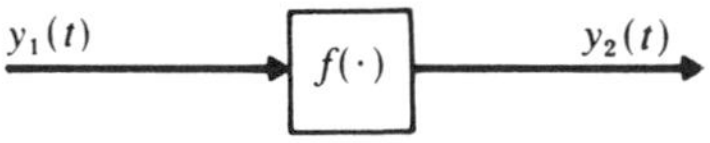	$y_2(t) = f(y_1(t))$

Example 9.4

One possible simulation diagram for $\ddot{y} + a\dot{y} + by = u + c\dot{u}$ is given in Fig. 9.5. Selecting the outputs of the integrators as x_1 and x_2 leads to the same state equations given earlier for this system.

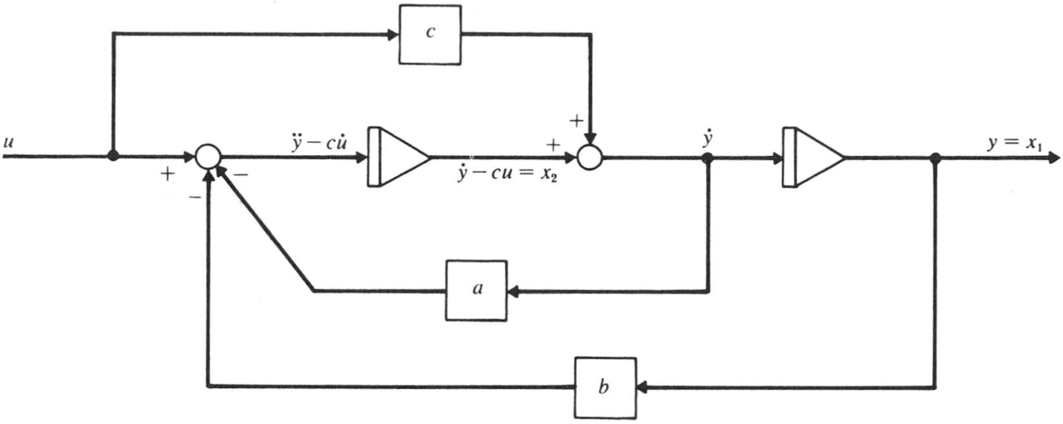

Figure 9.5 ■

Use of the simulation diagram can require a degree of ingenuity in more complicated systems unless a systematic approach is used. One such approach which is easy to use with constant coefficient systems is as follows.

1. Solve each differential equation for its highest derivative term. (Assume that it is an nth order derivative for discussion purposes only.)
2. Formally integrate each equation as many times as the order of the highest derivative n. This gives an expression for one output as a function of several integral terms.
3. Group terms on the right-hand side in the form of a nested sequence of integrations.
4. Draw the simulation diagram by inspection, using the fact that terms inside an integral sign constitute inputs to an integrator and terms already containing an integral are the outputs of other integrators.
5. Select the output of each integrator as a state variable. Write down state differential equations.

This procedure leads to what is called the "observable canonical form" of the state equations. The reason for this terminology will be more apparent in Chapter 12. For now, just note that it gives the simplest possible form for the **C** matrix, namely all zeros or ones.

Example 9.5

Consider $\dfrac{d^n y}{dt^n} + a_{n-1}\dfrac{d^{n-1} y}{dt^{n-1}} + \cdots + a_1\dfrac{dy}{dt} + a_0 y = \beta_0 u + \beta_1\dfrac{du}{dt} + \cdots + \beta_m\dfrac{d^m u}{dt^m}.$

The coefficients a_i and β_i are constant. Assume $m = n$. If $m < n$, then some of the β_i terms can be set to zero after the final result is obtained.

(1), (2) $$y(t) = \underbrace{\int\int\cdots\int}_{n \text{ integrals}} \left\{\beta_n \frac{d^n u}{dt^n} + \left(\beta_{n-1}\frac{d^{n-1}u}{dt^{n-1}} - a_{n-1}\frac{d^{n-1}y}{dt^{n-1}}\right) + \cdots + \left(\beta_1 \frac{du}{dt} - a_1 \frac{dy}{dt}\right) + (\beta_0 u - a_0 y)\right\} dt \ldots dt'$$

(3) $$y(t) = \beta_n u + \int \left\{(\beta_{n-1}u - a_{n-1}y) + \int \left[\beta_{n-2}u - a_{n-2}y + \int \left(\cdots + \int \left\{\beta_1 u - a_1 y + \int (\beta_0 u - a_0 y)\, dt\right\} \cdots\right) dt'\right] dt''\right\} dt'''$$

(4) The simulation diagram is shown in Fig. 9.6.

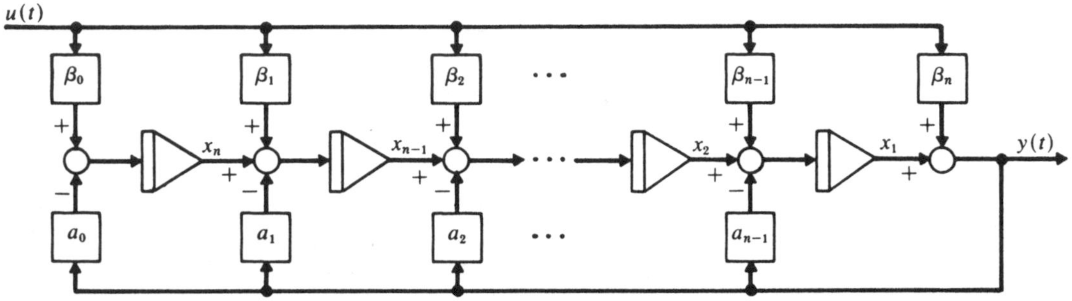

Figure 9.6

(5) Numbering outputs of integrators from the right gives

$$\dot{x}_1 = -a_{n-1}[x_1 + \beta_n u] + \beta_{n-1}u + x_2$$
$$= -a_{n-1}x_1 + x_2 + (\beta_{n-1} - a_{n-1}\beta_n)u$$
$$\dot{x}_2 = -a_{n-2}x_1 + x_3 + (\beta_{n-2} - a_{n-2}\beta_n)u$$
$$\vdots$$
$$\dot{x}_{n-1} = -a_1 x_1 + x_n + (\beta_1 - a_1\beta_n)u$$
$$\dot{x}_n = -a_0 x_1 + (\beta_0 - a_0\beta_n)u$$

The output equation is

$$y = x_1 + \beta_n u = [1 \quad 0 \quad 0 \quad \cdots \quad 0]\mathbf{x} + \beta_n u$$

This represents the observable canonical form of the state equations. ■

The above method works equally well when the input is not differentiated, or when coupled simultaneous equations must be considered.

State Equations from Transfer Functions

The discussion is restricted to single-input, single-output constant coefficient systems described by a transfer function:

$$\frac{Y(s)}{u(s)} = T(s) = \frac{\beta_m s^m + \beta_{m-1}s^{m-1} + \cdots + \beta_1 s + \beta_0}{s^n + a_{n-1}s^{n-1} + \cdots + a_1 s + a_0}$$

Multivariable systems and transfer matrices are considered in Chapter 15. The above transfer function gives the input-output relationship for the differential equation in Example 9.5. Therefore, the previous methods can be used to select state variables. The reason for considering this system further from the transfer function point of view is that several alternative forms for the state equations can easily be illustrated. Two forms of the state equations will be derived from the transfer function in its originally given form. These direct realizations are the observable canonical form already introduced and the "controllable canonical form." Then two alternate forms of the transfer function [equations (*9.15*) and (*9.16*)] will be used to obtain cascade and parallel realizations.

The transfer function can be written in factored form, equation (*9.15*). If all poles p_i are distinct and if $m = n$, the partial fraction expansion form of equation (*9.16*) can be obtained:

$$T(s) = \frac{\beta_m(s + z_1)(s + z_2)\cdots(s + z_m)}{(s + p_1)(s + p_2)\cdots(s + p_n)} \tag{9.15}$$

$$T(s) = b_0 + \frac{b_1}{s + p_1} + \frac{b_2}{s + p_2} + \cdots + \frac{b_n}{s + p_n} \tag{9.16}$$

If some poles are repeated, the partial fraction expansion will contain additional terms involving powers of the multiple pole $(s + p_i)$ in the denominator (see Problems 9.5 and 9.16). If $m < n$, $b_0 = 0$. For a dynamical system, m can never exceed n.

Equation (*9.15*) indicates that $T(s)$ can be written as the product of simple factors $1/(s + p_i)$ or $(s + z_j)/(s + p_i)$. Quadratic terms could be considered as well. In any case the system can be represented as a series (cascade) connection of simple terms shown in Fig. 9.7.

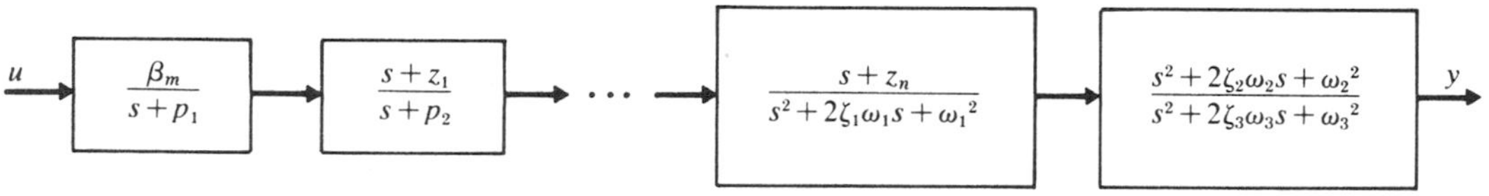

Figure 9.7

The overall simulation diagram is also a series connection of several simple simulation diagrams, such as those found in Problem 9.1.

Equation (*9.16*) indicates that $T(s)$ can be represented as a parallel connection of simple terms (Fig. 9.8).

The diagram of Fig. 9.8 assumes p_{n-1} is a double pole. The system simulation diagram is also a parallel connection of the individual terms.

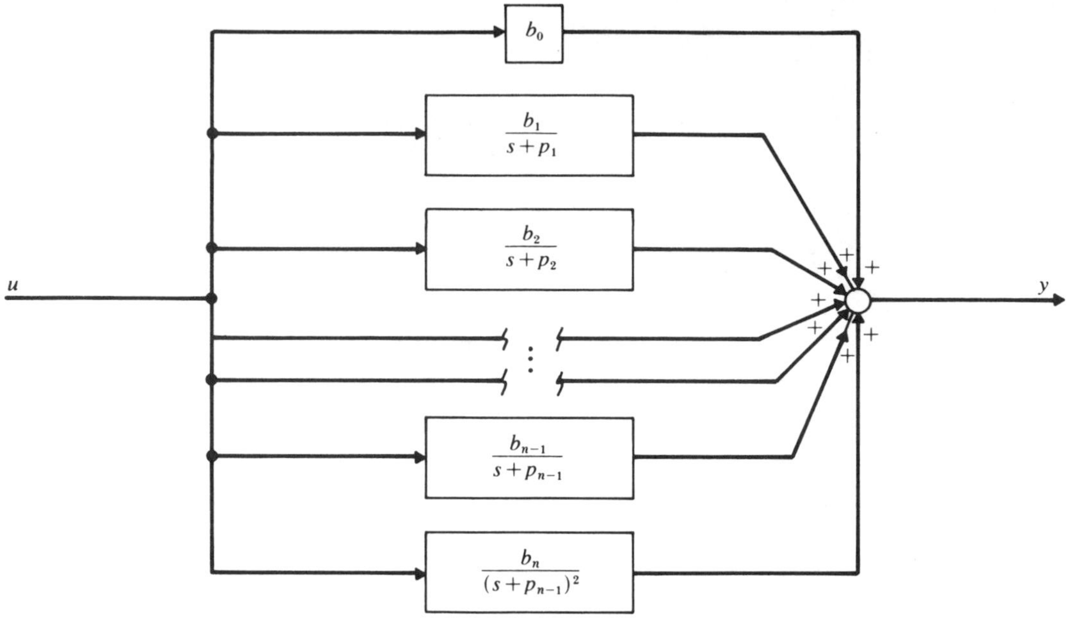

Figure 9.8

Example 9.6

Select a suitable set of state variables for the system whose transfer function is

$$T(s) = \frac{s+3}{s^3 + 9s^2 + 24s + 20}$$

Note that in factored form

$$T(s) = \frac{s+3}{(s+2)^2(s+5)}$$

and, using a partial fraction expansion,

$$T(s) = \frac{2/9}{s+2} + \frac{1/3}{(s+2)^2} + \frac{-2/9}{s+5}$$

Using the original form of the transfer function will lead to one possible direct realization. First, it is noted that

$$\dddot{y} = -9\ddot{y} - 24\dot{y} - 20y + \dot{u} + 3u$$

or

$$y = \int \left\{ -9y + \int \left[-24y + u + \int (-20y + 3u)\, dt \right] dt' \right\} dt''$$

from which the simulation diagram of Fig. 9.9 is obtained.

Using Fig. 9.9, the observable canonical form of the state equations is

$$\begin{bmatrix} \dot{x}_1 \\ \dot{x}_2 \\ \dot{x}_3 \end{bmatrix} = \begin{bmatrix} -9 & 1 & 0 \\ -24 & 0 & 1 \\ -20 & 0 & 0 \end{bmatrix} \begin{bmatrix} x_1 \\ x_2 \\ x_3 \end{bmatrix} + \begin{bmatrix} 0 \\ 1 \\ 3 \end{bmatrix} u$$

and

$$y = [1 \quad 0 \quad 0]\mathbf{x}$$

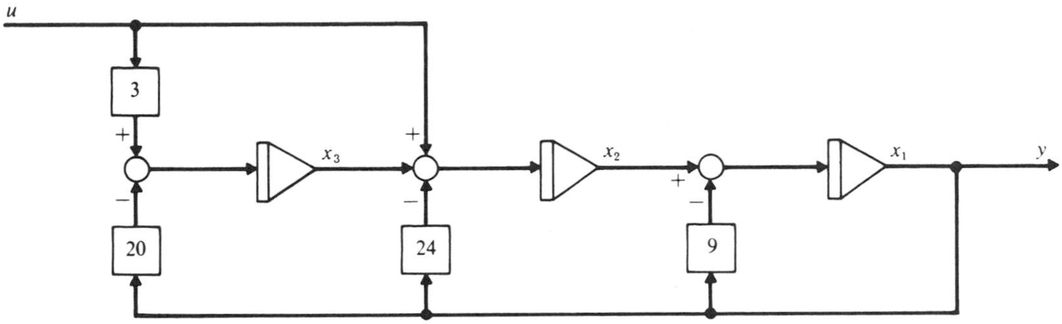

Figure 9.9

A second direct form of the transfer function will now be developed. The numerator and denominator of the transfer function are separated, as shown in Fig. 9.10. The inter-

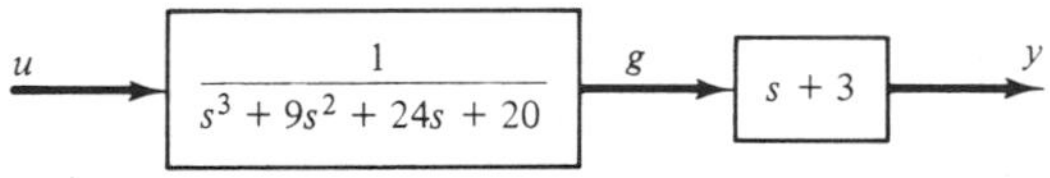

Figure 9.10

mediate variable thus created is labeled g for convenience. The relationship between u and g can be represented in exactly the same way as the system at the beginning of this section since there are no numerator dynamics.

$$\dddot{g} + 9\ddot{g} + 24\dot{g} + 20g = u$$

This means that the **A** matrix will be in companion form and the **B** matrix will assume its simplest possible form—all zeros or ones.

The relationship between the fictitious g and the output y depends only on the numerator of $T(s)$. In this simple case

$$y = \dot{g} + 3g$$

Noting that since the s variable indicates differentiation, y is seen to be a linear combination of g and its various derivatives. These derivatives are available as inputs to the various integrators. Using this reasoning leads to the simulation diagram of Fig. 9.11. Then, by

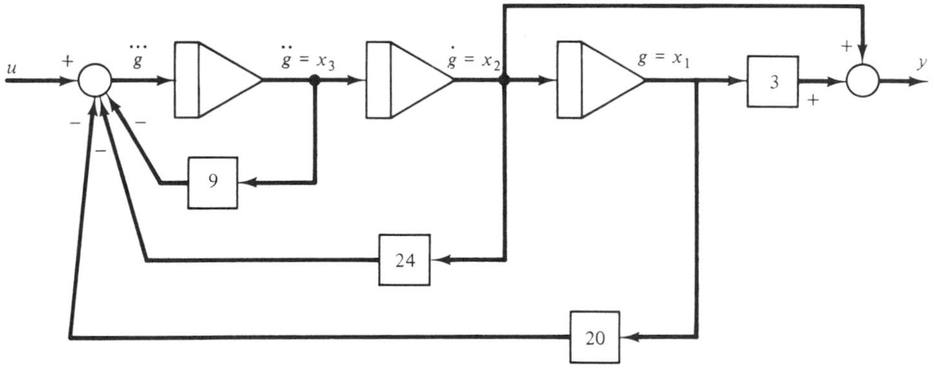

Figure 9.11

picking outputs of integrators as states, the controllable canonical form of the state equations is as follows:

$$\dot{\mathbf{x}} = \begin{bmatrix} 0 & 1 & 0 \\ 0 & 0 & 1 \\ -20 & -24 & -9 \end{bmatrix}\mathbf{x} + \begin{bmatrix} 0 \\ 0 \\ 1 \end{bmatrix}u$$

$$y = [3 \quad 1 \quad 0]\mathbf{x} + [0]u$$

Using the factored form of $T(s)$, the simulation diagram of Fig. 9.12 is obtained.

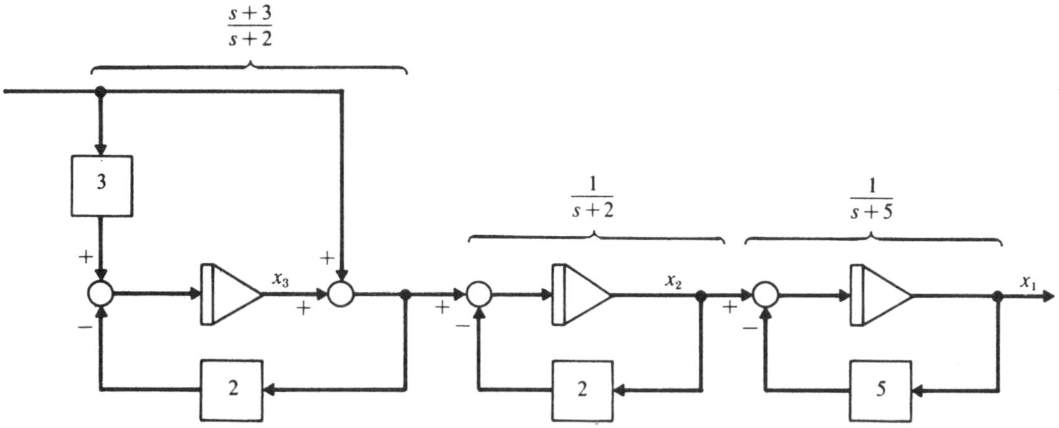

Figure 9.12

From Fig. 9.12,

$$\begin{bmatrix} \dot{x}_1 \\ \dot{x}_2 \\ \dot{x}_3 \end{bmatrix} = \begin{bmatrix} -5 & 1 & 0 \\ 0 & -2 & 1 \\ 0 & 0 & -2 \end{bmatrix}\begin{bmatrix} x_1 \\ x_2 \\ x_3 \end{bmatrix} + \begin{bmatrix} 0 \\ 1 \\ 1 \end{bmatrix}u \quad \text{and} \quad y = [1 \quad 0 \quad 0]\mathbf{x}$$

Using the partial fraction expansion, the simulation diagram of Fig. 9.13 is obtained.

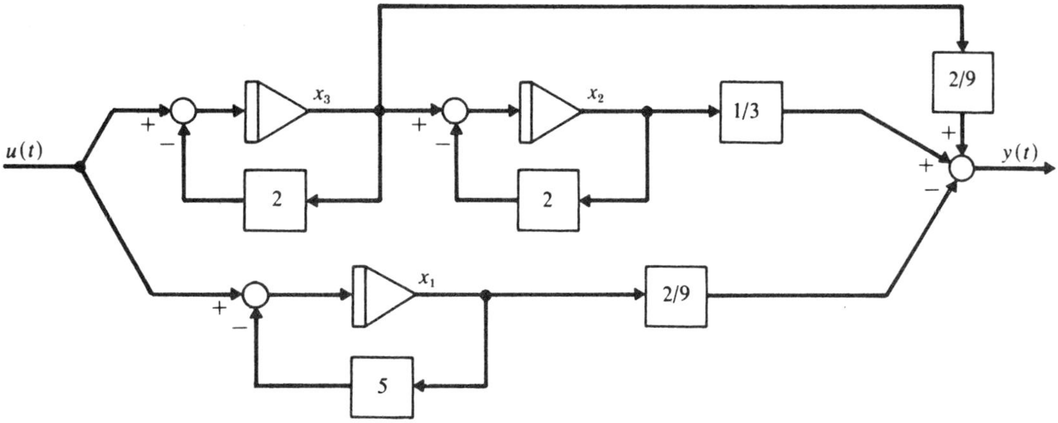

Figure 9.13

From Fig. 9.13,

$$\begin{bmatrix} \dot{x}_1 \\ \dot{x}_2 \\ \dot{x}_3 \end{bmatrix} = \begin{bmatrix} -5 & 0 & 0 \\ 0 & -2 & 1 \\ 0 & 0 & -2 \end{bmatrix} \begin{bmatrix} x_1 \\ x_2 \\ x_3 \end{bmatrix} + \begin{bmatrix} 1 \\ 0 \\ 1 \end{bmatrix} u \quad \text{and} \quad y = [-2/9 \quad 1/3 \quad 2/9]\mathbf{x}$$

■

Four different sets of valid state equations have been derived for this system. In the third and fourth forms, the diagonal terms of the matrix **A** are the same, the system poles (and the eigenvalues of **A**). The fourth form gives **A** in Jordan canonical form. All four representations are different, but all have the same number of state variables, and this is the order of the system. They also all have the same eigenvalues (poles). Note that if separate parallel paths had been used for the terms $1/(s + 2)^2$ and $1/(s + 2)$, one unnecessary integrator would have been used. This must be avoided if integrator outputs are to be used as state variables.

State Equations Directly from the System's Linear Graph [17, 78]

Linear graphs were used in Chapter 1 in connection with system modeling. Recall that linear graph techniques are not restricted to linear or constant coefficient systems. A method of obtaining state equations directly from the system graph is now presented. This is a powerful method because it avoids many of the intermediate manipulations with transfer functions and input-output differential equations which are restricted to linear, constant systems. Furthermore, the linear graph technique often gives greater engineering insight because the state variables thus obtained are usually related to the energy stored in the system. Before describing the method, a few additional definitions regarding linear graphs are required.

A *tree* is a set of branches of the graph that (1) contains every node of the graph, (2) is connected, and (3) contains no loops. A tree is formed from a graph by removing certain branches. A branch of the graph included in the tree is called a tree branch. Those branches which were deleted while forming a tree are called *links*. Each time a link is added to a tree, one loop is formed. A loop consisting of one link plus a number of tree branches is called the *fundamental loop* associated with that link. For a given tree, if any one tree branch is cut, the tree is separated into two parts. A *fundamental cutset* of a given tree branch consists of that one cut tree branch plus all links that connect between nodes of the two halves of the severed tree. In other words, if a line is drawn through the original graph in such a way as to (1) divide the graph into two parts and (2) cut only one tree branch, then that branch plus all links cut by the dividing line form a fundamental cutset.

The following procedure is a systematic method of obtaining state equations from a linear graph:

1. *Form a tree from the graph which includes:* (*a*) all across variable (voltage) sources; (*b*) as many elements as possible which store energy by virtue of their across variable (capacitors or the analogous elements in other disciplines), that

is, elements whose elemental equation has the across variable differentiated; (*c*) elements with algebraic elemental equations (resistors and their analogs); (*d*) as few elements as possible which store energy by virtue of their through variable and have the through variable differentiated in their elemental equation (inductors and their analogs); (*e*) no through variable (current) sources are included in the tree; (*f*) if ideal transformers are included in the graph, one side of the transformer should be treated like a through variable source and the other side like an across variable source. There will usually be several trees which satisfy these rules.

2. *Choose as state variables the across variables of all capacitor-like elements included in the tree and the through variables of all inductor-like elements not included in the tree.*
3. *The elemental equations for elements involving the selected state variables will be of the form*

$$\dot{x}_i = \text{function of through or across variables and inputs}$$

Use the compatibility laws (Kirchhoff's voltage laws) around the fundamental loops and the conservation laws (Kirchhoff's current laws) into the fundamental cutsets to eliminate nonstate variables from these functional equations.

Example 9.7

Consider the linear graph shown in Fig. 9.14. The symbols L, C, and R_i are used to indicate the type elements in each branch, although they need not be of an electrical nature.

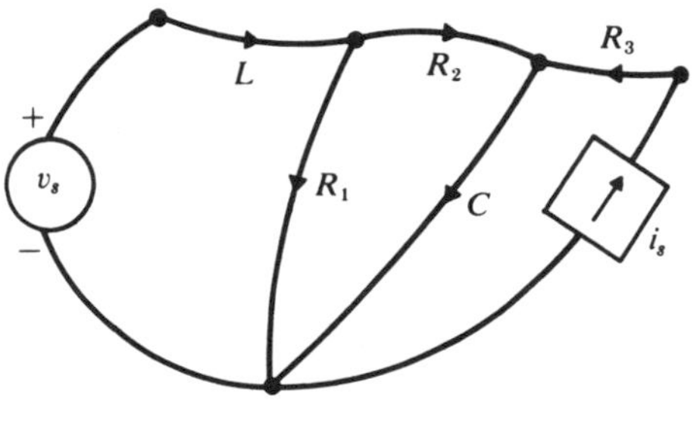

Figure 9.14

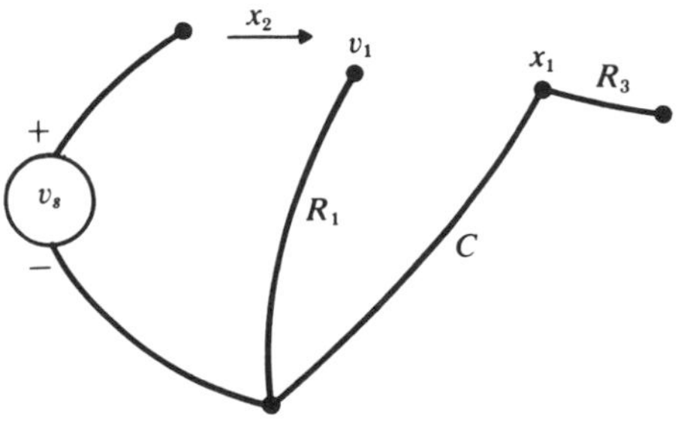

Figure 9.15

Using the given rules, the tree of Fig. 9.15 is selected. The state variables are the voltage x_1 across C and the current x_2 through L. It is assumed that the elemental equations are nonlinear,

$$C\dot{x}_1 = f_1(i_C) \qquad \text{and} \qquad L\dot{x}_2 = f_2(v_s - v_1)$$

To complete the state description, i_C and v_1 must be expressed as functions of x_1, x_2, v_s, and i_s.

The fundamental cutset associated with branch C is given in Fig. 9.16 and thus $i_C = i_2 + i_s$. The cutset associated with R_1 is given in Fig. 9.17, from which $i_1 = x_2 - i_2$. The fundamental loop formed with link R_2 gives the compatibility equation $v_1 - i_2 R_2 = x_1$, assuming R_2 is a linear resistor. If R_1 is also linear, $v_1 = R_1 i_1$.

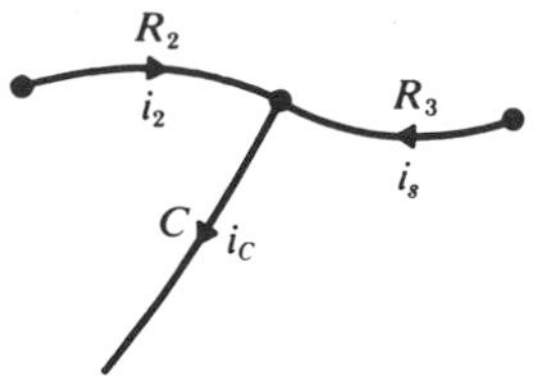

Figure 9.16

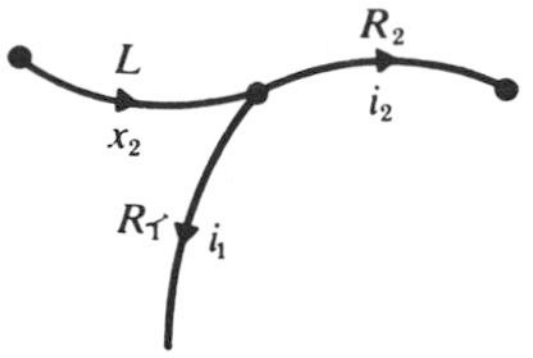

Figure 9.17

These equations are solved simultaneously to give

$$i_C = i_s + \frac{R_1 x_2 - x_1}{R_1 + R_2}$$

$$v_1 = \frac{R_1 R_2 x_2 + R_1 x_1}{R_1 + R_2}$$

so that the state equations are

$$\dot{x}_1 = \frac{1}{C} f_1\left(i_s + \frac{R_1 x_2 - x_1}{R_1 + R_2}\right)$$

$$\dot{x}_2 = \frac{1}{L} f_2\left(v_s - \frac{R_1 R_2 x_2 + R_1 x_1}{R_1 + R_2}\right)$$

If L and C are also linear, then letting $u_1 = v_s$ and $u_2 = i_s$ reduces the equations to

$$\begin{bmatrix} \dot{x}_1 \\ \dot{x}_2 \end{bmatrix} = \begin{bmatrix} -\dfrac{1}{C(R_1 + R_2)} & \dfrac{R_1}{C(R_1 + R_2)} \\ -\dfrac{R_1}{L(R_1 + R_2)} & -\dfrac{R_1 R_2}{L(R_1 + R_2)} \end{bmatrix} \begin{bmatrix} x_1 \\ x_2 \end{bmatrix} + \begin{bmatrix} 0 & 1/C \\ 1/L & 0 \end{bmatrix} \begin{bmatrix} u_1 \\ u_2 \end{bmatrix}$$

If the voltage across R_2 is considered to be the output y, then

$$y = v_1 - x_1 = \begin{bmatrix} -\dfrac{R_2}{R_1 + R_2} & \dfrac{R_1 R_2}{R_1 + R_2} \end{bmatrix} \mathbf{x}$$ ■

9.5 COMMENTS ON THE STATE SPACE REPRESENTATION

The selection of state variables is *not* a unique process. Various sets of state variables can be used. It is usually advantageous to use variables which have physical significance and, if possible, can be measured. Several methods of selecting state variables have been illustrated. The form of the information available for a system will often dictate which method should be used. In some cases, for example, a transfer function is obtained experimentally and must be used as the starting point. In any case the advantages of the state space method include the following:

1. It provides a convenient, compact notation and allows the application of the powerful vector-matrix theory.
2. The uniform notation for all systems makes possible a uniform set of solution techniques.
3. The state space representation is in an ideal format for computer solution, either

analog or digital. This is important because computers are frequently required for the analysis of complex systems.

4. The state space method gives a more complete description of a system than does the input-output approach. This will become more evident when the concepts of controllability and observability are considered in Chapters 12 and 15.

The state space method is the central theme of modern control theory. However, the ideas have been with us for some time. The generalized coordinates and momenta of Hamiltonian and Lagrangian mechanics are, in fact, a set of state variables. For those familiar with these subjects, an alternative method of state variable selection is available [44].

ILLUSTRATIVE PROBLEMS

9.1 Four input-output transfer functions $y(s)/u(s)$ are listed below. Describe the systems they represent in state variable form:

a. $1/(s+\alpha)$

b. $(s+\beta)/(s+\alpha)$

c. $(s+\beta)/(s^2+2\zeta\omega s+\omega^2)$

d. $(s^2+2\zeta_1\omega_1 s+\omega_1^2)/(s^2+2\zeta_2\omega_2 s+\omega_2^2)$

The solutions are obtained by writing the input-output differential equation, drawing the simulation diagram, and selecting the integrator outputs as state variables.

a. The differential equation is $\dot{y}+\alpha y=u$. This is simulated in Fig. 9.18*a*, from which $\dot{x}=-\alpha x+u$ and $y=x$.

b. The differential equation is $\dot{y}+\alpha y=\beta u+\dot{u}$. This is simulated in Fig. 9.18*b*, from which $\dot{x}=-\alpha x+(\beta-\alpha)u$ and $y=x+u$.

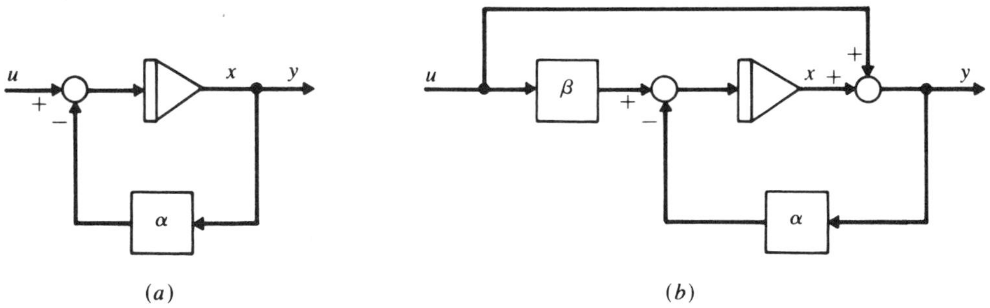

Figure 9.18*a*, *b*

c. The differential equation is $\ddot{y} + 2\zeta\omega\dot{y} + \omega^2 y = \beta u + \dot{u}$, and is simulated in Fig. 9.18*c*. The state equations are $\dot{x}_1 = -2\zeta\omega x_1 + x_2 + u$, $\dot{x}_2 = -\omega^2 x_1 + \beta u$, and $y = [1 \quad 0]\mathbf{x}$.

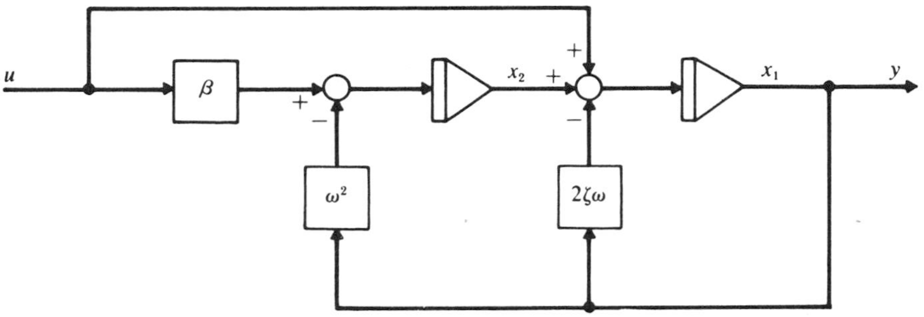

Figure 9.18*c*

d. The differential equation is $\ddot{y} + 2\zeta_2\omega_2\dot{y} + \omega_2^2 y = \ddot{u} + 2\zeta_1\omega_1\dot{u} + \omega_1^2 u$. Integrating twice allows this to be written as

$$y = u + \int \left\{ 2\zeta_1\omega_1 u - 2\zeta_2\omega_2 y + \int [\omega_1^2 u - \omega_2^2 y]\, dt \right\} dt'$$

The simulation diagram of Fig. 9.18*d* gives $\dot{x}_1 = -2\zeta_2\omega_2 x_1 + x_2 + (2\zeta_1\omega_1 - 2\zeta_2\omega_2)u$, $\dot{x}_2 = -\omega_2^2 x_1 + (\omega_1^2 - \omega_2^2)u$, and $y = [1 \quad 0]\mathbf{x} + u$.

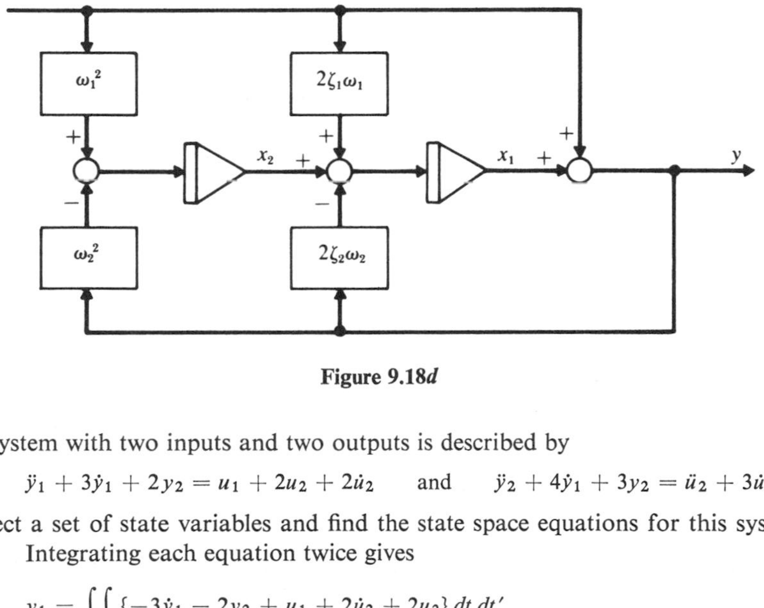

Figure 9.18*d*

9.2 A system with two inputs and two outputs is described by

$$\ddot{y}_1 + 3\dot{y}_1 + 2y_2 = u_1 + 2u_2 + 2\dot{u}_2 \qquad \text{and} \qquad \ddot{y}_2 + 4\dot{y}_1 + 3y_2 = \ddot{u}_2 + 3\dot{u}_2 + u_1$$

Select a set of state variables and find the state space equations for this system.

Integrating each equation twice gives

$$y_1 = \iint \{-3\dot{y}_1 - 2y_2 + u_1 + 2\dot{u}_2 + 2u_2\}\, dt\, dt'$$

$$y_2 = \iint \{-4\dot{y}_1 - 3y_2 + \ddot{u}_2 + 3\dot{u}_2 + u_1\}\, dt\, dt'$$

or

$$y_1 = \int \left\{ -3y_1 + 2u_2 + \int [-2y_2 + u_1 + 2u_2]\, dt \right\} dt'$$

$$y_2 = u_2 + \int \left\{ -4y_1 + 3u_2 + \int [-3y_2 + u_1]\, dt \right\} dt'$$

The simulation diagram of Fig. 9.19 can now be drawn.

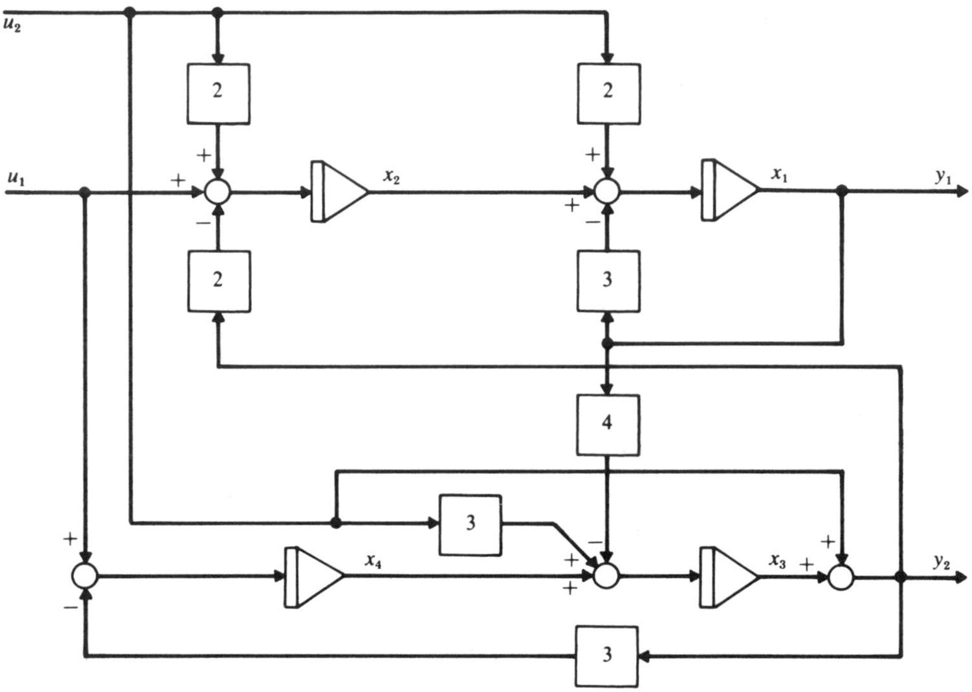

Figure 9.19

The state equations are

$$\left.\begin{aligned} \dot{x}_1 &= -3x_1 + x_2 + 2u_2 \\ \dot{x}_2 &= -2x_3 + u_1 \\ \dot{x}_3 &= -4x_1 + x_4 + 3u_2 \\ \dot{x}_4 &= -3x_3 + u_1 - 3u_2 \end{aligned}\right\} \quad \text{or} \quad \dot{\mathbf{x}} = \begin{bmatrix} -3 & 1 & 0 & 0 \\ 0 & 0 & -2 & 0 \\ -4 & 0 & 0 & 1 \\ 0 & 0 & -3 & 0 \end{bmatrix} \mathbf{x} + \begin{bmatrix} 0 & 2 \\ 1 & 0 \\ 0 & 3 \\ 1 & -3 \end{bmatrix} \mathbf{u}$$

and $\mathbf{y} = \begin{bmatrix} y_1 \\ y_2 \end{bmatrix} = \begin{bmatrix} 1 & 0 & 0 & 0 \\ 0 & 0 & 1 & 0 \end{bmatrix} \mathbf{x} + \begin{bmatrix} 0 & 0 \\ 0 & 1 \end{bmatrix} \mathbf{u}.$

9.3 A single-input, single-out system has the transfer function

$$\frac{y(s)}{u(s)} = \frac{1}{s^3 + 10s^2 + 27s + 18} = T(s)$$

Find three different state variable representations.

a. The transfer function represents the differential equation $\dddot{y} + 10\ddot{y} + 27\dot{y} + 18y$

$= u$. Setting $x_1 = y$, $x_2 = \dot{y}$, $x_3 = \ddot{y}$ gives

$$\begin{bmatrix} \dot{x}_1 \\ \dot{x}_2 \\ \dot{x}_3 \end{bmatrix} = \begin{bmatrix} 0 & 1 & 0 \\ 0 & 0 & 1 \\ -18 & -27 & -10 \end{bmatrix} \begin{bmatrix} x_1 \\ x_2 \\ x_3 \end{bmatrix} + \begin{bmatrix} 0 \\ 0 \\ 1 \end{bmatrix} u \quad \text{and} \quad y = [1 \quad 0 \quad 0]\mathbf{x}$$

b. In factored form $T(s) = 1/[(s+6)(s+1)(s+3)]$, and a simulation diagram is given in Fig. 9.20.

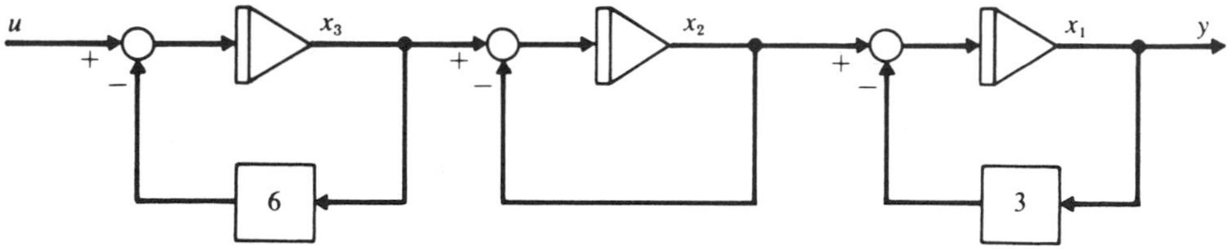

Figure 9.20

$$\text{Then} \quad \begin{bmatrix} \dot{x}_1 \\ \dot{x}_2 \\ \dot{x}_3 \end{bmatrix} = \begin{bmatrix} -3 & 1 & 0 \\ 0 & -1 & 1 \\ 0 & 0 & -6 \end{bmatrix} \begin{bmatrix} x_1 \\ x_2 \\ x_3 \end{bmatrix} + \begin{bmatrix} 0 \\ 0 \\ 1 \end{bmatrix} u \quad \text{and} \quad y = [1 \quad 0 \quad 0]\mathbf{x}.$$

c. Using partial fractions, $T(s) = a/(s+6) + b/(s+1) + c/(s+3)$, where

$a = (s+6)T(s)|_{s=-6} = 1/15 \qquad c = (s+3)T(s)|_{s=-3} = -1/6$

$b = (s+1)T(s)|_{s=-1} = 1/10$

so the simulation diagram of Fig. 9.21 is obtained.

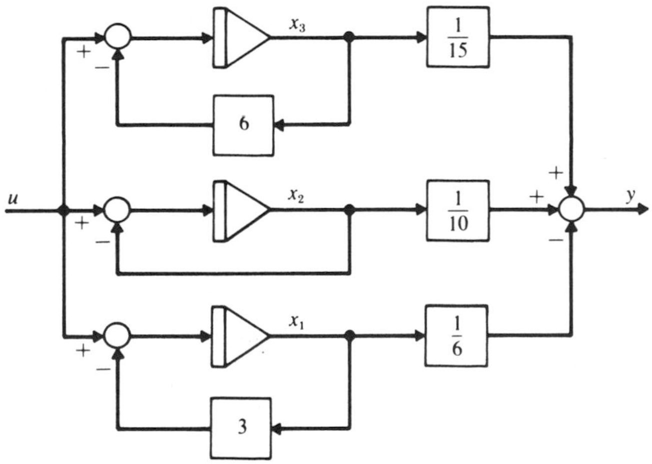

Figure 9.21

$$\text{The state equations are} \quad \dot{\mathbf{x}} = \begin{bmatrix} -3 & 0 & 0 \\ 0 & -1 & 0 \\ 0 & 0 & -6 \end{bmatrix} \mathbf{x} + \begin{bmatrix} 1 \\ 1 \\ 1 \end{bmatrix} u$$

$$y = [-1/6 \quad 1/10 \quad 1/15]\mathbf{x}$$

Note that since Fig. 9.22*a* and *b* are equivalent, an alternative form for the matrices **B** and **C** is $\mathbf{B} = [-1/6 \quad 1/10 \quad 1/15]^T$ and $\mathbf{C} = [1 \quad 1 \quad 1]$.

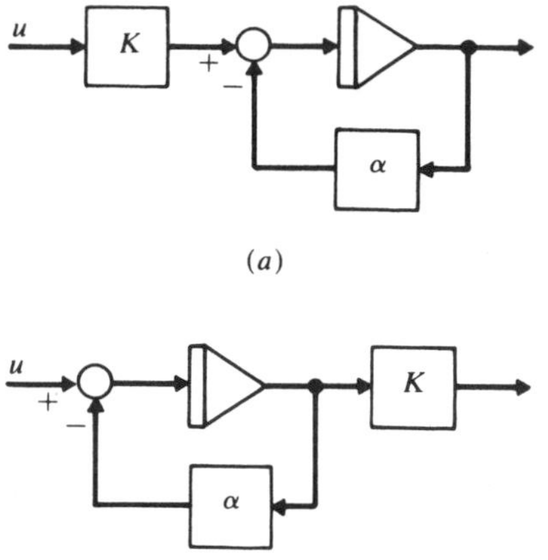

(*a*)

(*b*)

Figure 9.22

9.4 Obtain a state variable representation for the linear system with time-varying coefficients $\ddot{y} + e^{-t^2}\dot{y} + e^t y = u$.

Let $x_1 = y$, $x_2 = \dot{y}$, then $\dot{x}_2 = \ddot{y}$, so

$$\begin{bmatrix} \dot{x}_1 \\ \dot{x}_2 \end{bmatrix} = \begin{bmatrix} 0 & 1 \\ -e^t & -e^{-t^2} \end{bmatrix} \begin{bmatrix} x_1 \\ x_2 \end{bmatrix} + \begin{bmatrix} 0 \\ 1 \end{bmatrix} u \quad \text{and} \quad y = [1 \quad 0]\mathbf{x}$$

9.5 A system input-output transfer function is $T(s) = 1/[s^2(s+3)^3(s+1)]$. Find a state variable representation, using the partial fraction expansion of $T(s)$.

The expansion is

$$T(s) = a_1/s^2 + a_2/s + a_3/(s+3)^3 + a_4/(s+3)^2 + a_5/(s+3) + a_6/(s+1)$$

where

$$a_1 = s^2 T(s)|_{s=0} = 1/27 \qquad a_4 = \frac{d}{ds}\{(s+3)^3 T(s)\}|_{s=-3} = -7/108$$

$$a_2 = \frac{d}{ds}\{s^2 T(s)\}|_{s=0} = -2/27 \qquad a_5 = \frac{1}{2!}\frac{d^2}{ds^2}\{(s+3)^3 T(s)\}|_{s=-3} = -11/216$$

$$a_3 = (s+3)^3 T(s)|_{s=-3} = -1/18 \qquad a_6 = (s+1)T(s)|_{s=-1} = 1/8$$

Using this expansion gives the simulation diagram of Fig. 9.23.

From the diagram,

$$\dot{\mathbf{x}} = \left[\begin{array}{cc:ccc:c} 0 & 1 & 0 & 0 & 0 & 0 \\ 0 & 0 & 0 & 0 & 0 & 0 \\ \hdashline 0 & 0 & -3 & 1 & 0 & 0 \\ 0 & 0 & 0 & -3 & 1 & 0 \\ 0 & 0 & 0 & 0 & -3 & 0 \\ \hdashline 0 & 0 & 0 & 0 & 0 & -1 \end{array}\right] \mathbf{x} + \begin{bmatrix} 0 \\ 1 \\ 0 \\ 0 \\ 1 \\ 1 \end{bmatrix} u \qquad \text{(note } \mathbf{A} \text{ is in Jordan form)}$$

$$y = [1/27 \quad -2/27 \quad -1/18 \quad -7/108 \quad -11/216 \quad 1/8]\mathbf{x}$$

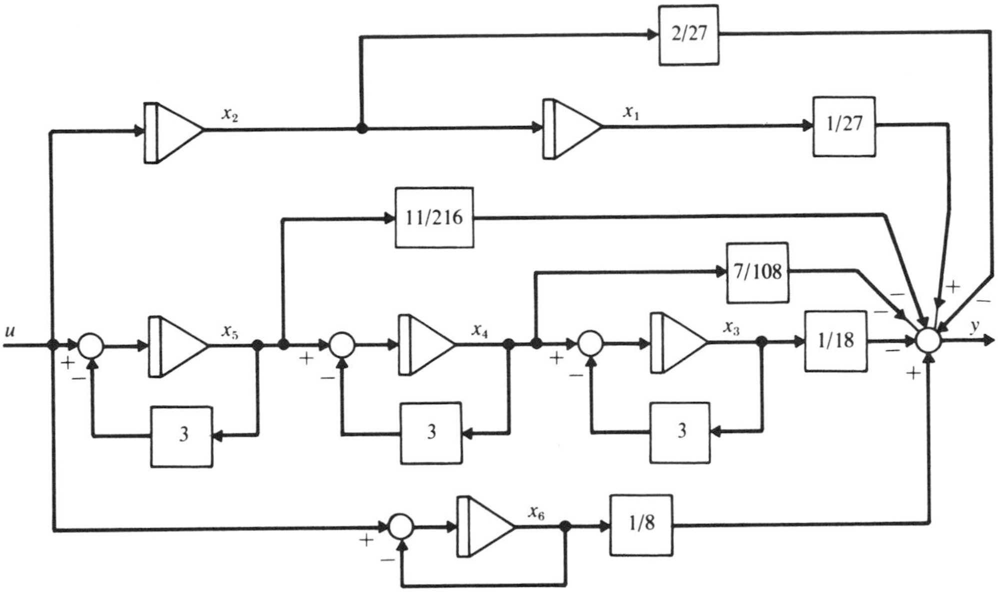

Figure 9.23

9.6 Find the state space representation for a system described by $T(s) = (s + 1)/(s^2 + 7s + 6)$.

Even though $T(s) = (s + 1)/[(s + 1)(s + 6)]$, the common factor should not be cancelled, or the system will be mistaken for a first-order system. Rather, use $\ddot{y} + 7\dot{y} + 6y = \dot{u} + u$, from which $\begin{bmatrix} \dot{x}_1 \\ \dot{x}_2 \end{bmatrix} = \begin{bmatrix} -7 & 1 \\ -6 & 0 \end{bmatrix} \begin{bmatrix} x_1 \\ x_2 \end{bmatrix} + \begin{bmatrix} 1 \\ 1 \end{bmatrix} u$ and $y = [1 \quad 0]\mathbf{x}$.

9.7 Write the differential equations for the circuit of Fig. 9.24 in state variable form. Consider the voltage across R_3 as the output.

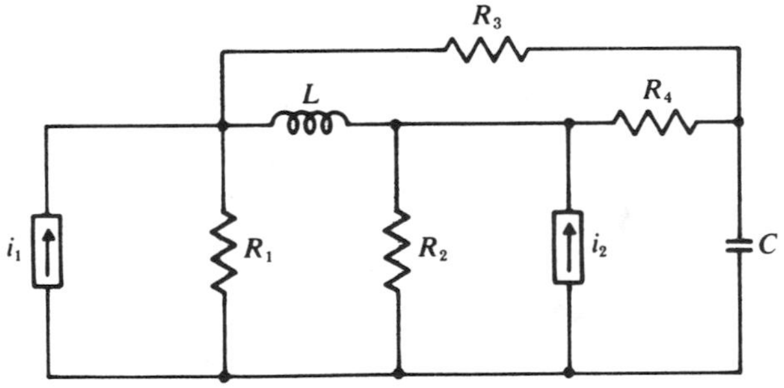

Figure 9.24

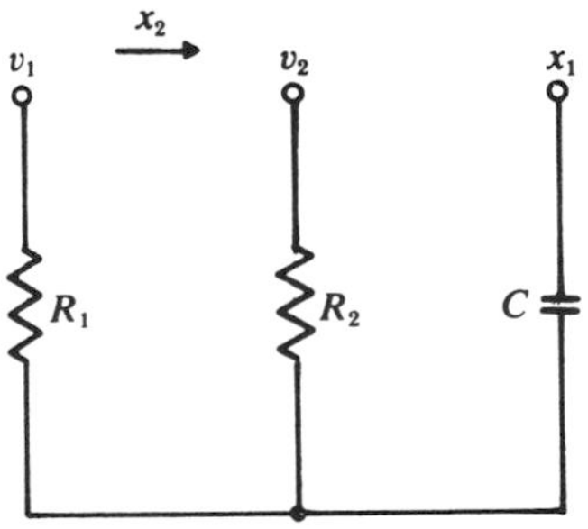

Figure 9.25

The tree selected is shown in Fig. 9.25.

The state variables are selected as the capacitor voltage x_1 and the inductor current x_2, so that $\dot{x}_1 = (1/C)i_C$, $\dot{x}_2 = (1/L)(v_1 - v_2)$. To express i_C, v_1, and v_2 in terms of x_1, x_2 and inputs, use

$$\begin{aligned}
i_C &= i_{R_4} + i_{R_3} && \text{(cutset equation for tree branch } C\text{)}\\
i_{R_1} &= i_1 - i_{R_3} - x_2 && \text{(cutset equation for tree branch } R_1\text{)}\\
i_{R_2} &= x_2 + i_2 - i_{R_4} && \text{(cutset equation for tree branch } R_2\text{)}\\
R_3 i_{R_3} &= v_1 - x_1 && \text{(fundamental loop equation using link } R_3\text{)}\\
R_4 i_{R_4} &= v_2 - x_1 && \text{(fundamental loop equation using link } R_4\text{)}\\
v_1 &= R_1 i_{R_1}\\
v_2 &= R_2 i_{R_2}
\end{aligned}$$

These are seven equations with seven unknowns. Solving for all the resistor currents gives

$$i_{R_1} = \frac{R_3}{R_1 + R_3} i_1 + \frac{x_1}{R_1 + R_3} - \frac{R_3 x_2}{R_1 + R_3} \qquad i_{R_3} = -\frac{x_1}{R_1 + R_3} - \frac{R_1 x_2}{R_1 + R_3} + \frac{R_1 i_1}{R_1 + R_3}$$

$$i_{R_2} = \frac{x_1}{R_2 + R_4} + \frac{R_4 x_2}{R_2 + R_4} + \frac{R_4 i_2}{R_2 + R_4} \qquad i_{R_4} = -\frac{x_1}{R_2 + R_4} + \frac{R_2 x_2}{R_2 + R_4} + \frac{R_2 i_2}{R_2 + R_4}$$

Using these to determine i_C, v_1, and v_2 gives

$$\begin{bmatrix} \dot{x}_1 \\ \dot{x}_2 \end{bmatrix} = \begin{bmatrix} \dfrac{-(R_1 + R_2 + R_3 + R_4)}{C(R_1 + R_3)(R_2 + R_4)} & \dfrac{R_2 R_3 - R_1 R_4}{C(R_2 + R_4)(R_1 + R_3)} \\ \dfrac{(R_1 R_4 - R_2 R_3)}{L(R_1 + R_3)(R_2 + R_4)} & -\dfrac{1}{L}\left[\dfrac{R_1 R_3}{R_1 + R_3} + \dfrac{R_2 R_4}{R_2 + R_4}\right] \end{bmatrix} \begin{bmatrix} x_1 \\ x_2 \end{bmatrix}$$

$$+ \begin{bmatrix} \dfrac{R_1}{C(R_1 + R_3)} & \dfrac{R_2}{C(R_2 + R_4)} \\ \dfrac{R_1 R_3}{L(R_1 + R_3)} & \dfrac{-R_2 R_4}{L(R_2 + R_4)} \end{bmatrix} \begin{bmatrix} u_1 \\ u_2 \end{bmatrix}$$

where $u_1 = i_1$, $u_2 = i_2$ are the inputs. The output is

$$y = R_3 i_{R_3} = \left[\frac{-R_3}{R_1 + R_3} \quad \frac{-R_1 R_3}{R_1 + R_2}\right]\mathbf{x} + \left[\frac{R_1 R_3}{R_1 + R_3} \quad 0\right]\mathbf{u}$$

9.8 A schematic of a motor-generator system driving an inertia load J, with viscous damping b, at an angular velocity Ω is shown in Fig. 9.26. Derive a state space model of this system.

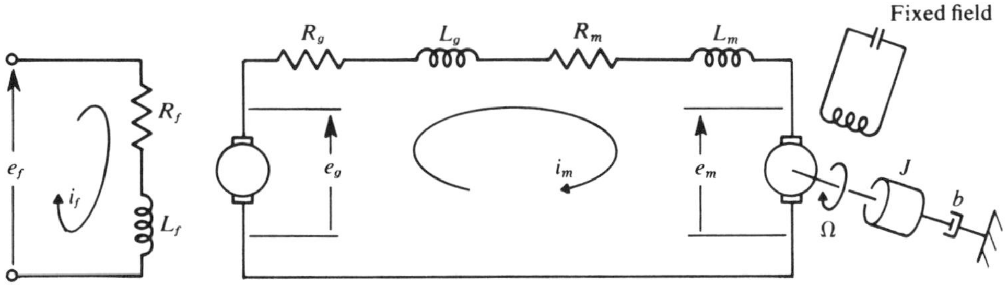

Figure 9.26

The pertinent equations are

1. $L_f \dfrac{di_f}{dt} + R_f i_f = e_f$
2. $e_g = f(i_f)$
3. $e_g - e_m = (R_g + R_m) i_m + (L_g + L_m) \dfrac{di_m}{dt}$
4. $T = K_m i_m$
5. $T = J\dot{\Omega} + b\Omega$
6. $e_m = K_m \Omega$

The simulation diagram of Fig. 9.27 can be constructed directly from these equations.

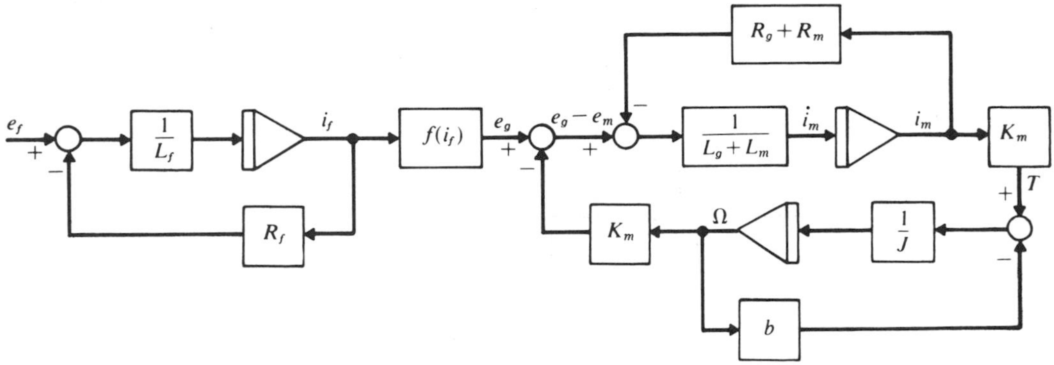

Figure 9.27

Selecting integrator outputs as state variables, $x_1 = \Omega$ (indicates kinetic energy in load), $x_2 = i_m$ (indicates magnetic energy in motor inductance), and $x_3 = i_f$ (indicates magnetic energy in the generator field), and letting the input be $e_f = u(t)$ gives

$$\begin{bmatrix} \dot{x}_1 \\ \dot{x}_2 \\ \dot{x}_3 \end{bmatrix} = \begin{bmatrix} -\frac{b}{J}x_1 + \frac{K_m}{J}x_2 \\ -\frac{K_m}{L_g + L_m}x_1 - \frac{R_g + R_m}{L_g + L_m}x_2 + \frac{f(x_3)}{L_g + L_m} \\ -\frac{R_f}{L_f}x_3 + \frac{u}{L_f} \end{bmatrix}$$

If it can be assumed that the generator characteristics are linear, i.e., if $e_g = K_g i_f$, then the above equations can be written in the standard linear form $\dot{\mathbf{x}} = \mathbf{Ax} + \mathbf{B}u$. If the speed Ω is considered to be the output y, then $y = [1 \quad 0 \quad 0]\mathbf{x}$.

9.9 Develop a state space model of a rocket vehicle (Fig. 9.28) moving vertically above the

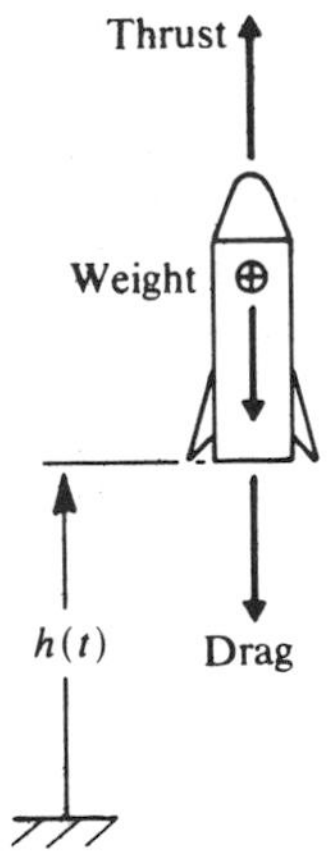

Figure 9.28

earth. The vehicle thrust is $T = K\dot{m}$, where $\dot{m}$ is the rate of mass expulsion and can be controlled. Assume a drag force is given as a nonlinear function of velocity.

Letting the instantaneous mass of the vehicle be $m(t)$ and letting $D = f(\dot{h})$ be the drag, Newton's second law gives the dynamic force balance $m\ddot{h} = T(t) - f(\dot{h}) - [m(t)k^2 g_0/(k + h)^2]$, where an inverse square gravity law has been assumed. A simulation diagram is given in Fig. 9.29.

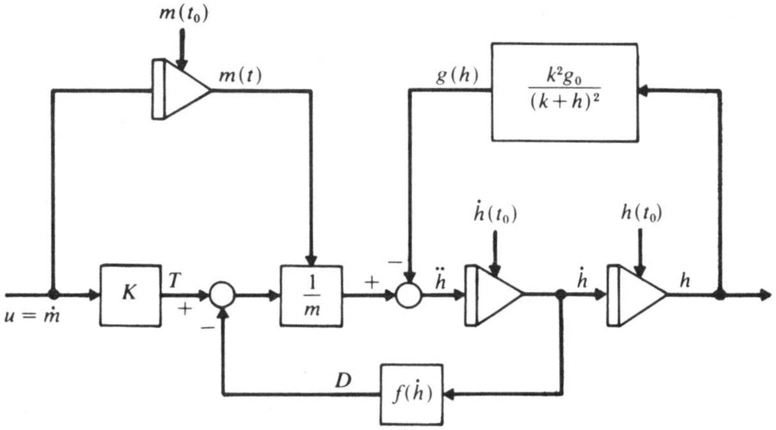

Figure 9.29

This is a nonlinear, time-variable system, but as before, outputs of the integrators are selected as state variables, $x_1 = h$, $x_2 = \dot{h}$, $x_3 = m$. Then

$$\begin{bmatrix} \dot{x}_1 \\ \dot{x}_2 \\ \dot{x}_3 \end{bmatrix} = \begin{bmatrix} x_2 \\ \dfrac{Ku - f(x_2)}{x_3} - g(x_1) \\ u \end{bmatrix}$$

In this case it is not possible to obtain the linear form $\dot{\mathbf{x}} = \mathbf{A}\mathbf{x} + \mathbf{B}\mathbf{u}$, but the above result is of the more general form $\dot{\mathbf{x}} = \mathbf{f}(\mathbf{x}, \mathbf{u}, t)$.

9.10 Derive the equations of motion for a satellite rotating in free space under the influence of gas jets mounted along three mutually orthogonal body-fixed axes.

Let $\omega = [\omega_x \quad \omega_y \quad \omega_z]^T$ be the three components of angular velocity expressed with respect to the body-fixed axes. Let $\mathbf{T} = [T_x \quad T_y \quad T_z]^T$ be the three components of input control torques. Newton's second law, as applied to a rotating body, states that $d\mathbf{H}/dt = \mathbf{T}$, where $\mathbf{H}$ is the angular momentum vector, and the time rate of change d/dt is with respect to a fixed inertial reference. The vector $\mathbf{H}$ can be expressed in body coordinates as $\mathbf{H} = [J_x\omega_x \quad J_y\omega_y \quad J_z\omega_z]^T$, where the constants J_i are moments of inertia of the body and x, y, z are assumed to be principal axes of inertia. The inertial rate of change $d\mathbf{H}/dt$ is related to the apparent rate $[\dot{\mathbf{H}}]$ as seen by an observer moving with the body by $d\mathbf{H}/dt = [\dot{\mathbf{H}}] + \boldsymbol{\omega} \times \mathbf{H}$ (see Problem 6.30, page 172). Therefore,

$$T_x = J_x\dot{\omega}_x + (J_z - J_y)\omega_y\omega_z, \qquad T_y = J_y\dot{\omega}_y + (J_x - J_z)\omega_x\omega_z,$$

$$T_z = J_z\dot{\omega}_z + (J_y - J_x)\omega_x\omega_y$$

These equations are often referred to as Euler's dynamical equations. Rearranging gives

$$\begin{bmatrix} \dot{\omega}_x \\ \dot{\omega}_y \\ \dot{\omega}_z \end{bmatrix} = \begin{bmatrix} \dfrac{J_y - J_z}{J_x}\omega_y\omega_z \\ \dfrac{J_z - J_x}{J_y}\omega_x\omega_z \\ \dfrac{J_x - J_y}{J_z}\omega_x\omega_y \end{bmatrix} + \begin{bmatrix} \dfrac{T_x}{J_x} \\ \dfrac{T_y}{J_y} \\ \dfrac{T_z}{J_z} \end{bmatrix}$$

This obviously is in the state variable form, and is linear in the control variables T_i, but nonlinear in the state variables ω_i due to the products $\omega_i\omega_j$.

9.11 Apply the results of the preceding problem to the satellite of Fig. 9.30, which is spin stabilized about the x axis. That is, $\omega_x = S$ is a large value, ω_y and $\omega_z \ll S$ represent

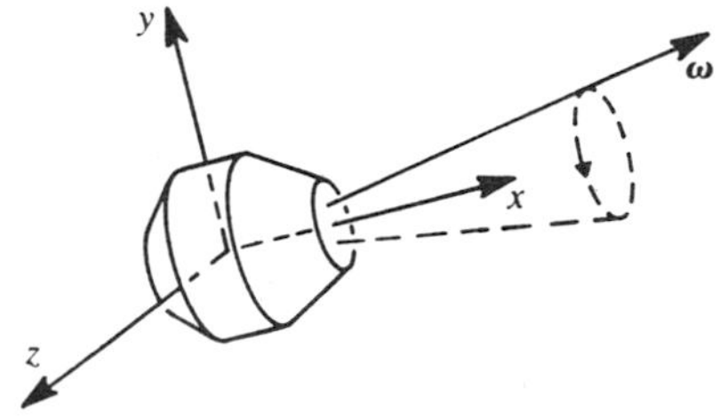

Figure 9.30

small wobbling errors. Assume the satellite is rotationally symmetric about the x axis. Find a linear approximation for the state equation.

Due to symmetry $J_y = J_z$, so $\dot{\omega}_x = T_x/J_x$. If $T_x = 0$, in the absence of any other torques, $\omega_x =$ constant. If the definition $[(J_x - J_y)/J_z]S \triangleq \Omega$ is introduced, then the linear relations between the control torques T_y, T_z and the small wobbling errors ω_y, ω_z are

$$\begin{bmatrix} \dot{\omega}_y \\ \dot{\omega}_z \end{bmatrix} = \begin{bmatrix} 0 & -\Omega \\ \Omega & 0 \end{bmatrix} \begin{bmatrix} \omega_y \\ \omega_z \end{bmatrix} + \frac{1}{J_y}\begin{bmatrix} T_y \\ T_z \end{bmatrix}$$

9.12 Describe the hydraulic system of Problem 1.4, page 14, in state variable form.

The linear graph is redrawn as Fig. 9.31 using the analogous RLC symbols for the elements.

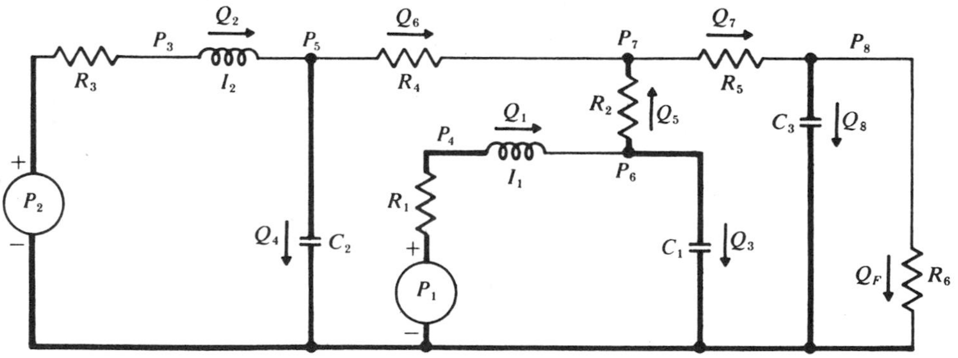

Figure 9.31

The tree is shown in heavy lines. The state variables are chosen as the pressures $x_1 = P_6$, $x_2 = P_5$, $x_3 = P_8$ and the flow rates $x_4 = Q_2$, $x_5 = Q_1$. Then $\dot{x}_1 = (1/C_1)Q_3$, $\dot{x}_2 = (1/C_2)Q_4$, $\dot{x}_3 = (1/C_3)Q_8$, $\dot{x}_4 = (1/I_2)P_{35}$, $\dot{x}_5 = (1/I_1)P_{46}$.

The last two are the easiest to complete and this is done first: $P_{35} = P_2 - R_3x_4 - x_2$, $P_{46} = P_1 - R_1x_5 - x_1$.

In order to express Q_3, Q_4, and Q_8 in terms of the state variables, the following simultaneous equations must be solved:

$$\left[\begin{array}{ccc|ccc} 1 & 0 & 0 & 1 & 0 & 0 \\ 0 & 1 & 0 & 0 & 1 & 0 \\ 0 & 0 & 1 & 0 & 0 & -1 \\ \hline 0 & 0 & 0 & 1 & 1 & -1 \\ 0 & 0 & 0 & 0 & R_4 & R_5 \\ 0 & 0 & 0 & -R_2 & R_4 & 0 \end{array}\right] \begin{bmatrix} Q_3 \\ Q_4 \\ Q_8 \\ \hdashline Q_5 \\ Q_6 \\ Q_7 \end{bmatrix} = \begin{bmatrix} x_5 \\ x_4 \\ -x_3/R_6 \\ \hdashline 0 \\ x_2 - x_3 \\ x_2 - x_1 \end{bmatrix}$$

The required solutions are

$$Q_3 = x_5 + \frac{1}{\Delta}[-R_4(x_2 - x_3) + (R_4 + R_5)(x_2 - x_1)]$$

$$Q_4 = x_4 + \frac{1}{\Delta}[-R_2(x_2 - x_3) - R_5(x_2 - x_1)]$$

$$Q_8 = -\frac{x_3}{R_6} + \frac{1}{\Delta}[(R_4 + R_2)(x_2 - x_3) - R_4(x_2 - x_1)]$$

where $\Delta = R_2R_4 + R_4R_5 + R_2R_5$. Using these relations allows the system equations to be put in the form $\dot{\mathbf{x}} = \mathbf{A}\mathbf{x} + \mathbf{B}\mathbf{u}$, where $\mathbf{u} = [P_1 \quad P_2]^T$. The output is $Q_F = y$ and is given by $y = [0 \quad 0 \quad 1/R_6 \quad 0 \quad 0]\mathbf{x}$.

PROBLEMS

9.13 Convince yourself that both Fig. 9.32*a* and *b* represent the system described by

$$\ddot{y} + a\dot{y} + by = u$$

and find the matrices **A**, **B**, **C**, and **D** for each case.

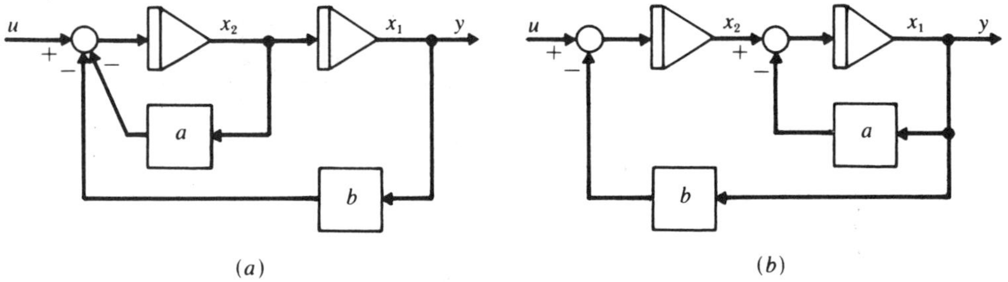

Figure 9.32

(9.14) Find a state space representation for the system described by

$$\dot{y}_1 + 3(y_1 + y_2) = u_1$$

$$\ddot{y}_2 + 4\dot{y}_2 + 3y_2 = u_2$$

9.15 Find a state space representation for the system described by

$$\ddot{y}_1 + 3\dot{y}_1 + 2(y_1 - y_2) = u_1 + \dot{u}_2$$
$$\dot{y}_2 + 3(y_2 - y_1) = u_2 + 2\dot{u}_1$$

9.16 Find the state variable equations for a system described by

$$T(s) = 1/(s^3 + 8s^2 + 13s + 6)$$

using the partial fraction expansion.

9.17 Describe the circuit of Fig. 9.33 in state variable form.

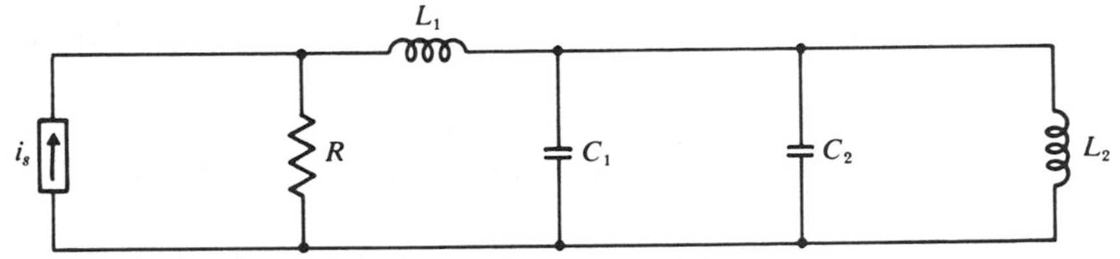

Figure 9.33

9.18 Represent the circuit of Fig. 9.34 in state variable form. Assume that the amplifier is an ideal voltage amplifier, that is $v_5 = Kv_4$ and the amplifier draws no current. The transformer ratio is N. The output is the voltage across C_2.

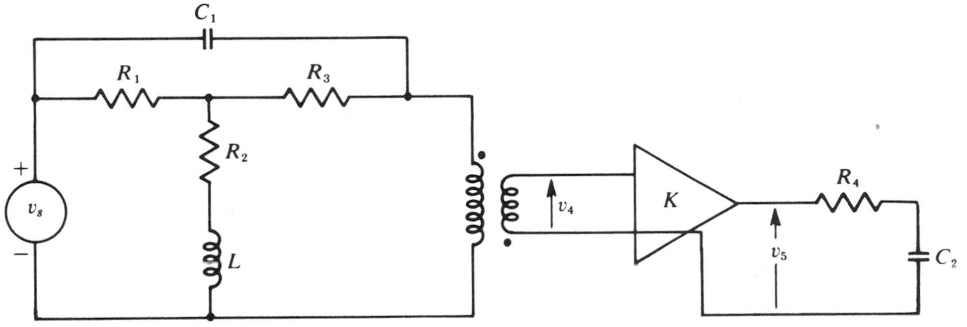

Figure 9.34

9.19 Use as state variables the voltages x_1 and x_2 and the current x_3 as shown in Fig. 9.35. Derive the state equations, letting $v_s = u_1$ and $i_s = u_2$.

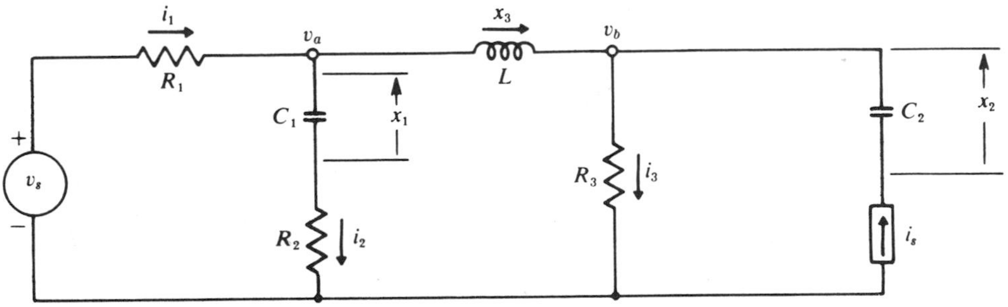

Figure 9.35

Chapter 10

ANALYSIS OF THE STATE EQUATIONS FOR CONTINUOUS-TIME LINEAR SYSTEMS

10.1 INTRODUCTION

The task of approximating a real system by a simplified physical model and then describing the model mathematically was discussed in Chapter 1. Methods of representing the system model in state space form were presented in Chapter 9. The most general form of the state space equations for a lumped-parameter, linear, continuous-time system consists of the pair of equations (*10.1*) and (*10.2*):

$$\dot{\mathbf{x}} = \mathbf{A}(t)\mathbf{x} + \mathbf{B}(t)\mathbf{u}(t) \tag{10.1}$$

$$\mathbf{y}(t) = \mathbf{C}(t)\mathbf{x} + \mathbf{D}(t)\mathbf{u}(t) \tag{10.2}$$

A typical control problem is to decide what the input $\mathbf{u}(t)$ should be so that the state $\mathbf{x}(t)$ and/or the output $\mathbf{y}(t)$ behave in an acceptable fashion while achieving the goal of the system. The determination of the control law is considered in Chapters 16 and 17. In this chapter the response of the system is analyzed, in terms of $\mathbf{x}(t)$ and $\mathbf{y}(t)$, when the input is a known function. Since $\mathbf{y}(t)$ is related to $\mathbf{x}$ and $\mathbf{u}$ in an algebraic fashion, the major effort is spent in solving the differential equations for $\mathbf{x}(t)$.

10.2 FIRST-ORDER SCALAR DIFFERENTIAL EQUATIONS

The familiar scalar differential equation

$$\dot{x} = a(t)x(t) + b(t)u(t) \tag{10.3}$$

is reviewed before considering the nth order matrix case. When the input $u(t)$ is zero, the differential equation for x is said to be homogeneous. In this case,

$$\frac{dx}{dt} = a(t)x(t) \qquad \text{or} \qquad \frac{dx}{x} = a(t)\,dt$$

In the latter form the dependent variable x and the independent variable t are separated so that both sides of the equation represent an exact differential and can be integrated:

$$\int_{x(t_0)}^{x(t)} \frac{dx}{x} = \ln(x)\Big|_{x(t_0)}^{x(t)} = \int_{t_0}^{t} a(\tau)\,d\tau$$

or

$$\ln x(t) - \ln x(t_0) = \int_{t_0}^{t} a(\tau)\,d\tau$$

Using $\ln x(t) - \ln x(t_0) = \ln[x(t)/x(t_0)]$ and the fact that $e^{\ln x} = x$ gives

$$x(t) = x(t_0)e^{\int_{t_0}^{t} a(\tau)\,d\tau} \tag{10.4}$$

In the particular case where $a(t) = a$ is constant, this reduces to

$$x(t) = x(t_0)e^{a(t-t_0)} \tag{10.5}$$

As expected, the initial condition $x(t_0)$ must be specified before a unique solution $x(t)$ can be determined.

When the nonhomogeneous equation (*10.3*) is considered, a solution can still be obtained by reducing the equation to a form which can be easily integrated. One extra step is required first. Consider

$$\frac{d}{dt}[k(t)x(t)] = k(t)\dot{x}(t) + \dot{k}(t)x(t)$$

If equation (*10.3*) is multiplied by $k(t)$ and rearranged, the result is

$$k(t)\dot{x}(t) - k(t)a(t)x(t) = k(t)b(t)u(t)$$

The left-hand side can be made an exact differential provided a function $k(t)$ is selected which satisfies $\dot{k}(t) = -k(t)a(t)$. This requirement on $k(t)$ represents a first-order homogeneous equation of the type just considered. Its solution is $k(t) = e^{-\int_{t_0}^{t} a(\tau)\,d\tau}$. Using this, the nonhomogeneous equation for $x(t)$ can be written in terms of exact differentials,

$$d[k(t)x(t)] = k(t)b(t)u(t)\,dt$$

Carrying out the integration of both sides and solving for $x(t)$ gives

$$x(t) = \left[\frac{k(t_0)}{k(t)}\right] x(t_0) + \int_{t_0}^{t} \frac{k(\tau)}{k(t)} b(\tau)u(\tau)\,d\tau$$

Using the agreed upon form for $k(t)$, the general solution becomes

$$x(t) = \{e^{\int_{t_0}^{t} a(\tau)\,d\tau}\}\,x(t_0) + \int_{t_0}^{t} e^{\int_{\tau}^{t} a(\zeta)\,d\zeta}\, b(\tau)u(\tau)\,d\tau \tag{10.6}$$

If the coefficient a is constant, this reduces to

$$x(t) = e^{a(t-t_0)}x(t_0) + \int_{t_0}^{t} e^{a(t-\tau)}b(\tau)u(\tau)\,d\tau \tag{10.7}$$

The last result can be derived directly using Laplace transforms and the convolution theorem (see Problem 10.1). A direct verification that this represents a solution of the differential equation is given in Problem 10.2.

When b is constant as well as a, equation (*10.7*) indicates that $x(t)$ depends only on the time difference $t - t_0$ and so the starting time t_0 is often replaced by 0 for simplicity. When both a and b are constant, the system is said to be time-invariant because the response due to a given input is always the same regardless of the label attached to the starting time.

10.3 THE CONSTANT COEFFICIENT MATRIX CASE

The homogeneous set of n state equations

$$\dot{\mathbf{x}} = \mathbf{A}\mathbf{x}, \qquad \mathbf{x}(t_0) \text{ given, } \mathbf{A} \text{ constant} \tag{10.8}$$

has a solution which is completely analogous to the scalar result of equation (*10.5*):

$$\mathbf{x}(t) = e^{\mathbf{A}(t-t_0)}\mathbf{x}(t_0) \tag{10.9}$$

There are several methods of verifying that this is a solution to the state equations (see Problems 10.3 and 10.4). First, note that the initial conditions are satisfied. That is,

$$\mathbf{x}(t_0) = e^{\mathbf{A}(t_0-t_0)}\mathbf{x}(t_0) = e^{[0]}\mathbf{x}(t_0) = \mathbf{x}(t_0)$$

Differentiating both sides of equation (*10.9*) and using the result of Example 8.3, it is easily verified that $\dot{\mathbf{x}} = \mathbf{A}\mathbf{x}(t)$. Since equation (*10.9*) satisfies the initial conditions and the differential equation, it represents a unique solution (see page 20 of Reference 22) of equation (*10.8*).

The nonhomogeneous set of state equations is now considered. The system matrix A is still constant, but $\mathbf{B}(t)$ may be time-varying. Components of $\mathbf{Bu}(t)$ are assumed to be piecewise continuous to guarantee a unique solution (see page 74 of Reference 22):

$$\dot{\mathbf{x}} = \mathbf{Ax} + \mathbf{B}(t)\mathbf{u}(t), \qquad \mathbf{x}(t_0) \text{ given} \tag{10.10}$$

The technique used in solving the scalar equation is repeated with only minor dimensional modifications. Let $\mathbf{K}(t)$ be an $n \times n$ matrix. Premultiplying equation (*10.10*) by $\mathbf{K}(t)$ and rearranging gives

$$\mathbf{K}(t)\dot{\mathbf{x}}(t) - \mathbf{K}(t)\mathbf{Ax}(t) = \mathbf{K}(t)\mathbf{B}(t)\mathbf{u}(t)$$

Since $d[\mathbf{K}(t)\mathbf{x}(t)]/dt = \mathbf{K}\dot{\mathbf{x}} + \dot{\mathbf{K}}\mathbf{x}$, the left-hand side can be written as an exact (vector) differential provided $\dot{\mathbf{K}} = -\mathbf{K}(t)\mathbf{A}$. One such matrix is $\mathbf{K}(t) = e^{-\mathbf{A}(t-t_0)}$. Agreeing that this is the $\mathbf{K}$ matrix to be used, the differential equation can be written

$$d[\mathbf{K}(t)\mathbf{x}(t)] = \mathbf{K}(t)\mathbf{B}(t)\mathbf{u}(t)\,dt$$

Integration gives

$$\mathbf{K}(t)\mathbf{x}(t) - \mathbf{K}(t_0)\mathbf{x}(t_0) = \int_{t_0}^{t} \mathbf{K}(\tau)\mathbf{B}(\tau)\mathbf{u}(\tau)\,d\tau$$

The selected form for $\mathbf{K}$ always has an inverse, so

$$\mathbf{x}(t) = \mathbf{K}^{-1}(t)\mathbf{K}(t_0)\mathbf{x}(t_0) + \int_{t_0}^{t} \mathbf{K}^{-1}(t)\mathbf{K}(\tau)\mathbf{B}(\tau)\mathbf{u}(\tau)\,d\tau$$

or

$$\mathbf{x}(t) = e^{\mathbf{A}(t-t_0)}\mathbf{x}(t_0) + \int_{t_0}^{t} e^{\mathbf{A}(t-\tau)}\mathbf{B}(\tau)\mathbf{u}(\tau)\,d\tau \tag{10.11}$$

This represents the solution for any system equation in the form of equation (*10.10*). Note that it is composed of a term depending only on the intial state and a convolution integral involving the input but not the initial state. These two terms are known by various names such as the homogeneous solution and the particular integral, the force-free response and the forced response, the zero input response and the zero state response, etc.

10.4 SYSTEM MODES AND MODAL DECOMPOSITION [28, 117]

Equation (*10.10*) is considered again. It is emphasized that the matrix A is constant, but $\mathbf{B}(t)$ may be time-varying. Assume that the n eigenvalues λ_i and n independent vectors, either eigenvectors or generalized eigenvectors, have been found for the matrix $\mathbf{A}$. These vectors are denoted by $\boldsymbol{\xi}_i$ to avoid confusion with the state vector $\mathbf{x}$. Since the set $\{\boldsymbol{\xi}_i\}$ is linearly independent, it can be used as a basis for the space Σ. Thus at any given time t, the state $\mathbf{x}(t)$ can be expressed as

$$\mathbf{x}(t) = q_1(t)\boldsymbol{\xi}_1 + q_2(t)\boldsymbol{\xi}_2 + \cdots + q_n(t)\boldsymbol{\xi}_n \tag{10.12}$$

The time variation of $\mathbf{x}$ is contained in the expansion coefficients q_i since A, and hence the $\boldsymbol{\xi}_i$, are constant. At any given time t, the vector $\mathbf{B}(t)\mathbf{u}(t) \in \Sigma$ and therefore it, too, can be expanded as

$$\mathbf{B}(t)\mathbf{u}(t) = \beta_1(t)\boldsymbol{\xi}_1 + \beta_2(t)\boldsymbol{\xi}_2 + \cdots + \beta_n(t)\boldsymbol{\xi}_n$$

In fact, $\beta_i(t) = \langle \mathbf{r}_i, \mathbf{B}(t)\mathbf{u}(t)\rangle$, where $\{\mathbf{r}_i\}$ is the set of reciprocal basis vectors.

Using the above expansion, equation (*10.10*) becomes

$$\dot{q}_1\boldsymbol{\xi}_1 + \dot{q}_2\boldsymbol{\xi}_2 + \cdots + \dot{q}_n\boldsymbol{\xi}_n = q_1\mathbf{A}\boldsymbol{\xi}_1 + q_2\mathbf{A}\boldsymbol{\xi}_2 + \cdots + q_n\mathbf{A}\boldsymbol{\xi}_n + \beta_1\boldsymbol{\xi}_1 + \beta_2\boldsymbol{\xi}_2 + \cdots + \beta_n\boldsymbol{\xi}_n$$

Assume for the moment that $\mathbf{A}$ is normal (see Sec. 6.6, page 159, and Problem 7.24, page 193) so that all $\boldsymbol{\xi}_i$ are eigenvectors rather than generalized eigenvectors. Then $\mathbf{A}\boldsymbol{\xi}_i = \lambda_i\boldsymbol{\xi}_i$, so that

$$(\dot{q}_1 - \lambda_1 q_1 - \beta_1)\boldsymbol{\xi}_1 + (\dot{q}_2 - \lambda_2 q_2 - \beta_2)\boldsymbol{\xi}_2 + \cdots + (\dot{q}_n - \lambda_n q_n - \beta_n)\boldsymbol{\xi}_n = \mathbf{0}$$

Since the set $\{\boldsymbol{\xi}_i\}$ is linearly independent, this requires that

$$\dot{q}_i = \lambda_i q_i + \beta_i \qquad \text{for } i = 1, 2, \ldots, n$$

This demonstrates that when $\mathbf{A}$ is constant and has a full set of eigenvectors, the system is completely described by a set of n uncoupled scalar equations whose solutions are of the form

$$q_i(t) = e^{\lambda_i(t-t_0)}q_i(t_0) + \int_{t_0}^{t} e^{\lambda_i(t-\tau)}\beta_i(\tau)\,d\tau$$

Of couse, $q_i(t_0) = \langle \mathbf{r}_i, \mathbf{x}(t_0)\rangle$. The state vector is given by

$$\mathbf{x}(t) = q_1(t)\boldsymbol{\xi}_1 + q_2(t)\boldsymbol{\xi}_2 + \cdots + q_n(t)\boldsymbol{\xi}_n$$

The terms in this sum are called the system modes. The general response of a complicated system can be broken down into the sum of n simple modal responses.

It should be recognized that equation (*10.12*) can be written in terms of the modal matrix $\mathbf{M} = [\boldsymbol{\xi}_1 \ldots \boldsymbol{\xi}_n]$ as $\mathbf{x} = \mathbf{M}\mathbf{q}$, and as such, represents a change of basis. Using this notation, equation (*10.10*) is considered again:

$\dot{\mathbf{x}}$ becomes $\mathbf{M}\dot{\mathbf{q}}$ since $\mathbf{M}$ is constant
$\mathbf{A}\mathbf{x}$ becomes $\mathbf{A}\mathbf{M}\mathbf{q}$ and $\mathbf{B}\mathbf{u}$ remains unchanged

Therefore,

$$\mathbf{M\dot{q}} = \mathbf{AMq} + \mathbf{Bu}$$

or

$$\mathbf{\dot{q}} = \mathbf{M}^{-1}\mathbf{AMq} + \mathbf{M}^{-1}\mathbf{Bu} = \mathbf{Jq} + \mathbf{B}_n\mathbf{u}$$

where $\mathbf{B}_n \triangleq \mathbf{M}^{-1}\mathbf{B}$ and the assumption regarding a full set of eigenvectors is dropped. $\mathbf{J}$ is the Jordan canonical form (or the diagonal matrix $\mathbf{\Lambda}$ in many cases). If the same change of basis is used in the output equation (*10.2*), then the system is described by the pair of *normal form* equations

$$\mathbf{\dot{q}} = \mathbf{Jq} + \mathbf{B}_n\mathbf{u} \qquad (10.13)$$

$$\mathbf{y} = \mathbf{C}_n\mathbf{q} + \mathbf{Du} \qquad (10.14)$$

where $\mathbf{C}_n \triangleq \mathbf{CM}$. One advantage of the normal form is that the state equations are as nearly uncoupled as possible. Each component of $\mathbf{q}$ is coupled to at most one other component because of the nature of the Jordan form matrix $\mathbf{J}$. The solution for equation (*10.13*) can be written as

$$\mathbf{q}(t) = e^{\mathbf{J}(t-t_0)}\mathbf{q}(t_0) + \int_{t_0}^{t} e^{\mathbf{J}(t-\tau)}\mathbf{B}_n(\tau)\mathbf{u}(\tau)\,d\tau$$

Relating this to the original state vector gives

$$\mathbf{x}(t) = \mathbf{Mq}(t) = \mathbf{M}e^{\mathbf{J}(t-t_0)}\mathbf{M}^{-1}\mathbf{x}(t_0) + \int_{t_0}^{t} \mathbf{M}e^{\mathbf{J}(t-\tau)}\mathbf{M}^{-1}\mathbf{B}(\tau)\mathbf{u}(\tau)\,d\tau \qquad (10.15)$$

The above equation used the fact that $\mathbf{M}^{-1}$ is the matrix of transposed reciprocal basis vectors, which means that

$$\mathbf{q}(t_0) = \mathbf{M}^{-1}\mathbf{x}(t_0)$$

Comparing equations (*10.15*) and (*10.11*) shows that

$$e^{\mathbf{A}(t-t_0)} = \mathbf{M}e^{\mathbf{J}(t-t_0)}\mathbf{M}^{-1}$$

a result given earlier in Chapter 8.

Modal decomposition is useful because of the insight it gives regarding the intrinsic properties of the system. The properties of controllability and observability (Chapter 12) are more easily understood and evaluated. The stability properties (Chapter 14) of the system are also more clearly revealed. Modal decomposition provides a simple geometrical picture for the motion of the state vector versus time. By retaining only the dominant modes, a high-order system can be approximated by a lower-order system.

It should be kept in mind that the modal decomposition technique is only useful if **A**, and thus ξ_i, λ_i are constant. It is the invariance of the vector parts of $\mathbf{x}(t)$, that is, the ξ_i terms, that gives value to the method. If the modal matrix were time-varying and had to be continually reevaluated, the advantages of modal decomposition would be lost.

Example 10.1

A system is described by

$$\begin{bmatrix} \dot{x}_1 \\ \dot{x}_2 \end{bmatrix} = \begin{bmatrix} 0 & 1 \\ 8 & -2 \end{bmatrix} \begin{bmatrix} x_1 \\ x_2 \end{bmatrix} + \begin{bmatrix} 1 \\ 1 \end{bmatrix} u$$

$$y(t) = [4 \quad 1]\mathbf{x}(t)$$

The initial conditions are $\mathbf{x}(0) = [1 \quad -4]^T$. Assume that $u(t) = 0$ and analyze this system.

With $\mathbf{A} = \begin{bmatrix} 0 & 1 \\ 8 & -2 \end{bmatrix}$, the eigenvalues are $\lambda_1 = -4$, $\lambda_2 = 2$.

The eigenvectors are $\xi_1 = [1 \quad -4],^T$ $\xi_2 = [1 \quad 2]^T$.

The modal matrix and its inverse are $\mathbf{M} = \begin{bmatrix} 1 & 1 \\ -4 & 2 \end{bmatrix}$, $\mathbf{M}^{-1} = \frac{1}{6}\begin{bmatrix} 2 & -1 \\ 4 & 1 \end{bmatrix}$.

Any one of several methods gives

$$e^{\mathbf{A}t} = \frac{1}{6}\begin{bmatrix} 2e^{-4t} + 4e^{2t} & -e^{-4t} + e^{2t} \\ -8e^{-4t} + 8e^{2t} & 4e^{-4t} + 2e^{2t} \end{bmatrix}$$

so the homogeneous solution is $\mathbf{x}(t) = e^{\mathbf{A}t}\mathbf{x}(0) = [e^{-4t} \quad -4e^{-4t}]^T$, and the output is $y(t) = 4e^{-4t} - 4e^{-4t} = 0$ for all t. ■

Example 10.2

Modal decomposition is now applied in an attempt to gain insight into the unusual result of Example 10.1.

Since the eigenvalues are distinct, $\mathbf{M}^{-1}\mathbf{A}\mathbf{M} = \mathbf{\Lambda}$ for this system, and equation (*10.13*) becomes

$$\begin{bmatrix} \dot{q}_1 \\ \dot{q}_2 \end{bmatrix} = \begin{bmatrix} -4 & 0 \\ 0 & 2 \end{bmatrix} \begin{bmatrix} q_1 \\ q_2 \end{bmatrix} + \begin{bmatrix} 1/6 \\ 5/6 \end{bmatrix} u$$

The initial conditions are $\mathbf{q}(0) = \mathbf{M}^{-1}\mathbf{x}(0) = [1 \quad 0]^T$. Equation (*10.14*) becomes $y = [0 \quad 6]\mathbf{q}$. The state vector $\mathbf{x}(t)$ can be written as the sum of two modes,

$$\mathbf{x}(t) = q_1(0)e^{-4t}\begin{bmatrix} 1 \\ -4 \end{bmatrix} + q_2(0)\, e^{2t}\begin{bmatrix} 1 \\ 2 \end{bmatrix}$$

The particular initial condition selected here has no component along the direction of mode 2, as evidenced by $q_2(0) = 0$. Thus the second mode is not excited. The output of this system consists only of the second mode contribution, as evidenced by $\mathbf{C}_n = [0 \quad 6]$. Mode 1 contributes nothing to the output and mode 2 is not excited, so the output remains identically zero. ■

10.5 THE TIME-VARYING MATRIX CASE

The time-varying homogeneous state equations

$$\dot{\mathbf{x}} = \mathbf{A}(t)\mathbf{x} \tag{10.16}$$

are considered first. In order that this qualify as a valid state equation, it is required that there be a *unique* solution for every $\mathbf{x}(t_0) \in \Sigma$. This places some restriction on the kind of time variation allowed on the matrix A. A *sufficient* condition for the existence of unique solutions is to require that all elements $a_{ij}(t)$ of $\mathbf{A}(t)$ be continuous. Weaker conditions may be found in textbooks on differential equations [22, 75].

Since dim $(\Sigma) = n$, n linearly independent initial vectors $\mathbf{x}_i(t_0)$ can be found, and each one defines a unique solution of equation (*10.16*), called $\mathbf{x}_i(t)$, $t \geq t_0$. A particular set of n independent initial condition vectors is selected,

$$\mathbf{x}_1(t_0) = [1 \quad 0 \quad 0 \quad \cdots \quad 0]^T, \qquad \mathbf{x}_2(t_0) = [0 \quad 1 \quad 0 \quad \cdots \quad 0]^T, \qquad \ldots,$$
$$\mathbf{x}_n(t_0) = [0 \quad 0 \quad 0 \quad \cdots \quad 1]^T$$

That is, the ith vector has one as its ith component and all other components are zero. The n solutions corresponding to these initial conditions are used as the columns in forming an $n \times n$ matrix $\mathbf{U}(t) = [\mathbf{x}_1(t) \quad \mathbf{x}_2(t) \quad \ldots \quad \mathbf{x}_n(t)]$. The matrix $\mathbf{U}(t)$ is called the *fundamental solution matrix*. Notice that $\mathbf{U}(t_0) = \mathbf{I}_n$ and

$$\dot{\mathbf{U}}(t) = \mathbf{A}(t)\mathbf{U}(t) \tag{10.17}$$

Assuming that the fundamental solution matrix is available, the solution to equation (*10.16*) with an arbitrary initial condition vector $\mathbf{x}(t_0)$ is

$$\mathbf{x}(t) = \mathbf{U}(t)\mathbf{x}(t_0) \tag{10.18}$$

This is easily verified. Checking initial conditions,

$$\mathbf{x}(t_0) = \mathbf{U}(t_0)\mathbf{x}(t_0) = \mathbf{I}_n\mathbf{x}(t_0) = \mathbf{x}(t_0)$$

Checking to see that this solution satisfies the differential equation,

$$\dot{\mathbf{x}}(t) = \dot{\mathbf{U}}(t)\mathbf{x}(t_0) = \mathbf{A}(t)\mathbf{U}(t)\mathbf{x}(t_0) = \mathbf{A}(t)\mathbf{x}(t)$$

Both the initial conditions and the differential equation are satisfied, so this represents the unique solution to the homogeneous problem.

The nonhomogeneous time-varying state equation is solved in an analogous manner to the scalar and constant matrix cases. That is, the equation is reduced to exact differentials so that it can be integrated. Preliminary to this, it is noted that $\mathbf{U}^{-1}(t)$ can be shown to exist for all $t \geq t_0$ and that

$$\mathbf{U}(t)\mathbf{U}^{-1}(t) = \mathbf{I}_n \qquad \text{so that} \qquad \frac{d}{dt}(\mathbf{U}(t)\mathbf{U}^{-1}(t)) = [\mathbf{0}]$$

or

$$\frac{d\mathbf{U}}{dt}\mathbf{U}^{-1} + \mathbf{U}\frac{d\mathbf{U}^{-1}}{dt} = [\mathbf{0}] \quad \text{or} \quad \frac{d\mathbf{U}^{-1}}{dt} = -\mathbf{U}^{-1}\frac{d\mathbf{U}}{dt}\mathbf{U}^{-1}$$

Therefore,

$$\frac{d\mathbf{U}^{-1}}{dt} = -\mathbf{U}^{-1}(t)\mathbf{A}(t) \tag{10.19}$$

Premultiplying equation (*10.1*) by $\mathbf{U}^{-1}(t)$, postmultiplying (*10.19*) by $\mathbf{x}(t)$, and adding the results gives

$$\mathbf{U}^{-1}(t)\dot{\mathbf{x}} + \frac{d\mathbf{U}^{-1}}{dt}\mathbf{x}(t) = \mathbf{U}^{-1}(t)\mathbf{B}(t)\mathbf{u}(t)$$

or

$$\frac{d}{dt}[\mathbf{U}^{-1}(t)\mathbf{x}(t)] = \mathbf{U}^{-1}(t)\mathbf{B}(t)\mathbf{u}(t)$$

The nonhomogeneous solution is obtained by integrating both sides from t_0 to t, that is,

$$\mathbf{U}^{-1}(t)\mathbf{x}(t) - \mathbf{U}^{-1}(t_0)\mathbf{x}(t_0) = \int_{t_0}^{t} \mathbf{U}^{-1}(\tau)\mathbf{B}(\tau)\mathbf{u}(\tau)\, d\tau$$

or

$$\mathbf{x}(t) = \mathbf{U}(t)\mathbf{U}^{-1}(t_0)\mathbf{x}(t_0) + \int_{t_0}^{t} \mathbf{U}(t)\mathbf{U}^{-1}(\tau)\mathbf{B}(\tau)\mathbf{u}(\tau)\, d\tau \tag{10.20}$$

The result again takes the form of a term depending on the initial state and a convolution integral involving the input function. In fact, since $\mathbf{U}(t_0) = \mathbf{I}_n = \mathbf{U}^{-1}(t_0)$, the first term is the same homogeneous solution given by equation (*10.18*).

• 10.6 THE TRANSITION MATRIX

The preceding results prompt the definition of an important matrix which can be associated with any linear system, namely the *transition matrix:*

$$\mathbf{\Phi}(t, \tau) \triangleq \mathbf{U}(t)\mathbf{U}^{-1}(\tau) \tag{10.21}$$

This $n \times n$ matrix is a linear transformation or mapping of Σ onto itself. That is, in the absence of any input $\mathbf{u}(t)$, given the state $\mathbf{x}(\tau)$ at any time τ, the state at any other time t is given by the mapping

$$\mathbf{x}(t) = \mathbf{\Phi}(t, \tau)\mathbf{x}(\tau)$$

The mapping of $\mathbf{x}(\tau)$ into itself requires that

$$\mathbf{\Phi}(\tau, \tau) = \mathbf{I}_n \quad \text{for any } \tau \tag{10.22}$$

This is obviously true from equation (*10.21*). Differentiating $\mathbf{\Phi}(t, \tau)$ with respect to its first argument t gives

$$\frac{d\mathbf{\Phi}(t,\tau)}{dt} = \frac{d\mathbf{U}(t)}{dt}\mathbf{U}^{-1}(\tau) = \mathbf{A}(t)\mathbf{U}(t)\mathbf{U}^{-1}(\tau)$$

so

$$\frac{d\mathbf{\Phi}(t,\tau)}{dt} = \mathbf{A}(t)\mathbf{\Phi}(t,\tau) \tag{10.23}$$

The set of differential equations (*10.23*), along with the initial condition, equation (*10.22*), is often considered as the definition for $\mathbf{\Phi}(t, \tau)$

Two other important properties of the transition matrix are the semigroup property, mentioned in Chapter 9 while defining "state,"

$$\mathbf{\Phi}(t_2, t_0) = \mathbf{\Phi}(t_2, t_1)\mathbf{\Phi}(t_1, t_0) \qquad \text{for any } t_0, t_1, t_2$$

and the relationship between $\mathbf{\Phi}^{-1}$ and $\mathbf{\Phi}$:

$$\mathbf{\Phi}^{-1}(t, t_0) = \mathbf{\Phi}(t_0, t) \qquad \text{for any } t_0, t$$

Both of these properties are immediately obvious if the definition of equation (*10.21*) is considered.

Methods of Computing the Transition Matrix

If the matrix $\mathbf{A}$ is constant, then

$$\mathbf{\Phi}(t, \tau) = e^{\mathbf{A}(t-\tau)} \qquad \text{(Compare equations (10.11) and (10.20).)}$$

Therefore, all of the methods of Chapter 8 are applicable for finding $\mathbf{\Phi}$, including:

1. $\mathbf{\Phi}(t, 0) = \mathcal{L}^{-1}\{[\mathbf{I}s - \mathbf{A}]^{-1}\}$. $\mathbf{\Phi}(t, \tau)$ is then found by replacing t by $t - \tau$, since $\mathbf{\Phi}(t, \tau) = \mathbf{\Phi}(t - \tau, 0)$ when $\mathbf{A}$ is constant.
2. $\mathbf{\Phi}(t, \tau) = \alpha_0\mathbf{I} + \alpha_1\mathbf{A} + \cdots + \alpha_{n-1}\mathbf{A}^{n-1}$, where $e^{\lambda_i(t-\tau)} = \alpha_0 + \alpha_1\lambda_i + \cdots + \alpha_{n-1}\lambda_i^{n-1}$ and, if some eigenvalues are repeated, derivatives of the above expression with respect to λ must be used.
3. $\mathbf{\Phi}(t, \tau) = \mathbf{M}e^{\mathbf{J}(t-\tau)}\mathbf{M}^{-1}$, where $\mathbf{J}$ is the Jordan form (or the diagonal matrix $\mathbf{\Lambda}$), and $\mathbf{M}$ is the modal matrix.
4. $\mathbf{\Phi}(t, \tau) = \sum_{i=1}^{n} e^{\lambda_i(t-\tau)}\mathbf{Z}_i(\lambda)$, where the $n \times n$ matrices $\mathbf{Z}_i$ are defined in Problem 8.22.
5. $\mathbf{\Phi}(t, \tau) \cong \mathbf{I} + \mathbf{A}(t - \tau) + \frac{1}{2}\mathbf{A}^2(t - \tau)^2 + \frac{1}{3!}\mathbf{A}^3(t - \tau)^3 + \cdots$. This infinite series can be truncated after a finite number of terms to obtain an approximation for the transition matrix. See Problem 10.11 for a more efficient computational form of this series.

A modification of method 1, using signal flow graphs to avoid the matrix inversion, can also be used. Since $\phi_{ij}(s) \triangleq \mathcal{L}\{\phi_{ij}(t, 0)\}$ is the transfer function from the input to the jth integrator to the output of the ith integrator, that is, the ith state variable x_i, Mason's gain rule [30] can be used to write the components $\phi_{ij}(s)$ directly. Inverse Laplace transformations then give the elements of $\mathbf{\Phi}(t, 0)$.

When $\mathbf{A}(t)$ is time-varying, the choices for finding $\mathbf{\Phi}(t, \tau)$ are more restricted:

1. *Computer solution of* $\dot{\mathbf{\Phi}} = \mathbf{A}(t)\mathbf{\Phi}$ *with* $\mathbf{\Phi}(\tau, \tau) = \mathbf{I}$. This is expensive in terms of computer time if the transition matrix is required for all t and τ. It means solving the matrix differential equation many times, using a large set of different τ values as initial times.
2. *Let* $\mathbf{B}(t, \tau) = \int_\tau^t \mathbf{A}(\zeta)\, d\zeta$. Unlike the scalar case, $\mathbf{\Phi}(t, \tau) \neq e^{\mathbf{B}(t,\tau)}$ unless $\mathbf{B}(t, \tau)$ and $\mathbf{A}(t)$ commute. Unfortunately, they generally do not commute, but two cases for which they do are when $\mathbf{A}$ is constant and when $\mathbf{A}$ is diagonal. Whenever $\mathbf{BA} = \mathbf{AB}$, any method may be used for computing $\mathbf{\Phi}(t, \tau) = e^{\mathbf{B}(t,\tau)}$.
3. *Successive approximations may be used to obtain an approximate transition matrix, as derived in Problem 10.5:*

$$\mathbf{\Phi}(t, t_0) = \mathbf{I}_n + \int_{t_0}^t \mathbf{A}(\tau_0)\, d\tau_0 + \int_{t_0}^t \mathbf{A}(\tau_0) \int_{t_0}^{\tau_0} \mathbf{A}(\tau_1)\, d\tau_1\, d\tau_0 + \int_{t_0}^t \mathbf{A}(\tau_0) \int_{t_0}^{\tau_0} \mathbf{A}(\tau_1) \int_{t_0}^{\tau_1} \mathbf{A}(\tau_2)\, d\tau_2\, d\tau_1\, d\tau_0 + \cdots$$

4. *In some special cases closed form solutions to the equations may be possible.*

In some cases it may be necessary or desirable to select a set of discrete time points, t_k, such that $\mathbf{A}(t)$ can be approximated by a constant matrix over each interval $[t_k, t_{k+1}]$. Then a set of difference equations can be used to describe the state of the system at these discrete times. The approximating difference equation is derived in Problems 10.10, 10.11, and 10.26. Solutions of this type of equation are discussed in the next chapter.

10.7 CLOSURE

The most general state space description of a linear system is given by

$$\dot{\mathbf{x}}(t) = \mathbf{A}(t)\mathbf{x}(t) + \mathbf{B}(t)\mathbf{u}(t) \qquad (10.1)$$

$$\mathbf{y}(t) = \mathbf{C}(t)\mathbf{x}(t) + \mathbf{D}(t)\mathbf{u}(t) \qquad (10.2)$$

The form of the solution for equation (*10.1*) has been shown to be

$$\mathbf{x}(t) = \mathbf{\Phi}(t, t_0)\mathbf{x}(t_0) + \int_{t_0}^t \mathbf{\Phi}(t, \tau)\mathbf{B}(\tau)\mathbf{u}(\tau)\, d\tau \qquad (10.24)$$

Equation (*10.24*) is the explicit form of the transformation

$$\mathbf{x}(t) = \mathbf{g}(\mathbf{x}(t_0), \mathbf{u}(t), t_0, t)$$

introduced in Chapter 9 when defining state. When the system matrix **A** is constant, the transition matrix can always be found in closed form, although it may be tedious to do so for high-order systems. In the time-varying case numerical solutions or approximations must be relied upon. When considering certain questions, it is valuable to know that a solution exists in the stated form, even if it cannot be easily computed.

When the solution for $\mathbf{x}(t)$ is used, the output equation (*10.2*) becomes

$$\mathbf{y}(t) = \mathbf{C}(t)\mathbf{\Phi}(t, t_0)\mathbf{x}(t_0) + \int_{t_0}^{t} \mathbf{C}(t)\mathbf{\Phi}(t, \tau)\mathbf{B}(\tau)\mathbf{u}(\tau)\, d\tau + \mathbf{D}(t)\mathbf{u}(t)$$

or

$$\mathbf{y}(t) = \mathbf{C}(t)\mathbf{\Phi}(t, t_0)\mathbf{x}(t_0) + \int_{t_0}^{t} [\mathbf{C}(t)\mathbf{\Phi}(t, \tau)\mathbf{B}(\tau) + \delta(t - \tau)\mathbf{D}(\tau)]\mathbf{u}(\tau)\, d\tau$$

The term inside the integral is an explicit expression for the weighting matrix $\mathbf{W}(t, \tau)$ used in the integral form of the input-output description for the system:

$$\mathbf{y}(t) = \int_{t_0}^{t} \mathbf{W}(t, \tau)\mathbf{u}(\tau)\, d\tau$$

It is seen that this input-output description is only valid when $\mathbf{x}(t_0) = \mathbf{0}$, the so-called zero state response case. This difficulty can be overcome by considering the initial state part of $\mathbf{y}(t)$ as having arisen because of some input between $t = -\infty$ and $t = t_0$. Then

$$\mathbf{y}(t) = \int_{-\infty}^{t} \mathbf{W}(t, \tau)\mathbf{u}(\tau)\, d\tau$$

ILLUSTRATIVE PROBLEMS

Derivation and Verification of Solutions

10.1 Use Laplace transforms to solve $\dot{x} = ax(t) + b(t)u(t)$, with the initial condition $x(0)$, and a is constant.

Transforming gives

$$sx(s) - x(0) = ax(s) + \mathcal{L}\{b(t)u(t)\} \qquad \text{or} \qquad x(s) = \frac{x(0)}{s - a} + \frac{\mathcal{L}\{b(t)u(t)\}}{s - a}$$

The inverse transform gives

$$x(t) = \mathcal{L}^{-1}\{x(s)\} = x(0)e^{at} + \mathcal{L}^{-1}\left\{\frac{\mathcal{L}\{b(t)u(t)\}}{s - a}\right\}.$$

Using the convolution theorem $\mathcal{L}^{-1}\{g_1(s)g_2(s)\} = \int_0^t g_1(t-\tau)g_2(\tau)\,d\tau$ on the last term gives

$$\mathcal{L}^{-1}\left\{\frac{\mathcal{L}\{b(t)u(t)\}}{s-a}\right\} = \int_0^t e^{a(t-\tau)}b(\tau)u(\tau)\,d\tau$$

so that

$$x(t) = e^{at}x(0) + \int_{t_0}^t e^{a(t-\tau)}b(\tau)u(\tau)\,d\tau$$

If b is also constant, the system is time-invariant. The solution due to any other initial condition $x(t_0)$ at time t_0 is

$$x(t) = e^{a(t-t_0)}x(t_0) + \int_{t_0}^t e^{a(t-\tau)}b(\tau)u(\tau)\,d\tau$$

10.2 Verify that equation (*10.6*) is the solution of equation (*10.3*).

Verification requires showing that the postulated solution $x(t)$ satisfies the initial condition and the differential equation.

Initial condition check: With $t = t_0$,

$$x(t_0) = \{e^{\int_{t_0}^{t_0} a(\tau)\,d\tau}\}\,x(t_0) + \int_{t_0}^{t_0} e^{\int_\tau^{t_0} a(\zeta)\,d\zeta}\,b(\tau)u(\tau)\,d\tau$$

Since $a(t)$ is continuous, $\int_{t_0}^{t_0} a(\tau)\,d\tau = 0$ so $e^{\int_{t_0}^{t_0} a(\tau)\,d\tau} = 1$.

Likewise, $\int_{t_0}^{t_0} e^{\int a(\zeta)\,d\zeta}\,b(\tau)u(\tau)\,d\tau = 0$ provided that $b(t)$ and $u(t)$ remain finite. Therefore, $x(t_0) = x(t_0)$.

Differential equation check: Differentiating the postulated solution gives

$$\dot{x}(t) = \frac{d}{dt}\left[\int_{t_0}^t a(\tau)\,d\tau\right]e^{\int_{t_0}^t a(\tau)\,d\tau}\,x(t_0) + \frac{d}{dt}\left[\int_{t_0}^t e^{\int_\tau^t a(\zeta)\,d\zeta}\,b(\tau)u(\tau)\,d\tau\right]$$

Using the general formula for differentiating an integral term,

$$\frac{d}{dt}\left[\int_{f(t)}^{g(t)} h(t,\tau)\,d\tau\right] = \int_{f(t)}^{g(t)} \frac{\partial h(t,\tau)}{\partial t}\,d\tau + h(t,g(t))\frac{dg}{dt} - h(t,f(t))\frac{df}{dt}$$

gives

$$\frac{d}{dt}\left[\int_{t_0}^t a(\tau)\,d\tau\right] = a(t)$$

$$\frac{d}{dt}\left[\int_{t_0}^t e^{\int_\tau^t a(\zeta)\,d\zeta}\,b(\tau)u(\tau)\,d\tau\right] = e^{\int_t^t a(\zeta)\,d\zeta}\,b(t)u(t) + \int_{t_0}^t \frac{\partial}{\partial t}\,[e^{\int_\tau^t a(\zeta)\,d\zeta}]\,b(\tau)u(\tau)\,d\tau$$

$$= b(t)u(t) + a(t)\int_{t_0}^t e^{\int_\tau^t a(\zeta)\,d\zeta}\,b(\tau)u(\tau)\,d\tau$$

so that

$$\dot{x}(t) = a(t)\Big[e^{\int_{t_0}^t a(\tau)\,d\tau}\,x(t_0) + \int_{t_0}^t e^{\int_\tau^t a(\zeta)\,d\zeta}\,b(\tau)u(\tau)\,d\tau\Big] + b(t)u(t)$$

The term in brackets is the postulated solution for $x(t)$, so $\dot{x} = a(t)x(t) + b(t)u(t)$ and the equation is satisfied.

10.3 Solve $\dot{\mathbf{x}} = \mathbf{Ax} + \mathbf{B}(t)\mathbf{u}(t)$ using Laplace transforms.

Let the vector $\mathbf{B}(t)\mathbf{u}(t) = \mathbf{f}(t)$ for convenience. $s\mathbf{x}(s) - \mathbf{x}(0) = \mathbf{Ax}(s) + \mathbf{f}(s)$ or $[s\mathbf{I} - \mathbf{A}]\mathbf{x}(s) = \mathbf{x}(0) + \mathbf{f}(s)$ so that $\mathbf{x}(s) = [s\mathbf{I} - \mathbf{A}]^{-1}\mathbf{x}(0) + [s\mathbf{I} - \mathbf{A}]^{-1}\mathbf{f}(s)$. Taking the inverse transform gives $\mathbf{x}(t) = \mathcal{L}^{-1}\{[s\mathbf{I} - \mathbf{A}]^{-1}\}\mathbf{x}(0) + \mathcal{L}^{-1}\{[s\mathbf{I} - \mathbf{A}]^{-1}\} * \mathbf{f}(t)$, where $g(t) * f(t)$ is used to indicate convolution. Since $\mathcal{L}^{-1}\{[s\mathbf{I} - \mathbf{A}]^{-1}\} = e^{\mathbf{A}t}$,

$$\mathbf{x}(t) = e^{\mathbf{A}t}\mathbf{x}(0) + \int_0^t e^{\mathbf{A}(t-\tau)}\mathbf{f}(\tau)\,d\tau = e^{\mathbf{A}t}\mathbf{x}(0) + \int_0^t e^{\mathbf{A}(t-\tau)}\,\mathbf{B}(\tau)\mathbf{u}(\tau)\,d\tau$$

10.4 Verify that equation *(10.11)* is the solution of equation *(10.10)*.

Setting $t = t_0$ gives $\mathbf{x}(t_0) = e^{[0]}\mathbf{x}(t_0) = \mathbf{x}(t_0)$, assuming that $\mathbf{B}(t)\mathbf{u}(t)$ remains finite, i.e., contains no impulse functions.

Differentiating equation *(10.11)* gives

$$\dot{\mathbf{x}} = \mathbf{A}e^{\mathbf{A}(t-t_0)}\mathbf{x}(t_0) + \mathbf{B}(t)\mathbf{u}(t) + \mathbf{A}\int_{t_0}^t e^{\mathbf{A}(t-\tau)}\mathbf{B}(\tau)\mathbf{u}(\tau)\,d\tau$$

Using equation *(10.11)* reduces this to $\dot{\mathbf{x}} = \mathbf{Ax}(t) + \mathbf{B}(t)\mathbf{u}(t)$, indicating that equation *(10.11)* does satisfy the differential equation.

10.5 Use a sequence of approximations for the solution of $\dot{\mathbf{x}} = \mathbf{A}(t)\mathbf{x}(t)$, $\mathbf{x}(t_0)$ given, and derive an approximation for the transition matrix $\mathbf{\Phi}(t, t_0)$.

As the zeroth approximation, let $\mathbf{x}^{(0)}(t) = \mathbf{x}(t_0)$. Then use the differential equation to find the next approximation $\mathbf{x}^{(1)}(t)$ by solving $\dot{\mathbf{x}}^{(1)}(t) = \mathbf{A}(t)\mathbf{x}^{(0)}(t)$. The solution is

$$\dot{\mathbf{x}}^{(1)}(t) = \mathbf{x}(t_0) + \int_{t_0}^t \dot{\mathbf{x}}^{(1)}(\tau)\,d\tau = \left[\mathbf{I} + \int_{t_0}^t \mathbf{A}(\tau_0)\,d\tau_0\right]\mathbf{x}(t_0)$$

Let $\dot{\mathbf{x}}^{(2)} = \mathbf{A}(t)\mathbf{x}^{(1)}(t)$. Then

$$\begin{aligned}\mathbf{x}^{(2)}(t) &= \mathbf{x}(t_0) + \int_{t_0}^t \dot{\mathbf{x}}^{(2)}(\tau)\,d\tau \\ &= \mathbf{x}(t_0) + \left[\int_{t_0}^t \mathbf{A}(\tau_0)\,d\tau_0 + \int_{t_0}^t \mathbf{A}(\tau_0)\int_{t_0}^{\tau_0} \mathbf{A}(\tau_1)\,d\tau_1\,d\tau_0\right]\mathbf{x}(t_0) \\ &= \left[\mathbf{I} + \int_{t_0}^t \mathbf{A}(\tau_0)\,d\tau_0 + \int_{t_0}^t \mathbf{A}(\tau_0)\int_{t_0}^{\tau_0} \mathbf{A}(\tau_1)\,d\tau_1\,d\tau_0\right]\mathbf{x}(t_0)\end{aligned}$$

Continuing this procedure with $\dot{\mathbf{x}}^{(k+1)}(t) = \mathbf{A}(t)\mathbf{x}^{(k)}(t)$ leads to

$$\begin{aligned}\mathbf{x}(t) \cong \Big[\mathbf{I} &+ \int_{t_0}^t \mathbf{A}(\tau_0)\,d\tau_0 + \int_{t_0}^t \mathbf{A}(\tau_0)\int_{t_0}^{\tau_0} \mathbf{A}(\tau_1)\,d\tau_1\,d\tau_0 \\ &+ \int_{t_0}^t \mathbf{A}(\tau_0)\int_{t_0}^{\tau_0} \mathbf{A}(\tau_1)\int_{t_0}^{\tau_1} \mathbf{A}(\tau_2)\,d\tau_2\,d\tau_1\,d\tau_0 + \cdots\Big]\mathbf{x}(t_0)\end{aligned}$$

Truncating the series in the brackets after a finite number of terms gives an approximation for $\mathbf{\Phi}(t, t_0)$.

Miscellaneous Applications

10.6 The satellite of Problem 9.11 is considered. If the two input torques are programmed to give $u_1(t) = (1/J_y)T_y(t) = C\sin\alpha t$, and $u_2(t) = (1/J_y)T_z(t) = C\cos\alpha t$, find the resultant time history of the state $\mathbf{x}(t) = [\omega_y \quad \omega_z]^T$. Use arbitrary initial conditions at time $t = 0$.

The state equations are $\dot{\mathbf{x}} = \begin{bmatrix} 0 & -\Omega \\ \Omega & 0 \end{bmatrix}\mathbf{x} + \begin{bmatrix} u_1(t) \\ u_2(t) \end{bmatrix}$. From Problem 8.17, the transition matrix is $\mathbf{\Phi}(t, 0) = \begin{bmatrix} \cos \Omega t & -\sin \Omega t \\ \sin \Omega t & \cos \Omega t \end{bmatrix}$ and

$$\mathbf{x}(t) = \mathbf{\Phi}(t, 0)\mathbf{x}(0) + \int_0^t \mathbf{\Phi}(t, \tau)\mathbf{u}(\tau)\, d\tau$$

The transition matrix properties can be used to write $\mathbf{\Phi}(t, \tau) = \mathbf{\Phi}(t, 0)\mathbf{\Phi}(0, \tau)$ and $\mathbf{\Phi}(0, \tau) = \mathbf{\Phi}(-\tau, 0)$, so

$$\mathbf{x}(t) = \mathbf{\Phi}(t, 0)\left\{\mathbf{x}(0) + C\int_0^t \begin{bmatrix} \cos \Omega\tau \sin \alpha\tau + \sin \Omega\tau \cos \alpha\tau \\ -\sin \Omega\tau \sin \alpha\tau + \cos \Omega\tau \cos \alpha\tau \end{bmatrix} d\tau\right\}$$

The trigonometric identities $\cos a \sin b + \sin a \cos b = \sin (a + b)$ and $\cos a \cos b - \sin a \sin b = \cos (a + b)$ are used inside the integral to give

$$\mathbf{x}(t) = \mathbf{\Phi}(t, 0)\left\{\mathbf{x}(0) + \frac{C}{\Omega + \alpha}\begin{bmatrix} 1 - \cos (\Omega + \alpha)t \\ \sin (\Omega + \alpha)t \end{bmatrix}\right\}$$

10.7 The input to the circuit of Fig. 10.1 is an ideal current source $u(t)$. The output (and also the state) is the voltage across the capacitor $x(t)$. If

$$u(t) = e^{t/RC}\frac{10 - e^{-t_f/RC}x_0}{R \sinh (t_f/RC)}$$

and if $x(0) = x_0$, find the output $x(t_f)$ at some final time $t = t_f$.

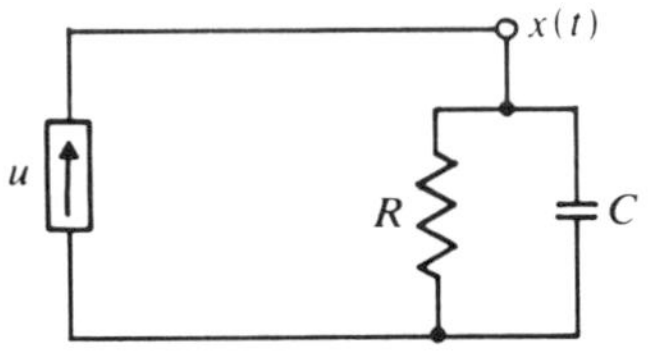

Figure 10.1

The state equation is $\dot{x} = -x/RC + u/C$. The solution is

$$x(t) = e^{-t/RC}\, x_0 + \frac{1}{C}\int_0^t e^{-(t-\tau)/RC}\, u(\tau)\, d\tau$$

Letting $u(t) = Ke^{t/RC}$ for simplicity gives

$$x(t) = e^{-t/RC}\, x_0 + \frac{K}{C}\, e^{-t/RC}\int_0^t e^{2\tau/RC}\, d\tau$$

$$= e^{-t/RC}\, x_0 + \frac{K}{C}\, e^{-t/RC}\left\{\frac{RC}{2}[e^{2t/RC} - 1]\right\} = e^{-t/RC}\, x_0 + KR \sinh\left(\frac{t}{RC}\right)$$

Using $K = [10 - e^{-t_f/RC}x_0]/[R \sinh (t_f/RC)]$ and evaluating at $t = t_f$ gives $x(t_f) = 10$.

Although not proven here, the specified input $u(t)$ is the one which charges the capacitor from $x(0) = x_0$ to $x(t_f) = 10$ while minimizing the energy dissipated in R.

10.8 A system is described by $\begin{bmatrix} \dot{x}_1 \\ \dot{x}_2 \\ \dot{x}_3 \end{bmatrix} = \begin{bmatrix} -2 & -2 & 0 \\ 0 & 0 & 1 \\ 0 & -3 & -4 \end{bmatrix}\begin{bmatrix} x_1 \\ x_2 \\ x_3 \end{bmatrix} + \begin{bmatrix} 1 & 0 \\ 0 & 1 \\ 1 & 1 \end{bmatrix}\begin{bmatrix} u_1(t) \\ u_2(t) \end{bmatrix}$

a. Find the change of variables $\mathbf{x} = \mathbf{M}\mathbf{q}$ which uncouples this system.
b. If $\mathbf{x}(0) = [10 \quad 5 \quad 2]^T$ and if $\mathbf{u}(t) = [t \quad 1]^T$, find $\mathbf{x}(t)$.

a. The modal matrix $\mathbf{M}$ must be found. $|\mathbf{A} - \mathbf{I}\lambda| = (\lambda + 1)(\lambda + 2)(\lambda + 3)$ so the eigenvalues are $\lambda_i = -1, -2, -3$:

$$\text{Adj}\,[\mathbf{A} - \mathbf{I}\lambda] = \begin{bmatrix} \lambda^2 + 4 + 3 & -2(\lambda + 4) & -2 \\ 0 & (\lambda + 2)(\lambda + 4) & 2 + \lambda \\ 0 & -3(2 + \lambda) & \lambda(2 + \lambda) \end{bmatrix}$$

From this, the eigenvectors are $\boldsymbol{\xi}_1 = [-2 \quad 1 \quad -1]^T$, $\boldsymbol{\xi}_2 = [1 \quad 0 \quad 0]^T$, $\boldsymbol{\xi}_3 = [-2 \quad -1 \quad 3]^T$ so that $\mathbf{x} = \begin{bmatrix} -2 & 1 & -2 \\ 1 & 0 & -1 \\ -1 & 0 & 3 \end{bmatrix}\mathbf{q}$ is the decoupling transformation.

Using this substitution along with $\mathbf{M}^{-1} = \begin{bmatrix} 0 & 3/2 & 1/2 \\ 1 & 4 & 2 \\ 0 & 1/2 & 1/2 \end{bmatrix}$ leads to

$$\dot{\mathbf{q}} = \begin{bmatrix} -1 & 0 & 0 \\ 0 & -2 & 0 \\ 0 & 0 & -3 \end{bmatrix}\mathbf{q} + \begin{bmatrix} 1/2 & 2 \\ 3 & 6 \\ 1/2 & 1 \end{bmatrix}\begin{bmatrix} u_1 \\ u_2 \end{bmatrix}$$

b. The three uncoupled equations and their solutions are

$$\dot{q}_1 = -q_1 + \tfrac{1}{2}t + 2 \Rightarrow q_1(t) = e^{-t}q_1(0) + \tfrac{1}{2}t + \tfrac{3}{2}(1 - e^{-t})$$

$$\dot{q}_2 = -2q_2 + 3t + 6 \Rightarrow q_2(t) = e^{-2t}q_2(0) + \tfrac{3}{2}t + \tfrac{9}{4}(1 - e^{-2t})$$

$$\dot{q}_3 = -3q_3 + \tfrac{1}{2}t + 1 \Rightarrow q_3(t) = e^{-3t}q_3(0) + \tfrac{1}{6}t + \tfrac{5}{18}(1 - e^{-3t})$$

Since $\mathbf{q}(0) = \mathbf{M}^{-1}\mathbf{x}(0) = [17/2 \quad 34 \quad 7/2]^T$, and since $\mathbf{x}(t) = \mathbf{M}\mathbf{q}(t)$, the solution is

$$\mathbf{x}(t) = \begin{bmatrix} -14e^{-t} + (127/4)e^{-2t} - (58/9)e^{-3t} + (1/6)t - 47/36 \\ 7e^{-t} - (29/9)e^{-3t} + (1/3)t + 11/9 \\ -7e^{-t} + (29/3)e^{-3t} - 2/3 \end{bmatrix}$$

10.9 The motor-generator system of Problem 9.8, page 250, has been driving the load at a constant speed $\Omega = 100$ rad/sec for some time. At time $t = 0$ the input voltage $e_f(t)$ is suddenly removed, that is, $e_f(t) = 0$ for $t \geq 0$. Find the resulting motion of the system. Assume the linear relation $e_g = K_g i_f$ and use the parameter values $b/J = 1$, $K_m/J = 2$, $K_m/(L_g + L_m) = 2.5$, $(R_g + R_m)/(L_g + L_m) = 7$, $K_g/(L_g + L_m) = 4$, $R_f/L_f = 5$, $L_f = 1$.

The state equations are $\begin{bmatrix} \dot{x}_1 \\ \dot{x}_2 \\ \dot{x}_3 \end{bmatrix} = \begin{bmatrix} -1 & 2 & 0 \\ -2.5 & -7 & 4 \\ 0 & 0 & -5 \end{bmatrix}\begin{bmatrix} x_1 \\ x_2 \\ x_3 \end{bmatrix} + \begin{bmatrix} 0 \\ 0 \\ 1 \end{bmatrix}u(t).$

The desired solution is $\mathbf{x}(t) = \boldsymbol{\Phi}(t, 0)\mathbf{x}(0) = e^{\mathbf{A}t}\mathbf{x}(0)$. The initial value of $\Omega = x_1(0)$ is 100. The initial values of the other state variables can be determined from the fact that the system was initially in steady-state, $\dot{\Omega} = 0$ and $T(0) = b\Omega(0) = K_m i_m(0)$. From this, $x_2(0) = i_m(0) = (b/K_m)\Omega(0) = (b/J)(J/K_m)\Omega(0) = 50$. Also $di_m/dt = 0$ for $t \leq 0$, so $e_g - e_m = (R_g + R_m)i_m$ at $t = 0$. Therefore, $e_g(0) = e_m(0) + (R_g + R_m)i_m(0)$ and $i_f(0) = x_3(0) = e_g(0)/K_g$. $x_3(0) = [K_m\Omega(0) + (R_g + R_m)i_m(0)]/K_g = 2.5(100)/4 + 7(50)/4 = 150$ or $\mathbf{x}(0) = [100 \quad 50 \quad 150]^T$.

To find $\boldsymbol{\Phi}(t, 0)$, the eigenvalues of $\mathbf{A}$ are found.

$$|\mathbf{A} - \mathbf{I}\lambda| = (-\lambda - 5)\begin{vmatrix} -\lambda - 1 & 2 \\ -2.5 & -\lambda - 7 \end{vmatrix} = (-\lambda - 5)(\lambda + 2)(\lambda + 6);$$

$$\lambda_1 = -2,\ \lambda_2 = -5,\ \lambda_3 = -6$$

Using the Cayley-Hamilton remainder technique,

$$e^{\mathbf{A}t} = \alpha_0\mathbf{I} + \alpha_1\mathbf{A} + \alpha_2\mathbf{A}^2 = \begin{bmatrix} \alpha_0 - \alpha_1 - 4\alpha_2 & 2\alpha_1 - 16\alpha_2 & 8\alpha_2 \\ -2.5\alpha_1 + 20\alpha_2 & \alpha_0 - 7\alpha_1 + 44\alpha_2 & 4\alpha_1 - 48\alpha_2 \\ 0 & 0 & \alpha_0 - 5\alpha_1 + 25\alpha_2 \end{bmatrix}$$

where

$$\left.\begin{aligned} e^{-2t} &= \alpha_0 - 2\alpha_1 + 4\alpha_2 \\ e^{-5t} &= \alpha_0 - 5\alpha_1 + 25\alpha_2 \\ e^{-6t} &= \alpha_0 - 6\alpha_1 + 36\alpha_2 \end{aligned}\right\} \Rightarrow \left\{\begin{aligned} \alpha_0 &= (5/2)e^{-2t} - 4e^{-5t} + (5/2)e^{-6t} \\ \alpha_1 &= (11/12)e^{-2t} - (8/3)e^{-5t} + (7/4)e^{-6t} \\ \alpha_2 &= (1/12)e^{-2t} - (1/3)e^{-5t} + (1/4)e^{-6t} \end{aligned}\right.$$

Using these gives

$$\mathbf{\Phi}(t, 0) = \begin{bmatrix} \frac{5}{4}e^{-2t} - \frac{1}{4}e^{-6t} & \frac{1}{2}e^{-2t} - \frac{1}{2}e^{-6t} & \frac{2}{3}e^{-2t} - \frac{8}{3}e^{-5t} + 2e^{-6t} \\ -\frac{5}{8}e^{-2t} + \frac{5}{8}e^{-6t} & -\frac{1}{4}e^{-2t} + \frac{5}{4}e^{-6t} & -\frac{1}{3}e^{-2t} + \frac{16}{3}e^{-5t} - 5e^{-6t} \\ 0 & 0 & e^{-5t} \end{bmatrix}$$

Then $\mathbf{x}(t) = \mathbf{\Phi}(t, 0)\mathbf{x}(0)$; and since $\Omega(t) = y = [1 \quad 0 \quad 0]\mathbf{x}(t)$, $\Omega(t) = 250e^{-2t} - 400e^{-5t} + 250e^{-6t}$.

Discrete Approximations for Continuous-Time State Equations

10.10 Consider $\dot{\mathbf{x}}(t) = \mathbf{A}(t)\mathbf{x}(t) + \mathbf{B}(t)\mathbf{u}(t)$, $\mathbf{x}(t_0)$ given. Assume $\{t_0, t_1, t_2, \ldots, t_k, t_{k+1}, \ldots\}$ is a set of discrete-time points which are sufficiently close together so that during any interval $[t_k, t_{k+1}]$ the following piecewise constant approximations can be made for the input: $\mathbf{u}(t_k) = \mathbf{u}(t)$. Develop an approximate difference equation for this system.

For a typical interval, the differential equation is $\dot{\mathbf{x}}(t) = \mathbf{A}(t)\mathbf{x}(t) + \mathbf{B}(t)\mathbf{u}(t_k)$ and its solution, evaluated at t_{k+1}, is

$$\mathbf{x}(t_{k+1}) = \mathbf{\Phi}(t_{k+1}, t_k)\mathbf{x}(t_k) + \int_{t_k}^{t_{k+1}} \mathbf{\Phi}(t_{k+1}, \tau)\mathbf{B}(\tau)\, d\tau\, \mathbf{u}(t_k)$$

This represents an approximating difference equation.

If, in addition, $\mathbf{A}(t)$ can be approximated by the constant matrix $\mathbf{A}(t_k)$ during the interval, then

$$\mathbf{\Phi}(t_{k+1}, t_k) = e^{\mathbf{A}(t_k)\Delta t_k} \qquad \text{where } \Delta t_k \triangleq t_{k+1} - t_k$$

If $\mathbf{B}(t)$ can also be approximated by $\mathbf{B}(t_k)$, then

$$\int_{t_k}^{t_{k+1}} \mathbf{\Phi}(t_{k+1}, \tau)\mathbf{B}(\tau)\, d\tau = e^{\mathbf{A}(t_k)t_{k+1}} \int_{t_k}^{t_{k+1}} e^{-\mathbf{A}(t_k)\tau}\, d\tau\, \mathbf{B}(t_k)$$

If $\mathbf{A}(t_k)^{-1}$ exists, a result similar to that of Example 8.4 gives

$$\mathbf{x}(t_{k+1}) = \mathbf{A}_1(t_k)\mathbf{x}(t_k) + \mathbf{B}_1(t_k)\mathbf{u}(t_k)$$

where

$$\mathbf{A}_1(t_k) \triangleq e^{\mathbf{A}(t_k)\Delta t_k}, \qquad \mathbf{B}_1(t_k) \triangleq [\mathbf{A}_1(t_k) - \mathbf{I}]\mathbf{A}^{-1}\mathbf{B}(t_k)$$

10.11 A system is described by $\dot{\mathbf{x}} = \mathbf{A}\mathbf{x} + \mathbf{B}\mathbf{u}$, with $\mathbf{A}$ and $\mathbf{B}$ constant. Develop an efficient computational procedure for finding $\mathbf{x}(T)$, assuming $\mathbf{u}(t)$ is constant over $[0, T]$.

Setting $t_{k+1} = T$ and $t_k = 0$ in Problem 10.10 gives the form of $\mathbf{x}(T)$

$$\mathbf{x}(T) = \mathbf{\Phi}(T, 0)\mathbf{x}(0) + \int_0^T \mathbf{\Phi}(T, \tau)\, d\tau\, \mathbf{B}\mathbf{u}$$

It is known that $\mathbf{\Phi}(T, 0) = e^{\mathbf{A}T}$ and the series form is

$$\begin{aligned}\mathbf{\Phi}(T, 0) &= \mathbf{I} + \mathbf{A}T + (\mathbf{A}T)^2/2 + (\mathbf{A}T)^3/3! + (\mathbf{A}T)^4/4! + \cdots \\ &= \mathbf{I} + \mathbf{A}T\{\mathbf{I} + \mathbf{A}T/2 + (\mathbf{A}T)^2/3! + (\mathbf{A}T)^3/4! + \cdots\} \\ &\;\;\vdots \\ &= \mathbf{I} + \mathbf{A}T\{\mathbf{I} + \mathbf{A}T/2[\mathbf{I} + \mathbf{A}T/3(\mathbf{I} + \mathbf{A}T/4)(\mathbf{I} + \cdots (\mathbf{I} + \mathbf{A}T/N)))]\}\end{aligned}$$

This nested form for $\mathbf{\Phi}(T, 0)$ does not require the direct computation of increasingly high powers of $\mathbf{A}T$ and therefore avoids many overflow and underflow problems. How many terms need to be retained depends upon $|\lambda_{max}|T$, where $|\lambda_{max}|$ is the largest magnitude eigenvalue of $\mathbf{A}$. This test is not normally used, however. On the Nth step in the nested sequence, the first neglected term would be $(\mathbf{A}T)^2/[(N)(N+1)]$, and this should be acceptably small compared with $\mathbf{A}T/N$.

The previous problem gave a result for $\mathbf{B}_1$ which depends on the existence of $\mathbf{A}^{-1}$. That is too restrictive in many cases, so another form which is better suited to machine computation is sought. Clearly,

$$\mathbf{B}_1 = \int_0^T e^{\mathbf{A}(T-\tau)}\, d\tau\, \mathbf{B} = -\int_T^0 e^{\mathbf{A}\xi}\, d\xi\, \mathbf{B} = \int_0^T e^{\mathbf{A}\xi}\, d\xi\, \mathbf{B}$$

Direct term-by-term integration of the exponential matrix gives

$$\begin{aligned}\mathbf{B}_1 &= \{\mathbf{I}T + \mathbf{A}T^2/2 + \mathbf{A}^2T^3/3! + \cdots\}\mathbf{B} \\ &= T\{\mathbf{I} + \mathbf{A}T/2 + (\mathbf{A}T)^2/3! + \cdots\}\mathbf{B}\end{aligned}$$

Note that the series inside { } is the same as the one which appeared in the calculation of $\mathbf{\Phi}(T, 0)$. Therefore, the same nested form is possible. Define this part of the solution as $\mathbf{\Psi}$. Then

$$\mathbf{\Psi} = \mathbf{I} + \mathbf{A}T/2[\mathbf{I} + \mathbf{A}T/3(\mathbf{I} + \mathbf{A}T/4)(\mathbf{I} + \ldots (\mathbf{I} + \mathbf{A}T/N)))]$$

This partial result is then used to obtain the desired approximations

$$\begin{aligned}\mathbf{B}_1 &= T\mathbf{\Psi}\mathbf{B} \\ \mathbf{\Phi} &= \mathbf{I} + \mathbf{A}T\mathbf{\Psi}\end{aligned}$$

These are widely used in obtaining discrete approximations to continuous-time systems.

10.12 A second-order system is described by $\ddot{x} + 2\dot{x} + 4x = u(t)$. Using $x = x_1$ and $x_2 = \dot{x}$ as states, find the state equations and evaluate the exact transition matrix $\mathbf{\Phi}(T, 0)$ and input matrix $\mathbf{B}_1$ using the results of Problem 10.10 with $T = 0.2$. Then use results of Problem 10.11 to obtain approximate numerical results. Compare these.

The state equation is

$$\dot{\mathbf{x}} = \begin{bmatrix} 0 & 1 \\ -4 & -2 \end{bmatrix}\mathbf{x} + \begin{bmatrix} 0 \\ 1 \end{bmatrix} u$$

The exact state transition matrix is found to be

$$\mathbf{\Phi}(T, 0) = \begin{bmatrix} C + S & S \\ -4S & C - S \end{bmatrix}, \text{ where } C = e^{-T}\cos(\sqrt{3}\,T) \text{ and } S = e^{-T}\sin(\sqrt{3}\,T)/\sqrt{3}$$

Using $T = 0.2$ gives $\mathbf{\Phi} = \begin{bmatrix} 0.9306 & 0.1605 \\ -0.6420 & 0.6096 \end{bmatrix}$. Since $\mathbf{A}$ is nonsingular here, the results of Problem 10.10 can be used to find

$$\mathbf{B}_1 = \mathbf{A}^{-1}[\mathbf{\Phi}(T, 0) - \mathbf{I}]\mathbf{B} = \begin{bmatrix} 0.017 \\ 0.160 \end{bmatrix}$$

The truncated series defined as $\mathbf{\Psi}$ in Problem 10.11 is now used. Note that for an Nth order approximation in T for $\mathbf{\Phi}$ and $\mathbf{B}_1$, an $(N-1)$th order approximation in $\mathbf{\Psi}$ is used.

Highest power of T	$\mathbf{\Psi}$	$\mathbf{\Phi}(T, 0)$	$\mathbf{B}_1$
1	$\begin{bmatrix} 1 & 0 \\ 0 & 1 \end{bmatrix}$	$\begin{bmatrix} 1 & 0.2 \\ -0.8 & 0.6 \end{bmatrix}$	$\begin{bmatrix} 0 \\ 0.2 \end{bmatrix}$
2	$\begin{bmatrix} 1 & 0.1 \\ -0.4 & 0.8 \end{bmatrix}$	$\begin{bmatrix} 0.92 & 0.16 \\ -0.64 & 0.60 \end{bmatrix}$	$\begin{bmatrix} 0.02 \\ 0.16 \end{bmatrix}$
3	$\begin{bmatrix} 0.9733 & 0.0867 \\ -0.3467 & 0.8 \end{bmatrix}$	$\begin{bmatrix} 0.931 & 0.160 \\ -0.64 & 0.611 \end{bmatrix}$	$\begin{bmatrix} 0.017 \\ 0.160 \end{bmatrix}$

Depending on the application, the second- or third-order approximation may suffice. The first-order approximation probably would not, because very large differences between $(1)^k$ and $(0.93)^k$ will quickly appear in $\mathbf{\Phi}(k, 0)$ as the approximate difference equations are solved over k time steps.

10.13 Find a discrete-time approximate model for the system of Fig. 10.2. Use $t_{k+1} - t_k = \Delta t = 1$ and approximate u_1 and u_2 as piecewise constant functions.

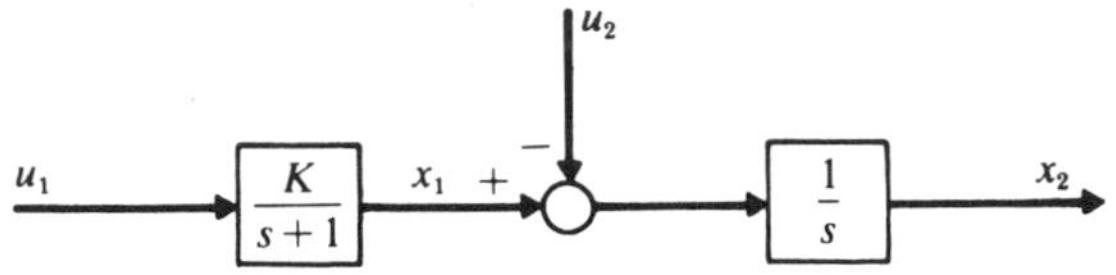

Figure 10.2

The continuous-time state equations are

$$\begin{bmatrix} \dot{x}_1 \\ \dot{x}_2 \end{bmatrix} = \begin{bmatrix} -1 & 0 \\ 1 & 0 \end{bmatrix} \begin{bmatrix} x_1 \\ x_2 \end{bmatrix} + \begin{bmatrix} K & 0 \\ 0 & -1 \end{bmatrix} \begin{bmatrix} u_1 \\ u_2 \end{bmatrix}$$

and the transition matrix is

$$\mathbf{\Phi}(t, 0) = e^{\mathbf{A}t} = \mathcal{L}^{-1}\{[s\mathbf{I} - \mathbf{A}]^{-1}\}, \qquad \mathbf{\Phi}(s) = \frac{\begin{bmatrix} s & 0 \\ 1 & s+1 \end{bmatrix}}{s(s+1)}$$

or

$$\mathbf{\Phi}(t, 0) = \begin{bmatrix} e^{-t} & 0 \\ (1 - e^{-t}) & 1 \end{bmatrix}$$

The state at time $t_{k+1} = t_k + \Delta t$ can be written as

$$\mathbf{x}(t_{k+1}) = \mathbf{\Phi}(t_{k+1}, t_k)\mathbf{x}(t_k) + \int_{t_k}^{t_{k+1}} \mathbf{\Phi}(t_{k+1}, \tau)\, d\tau \begin{bmatrix} K & 0 \\ 0 & -1 \end{bmatrix} \begin{bmatrix} u_1(t_k) \\ u_2(t_k) \end{bmatrix}$$

But

$$\mathbf{\Phi}(t_{k+1}, t_k) = \mathbf{\Phi}(t_{k+1} - t_k, 0) = \begin{bmatrix} e^{-1} & 0 \\ 1 - e^{-1} & 1 \end{bmatrix} = \begin{bmatrix} 0.368 & 0 \\ 0.632 & 1 \end{bmatrix}$$

and

$$\int_{t_k}^{t_{k+1}} \mathbf{\Phi}(t_{k+1}, \tau)\, d\tau = \begin{bmatrix} 1 - e^{-1} & 0 \\ e^{-1} & 1 \end{bmatrix} = \begin{bmatrix} 0.632 & 0 \\ 0.368 & 1 \end{bmatrix}$$

The approximating difference equation is

$$\begin{bmatrix} x_1(t_{k+1}) \\ x_2(t_{k+1}) \end{bmatrix} = \begin{bmatrix} 0.368 & 0 \\ 0.632 & 1 \end{bmatrix} \begin{bmatrix} x_1(t_k) \\ x_2(t_k) \end{bmatrix} + \begin{bmatrix} 0.632K & 0 \\ 0.368K & -1 \end{bmatrix} \begin{bmatrix} u_1(t_k) \\ u_2(t_k) \end{bmatrix}$$

10.14 The system of Fig. 10.2 represents a simple model of a production and inventory control system. The input $u_1(t)$ represents the scheduled production rate, $x_1(t)$ represents the actual production rate, $u_2(t)$ represents the sales rate, and $x_2(t)$ represents the current inventory level. Suppose that the production schedule is selected as $u_1(t) = c - x_2(t)$, where c is the desired inventory level. This is a feedback control policy. The system is originally in equilibrium with $x_1(0)$ equal to the sales rate and $x_2(0) = c$. At time $t = 0$ the sales rate suddenly increases by 10%. That is, $u_2(t) = 1.1x_1(0)$ for $t \geq 0$. Find the resulting system response. Use $K = 3/16$.

The simulation diagram for the feedback system is shown in Fig. 10.3.

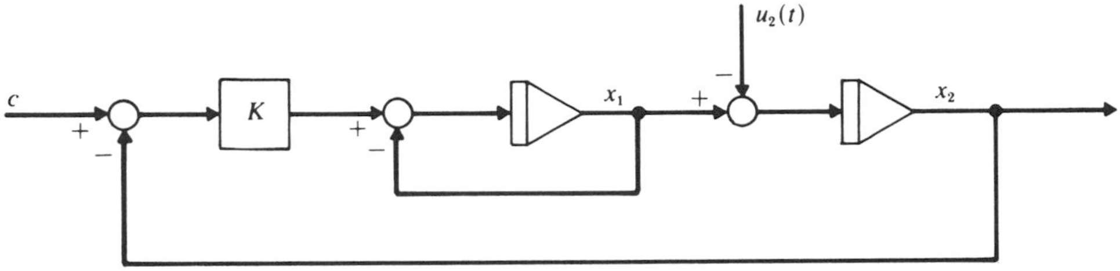

Figure 10.3

The state equations are $\dot{\mathbf{x}} = \begin{bmatrix} -1 & -K \\ 1 & 0 \end{bmatrix}\mathbf{x} + \begin{bmatrix} K & 0 \\ 0 & -1 \end{bmatrix}\begin{bmatrix} c \\ u_2 \end{bmatrix}$. The eigenvalues are determined from $|\mathbf{A} - \mathbf{I}\lambda| = \lambda^2 + \lambda + K = 0$ so that, with $K = 3/16$, $\lambda_1 = -1/4$ and $\lambda_2 = -3/4$.

The transition matrix is

$$\mathbf{\Phi}(t, 0) = e^{\mathbf{A}t} = \alpha_0 \mathbf{I} + \alpha_1 \mathbf{A} = \begin{bmatrix} \alpha_0 - \alpha_1 & -3\alpha_1/16 \\ \alpha_1 & \alpha_0 \end{bmatrix}$$

Using the eigenvalues to solve for α_0 and α_1 gives

$$\mathbf{\Phi}(t, 0) = \begin{bmatrix} -\frac{1}{2}e^{-t/4} + \frac{3}{2}e^{-3t/4} & -\frac{3}{8}(e^{-t/4} - e^{-3t/4}) \\ 2(e^{-t/4} - e^{-3t/4}) & \frac{3}{2}e^{-t/4} - \frac{1}{2}e^{-3t/4} \end{bmatrix}$$

The solution is

$$\mathbf{x}(t) = \mathbf{\Phi}(t, 0)\begin{bmatrix} x_1(0) \\ c \end{bmatrix} + \int_0^t \mathbf{\Phi}(t, \tau)\, d\tau\, \mathbf{B}\begin{bmatrix} c \\ 1.1x_1(0) \end{bmatrix}$$

Using $\mathbf{\Phi}(t, \tau) = \mathbf{\Phi}(t - \tau, 0)$, carrying out the integration, and simplifying give

$$x_1(t) = x_1(0)\{1.1 - (4.3/2)e^{-t/4} + (4.1/2)e^{-3t/4}\}$$

$$x_2(t) = c + x_1(0)\{-17.6/3 + 8.6e^{-t/4} - (8.2/3)e^{-3t/4}\}$$

Additional Properties of the Transition Matrix

10.15 Assume that the eigenvectors of the constant system matrix **A** form a basis and show that $\mathbf{\Phi}(t, t_0) = \sum_{i=1}^{n} e^{\lambda_i(t-t_0)}\boldsymbol{\xi}_i\rangle\langle\mathbf{r}_i$, where λ_i, $\boldsymbol{\xi}_i$, and $\mathbf{r}_i$ are eigenvalues, eigenvectors, and reciprocal basis vectors of **A**, respectively.

The modal decomposition developed in Sec. 10.4 led to the expression

$$\mathbf{x}(t) = q_1(t)\boldsymbol{\xi}_1 + q_2(t)\boldsymbol{\xi}_2 + \cdots + q_n(t)\boldsymbol{\xi}_n = \sum_{i=1}^{n} q_i(t)\boldsymbol{\xi}_i$$

For the homogeneous system $\dot{\mathbf{x}} = \mathbf{A}\mathbf{x}$,

$$q_i(t) = e^{\lambda_i(t-t_0)}q_i(t_0) = e^{\lambda_i(t-t_0)}\langle\mathbf{r}_i, \mathbf{x}(t_0)\rangle$$

so that

$$\mathbf{x}(t) = \sum_{i=1}^{n} e^{\lambda_i(t-t_0)}\langle\mathbf{r}_i, \mathbf{x}(t_0)\rangle\boldsymbol{\xi}_i = \left[\sum_{i=1}^{n} e^{\lambda_i(t-t_0)}\boldsymbol{\xi}_i\rangle\langle\mathbf{r}_i\right]\mathbf{x}(t_0)$$

Comparing this with the known solution $\mathbf{x}(t) = \mathbf{\Phi}(t, t_0)\mathbf{x}(t_0)$ gives the desired result.

10.16 For fixed times t_0 and t, the transition matrix is a transformation of the state space Σ onto itself. A linear transformation which possesses a full set of n linearly independent eigenvectors has a spectral representation

$$\mathbf{\Phi}(t, t_0) = \sum_{i=1}^{n} \gamma_i\boldsymbol{\eta}_i\rangle\langle\mathbf{v}_i$$

where γ_i, $\boldsymbol{\eta}_i$ are the eigenvalues and eigenvectors of $\mathbf{\Phi}(t, t_0)$ and $\mathbf{v}_i$ are reciprocal to $\boldsymbol{\eta}_i$. This is the result of equation (*7.4*), page 182, with notational changes. Comparing this with the previous problem, draw conclusions about the relationships between eigenvalues and eigenvectors of **A** and of $\mathbf{\Phi}(t, t_0)$.

The indicated comparison suggests the following relationship between eigenvalues, a result known as Frobenius' theorem. If $\lambda_1, \lambda_2, \ldots, \lambda_n$ are eigenvalues of the $n \times n$ matrix **A**, and if $f(x)$ is a function which is analytic inside a circle in the complex plane which contains all the λ_i, then $f(\lambda_1), f(\lambda_2), \ldots, f(\lambda_n)$ are the eigenvalues of the matrix function $f(\mathbf{A})$. In the present case the eigenvalues of $\mathbf{\Phi}(t, t_0) = e^{\mathbf{A}(t-t_0)}$ are $\gamma_i = e^{\lambda_i(t-t_0)}$. Furthermore, it can be verified that the eigenvectors of **A** and of $\mathbf{\Phi}(t, t_0)$ are the same, that is, $\boldsymbol{\xi}_i = \boldsymbol{\eta}_i$.

The Adjoint Equations

10.17 The formal adjoint of the differential equation $\dot{\mathbf{x}} = \mathbf{A}\mathbf{x}$ is $\dot{\mathbf{y}} = -\mathbf{A}^T\mathbf{y}$ (see Problem 6.18) Let the transition matrix for the adjoint equation be $\mathbf{\Theta}(t, \tau)$. Show that $\mathbf{\Theta}(t, \tau) = \mathbf{\Phi}^T(\tau, t)$, where $\mathbf{\Phi}(t, \tau)$ is the transition matrix for the equation in **x**.

Since $\mathbf{\Theta}(t, \tau)$ is the transition matrix, it satisfies

$$\frac{d}{dt}[\mathbf{\Theta}(t, \tau)] = -\mathbf{A}^T\mathbf{\Theta}(t, \tau), \qquad \mathbf{\Theta}(\tau, \tau) = \mathbf{I}$$

The desired result is established by showing that $\mathbf{\Phi}^T(\tau, t)$ satisfies the same differential equation and initial conditions, since these equations define a unique solution. Since

$\mathbf{\Phi}(\tau, t) = \mathbf{\Phi}^{-1}(t, \tau)$,

$$\frac{d}{dt}[\mathbf{\Phi}(\tau, t)] = \frac{d}{dt}[\mathbf{\Phi}^{-1}(t, \tau)] = -\mathbf{\Phi}^{-1}(t, \tau)\frac{d}{dt}[\mathbf{\Phi}(t, \tau)\mathbf{\Phi}^{-1}(t, \tau)]$$

But $\frac{d}{dt}[\mathbf{\Phi}(t, \tau)] = \mathbf{A}\mathbf{\Phi}(t, \tau)$. Therefore, $\frac{d}{dt}[\mathbf{\Phi}(\tau, t) = -\mathbf{\Phi}^{-1}(t, \tau)\mathbf{A} = -\mathbf{\Phi}(\tau, t)\mathbf{A}$.

Transposing shows that $\frac{d}{dt}[\mathbf{\Phi}(\tau, t)]^T = -\mathbf{A}^T[\mathbf{\Phi}(\tau, t)]^T$. This, plus the fact that $\mathbf{\Phi}(\tau, \tau) = \mathbf{I}$, establishes that

$$\mathbf{\Theta}(t, \tau) = \mathbf{\Phi}^T(\tau, t)$$

It follows that the adjoint transition matrix can be expressed in terms of the fundamental matrix $\mathbf{U}(t)$ as $\mathbf{\Theta}^T(t, \tau) = \mathbf{U}(\tau)\mathbf{U}^{-1}(t)$.

10.18 Suppose a simulation of the system $\dot{\mathbf{x}} = \mathbf{A}\mathbf{x}$ is available, as well as a simulation of the adjoint system $\dot{\mathbf{y}} = -\mathbf{A}^T\mathbf{y}$. Interpret the meaning of the vector functions $\mathbf{x}(t)$ and $\mathbf{y}(t)$ which are obtained if the initial conditions used are $\mathbf{x}(t_0) = [1 \quad 0 \quad 0 \quad \cdots \quad 0]^T$ and $\mathbf{y}(t_0) = [1 \quad 0 \quad 0 \quad \cdots \quad 0]^T$.

Since the solutions are $\mathbf{x}(t) = \mathbf{\Phi}(t, t_0)\mathbf{x}(t_0)$ and $\mathbf{y}(t) = \mathbf{\Theta}(t, t_0)\mathbf{y}(t_0)$, the first simulation generates the first column of $\mathbf{\Phi}(t, t_0)$ for all $t \geq t_0$ and the second generates the first column of $\mathbf{\Theta}(t, t_0)$ for all $t \geq t_0$. But the first column of $\mathbf{\Theta}(t, t_0)$ equals the first *row* of $\mathbf{\Phi}(t_0, t)$.

The adjoint system simulation provides a means of reversing the roles of t_0 and t. In a sense, a reversed time impulse response for the original system can be generated. The complete matrix $\mathbf{\Phi}(t_0, t)$ can be obtained, one row at a time, by modifying the initial conditions for the adjoint simulation. This property has several uses (see pages 379–394 of Reference 28).

State Equations as Linear Transformations

10.19 Consider the operator $\mathcal{A}(\mathbf{x}) \triangleq [\mathbf{I}(d/dt) - \mathbf{A}]\mathbf{x}$ as a transformation on infinite dimensional function spaces. Discuss the form of the solution for $\mathcal{A}(\mathbf{x}) = \mathbf{B}\mathbf{u}(t)$ given by equation (*10.24*) in terms of the results of Problem 6.9, page 165.)

Problem 6.9 indicates that if $\mathbf{x}_1$ is a nonzero solution of the homogeneous equation $\mathcal{A}(\mathbf{x}) = \mathbf{0}$, then the most general solution of $\mathcal{A}(\mathbf{x}) = \mathbf{B}\mathbf{u}$ takes the form $\mathbf{x} + \mathbf{x}_1$, where $\mathbf{x}$ is a solution to the nonhomogeneous equation. This is precisely the form of equation (*10.24*), with $\mathbf{x}_1 = \mathbf{\Phi}(t, t_0)\mathbf{x}(t_0)$ being the homogeneous solution. A unique solution is not possible without specifying initial conditions.

10.20 Discuss the implications of Problem 6.10, page 166, in the context of linear state equations.

Problem 6.10 indicates that $\mathcal{A}(\mathbf{x}) = \mathbf{B}\mathbf{u}$ will have a solution for all $\mathbf{B}\mathbf{u}(t)$ if and only if the only solution to $\mathcal{A}^*(\mathbf{y}) = \mathbf{0}$ is the trivial solution. This poses an *apparent* contradiction, since nontrivial solutions to the adjoint equation have been discussed in Problem 10.16 and since solutions to $\mathcal{A}(\mathbf{x}) = \mathbf{B}\mathbf{u}(t)$ have been explicitly displayed in equation (*10.24*). The difficulty arises because of the differences between the adjoint transformation of Chapter 6 and the *formal* adjoint as used in this chapter.

A heuristic reconciliation is provided in a nonrigorous, formal manner. In Problem 6.18, page 168, it is shown that

$$\langle \mathbf{y}(t), \mathcal{A}(x(t))\rangle = \mathbf{y}^T(t_f)\mathbf{x}(t_f) - \mathbf{y}^T(t_0)\mathbf{x}(t_0) - \left\langle \frac{d\mathbf{y}}{dt} + \mathbf{A}^T\mathbf{y}, \mathbf{x}(t)\right\rangle$$

The *formal* adjoint was defined by dropping the two boundary terms. These two terms automatically cancel if $\mathbf{u}(t) = \mathbf{0}$ (see Problem 6.19). Generally, they are nonzero and can be included as follows:

$$\mathbf{y}^T(t_f)\mathbf{x}(t_f) - \mathbf{y}^T(t_0)\mathbf{x}(t_0) - \int_{t_0}^{t_f} \left[\frac{d\mathbf{y}^T(\tau)}{dt} + \mathbf{y}^T(\tau)\mathbf{A}\right] \mathbf{x}(\tau)\, d\tau$$

$$= \int_{t_0}^{t_f} \left\{-\frac{d\mathbf{y}^T(\tau)}{dt} - \mathbf{y}^T(\tau)\mathbf{A} + \mathbf{y}^T(t_f)\delta(t_f - \tau) - \mathbf{y}^T(t_0)\delta(\tau - t_0)\right\} \mathbf{x}(\tau)\, d\tau$$

The term in brackets is the transpose of the adjoint transformation, that is,

$$\mathcal{A}^*(\mathbf{y}) = -\frac{d\mathbf{y}}{dt} - \mathbf{A}^T\mathbf{y}(t) + \mathbf{y}(t_f)\delta(t_f - t) - \mathbf{y}(t_0)\delta(t - t_0) \tag{1}$$

Then treating the two impulse terms as forcing terms, the solution of equation (*1*) is

$$\mathbf{y}(t) = \mathbf{\Theta}(t, t_0)\mathbf{y}(t_0) + \int_{t_0}^{t} \mathbf{\Theta}(t, \tau)[\mathbf{y}(t_f)\delta(t_f - \tau) - \mathbf{y}(t_0)\delta(\tau - t_0)]\, d\tau$$

But

$$\int_{t_0}^{t} \mathbf{\Theta}(t, \tau)\mathbf{y}(t_f)\delta(t_f - \tau)\, d\tau = \mathbf{0} \qquad \text{for all } t < t_f$$

and

$$\int_{t_0}^{t} \mathbf{\Theta}(t, \tau)\mathbf{y}(t_0)\delta(\tau - t_0)\, d\tau = \mathbf{\Theta}(t, t_0)\mathbf{y}(t_0)$$

because of the sifting property of the impulse function. Therefore, the solution is $\mathbf{y}(t) = \mathbf{0}$ for all $t < t_f$. At the final time an identity is obtained, $\mathbf{y}(t_f) = \mathbf{y}(t_f)$. The solution $\mathbf{y}(t)$ is therefore zero for all t except possibly at the single time $t = t_f$. Such a function will be considered $\mathbf{0}$, since its norm is zero (the Hilbert inner product norm, for example). Thus the only solution to $\mathcal{A}^*(\mathbf{y}) = \mathbf{0}$ is the trivial solution, and the results of Problem 6.10 are still true and do not lead to a contradiction.

PROBLEMS

10.21 A system has two inputs $\mathbf{u} = [u_1 \quad u_2]^T$, and two outputs $\mathbf{y} = [y_1 \quad y_2]^T$. The input-output equations are $\dot{y}_1 + 3(y_1 + y_2) = u_1$ and $\ddot{y}_2 + 4\dot{y}_2 + 3y_2 = u_2$. Find $\mathbf{y}(t)$ if $y_1(0) = 1$, $y_2(0) = 2$, $\dot{y}_2(0) = 1$, and $\mathbf{u}(t) = \mathbf{0}$.

10.22 A system is described by the coupled input-output equations $\dot{y}_1 + 2(y_1 + y_2) = u_1$ and $\ddot{y}_2 + 4\dot{y}_2 + 3y_2 = u_2$. Find the output $\mathbf{y}(t) = [y_1(t) \quad y_2(t)]^T$ if $y_1(0) = 1$, $y_2(0) = 2$, $\dot{y}_2(0) = 0$, $u_1(t) = 0$, $u_2(t) = \delta(t)$ (i.e., an impulse at $t = 0$).

10.23 A system is described by $\begin{bmatrix} \dot{x}_1 \\ \dot{x}_2 \end{bmatrix} = \begin{bmatrix} -2 & 1 \\ -1 & 0 \end{bmatrix} \begin{bmatrix} x_1 \\ x_2 \end{bmatrix} + \begin{bmatrix} 3 \\ 1 \end{bmatrix} u(t)$. If $\mathbf{x}(0) = [10 \quad 1]^T$ and if $u(t) = 0$, find $\mathbf{x}(t)$.

10.24 If the input to the system of the previous problem is $u(t) = e^{2t}$, what is $\mathbf{x}(t)$?

10.25 The wobbling satellite of Problems 9.11, page 253, and 10.6, page 269, has the initial state $\mathbf{x}(0) = [\omega_y(0) \quad \omega_z(0)]^T$. If the input torques are programmed as

$$u_1(t) = -\frac{1}{t_f}[\omega_y(0) \cos \Omega t - \omega_z(0) \sin \Omega t]$$

and

$$u_2(t) = -\frac{1}{t_f}[\omega_y(0) \sin \Omega t + \omega_z(0) \cos \Omega t]$$

find the state (wobble) $\mathbf{x}$ at time $t = t_f$.

10.26 Show that the approximate numerical solution of

$$\dot{\mathbf{x}} = \mathbf{A}\mathbf{x} + \mathbf{B}\mathbf{u} \qquad (1)$$

at time T, expressed as

$$\dot{\mathbf{x}}(T) = \mathbf{x}(0) + \dot{\mathbf{x}}(0)T + \ddot{\mathbf{x}}(0)T^2/2 + \dddot{\mathbf{x}}(0)T^3/3! + \cdots$$

leads to exactly the same series representation for $\mathbf{\Phi}(T, 0)$ and $\mathbf{B}_1$ as found in Problem 10.11. Hint: Repeatedly use equation (*1*) and its derivatives to express all derivatives of $\mathbf{x}$ in terms of $\mathbf{x}$ and $\mathbf{u}$. Treat $\mathbf{A}$, $\mathbf{B}$, and $\mathbf{u}$ as constants. Notice that the first-order approximation is just rectangular integration of $\dot{\mathbf{x}}$, the second-order approximation is trapezoidal integration of $\dot{\mathbf{x}}$, etc.

10.27 Find the transition matrix $\mathbf{\Phi}(t, 0)$ for the feedback system of Problem 10.14 if K is increased to 2.5.

10.28 Let $\mathbf{A}$ be a constant $n \times n$ matrix with n linearly independent eigenvectors. Use the Cayley-Hamilton remainder form for $\mathbf{\Phi}(t, t_0) = e^{\mathbf{A}(t-t_0)}$ to verify the results stated in Problem 10.16. That is, show that $\mathbf{\Phi}(t, t_0)\boldsymbol{\xi}_i = e^{\lambda_i(t-t_0)}\boldsymbol{\xi}_i$, where λ_i and $\boldsymbol{\xi}_i$ are eigenvalues and eigenvectors of $\mathbf{A}$.

Chapter 11

ANALYSIS OF DISCRETE-TIME LINEAR SYSTEMS

11.1 INTRODUCTION

Dynamic systems which are described by difference equations, rather than by differential equations, are referred to as discrete-time systems. These systems can arise as a result of approximating a continuous system, perhaps for the purpose of simulation on a digital computer. In other cases discrete system equations arise because of the introduction of sampling (sampled-data systems) [39, 69, 103] or because certain parts of the equipment must be time-shared with several other systems. In still other cases, the system is inherently discrete in nature. Digital electronic circuits fall into this category. The purely resistive ladder network of Problem 1.6, page 17, is another example of a system which is described by difference equations. In that example the discrete index refers to a particular circuit loop rather than a discrete time point. In this book it will be assumed that the discrete events of interest can be indexed by a sequence of real numbers $t_0, t_1, t_2, \ldots, t_k, \ldots$ which will be thought of as discrete time points. For convenience, a general index t_k will often be referred to simply as k.

The notion of state, as introduced in Chapter 9, is sufficiently general to include discrete-time systems, and the state equations for these systems take the form given in equations (*9.9*) and (*9.10*), page 230. In this chapter the linear versions of these equations, given in equations (*9.13*) and (*9.14*), page 231, are considered. Methods of expressing discrete-time systems in state variable form are first presented. Then solution techniques for these equations are considered, in a manner which closely parallels the treatment of continuous systems in Chapter 10.

11.2 STATE SPACE REPRESENTATION OF DISCRETE-TIME SYSTEMS

The most appropriate method of selecting state variables for a discrete-time system depends upon the form in which the given information is available. First consider the case where the discrete system model is to be derived as an approximation to a continuous system model for which state equations are available. In this situation the discrete-time state variables can be chosen as the same state variables which represent the continuous system, but with $\mathbf{x}(t)$ restricted to discrete times t_k. Obtaining the discrete state equations is then a matter of solving, at least approximately, for $\mathbf{x}(k+1)$ in terms of $\mathbf{x}(k)$ and a discrete approximation of the input $\mathbf{u}(k)$. Problems 10.10 through 10.14 illustrate this approach.

If the continuous-time system is described in transfer function form, several options are available. Continuous-time state equations could be selected by any of the methods in Chapter 9. Then discrete-time state equations could be selected, as discussed in the previous paragraph. Alternatively, a discrete system transfer function (Z-transforms) could be derived from the s-domain transfer function. There are several ways of doing this s-domain to z-domain transformation (see Problems 2.19 and 2.20). Then discrete-time state equations can be derived from the z-domain transfer functions, in ways that are exact analogs of those developed in Sec. 9.4. The one significant difference in using the simulation diagram approach for discrete systems is that an ideal delay element (see Table 9.1) will be the only element with memory. Delays replace integrators. The close analogy between these two elements is illustrated in Fig. 11.1, both in the time domain and transfer function representations. As might be expected, the discrete state variables will be selected as the outputs of the delay elements in the system simulation diagram.

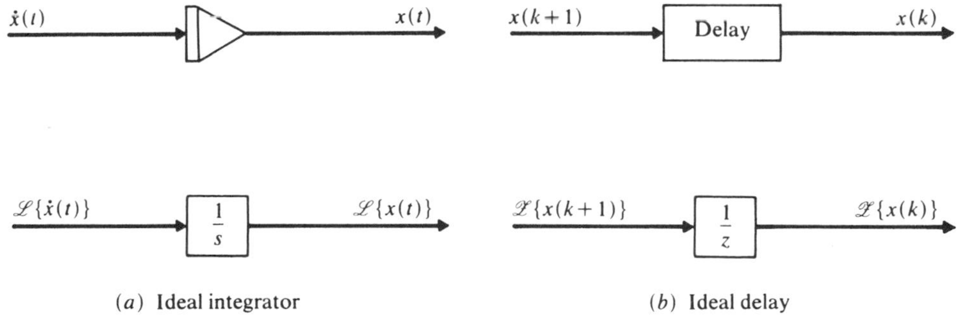

Figure 11.1

When the original system description is in terms of one or more coupled difference equations, simulation diagrams are again very useful in selecting states and determining the state equations.

Example 11.1

A system is described by the input-output equation

$$y(k+3) + 2y(k+2) + 4y(k+1) + y(k) = u(k)$$

This case is analogous to the simplest continuous-time problem where the input is not differentiated. Consequently, state variables can be selected as the output $y(k)$ and the output advanced by one and by two time steps. That is,

$$x_1(k) = y(k), \qquad x_2(k) = y(k+1), \qquad x_3(k) = y(k+2)$$

Then

$$x_1(k+1) = x_2(k), \qquad x_2(k+1) = x_3(k)$$

The final component of the state vector equation comes from the original difference equation and is

$$x_3(k+1) = -2x_3(k) - 4x_2(k) - x_1(k) + u(k)$$

Although its use is unnecessary in this simple problem, a possible simulation diagram is shown in Fig. 11.2.

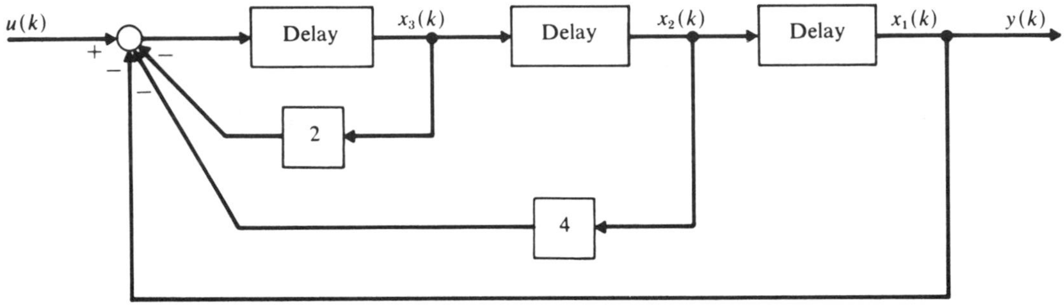

Figure 11.2

The state equations for this example are

$$\begin{bmatrix} x_1(k+1) \\ x_2(k+1) \\ x_3(k+1) \end{bmatrix} = \begin{bmatrix} 0 & 1 & 0 \\ 0 & 0 & 1 \\ -1 & -4 & -2 \end{bmatrix} \begin{bmatrix} x_1(k) \\ x_2(k) \\ x_3(k) \end{bmatrix} + \begin{bmatrix} 0 \\ 0 \\ 1 \end{bmatrix} u(k)$$

$$y(k) = [1 \quad 0 \quad 0]\mathbf{x}(k) + 0u(k)$$

■

When the input-output equation involves not only $u(k)$ but advanced terms such as $u(k+1), u(k+2), \ldots$, then a technique similar to the one presented for continuous-time systems with input derivatives can be used.

Example 11.2

(*Compare with Example 9.5.*)

A system is described by the following equation:

$$y(k+n) + a_{n-1}y(k+n-1) + \cdots + a_1 y(k+1) + a_0 y(k) = \beta_0 u(k) + \beta_1 u(k+1) + \cdots + \beta_m u(k+m)$$

First solve for the most advanced output term (or terms if coupled equations are involved):

$$y(k+n) = -a_{n-1}y(k+n-1) - \cdots - a_1 y(k+1) - a_0 y(k) + \beta_0 u(k) + \beta_1 u(k+1) + \cdots + \beta_m u(k+m)$$

Then delay every term in the equation n times so that $y(k)$ is obtained on the left-hand side. The symbol $\mathfrak{D}$ will be used to represent the delay operation. As in Example 9.5, it is assumed that $m = n$ for convenience. If $m < n$, then some of the coefficients β_i can be set to zero. It is impossible that $m > n$ for physically realizable systems:

$$y(k) = -a_{n-1}\mathfrak{D}(y(k)) - \cdots - a_1\mathfrak{D}^{n-1}(y(k)) - a_0\mathfrak{D}^n(y(k)) + \beta_0\mathfrak{D}^n(u(k)) + \beta_1\mathfrak{D}^{n-1}(u(k)) + \cdots + \beta_n u(k)$$

Rearrange this expression as a nested sequence of delayed terms:

$$y(k) = \beta_n u(k) + \mathfrak{D}\{-a_{n-1}y(k) + \beta_{n-1}u(k) + \mathfrak{D}[-a_{n-2}y(k) + \beta_{n-2}u(k) + \mathfrak{D}(\cdots + \mathfrak{D}\{-a_0 y(k) + \beta_0 u(k)\})]\}$$

The simulation diagram of Fig. 11.3 can now be drawn, noting that everything which is operated upon by the delay operator $\mathfrak{D}$ forms the input to that delay.

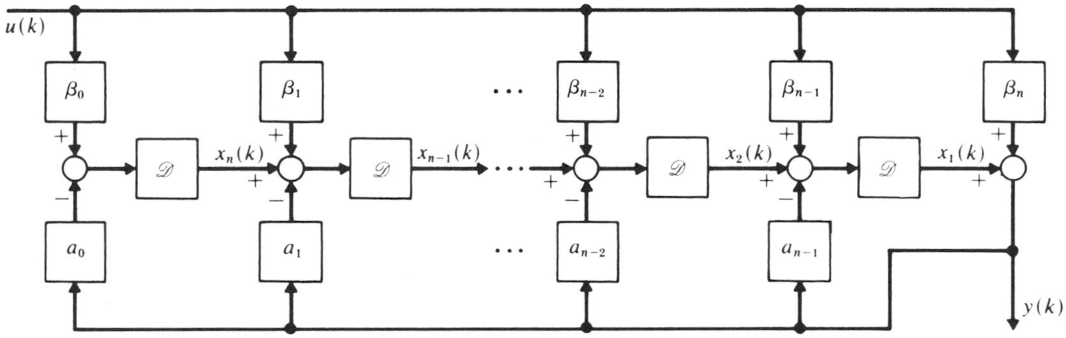

Figure 11.3

This diagram is exactly like the one of Fig. 9.6 except that the integrators have been replaced by delay elements. The state equations can be written down from the simulation diagram by using the fact that if the output of a delay is $x_i(k)$, then its input signal must be $x_i(k+1)$:

$$\begin{aligned}
x_1(k+1) &= -a_{n-1}[x_1(k) + \beta_n u(k)] + x_2(k) + \beta_{n-1}u(k) \\
&= -a_{n-1}x_1(k) + x_2(k) + [\beta_{n-1} - a_{n-1}\beta_n]u(k) \\
x_2(k+1) &= -a_{n-2}x_1(k) + x_3(k) + [\beta_{n-2} - a_{n-2}\beta_n]u(k) \\
&\vdots \\
x_{n-1}(k+1) &= -a_1 x_1(k) + x_n(k) + [\beta_1 - a_1\beta_n]u(k) \\
x_n(k+1) &= -a_0 x_1(k) + [\beta_0 - a_0\beta_n]u(k)
\end{aligned}$$

The output equation is

$$y(k) = x_1(k) + \beta_n u(k) = [1 \quad 0 \quad 0 \quad \cdots \quad 0]\mathbf{x}(k) + \beta_n u(k)$$

■

The preceding method is immediately applicable to sets of coupled linear constant coefficient difference equations. The general idea of using simulation diagrams to select state variables also applies to nonlinear or time-varying difference equations. In this case the step-by-step method of drawing the simulation diagram will not apply. However, once a simulation diagram is obtained, regardless of how it is obtained, the state variables can be selected as the outputs of the delay elements. This will result in a set of state equations of the form [28, 117]

$$\mathbf{x}(k+1) = \mathbf{f}(\mathbf{x}(k), \mathbf{u}(k), k) \tag{11.1}$$

$$\mathbf{y}(k) = \mathbf{h}(\mathbf{x}(k), \mathbf{u}(k), k) \tag{11.2}$$

After developing methods of obtaining state equations from transfer functions, the rest of this chapter is devoted to the analysis of the linear form of these equations,

$$\mathbf{x}(k+1) = \mathbf{A}(k)\mathbf{x}(k) + \mathbf{B}(k)\mathbf{u}(k) \tag{11.3}$$

$$\mathbf{y}(k) = \mathbf{C}(k)\mathbf{x}(k) + \mathbf{D}(k)\mathbf{u}(k) \tag{11.4}$$

Obtaining State Equations from Z-Transfer Functions

Consider a single-input, single-output discrete-time system described by the transfer function

$$\frac{y(z)}{u(z)} = T(z) = \frac{\beta_m z^m + \beta_{m-1} z^{m-1} + \cdots + \beta_i z + \beta_0}{z^n + a_{n-1} z^{n-1} + \cdots + a_1 z + a_0}$$

For any physical system, causality (physical realizability) requires that $m < n$. Otherwise, the ouptut $y(k)$ at time t_k would depend upon future inputs $u(j)$ at times t_j with $j > k$. The transfer function can also be written in terms of negative powers of z:

$$T(z) = \frac{z^{m-n}[\beta_m + \beta_{m-1} z^{-1} + \cdots + \beta_1 z^{-(m-1)} + \beta_0 z^{-m}]}{1 + a_{n-1} z^{-1} + a_{n-2} z^{-2} + \cdots + a_1 z^{-(n-1)} + a_0 z^{-n}}$$

Since z^{-1} provides a delay of one sample period, this might be called the delay operator form. It is clear that there is a delay of $n - m$ sample periods from input to output.

Several forms of the state equations can be derived, depending on whether the transfer function is used directly in the given polynomial form, or if it is first factored, or if it is factored and then expanded using partial fractions. Four commonly used variations will be demonstrated by way of examples: two direct realizations, called the observable canonical form and the controllable canonical form, and a cascade realization and a parallel realization. Within each of these categories there remains a certain amount of freedom of choice, so there is a very large number of possibilities. Actually, the possibilities are infinite because if $\mathbf{x}$ is any valid state vector, then so is $\mathbf{Sx}$ for any nonsingular transformation matrix $\mathbf{S}$.

Example 11.3 Direct Realization, Observable Canonical Form

The delay operator form of the transfer function converts immediately to the difference equation considered in Example 11.2. The result of the previous example is a set of observable canonical form state equations

$$\mathbf{x}(k+1) = \begin{bmatrix} -a_{n-1} & 1 & 0 & 0 & \cdots & 0 & 0 \\ -a_{n-2} & 0 & 1 & 0 & \cdots & 0 & 0 \\ \cdot & & & & & & \cdot \\ \cdot & & & & & & \cdot \\ \cdot & & & & & & \cdot \\ -a_1 & 0 & 0 & 0 & \cdots & 0 & 1 \\ -a_0 & 0 & 0 & 0 & \cdots & 0 & 0 \end{bmatrix} \mathbf{x}(k) + \begin{bmatrix} \beta_{n-1} - a_{n-1}\beta_n \\ \beta_{n-2} - a_{n-2}\beta_n \\ \cdot \\ \cdot \\ \cdot \\ \beta_1 - a_1\beta_n \\ \beta_0 - a_0\beta_n \end{bmatrix} u(k)$$

$$y(k) = [1 \quad 0 \quad 0 \quad 0 \quad \cdots \quad 0 \quad 0]\mathbf{x}(k) + \beta_n u(k)$$

If the system treated in Example 9.6, along with a zero-order hold, is Z-transformed using a sample period of 0.2, the resulting discrete transfer is

$$T(z) = \frac{0.013667z^2 + 0.00167z - 0.0050}{z^3 - 1.7085z^2 + 0.9425z - 0.1653}$$

A specific example of the observable canonical state equations is therefore

$$\mathbf{x}(k+1) = \begin{bmatrix} 1.7085 & 1 & 0 \\ -0.9425 & 0 & 1 \\ 0.1653 & 0 & 0 \end{bmatrix} \mathbf{x}(k) + \begin{bmatrix} 0.01361 \\ 0.00167 \\ -0.0050 \end{bmatrix} u(k)$$

$$y(k) = [1 \quad 0 \quad 0]\mathbf{x}(k)$$

■

Example 11.4 Direct Realization, Controllable Canonical Form

The previous transfer function is artificially split into two parts with the fictitious variable $g(k)$ in between, as shown in Fig. 11.4. The simulation diagram for determining $g(k)$

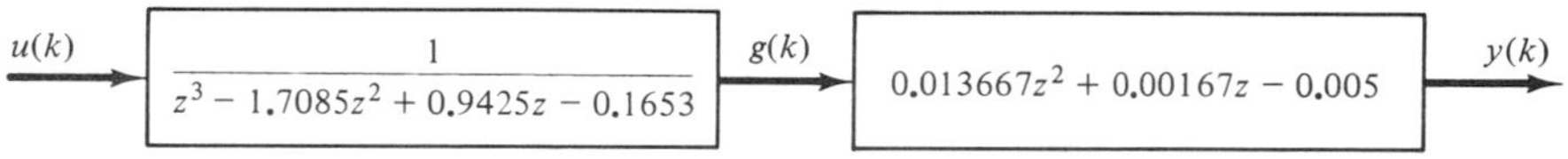

Figure 11.4

is first developed using

$$g(k+3) = 1.7085g(k+2) - 0.9425g(k+1) + 0.1653g(k) + u(k)$$

Then three successive delay elements give $g(k+2)$, $g(k+1)$, and $g(k)$. This constitutes part of the diagram in Fig. 11.5. The second transfer function in Fig. 11.4 states that

$$y(k) = -0.0050g(k) + 0.00167g(k+1) + 0.013667g(k+2)$$

This relationship constitutes the rest of Fig.11.5. Numbering the outputs of the delays as states in the order shown immediately gives a controllable canonical form of the state equations

$$\mathbf{x}(k+1) = \begin{bmatrix} 0 & 1 & 0 \\ 0 & 0 & 1 \\ 0.1653 & -0.9425 & 1.7085 \end{bmatrix} \mathbf{x}(k) + \begin{bmatrix} 0 \\ 0 \\ 1 \end{bmatrix} u(k)$$

$$y(k) = [-0.0050 \quad 0.00167 \quad 0.013667]\mathbf{x}(k)$$

■

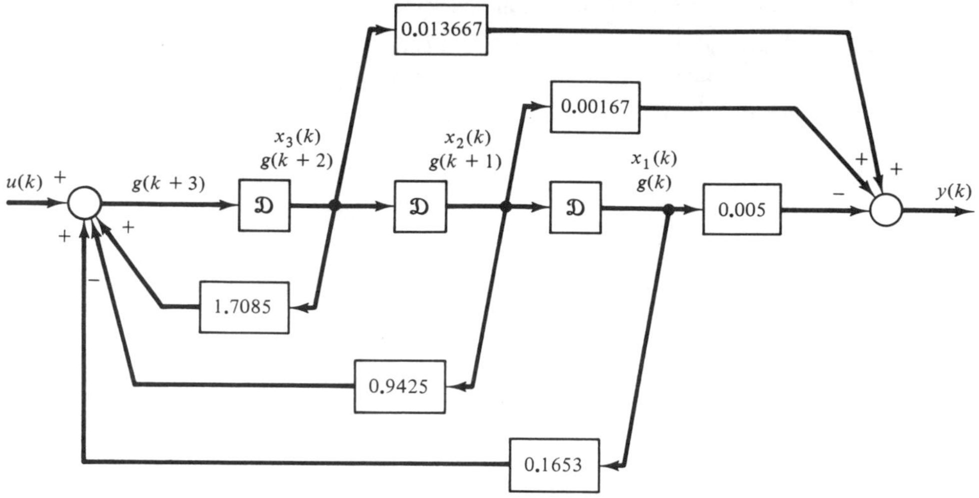

Figure 11.5

Example 11.5 Cascade Realization

The same transfer function is again considered, but this time in factored form (and grouped in the way terms will be cascaded together):

$$T(z) = \left[\frac{1}{z - 0.3679}\right]\left[\frac{z - 0.5488}{z - 0.6708}\right]\left[\frac{z + 0.0714}{z - 0.6703}\right][0.013667]$$

There are at least two valid simulation diagrams for factors like $\frac{z+a}{z-b}$, as shown in Fig. 11.6*a* and *b*. Using the second form, the total simulation diagram for a cascade realization

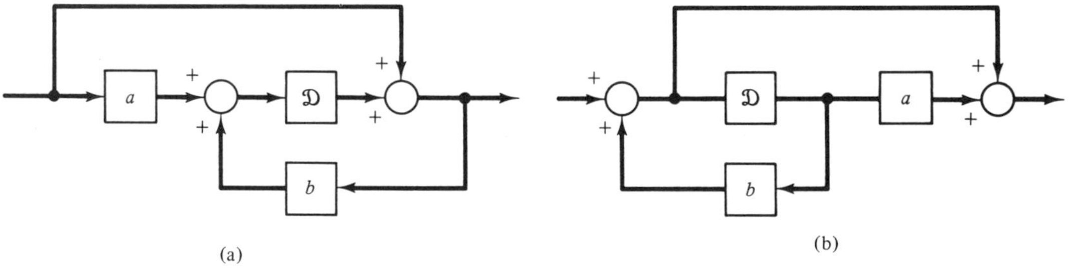

Figure 11.6*a*, *b*

is as shown in Fig. 11.7. Using state variables numbered as shown, the state equations are

$$\mathbf{x}(k+1) = \begin{bmatrix} 0.6703 & 0.1215 & 1 \\ 0 & 0.6703 & 1 \\ 0 & 0 & 0.3679 \end{bmatrix}\mathbf{x}(k) + \begin{bmatrix} 0 \\ 0 \\ 1 \end{bmatrix}u(k)$$

$$y(k) = 0.013667[1.3417 \quad 0.1215 \quad 1]\mathbf{x}(k)$$

$$= [0.01834 \quad 0.00166 \quad 0.013667]\mathbf{x}(k)$$

■

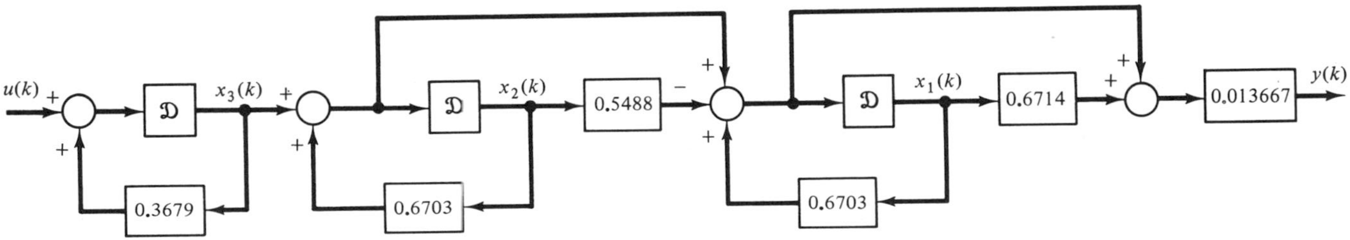

Figure 11.7

Example 11.6 Parallel Realization

A parallel realization of the same transfer function is now developed using partial fraction expansions. If $T(z)/z$ is expanded and that result is then multiplied by z, the result is

$$T(z) = 0.0305 - 0.7637\frac{z}{(z - 0.3679)} + 0.011\frac{z}{(z - 0.6703)^2} + 0.0459\frac{z}{(z - 0.6703)}$$

Although there is often good reason for going through the extra steps to put an expanded Z-transfer function into this form, there is no reason to do so in the present context. A direct expansion of $T(z)$ in the usual manner gives

$$T(z) = \frac{0.0074}{(z - 0.6703)^2} + \frac{0.0418}{(z - 0.6703)} - \frac{0.0281}{(z - 0.3675)}$$

A simulation diagram of this is shown in Fig. 11.8. The state equations are written directly from this as

$$\mathbf{x}(k + 1) = \begin{bmatrix} 0.6703 & 1 & 0 \\ 0 & 0.6703 & 0 \\ 0 & 0 & 0.3675 \end{bmatrix}\mathbf{x}(k) + \begin{bmatrix} 0 \\ 1 \\ 1 \end{bmatrix}u(k)$$

$$y(k) = [0.0074 \quad 0.0418 \quad -0.0281]\mathbf{x}(k)$$

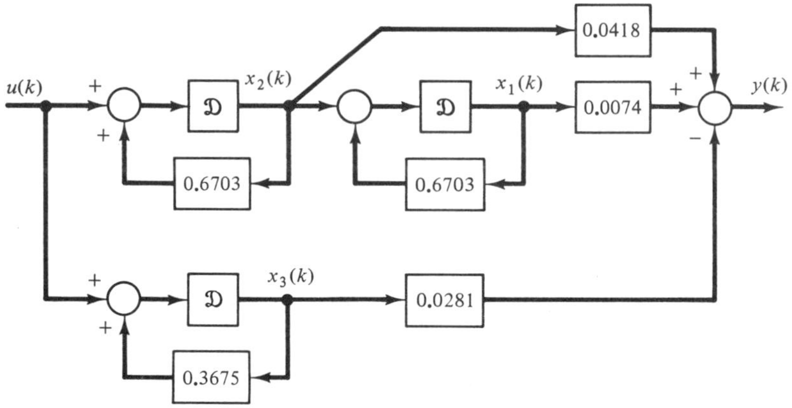

Figure 11.8

As in the continuous-time case, the partial fraction procedure leads to an **A** matrix in Jordan canonical form.

The alternative form of the partial fraction expansion equation leads to exactly the same final state equations. It would be a good exercise for the reader to verify this. The *apparent* direct path from u to y through the gain of 0.0305 will cancel out in the manipulations, and a zero D term does in fact result. There is no path from input to output with less than a one sample period delay. ■

11.3 CONSTANT COEFFICIENT DISCRETE-TIME STATE EQUATIONS

Assume that the system matrix $\mathbf{A}(k)$ is constant, so the index k can be omitted. The homogeneous case is first considered:

$$\mathbf{x}(k + 1) = \mathbf{A}\mathbf{x}(k)$$

The initial conditions $\mathbf{x}(0)$ are assumed known, so that $\mathbf{x}(1) = \mathbf{A}\mathbf{x}(0)$. Using this in the difference equation gives $\mathbf{x}(2) = \mathbf{A}\mathbf{x}(1) = \mathbf{A}^2\mathbf{x}(0)$. Continuing this process, the solution at a general time t_k is expressed in terms of $\mathbf{x}(0)$ as

$$\mathbf{x}(k) = \mathbf{A}^k\mathbf{x}(0) \tag{11.5}$$

The methods of Chapter 8 can be used to determine $\mathbf{A}^k$ as a general function of k, so that repeated matrix multiplications are unnecessary.

The nonhomogeneous case is now considered. A sequence of input vectors $\mathbf{u}(0), \mathbf{u}(1), \mathbf{u}(2), \ldots$ is given, as well as the initial conditions $\mathbf{x}(0)$. Then

$$\mathbf{x}(1) = \mathbf{A}\mathbf{x}(0) + \mathbf{B}(0)\mathbf{u}(0)$$
$$\mathbf{x}(2) = \mathbf{A}\mathbf{x}(1) + \mathbf{B}(1)\mathbf{u}(1) = \mathbf{A}^2\mathbf{x}(0) + \mathbf{A}\mathbf{B}(0)\mathbf{u}(0) + \mathbf{B}(1)\mathbf{u}(1)$$
$$\mathbf{x}(3) = \mathbf{A}\mathbf{x}(2) + \mathbf{B}(2)\mathbf{u}(2) = \mathbf{A}^3\mathbf{x}(0) + \mathbf{A}^2\mathbf{B}(0)\mathbf{u}(0) + \mathbf{A}\mathbf{B}(1)\mathbf{u}(1) + \mathbf{B}(2)\mathbf{u}(2)$$

At a general time t_k, this leads to

$$\mathbf{x}(k) = \mathbf{A}^k\mathbf{x}(0) + \sum_{j=0}^{k-1} \mathbf{A}^{k-1-j}\mathbf{B}(j)\mathbf{u}(j) \tag{11.6}$$

A change in the dummy summation index allows this result to be written in the alternative form

$$\mathbf{x}(k) = \mathbf{A}^k\mathbf{x}(0) + \sum_{j=1}^{k} \mathbf{A}^{k-j}\mathbf{B}(j-1)\mathbf{u}(j-1) \tag{11.7}$$

Either of these forms may be used. The close analogy with the continuous-time system results is made more apparent by using the definition for the discrete system transition matrix. Whenever $\mathbf{A}$ is constant, the discrete transition matrix is given by

$$\mathbf{\Phi}(k, j) = \mathbf{A}^{k-j}$$

Then

$$\mathbf{x}(k) = \mathbf{\Phi}(k, 0)\mathbf{x}(0) + \sum_{j=1}^{k} \mathbf{\Phi}(k, j)\mathbf{B}(j-1)\mathbf{u}(j-1) \tag{11.8}$$

and the only difference from the continuous result is the replacement of the convolution integration by a discrete summation.

11.4 MODAL DECOMPOSITION

The system matrix $\mathbf{A}$ is still considered constant, and its eigenvalues and eigenvectors (or generalized eigenvectors) are λ_i and ξ_i, respectively. Then, if the change of basis $\mathbf{x}(k) = \mathbf{M}\mathbf{q}(k)$ is used, where $\mathbf{M} = [\xi_1 \quad \xi_2 \quad \cdots \quad \xi_n]$, the state equations reduce to

$$\mathbf{M}\mathbf{q}(k+1) = \mathbf{A}\mathbf{M}\mathbf{q}(k) + \mathbf{B}(k)\mathbf{u}(k)$$

or

$$\mathbf{q}(k+1) = \mathbf{J}\mathbf{q}(k) + \mathbf{B}_n(k)\mathbf{u}(k) \tag{11.9}$$

and

$$\mathbf{y}(k) = \mathbf{C}(k)\mathbf{M}\mathbf{q}(k) + \mathbf{D}(k)\mathbf{u}(k)$$

or

$$\mathbf{y}(k) = \mathbf{C}_n(k)\mathbf{q}(k) + \mathbf{D}(k)\mathbf{u}(k) \tag{11.10}$$

where $\mathbf{J} = \mathbf{M}^{-1}\mathbf{A}\mathbf{M}$, $\mathbf{B}_n(k) = \mathbf{M}^{-1}\mathbf{B}(k)$, and $\mathbf{C}_n(k) = \mathbf{C}(k)\mathbf{M}$. Just as in the continuous case, the equations in $\mathbf{q}$ are as nearly uncoupled as possible and provide the same advantages. When $\mathbf{A}$ has a full set of eigenvectors, then $\mathbf{J}$ will be the diagonal matrix $\mathbf{\Lambda}$. A typical equation for the q_i components then takes the form

$$q_i(k+1) = \lambda_i q_i(k) + \langle \mathbf{r}_i, \mathbf{B}(k)\mathbf{u}(k)\rangle$$

The solution is

$$q_i(k) = \lambda_i^k q_i(0) + \sum_{j=1}^{k} \lambda_i^{k-j}\langle \mathbf{r}_i, \mathbf{B}(j-1)\mathbf{u}(j-1)\rangle$$

so that

$$\mathbf{q}(k) = \mathbf{\Lambda}^k\mathbf{q}(0) + \sum_{j=1}^{k} \mathbf{\Lambda}^{k-j}\mathbf{M}^{-1}\mathbf{B}(j-1)\mathbf{u}(j-1)$$

Using $\mathbf{x}(k) = \mathbf{M}\mathbf{q}(k)$ and $\mathbf{q}(0) = \mathbf{M}^{-1}\mathbf{x}(0)$ gives

$$\mathbf{x}(k) = \mathbf{M}\mathbf{\Lambda}^k\mathbf{M}^{-1}\mathbf{x}(0) + \sum_{j=1}^{k} \mathbf{M}\mathbf{\Lambda}^{k-j}\mathbf{M}^{-1}\mathbf{B}(j-1)\mathbf{u}(j-1)$$

This demonstrates again that $\mathbf{A}^k = \mathbf{M}\mathbf{\Lambda}^k\mathbf{M}^{-1}$ and provides a means of computing the transition matrix, provided $\mathbf{A}$ has a full set of eigenvectors.

The modal decomposition technique provides geometrical insight into the system's structure. The behavior of $\mathbf{x}(k)$ versus the time index k can be represented as the vector sum of the eigenvectors $\boldsymbol{\xi}_i$ multiplied by the easily evaluated time-variable coefficients $q_i(k)$. That is,

$$\mathbf{x}(k) = \mathbf{M}\mathbf{q}(k) = \boldsymbol{\xi}_1 q_1(k) + \boldsymbol{\xi}_2 q_2(k) + \cdots + \boldsymbol{\xi}_n q_n(k)$$

11.5 TIME-VARIABLE COEFFICIENTS

When $\mathbf{A}(k)$ is a time-variable matrix, then the solution technique of Sec. 11.3 must be modified slightly. Rather than the powers of $\mathbf{A}$, products of $\mathbf{A}$ evaluated at successive time points k are obtained. That is, the solution for $\mathbf{x}(k)$ at a general time t_k becomes

$$\mathbf{x}(k) = \mathbf{A}(k-1)\mathbf{A}(k-2)\cdots\mathbf{A}(0)\mathbf{x}(0) + \sum_{j=1}^{k}\left[\prod_{p=j}^{k-1}\mathbf{A}(p)\right]\mathbf{B}(j-1)\mathbf{u}(j-1) \tag{11.11}$$

In equation (*11.11*), the notation $\prod_{p=j}^{k-1}\mathbf{A}(p)$ indicates the product $\mathbf{A}(k-1)\mathbf{A}(k-2)\ldots\mathbf{A}(j+1)\mathbf{A}(j)$. It is understood that if $j = k-1$, the product is just $\mathbf{A}(k-1)$ and if $j = k$, then $\prod_{p=k}^{k-1}\mathbf{A}(p) \triangleq \mathbf{I}_n$. The transition matrix for the time-varying case is given by

$$\mathbf{\Phi}(k, j) = \prod_{p=j}^{k-1}\mathbf{A}(p) \tag{11.12}$$

When this definition is used, the solution for the time-variable case, equation (*11.11*), is exactly that given in equation (*11.8*). Evaluation of the transition matrix is much more cumbersome for the time-variable case, however.

11.6 THE DISCRETE-TIME TRANSITION MATRIX

The discrete-time transition matrix has been defined and used in the previous sections. The principal properties of this important matrix are summarized here. For the most part, the same properties hold for both the continuous-time and discrete-time transition matrices. In particular, the transition matrix $\mathbf{\Phi}(k, j)$ represents the mapping of the state at time t_j into the state at time t_k provided the input sequence $\mathbf{u}$ is zero in that interval. It completely describes the unforced behavior of the state vector.

The semigroup property applies, that is, $\mathbf{\Phi}(k, m)\mathbf{\Phi}(m, j) = \mathbf{\Phi}(k, j)$ for any k, m, j satisfying $j \le m \le k$. The identity property holds, that is, $\mathbf{\Phi}(k, k) = \mathbf{I}_n$ for any time index k.

One major difference for the discrete-time transition matrix is that its inverse *need not* exist. When the inverse does exist, then the reversed time property holds:

$$\mathbf{\Phi}^{-1}(k, j) = \mathbf{\Phi}(j, k)$$

The inverse will exist if the discrete system is correctly derived as an approximation to a continuous system, since then $\mathbf{A}(k) = \mathbf{\Phi}(t_{k+1}, t_k)$ and $\mathbf{\Phi}(k, j) = \mathbf{\Phi}(t_k, t_j)$ and the continuous system transition matrix is always nonsingular.

In Problem 6.35, page 172, the formal adjoint for the discrete-time system operator was shown to be

$$\mathbf{w}(k-1) = \mathbf{A}^T(k)\mathbf{w}(k)$$

Notice the backward time indexing. If the transition matrix for this adjoint system is defined as $\mathbf{\Theta}(k, j)$, then many (but not all) of the relationships existing between $\mathbf{\Phi}$ and $\mathbf{\Theta}$ in the continuous case will also be true in the discrete case. These properties are less useful in the discrete case and are not presented.

11.7 CLOSURE

The most general solution for the linear discrete-time state equation (*11.3*) is given by equation (*11.8*), repeated here:

$$\mathbf{x}(k) = \mathbf{\Phi}(k, 0)\mathbf{x}(0) + \sum_{j=1}^{k} \mathbf{\Phi}(k, j)\mathbf{B}(j-1)\mathbf{u}(j-1) \tag{11.8}$$

The output is given by

$$\mathbf{y}(k) = \mathbf{C}(k)\mathbf{\Phi}(k, 0)\mathbf{x}(0) + \sum_{j=1}^{k} \mathbf{C}(k)\mathbf{\Phi}(k, j)\mathbf{B}(j-1)\mathbf{u}(j-1) + \mathbf{D}(k)\mathbf{u}(k)$$

When the system matrix $\mathbf{A}$ is constant, then the transition matrix $\mathbf{\Phi}(k, j) = \mathbf{A}^{k-j}$ can be computed by any one of the several methods presented in Chapter 8. When $\mathbf{A}(k)$ is time-varying, no simple method exists for evalutaing $\mathbf{\Phi}$ other than the direct calculation of the products indicated in equation (*11.12*). Hopefully, a digital computer would be available for this task.

ILLUSTRATIVE PROBLEMS

Obtaining Discrete-Time State Equations

11.1 A system has two inputs $u_1(k)$ and $u_2(k)$ and three outputs $y_1(k)$, $y_2(k)$, and $y_3(k)$. The input-output difference equations are

$$y_1(k+3) + 6[y_1(k+2) - y_3(k+2)] + 2y_1(k+1) + y_2(k+1) + y_1(k) - 2y_3(k) = u_1(k) + u_2(k+1) \tag{1}$$

$$y_2(k+2) + 3y_2(k+1) - y_1(k+1) + 5y_2(k) + y_3(k) = u_1(k) + u_2(k) + u_3(k) \tag{2}$$

$$y_3(k+1) + 2y_3(k) - y_2(k) = u_3(k) - u_2(k) + 7u_3(k+1) \tag{3}$$

Draw a simulation diagram, select state variables, and write the matrix state equations.

Delaying each term in equation (*1*) three times gives

$$y_1(k) = \mathfrak{D}\{-6[y_1(k) - y_3(k)] + \mathfrak{D}[u_2(k) - 2y_1(k) - y_2(k) + \mathfrak{D}[u_1(k) - y_1(k) + 2y_3(k)]]\}$$

Delaying each term in equation (*2*) twice gives

$$y_2(k) = \mathfrak{D}\{-3y_2(k) + y_1(k) + \mathfrak{D}[u_1(k) + u_2(k) + u_3(k) - 5y_2(k) - y_3(k)]\}$$

and from equation (*3*), delayed once,

$$y_3(k) = 7u_3(k) + \mathfrak{D}\{u_3(k) - u_2(k) - 2y_3(k) + y_2(k)\}$$

The simulation diagram can be represented as shown in Fig. 11.9.

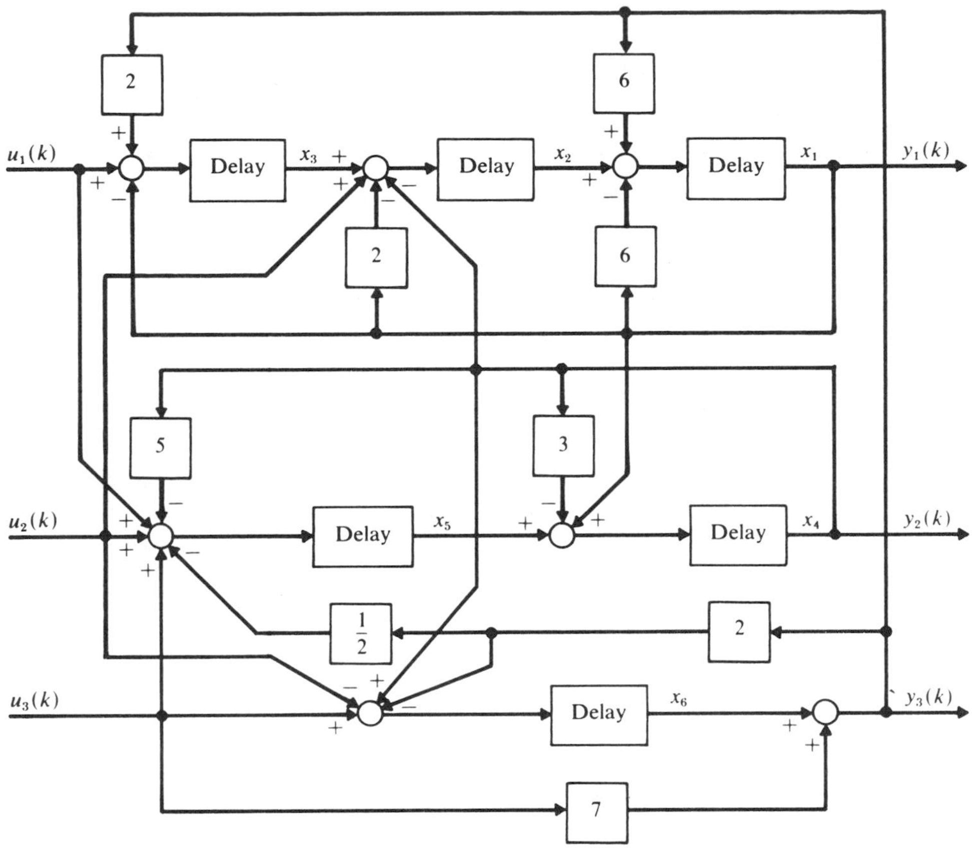

Figure 11.9

Labeling x_1 through x_6 as shown in Fig. 11.9 gives

$$\mathbf{x}(k+1) = \begin{bmatrix} -6 & 1 & 0 & 0 & 0 & 6 \\ -2 & 0 & 1 & -1 & 0 & 0 \\ -1 & 0 & 0 & 0 & 0 & 2 \\ 1 & 0 & 0 & -3 & 1 & 0 \\ 0 & 0 & 0 & -5 & 0 & -1 \\ 0 & 0 & 0 & 1 & 0 & -2 \end{bmatrix} \mathbf{x}(k) + \begin{bmatrix} 0 & 0 & 42 \\ 0 & 1 & 0 \\ 1 & 0 & 14 \\ 0 & 0 & 0 \\ 1 & 1 & -6 \\ 0 & -1 & -13 \end{bmatrix} \begin{bmatrix} u_1(k) \\ u_2(k) \\ u_3(k) \end{bmatrix}$$

and

$$\mathbf{y}(k) = \begin{bmatrix} 1 & 0 & 0 & 0 & 0 & 0 \\ 0 & 0 & 0 & 1 & 0 & 0 \\ 0 & 0 & 0 & 0 & 0 & 1 \end{bmatrix} \mathbf{x}(k) + \begin{bmatrix} 0 & 0 & 0 \\ 0 & 0 & 0 \\ 0 & 0 & 7 \end{bmatrix} \begin{bmatrix} u_1(k) \\ u_2(k) \\ u_3(k) \end{bmatrix}$$

11.2 Draw a simulation diagram and express the following nonlinear, time-varying system in state space form:

$$y(k+3) + y(k+2)y(k+1) + \alpha \sin(\omega k)y^2(k) + y(k) = u(k) - 3u(k+1)$$

Rewriting the equation as

$$y(k+3) + 3u(k+1) = u(k) - y(k+2)y(k+1) - \alpha \sin(\omega k) y^2(k) - y(k)$$

allows the simulation diagram to be drawn as shown in Fig. 11.10.

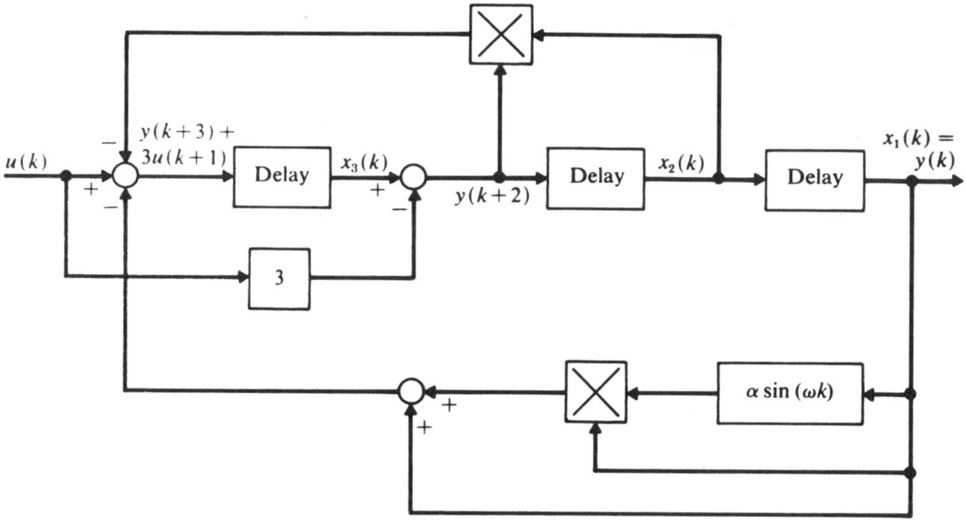

Figure 11.10

Labeling x_1 through x_3 as shown, the state equations are

$$x_1(k+1) = x_2(k)$$

$$x_2(k+1) = x_3(k) - 3u(k)$$

$$x_3(k+1) = -x_2(k)x_3(k) + 3x_2(k)u(k) - \alpha \sin(\omega k) x_1^2(k) - x_1(k) + u(k)$$

and the output equation is $y(k) = [1 \quad 0 \quad 0]\mathbf{x}(k)$.

Solution of Linear Discrete-Time State Equations

11.3 Solve for $\mathbf{x}(k)$ if

$$x_1(k+1) = \tfrac{1}{2}x_1(k) - \tfrac{1}{2}x_2(k) + x_3(k) \qquad x_3(k+1) = \tfrac{1}{2}x_3(k)$$

$$x_2(k+1) = \tfrac{1}{2}x_2(k) + 2x_3(k) \qquad x(0) = [2 \quad 4 \quad 6]^T$$

When put in matrix form $\mathbf{x}(k+1) = \mathbf{A}\mathbf{x}(k)$, the system matrix is $\mathbf{A} = \begin{bmatrix} \frac{1}{2} & -\frac{1}{2} & 1 \\ 0 & \frac{1}{2} & 2 \\ 0 & 0 & \frac{1}{2} \end{bmatrix}$. The solution for this homogeneous system is $\mathbf{x}(k) = \mathbf{A}^k\mathbf{x}(0)$. The matrix $\mathbf{A}^k$ was found in Problem 8.30, page 224. Using that result,

$$x_1(k) = 2(\tfrac{1}{2})^k - 4k(\tfrac{1}{2})^k + 6k(2-k)(\tfrac{1}{2})^{k-1}$$

$$x_2(k) = 4(\tfrac{1}{2})^k + 6k(\tfrac{1}{2})^{k-2}$$

$$x_3(k) = 6(\tfrac{1}{2})^k$$

11.4 Write the solution for the homogeneous discrete-time system

$$\begin{bmatrix} x_1(k+1) \\ x_2(k+1) \end{bmatrix} = \frac{1}{12}\begin{bmatrix} 5 & 1 \\ 1 & 5 \end{bmatrix}\begin{bmatrix} x_1(k) \\ x_2(k) \end{bmatrix}, \qquad \mathbf{x}(0) = \begin{bmatrix} 2 \\ 1 \end{bmatrix}$$

in the modal expansion form $\mathbf{x}(k) = \sum_{i=1}^{2} \langle \mathbf{r}_i, \mathbf{x}(0)\rangle \lambda_i^k \xi_i$.

The matrix $\mathbf{A} = \frac{1}{12}\begin{bmatrix} 5 & 1 \\ 1 & 5 \end{bmatrix}$ has as its eigenvalues $\lambda_1 = 1/2$, $\lambda_2 = 1/3$. The eigenvectors are $\xi_1 = [1 \quad 1]^T$ and $\xi_2 = [-1 \quad 1]^T$. Thus $\mathbf{M} = \begin{bmatrix} 1 & -1 \\ 1 & 1 \end{bmatrix}$ and $\mathbf{M}^{-1} = \frac{1}{2}\begin{bmatrix} 1 & 1 \\ -1 & 1 \end{bmatrix}$.

The rows of $\mathbf{M}^{-1}$ give the reciprocal basis vectors $\mathbf{r}_1 = [1/2 \quad 1/2]^T$, $\mathbf{r}_2 = [-1/2 \quad 1/2]^T$. Since $\langle \mathbf{r}_1, \mathbf{x}(0)\rangle = 3/2$, $\langle \mathbf{r}_2, \mathbf{x}(0)\rangle = -1/2$, the solution is

$$\mathbf{x}(k) = \frac{3}{2}\left(\frac{1}{2}\right)^k \begin{bmatrix} 1 \\ 1 \end{bmatrix} - \frac{1}{2}\left(\frac{1}{3}\right)^k \begin{bmatrix} -1 \\ 1 \end{bmatrix}$$

11.5 Consider the discrete-time system

$$\begin{bmatrix} x_1(k+1) \\ x_2(k+1) \end{bmatrix} = \begin{bmatrix} 1/2 & 1/8 \\ 1/8 & 1/2 \end{bmatrix}\begin{bmatrix} x_1(k) \\ x_2(k) \end{bmatrix} + \begin{bmatrix} 1 & 0 \\ 0 & 1 \end{bmatrix}\begin{bmatrix} u_1(k) \\ u_2(k) \end{bmatrix}$$

$$y(k) = x_1(k) + 2x_2(k)$$

Find $y(k)$ if $x_1(0) = -1$, $x_2(0) = 3$. The input $u_1(k)$ is obtained by sampling the ramp function t at times $t_0 = 0, t_1 = 1, \ldots, t_k = k$, and $u_2(k)$ is obtained by sampling e^{-t} at the same set of discrete times.

The transition matrix is first found:

$$\mathbf{A}^k = \alpha_0 \mathbf{I} + \alpha_1 \mathbf{A} = \begin{bmatrix} \alpha_0 + \frac{1}{2}\alpha_1 & \frac{1}{8}\alpha_1 \\ \frac{1}{8}\alpha_1 & \alpha_0 + \frac{1}{2}\alpha_1 \end{bmatrix}$$

The eigenvalues are required. $|\mathbf{A} - \mathbf{I}\lambda| = \lambda^2 - \lambda + 15/64$ so $\lambda_1 = 3/8$, $\lambda_2 = 5/8$. Solving for α_0 and α_1,

$$\left.\begin{matrix} (\frac{3}{8})^k = \alpha_0 + (\frac{3}{8})\alpha_1 \\ (\frac{5}{8})^k = \alpha_0 + (\frac{5}{8})\alpha_1 \end{matrix}\right\} \Longrightarrow \left\{\begin{matrix} 4[(\frac{5}{8})^k - (\frac{3}{8})^k] = \alpha_1 \\ (\frac{5}{2})(\frac{3}{8})^k - (\frac{3}{2})(\frac{5}{8})^k = \alpha_0 \end{matrix}\right.$$

Thus

$$\mathbf{A}^k = \mathbf{\Phi}(k, 0) = \begin{bmatrix} \frac{1}{2}[(\frac{5}{8})^k + (\frac{3}{8})^k] & \frac{1}{2}[(\frac{5}{8})^k - (\frac{3}{8})^k] \\ \frac{1}{2}[(\frac{5}{8})^k - (\frac{3}{8})^k] & \frac{1}{2}[(\frac{5}{8})^k + (\frac{3}{8})^k] \end{bmatrix}$$

Using this and the fact that $\mathbf{\Phi}(k, j) = \mathbf{A}^{k-j}$ gives

$$\mathbf{x}(k) = \begin{bmatrix} (\frac{5}{8})^k - 2(\frac{3}{8})^k \\ (\frac{5}{8})^k + 2(\frac{3}{8})^k \end{bmatrix} + \sum_{j=1}^{k} \begin{bmatrix} \frac{1}{2}[(\frac{5}{8})^{k-j} + (\frac{3}{8})^{k-j}] & \frac{1}{2}[(\frac{5}{8})^{k-j} - (\frac{3}{8})^{k-j}] \\ \frac{1}{2}[(\frac{5}{8})^{k-j} - (\frac{3}{8})^{k-j}] & \frac{1}{2}[(\frac{5}{8})^{k-j} + (\frac{3}{8})^{k-j}] \end{bmatrix}\begin{bmatrix} j-1 \\ e^{1-j} \end{bmatrix}$$

The output is $y(k) = [1 \quad 2]\mathbf{x}(k)$.

11.6 Express the following discrete-time state equations in normal form using the modal matrix in a change of basis:

$$\begin{bmatrix} x_1(k+1) \\ x_2(k+1) \\ x_3(k+1) \end{bmatrix} = \begin{bmatrix} \frac{1}{2} & \frac{1}{2} & 0 \\ 0 & 1 & 0 \\ \frac{5}{6} & -\frac{13}{6} & -\frac{1}{3} \end{bmatrix} \begin{bmatrix} x_1(k) \\ x_2(k) \\ x_3(k) \end{bmatrix} + \begin{bmatrix} 3 & 1 \\ 2 & 0 \\ -1 & 1 \end{bmatrix} \begin{bmatrix} u_1(k) \\ u_2(k) \end{bmatrix}$$

$$\begin{bmatrix} y_1(k) \\ y_2(k) \end{bmatrix} = \begin{bmatrix} -1 & 3 & 1 \\ 0 & 1 & 1 \end{bmatrix} \begin{bmatrix} x_1(k) \\ x_2(k) \\ x_3(k) \end{bmatrix}$$

When the equations are expressed as in equations (*11.9*) and (*11.10*), they are said to be in normal form. To put the equations in this form, the eigenvalues and eigenvectors must be determined:

$$|\mathbf{A} - \mathbf{I}\lambda| = (1-\lambda)(\tfrac{1}{2}-\lambda)(-\tfrac{1}{3}-\lambda)$$

The eigenvectors are $\xi_1 = [1 \quad 1 \quad -1]^T$, $\xi_2 = [1 \quad 0 \quad 1]^T$, $\xi_3 = [0 \quad 0 \quad 1]^T$, so that

$$\mathbf{M} = \begin{bmatrix} 1 & 1 & 0 \\ 1 & 0 & 0 \\ -1 & 1 & 1 \end{bmatrix} \quad \text{and} \quad \mathbf{M}^{-1} = \begin{bmatrix} 0 & 1 & 0 \\ 1 & -1 & 0 \\ -1 & 2 & 1 \end{bmatrix}$$

$$\mathbf{\Lambda} = \mathbf{M}^{-1}\mathbf{A}\mathbf{M} = \begin{bmatrix} 1 & 0 & 0 \\ 0 & \frac{1}{2} & 0 \\ 0 & 0 & -\frac{1}{3} \end{bmatrix}, \qquad \mathbf{B}_n = \mathbf{M}^{-1}\mathbf{B} = \begin{bmatrix} 2 & 0 \\ 1 & 1 \\ 0 & 0 \end{bmatrix},$$

$$\mathbf{C}_n = \mathbf{C}\mathbf{M} = \begin{bmatrix} 1 & 0 & 1 \\ 0 & 1 & 1 \end{bmatrix}$$

Collecting and using these results in $\mathbf{q}(k+1) = \mathbf{\Lambda}\mathbf{q}(k) + \mathbf{B}_n\mathbf{u}(k)$ and $\mathbf{y}(k) = \mathbf{C}_n\mathbf{q}(k)$ gives the normal form equations.

11.7 Find an expression for $\mathbf{y}(k)$, valid for all time t_k, for the system of Problem 11.6. Use $\mathbf{x}(0) = [1 \quad 2 \quad 1]^T$, $\mathbf{u}(k) = [k \quad -1-k]$.

Using the normal form equations,

$$\mathbf{q}(0) = \mathbf{M}^{-1}\mathbf{x}(0) = [2 \quad -1 \quad 4]^T$$

$$q_1(k) = q_1(0) + \sum_{j=1}^{k} 2u_1(j-1) = 2 + k(k-1)$$

$$q_2(k) = (\tfrac{1}{2})^k q_2(0) + \sum_{j=1}^{k} (\tfrac{1}{2})^{k-j}[u_1(j-1) + u_2(j-1)] = -(\tfrac{1}{2})^k\left[1 + \sum_{j=1}^{k} 2^j\right]$$

$$q_3(k) = (-\tfrac{1}{3})^k q_3(0) = 4(-\tfrac{1}{3})^k$$

Using the output equation $\mathbf{y}(k) = \mathbf{C}_n\mathbf{q}(k)$ yields

$$y_1(k) = q_1(k) + q_3(k) = 2 + k(k-1) + 4(-\tfrac{1}{3})^k$$

$$y_2(k) = q_2(k) + q_3(k) = (-\tfrac{1}{2})^k\left[1 + \sum_{j=1}^{k} 2^j\right] + 4(-\tfrac{1}{3})^k$$

11.8 Consider a homogeneous discrete-time system described by $\mathbf{x}(k+1) = \mathbf{A}\mathbf{x}(k)$.

a. Show that if a nontrivial steady-state (constant) solution is to exist, the matrix $\mathbf{A}$ must have unity as an eigenvalue.

b. Construct a 2×2 nondiagonal, symmetric matrix with this property and find the steady-state solution.

a. A constant steady-state solution implies that for k sufficiently large, $\mathbf{x}(k+1) = \mathbf{x}(k)$. Call this solution $\mathbf{x}_e$. Then the difference equation requires that $\mathbf{x}_e = \mathbf{A}\mathbf{x}_e$. But this is just the eigenvalue equation $\mathbf{A}\mathbf{x}_e = \lambda\mathbf{x}_e$, with $\lambda = 1$.

b. Let $\mathbf{A} = \begin{bmatrix} a_{11} & a_{12} \\ a_{12} & a_{22} \end{bmatrix}$. Then

$$|\mathbf{A} - \mathbf{I}\lambda| = \lambda^2 - (a_{11} + a_{22})\lambda + (a_{11}a_{22} - a_{12}^2) = 0$$

The roots are

$$\lambda_{1,2} = \tfrac{1}{2}(a_{11} + a_{22}) \pm \sqrt{[\tfrac{1}{2}(a_{11} + a_{22})]^2 - (a_{11}a_{22} - a_{12}^2)}$$

There are many possible solutions. One is obtained by arbitrarily setting $a_{11} = a_{22} = 2$. Then $\lambda_{1,2} = 2 \pm \sqrt{4 - 4 + a_{12}^2}$. If a root is to be $\lambda = 1$, then $a_{12} = 1$. Using $\mathbf{A} = \begin{bmatrix} 2 & 1 \\ 1 & 2 \end{bmatrix}$, the steady-state solution $\mathbf{x}_e$ is just the eigenvector associated with the root $\lambda = 1$:

$$\text{Adj}\,[\mathbf{A} - \mathbf{I}\lambda]|_{\lambda=1} = \begin{bmatrix} 2-\lambda & -1 \\ -1 & 2-\lambda \end{bmatrix}\Bigg|_{\lambda=1} = \begin{bmatrix} 1 & -1 \\ -1 & 1 \end{bmatrix}$$

The steady-state solution will be proportional to $\mathbf{x}_e = \begin{bmatrix} 1 \\ -1 \end{bmatrix}$.

11.9 Assume that a system is described by equations (*11.3*) and (*11.4*), with **A**, **B**, **C**, and **D** constant. Use Z-transforms to find the transforms of $\mathbf{x}(k)$ and $\mathbf{y}(k)$. Then use the inverse Z-transform to find $\mathbf{x}(k)$. The initial condition $\mathbf{x}(0)$ is known and the input $\mathbf{u}(k)$ is zero for $k < 0$.

The Z-transform introduced in Chapters 1 and 2 applies to vector-matrix equations on an obvious component-by-component basis. If $Z\{\mathbf{x}(k)\} \triangleq \mathbf{X}(z)$, then $Z\{\mathbf{x}(k+1)\} = z\mathbf{X}(z) - z\mathbf{x}(0)$. Note the z multiplier on $\mathbf{x}(0)$, which deviates from the analogy expected from the Laplace transform of a derivative term. With this result and the linearity of the Z-transform operator, equation (*11.3*) gives

$$z\mathbf{X}(z) - z\mathbf{x}(0) = \mathbf{A}\mathbf{X}(z) + \mathbf{B}\mathbf{U}(z)$$

or

$$\mathbf{X}(z) = [z\mathbf{I} - \mathbf{A}]^{-1}\{\mathbf{B}\mathbf{U}(z) + z\mathbf{x}(0)\} \qquad (1)$$

Since the form of the initial condition response is known from earlier time-domain analysis, it is noted that

$$\mathbf{\Phi}(k, 0) \triangleq Z^{-1}\{[z\mathbf{I} - \mathbf{A}]^{-1}z\}$$

Using this definition allows the forcing function term to be written as

$$[z\mathbf{I} - \mathbf{A}]^{-1}\mathbf{B}\mathbf{U}(z) = [z\mathbf{I} - \mathbf{A}]^{-1}zz^{-1}\mathbf{B}\mathbf{U}(z) = Z\{\mathbf{\Phi}(k,0)\}Z\{\mathbf{B}\mathbf{u}(k-1)\}$$

Then the convolution theorem of Z-transforms gives the inverse as

$$Z^{-1}\{[z\mathbf{I} - \mathbf{A}]^{-1}\mathbf{B}\mathbf{U}(z)\} = \sum_{j=0}^{k} \mathbf{\Phi}(k, j)\mathbf{B}\mathbf{u}(j-1)$$

Since $\mathbf{u}(j-1) = \mathbf{0}$ for $j \le 0$, the lower summation limit is changed to $j = 1$. The solution for $\mathbf{x}(k)$ is then exactly equation (*11.8*). Note that the assumption that **B** was constant is unnecessary. From the transform of equation (*11.4*),

$$\mathbf{Y}(z) = \mathbf{C}\mathbf{X}(z) + \mathbf{D}\mathbf{U}(z) \qquad (2)$$

Combined with (*1*) this gives

$$\mathbf{Y}(z) = \{\mathbf{C}[z\mathbf{I} - \mathbf{A}]^{-1}\mathbf{B} + \mathbf{D}\}\mathbf{U}(z) + [z\mathbf{I} - \mathbf{A}]^{-1}z\mathbf{x}(0) \qquad (3)$$

Note that a useful formula for computing input-output transfer functions has been found, namely,

$$T(z) = \mathbf{C}[z\mathbf{I} - \mathbf{A}]^{-1}\mathbf{B} + \mathbf{D}$$

Finally, the sequence $\mathbf{y}(k)$ could be computed by inverse transforming (*3*) or by first finding $\mathbf{x}(k)$ as the inverse transform of (*1*) and then using that result in equation (*11.4*).

Recursive Weighted Least Squares with Discrete-Time Systems

11.10 A homogeneous linear discrete-time system is described by $\mathbf{x}(k+1) = \mathbf{A}(k)\mathbf{x}(k)$. Measured outputs are given by $\mathbf{y}(k) = \mathbf{C}(k)\mathbf{x}(k) + \mathbf{e}(k)$, where $\mathbf{e}(k)$ is an error vector due to imperfect measuring devices. The precise initial conditions for this system, $\mathbf{x}(0)$, are not known, although an estimate, $\hat{\mathbf{x}}(0)$, is available.* Use the recursive weighted least-squares technique of Chapter 5 and develop a scheme for improving on the estimate $\hat{\mathbf{x}}(0)$ each time a new measurement $\mathbf{y}(k)$ becomes available.

In order to place this problem in the framework developed in Sec. 5.8, page 134, all that is required is a few notational changes and the elimination of the difference equation. The solution to the difference equation is $\mathbf{x}(k) = \mathbf{\Phi}(k, 0)\mathbf{x}(0)$, so that the measurement equations can be written as

$$\mathbf{y}(k) = \mathbf{C}(k)\mathbf{\Phi}(k, 0)\mathbf{x}(0) + \mathbf{e}(k)$$

This is in the form used in Sec. 5.8, where $\mathbf{C}(k)\mathbf{\Phi}(k, 0)$ corresponds to $\mathbf{H}_k$ used in that earlier section. In Sec. 5.8 the index k referred to the kth estimate of a constant vector $\mathbf{x}$. To avoid confusion here, $\hat{\mathbf{x}}(0 \mid k)$ will be used to indicate the estimate of $\mathbf{x}(0)$ based on all measurements up to and including $\mathbf{y}(k)$. Assuming that the original estimate $\hat{\mathbf{x}}(0)$ is to be weighted by a nonsingular $n \times n$ matrix $\mathbf{P}_0^{-1}$ and that each succeeding measurement is weighted by the $m \times m$ nonsingular matrix $\mathbf{R}_k^{-1}$ the results of Sec. 5.8 give

$$\hat{\mathbf{x}}(0 \mid 1) = \hat{\mathbf{x}}(0) + \mathbf{K}_0[\mathbf{y}(1) - \mathbf{C}(1)\mathbf{\Phi}(1, 0)\hat{\mathbf{x}}(0)]$$

The gain matrix $\mathbf{K}_0$ is given by

$$\mathbf{K}_0 = \mathbf{P}_0\mathbf{\Phi}^T(1, 0)\mathbf{C}^T(1)[\mathbf{C}(1)\mathbf{\Phi}(1, 0)\mathbf{P}_0\mathbf{\Phi}^T(1, 0)\mathbf{C}^T(1) + \mathbf{R}_1]^{-1}$$

The estimate of $\mathbf{x}(0)$ may be further refined each time a new measurement $\mathbf{y}(k)$ is taken by using the recursive relations

$$\hat{\mathbf{x}}(0 \mid k+1) = \hat{\mathbf{x}}(0 \mid k) + \mathbf{K}_k[\mathbf{y}(k+1) - \mathbf{C}(k+1)\mathbf{\Phi}(k+1, 0)\hat{\mathbf{x}}(0 \mid k)]$$

where

$$K_k = \mathbf{P}_k\mathbf{\Phi}^T(k+1, 0)\mathbf{C}^T(k+1)[\mathbf{C}(k+1)\mathbf{\Phi}(k+1, 0) \times \mathbf{P}_k\mathbf{\Phi}^T(k+1, 0)\mathbf{C}^k(k+1) + \mathbf{R}_{k+1}]^{-1}$$

and where $\mathbf{P}_k$ is computed recursively using equation (*5.7*), page 135, which can be written as

$$\mathbf{P}_{k+1} = \mathbf{P}_k - \mathbf{K}_k\mathbf{C}(k+1)\mathbf{\Phi}(k+1, 0)\mathbf{P}_k$$

If the error vector $\mathbf{e}(k)$ and the weighting matrices $\mathbf{P}_k$ and $\mathbf{R}_k$ are given the proper statistical interpretation, the above technique constitutes a simple example of the fixed-point smoothing algorithm [80]. A block diagram of the procedure is given in Fig. 11.11.

*The circumflex ^ is used in this and the next problem to indicate an estimated quantity. It should not be confused with the notation for a unit vector in Chapter 4.

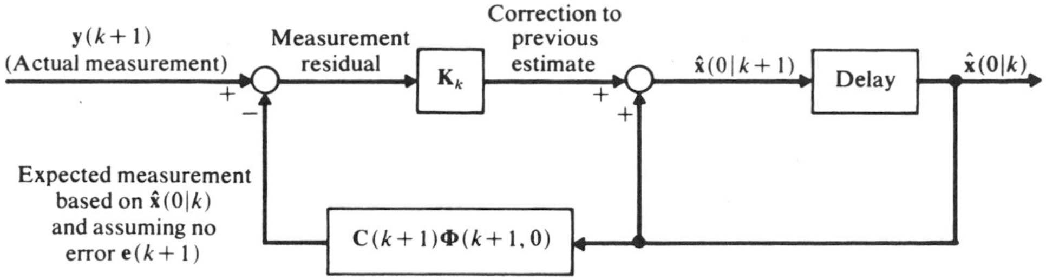

Figure 11.11

11.11 Consider the results of Problem 11.10. Instead of reestimating $\mathbf{x}(0)$ after each measurement, an estimate of the current state is now desired. Let $\hat{\mathbf{x}}(k \mid k)$ be the estimate of $\mathbf{x}(k)$ based on all measurements up to and including $\mathbf{y}(k)$. For this estimate, use the intuitively reasonable relation

$$\hat{\mathbf{x}}(k+1 \mid k+1) = \mathbf{\Phi}(k+1, 0)\hat{\mathbf{x}}(0 \mid k+1)$$

Modify the previous block diagram so that $\hat{\mathbf{x}}(k \mid k)$ is the output.

If the transition matrix $\mathbf{\Phi}(k+1, 0)$ is inserted into the diagram of Fig. 11.11 before the delay, then the output will be $\hat{\mathbf{x}}(k \mid k)$ as shown in Fig. 11.12. To maintain the correct relations in the rest of the diagram, the term $\hat{\mathbf{x}}(0 \mid k) = \mathbf{\Phi}(0, k)\hat{\mathbf{x}}(k \mid k)$ is needed, so $\mathbf{\Phi}(0, k)$ is inserted in the feedback path as shown in Fig. 11.12.

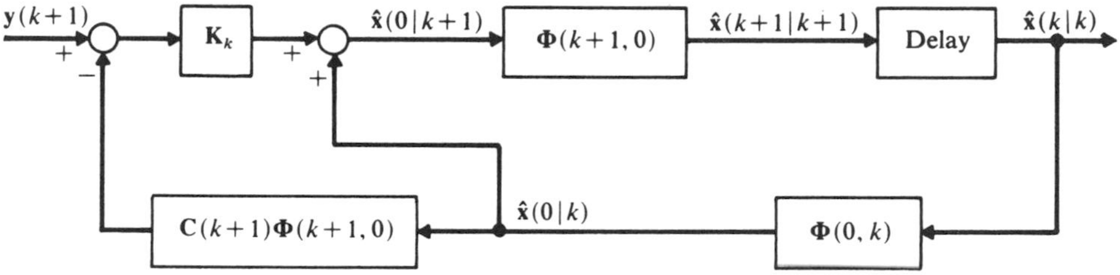

Figure 11.12

Using standard matrix block diagram manipulations, $\mathbf{\Phi}(k+1, 0)$ is moved past the summing junction, into both paths. A new gain matrix $\mathbf{K}'_{k+1} = \mathbf{\Phi}(k+1, 0)\mathbf{K}_k$ is defined. The other $\mathbf{\Phi}(k+1, 0)$ term is shifted into the feedback path and combined with $\mathbf{\Phi}(0, k)$ to give $\mathbf{\Phi}(k+1, k) = \mathbf{\Phi}(k+1, 0)\mathbf{\Phi}(0, k)$. This also removes the $\mathbf{\Phi}(k+1, 0)$ term multiplying $\mathbf{C}(k+1)$. This results in the most commonly used form of the discrete Kalman filter [80], shown in Fig. 11.13.

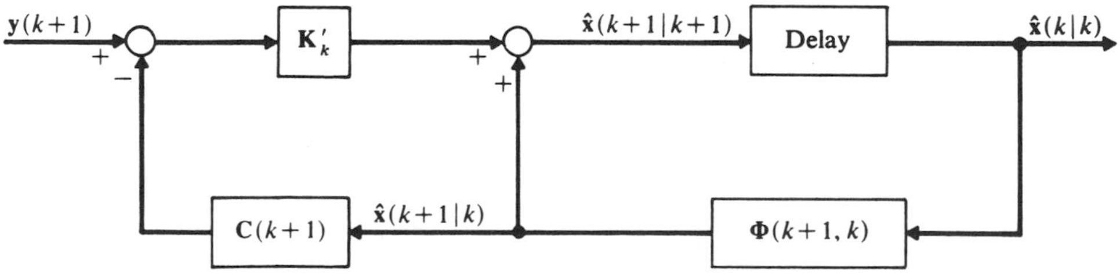

Figure 11.13

For easy reference the five equations which constitute the discrete Kalman filter are summarized below. An additive deweighting matrix $\mathbf{Q}_k$ (see Sec. 5.8) is included. To appreciate these results fully some knowledge of random processes is required [57]. Lacking this, the procedure can still be interpreted and used successfully as a recursive least-squares algorithm with ad hoc additive deweighting included.

To **find** the gain $\mathbf{K}'_k$, recursively compute

$$\mathbf{M}_k = \mathbf{\Phi}(k, k-1)\mathbf{P}_{k-1}\mathbf{\Phi}^T(k, k-1) + \mathbf{Q}_{k-1} \tag{1}$$

$$K'_k = \mathbf{M}_k\mathbf{C}^T(k)[\mathbf{C}(k)\mathbf{M}_k\mathbf{C}^T(k) + \mathbf{R}(k)]^{-1} \tag{2}$$

$$\mathbf{P}_k = [\mathbf{I} - \mathbf{K}'_k\mathbf{C}(k)]\mathbf{M}_k \tag{3}$$

To **use** the gain to estimate $\mathbf{x}$, recursively compute

$$\hat{\mathbf{x}}(k+1 \mid k) = \mathbf{\Phi}(k+1, k)\hat{\mathbf{x}}(k \mid k) \tag{4}$$

$$\hat{\mathbf{x}}(k+1 \mid k+1) = \hat{\mathbf{x}}(k+1 \mid k) + \mathbf{K}'_{k+1}[\mathbf{y}(k+1) - \mathbf{C}(k+1)\hat{\mathbf{x}}(k+1 \mid k)] \tag{5}$$

To initialize the procedure, there are two possibilities:

i. If $\hat{\mathbf{x}}(k+1 \mid k)$ and $\mathbf{M}_{k+1}$ are given, then start by using equation (*2*) to find $\mathbf{K}'_{k+1}$, then use (*5*), along with the measurement $\mathbf{y}(k+1)$, to find $\mathbf{x}(k+1 \mid k+1)$. To get ready for the next cycle, use (*3*), (*4*), and (*1*).
ii. If $\hat{\mathbf{x}}(k \mid k)$ and $\mathbf{P}_k$ are given, then start by using (*1*), then (*2*) to find $\mathbf{K}'_{k+1}$. Then use (*4*), followed by (*5*). To complete the first cycle and get ready for the next, (*3*) is then used.

Note that the above algorithm reduces to those at the end of Sec. 5.8 when $\mathbf{\Phi}(k+1, k) = \mathbf{I}$, that is, when $\mathbf{x}(k)$ is just a constant. Also be aware that the above algorithm can be written in several other forms which are *algebraically* equivalent, but which may have different numerical behavior on a finite word-length computer.

Approximation of a Continuous-Time System

11.12 A simple scalar system is described by $\dot{x} = -x + u$, $x(0) = 10$.

a. Solve for $x(t)$ if $u(t) = e^t$.
b. Derive a discrete approximation for the above system, using $t_{k+1} - t_k = 1$. Solve this discrete system and compare the results with the continuous solution.

a. By inspection, $\phi(t, \tau) = e^{-(t-\tau)}$, so

$$x(t) = 10e^{-t} + \int_0^t e^{-(t-\tau)}e^{\tau}\, d\tau = 10e^{-t} + \sinh t$$

b. Using the scalar transition matrix, the relation between $x(k+1)$ and $x(k)$ is

$$x(k+1) = e^{-1}x(k) + \int_{t_k}^{t_{k+1}} e^{-(t_{k+1}-\tau)}u(\tau)\, d\tau$$

Assuming $u(\tau)$ is constant over the interval $[t_k, t_{k+1}]$ and carrying out the integration gives

$$x(k+1) = e^{-1}x(k) + [1 - e^{-1}]u(k) \cong 0.368x(k) + 0.632u(k)$$

The solution for the discrete system is

$$x(k) = 10(0.368)^k + \sum_{j=1}^{k} (0.368)^{k-j}[0.632u(j-1)]$$

The initial condition term is the same for both the continuous and the discrete cases. A comparison of the forced response for the first five sampling periods is given in Table 11.1. The two separate approximations of Fig. 11.14 are used for $u(k)$.

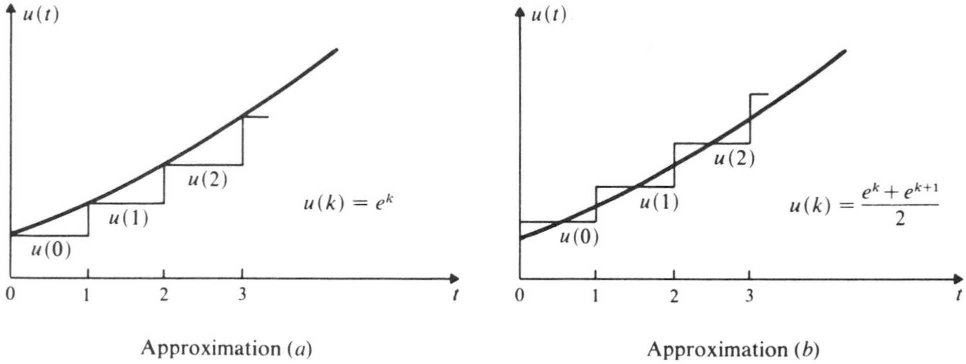

Figure 11.14

TABLE 11.1

		$t=k=0$	$t=k=1$	$t=k=2$	$t=k=3$	$t=k=4$
Continuous result: $\sinh t$		0	1.1752	3.6269	10.018	27.290
Discrete result: $\sum_{j=1}^{k} (0.368)^{k-j}[0.632u(j-1)]$	input (*a*)	0	0.632	1.951	5.388	14.677
	input (*b*)	0	1.175	3.626	10.016	27.286

The accuracy of the discrete approximation depends on how closely the piecewise constant discrete input matches the continuous input. The accuracy can be good or poor as demonstrated in the example, but improves as the discrete step-size is decreased.

11.13 Apply the method of Problem 10.11 to the observable canonical form of the continuous-time system in Example 9.6 to obtain discrete state equations. Recall that this same system is treated in Examples 11.3 through 11.6.

To facilitate comparisons, $T = 0.2$ is again used. Keeping fifth-order terms leads to

$$\mathbf{\Phi}(T, 0) = \begin{bmatrix} 0.0056 & 0.0793 & 0.1100 \\ -2.1224 & 0.7194 & 0.1779 \\ -1.5861 & -0.2191 & 0.9823 \end{bmatrix}, \qquad \mathbf{B}_1 = \begin{bmatrix} 0.0136 \\ 0.2347 \\ 0.5797 \end{bmatrix}$$

The output matrix is still $\mathbf{C} = [1 \quad 0 \quad 0]$, and $\mathbf{D} = 0$. C agrees with Example 11.3, meaning that both systems have at least the first state variable the same and equal to the output $\mathbf{y}$. However, the rest of the model terms are totally different. Still another valid state variable representation of this system has been found! If the above model, and any of those in Examples 11.3–11.6 as well, is used in the equation for the transfer function found in Problem 11.9, with $\mathbf{A} = \mathbf{\Phi}$ and $\mathbf{B}_1 = \mathbf{B}$, of course, it can be verified that the correct transfer function $T(z)$ is recovered, to within small round-off error.

PROBLEMS

11.14 A system has two inputs and two outputs. The input-output equations are

$$y_1(k+2) + 10y_1(k+1) - y_2(k+1) + 3y_1(k) + 2y_2(k) = u_1(k) + 2u_1(k+1)$$

$$y_2(k+1) + 4[y_2(k) - y_1(k)] = 2u_2(k) - u_1(k)$$

Select state variables and write the vector matrix state equations.

11.15 Draw two different simulation diagrams and obtain two different state variable representations for the system described by

$$y(k+2) + 3y(k+1) + 2y(k) = u(k)$$

11.16 Solve the following homogeneous difference equations:

$$x_1(k+1) = x_1(k) - x_2(k) + x_3(k)$$

$$x_2(k+1) = x_2(k) + x_3(k)$$

$$x_3(k+1) = x_3(k)$$

with $x_1(0) = 2$, $x_2(0) = 5$, $x_3(0) = 10$.

11.17 Find the time response of the discrete model developed in Problem 10.13 if $x_1(0) = 0$, $x_2 = 10$, $u_1(k) = 1/K$, and $u_2(k) = 1$.

11.18 A single-input, single-output system is described by

$$\mathbf{x}(k+1) = \begin{bmatrix} 1 & 0 \\ -1/2 & 1/2 \end{bmatrix}\mathbf{x}(k) + \begin{bmatrix} 1 \\ -1 \end{bmatrix}u(k) \quad \text{and} \quad y(k) = [5 \quad 1]\mathbf{x}(k)$$

Use a change of basis to determine the normal form equations.

11.19 If a system is described by $\mathbf{x}(k+1) = \begin{bmatrix} 3 & 2 & 3 \\ 2 & 1 & 1 \\ 1 & 1 & 2 \end{bmatrix}\mathbf{x}(k)$, is it true that $\mathbf{\Phi}(j, k) = \mathbf{\Phi}^{-1}(k, j)$?

11.20 A simplified model of a motor is given by the transfer function $\theta(s)/u(s) = K/[s(\tau s + 1)]$. Let $x_1 = \theta$, $x_2 = \dot{\theta}$ and develop the continuous state equations. Then determine the approximate discrete-time state equations, using time points separated by $t_{k+1} - t_k = \Delta t$.

11.21 The motor of Problem 11.20 is used in a sampled-data feedback system as shown in Fig. 11.15. The signal $u(k)$ is $e(t_k) = r(t_k) - \theta(k)$. Write the discrete state equations, using $r(t_k)$ as the input and $\theta(t_k)$ as the output.

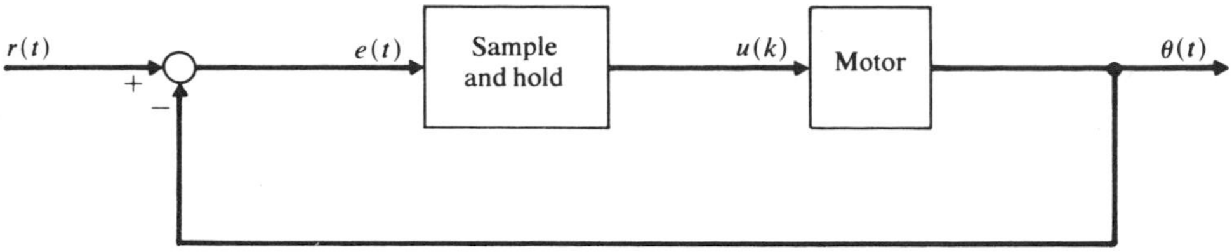

Figure 11.15

11.22 Apply the method of Problem 10.11 to the cascade realization of the continuous-time system in Example 9.6. Use $T = 0.2$ and keep terms through fifth order. After finding the approximate discrete models for **A** and **B**, verify that the transfer function of Example 11.3 is obtained from

$$T(z) = \mathbf{C}[z\mathbf{I} - \mathbf{A}]^{-1}\mathbf{B}$$

11.23 Find a discrete-time state variable model for a system with transfer function

$$T(z) = \frac{0.006745(z + 0.0672)(z + 1.2416)}{(z - 0.04979)(z - 0.22313)(z - 0.60653)}$$

Chapter
12

CONTROLLABILITY AND OBSERVABILITY FOR LINEAR SYSTEMS

12.1 INTRODUCTION

In this chapter the important properties of controllability and observability are defined. Three sets of criteria are presented for determining whether a linear system possesses these properties [43, 59, 117].

12.2 DEFINITIONS

It has been shown previously that the description of a linear system, either continuous-time or discrete-time, depends upon four matrices **A**, **B**, **C**, and **D**. Depending on the choice of state variables, or alternatively on the choice of the basis for the state space Σ, different matrices can be used to describe the same system. A particular set $\{\mathbf{A}, \mathbf{B}, \mathbf{C}, \mathbf{D}\}$ is called a *system representation* or *realization*. In some cases these matrices will be constant. In other cases they will depend on time, in either a continuous fashion $\{\mathbf{A}(t), \mathbf{B}(t), \mathbf{C}(t), \mathbf{D}(t)\}$ or in a discrete fashion $\{\mathbf{A}(k), \mathbf{B}(k), \mathbf{C}(k), \mathbf{D}(k)\}$. Both the continuous-time and discrete-time cases are considered simultaneously using the notation of Chapter 9. The times of interest will be referred to as the set of scalars $\mathfrak{J}$, where $\mathfrak{J}$ can be a continuous interval $[t_0, t_f]$, or a set of discrete points $[t_0, t_1, \ldots, t_N]$. At any particular time $t \in \mathfrak{J}$, the four system matrices are representations of transformations on the n-dimensional state space Σ, the r-dimensional input space $\mathcal{U}^r$, and the m-dimensional output space $\mathcal{Y}^m$. That is,

$\mathbf{A}: \Sigma \longrightarrow \Sigma$
$\mathbf{B}: \mathfrak{U}^r \longrightarrow \Sigma$
$\mathbf{C}: \Sigma \longrightarrow \mathfrak{Y}^m$
$\mathbf{D}: \mathfrak{U}^r \longrightarrow \mathfrak{Y}^m$

Figure 12.1 symbolizes these relationships.

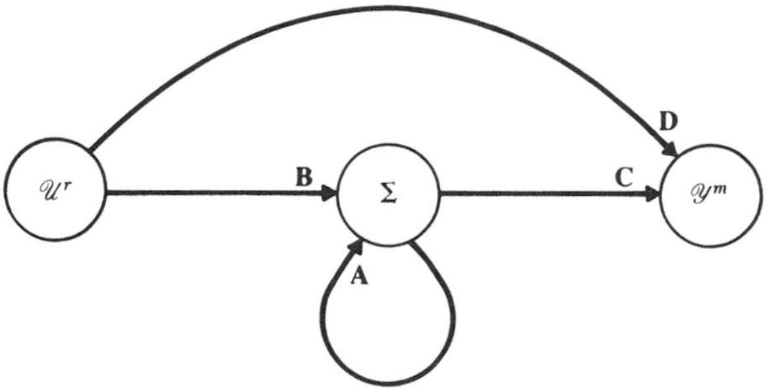

Figure 12.1

Controllability is a property of the coupling between the input and the state, and thus involves the matrices **A** and **B**.

Definition 12.1: A linear system is said to be *controllable* at t_0 if it is possible to find some input function (or sequence in the discrete case) $\mathbf{u}(t)$, defined over $t \in \mathfrak{I}$, which will transfer the initial state $\mathbf{x}(t_0)$ to the origin at some finite time $t_1 \in \mathfrak{I}$, $t_1 > t_0$. That is, there exists some input $\mathbf{u}_{[t_0,t_1]}$ which gives $\mathbf{x}(t_1) = 0$ for some finite $t_1 \in \mathfrak{I}$. If this is true for all initial times t_0 and all initial states $\mathbf{x}(t_0)$, the system is *completely controllable.*

Some authors define another kind of controllability involving the output $\mathbf{y}(t)$ [16]. The definition given above is referred to as state controllability. It is the most common definition, and is the only type used in this text, so the adjective "state" is omitted. The significance of complete controllability is easily grasped. If a system is not completely controllable, then for some initial states no input exists which can drive the system to the zero state. It would be meaningless to search for an *optimal control* in this case. A trivial example of an uncontrollable system arises when the matrix **B** is zero, because then the input is disconnected from the state.

Observability is a property of the coupling between the state and the output, and thus involves the matrices **A** and **C**.

Definition 12.2: A linear system is said to be *observable* at t_0 if $\mathbf{x}(t_0)$ can be determined from the output function $\mathbf{y}_{[t_0,t_1]}$ (or output sequence) for $t_0 \in \mathfrak{I}$ and $t_0 \leq t_1$, where t_1 is some *finite* time belonging to $\mathfrak{I}$. If this is true for all t_0 and $\mathbf{x}(t_0)$, the system is said to be *completely observable.*

If a system is not completely observable, then the initial state $\mathbf{x}(t_0)$ cannot be determined from the output, no matter how long the output is observed. The system of Example 10.1 gave an output which was identically zero for all time. That system is obviously not completely observable.

12.3 TIME-INVARIANT SYSTEMS WITH DISTINCT EIGENVALUES

Controllability and observability for time-invariant systems depend only on the constant matrices $\{\mathbf{A}, \mathbf{B}, \mathbf{C}\}$. No reference need be made to a particular interval $[t_0, t_1]$. In this section the n eigenvalues of $\mathbf{A}$ are assumed to be distinct. Then the Jordan form representation for continuous-time systems is

$$\dot{\mathbf{q}} = \mathbf{\Lambda q} + \mathbf{B}_n\mathbf{u}(t) \tag{12.1}$$

$$\mathbf{y}(t) = \mathbf{C}_n\mathbf{q}(t) + \mathbf{Du}(t) \tag{12.2}$$

For discrete-time systems,

$$\mathbf{q}(k+1) = \mathbf{\Lambda q}(k) + \mathbf{B}_n\mathbf{u}(k) \tag{12.3}$$

$$\mathbf{y}(k) = \mathbf{C}_n\mathbf{q}(k) + \mathbf{Du}(k) \tag{12.4}$$

The following definitions from earlier chapters are recalled. The modal matrix is $\mathbf{M}$, and $\mathbf{\Lambda} = \mathbf{M}^{-1}\mathbf{AM}$, $\mathbf{B}_n = \mathbf{M}^{-1}\mathbf{B}$, $\mathbf{C}_n = \mathbf{CM}$. Clearly, if any one row of $\mathbf{B}_n$ contains only zero elements, then the corresponding mode q_i is unaffected by the input. Then $\dot{q}_i = \lambda_i q_i$ or $q_i(k+1) = \lambda_i q_i(k)$. In this case the homogeneous solution for q_i may eventually approach zero as t (or k) $\longrightarrow \infty$, but there is no *finite* time at which this component of $\mathbf{q}$ will be zero. Thus there is no finite time at which $\mathbf{q}$, and consequently $\mathbf{x}$, can be driven to zero.

Controllability Criterion 1

The constant coefficient system, for which $\mathbf{A}$ has distinct eigenvalues, is completely controllable if and only if there are no zero rows of $\mathbf{B}_n = \mathbf{M}^{-1}\mathbf{B}$.

Example 12.1

The system of Problem 10.8, page 270, has $\mathbf{B}_n = \begin{bmatrix} 1/2 & 2 \\ 3 & 6 \\ 1/2 & 1 \end{bmatrix}$. Since $\mathbf{A}$ is constant and its eigenvalues are distinct, the above criterion applies. This system is completely controllable because no row of $\mathbf{B}_n$ is all zero. ■

Example 12.2

The system of Problem 11.6, page 295, has $\mathbf{B}_n = \begin{bmatrix} 2 & 0 \\ 1 & 1 \\ 0 & 0 \end{bmatrix}$. Since $\mathbf{A}$ is constant and has distinct eigenvalues, the above criterion indicates that this system is not completely controllable. Since the third row of $\mathbf{B}_n$ contains only zeros, no control can affect the third mode

$q_3(k)$ of this system. Since $q_3(k) = (-1/3)^k q_3(0)$, this component approaches zero only as $k \longrightarrow \infty$. ■

Equations (*12.2*) and (*12.4*) indicate that the output $\mathbf{y}$ will not be influenced by the ith system mode q_i if column i of $\mathbf{C}_n$ contains only zero elements. If this is true, then $q_i(t_0)$ could take on any arbitrary value without influencing $\mathbf{y}$. There is no possibility of determining $\mathbf{q}(t_0)$ or $\mathbf{x}(t_0)$ in this case.

Observability Criterion 1

The constant coefficient system, for which $\mathbf{A}$ has distinct eigenvalues, is completely observable if and only if there are no zero columns of $\mathbf{C}_n = \mathbf{CM}$.

Example 12.3

The system of Example 10.2 has $\mathbf{C}_n = [0 \quad 6]$. Since $\mathbf{A}$ is constant, with distinct eigenvalues, the above criterion applies. Column one of $\mathbf{C}_n$ is zero, so this system is not completely observable.

The discrete-time system of Problem 11.6, page 295, has $\mathbf{C}_n = \begin{bmatrix} 1 & 0 & 1 \\ 0 & 1 & 1 \end{bmatrix}$. Since no columns are all zero, this system is completely observable. ■

The criteria for complete controllability and observability given in this section (see also Problem 12.11) are useful because of the geometrical insight they give. They also make it possible to speak of individual system modes as being controllable or uncontrollable, observable or unobservable. However, these criteria are not the most useful because they are restricted to the distinct eigenvalue case (and not merely a diagonal Jordan form). Furthermore, since the modal matrix and its inverse are required, these tests are not the easiest to apply. In the next section a more general set of conditions, which is often easier to use, is given.

12.4 TIME-INVARIANT SYSTEMS WITH ARBITRARY EIGENVALUES

Controllability Criterion 2

A constant coefficient linear system with the representation $\{\mathbf{A}, \mathbf{B}, \mathbf{C}, \mathbf{D}\}$ is completely controllable if and only if the $n \times rn$ matrix of

$$\mathbf{P} \triangleq [\mathbf{B} \mid \mathbf{AB} \mid \mathbf{A}^2\mathbf{B} \mid \cdots \mid \mathbf{A}^{n-1}\mathbf{B}] \tag{12.5}$$

has rank n. See Problem 12.12.

Example 12.4

For the continuous-time system of Problem 10.8,

$$\mathbf{A} = \begin{bmatrix} -2 & -2 & 0 \\ 0 & 0 & 1 \\ 0 & -3 & -4 \end{bmatrix}, \qquad \mathbf{B} = \begin{bmatrix} 1 & 0 \\ 0 & 1 \\ 1 & 1 \end{bmatrix}$$

and

$$\mathbf{P} = \begin{bmatrix} 1 & 0 & \vdots & -2 & -2 & \vdots & 2 & 2 \\ 0 & 1 & \vdots & 1 & 1 & \vdots & -4 & -7 \\ 1 & 1 & \vdots & -4 & -7 & \vdots & 13 & 25 \end{bmatrix}$$

The rank of **P** is 3, since the determinant of the first three columns is nonzero (= −3). Therefore, this system is completely controllable. ■

Example 12.5

For the discrete-time system considered in Problem 11.6.

$$\mathbf{P} = \begin{bmatrix} 3 & 1 & \vdots & 5/2 & 1/2 & \vdots & 9/4 & 1/4 \\ 2 & 0 & \vdots & 2 & 0 & \vdots & 2 & 0 \\ -1 & 1 & \vdots & -3/2 & 1/2 & \vdots & -7/4 & 1/4 \end{bmatrix}$$

The rank of **P** is not 3, since subtracting row 3 from row 1 gives twice row 2. Since rank **P** $\neq n$, this system is not completely controllable. ■

Observability Criterion 2

A constant coefficient linear system is completely observable if and only if the $n \times mn$ matrix of equation (*12.6*) has rank n:

$$\mathbf{Q} \triangleq [\bar{\mathbf{C}}^T \mid \bar{\mathbf{A}}^T\bar{\mathbf{C}}^T \mid \bar{\mathbf{A}}^{2^T}\bar{\mathbf{C}}^T \mid \cdots \mid \bar{\mathbf{A}}^{n-1^T}\bar{\mathbf{C}}^T] \tag{12.6}$$

Example 12.6

The continuous-time system of Example 10.1 has $\mathbf{A} = \begin{bmatrix} 0 & 1 \\ 8 & -2 \end{bmatrix}$, $\mathbf{C} = [4 \quad 1]$. These are both real, so the complex conjugate in **Q** is unnecessary for this example, and $\mathbf{Q} = \begin{bmatrix} 4 & \vdots & 8 \\ 1 & \vdots & 2 \end{bmatrix}$. The second column is twice the first, so $r_Q = 1$. Since $r_Q \neq 2$, this system is not completely observable. Similar manipulations show that the system of Problem 11.6 is completely observable. ■

The forms of equations (*12.5*) and (*12.6*) are the same. Therefore, a computer algorithm for forming P and checking its rank will also form **Q** and check its rank. In forming **Q**, $\bar{\mathbf{A}}^T$ replaces **A** and $\bar{\mathbf{C}}^T$ replaces **B**. In the most common case where **A** and **B** are real, the complex conjugates are superfluous. In developing such a computer algorithm the crucial consideration is accurate determination of rank. If a row-reduced echelon technique is used, the notion of "machine zero" must be used to distinguish between legitimate nonzero but small divisors and divisors which would have been zero were it not for round-off error. The discussion of the GSE method in Sec. 5.5 and the SVD method in Problems 7.25 through 7.29 are of significance in this regard. These same considerations arise many other places as well, e.g.: Do nontrivial solutions exist for a set of homogeneous equations? See Reference 38 for a fuller discussion of the implications of the fact that rank of a matrix is a discontinuous function. The wrong rank can easily be computed due to very small computer errors, unless due precautions are taken.

12.5 TIME-VARYING LINEAR SYSTEMS

A continuous-time system with the representation $\{\mathbf{A}(t), \mathbf{B}(t), \mathbf{C}(t), \mathbf{D}(t)\}$ is considered. For a given input function $\mathbf{u}(t)$, the solution for the state at a fixed time t_1 is

$$\mathbf{x}(t_1) = \mathbf{\Phi}(t_1, t_0)\mathbf{x}(t_0) + \int_{t_0}^{t_1} \mathbf{\Phi}(t_1, \tau)\mathbf{B}(\tau)\mathbf{u}(\tau)\, d\tau$$

The vector defined by $\mathbf{x}_1 = \mathbf{x}(t_1) - \mathbf{\Phi}(t_1, t_0)\mathbf{x}(t_0)$ is a constant vector in Σ for any fixed time t_1. The notation of Chapter 6 is used to define the linear transformation

$$\mathcal{A}_c(\mathbf{u}) \triangleq \int_{t_0}^{t_1} \mathbf{\Phi}(t_1, \tau)\mathbf{B}(\tau)\mathbf{u}(\tau)\, d\tau$$

The transformation $\mathcal{A}_c$ maps functions in $\mathcal{U}$ into vectors in Σ. The question of complete controllability on $[t_0, t_1]$ reduces to asking whether $\mathcal{A}_c(\mathbf{u}) = \mathbf{x}_1$ has a solution $\mathbf{u}(t)$ for arbitrary $\mathbf{x}_1 \in \Sigma$. It was shown in Problem 6.10, page 166, that a necessary and sufficient condition for the existence of such a solution is that the null space of $\mathcal{A}_c^*$ contain only the zero element $\mathfrak{N}(\mathcal{A}_c^*) = \{0\}$. This is the requirement for complete controllability on $[t_0, t_1]$,but it can be put into a more useful form. Since $\mathcal{A}_c^*: \Sigma \longrightarrow \mathcal{U}$, the range of $\mathcal{A}_c^*$ is an infinite dimensional function space. The following lemma allows the use of a finite dimensional transformation.

Lemma 12.1: The null space of $\mathcal{A}_c^*$ is the same as the null space of $\mathcal{A}_c\mathcal{A}_c^*$. That is, $\mathfrak{N}(\mathcal{A}_c^*) = \mathfrak{N}(\mathcal{A}_c\mathcal{A}_c^*)$.

Proof: Let $\mathbf{v} \in \mathfrak{N}(\mathcal{A}_c^*)$. Then $\mathcal{A}_c^*(\mathbf{v}) = \mathbf{0}$. Therefore, $\mathcal{A}_c\mathcal{A}_c^*(\mathbf{v}) = \mathcal{A}_c(\mathbf{0}) = \mathbf{0}$, so $\mathbf{v} \in \mathfrak{N}(\mathcal{A}_c\mathcal{A}_c^*)$ also. Now assume that $\mathbf{v} \in \mathfrak{N}(\mathcal{A}_c\mathcal{A}_c^*)$. Then $\mathcal{A}_c\mathcal{A}_c^*(\mathbf{v}) = \mathbf{0}$. Therefore, $\langle \mathcal{A}_c\mathcal{A}_c^*(\mathbf{v}), \mathbf{v}\rangle = 0$ or $\langle \mathcal{A}_c^*(\mathbf{v}), \mathcal{A}_c^*(\mathbf{v})\rangle = 0$. But this indicates that $\| \mathcal{A}_c^*(\mathbf{v}) \|^2 = 0$, so that $\mathcal{A}_c^*(\mathbf{v}) = \mathbf{0}$. Thus for every $\mathbf{v} \in \mathfrak{N}(\mathcal{A}_c^*)$, $\mathbf{v}$ also belongs to $\mathfrak{N}(\mathcal{A}_c\mathcal{A}_c^*)$ and conversely. The two null spaces are therefore equal.

To use the lemma in developing the criterion for complete observability, an expression for the transformation $\mathcal{A}_c\mathcal{A}_c^*$ must be found:

$$\mathcal{A}_c(\mathbf{u}) = \int_{t_0}^{t_1} \mathbf{\Phi}(t_1, \tau)\mathbf{B}(\tau)\mathbf{u}(\tau)\, d\tau$$

Therefore,

$$\langle \mathbf{v}, \mathcal{A}_c(\mathbf{u})\rangle = \langle \mathcal{A}_c^*(\mathbf{v}), \mathbf{u}\rangle = \int_{t_0}^{t_1} \bar{\mathbf{v}}^T\mathbf{\Phi}(t_1, \tau)\mathbf{B}(\tau)\mathbf{u}(\tau)\, d\tau$$

so that $\mathcal{A}_c^*(\mathbf{v}) = \bar{\mathbf{B}}^T(t)\bar{\mathbf{\Phi}}^T(t_1, t)\mathbf{v}$. Then

$$\mathcal{A}_c\mathcal{A}_c^*(\mathbf{v}) = \int_{t_0}^{t_1} \mathbf{\Phi}(t_1, \tau)\mathbf{B}(\tau)\bar{\mathbf{B}}^T(\tau)\bar{\mathbf{\Phi}}^T(t_1, \tau)\, d\tau\, \mathbf{v}$$

The transformation $\mathcal{A}_c\mathcal{A}_c^*$ is just an $n \times n$ matrix, redefined as $\mathbf{G}(t_1, t_0)$,

$$\mathbf{G}(t_1, t_0) \triangleq \int_{t_0}^{t_1} \mathbf{\Phi}(t_1, \tau)\mathbf{B}(\tau)\bar{\mathbf{B}}^T(\tau)\bar{\mathbf{\Phi}}^T(t_1, \tau)\, d\tau \tag{12.7}$$

The null space of $\mathcal{A}_c\mathcal{A}_c^*$ will contain only the zero element if and only if $\mathbf{G}(t_1, t_0)$ does not have zero as an eigenvalue.

Controllability Criterion 3

The system described by $\dot{\mathbf{x}} = \mathbf{A}(t)\mathbf{x} + \mathbf{B}(t)\mathbf{u}(t)$ is completely controllable on the interval $[t_0, t_1]$ if any of the following equivalent conditions is satisfied:

(a) The matrix $\mathbf{G}(t_1, t_0)$ is positive definite.
(b) Zero is not an eigenvalue of $\mathbf{G}(t_1, t_0)$.
(c) $|\mathbf{G}(t_1, t_0)| \neq 0$.

The general form for the output $\mathbf{y}(t)$ is

$$\mathbf{y}(t) = \mathbf{C}(t)\mathbf{\Phi}(t, t_0)\mathbf{x}(t_0) + \int_{t_0}^{t_1} \mathbf{C}(t)\mathbf{\Phi}(t, \tau)\mathbf{B}(\tau)\mathbf{u}(\tau)\, d\tau + \mathbf{D}(t)\mathbf{u}(t)$$

Since the input $\mathbf{u}(t)$ is assumed known, the two terms containing the input could be combined with the output function $\mathbf{y}(t)$ to give a modified function $\mathbf{y}_1(t)$. Alternatively, only the unforced solution could be considered. In either case complete observability requires that a knowledge of $\mathbf{y}(t)$ (or $\mathbf{y}_1(t)$) be sufficient for the determination of $\mathbf{x}(t_0)$. Defining the linear transformation $\mathcal{A}_0(\mathbf{x}(t_0)) = \mathbf{C}(t)\mathbf{\Phi}(t, t_0)\mathbf{x}(t_0)$, the requirement for complete observability is that a unique $\mathbf{x}(t_0)$ can be associated with each output function $\mathbf{y}(t)$. This requires that $\mathfrak{N}(\mathcal{A}_0) = \{0\}$ (see Problem 6.9, page 165). Using only minor changes in the previous lemma, it can be shown that $\mathfrak{N}(\mathcal{A}_0^*\mathcal{A}_0) = \mathfrak{N}(\mathcal{A}_0)$. To find the adjoint transformation, consider

$$\langle \mathbf{w}(t), \mathcal{A}_0(\mathbf{x}(t_0))\rangle = \langle \mathcal{A}_0^*\mathbf{w}(t), \mathbf{x}(t_0)\rangle = \int_{t_0}^{t_1} \bar{\mathbf{w}}^T(\tau)\mathbf{C}(\tau)\mathbf{\Phi}(\tau, t_0)\, d\tau\, \mathbf{x}(t_0)$$

Thus

$$\mathcal{A}_0^*(\mathbf{w}) = \int_{t_0}^{t_1} \bar{\mathbf{\Phi}}^T(\tau, t_0)\bar{\mathbf{C}}^T(\tau)\mathbf{w}(\tau)\, d\tau$$

and

$$\mathcal{A}_0^*\mathcal{A}_0(\mathbf{x}(t_0)) = \int_{t_0}^{t_1} \bar{\mathbf{\Phi}}^T(\tau, t_0)\bar{\mathbf{C}}^T(\tau)\mathbf{C}(\tau)\mathbf{\Phi}(\tau, t_0)\, d\tau\, \mathbf{x}(t_0)$$

The transformation $\mathcal{A}_0^*\mathcal{A}_0 : \Sigma \longrightarrow \Sigma$ is just an $n \times n$ matrix, redefined as $\mathbf{H}(t_1, t_0)$,

$$\mathbf{H}(t_1, t_0) \triangleq \int_{t_0}^{t_1} \bar{\mathbf{\Phi}}^T(\tau, t_0)\bar{\mathbf{C}}^T(\tau)\mathbf{C}(\tau)\mathbf{\Phi}(\tau, t_0)\, d\tau \tag{12.8}$$

Observability Criterion 3

The system

$$\dot{\mathbf{x}} = \mathbf{A}(t)\mathbf{x}(t) + \mathbf{B}(t)\mathbf{u}(t)$$
$$\mathbf{y}(t) = \mathbf{C}(t)\mathbf{x}(t) + \mathbf{D}(t)\mathbf{u}(t)$$

is completely observable at t_0 if there exists some finite time t_1 for which any one of the following equivalent conditions holds:

(a) The matrix $\mathbf{H}(t_1, t_0)$ is positive definite.
(b) Zero is not an eigenvalue of $\mathbf{H}(t_1, t_0)$.
(c) $|\mathbf{H}(t_1, t_0)| \neq 0$.

The corresponding forms of the controllability and observability criteria 3 for discrete systems are derived in the same way. However, since the input and output spaces have sequences, rather than functions, as their elements, the appropriate inner product is a summation rather than an integral:

$$\langle \mathbf{w}(k), \mathbf{y}(k) \rangle = \sum_{k=0}^{N} \bar{\mathbf{w}}^T(k)\mathbf{y}(k)$$

The criteria may be stated as follows.

Controllability and Observability Criteria 3, Discrete Systems

The system

$$\mathbf{x}(k+1) = \mathbf{A}(k)\mathbf{x}(k) + \mathbf{B}(k)\mathbf{u}(k)$$
$$\mathbf{y}(k) = \mathbf{C}(k)\mathbf{x}(k) + \mathbf{D}(k)\mathbf{u}(k)$$

is completely controllable at $k = 0$ if and only if for some finite time index N, the $n \times n$ matrix

$$\sum_{k=0}^{N} \mathbf{\Phi}(N, k)\mathbf{B}(k)\bar{\mathbf{B}}^T(k)\bar{\mathbf{\Phi}}^T(N, k)$$

is positive definite (or does not have zero as an eigenvalue, or has a nonzero determinant). This system is completely observable at $k = 0$ if and only if there exists some finite index N such that the $n \times n$ matrix

$$\sum_{k=0}^{N} \bar{\mathbf{\Phi}}^T(k, 0)\bar{\mathbf{C}}^T(k)\mathbf{C}(k)\mathbf{\Phi}(k, 0)$$

is positive definite (or does not have zero as an eigenvalue, or has a nonzero determinant).

ILLUSTRATIVE PROBLEMS

Application of the Criteria

12.1 Is the following system completely controllable and completely observable?

$$\dot{\mathbf{x}} = \begin{bmatrix} -3/4 & -1/4 \\ -1/2 & -1/2 \end{bmatrix}\mathbf{x}(t) + \begin{bmatrix} 1 \\ 1 \end{bmatrix}u(t), \qquad y(t) = [4 \quad 2]x(t)$$

Using criteria 2, $\mathbf{P} = [\mathbf{B} \mid \mathbf{AB}] = \begin{bmatrix} 1 & -1 \\ 1 & -1 \end{bmatrix}$ has rank 1 and $\mathbf{Q} = [\mathbf{C}^T \mid \mathbf{A}^T\mathbf{C}^T] = \begin{bmatrix} 4 & -4 \\ 2 & -2 \end{bmatrix}$ has rank 1. Therefore, the system is neither completely controllable nor completely observable.

12.2 Is the following discrete-time system completely controllable and completely observable?

$$\mathbf{x}(k+1) = \begin{bmatrix} 1 & 0 \\ -1/2 & 1/2 \end{bmatrix}\mathbf{x}(k) + \begin{bmatrix} 1 \\ -1 \end{bmatrix}u(k), \qquad y(k) = [5 \quad 1]\mathbf{x}(k)$$

Using criteria 2, $\mathbf{P} = \begin{bmatrix} 1 & 1 \\ -1 & -1 \end{bmatrix}$, rank $\mathbf{P} = 1$ but $n = 2$. Therefore, the system is *not* completely controllable. $\mathbf{Q} = \begin{bmatrix} 5 & 9/2 \\ 1 & 1/2 \end{bmatrix}$ has rank 2. The system is completely observable.

12.3 Investigate the controllability and observability of the systems in Fig. 12.2(*a*) and (*b*) individually and when connected in series as in (*c*).

For system (*a*), $\dot{y}_1 + \beta y_1 = \dot{u}_1 + \alpha u_1$. Letting $x_1 = y_1 - u_1$ gives the state equation $\dot{x}_1 = -\beta x_1 + (\alpha - \beta)u_1$. This system is completely controllable if $\alpha \neq \beta$. It is always completely observable.

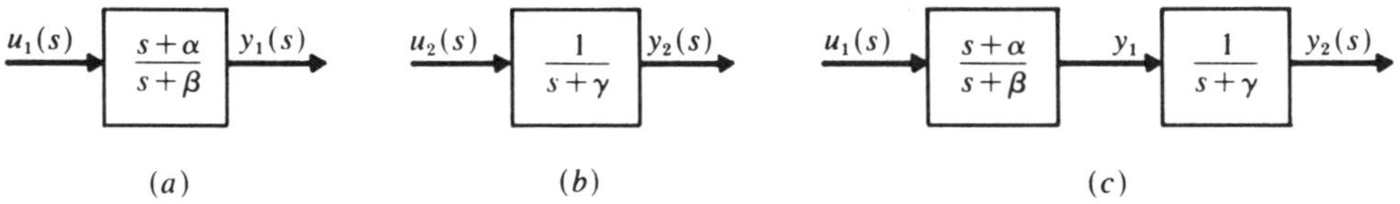

Figure 12.2

For system (*b*), $\dot{y}_2 + \gamma y_2 = u_2$. Letting $x_2 = y_2$ gives the state equation $\dot{x}_2 = -\gamma x_2 + u_2$. This system is completely controllable and observable.

Using the same definition for x_1 and x_2 in system (*c*), and noting that $y_1 = x_1 + u_1$ replaces u_2, the state equations are

$$\begin{bmatrix} \dot{x}_1 \\ \dot{x}_2 \end{bmatrix} = \begin{bmatrix} -\beta & 0 \\ 1 & -\gamma \end{bmatrix}\begin{bmatrix} x_1 \\ x_2 \end{bmatrix} + \begin{bmatrix} \alpha - \beta \\ 1 \end{bmatrix}u_1$$

The controllability matrix is

$$\mathbf{P} = \begin{bmatrix} \alpha - \beta & -\beta(\alpha - \beta) \\ 1 & \alpha - \beta - \gamma \end{bmatrix} \quad \text{and} \quad |\mathbf{P}| = (\alpha - \beta)(\alpha - \gamma)$$

The rank of **P** is 2 and system (*c*) is completely controllable, unless $\alpha = \beta$ or $\alpha = \gamma$. If either of these conditions is satisfied, the pole-zero cancellation leads to an uncontrollable system. The observability matrix is $\mathbf{Q} = \begin{bmatrix} 0 & 1 \\ 1 & -\gamma \end{bmatrix}$; and since its rank is 2, system (*c*) is completely observable.

12.4 Investigate the controllability and observability of the two systems shown in Fig. 12.3.

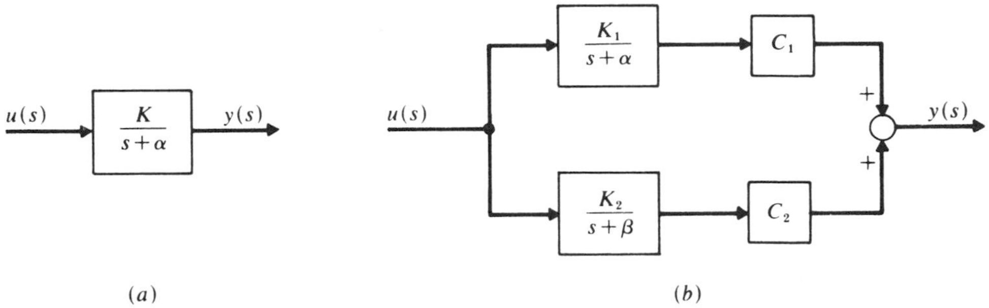

Figure 12.3

System (*a*) is described by $\dot{x}_1 = -\alpha x_1 + Ku$, $y = x_1$, and is completely controllable and observable.

System (*b*) can be described by

$$\begin{bmatrix} \dot{x}_1 \\ \dot{x}_2 \end{bmatrix} = \begin{bmatrix} -\alpha & 0 \\ 0 & -\beta \end{bmatrix} \begin{bmatrix} x_1 \\ x_2 \end{bmatrix} + \begin{bmatrix} K_1 \\ K_2 \end{bmatrix} u, \qquad y = [C_1 \quad C_2]\mathbf{x}$$

The controllability and observability matrices of criterion 2 are

$$\mathbf{P} = \begin{bmatrix} K_1 & -\alpha K_1 \\ K_2 & -\beta K_2 \end{bmatrix} \quad \text{and} \quad \mathbf{Q} = \begin{bmatrix} C_1 & -\alpha C_1 \\ C_2 & -\beta C_2 \end{bmatrix}$$

System (*b*) is completely controllable and observable except when $\alpha = \beta$.

12.5 Use the system of Problem 10.6 to draw conclusions regarding the observability of a discrete-time system obtained by sampling a completely observable continuous-time system. Assume that the output is $y = x_1$.

To consider observability, it is only necessary to consider the unforced system $\dot{\mathbf{x}} = \begin{bmatrix} 0 & -\Omega \\ \Omega & 0 \end{bmatrix}\mathbf{x}$, $\mathbf{y} = [1 \quad 0]\mathbf{x}$. This system is completely observable since $\mathbf{Q} = \begin{bmatrix} 1 & 0 \\ 0 & -\Omega \end{bmatrix}$ has rank 2.

Using the state transition matrix $\mathbf{\Phi}(t_{k+1}, t_k) = \mathbf{\Phi}(\Delta t, 0)$, with $\Delta t \triangleq t_{k+1} - t_k$, the discrete-time equations are

$$\mathbf{x}(k+1) = \begin{bmatrix} \cos \Omega\,\Delta t & -\sin \Omega\,\Delta t \\ \sin \Omega\,\Delta t & \cos \Omega\,\Delta t \end{bmatrix} \mathbf{x}(k), \qquad \mathbf{y}(k) = [1 \quad 0]\mathbf{x}(k)$$

The discrete observability matrix is $\mathbf{Q} = \begin{bmatrix} 1 & \cos \Omega\,\Delta t \\ 0 & -\sin \Omega\,\Delta t \end{bmatrix}$. The rank is 2 unless the sampling period Δt is an integer multiple of π/Ω. The property of complete observability is lost if an oscillatory system is sampled at its natural frequency.

12.6 Is the following time-variable system completely controllable?

$$\dot{\mathbf{x}} = \frac{1}{12}\begin{bmatrix} 5 & 1 \\ 1 & 5 \end{bmatrix}\mathbf{x} + e^{t/2}\begin{bmatrix} 1 \\ 1 \end{bmatrix}u(t)$$

Since $\mathbf{B}(t)$ is time-varying, criterion 3 is used. The controllability matrix of equation (*12.7*) can be written as

$$\mathbf{G}(t_1, t_0) = \mathbf{\Phi}(t_1, 0)\int_{t_0}^{t_1} \mathbf{\Phi}^{-1}(\tau, 0)\mathbf{B}(\tau)\mathbf{B}^T(\tau)[\mathbf{\Phi}^{-1}(\tau, 0)]^T\,d\tau\,\mathbf{\Phi}^T(t_0, 0)$$

The transition matrix $\mathbf{\Phi}(t, 0)$ can be found by any of the methods of Chapter 8, and then

$$\mathbf{\Phi}^{-1}(\tau, 0) = \mathbf{\Phi}(-\tau, 0) = \frac{1}{2}\begin{bmatrix} e^{-\tau/2} + e^{-\tau/3} & e^{-\tau/2} - e^{-\tau/3} \\ e^{-/\tau 2} - e^{-\tau/3} & e^{-\tau/2} + e^{-\tau/3} \end{bmatrix}$$

Therefore, $\mathbf{\Phi}^{-1}(\tau, 0)\mathbf{B}(\tau) = \begin{bmatrix} 1 \\ 1 \end{bmatrix}$ so that

$$|\mathbf{G}(t_1, t_0)| = |\mathbf{\Phi}(t_1, 0)|\left|\begin{bmatrix} t_1 - t_0 & t_1 - t_0 \\ t_1 - t_0 & t_1 - t_0 \end{bmatrix}\right||\mathbf{\Phi}^T(t_1, 0)| = 0$$

This is true for all t_0, t_1. The system is not completely controllable.

12.7 An approximate linear model of the lateral dynamics of an aircraft, for a particular set of flight conditions, has [1] the state and control vectors in the perturbation quantities

$$\mathbf{x} = [p \quad r \quad \beta \quad \phi]^T \qquad \text{and} \qquad \mathbf{u} = [\delta_a \quad \delta_r]^T$$

where p and r are incremental roll and yaw rates, β is an incremental sideslip angle, and ϕ is an incremental roll angle. The control inputs are the incremental changes in the aileron angle δ_a and in the rudder angle δ_r, respectively. These variables are shown in Fig. 12.4.

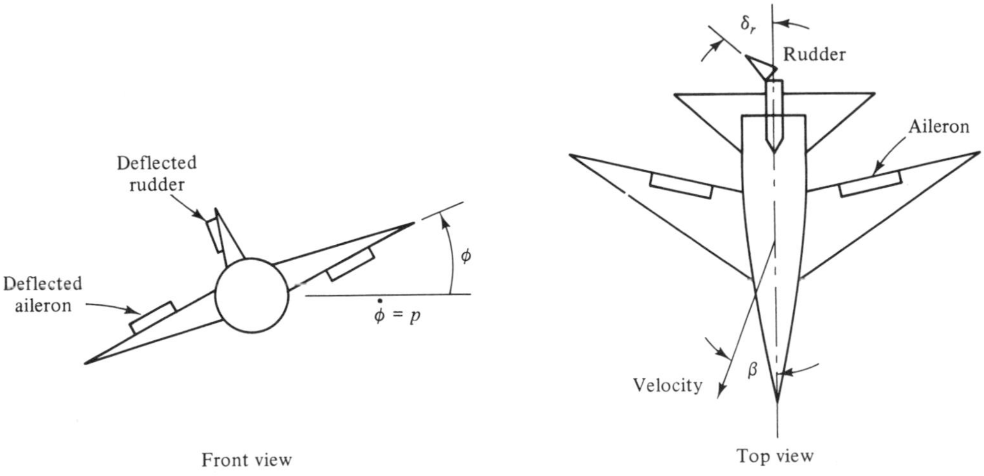

Figure 12.4

In a consistent set of units this linearized model has

$$\mathbf{A} = \begin{bmatrix} -10 & 0 & -10 & 0 \\ 0 & -0.7 & 9 & 0 \\ 0 & -1 & -0.7 & 0 \\ 1 & 0 & 0 & 0 \end{bmatrix} \qquad \mathbf{B} = \begin{bmatrix} 20 & 2.8 \\ 0 & -3.13 \\ 0 & 0 \\ 0 & 0 \end{bmatrix}$$

Suppose a malfunction prevents manipulation of the input δ_r. Is it possible to control the aircraft using only δ_a? Is the aircraft controllable with just δ_r? Verify that it is controllable with both inputs operable.

When δ_a is the only input, just the first column of the **B** matrix must be used in checking for controllability. The **P** matrix is determined to be

$$\mathbf{P} = \left[\begin{array}{c|c|c|c} 20 & -200 & 2000 & -2 \times 10^4 \\ 0 & 0 & 0 & 0 \\ 0 & 0 & 0 & 0 \\ 0 & 20 & -200 & 2 \times 10^3 \end{array}\right]$$

and its rank is $2 < n$. Thus it is not controllable. The aircraft can be made to roll using only the ailerons, but it cannot be made to turn, at least insofar as this linearized model is concerned.

With δ_r as the only input, column 2 of **B** is used to compute another **P** matrix:

$$\mathbf{P} = \left[\begin{array}{c|c|c|c} 2.8 & -28 & 248.7 & -2443.18 \\ -3.13 & 2.191 & 26.636 & -58.08 \\ 0 & 3.13 & -4.382 & -23.57 \\ 0 & 2.8 & -28 & 248.7 \end{array}\right]$$

The rank is now 4, and the system is controllable. The controllability index (the number of partitions in **P** that are required before full rank is achieved) is 4. The maneuverability of the aircraft would be greatly degraded, and a very sloppy flight profile would be expected under such conditions. Controllability does not guarantee a high-quality control, it just guarantees that all the states can be manipulated to zero in some fashion in some finite time. Adding the second input only adds more columns to **P** so its rank is also 4. Now, however, a rank 4 matrix can be formed from just the first two partitions of **P**. The controllability index is 2, indicating a stronger degree of controllability in some sense.

12.8 If the only output is a measurement of the roll rate p (provided by a rate gyro) in the previous problem, is the system observable?

The output matrix is $\mathbf{C} = [1 \quad 0 \quad 0 \quad 0]$. Using this, the **Q** matrix is

$$\mathbf{Q} = \left[\begin{array}{c|c|c|c} 1 & -10 & 100 & -1000 \\ 0 & 0 & 10 & -114 \\ 0 & -10 & 107 & -984.9 \\ 0 & 0 & 0 & 0 \end{array}\right]$$

The rank is $3 < 4$, so the system is not observable. Measurements of roll rate allow the *change* in roll angle to be monitored, but will never allow determination of the roll angle itself because its initial value is unknown. In fact, if any one or all of the states except ϕ are measured outputs, this system remains unobservable. A bank indicator or some other means of measuring ϕ is required in order to obtain an observable system. If ϕ is the only measurement, then $\mathbf{C} = [0 \quad 0 \quad 0 \quad 1]$, which leads to

$$\mathbf{Q} = \begin{bmatrix} 0 & 1 & -10 & 100 \\ 0 & 0 & 0 & 10 \\ 0 & 0 & -10 & 107 \\ 1 & 0 & 0 & 0 \end{bmatrix}$$

This has rank 4, so the system is observable. The observability index is 4, meaning that all four partitions in $\mathbf{Q}$ are required to give a rank 4 result. If other states can also be measured and combined with ϕ to form a vector rather than scalar output, the observability index will improve (decrease).

12.9 a. Show that a system governed by

$$\mathbf{x}(k+1) = \mathbf{x}(k)$$
$$\mathbf{y}(k) = \mathbf{C}\mathbf{x}(k)$$

with $\mathbf{C}$ constant, is never observable unless $\text{rank}(\mathbf{C}) = n$.

b. Show that if $\mathbf{C}(k)$ is time-varying, observability criterion 3 leads to the condition that the normal equations of least squares must eventually become invertible for some finite time N if the system is to be observable.

a. With $\mathbf{A} = \mathbf{I}$, criterion 2 gives

$$\mathbf{Q} = [\mathbf{C}^T \mid \mathbf{C}^T \mid \cdots \mid \mathbf{C}^T]$$

Therefore, $\text{rank}(\mathbf{Q}) = \text{rank}(\mathbf{C}^T) = \text{rank}(\mathbf{C})$. Observability requires that $\text{rank}(\mathbf{C}) = n$, where n is the number of components in $\mathbf{x}$.

b. Again, $\mathbf{A} = \mathbf{I}$, so the observability matrix of criterion 3 becomes

$$\mathbf{Q}' = \sum_{k=0}^{N} \mathbf{C}^T(k)\mathbf{C}(k) = [\mathbf{C}^T(0) \mid \mathbf{C}^T(1) \mid \cdots \mid \mathbf{C}^T(N)] \begin{bmatrix} \mathbf{C}(0) \\ \mathbf{C}(1) \\ \vdots \\ \mathbf{C}(N) \end{bmatrix}$$

Define the stacked-up measurement vector and measurement matrix as

$$\mathbf{Y} = \begin{bmatrix} \mathbf{y}(0) \\ \mathbf{y}(1) \\ \vdots \\ \mathbf{y}(N) \end{bmatrix}, \qquad \mathcal{H} = \begin{bmatrix} \mathbf{C}(0) \\ \mathbf{C}(1) \\ \vdots \\ \mathbf{C}(N) \end{bmatrix}$$

Then the entire group of measurements, as it would be processed in batch least squares, is $\mathbf{Y} = \mathcal{H}\mathbf{x}(0)$, and $\mathbf{Q}' = \mathcal{H}^T\mathcal{H}$. The normal equation is $\mathcal{H}^T\mathcal{H}\mathbf{x}(0) = \mathcal{H}^T\mathbf{Y}$, and is invertible if and only if $\mathcal{H}$ has rank n. This is the same as requiring that $\mathbf{Q}'$ have rank n. Thus, observability is seen to be the same as having a *unique* least-squares solution in this case of a constant state vector.

Extensions and Proofs

12.10 Show that if a continuous-time linear system is completely controllable at t_0, then any initial state $\mathbf{x}(t_0)$ can be transferred to any other state $\mathbf{x}(t_1)$ at some finite time t_1.

Complete controllability means that any $\mathbf{x}(t_0)$ can be transferred to the origin

$\mathbf{x}(t_1) = \mathbf{0}$. The solution for a given input is of the form

$$\mathbf{x}(t_1) = \mathbf{\Phi}(t_1, t_0)\mathbf{x}(t_0) + \int_{t_0}^{t_1} \mathbf{\Phi}(t_1, \tau)\mathbf{B}(\tau)\mathbf{u}(\tau)\, d\tau \tag{1}$$

This could be written as

$$\mathbf{0} = \mathbf{\Phi}(t_1, t_0)[\mathbf{x}(t_0) - \mathbf{\Phi}(t_0, t_1)\mathbf{x}(t_1)] + \int_{t_0}^{t_1} \mathbf{\Phi}(t_1, \tau)\mathbf{B}(\tau)\mathbf{u}(\tau)\, d\tau \tag{2}$$

Since $\mathbf{x}' \triangleq \mathbf{x}(t_0) - \mathbf{\Phi}(t_0, t_1)\mathbf{x}(t_1)$ belongs to Σ, it is a possible initial state. Complete controllability at t_0 guarantees that $\mathbf{x}'$ can be driven to the origin (equation (*2*)), which means that $\mathbf{x}(t_0)$ can be driven to any arbitrary $\mathbf{x}(t_1)$ (equation (*1*)).

12.11 The arguments in Sec. 12.3 establish the necessity of controllability criterion 1. Show that this criterion is sufficient by assuming no zero rows of $\mathbf{B}_n$ and deriving the control which drives an arbitrary initial state to the origin.

The normal form description of the system is $\dot{\mathbf{q}} = \mathbf{\Lambda q} + \mathbf{B}_n\mathbf{u}$ and the solution at t_1 is

$$\mathbf{q}(t_1) = e^{\mathbf{\Lambda}t_1}\mathbf{q}(0) + \int_0^{t_1} e^{\mathbf{\Lambda}t_1}e^{-\mathbf{\Lambda}\tau}\mathbf{B}_n\mathbf{u}(\tau)\, d\tau$$

Let the ith row of $\mathbf{B}_n$ define the *row* vector $\mathbf{b}_i$, and let $\mathbf{u}(t) = \sum_{j=1}^{n} \beta_j e^{-\bar{\lambda}_j t}\bar{\mathbf{b}}_j^T$. The coefficients β_j are unknown constants. It is to be shown that these constants can be selected in such a way that $\mathbf{q}(t_1) = \mathbf{0}$ if none of the rows $\mathbf{b}_i$ are identically zero. Using the assumed form for $\mathbf{u}(t)$, a typical component of $\mathbf{q}(t_1)$ is

$$q_i(t_1) = e^{\lambda_i t_1}q_i(0) = \sum_{j=1}^{n} e^{\lambda_i t_1} \int_0^{t_1} e^{-\lambda_i \tau}\mathbf{b}_i\bar{\mathbf{b}}_j^T e^{-\bar{\lambda}_j \tau}\, d\tau\, \beta_j$$

or

$$e^{-\lambda_i t_1}q_i(t_1) - q_i(0) = \sum_{j=1}^{n} \langle \boldsymbol{\theta}_i(\tau), \boldsymbol{\theta}_j(\tau)\rangle \beta_j$$

The integral inner product (Problem 4.20, page 112) of the functions $\boldsymbol{\theta}_j(\tau) = e^{-\bar{\lambda}_j\tau}\bar{\mathbf{b}}_j^T$ is used. The unknown coefficients can be obtained by solving n simultaneous equations, and are given by

$$\begin{bmatrix} \beta_1 \\ \beta_2 \\ \vdots \\ \beta_n \end{bmatrix} = [\langle \boldsymbol{\theta}_i(\tau), \boldsymbol{\theta}_j(\tau)\rangle]^{-1}[e^{-\mathbf{\Lambda}t_1}\mathbf{q}(t_1) - \mathbf{q}(0)]$$

The indicated inverse is guaranteed to exist if $\mathbf{b}_i \neq \mathbf{0}$ for all i and if all the λ_i are distinct. This is true because under these conditions the set of functions $\{\boldsymbol{\theta}_j(\tau)\}$ is linearly independent over every finite interval $[0, t_1]$. The matrix $[\langle \boldsymbol{\theta}_i(\tau), \boldsymbol{\theta}_j(\tau)\rangle]$, which can also be written as $\int_0^{t_1} e^{-\mathbf{\Lambda}\tau}\, \mathbf{B}_n\bar{\mathbf{B}}_n^T e^{-\bar{\mathbf{\Lambda}}\tau}\, d\tau$, is the Grammian matrix and is nonsingular. The conditions of controllability criterion 1 are sufficient to guarantee that any $\mathbf{q}(0)$ can be driven to any $\mathbf{q}(t_1)$, including $\mathbf{q}(t_1) = \mathbf{0}$. An input function which drives $\mathbf{q}_1(0)$ to the origin at t_1 is

$$\mathbf{u}(t) = \bar{\mathbf{B}}_n^T e^{\bar{\mathbf{\Lambda}}t}\boldsymbol{\beta} = -\bar{\mathbf{B}}_n^T e^{-\bar{\mathbf{\Lambda}}t}\left[\int_0^{t_1} e^{-\mathbf{\Lambda}\tau}\mathbf{B}_n\bar{\mathbf{B}}_n^T e^{-\bar{\mathbf{\Lambda}}\tau}\, d\tau\right]^{-1}\mathbf{q}(0)$$

12.12 Assume that the time-invariant system $\dot{\mathbf{x}} = \mathbf{Ax} + \mathbf{Bu}$ is completely controllable. Prove that the controllability criterion 2 is a necessary condition.

Complete controllability means that for every $\mathbf{x}_0$ there is some finite time t_1 and some input function $\mathbf{u}(t)$ such that

$$\mathbf{0} = e^{\mathbf{A}t_1}x_0 + \int_0^{t_1} e^{\mathbf{A}(t_1-\tau)}\mathbf{Bu}(\tau)\,d\tau \qquad \text{or} \qquad -\mathbf{x}_0 = \int_0^{t_1} e^{-\mathbf{A}\tau}\mathbf{Bu}(\tau)\,d\tau$$

Using the remainder form from the matrix exponential,

$$e^{-\mathbf{A}\tau} = \alpha_0(\tau)\mathbf{I} + \alpha_1(\tau)\mathbf{A} + \alpha_2(\tau)\mathbf{A}^2 + \cdots + \alpha_{n-1}(\tau)\mathbf{A}^{n-1}$$

gives

$$-\mathbf{x}_0 = \sum_{j=0}^{n-1} \mathbf{A}^j\mathbf{B}\int_0^{t_1} \alpha_j(\tau)\mathbf{u}(\tau)\,d\tau$$

Each integral term is an $r \times 1$ constant vector, defined as

$$\mathbf{v}_j = \int_0^{t_1} \alpha_j(\tau)\mathbf{u}(\tau)\,d\tau$$

Then

$$-\mathbf{x}_0 = [\mathbf{B} \mid \mathbf{AB} \mid \mathbf{A}^2\mathbf{B} \mid \cdots \mid \mathbf{A}^{n-1}\mathbf{B}]\begin{bmatrix} \mathbf{v}_0 \\ \hline \mathbf{v}_1 \\ \hline \vdots \\ \hline \mathbf{v}_{n-1}\end{bmatrix}$$

This result states that every *vector* $-\mathbf{x}_0$ can be expressed as some linear combination of the columns of $\mathbf{P} = [\mathbf{B} \mid \mathbf{AB} \mid \cdots \mid \mathbf{A}^{n-1}\mathbf{B}]$. These columns must span the n-dimensional state space Σ, that is, it is necessary that rank $\mathbf{P} = n$. The necessity of the observability condition 2 can be established in a similar manner.

12.13 Assume that the following system is completely controllable and completely observable over the interval $[t_0, t_1]$:

$$\dot{\mathbf{x}} = \mathbf{A}(t)\mathbf{x}(t) + \mathbf{B}(t)\mathbf{u}(t), \qquad \mathbf{y}(t) = \mathbf{C}(t)\mathbf{x}(t) + \mathbf{D}(t)\mathbf{u}(t)$$

a. Derive an explicit expression for an input which transfers the state from $\mathbf{x}(t_0)$ to $\mathbf{x}(t_1)$.

b. If the input is zero, find an explicit expression for $\mathbf{x}(t_0)$ in terms of the output function $\mathbf{y}(t)$, $t_0 \le t \le t_1$.

a. The solution for the state at t_1 can be written in terms of the transformation $\mathcal{A}_c : \mathcal{U} \longrightarrow \Sigma$,

$$\mathbf{x}(t_1) - \mathbf{\Phi}(t_1, t_0)\mathbf{x}(t_0) = \mathcal{A}_c(\mathbf{u})$$

Let $\mathbf{u}(t) = \mathcal{A}_c^*(\mathbf{w})$, where $\mathbf{w}$ is an unknown vector in Σ. Then $\mathbf{x}(t_1) - \mathbf{\Phi}(t_1, t_0)\mathbf{x}(t_0) = \mathcal{A}_c\mathcal{A}_c^*(\mathbf{w})$. The condition for complete controllability is that the $n \times n$ matrix $\mathcal{A}_c\mathcal{A}_c^*$ has an inverse. Inverting this matrix to solve for $\mathbf{w}$ leads to

$$\begin{aligned}\mathbf{u}(t) &= \mathcal{A}_c^*(\mathcal{A}_c\mathcal{A}_c^*)^{-1}[\mathbf{x}(t_1) - \mathbf{\Phi}(t_1, t_0)\mathbf{x}(t_0)] \\ &= \bar{\mathbf{B}}^T(t)\bar{\mathbf{\Phi}}^T(t_1, t)\left[\int_{t_0}^{t_1} \mathbf{\Phi}(t_1, \tau)\mathbf{B}(\tau)\bar{\mathbf{B}}^T(\tau)\bar{\mathbf{\Phi}}^T(t_1, \tau)\,d\tau\right]^{-1}[\mathbf{x}(t_1) - \mathbf{\Phi}(t_1, t_0)\mathbf{x}(t_0)]\end{aligned}$$

b. In terms of the transformation $\mathcal{A}_0$, the output of the unforced system is $\mathbf{y}(t) = \mathcal{A}_0(\mathbf{x}(t_0))$. Operating on both sides with the adjoint transformation $\mathcal{A}_0^*$ gives $\mathcal{A}_0^*(\mathbf{y}(t)) = \mathcal{A}_0^*\mathcal{A}_0(\mathbf{x}(t_0))$. The criterion for complete observability ensures that the matrix $\mathcal{A}_0^*\mathcal{A}_0$ has an inverse, so

$$\mathbf{x}(t_0) = (\mathcal{A}_0^*\mathcal{A}_0)^{-1}\mathcal{A}_0^*(\mathbf{y}(t))$$

$$= \left[\int_{t_0}^{t_1} \bar{\mathbf{\Phi}}^T(\tau, t_0)\bar{\mathbf{C}}^T(\tau)\mathbf{C}(\tau)\mathbf{\Phi}(\tau, t_0)\, d\tau\right]^{-1} \int_{t_0}^{t_1} \bar{\mathbf{\Phi}}(\tau, t_0)\bar{\mathbf{C}}^T(\tau)\mathbf{y}(\tau)\, d\tau$$

Decomposition

12.14 Show that at any time $t_1 > t_0$, the state space Σ can be expressed as the direct sum $\Sigma = \mathcal{X}_1 \oplus \mathcal{X}_2$, where $\mathcal{X}_1$ is the subspace which contains all the controllable initial states and $\mathcal{X}_2$ is the null space of $\mathbf{G}'(t_1, t_0)$ of Problem 12.26.

Let $\mathbf{x}(t_0)$ be a controllable initial state. Then by definition, there exists an input such that

$$-\mathbf{x}(t_0) = \int_{t_0}^{t_1} \mathbf{\Phi}(t_0, \tau)\mathbf{B}(\tau)\mathbf{u}(\tau)\, d\tau \triangleq = \mathcal{A}(\mathbf{u})$$

All controllable initial states belong to $\mathfrak{R}(\mathcal{A}) \triangleq \mathcal{X}_1$. Since $\mathcal{A}: \mathcal{U} \longrightarrow \Sigma$, the results of Problem 6.8 give $\Sigma = \mathfrak{R}(\mathcal{A}) \oplus \mathfrak{N}(\mathcal{A}^*)$. From Sec. 12.5, $\mathfrak{N}(\mathcal{A}^*) = \mathfrak{N}(\mathcal{A}\mathcal{A}^*)$. For this example, $\mathcal{A}^* = \bar{\mathbf{B}}^T(t)\bar{\mathbf{\Phi}}^T(t_0, t)$ so that $\mathcal{A}\mathcal{A}^* = \mathbf{G}'(t_1, t_0)$. Therefore, $\mathfrak{N}(\mathcal{A}^*) = \mathfrak{N}(\mathbf{G}'(t_1, t_0))$. Note that this result implies that for each $\mathbf{x}(t_0) \in \Sigma$, $\mathbf{x}(t_0) = \mathbf{x}_1 + \mathbf{x}_2$ with $\mathbf{x}_1 \in \mathcal{X}_1$ (controllable) and $\mathbf{x}_2 \in \mathcal{X}_2$ (not controllable). However, it does not imply that the set of all uncontrollable states is the subspace $\mathcal{X}_2$.

Let $\mathbf{x}_a(t_0) = \mathbf{x}_1 + \mathbf{x}_2$ and $\mathbf{x}_b(t_0) = \mathbf{x}_1 - \mathbf{x}_2$. Since $\mathbf{x}_a(t_0)$ and $\mathbf{x}_b(t_0)$ do not belong to $\mathcal{X}_1$ they are not controllable. They do not belong to $\mathcal{X}_2$ either. Still, $\mathbf{x}_a(t_0) + \mathbf{x}_b(t_0) = 2\mathbf{x}_1$ belongs to $\mathcal{X}_1$ and is therefore controllable. The set of all uncontrollable initial states is *not* a subspace.

12.15 Let $\mathcal{A}_0: \Sigma \longrightarrow \mathcal{Y}$ be the output transformation defined in Sec. 12.5. Then the results of Problem 6.8 allow the decomposition $\Sigma = \mathfrak{N}(\mathcal{A}_0) \oplus \mathfrak{R}(\mathcal{A}_0^*)$. It has been shown that $\mathfrak{N}(\mathcal{A}_0) = \mathfrak{N}(\mathcal{A}_0^*\mathcal{A}_0) = \mathfrak{N}(\mathbf{H}(t_1, t_0))$. Define this null space as $\mathcal{X}_3$. Then every $\mathbf{x}(t_0) \in \mathcal{X}_3$ contributes nothing to the output $\mathbf{y}(t)$, and these are referred to as unobservable states. Use this and the results of Problem 12.14 to show that for all $\mathbf{x}(t_0) \in \Sigma$, $\mathbf{x}(t_0) = \mathbf{x}_a + \mathbf{x}_b + \mathbf{x}_c + \mathbf{x}_d$, where $\mathbf{x}_a$ is controllable but unobservable, $\mathbf{x}_b$ is controllable and observable, $\mathbf{x}_c$ is uncontrollable but observable, and $\mathbf{x}_d$ is uncontrollable and unobservable.

Define $\mathcal{X}_4 = \mathcal{X}_3^{\perp} = \mathfrak{R}(\mathcal{A}_0^*)$. Each $\mathbf{x}(t_0) \in \mathcal{X}_4$ is observable in the sense that a unique $\mathbf{x}(t_0) \in \mathcal{X}_4$ can be associated with a given unforced output record $\mathbf{y}(t)$. (Of course, $\mathbf{x}'(t_0) = \mathbf{x}(t_0) + \mathbf{x}_3$ will give the same $\mathbf{y}(t)$ if $\mathbf{x}_3 \in \mathcal{X}_3$, so it is not possible to determine whether $\mathbf{x}(t_0)$ or $\mathbf{x}'(t_0)$ is the actual initial state.)

Every $\mathbf{x}(t_0)$ can be written as $\mathbf{x}(t_0) = \mathbf{x}_1 + \mathbf{x}_2$ with $\mathbf{x}_1 \in \mathcal{X}_1$, $\mathbf{x}_2 \in \mathcal{X}_2$. The orthogonal projection of $\mathbf{x}_1$ into $\mathcal{X}_3$ gives $\mathbf{x}_a$. The projection into $\mathcal{X}_4$ gives $\mathbf{x}_b$. Similarly, projecting $\mathbf{x}_2$ into $\mathcal{X}_4$ gives $\mathbf{x}_c$ and projecting $\mathbf{x}_2$ into $\mathcal{X}_3$ gives $\mathbf{x}_d$.

12.16 Indicate how a time-invariant linear system with distinct eigenvalues can be decomposed into four possible subsystems with the respective properties (1) controllable but unobservable, (2) controllable and observable, (3) uncontrollable but observable, and (4) uncontrollable and unobservable.

The system can be put into normal form, giving $\mathbf{x}(t_0) = q_1(t_0)\boldsymbol{\xi}_1 + q_2(t_0)\boldsymbol{\xi}_2 + \cdots + q_n(t_0)\boldsymbol{\xi}_n$.

For this class of systems, controllability and observability criteria 1 apply. The controllability and observability can be ascertained for each mode individually. The modes are each assigned to one of the four categories. The resulting decomposition is illustrated in Fig. 12.5.

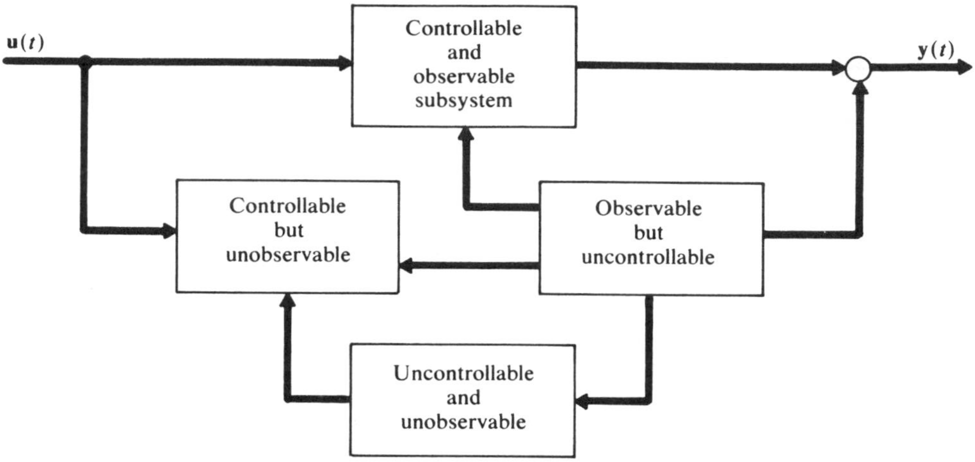

Figure 12.5

Notice that there is no signal path from the input to an uncontrollable subsystem, either directly or through other subsystems. Also, there is no signal path from an unobservable subsystem to the output. The decomposition of Fig. 12.5 can be accomplished for any linear system, but the process is not always this simple [59].

12.17 Subdivide the following system into subsystems as discussed in Problem 12.16:

$$\dot{\mathbf{x}} = \begin{bmatrix} -7 & -2 & 6 \\ 2 & -3 & -2 \\ -2 & -2 & 1 \end{bmatrix}\mathbf{x} + \begin{bmatrix} 1 & 1 \\ 1 & -1 \\ 1 & 0 \end{bmatrix}\mathbf{u}, \qquad \mathbf{y} = \begin{bmatrix} -1 & -1 & 2 \\ 1 & 1 & -1 \end{bmatrix}\mathbf{x}$$

The eigenvalues of $\mathbf{A}$ are $\lambda_i = -1, -3$, and -5. The Jordan normal form will be used, since the controllability and observability criteria 1 apply. The modal matrix containing the eigenvectors is

$$\mathbf{M} = \begin{bmatrix} 1 & 1 & 1 \\ 0 & 1 & -1 \\ 1 & 1 & 0 \end{bmatrix} \quad \text{and} \quad \mathbf{M}^{-1} = \begin{bmatrix} -1 & -1 & 2 \\ 1 & 1 & -1 \\ 1 & 0 & -1 \end{bmatrix}$$

so that

$$\dot{\mathbf{q}} = \begin{bmatrix} -1 & 0 & 0 \\ 0 & -3 & 0 \\ 0 & 0 & -5 \end{bmatrix}\mathbf{q} + \begin{bmatrix} 0 & 0 \\ 1 & 0 \\ 0 & 1 \end{bmatrix}\mathbf{u}, \qquad \mathbf{y} = \begin{bmatrix} 1 & 0 & 0 \\ 0 & 1 & 0 \end{bmatrix}\mathbf{q}$$

The first mode is uncontrollable and the third mode is unobservable. The second mode is both controllable and observable. There is no mode which is both uncontrollable and unobservable. Figure 12.6 illustrates the three subsystems.

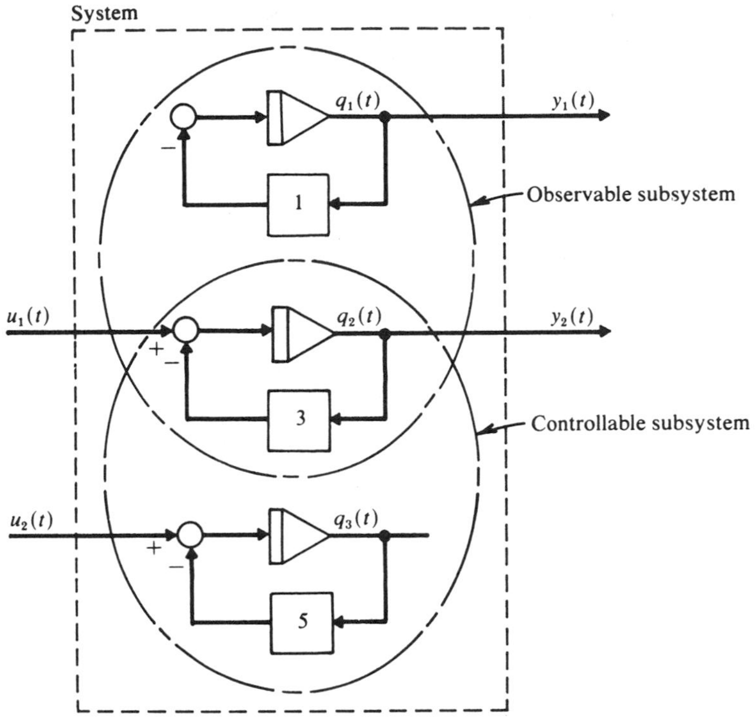

Figure 12.6

PROBLEMS

12.18 A continuous-time system is represented by $\mathbf{A} = \begin{bmatrix} 2 & -5 \\ -4 & 0 \end{bmatrix}$, $\mathbf{B} = \begin{bmatrix} 1 \\ -1 \end{bmatrix}$, $\mathbf{C} = [1 \quad 1]$. Is this system completely controllable and completely observable?

12.19 Investigate the controllability properties of time-invariant systems $\dot{\mathbf{x}} = \mathbf{A}\mathbf{x} + \mathbf{B}u$ if $u(t)$ is a scalar, and

a. $\mathbf{A} = \begin{bmatrix} -5 & 1 \\ 0 & 4 \end{bmatrix}$, $\mathbf{B} = \begin{bmatrix} 1 \\ 1 \end{bmatrix}$ b. $\mathbf{A} = \begin{bmatrix} 0 & 1 & 0 \\ 0 & 0 & 1 \\ 0 & 0 & 0 \end{bmatrix}$, $\mathbf{B} = \begin{bmatrix} 0 \\ 0 \\ 1 \end{bmatrix}$

c. $\mathbf{A} = \begin{bmatrix} 3 & 3 & 6 \\ 1 & 1 & 2 \\ 2 & 2 & 4 \end{bmatrix}$, $\mathbf{B} = \begin{bmatrix} 0 \\ 0 \\ 1 \end{bmatrix}$

12.20 Investigate the controllability and observability of the following systems. Note the results are unaffected by whether or not the system has a nonzero **D** matrix.

a.

$$\mathbf{A} = \begin{bmatrix} -1 & 0 & 0 & 0 & 0 & 0 \\ 0 & -2 & 1 & 0 & 0 & 0 \\ 0 & 0 & -1 & 0 & 0 & 0 \\ 0 & 0 & 0 & -3 & 0 & 0 \\ 0 & 0 & 0 & 0 & -3 & 1 \\ 0 & 0 & 0 & 0 & 0 & -1 \end{bmatrix}, \quad \mathbf{B} = \begin{bmatrix} 1 & 0 \\ 0 & 0 \\ 0 & 1 \\ 0 & 1 \\ 0 & 0 \\ 1 & 0 \end{bmatrix}, \quad \mathbf{C}^T = \begin{bmatrix} 1 & 0 \\ 2 & 0 \\ 0 & 0 \\ 0 & 1 \\ 0 & 1 \\ 0 & 0 \end{bmatrix}$$

b.

$$\mathbf{A} = \begin{bmatrix} -6 & 1 & 0 \\ -11 & 0 & 1 \\ -6 & 0 & 0 \end{bmatrix}, \quad \mathbf{B} = \begin{bmatrix} 1 \\ 6 \\ 5 \end{bmatrix}, \quad \mathbf{C} = [1 \quad 0 \quad 0]$$

c.

$$\mathbf{A} = \begin{bmatrix} -1 & 3 & 0 & 0 \\ -3 & -1 & 0 & 0 \\ 0 & 0 & -5 & 0 \\ 0 & 0 & 0 & -5 \end{bmatrix}, \quad \mathbf{B} = \begin{bmatrix} 1 & 0 \\ 0 & 0 \\ 1 & 0 \\ 0 & -1 \end{bmatrix}, \quad \mathbf{C} = \frac{1}{25}\begin{bmatrix} -3 & -4 & 3 & 250 \\ 4 & -3 & -4 & 100 \end{bmatrix}$$

12.21 A factor in determining useful life of a flexible structure, such as a ship, a tall building, or a large airplane, is the possibility of fatigue failures due to structural vibrations. Each vibration mode is described by an equation of the form $m\ddot{x} + kx = u(t)$, where $u(t)$ is the input force. Is it possible to find an input which will drive both the deflection $x(t)$ and the velocity $\dot{x}(t)$ to zero in finite time for arbitrary initial conditions?

12.22 Investigate the controllability and observability of the mechanical system of Fig. 12.7. Use x_1 and x_2 as state variables, $u(t)$ as the input force, and $y(t) = x_1(t)$ as the output. Assume the masses m_1 and m_2 are negligible [31].

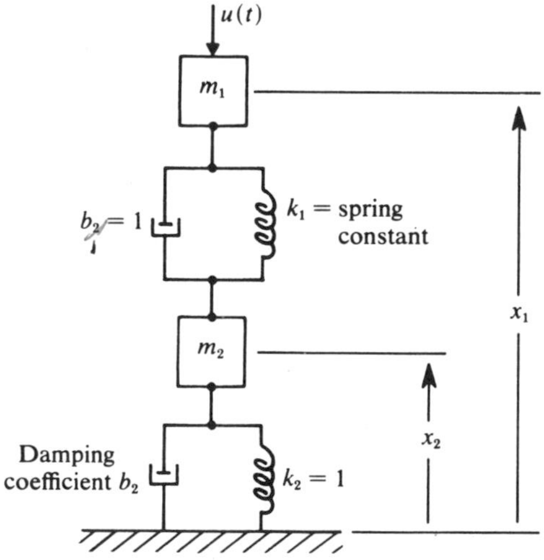

Figure 12.7

12.23 Is the system of Fig. 12.8 completely controllable and completely observable?

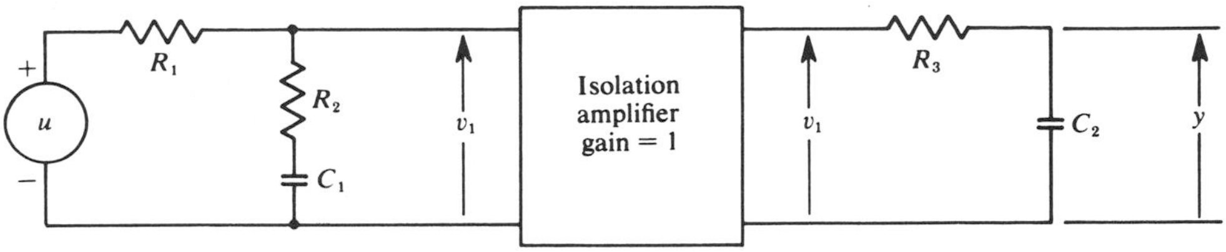

Figure 12.8

12.24 Determine whether the circuits of Fig. 12.9 are completely controllable and completely observable.

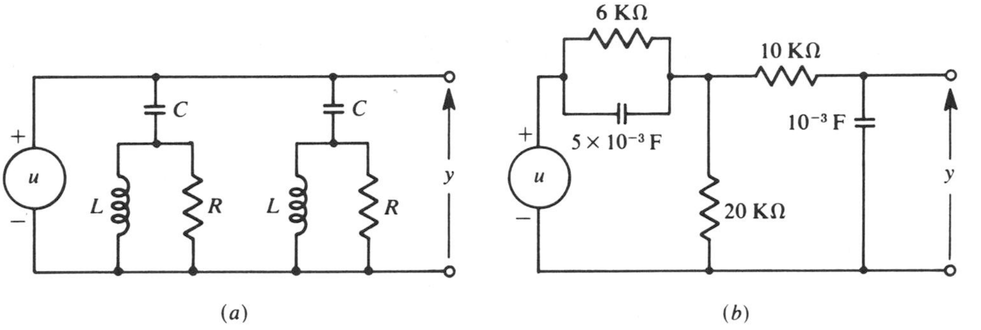

Figure 12.9

12.25 An nth order system has a time-invariant representation $\{\mathbf{A}, \mathbf{B}, \mathbf{C}, \mathbf{D}\}$.

a. Show that rank $\mathbf{P} = n \Leftrightarrow \mathbf{G}(t_1, t_0)$ is positive definite for all $t_1 > t_0$. Use the definitions in equations (*12.5*) and (*12.7*).

b. Show that rank $\mathbf{Q} = n \Leftrightarrow \mathbf{H}(t_1, t_0)$ is positive definite for all $t_1 > t_0$. Use the definitions in equations (*12.6*) and (*12.8*).

12.26 The criterion for complete controllability is often stated as the requirement that

$$\mathbf{G}'(t_1, t_0) = \int_{t_0}^{t_1} \mathbf{\Phi}(t_0, t)\mathbf{B}(t)\mathbf{B}^T(t)\mathbf{\Phi}^T(t_0, t)\, dt$$

be positive definite for some finite $t_1 > t_0$. Show that this is equivalent to the controllability criterion 3, provided that the system matrices $\mathbf{A}$ and $\mathbf{B}$ are real.

Chapter
13

NONLINEAR EQUATIONS AND PERTURBATION THEORY

13.1 INTRODUCTION

The formulation of the state equations in Chapter 9 allowed for the possibility of nonlinear equations. However, most of the work in intervening chapters was restricted to linear systems. The reason for this is that a general theory is available for the solution of linear equations. This is true for algebraic equations, differential equations, and difference equations. It is not true in the nonlinear case. Except in some special cases, complete analytical solutions to systems of nonlinear equations are not possible. Computer solution, either analog or digital or a hybrid combination, can be used to obtain the solution for a specific input and specific initial conditions.

This chapter presents some techniques which are useful for (1) obtaining approximate solutions for nonlinear equations, (2) understanding the behavior of a nonlinear system when it is perturbed slightly from a known nominal solution, (3) developing iterative correction techniques which are often required when controlling nonlinear systems, even if a computer is available.

13.2 THE GRADIENT OPERATOR AND DIFFERENTIATION WITH RESPECT TO A VECTOR

Let $f(x_1, x_2, \ldots, x_n)$ be a scalar-valued function of n variables x_i. The variables may be, but need not be, state variables in the present discussion. For notational convenience the dependence on n variables x_i is written as $f(\mathbf{x})$, with $\mathbf{x}$ being a vector with components x_i. The n partial derivatives of $f(\mathbf{x})$, $\partial f/\partial x_i$, will be used frequently. It is convenient to group these partials into an array and give the array a special symbol. A single-rowed array, a row vector, could be, and frequently is, used for this purpose. Here the array is arranged as a single column, that is, a column vector. This convention is an arbitrary choice. Both forms are used in the literature, and both are referred to as the *gradient vector* [28]. Three different symbols are frequently used to identify the gradient of $f(\mathbf{x})$, $\nabla_{\mathbf{x}} f = \text{grad}_{\mathbf{x}} f = df/d\mathbf{x}$. The meaning of these symbols is given by

$$\nabla_{\mathbf{x}} f = \begin{bmatrix} \dfrac{\partial f}{\partial x_1} & \dfrac{\partial f}{\partial x_2} & \cdots & \dfrac{\partial f}{\partial x_n} \end{bmatrix}^T \tag{13.1}$$

The only differences which arise between the row and column vector definitions are the presence or absence of the transpose in various algebraic manipulations. Conformability requirements for matrix multiplication must always be satisfied and can be used to determine whether a row or column definition is implied.

Example 13.1

Let $f_1(\mathbf{x}) = \mathbf{x}^T\mathbf{A}\mathbf{y}$, a bilinear function. Expanding this in terms of individual components gives $f_1(\mathbf{x}) = \sum_i \sum_j a_{ij}x_i y_j$. A typical component of the gradient is $\partial f_1/\partial x_k = \sum_i \sum_j a_{ij}(\partial x_i/\partial x_k)y_j$.

Using the independence of the x_i components gives $\partial x_i/\partial x_k = \delta_{ik} = \begin{cases} 1 \text{ if } i = k \\ 0 \text{ if } i \neq k \end{cases}$. This means that only one term in the summation over i is nonzero, so $\partial f_1/\partial x_k = \sum_j a_{kj}y_j$. This is just the kth component of the matrix product $\mathbf{Ay}$, so the gradient vector for this example is $\nabla_{\mathbf{x}}(\mathbf{x}^T\mathbf{A}\mathbf{y}) = \mathbf{Ay}$. ■

Example 13.2

If $f_2(\mathbf{x}) = \mathbf{y}^T\mathbf{A}\mathbf{x}$, then $\nabla_{\mathbf{x}} f_2 \neq \mathbf{y}^T\mathbf{A}$. The gradient operator *is not* simply a cancelling of the $\mathbf{x}$ vector as might be inferred from Example 13.1. By convention, the gradient is a column vector, so it cannot be equal to the row vector $\mathbf{y}^T\mathbf{A}$. The correct expression for the gradient is $\nabla_{\mathbf{x}} f_2 = \mathbf{A}^T\mathbf{y}$. ■

Example 13.3

Let $f_3(\mathbf{x}) = \mathbf{x}^T\mathbf{A}\mathbf{x}$, a quadratic form. In summation notation,

$$f_3(\mathbf{x}) = \sum_i \sum_j a_{ij}x_i x_j \quad \text{and} \quad \frac{\partial f_3}{\partial x_k} = \sum_i \sum_j a_{ij} \left\{ \frac{\partial x_i}{\partial x_k} x_j + x_i \frac{\partial x_j}{\partial x_k} \right\} = \sum_j a_{kj}x_j + \sum_i a_{ik}x_i$$

Returning to matrix notation $\nabla_{\mathbf{x}}(\mathbf{x}^T\mathbf{A}\mathbf{x}) = \mathbf{Ax} + \mathbf{A}^T\mathbf{x}$. If $\mathbf{A} = \mathbf{A}^T$, as is usual when dealing with quadratic forms, then $\nabla_{\mathbf{x}}(\mathbf{x}^T\mathbf{A}\mathbf{x}) = 2\mathbf{Ax}$. ■

The geometrical interpretation of the gradient is often useful. To aid in visualization, the vector $\mathbf{x}$ is restricted to two components. Then for each point $\mathbf{x}$ in the plane, the function $f(\mathbf{x})$ has some prescribed value. Figure 13.1 shows such a function.

The equation $f(\mathbf{x}) = c$, with c constant, specifies a locus of points in the plane. Figure 13.2 shows the locus of points in the plane for several different values of c.

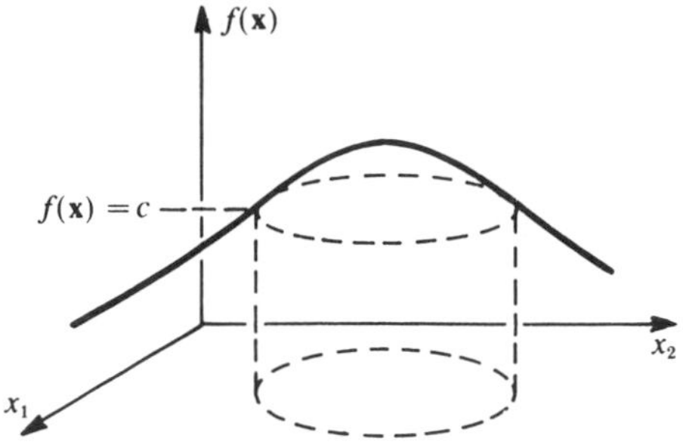

Figure 13.1

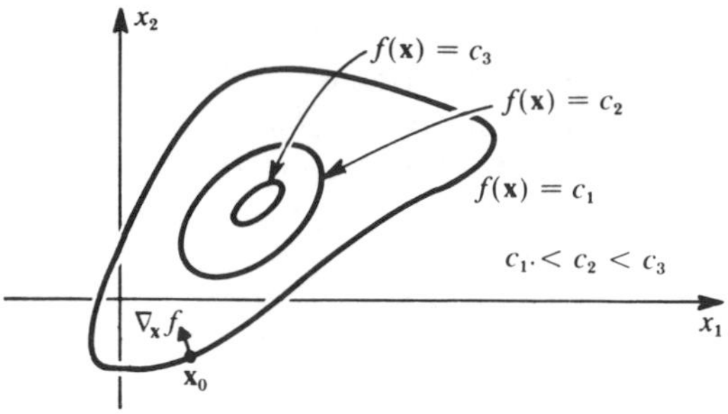

Figure 13.2

At a given point such as $\mathbf{x}_0$ in Fig. 13.2, $\nabla_{\mathbf{x}} f$ is a vector normal to the curve $f(\mathbf{x}) = c$, and it points in the direction of increasing values of $f(\mathbf{x})$. The gradient defines the direction of maximum increase of the function $f(\mathbf{x})$ [111]. This normal can be used to define a tangent line, plane or hyperplane (depending on the dimension of $\mathbf{x}$) at the point $\mathbf{x}_0$. Using the notation $\nabla_{\mathbf{x}_0} f$ to indicate the gradient evaluated at $\mathbf{x}_0$, the tangent line, plane, or hyperplane is the locus of points $\mathbf{x}$ satisfying $\langle \nabla_{\mathbf{x}_0} f, \mathbf{x} \rangle = k$, where the constant k is given by $\langle \nabla_{\mathbf{x}_0} f, \mathbf{x}_0 \rangle$. These concepts are useful in describing the behavior of $f(\mathbf{x})$ near a given point $\mathbf{x}_0$. The determination of the points at which $f(\mathbf{x})$ attains relative maximum or minimum values is also facilitated by use of the gradient.

The derivative of a scalar function with respect to a vector yields a vector, the gradient vector. If a vector-valued function of a vector, $\mathbf{f}(\mathbf{x}) = [f_1(\mathbf{x}) f_2(\mathbf{x}) \cdots f_m(\mathbf{x})]^T$, is considered, the gradient of each component $f_i(\mathbf{x})$ is a column vector with the same dimension as $\mathbf{x}$. If the function $\mathbf{f}(\mathbf{x})$ is transposed to a row vector, then

$$\nabla_{\mathbf{x}} \mathbf{f}^T(\mathbf{x}) = [\nabla_{\mathbf{x}} f_1 \quad \nabla_{\mathbf{x}} f_2(\mathbf{x}) \quad \cdots \quad \nabla_{\mathbf{x}} f_m(\mathbf{x})] \tag{13.2}$$

is an $n \times m$ matrix whose columns are gradients. The transpose of this matrix will be denoted by the symbols $\nabla_{\mathbf{x}} \mathbf{f}(x)$ or simply $d\mathbf{f}/d\mathbf{x}$. That is, $d\mathbf{f}/d\mathbf{x} \triangleq [\partial f_i/\partial x_j]$ and this $m \times n$ matrix is the *Jacobian matrix*. Note that the symbol $d\mathbf{f}/d\mathbf{x}$ is just a suggestive name attached to the prescribed array of first partial derivatives.

The second partial derivatives of a function of a vector also arise on occasion. When $f(\mathbf{x})$ is a scalar-valued function, the matrix of all second partial derivatives, called the *Hessian matrix*, will be denoted by

$$\frac{d^2 f}{d\mathbf{x}^2} = \begin{bmatrix} \dfrac{\partial^2 f}{\partial x_1^2} & \dfrac{\partial^2 f}{\partial x_1 \partial x_2} & \cdots & \dfrac{\partial^2 f}{\partial x_1 \partial x_n} \\ \dfrac{\partial^2 f}{\partial x_2 \partial x_1} & \dfrac{\partial^2 f}{\partial x_2^2} & \cdots & \dfrac{\partial^2 f}{\partial x_2 \partial x_n} \\ \vdots & & & \vdots \\ \dfrac{\partial^2 f}{\partial x_n \partial x_1} & \dfrac{\partial^2 f}{\partial x_n \partial x_2} & \cdots & \dfrac{\partial^2 f}{\partial x_n^2} \end{bmatrix} \tag{13.3}$$

It becomes notationally awkward to continue these definitions to the second derivative of a vector function **f** with respect to a vector. This would require a three-dimensional array, with a typical element being $\partial^2 f_i/\partial x_j \partial x_k$.

13.3 GENERALIZED TAYLOR SERIES

The Taylor series expansion is one of the most useful formulas in the analysis of nonlinear equations. The expansion of a scalar-valued function of a scalar is recalled [111]:

$$f(x_0 + \delta x) = f(x_0) + \left.\frac{df}{dx}\right|_0 \delta x + \frac{1}{2!}\left.\frac{d^2f}{dx^2}\right|_0 \delta x^2 + \cdots \tag{13.4}$$

The notation $\left.\frac{df}{dx}\right|_0$ indicates that all derivatives are evaluated at the point x_0.

The Taylor expansion of a function of two variables x and y is

$$f(x_0 + \delta x, y_0 + \delta y) = f(x_0, y_0) + \left.\frac{\partial f}{\partial x}\right|_0 \delta x + \left.\frac{\partial f}{\partial y}\right|_0 \delta y$$
$$+ \frac{1}{2!}\left[\left.\frac{\partial^2 f}{\partial x^2}\right|_0 \delta x^2 + 2\left.\frac{\partial^2 f}{\partial x \partial y}\right|_0 \delta x\, \delta y + \left.\frac{\partial^2 f}{\partial y^2}\right|_0 \delta y^2\right] + \cdots \tag{13.5}$$

If the two variables x and y are used to define the vector $\mathbf{x} = [x \quad y]^T$, the above expansion is more compactly written as

$$f(\mathbf{x}_0 + \delta\mathbf{x}) = f(\mathbf{x}_0) + (\nabla_{\mathbf{x}} f|_0)^T \delta\mathbf{x} + \frac{1}{2!}\delta\mathbf{x}^T \left.\frac{d^2 f}{d\mathbf{x}^2}\right|_0 \delta\mathbf{x} + \cdots \tag{13.6}$$

This generalized form of the Taylor expansion is valid for any number of components of the vector **x**.

An m component vector-valued function **f**(**x**) can be viewed as m separate scalar functions. The Taylor expansion through the first two terms can be written as

$$\mathbf{f}(\mathbf{x}_0 + \delta\mathbf{x}) = \mathbf{f}(\mathbf{x}_0) + \nabla_{\mathbf{x}}\mathbf{f}|_{\mathbf{x}_0} \delta\mathbf{x} + \cdots \tag{13.7}$$

The slight discrepancy in the gradient terms of equations (*13.6*) and (*13.7*) is due to the definition of the gradient as a column vector and is the reason why the row definition is preferred by some authors. Higher terms in equation (*13.7*) cannot be written conveniently in matrix notation. However, the first-order terms in $\delta\mathbf{x}$ are frequently all that are used, and a good approximation results if all components of $\delta\mathbf{x}$ are sufficiently small.

13.4 LINEARIZATION OF NONLINEAR EQUATIONS

The equations for a continuous-time nonlinear dynamic system take the form $\dot{\mathbf{x}}(t) = \mathbf{f}(\mathbf{x}(t), \mathbf{u}(t), t)$. Suppose that a nominal input $\mathbf{u}_n(t)$ and the resulting nominal state $\mathbf{x}_n(t)$ are known, perhaps from computer solution. That is, $\mathbf{u}_n$ and $\mathbf{x}_n$ satisfy

$\dot{\mathbf{x}}_n = \mathbf{f}(\mathbf{x}_n, \mathbf{u}_n, t)$. A solution is desired for another input $\mathbf{u}(t)$ which is slightly different from $\mathbf{u}_n(t)$, $\mathbf{u}(t) = \mathbf{u}_n(t) + \delta\mathbf{u}(t)$. Let the resulting state be $\mathbf{x}(t) = \mathbf{x}_n(t) + \delta\mathbf{x}$. If $\delta\mathbf{x}(t)$ can be determined approximately, then an approximate solution for $\mathbf{x}(t)$ will have been found. This is accomplished as follows:

$$\dot{\mathbf{x}}(t) = \dot{\mathbf{x}}_n(t) + \delta\dot{\mathbf{x}}(t) = \mathbf{f}(\mathbf{x}_n + \delta\mathbf{x}, \mathbf{u}_n + \delta\mathbf{u}, t)$$

The Taylor series expansion of the nonlinear function $\mathbf{f}$ can be utilized provided the indicated derivatives exist:

$$\mathbf{f}(\mathbf{x}_n + \delta\mathbf{x}, \mathbf{u}_n + \delta\mathbf{u}, t) \simeq \mathbf{f}(\mathbf{x}_n, \mathbf{u}_n, t) + \nabla_{\mathbf{x}}\mathbf{f}\,|_n\,\delta\mathbf{x} + \nabla_{\mathbf{u}}\mathbf{f}\,|_n\,\delta\mathbf{u}$$

This approximation assumes $\delta\mathbf{x}$ and $\delta\mathbf{u}$ are sufficiently small so that squared terms and product terms δx_i^2, δu_j^2, $\delta x_i \delta u_j$ are negligible. By assumption, $\dot{\mathbf{x}}_n = \mathbf{f}(\mathbf{x}_n, \mathbf{u}_n, t)$. Cancelling these terms leaves a *linear* differential equation for $\delta\mathbf{x}(t)$:

$$\delta\dot{\mathbf{x}} = \nabla_{\mathbf{x}}\mathbf{f}\,|_n\,\delta\mathbf{x} + \nabla_{\mathbf{u}}\mathbf{f}\,|_n\,\delta\mathbf{u}$$

or (*13.8*)

$$\delta\dot{\mathbf{x}} = \mathbf{A}(t)\,\delta\mathbf{x} + \mathbf{B}(t)\,\delta\mathbf{u}$$

where $\mathbf{A}(t)$ and $\mathbf{B}(t)$ are the matrices defined from the gradients and are evaluated with the nominal arguments $\mathbf{x}_n(t)$ and $\mathbf{u}_n(t)$. The linear equations (*13.8*) may still not be easy to solve because the matrices $\mathbf{A}$ and $\mathbf{B}$ will almost always be time variable. However, the *form* of the solutions is well known and has been discussed in Chapter 10. Although solving equation (*13.8*) may be a nontrivial task, it is certainly easier than solving the original nonlinear equations.

The perturbation equations (*13.8*) provide the most common approach to studying the properties and behavior of a nonlinear system in the neighborhood of a known nominal solution [14]. They also provide a means of obtaining approximate solutions to nonlinear equations. Exactly the same technique can be applied to nonlinear difference equations which describe discrete-time systems.

13.5 QUASILINEARIZATION

The linearization technique of Sec. 13.4 requires knowledge of the nominal input $\mathbf{u}_n(t)$ and state $\mathbf{x}_n(t)$. A brief introduction to a method known as quasilinearization is given here. It is an iterative method of obtaining approximate solutions to nonlinear equations, and it does not require advance knowledge of a nominal solution $\mathbf{x}_n(t)$. The input $\mathbf{u}(t)$ must be known in all cases. It makes little sense to seek solutions if the input is unknown.

Only the simplest application of quasilinearization is discussed here. Additional material is contained in the literature [8, 79, 100]. Consider the equations

$$\dot{\mathbf{x}}(t) = \mathbf{f}(\mathbf{x}, \mathbf{u}, t), \qquad \mathbf{x}(t_0) \text{ known}$$

As a first guess at the solution, set $\mathbf{x}^{(1)}(t) = \mathbf{x}(t_0)$; this constant vector at least satisfies the boundary conditions, but is otherwise a poor approximation. The next approximation is called $\mathbf{x}^{(2)}(t)$, and the difference $\delta\mathbf{x}^{(1)}$ is defined by $\mathbf{x}^{(2)}(t) = \mathbf{x}^{(1)}(t) + \delta\mathbf{x}^{(1)}(t)$. Then $\dot{\mathbf{x}}^{(2)} = \mathbf{f}(\mathbf{x}^{(2)}, \mathbf{u}, t) = \mathbf{f}(\mathbf{x}^{(1)} + \delta\mathbf{x}^{(1)}, \mathbf{u}, t)$. The Taylor series expansion gives

$$\dot{\mathbf{x}}^{(2)} = \mathbf{f}(\mathbf{x}^{(1)}, \mathbf{u}, t) + \nabla_{\mathbf{x}}\mathbf{f}^{(1)}\,\delta\mathbf{x}^{(1)}$$

Eliminating $\delta\mathbf{x}^{(1)}$ gives

$$\dot{\mathbf{x}}^{(2)} = \nabla_{\mathbf{x}}\mathbf{f}^{(1)}\mathbf{x}^{(2)} + [\mathbf{f}(\mathbf{x}^{(1)}, \mathbf{u}, t) - \nabla_{\mathbf{x}}\mathbf{f}^{(1)}\mathbf{x}^{(1)}(t)] = \mathbf{A}(t)\mathbf{x}^{(2)} + \mathbf{v}^{(1)}(t)$$

Clearly the bracketed term, which is defined as $\mathbf{v}^{(1)}(t)$, is a known function of time and can be treated as an input. This allows $\mathbf{x}^{(2)}(t)$ to be obtained by solving a nonhomogeneous *linear* equation. Then $\mathbf{x}^{(2)}(t)$ is used to solve another equation for $\mathbf{x}^{(3)}(t)$, and so on. The $(k+1)$th approximation is obtained by solving

$$\dot{\mathbf{x}}^{(k+1)}(t) = \nabla_{\mathbf{x}}\mathbf{f}^{(k)}\mathbf{x}^{(k+1)}(t) + [\mathbf{f}(\mathbf{x}^{(k)}(t), \mathbf{u}(t), t) - \nabla_{\mathbf{x}}\mathbf{f}^{(k)}\mathbf{x}^{(k)}(t)]$$

The iterative process is continued until $\mathbf{x}^{(k+1)}(t)$ and $\mathbf{x}^{(k)}(t)$ are sufficiently close to each other. At that point, $\mathbf{x}^{(k+1)}(t)$ represents a good approximation to the original nonlinear equation.

The iteration procedure described above is best suited to a computer because each step requires the solution of a set of linear but time-varying nonhomogeneous differential equations. It would be fair to ask, "Why not solve the original nonlinear equation numerically in the first place?" The answer rests with the boundary conditions. If all components of $\mathbf{x}(t_0)$ are known, then it is true that numerical integration gives $\mathbf{x}(t)$ quite easily. As will be seen in Chapter 17, many optimization problems lead to a two-point boundary value problem. Some boundary conditions are given at the initial time t_0 and others at a final time t_1. If the differential equations are linear, this does not present much difficulty (see Problem 5.24). It is the combination of nonlinear equations and split boundary conditions that makes solutions difficult to obtain. The worth of quasilinearization is fully demonstrated with this class of problem [8, 79, 100].

13.6 CONDITIONS FOR EXTREMALS OF FUNCTIONS

Another common use of the Taylor expansion is in deriving conditions which a function must satisfy at a point of relative maximum or relative minimum. Figure 13.3 shows a function of a single variable $f(x)$ with $a \leq x \leq b$.

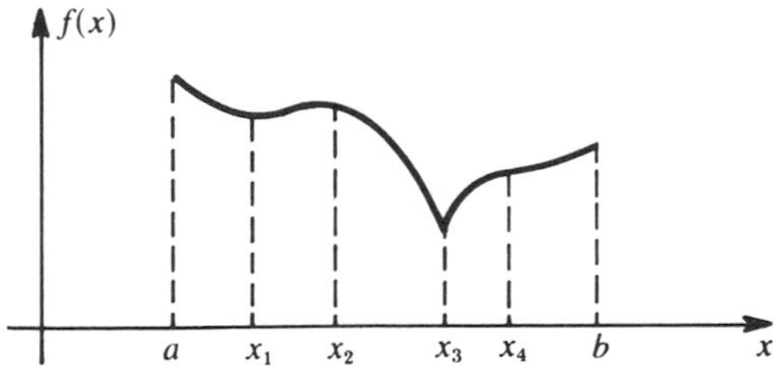

Figure 13.3

A maximum or minimum, i.e., an extremal, can occur at only three types of points:

1. Boundary points, such as $x = a$ or $x = b$.
2. Points where $df/dx = 0$, for example, at x_1, x_2, or x_4.
3. Points where df/dx is not uniquely defined, as at x_3.

In Fig. 13.3, $x = a$ is the overall or *global maximum*, x_3 is the *global minimum*, x_1 is a *relative minimum*, and x_2 and $x = b$ are *relative maxima*. The point x_4 is a *point of inflection.*

For simplicity, this section considers only functions defined on open domains so that boundary points need not be considered. Functions are assumed to possess continuous derivatives so that points like x_3 need not be considered.

Conditions for a relative minimum will be developed. Relative maximums are treated in an analogous manner. If x_1 is a point of relative minimum, then $f(x_1 + \delta x) - f(x_1) \geq 0$ for all δx sufficiently small. Using Taylor's series, the requirement is that

$$f(x_1) + \frac{df}{dx}\bigg|_{x_1} \delta x + \frac{1}{2}\frac{d^2f}{dx^2}\bigg|_{x_1} \delta x^2 + \cdots - f(x_1) \geq 0$$

The $f(x_1)$ terms cancel, and for δx sufficiently small, the sign is dominated by the linear term in δx. Since the sign of δx is not restricted, $\frac{df}{dx}\Big|_{x_1} \delta x \geq 0$ for all small δx requires that $df/dx = 0$ at a point of relative minimum. When this is true, the first nonzero term in the series is $\frac{1}{2}\frac{d^2f}{dx^2}\Big|_{x_1} \delta x^2$ and this term dominates the sign of the series. Therefore, at a point of relative minimum, $df/dx = 0$ and $d^2f/dx^2 > 0$. At a relative maximum, $df/dx = 0$ and $d^2f/dx^2 < 0$. At a point of inflection, $df/dx = 0$ and $d^2f/dx^2 = 0$. These well-known results are now generalized to scalar-valued functions of n variables. If the vector $\mathbf{x}_1$ is a point of relative minimum, then $f(\mathbf{x}_1 + \boldsymbol{\delta}\mathbf{x}) - f(\mathbf{x}_1) \geq 0$ for all perturbation vectors $\boldsymbol{\delta}\mathbf{x}$ which are sufficiently small. Using the vector version of the Taylor expansion reduces this to

$$[\nabla_{\mathbf{x}} f|_{\mathbf{x}_1}]^T \,\boldsymbol{\delta}\mathbf{x} + \frac{1}{2}\,\boldsymbol{\delta}\mathbf{x}^T\left[\frac{d^2f}{d\mathbf{x}^2}\bigg|_{\mathbf{x}_1}\right]\boldsymbol{\delta}\mathbf{x} + \cdots \geq 0$$

Unless the gradient is zero, the sign is dominated by the first term. To preserve the positive sign for all $\boldsymbol{\delta}\mathbf{x}$ it is necessary that $\nabla_{\mathbf{x}} f = \mathbf{0}$. If this is true and if the Hessian matrix $d^2f/d\mathbf{x}^2$ is positive definite at $\mathbf{x}_1$, then $\mathbf{x}_1$ is a point of relative minimum. If $d^2f/d\mathbf{x}^2$ is negative definite, $\mathbf{x}_1$ is a point of relative maximum.

Optimal control problems seek to maximize or minimize a performance criterion. Optimal control problems are more involved than those treated here because:

1. *The criterion to be maximized or minimized is not just a function of a vector* $\mathbf{x}$, *but a function of the control input function.* A function of a function is called a *functional.*

2. *There are additional side conditions or constraints that must be satisfied while searching for maximums or minimums.*
3. *The situation which is analogous to the boundary points $x = a$ and $x = b$ of Fig. 13.3 usually must be considered.* This often prevents the use of the simple necessary conditions for internal extremal points. Optimal control problems are considered in Chapter 17. Even though they are more complicated, extensive use will be made of the elementary reasoning presented in this section.

13.7 ITERATIVE SCHEMES FOR MINIMIZING OR MAXIMIZING A COST FUNCTION

The problem of finding the minimum or maximum of a scalar-valued cost function which depends on a vector of unknown parameters arises in many contexts. Nonlinear least squares (minimize $J(\mathbf{x})$) and maximum likelihood (maximize $L(\mathbf{x})$) are but two examples. Aside from direct enumeration, which is not feasible in most cases, the available methods for numerical solution divide into:

1. First-derivative-based schemes, known as steepest descent, hill climbing, or gradient methods.
2. Second-derivative-based schemes, known by a variety of names, including Newton-Raphson-type methods.

There are numerous variations on the above themes, involving how to select the step size, starting methods and stopping criteria, etc. In addition to these basic categories, three more can be mentioned.

3. Derivative-free methods (which often involve finite difference approximations to the above, by using just function evaluations) [13].
4. Combinations of the basic two categories. The Marquardt algorithm [12, 77, 101] is an example, as are the Fletcher-Powell-Davidon [37] methods.
5. Transformations of various types which lead to a simpler extremal problem to be solved. Sometimes the transformed problem is sufficiently simple to allow analytical maxima or minima to be found. Transforming from the product of variables to the sum of their logarithms is one simple example.

The central idea behind gradient methods is that if changes in the parameter estimates $\hat{\mathbf{x}}$ are made in the *direction* of the cost function gradient, say $\nabla_{\mathbf{x}} L$, and if the step size taken in that direction is not too large, then L ought to increase at each step. Hence the maximum should be found *eventually*. These schemes will usually converge if enough iterations are made. However, they are notoriously slow as the extremal point is approached. Rather large improvements in the cost function are achieved on the first few iterations, but the changes are very slow in the final stages. Clearly, changes in the unknown parameter vector should be made in the direction of the *negative* gradient if the minimum is being sought rather than the maximum. This is applicable to the least-squares problem.

Second-derivative Newton-type algorithms need a slightly more detailed discussion. Assume that an initial estimate $\hat{\mathbf{x}}_0$ is available, and expand $L(\mathbf{x})$ in a Taylor series through second-order terms, as in equation (*13.6*). Let the approximating quadratic function obtained by dropping all cubic and higher terms be called $\hat{L}(\mathbf{x})$.

$$\hat{L}(\mathbf{x}) = L(\hat{\mathbf{x}}_0 + \delta\mathbf{x}) = L(\hat{\mathbf{x}}_0) + [\nabla_{\mathbf{x}} L|_0]^T \delta\mathbf{x} + \frac{1}{2}\delta\mathbf{x}^T \frac{\mathbf{d}^2 L}{d\mathbf{x}^2}\bigg|_0 \delta\mathbf{x} \tag{13.9}$$

A depiction of the actual and approximate functions is presented in Fig. 13.4.

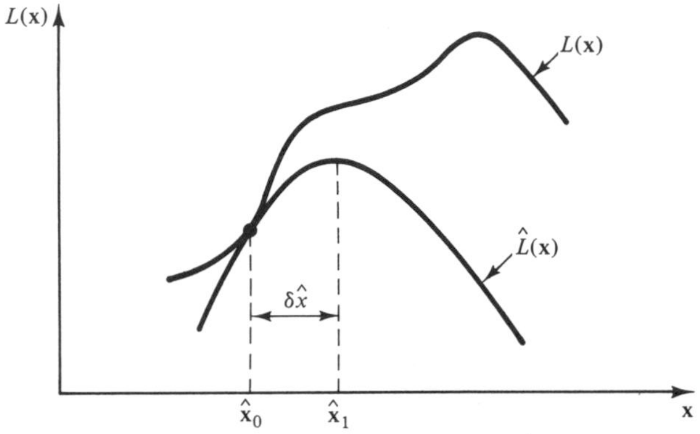

Figure 13.4

The operating principle in Newton-type methods is to select $\delta\hat{\mathbf{x}}$ so that the new estimate of $\hat{\mathbf{x}}$ is at the maximum (or mimimum) of the approximating function. Since the approximate function is a nice, smooth quadratic, its maximum is easily found by setting its first derivative with respect to $\mathbf{x}$ to zero.

$$\frac{\partial L(\mathbf{x})}{\partial \delta\boldsymbol{\theta}} = \nabla_{\mathbf{x}} L|_0 + \frac{d^2 L}{d\mathbf{x}^2}\bigg|_0 \delta\mathbf{x} = \mathbf{0}$$

or

$$\delta\hat{\mathbf{x}} = -\left[\frac{d^2 L}{d\mathbf{x}^2}\right]_0^{-1} \nabla_{\mathbf{x}} L|_0 \tag{13.10}$$

This correction is used to give the next estimate for $\hat{\mathbf{x}}_1$. The general step would be

$$\hat{\mathbf{x}}(k+1) = \hat{\mathbf{x}}(k) - \left[\frac{d^2 L}{d\mathbf{x}^2}\right]_k^{-1} \nabla_{\mathbf{x}} L|_k \tag{13.11}$$

The gradient vector $\mathbf{g}(k) = \nabla_{\mathbf{x}} L|_k$ and the Hessian matrix $\mathbf{S}(k) = \dfrac{d^2 L}{d\mathbf{x}^2}$ and the inverse of $\mathbf{S}$ must be evaluated at each step based upon the current estimate $\hat{\mathbf{x}}(k)$. Although the gradient vector is involved here, the direction in which changes in parameter estimate are made is not generally in the direction of the gradient.

ILLUSTRATIVE PROBLEMS

Linearizing Nonlinear Equations

13.1 The two-port nonlinear electrical device shown in Fig. 13.5 is characterized by the four quantities i_1, i_2, v_1, and v_2, Use Taylor series to develop a linear model for small signal variations about a nominal operating point.

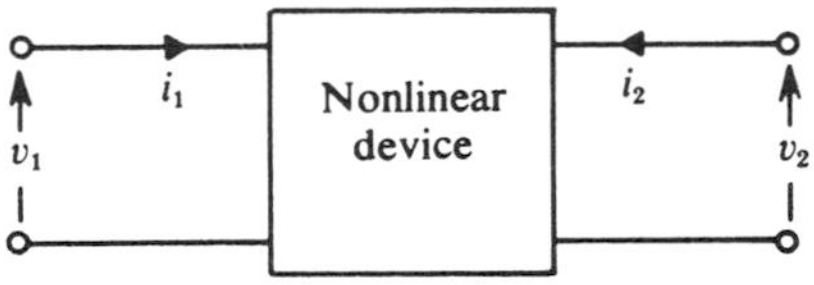

Figure 13.5

There are a variety of linear models that can be developed, depending on the form assumed for the functional relations among the four variables. There is one relationship at the input port and another at the output port. One possibility is the pair $v_1 = f_0(v_2, i_1, i_2)$ and $i_2 = g_0(i_1, v_2, v_1)$. These can be combined to yield $v_1 = f_0(v_2, i_1, g_0(i_1, v_2, v_1))$ and $i_2 = g_0(i_1, v_2, f_0(v_2, i_1, i_2))$. This shows that only two independent variables i_1 and v_2 suffice to determine v_1 and i_2. These relationships are rewritten more simply as $v_1 = f(i_1, v_2)$ and $i_2 = g(i_1, v_2)$. Let $v_2 = v_{2n} + \delta v_2$ and $i_1 = i_{1n} + \delta i_1$, where v_{2n} and i_{1n} define the nominal operating point and δv_2 and δi_1 are small variations from the nominal. Then Taylor series expansion gives

$$v_1 = f(i_{1n}, v_{2n}) + \frac{\partial f}{\partial i_1}\bigg|_n \delta i_1 + \frac{\partial f}{\partial v_2}\bigg|_n \delta v_2$$

$$i_2 = g(i_{1n}, v_{2n}) + \frac{\partial g}{\partial i_1}\bigg|_n \delta i_1 + \frac{\partial g}{\partial v_2}\bigg|_n \delta v_2$$

Obviously, $v_{1n} = f(i_{1n}, v_{2n})$ and $i_{2n} = g(i_{1n}, v_{2n})$ so that

$$\delta v_1 \triangleq v_1 - v_{1n} = h_{11}\,\delta i_1 + h_{12}\,\delta v_2$$

$$\delta i_2 \triangleq i_2 - i_{2n} = h_{21}\,\delta i_1 + h_{22}\,\delta v_2$$

where the h_{ij} terms are the partial derivatives evaluated at the nominal point. These are the hybrid or h parameters commonly used in small signal, linearized analysis of transistors and other nonlinear devices.

13.2 A tracking station measures the azimuth angle α, the elevation angle β, and the range r to an earth satellite as shown in Fig. 13.6.

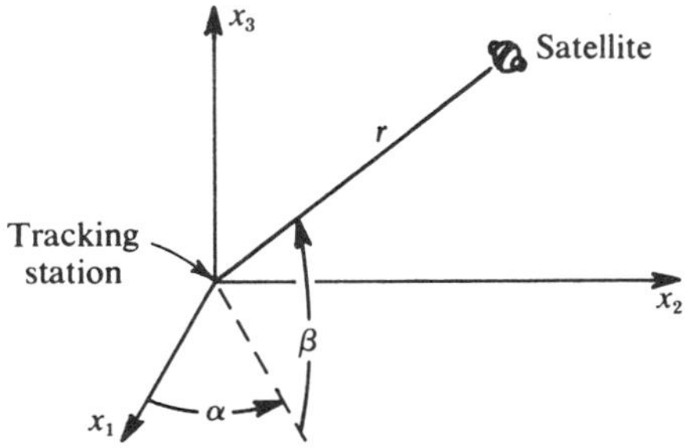

Figure 13.6

a. Derive the nonlinear equations which relate the satellite's relative position $[x_1 \quad x_2 \quad x_3]^T$ to the measured quantities.
b. Obtain linear equations which relate small perturbations in satellite location to small perturbations in the measurements.

a. The station-to-satellite range magnitude is $r = \sqrt{x_1^2 + x_2^2 + x_3^2}$ and the tracking antenna angles are $\alpha = \tan^{-1}(x_2/x_1)$ and $\beta = \tan^{-1}(x_3/\sqrt{x_1^2 + x_2^2})$.
b. Letting $\mathbf{x}(t) = \mathbf{x}_n(t) + \delta\mathbf{x}(t)$, $r(t) = r_n(t) + \delta r(t)$, $\alpha(t) = \alpha_n(t) + \delta\alpha(t)$, and $\beta(t) = \beta_n(t) + \delta\beta(t)$, the Taylor series expansion gives

$$\delta r = \sum_{i=1}^{3} \frac{\partial r}{\partial x_i}\,\delta x_i = \frac{1}{r_n}\mathbf{x}_n^T\,\delta\mathbf{x}$$

$$\delta\alpha = \sum_{i=1}^{3} \frac{\partial \alpha}{\partial x_i}\,\delta x_i = 1/(x_{1n}^2 + x_{2n}^2)[-x_{2n} \quad x_{1n} \quad 0]\delta\mathbf{x}$$

$$\delta\beta = \sum_{i=1}^{3} \frac{\partial \beta}{\partial x_i}\,\delta x_i = \frac{1}{r_n^2\sqrt{x_{1n}^2 + x_{2n}^2}}[-x_{1n}x_{3n} \quad -x_{2n}x_{3n} \quad x_{1n}^2 + x_{2n}^2]\,\delta\mathbf{x}$$

Letting $\delta\mathbf{y} = [\delta r \quad \delta\alpha \quad \delta\beta]^T$ allows the above results to be expressed as $\delta\mathbf{y}(t) = \mathbf{C}(t)\delta\mathbf{x}(t)$, where $\mathbf{C}$ is a 3×3 matrix.

13.3 Describe how linear control theory can be used to control a nonlinear system in the vicinity of a known nominal state trajectory $\mathbf{x}_n(t)$ [14].

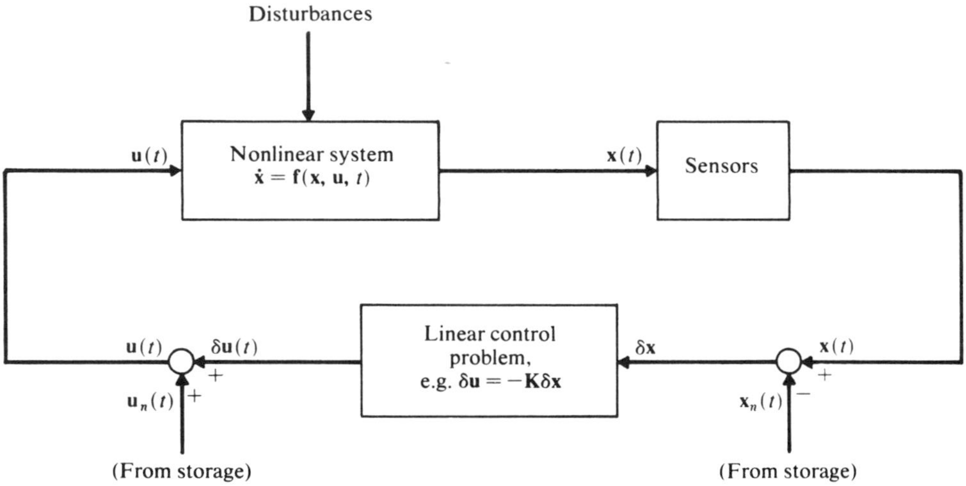

Figure 13.7

Let the system's state equations be $\dot{\mathbf{x}} = \mathbf{f}(\mathbf{x}, \mathbf{u}, t)$. It is assumed that $\mathbf{x}_n(t)$ and $\mathbf{u}_n(t)$ are known, and they satisfy $\dot{\mathbf{x}}_n = \mathbf{f}(\mathbf{x}_n, \mathbf{u}_n, t)$. Then $\delta\mathbf{x}(t) \triangleq \mathbf{x}(t) - \mathbf{x}_n(t)$ and $\delta\mathbf{u}(t) \triangleq \mathbf{u}(t) - \mathbf{u}_n(t)$ are related by the linear equation

$$\delta\dot{\mathbf{x}} = \left.\frac{d\mathbf{f}}{d\mathbf{x}}\right|_n \delta\mathbf{x} + \left.\frac{d\mathbf{f}}{d\mathbf{u}}\right|_n \delta\mathbf{u}$$

The determination of the $\delta\mathbf{u}(t)$ which drives $\delta\mathbf{x}(t)$ to a small value and keeps it near zero is a linear control problem. One form of the control function which is often applicable is the linear feedback law $\delta\mathbf{u}(t) = -\mathbf{K}\delta\mathbf{x}(t)$. This subject is discussed further in

Chapters 16 and 17, and methods of determining the matrix **K** can be found there. Omitting these details, a block diagram for controlling the nonlinear system is given in Fig. 13.7.

Error Analysis

13.4 A certain process is characterized by a set of parameters **x**. Measurements **y** can be made on this process, and they are related to **x** by a nonlinear algebraic equation, $\mathbf{y} = \mathbf{f}(\mathbf{x})$. The nominal values of **x** are $\mathbf{x}_n$. Describe a method of estimating the actual values of **x** based on the measurements **y**.

Let $\mathbf{x} = \mathbf{x}_n + \delta\mathbf{x}$. Then $\mathbf{y} = \mathbf{f}(\mathbf{x}_n + \delta\mathbf{x}) \cong \mathbf{f}(\mathbf{x}_n) + \left.\dfrac{d\mathbf{f}}{d\mathbf{x}}\right|_n \delta\mathbf{x}$.

Call $\mathbf{f}(\mathbf{x}_n) \triangleq \mathbf{y}_n$ and $\mathbf{y} - \mathbf{y}_n \triangleq \delta\mathbf{y}$. Since **y** is measured and since $\mathbf{y}_n$ can be computed from a knowledge of $\mathbf{x}_n$, $\delta\mathbf{y}$ is a known vector. The Jacobian matrix $\left.\dfrac{d\mathbf{f}}{d\mathbf{x}}\right|_n \triangleq \mathbf{A}$ can also be computed. The relation $\delta\mathbf{y} = \mathbf{A}\delta\mathbf{x}$ is of the form treated in Chapter 5. Because of measurement inaccuracies, redundant measurements and the least-squares technique are most commonly used to solve for $\delta\mathbf{x}$. Then the estimated parameter values are given by $\mathbf{x} = \mathbf{x}_n + \delta\mathbf{x}$.

13.5 Consider a coordinate transformation between two sets of orthogonal coordinates. The symbolic representation is shown in Fig. 13.8. For fixed inputs the outputs are nonlinear functions of the system parameters α and β. Assume that α and β are gimbal angles of a physical device, with nominal values $\alpha = 30°$, $\beta = 45°$. In order to check the alignment, a test input $u_1 = 1$, $u_2 = u_3 = 0$ is used and the outputs are found to be $f_1 = 0.850$, $f_2 = -0.375$, $f_3 = 0.370$. Determine the errors in α and β.

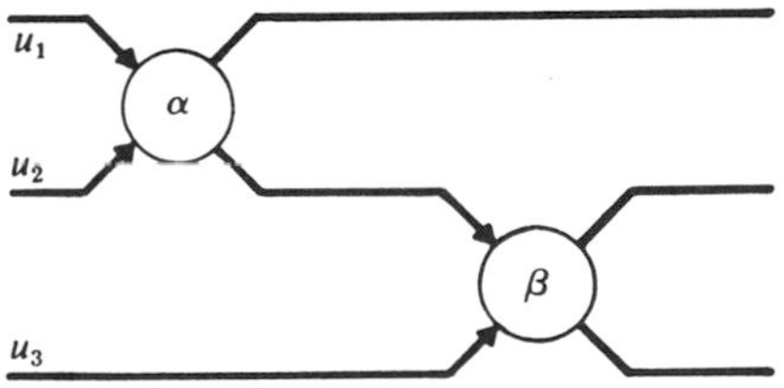

$f_1(\alpha, \beta) = u_1 \cos\alpha + u_2 \sin\alpha$

$f_2(\alpha, \beta) = -u_1 \sin\alpha\cos\beta + u_2\cos\alpha\cos\beta + u_3\sin\beta$

$f_3(\alpha, \beta) = u_1\sin\alpha\sin\beta - u_2\cos\alpha\sin\beta + u_3\cos\beta$

Figure 13.8

Let $\mathbf{y} = \mathbf{f}(\alpha, \beta)$. Then for small errors $\delta\alpha$ and $\delta\beta$,

$$\mathbf{y} = \mathbf{f}(\alpha_n, \beta_n) + \left[\frac{d\mathbf{f}}{d\alpha} \;\vdots\; \frac{d\mathbf{f}}{d\beta}\right]_n \begin{bmatrix} \delta\alpha \\ \delta\beta \end{bmatrix}$$

Define $\mathbf{y}_n = \mathbf{f}(\alpha_n, \beta_n)$ and $\delta\mathbf{y} = \mathbf{y} - \mathbf{y}_n$. Using nominal values gives $\mathbf{y}_n = [0.866 \quad -0.353 \quad 0.353]^T$. Combining this with the measured outputs gives $\delta\mathbf{y} = [-0.016 \quad -0.022 \quad 0.017]^T$. Using the given values of u_i leads to

$$\frac{\partial f_1}{\partial\alpha} = -\sin\alpha \qquad \frac{\partial f_2}{\partial\alpha} = -\cos\alpha\cos\beta \qquad \frac{\partial f_3}{\partial\alpha} = \cos\alpha\sin\beta$$

$$\frac{\partial f_1}{\partial\beta} = 0 \qquad \frac{\partial f_2}{\partial\beta} = \sin\alpha\sin\beta \qquad \frac{\partial f_3}{\partial\beta} = \sin\alpha\cos\beta$$

Therefore, the perturbation equation is $\begin{bmatrix} -0.016 \\ -0.022 \\ 0.017 \end{bmatrix} = \begin{bmatrix} -0.5 & 0 \\ -0.612 & 0.353 \\ 0.612 & 0.353 \end{bmatrix} \begin{bmatrix} \delta\alpha \\ \delta\beta \end{bmatrix}$. The least-squares solution is $\begin{bmatrix} \delta\alpha \\ \delta\beta \end{bmatrix} = [\mathbf{A}^T\mathbf{A}]^{-1}\mathbf{A}^T\,\delta\mathbf{y} = \begin{bmatrix} 0.032 \\ -0.007 \end{bmatrix}$. These are the misalignment angles, in radian measure. Expressed in degrees, $\delta\alpha = 1.83°$ and $\delta\beta = -0.40°$. The actual angles are $\alpha = 31.83°$ and $\beta = 44.60°$.

Newton-Raphson and Quasilinearization

13.6 Derive an iterative technique for determining the roots of the nonlinear algebraic equation $f(x) = 0$.

Let the initial estimate of a root be $x^{(0)}$. Normally this will not be a root, so an improved estimate $x^{(1)} = x^{(0)} + \delta x^{(0)}$ is sought. Keeping only linear terms in the Taylor expansion gives

$$f(x^{(0)} + \delta x^{(0)}) = f(x^{(0)}) + \left.\frac{df}{dx}\right|_0 \delta x$$

Setting this equal to zero and solving for δx gives

$$\delta x^{(0)} = -f(x^{(0)}) \Big/ \left.\frac{df}{dx}\right|_0 \qquad \text{or} \qquad x^{(1)} = x^{(0)} - f(x^{(0)}) \Big/ \left.\frac{df}{dx}\right|_0$$

Figure 13.9 gives the graphical interpretation for an arbitrary function $f(x)$ in the vicinity of a root. The $(k+1)$th estimate is $x^{(k+1)} = x^{(k)} - f(x^{(k)}) \Big/ \left.\frac{df}{dx}\right|_k$. This formula constitutes the Newton-Raphson root-finding technique [53, 68]. For the function shown, $x^{(2)}$ is an improved estimate and $x^{(3)}$ would be quite close to the actual root.

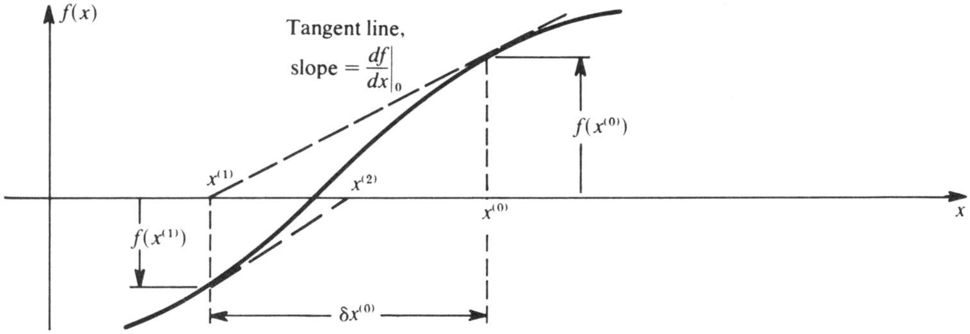

Figure 13.9

Because of the similarity between the Newton-Raphson technique and those used in quasilinearization of differential equations, the method of quasilinearization is sometimes referred to as a generalized Newton-Raphson technique [79].

13.7 Use the Newton-Raphson technique to determine roots of $f(x) = x^3 + 4x^2 + x - 6 = 0$.

The iterative equation for this case is

$$x^{(k+1)} = x^{(k)} - \left.\frac{x^3 + 4x^2 + x - 6}{3x^2 + 8x + 1}\right|_{x=x^{(k)}}$$

Results for three different initial estimates of $x^{(0)}$ are given in Table 13.1.

TABLE 13.1

k	$x^{(k)}$	$f(x^{(k)})$	$x^{(k)}$	$f(x^{(k)})$	$x^{(k)}$	$f(x^{(k)})$
0	10.0	1404.0	−10.0	−616.0	−1.5	−1.875
1	6.315	411.662	−7.213	−180.344	−1.941	−0.1832
2	3.910	118.821	−5.398	−52.121	−1.998	−0.006
3	2.389	32.857	−4.245	−14.665	−2.000	0.0
4	1.507	8.008	−3.550	−3.883		
5	1.104	1.319	−3.177	−0.873		
6	1.006	0.069	−3.0287	−0.119		
7	1.000	0.0	−3.001	−0.004		

For this simple case the exact roots are known to be 1, −2, and −3. The convergence to a particular root depends on the initial estimate. If $x^{(0)} = 100$ is used, the procedure again converges to $x = 1.0$, but it requires 14 iterations. A rough sketch of the function is useful in determining the initial estimates.

This simple procedure is not able to determine complex roots. If $f'(x^{(k)}) = 0$ or is very small, difficulties will also be encountered.

13.8 How might the Newton-Raphson method be applied to find the roots of a scalar-valued function of a vector $\mathbf{x}$?

Consider $f(\mathbf{x}) = 0$. Let $\mathbf{x}^{(0)}$ be the initial estimate and let $\mathbf{x}^{(1)} = \mathbf{x}^{(0)} + \delta\mathbf{x}^{(0)}$. Using the first two terms in the Taylor expansion gives

$$f(\mathbf{x}^{(1)}) = f(\mathbf{x}^{(0)} + \delta\mathbf{x}^{(0)}) = f(\mathbf{x}^{(0)}) + [\nabla_{\mathbf{x}} f|_0]^T \delta\mathbf{x}^{(0)}$$

Setting this equal to zero gives a single equation which is not sufficient for finding a unique $\delta\mathbf{x}^{(0)}$:

$$[\nabla_{\mathbf{x}} f|_0]^T \delta\mathbf{x}^{(0)} = -f(\mathbf{x}^{(0)})$$

In general, many solutions exist. The minimum norm solution, discussed in Chapters 5 and 6, is given by

$$\delta\mathbf{x}^{(0)} = -\nabla_{\mathbf{x}} f|_0[(\nabla_{\mathbf{x}} f|_0)^T \nabla_{\mathbf{x}} f|_0]^{-1} f(\mathbf{x}^{(0)}) = -\frac{\nabla_{\mathbf{x}} f|_0 f(\mathbf{x}^{(0)})}{\|\nabla_{\mathbf{x}} f|_0\|^2}.$$

This could be used to develop the iterative formula

$$\mathbf{x}^{(k+1)} = \mathbf{x}^{(k)} - \frac{\nabla_{\mathbf{x}} f|_k f(\mathbf{x}^{(k)})}{\|\nabla_{\mathbf{x}} f|_k\|^2}$$

13.9 Use quasilinearization to obtain an approximate solution for the equation $\dot{x} = \cos^2 x$ with $x(0) = \pi/4$. Use $x^{(0)}(t) = \pi/4$ as the zeroth estimate.

The nonlinear term is $f(x) = \cos^2 x$ and its gradient is $\nabla_x f = -2\cos x \sin x$. Therefore,

$$\dot{x}^{(k+1)} = [-2\cos x^{(k)} \sin x^{(k)}]x^{(k+1)} + \cos^2 x^{(k)} - [-2\cos x^{(k)} \sin x^{(k)}]\, x^{(k)}$$

Using $x^{(0)} = \pi/4$ gives $\dot{x}^{(1)} = -x^{(1)} + (0.5 + \pi/4)$. The solution for the first approximation is easily found to be $x^{(1)}(t) = \pi/4 + 0.5(1 - e^{-t})$. In order to find the second approximation, $x^{(1)}(t)$ must be used to form a time-variable coefficient linear differential equation. This equation for $x^{(2)}(t)$ is not easily solved analytically. Numerical integration was used to evaluate $x^{(2)}(t)$. The exact solution for this problem is known to be $x(t) = \tan^{-1}(t + 1)$. A comparison of $x(t)$ with $x^{(1)}(t)$ and $x^{(2)}(t)$ is given in Table 13.2. Further iterations with the quasilinearization scheme would continue to reduce the error.

TABLE 13.2

t	$x(t)$	$x^{(1)}(t)$	$x^{(2)}(t)$
0	0.78540	0.78540	0.78540
1	1.1071	1.1015	1.1071
2	1.2490	1.2177	1.2488
3	1.3258	1.2605	1.3241
4	1.3734	1.2762	1.3678
5	1.4056	1.2820	1.3939
6	1.4289	1.2842	1.4096
7	1.4464	1.2849	1.4188
8	1.4601	1.2852	1.4243
9	1.4711	1.2853	1.4275
10	1.4801	1.2854	1.4294

Maxima and Minima

13.10 Consider the nonlinear function of $\mathbf{x} = [x_1 \quad x_2]^T$:

$$f(\mathbf{x}) = \mathbf{x}^T \begin{bmatrix} -4/(3\pi) & -2/(3\pi) \\ -2/(3\pi) & -4/(3\pi) \end{bmatrix} \mathbf{x} + \sin^2 x_1 - \cos^2 x_2$$

a. Is the point $\mathbf{x} = \mathbf{0}$ a relative minimum or maximum?
b. Is the point $\mathbf{x} = [\pi/4 \quad \pi/4]^T$ a relative minimum or maximum or neither?
c. What can be said about the point $\mathbf{x} = [-\pi/4 \quad -\pi/4]^T$?

a. The gradient is $\nabla_{\mathbf{x}} f = 2\mathbf{A}\mathbf{x} + [2 \sin x_1 \cos x_1 \quad 2 \sin x_2 \cos x_2]^T$, where $\mathbf{A}$ is the 2×2 coefficient matrix. Obviously, $\mathbf{x} = \mathbf{0}$ yields $\nabla_{\mathbf{x}} f = \mathbf{0}$, so this point satisfies the necessary condition to be a maximum or a minimum. The Hessian matrix is

$$\frac{d^2 f}{d\mathbf{x}^2} = 2\mathbf{A} + \begin{bmatrix} 2 \cos 2x_1 & 0 \\ 0 & 2 \cos 2x_2 \end{bmatrix}$$

With $\mathbf{x} = \mathbf{0}$, $d^2f/d\mathbf{x}^2 = 2[\mathbf{A} + \mathbf{I}_2] = 2\begin{bmatrix} 1 - 4/(3\pi) & -2/(3\pi) \\ -2/(3\pi) & 1 - 4/(3\pi) \end{bmatrix}$. This matrix is positive definite, as is verified using the principal minor test of Sec. 7.7. Therefore, $\mathbf{x} = \mathbf{0}$ is a point of relative minimum for the function $f(\mathbf{x})$.

b. Using $x_1 = x_2 = \pi/4$, $\nabla_{\mathbf{x}} f = \mathbf{0}$. The Hessian matrix evaluated at this point is $d^2f/d\mathbf{x}^2 = 2\mathbf{A}$. This matrix is negative definite, so $[\pi/4 \quad \pi/4]^T$ is a point of relative maximum.

c. Using $x_1 = x_2 = -\pi/4$ again gives $\nabla_{\mathbf{x}} f = \mathbf{0}$ and the Hessian is equal to $2\mathbf{A}$. This is a point of relative maximum.

13.11 Find the gradient of $f(\mathbf{x}) = x_1^3 + x_1x_4 - x_2x_3 + x_3^2$.

$$\nabla_{\mathbf{x}} f = \left[\frac{\partial f}{\partial x_1} \quad \frac{\partial f}{\partial x_2} \quad \frac{\partial f}{\partial x_3} \quad \frac{\partial f}{\partial x_4}\right]^T = [3x_1^2 + x_4 \quad -x_3 \quad 2x_3 - x_2 \quad x_1]^T$$

13.12 a. Find all stationary points for $f(\mathbf{x})$ of Problem 13.11. That is, find all points satisfying $\nabla_{\mathbf{x}} f = \mathbf{0}$.

b. Determine whether these points correspond to a relative minimum, maximum, or neither.

a. $\nabla_{\mathbf{x}} f = \mathbf{0}$ requires that each component of the gradient be zero. The second and fourth components give $x_1 = x_3 = 0$. This, along with the other components, requires that $x_2 = x_4 = 0$ so $\mathbf{x} = \mathbf{0}$ is the only stationary point.

b. The Hessian matrix is

$$\frac{d^2 f}{d\mathbf{x}^2} = \begin{bmatrix} 6x_1 & 0 & 0 & 1 \\ 0 & 0 & -1 & 0 \\ 0 & -1 & 2 & 0 \\ 1 & 0 & 0 & 0 \end{bmatrix}$$

Evaluating this at $\mathbf{x} = \mathbf{0}$ gives a symmetric matrix with principal minors $\Delta_1 = 0$, $\Delta_2 = 0$, $\Delta_3 = 0$, $\Delta_4 = 1$. Therefore, the Hessian is neither positive nor negative definite and no conclusion can be drawn from this test regarding a maximum or minimum. Direct inspection of $f(\mathbf{x})$ reveals that $\mathbf{x} = \mathbf{0}$ is not a point of relative maximum or minimum, since $f(\mathbf{0}) = \mathbf{0}$ and $f(\mathbf{x})$ can be either positive or negative near $\mathbf{x} = \mathbf{0}$, depending on the sign of x_i.

13.13 A truck is to be driven at constant speed between two warehouses. Find the most economical speed if the driver is paid \$14.50/hr. Gasoline costs \$1.50/gal, and over the usual range of operating speeds, fuel consumption in gal/mi is $0.1 + 0.003V$, where V is the speed in miles per hour.

Let the distance traveled be D miles. Travel time is $T = D/V$ and the driver's wages are $14.5D/V$. Operating costs are [\$1.50/gal][$0.1 + 0.003V$](gal/mi)[Dmi]. The total costs are $f(V) = D[14.5/V + 0.15 + 0.0045V]$. For minimum cost, $df/dV = 0 = D[-14.5/V^2 + 0.0045]$. Therefore, $V^2 = 3222.22$ or $V = 56.76$ mph.

To verify that this is a minimum, the second derivative is evaluated, $d^2f/dV^2 = 29D/V^3$. This is positive for positive V, so the stationary point is a minimum.

13.14 A factory buys raw material and sells its finished products at locations $1, 2, \ldots, N$. The cost of shipping materials in and finished products out is assumed proportional to the shipping distance. Each commodity has its own shipping rate c_i, so that the total shipping cost is $\sum_{i=1}^{N} c_i\sqrt{(x - x_i)^2 + (y - y_i)^2}$, where point i has coordinates x_i, y_i and the factory is at x, y. The factory is to be moved to a new location. Find the direction it should be moved to give the maximum reduction in shipping costs.

Let $f(\mathbf{x})$ = total shipping costs, where $\mathbf{x} = [x \quad y]^T$.

The maximum rate of decrease in $f(\mathbf{x})$ occurs when the change in $\mathbf{x}$ is along $-\nabla_{\mathbf{x}} f$. Evaluating the gradient shows that $\nabla_{\mathbf{x}} f = [\hat{\mathbf{v}}_1 \quad \hat{\mathbf{v}}_2 \quad \cdots \quad \hat{\mathbf{v}}_N]\mathbf{c}$, where $\mathbf{v}_i = \mathbf{0}$ if $\mathbf{x} = \mathbf{x}_i$. Otherwise, $\hat{\mathbf{v}}_i$ is a unit vector from the current location towards point $\mathbf{x}_i$ and $\mathbf{c} \triangleq [c_1 \quad c_2 \quad \cdots \quad c_N]^T$.

The best *direction* to consider moving is given by $-\sum_{i=1}^{N} c_i\hat{\mathbf{v}}_i$. How far the factory should be moved along this line is unknown, and the optimum location need not be on this line. An iterative gradient scheme can be used to find the best final location. A move of a certain distance is assumed. The gradient is then re-evaluated at that point. A new direction is then computed. This can be continued until $\nabla_{\mathbf{x}} f = \mathbf{0}$ at the optimal location. If all c_i are equal, the optimum point is the geometric mean of the N shipping points. With unequal c_i, it is a weighted mean.

13.15 Use a gradient adjustment scheme to find the approximate location of the point $[x, y]$ which minimizes

$$f(x, y) = 6\sqrt{(x-3)^2 + (y-5)^2} + 2\sqrt{x^2 + y^2} + 3\sqrt{(x+1)^2 + (y-2)^2} + 4\sqrt{(x-2)^2 + (y+3)^2}$$

Applying the results of Problem 13.14 to this specific problem gives

$$\nabla_{\mathbf{x}} f = \begin{bmatrix} (x-3)/d_1 & x/d_2 & (x+1)/d_3 & (x-2)/d_4 \\ (y-5)/d_1 & y/d_2 & (y-2)/d_3 & (y+3)/d_4 \end{bmatrix} \begin{bmatrix} 6 \\ 2 \\ 3 \\ 4 \end{bmatrix}$$

where

$$d_1 = [(x-3)^2 + (y-5)^2]^{1/2} \qquad d_2 = [x^2 + y^2]^{1/2}$$
$$d_3 = [(x+1)^2 + (y-2)^2]^{1/2} \qquad d_4 = [(x-2)^2 + (y+3)^2]^{1/2}$$

An intentionally poor initial guess of $\mathbf{x}_0 = (x_0, y_0) = (10, 10)$ and the adjustment scheme $\mathbf{x}_{k+1} = \mathbf{x}_k - 0.5\nabla_{\mathbf{x}} f$ were implemented on a digital computer. The results are given in Table 13.3.

TABLE 13.3

k	x_k	y_k	$f(\mathbf{x}_k)$	$\left.\frac{\partial f}{\partial x}\right\|_k$	$\left.\frac{\partial f}{\partial y}\right\|_k$
0	10.0000	10.0000	181.760	10.81921	10.07279
1	4.5904	4.9636	75.546	11.24425	6.54003
2	−1.0317	1.6936	58.525	−8.15926	−1.72076
3	3.0479	2.5540	57.497	5.36434	−0.37689
4	0.3657	2.7424	54.894	−2.75065	3.35809
5	1.7411	1.0634	53.855	2.46364	−1.65054
6	0.5092	1.8886	52.809	−1.40392	0.85237
7	1.2112	1.4625	52.536	0.78700	−0.58377
8	0.8177	1.7543	52.437	−0.49529	0.31361
9	1.0653	1.5975	52.403	0.29162	−0.20582
10	0.9195	1.7005	52.391	−0.18080	0.11969
11	1.0099	1.6406	52.386	0.10894	−0.07514
12	0.9555	1.6782	52.385	−0.06690	0.04502
15	0.9810	1.6611	52.384	0.01513	−0.01030
20	0.9759	1.6646	52.384	−0.00128	0.00087
25	0.9764	1.6643	52.3835	0.00010	−0.00007
30	0.9763	1.6643	52.3835	−0.00001	0.00000

These results are typical of many gradient schemes. The reduction in $f(\mathbf{x})$ is rapid at first and then very slow as the minimum is approached. Fewer iterations would be required if a better initial guess were used. The step size of 0.5 also affects convergence. A smaller value would reduce the oscillations about the final answer and may require fewer iterations. More iterations will be required if the step size is too small. If it is too large, the scheme may not converge at all.

13.16 Consider finding the minimum of the function

$$J(x, y) = (x - 3)^2 + 10(y - 3)^2$$

using the methods of Sec. 13.7.

In this form the minimum is obviously at $(x, y) = (3, 3)$. If the problem is "disguised" by rotating the axes by 45° and dropping the constants, then J becomes (in the rotated θ_1, θ_2 coordinates)

$$J(\boldsymbol{\theta}) = 5.5\theta_1^2 + 5.5\theta_2^2 + 9\theta_1\theta_2 - 46.662\theta_1 - 38.178\theta_2$$

Because of the origin of the problem, the exact point of minimum is known to be $\boldsymbol{\theta} = [4.242 \quad 0]^T$. Let us investigate the problem in terms of the gradient and Newton methods. Assume that the current estimate of $\hat{\boldsymbol{\theta}}_0$ is $[0 \quad 0]^T$. The gradient at that point is

$$\mathbf{g} = \nabla_{\boldsymbol{\theta}} J = \begin{bmatrix} 11 & +9 & -46.662 \\ 11 & +9 & -38.178 \end{bmatrix} = \begin{bmatrix} -46.662 \\ -38.178 \end{bmatrix} \simeq \begin{bmatrix} 1.222 \\ 1 \end{bmatrix}$$

The Hessian for this quadratic takes on the same value at every point:

$$\mathbf{S} = \frac{d^2 J}{d\boldsymbol{\theta}^2} = \begin{bmatrix} 11 & 9 \\ 9 & 11 \end{bmatrix} \quad \text{and} \quad \mathbf{S}^{-1} = \begin{bmatrix} 11 & -9 \\ -9 & 11 \end{bmatrix} \Big/ 40$$

Therefore,

$$\mathbf{S}^{-1}\mathbf{g} = \begin{bmatrix} -4.242 \\ 0 \end{bmatrix} \quad \text{and} \quad \hat{\boldsymbol{\theta}}(1) = \epsilon \begin{bmatrix} 1.222 \\ 1 \end{bmatrix} \quad \text{(with the gradient)}$$

and

$$\hat{\boldsymbol{\theta}}(1) = \begin{bmatrix} 4.242 \\ 0 \end{bmatrix} \quad \text{(with Newton)}$$

In the gradient method the step size ϵ is still to be selected. Its direction is shown in Fig. 13.10. If too small a step is taken, J will decrease, but not very much. If too large

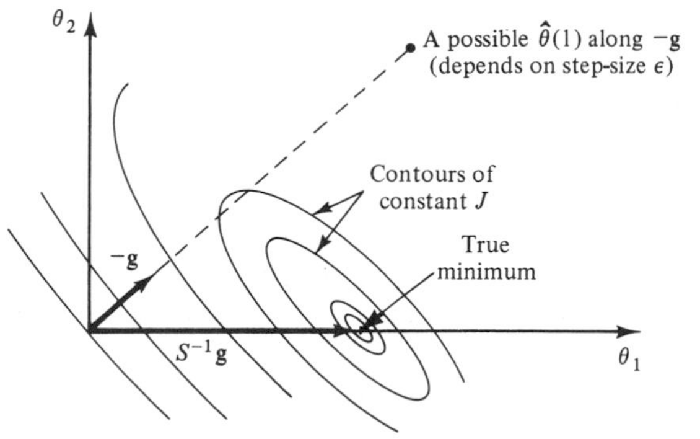

Figure 13.10

a step is taken, the estimate could get farther from the minimum. It is apparent that no step size will give as good an estimate as does Newton's method (in this example) because the gradient is normal to the local slope and does not point towards the true minimum. These two correction vectors generally do not have the same directions in parameter space.

13.17 A set of measurements $\mathbf{Y}$ are related to an unknown parameter vector $\boldsymbol{\theta}$ in a nonlinear way,

$$\mathbf{Y} = \mathbf{h}(\boldsymbol{\theta}) + \mathbf{V}$$

where $\mathbf{V}$ consists of measurement errors. The parameters $\boldsymbol{\theta}$ are to be selected to minimize

$$J(\boldsymbol{\theta}) = [\mathbf{Y} - \mathbf{h}(\boldsymbol{\theta})]^T \mathbf{W} [\mathbf{Y} - \mathbf{h}(\boldsymbol{\theta})]$$

with $\mathbf{W}$ being a given positive definite weighting matrix. A closed-form solution will not generally be possible, so apply the methods of Sec. 13.7 to develop an iterative algorithm.

When J is expanded out

$$J(\boldsymbol{\theta}) = \mathbf{Y}^T\mathbf{W}\mathbf{Y} + \mathbf{h}(\boldsymbol{\theta})^T\mathbf{W}\mathbf{h}(\boldsymbol{\theta}) - 2\mathbf{h}(\boldsymbol{\theta})^T\mathbf{W}\mathbf{Y}$$

so that

$$\mathbf{g} = \nabla_{\boldsymbol{\theta}} J = \left[\frac{d\mathbf{h}}{d\boldsymbol{\theta}}\right]^T \nabla_{\mathbf{h}} J = -2\left[\frac{d\mathbf{h}}{d\boldsymbol{\theta}}\right]\mathbf{W}[\mathbf{Y} - \mathbf{h}(\boldsymbol{\theta})]$$

Taking second derivatives gives

$$\frac{d^2 J}{d\boldsymbol{\theta}^2} = \frac{d}{d\boldsymbol{\theta}}\left\{-2\left[\frac{d\mathbf{h}}{d\boldsymbol{\theta}}\right]^T \mathbf{W}[\mathbf{Y} - \mathbf{h}(\boldsymbol{\theta})]\right\}$$

$$= 2\left[\frac{d\mathbf{h}}{d\boldsymbol{\theta}}\right]^T \mathbf{W}\left[\frac{d\mathbf{h}}{d\boldsymbol{\theta}}\right] - 2\frac{d}{d\boldsymbol{\theta}}\left\{\frac{d\mathbf{h}}{d\boldsymbol{\theta}}\right\}^T \mathbf{W}[\mathbf{Y} - \mathbf{h}(\boldsymbol{\theta})]$$

Note that if $\mathbf{h}(\boldsymbol{\theta}) = \mathcal{H}\theta$ (i.e., linear in the parameters), then $d\mathbf{h}/d\boldsymbol{\theta} = \mathcal{H}$. In this case the second term in the above equation is zero. Even in the nonlinear case the second term is usually neglected because

a. It should be small near the extremal point (due to $\mathbf{Y} \cong \mathbf{h}(\boldsymbol{\theta})$) so the first term will dominate.

b. The calculation of the second term involves all the second derivatives of $\mathbf{h}$ with respect to $\boldsymbol{\theta}$, and thus represents an undesirable computational burden.

When the second term is neglected, the iterative algorithm for the nonlinear least-squares problem becomes

$$\hat{\boldsymbol{\theta}}(k+1) = \hat{\boldsymbol{\theta}}(k) + \left\{\left[\frac{d\mathbf{h}}{d\boldsymbol{\theta}}\right]^T \mathbf{W}\left[\frac{d\mathbf{h}}{d\boldsymbol{\theta}}\right]\right\}^{-1}\left[\frac{d\mathbf{h}}{d\boldsymbol{\theta}}\right]^T \mathbf{W}[\mathbf{Y} - \mathbf{h}(\hat{\boldsymbol{\theta}}(k))]$$

13.18 Investigate and discuss the Marquardt algorithm [77] for iteratively solving nonlinear equations and nonlinear maximization or minimization problems.

First of all, note that solving nonlinear algebraic equations of the form

$$\mathbf{f}(\mathbf{x}) = \mathbf{0}$$

is equivalent to solving for the extremal of a scalar function. In the later case the function to be solved is the gradient vector set equal to zero.

The Marquardt algorithm could be viewed as a heuristic blending of gradient search and Newton-type methods. Actually, however, it was derived as the solution to a constrained optimization of a function. The magnitude of the step size is constrained, using a Lagrange multiplier approach [111]. The parameter μ in the algorithm below is in fact the Lagrange multiplier.

In general, the Newton-type algorithms have very rapid convergence once they get sufficiently near the minimum or maximum, or if the cost function is sufficiently quadratic-like. However, they frequently fail to converge due to poor initial guesses, or other reasons. The Marquardt algorithm (as well as some others) attempts to exploit the best features of both the gradient and the Newton algorithms. The algorithm is

$$\hat{\boldsymbol{\theta}}(k+1) = \hat{\boldsymbol{\theta}}(k) + \left[\mu_k \mathbf{I} + \left(\frac{d\mathbf{h}}{d\boldsymbol{\theta}}\bigg|_k\right)^T \mathbf{W}\left(\frac{d\mathbf{h}}{d\boldsymbol{\theta}}\bigg|_k\right)\right]^{-1} \left(\frac{d\mathbf{h}}{d\boldsymbol{\theta}}\bigg|_k\right)^T \mathbf{W}[\mathbf{Y} - \mathbf{h}(\hat{\boldsymbol{\theta}}(k))]$$

As can be seen, when $\mu_k = 0$, the Marquardt algorithm takes its corrective steps in the Newton direction. When μ_k is very large, the step direction approaches the gradient direction (or its negative, depending on the sign of μ_k). Thus if L is increasing (or J is decreasing), the algorithm is allowed to approach the Newton direction to take advantage of its quadratic convergence. If on a given step L decreases (or J increases), we have moved away from our objective, so μ_k is increased. This changes the next corrective direction toward the "safer" gradient direction.

Many ways of initializing and adjusting μ could be used. One commercially available software package [101] initializes μ to a very small value (10^{-8}). Then, if on any step $J[\hat{\boldsymbol{\theta}}(k+1)]$ is less than $J[\hat{\boldsymbol{\theta}}(k)]$ as is desired, μ_{k+1} is selected as $\mu_k/10$. On the other hand, if J increases on any step, then μ is increased by a factor of 10.

13.19 Investigate the neglected "second term" of Problem 13.17.

In order to investigate the nature of this term, let $\mathbf{W}[\mathbf{Y} - \mathbf{h}(\boldsymbol{\theta})]$ be defined as a vector $\boldsymbol{\eta}$. If this vector is treated as if it were not a function of $\boldsymbol{\theta}$, it can be brought inside the braces. The matrix-vector product gives a vector, which we can then differentiate.

$$\frac{d}{d\boldsymbol{\theta}}\left\{\left[\frac{d\mathbf{h}}{d\boldsymbol{\theta}}\right]^T\right\}\boldsymbol{\eta} = \frac{d}{d\boldsymbol{\theta}}\left\{\left[\frac{d\mathbf{h}}{d\boldsymbol{\theta}}\right]^T \boldsymbol{\eta}\right\}$$
$$= \frac{d}{d\boldsymbol{\theta}}\left\{\sum_{i=1}^{n} (\nabla_{\boldsymbol{\theta}} h_i)^T \eta_i\right\} = \sum_{i=1}^{n} \left[\frac{d^2 h_i}{d\boldsymbol{\theta}^2}\right]\eta_i$$

This term thus consists of the weighted sum of second-derivative matrices. To make the result more concrete, consider the two-component vector case, and let $\mathbf{W}$ be a diagonal matrix. Then

$$\frac{d}{d\boldsymbol{\theta}}\left\{\left[\frac{d\mathbf{h}}{d\boldsymbol{\theta}}\right]^T\right\}\mathbf{W}[\mathbf{Y} - \mathbf{h}(\boldsymbol{\theta})] = \begin{bmatrix} \dfrac{d^2 h_1}{d\theta_1^2} & \dfrac{d^2 h_1}{d\theta_1\, d\theta_2} \\ \dfrac{d^2 h_1}{d\theta_1\, d\theta_2} & \dfrac{d^2 h_1}{d\theta_2^2} \end{bmatrix} \mathbf{W}_{11}[y_1 - h_1(\boldsymbol{\theta})]$$
$$+ \begin{bmatrix} \dfrac{d^2 h_2}{d\theta_1^2} & \dfrac{d^2 h_2}{d\theta_1\, d\theta_2} \\ \dfrac{d^2 h_2}{d\theta_1\, d\theta_2} & \dfrac{d^2 h_2}{d\theta_2^2} \end{bmatrix} \mathbf{W}_{22}[y_2 - h_2(\boldsymbol{\theta})]$$

PROBLEMS

13.20 Pressure drop-flow rate relations through many devices are nonlinear. For an orifice the flow rate Q and the pressure drop $P_1 - P_2$ are related by $Q = c\sqrt{P_1 - P_2}$. Derive a linear relation for the flow out of an orifice at the bottom of a tank. The tank is nominally kept filled to a height h_n. The fluid density is ρ lb-sec²/ft⁴. Thus $P_1 = \rho g h$ and $P_2 = 0$.

13.21 A navigation scheme uses a sextant to measure the angle included between the directions to two known landmarks from the position of the sextant. Let the sextant position vector be $\mathbf{x}$. The landmark position vectors are $\mathbf{r}_1$ and $\mathbf{r}_2$.

a. Find the nonlinear expression for the measured angle θ.
b. Find the linear relation between small perturbations in θ and in $\mathbf{x}$.

13.22 Use the Newton-Raphson procedure to estimate a root of $f(x) = x^3 + 3x^2 + 5x +$

13.22 Use the Newton-Raphson procedure to estimate a root of $f(x) = x^3 + 3x^2 + 5x$ $2 = 0$. Note that $f(0) = 2$ and $f(-2) = -4$, so a root exists between 0 and -2.

13.23 Use the quasilinearization technique to find a first approximation for the solution of $\dot{x} = 3[\sinh^2 x + 1]^{-1/2}$ with $x(0) = 2$. Use $x^{(0)}(t) = 2$ as the zeroth approximation.

13.24 In linear programming problems the goal is to maximize or minimize a linear function $f(\mathbf{x}) = \mathbf{p}^T\mathbf{x}$, where $\mathbf{p}$ is a fixed vector. Can such a function have a relative maximum or minimum if $\mathbf{x}$ is allowed to take on any values whatsoever?

13.25 Apply the Marquardt algorithm to the minimization of f in Problem 13.15. Compare the convergence rates.

Chapter

14

STABILITY FOR LINEAR AND NONLINEAR SYSTEMS

14.1 INTRODUCTION

Stability of single-input, single-output linear time-invariant systems was discussed in Chapter 2. There the conditions for stability were given in terms of the pole locations of the input-output transfer functions. Classical methods of stability analysis were also presented.

The goal of this chapter is to extend the previous stability concepts to multivariable systems described in terms of state variables. The classical method of treating nonlinear systems depends heavily on graphical phase-plane methods. Since success of these methods has been largely limited to two-dimensional state space, they are not discussed here.

There are many different definitions of "stability." A few of the more common definitions are given, along with methods of investigating them. A sampling of the many works on stability may be found in References 9, 29, 62, 63, 71, 86, and 107.

14.2 EQUILIBRIUM POINTS AND STABILITY CONCEPTS

Consider the ball which is free to roll on the surface shown in Fig. 14.1. The ball could be made to rest at points A, E, F, and G and anywhere between points B and D, such as at C. Each of these points is an *equilibrium point* of the system. An infinitesimal perturbation away from A or F will cause the ball to diverge from these points. Thus A and F are *unstable* equilibrium points. After small perturbations away from E or G, the ball will eventually return to these points of *stable* equilibrium. If the ball is perturbed slightly away from point C, it will remain at the new position. Points like C are sometimes said to be *neutrally* stable.

Assume that the shape of the surface in Fig. 14.1 changes with time. Specifically,

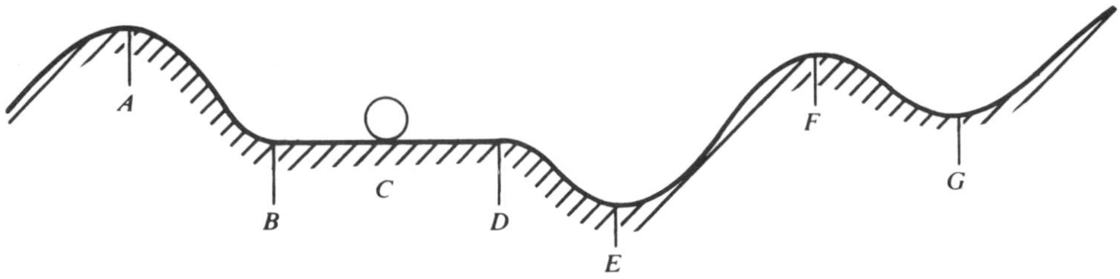

Figure 14.1

assume that point E moves vertically so that the slope at that point is always zero, but the surface is sometimes concave upward (as shown) and sometimes concave downward. Point E is still an equilibrium point, but whether it is stable or not now depends upon time.

Thus far only *local* stability has been considered, since the perturbations were assumed to be small. If the ball were displaced sufficiently far from point G, it would not return to that point. Stability therefore depends on the size of the original perturbation and on the nature of any disturbances which may be acting. These intuitive notions are now developed for dynamical systems.

A particular point $\mathbf{x}_e \in \Sigma$ is an equilibrium point of a dynamical system if the system's state at t_0 is $\mathbf{x}_e$ and $\mathbf{x}(t) = \mathbf{x}_e$ for all $t \geq t_0$ in the absence of inputs or disturbances. For the continuous-time system $\dot{\mathbf{x}} = \mathbf{f}(\mathbf{x}(t), \mathbf{u}(t), t)$, this means that $\mathbf{f}(\mathbf{x}_e, \mathbf{0}, t) = \mathbf{0}$ for $t \geq t_0$. For the discrete-time system $\mathbf{x}(k+1) = \mathbf{f}(\mathbf{x}(k), \mathbf{u}(k), k)$, this means that $\mathbf{f}(\mathbf{x}_e, \mathbf{0}, k) = \mathbf{x}_e$ for all $k > 0$.

The origin of the state space is always an equilibrium point for linear systems, although it need not be the only one. In the continuous-time case, if the system matrix $\mathbf{A}$ has a zero eigenvalue, then there is an infinity of vectors (eigenvectors) satisfying $\mathbf{A}\mathbf{x}_e = \mathbf{0}$. In the discrete-time case a unity eigenvalue of $\mathbf{A}$ means there is an infinity of vectors satisfying $\mathbf{A}\mathbf{x}_e = \mathbf{x}_e$. These points loosely correspond to points between B and D of Fig. 14.1 Only *isolated equilibrium points* will be considered in this text, and for linear systems the only isolated equilibrium point is the origin.

Any isolated singular point can be transferred to the origin by a change of variables, $\mathbf{x}' = \mathbf{x} - \mathbf{x}_e$. For this reason it is assumed in the sequel that $\mathbf{x}_e = \mathbf{0}$.

Stability deals with the following questions. If at time t_0 the state is perturbed from its equilibrium point, does the state return to $\mathbf{x}_e$, or remain close to $\mathbf{x}_e$, or diverge from it? Similar questions could be raised if system inputs or disturbances are allowed. Another class of stability questions deals with the state trajectories of an unperturbed system and of a perturbed system. Let the solution to $\dot{\mathbf{x}}_1 = \mathbf{f}(\mathbf{x}_1(t), \mathbf{u}(t), t)$, with $\mathbf{x}_1(t_0)$ given, define the unperturbed trajectory $\mathbf{x}_1(t)$. Let the perturbed trajectory $\mathbf{x}_2(t)$ be defined by $\dot{\mathbf{x}}_2 = \mathbf{f}(\mathbf{x}_2(t), \mathbf{u}(t) + \mathbf{v}(t), t)$, where $\mathbf{x}_2(t_0) = \mathbf{x}_1(t_0) + \mathbf{e}(t_0)$. The initial state and control perturbations are $\mathbf{e}(t_0)$ and $\mathbf{v}(t)$, respectively. Does $\mathbf{x}_2(t)$ return to $\mathbf{x}_1(t)$, or remain close to it, or diverge from it? These questions can be studied by considering the difference $\mathbf{e}(t) = \mathbf{x}_2(t) - \mathbf{x}_1(t)$ which satisfies $\dot{\mathbf{e}} = \mathbf{f}(\mathbf{x}_1(t) + \mathbf{e}(t), \mathbf{u}(t) + \mathbf{v}(t), t) - \mathbf{f}(\mathbf{x}_1(t), \mathbf{u}(t), t)$ or simply $\dot{\mathbf{e}} = \mathbf{f}'(\mathbf{e}(t), \mathbf{v}(t), t)$ with $\mathbf{e}(t_0)$, $\mathbf{x}_1(t)$, and $\mathbf{u}(t)$ given. Now $\mathbf{e} = \mathbf{0}$

is an equilibrium point and the questions regarding the perturbed motion can be studied in terms of perturbations about the origin, as before.

Whether an equilibrium point is stable or not depends upon what is meant by "remaining close," the magnitude of state or input disturbances and their time of application. These qualifying conditions are the reasons for the existence of a variety of stability definitions.

14.3 STABILITY DEFINITIONS

For the continuous-time case with zero input

$$\dot{\mathbf{x}} = \mathbf{f}(\mathbf{x}, \mathbf{0}, t), \qquad \mathbf{x}(t_0) = \mathbf{x}_0$$

with the origin an equilibrium point, the following apply.

Definition 14.1: The origin is a *stable* equilibrium point if for any given value $\epsilon > 0$ there exists a number $\delta(\epsilon, t_0) > 0$ such that if $\|\mathbf{x}(t_0)\| < \delta$, then the resultant motion $\mathbf{x}(t)$ satisfies $\|\mathbf{x}(t)\| < \epsilon$ for all $t > t_0$.

This definition of stability is sometimes called *stability in the sense of Lyapunov*, abbreviated as stable i.s.L. If a system possesses this type of stability, then it is ensured that the state can be kept within ϵ, in norm, of the origin by restricting the initial perturbation to be less than δ, in norm. Note that it is necessarily true that $\delta \leq \epsilon$.

Definition 14.2: The origin is an *asymptotically stable* equilibrium point if
(*a*) it is stable, and if in addition,
(*b*) there exists a number $\delta'(t_0) > 0$ such that whenever $\|\mathbf{x}(t_0)\| < \delta'(t_0)$ the resultant motion satisfies $\lim_{t\to\infty} \|\mathbf{x}(t)\| = \mathbf{0}$.

These constitute the two basic definitions of stability for an unforced continuous-time system. If δ and δ' are not functions of t_0, then the origin is said to be *uniformly stable* and *uniformly asymptotically stable*, respectively. If $\delta'(t_0)$ in Definition 14.2 can be made arbitrarily large, i.e., if all $\mathbf{x}(t_0)$ converge to $\mathbf{0}$, then the origin is said to be *globally* asymptotically stable or asymptotically stable *in the large*.

Stability definitions for the discrete-time system with zero input

$$\mathbf{x}(k+1) = \mathbf{f}(\mathbf{x}(k), \mathbf{0}, k), \qquad \mathbf{x}(0) = \mathbf{x}_0$$

are identical to those given above, provided the discrete-time index k is used in place of t. As before, it is assumed that the coordinates have been chosen so that the origin is an equilibrium state.

When nonzero inputs $\mathbf{u}(t)$ or $\mathbf{u}(k)$ are considered, two additional types of stability are often used.

Definition 14.3: (Bounded input, bounded state stability.) If there is a fixed, finite constant K such that $\|\mathbf{u}\| \leq K$ for every t (or k), then the input is said to be bounded. If for every bounded input, and for arbitrary initial conditions $\mathbf{x}(t_0)$, there exists a scalar $0 < \delta(K, t_0, \mathbf{x}(t_0))$ such that the resultant state satisfies $\|\mathbf{x}\| \leq \delta$, then the system is *bounded input, bounded state* stable, abbreviated as BIBS stable.

All of the previous definitions of stability deal with the behavior of the state vector relative to an equilibrium state. Frequently, the main interest is in the system output behavior. This motivates the final stability definition.

Definition 14.4: (Bounded input, bounded output stability.) Let $\mathbf{u}$ be a bounded input with K_m as the least upper bound. If there exists a scalar α such that for every t (or k), the output satisfies $\|\mathbf{y}\| \leq \alpha K_m$, then the system is *bounded input, bounded output* stable, abbreviated as BIBO stable.

14.4 LINEAR SYSTEM STABILITY

The following linear continuous-time system is considered:

$$\begin{aligned} \dot{\mathbf{x}} &= \mathbf{A}(t)\mathbf{x}(t) + \mathbf{B}(t)\mathbf{u}(t) \\ \mathbf{y}(t) &= \mathbf{C}(t)\mathbf{x}(t) + \mathbf{D}(t)\mathbf{u}(t) \end{aligned} \tag{14.1}$$

The unforced case is treated first. With $\mathbf{u}(t) = \mathbf{0}$, the state vector is given by

$$\mathbf{x}(t) = \mathbf{\Phi}(t, t_0)\mathbf{x}(t_0) \tag{14.2}$$

The norm of $\mathbf{x}(t)$ is a measure of the distance of the state from the origin:

$$\|\mathbf{x}(t)\| = \|\mathbf{\Phi}(t, t_0)\mathbf{x}(t_0)\| \leq \|\mathbf{\Phi}(t, t_0)\| \, \|\mathbf{x}(t_0)\| \tag{14.3}$$

Suppose there exists a number $N(t_0)$, possibly depending on t_0, such that

$$\|\mathbf{\Phi}(t, t_0)\| \leq N(t_0) \qquad \text{for all } t \geq t_0 \tag{14.4}$$

Then the conditions of Definition 14.1 can be satisfied for any $\epsilon > 0$ by letting $\delta(t_0, \epsilon) = \epsilon/N(t_0)$. It follows from equation (*14.3*) that equation (*14.4*) is *sufficient* to ensure that the origin is stable in the sense of Lyapunov. It is easy to show that this condition is also *necessary*. The origin is asymptotically stable if and only if equation (*14.4*) holds, and if in addition, $\|\mathbf{\Phi}(t, t_0)\| \longrightarrow 0$ for $t \longrightarrow \infty$. Note that for a linear system, asymptotic stability does not depend on $\mathbf{x}(t_0)$. If a linear system is asymptotically stable, it is globally asymptotically stable.

The stability types which depend upon the input $\mathbf{u}(t)$ are now considered. For the linear continuous-time system, the state vector is given by

$$\mathbf{x}(t) = \mathbf{\Phi}(t, t_0)\mathbf{x}(t_0) + \int_{t_0}^{t} \mathbf{\Phi}(t, \tau)\mathbf{B}(\tau)\mathbf{u}(\tau)\, d\tau \tag{14.5}$$

BIBS stability requires that $\mathbf{x}(t)$ remain bounded for all bounded inputs. Since $\mathbf{u}(t) = \mathbf{0}$ is bounded, it is clear that stability i.s.L. is a necessary condition for BIBS stability. By taking the norm of both sides of equation (*14.5*) and using well-known properties of the norm, it is found that $\|\mathbf{x}(t)\|$ remains bounded, and thus the origin is BIBS stable, if equation (*14.4*) holds and if in addition there exists a number $N_1(t_0)$ such that

$$\int_{t_0}^{t} \|\mathbf{\Phi}(t, \tau)\mathbf{B}(\tau)\|\, d\tau \leq N_1(t_0) \qquad \text{for all } t \geq t_0 \tag{14.6}$$

Similar arguments show that a linear discrete-time system is BIBS stable if the discrete transition matrix satisfies equation (*14.4*) and if $\sum_{k=0}^{k_1} \|\mathbf{\Phi}(k_1, k)\mathbf{B}(k-1)\| \leq N_1$.

BIBO stability is investigated by considering the output of a linear system

$$\mathbf{y}(t) = \mathbf{C}(t)\mathbf{x}(t) + \mathbf{D}(t)\mathbf{u}(t) \tag{14.7}$$

Substitution of equation (*14.5*) into equation (*14.7*) and viewing the initial state $\mathbf{x}(t_0)$ as having arisen because of a bounded input over the interval $(-\infty, t_0)$ gives

$$\mathbf{y}(t) = \int_{-\infty}^{t} \mathbf{W}(t, \tau)\mathbf{u}(\tau)\, d\tau \tag{14.8}$$

Only bounded inputs are considered, that is,

$$\|\mathbf{u}(\tau)\| \leq K \qquad \text{for all } \tau \tag{14.9}$$

The output remains bounded in norm if there exists a constant $M > 0$ such that the impulse response or weighting matrix $\mathbf{W}(t, \tau)$ satisfies

$$\int_{-\infty}^{t} \|\mathbf{W}(t, \tau)\|\, d\tau \leq M \qquad \text{for all } t \tag{14.10}$$

Equation (*14.10*) is the necessary and sufficient condition for BIBO stability of continuous-time systems. The analogous result, with a summation replacing the integration, holds for discrete-time systems.

The matrix norm $\|\mathbf{\Phi}(t, t_0)\|$ plays a central role in the stability conditions. This norm can be defined in various ways, including the inner product definition

$$\|\mathbf{\Phi}(t, t_0)\|^2 = \max_{\mathbf{x}} \{\langle \mathbf{\Phi}(t, t_0)\mathbf{x}, \mathbf{\Phi}(t, t_0)\mathbf{x}\rangle \mid \langle \mathbf{x}, \mathbf{x}\rangle = 1\} \tag{14.11}$$

By introducing the adjoint of $\mathbf{\Phi}(t, t_0)$, it is found that equation (*14.11*) leads to

$$\|\mathbf{\Phi}(t, t_0)\|^2 = \text{max eigenvalue of } \bar{\mathbf{\Phi}}^T(t, t_0)\mathbf{\Phi}(t, t_0) \tag{14.12}$$

If $\mathbf{\Phi}(t, t_0)$ is normal, then $\bar{\mathbf{\Phi}}^T(t, t_0)\mathbf{\Phi}(t. t_0) = \mathbf{\Phi}(t, t_0)\bar{\mathbf{\Phi}}^T(t, t_0)$ and then

$$\|\mathbf{\Phi}(t, t_0)\| = \max_i |\alpha_i| \tag{14.13}$$

where α_i is an eigenvalue of $\mathbf{\Phi}(t, t_0)$. In all cases a useful lower bound on the norm is given by

$$\| \mathbf{\Phi}(t, t_0) \|^2 \geq |\alpha_i|^2 \quad \text{for any eigenvalue } \alpha_i \tag{14.14}$$

14.5 LINEAR CONSTANT SYSTEMS

Whenever the system under consideration has a constant system matrix **A**, the following results hold:

$$\mathbf{\Phi}(t, t_0) = \mathbf{e}^{\mathbf{A}(t-t_0)} \quad \text{(continuous-time)} \tag{14.15}$$

$$\mathbf{\Phi}(k, 0) = \mathbf{A}^k \quad \text{(discrete-time)} \tag{14.16}$$

By virtue of the Cayley-Hamilton theorem, Chapter 8, both of these results can be expressed as polynomials in **A**. Then by Frobenius' theorem (Problem 10.16, page 276) the eigenvalues α_i of $\mathbf{\Phi}$ are related to the eigenvalues λ_i of **A** by

$$\alpha_i = e^{\lambda_i(t-t_0)} \quad \text{or} \quad \alpha_i = \lambda_i^k$$

for the continuous-time and discrete-time cases, respectively. It is relatively simple to express the previous stability conditions in terms of the eigenvalues of the system matrix **A**. Letting these eigenvalues be $\lambda_i = \beta_i \pm j\omega_i$, the resulting conditions are summarized in Table 14.1.

TABLE 14.1 STABILITY CRITERIA FOR LINEAR CONSTANT SYSTEMS

(Eigenvalues of **A** are $\lambda_i = \beta_i \pm j\omega_i$)

	Continuous Time $\dot{\mathbf{x}} = \mathbf{A}\mathbf{x}$	Discrete Time $\mathbf{x}(k+1) = \mathbf{A}\mathbf{x}(k)$
Unstable	If $\beta_i > 0$ for any simple root or $\beta_i \geq 0$ for any repeated root	If $\|\lambda_i\| > 1$ for any simple root or $\|\lambda_i\| \geq 1$ for any repeated root
Stable i.s.L.	If $\beta_i \leq 0$ for all simple roots and $\beta_i < 0$ for all repeated roots	If $\|\lambda_i\| \leq 1$ for all simple roots and $\|\lambda_i\| < 1$ for all repeated roots
Asymptotically Stable	If $\beta_i < 0$ for all roots	$\|\lambda_i\| < 1$ for all roots

14.6 THE DIRECT METHOD OF LYAPUNOV*

Late in the nineteenth century the Russian mathematician A. M. Lyapunov developed an approach to stability analysis, now known as the direct method (or second method) of Lyapunov. Probably because of the language barrier, the technique was largely

*The alternate spelling, Liapunov, is frequently used. The difference arose during transliteration from Russian characters.

ignored by the English-speaking world for about half a century. At the present time, it is widely used for stability analysis of linear and nonlinear systems, both time-invariant and time-varying. In this brief treatment an intuitive discussion of the method is presented. Then a few of the main theorems are given without proof. Uses of the theorems are illustrated in the problems. Some other areas of current application are also mentioned [62, 63, 71, 86].

Energy concepts are widely used and easily understood by engineers. Lyapunov's direct method can be viewed as a generalized energy method. Consider a second-order system, such as the unforced *LC* circuit of Fig. 14.2*a* or the mass-spring system of Fig. 14.2*b*.

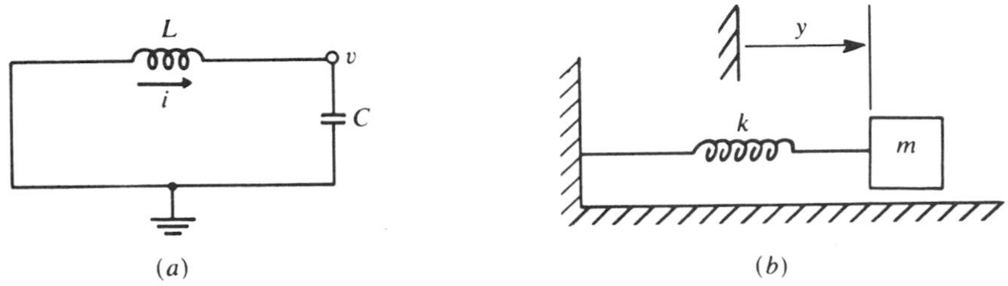

Figure 14.2

In the first case the capacitor voltage v and inductor current i can be used as state variables $\mathbf{x}$, and the total energy (magnetic plus electric) in the system at any time is $Li^2/2 + Cv^2/2$. In the second case the position y (measured from the free length of the spring) and the velocity $\dot{y}$ can be used as state variables $\mathbf{x}$, and the total energy (kinetic plus potential) at any time is $m\dot{y}^2/2 + ky^2/2$. In both cases the energy $\mathcal{E}$ is a quadratic function of the state variables. Thus $\mathcal{E}(\mathbf{x}) > 0$ if $\mathbf{x} \neq \mathbf{0}$, and $\mathcal{E} = 0$ if and only if $\mathbf{x} = \mathbf{0}$. If the time rate of change of the energy $\dot{\mathcal{E}}$ is always negative, except at $\mathbf{x} = \mathbf{0}$, then $\mathcal{E}$ will be continually decreasing and will eventually approach zero. Because of the nature of the energy function, $\mathcal{E} = 0$ implies $\mathbf{x} = \mathbf{0}$. Therefore, if $\dot{\mathcal{E}} < 0$ for all t, except when $\mathbf{x} = \mathbf{0}$, it is concluded that $\mathbf{x}(t) \longrightarrow \mathbf{0}$ for sufficiently large t. The close relationship of this conclusion to asymptotic stability, Definition 14.2, is obvious.

If the time rate of change of energy is never positive, that is, $\dot{\mathcal{E}} \leq 0$, then $\mathcal{E}$ can never increase, but it need not approach zero either. It can then be concluded that $\mathcal{E}$, and hence $\mathbf{x}$, remain bounded in some sense. This situation is related in an obvious way to stability i.s.L., Definition 14.1.

For both systems of Fig. 14.2, the energy can be expressed in the form $\mathcal{E}(\mathbf{x}) = \frac{1}{2}a_1x_1^2 + \frac{1}{2}a_2x_2^2$. Assuming for the present that the coefficients a_1 and a_2 are constant, the time rate of change is given by

$$\dot{\mathcal{E}} = a_1x_1\dot{x}_1 + a_2x_2\dot{x}_2 \tag{14.17}$$

knowledge of the *form* of the differential equations $\dot{x}_1 = f_1(\mathbf{x})$ and $\dot{x}_2 = f_2(\mathbf{x})$ allows both $\mathcal{E}$ and $\dot{\mathcal{E}}$ to be expressed as functions of the state $\mathbf{x}$. No knowledge of the *solutions* of the differential equations is required in order to draw conclusions regarding stability. Lyapunov's direct method is a generalization of these ideas.

Example 14.1

For the system of Fig. 14.2*a*, with $x_1 = i$ and $x_2 = v$, $\dot{x}_1 = -x_2/L$, $\dot{x}_2 = x_1/C$, so that $\dot{\mathcal{E}} = Lx_1(-x_2/L) + Cx_2(x_1/C) = 0$. Thus $\dot{\mathcal{E}} \equiv 0$ for all t and $\mathcal{E}$ is constant. This conservative system is stable i.s.L. but not asymptotically stable. Of course, this is a well-known result for this undamped oscillator. If the system is modified to include a positive resistance, then the system is dissipative and $\dot{\mathcal{E}}$ would be always negative, except when $\mathbf{x} = \mathbf{0}$. The system is then asymptotically stable. ■

Example 14.2

The system of Fig. 14.2*b* is assumed to have a nonlinear frictional force d acting between the mass and the supporting surface. This does not change the definition of the energy $\mathcal{E}$. Letting $x_1 = y$ and $x_2 = \dot{y}$, the differential equations are $\dot{x}_1 = x_2$, $\dot{x}_2 = (-kx_1 + d)/m$. Then

$$\dot{\mathcal{E}} = kx_1\dot{x}_1 + mx_2\dot{x}_2 = kx_1x_2 - kx_1x_2 + x_2d = x_2d$$

$\dot{\mathcal{E}}$ is nonpositive if the friction force d is always opposing the direction of the velocity x_2. In this case $\mathcal{E}$ can never increase, and the system is clearly stable i.s.L. It may, in fact, be asymptotically stable, as will become clear later. On the other hand, if d and x_2 have the same sign, then $\dot{\mathcal{E}} > 0$ and $\mathcal{E}$ may increase. Neither stability nor instability can be concluded without further analysis. ■

Lyapunov's direct method makes use of a *Lyapunov function* $V(\mathbf{x})$. This scalar function of the state may be thought of as a generalized energy. In many problems the energy function can serve as a Lyapunov function. In cases where a system model is described mathematically, it may not be clear what "energy" means. The conditions which $V(\mathbf{x})$ must satisfy in order to be a Lyapunov function are therefore based on mathematical rather than physical considerations.

A single-valued function $V(\mathbf{x})$ which is continuous and has continuous partial derivatives is said to be *positive definite* in some region Ω about the origin of the state space if (1) $V(\mathbf{0}) = 0$ and (2) $V(\mathbf{x}) > 0$ for all nonzero $\mathbf{x}$ in Ω. A special case of a positive definite function was the quadratic form discussed in Sec. 7.7, page 183. If condition (2) is relaxed to $V(\mathbf{x}) \geq 0$ for all $\mathbf{x} \in \Omega$, then $V(\mathbf{x})$ is said to be *positive semidefinite*. Reversing the inequalities leads to corresponding definitions of *negative definite* and *negative semidefinite* functions.

Consider the autonomous (i.e., unforced, no explicit time dependence) system

$$\dot{\mathbf{x}} = \mathbf{f}(\mathbf{x}) \tag{14.18}$$

The origin is assumed to be an equilibrium point, that is, $\mathbf{f}(\mathbf{0}) = \mathbf{0}$. The stability of this equilibrium point can be investigated by means of the following theorems.

Theorem 14.1: If a positive definite function $V(\mathbf{x})$ can be determined such that $\dot{V}(\mathbf{x}) \leq 0$ (negative semidefinite), then the origin is stable i.s.L.

A function $V(\mathbf{x})$ satisfying these requirements is called a Lyapunov function. The Lyapunov function is not unique; rather, many different Lyapunov functions may be found for a given system. Likewise, the inability to find a satisfactory Lyapunov function does not mean that the system is unstable.

Theorem 14.2: If a positive definite function $V(\mathbf{x})$ can be found such that $\dot{V}(\mathbf{x})$ is negative definite, then the origin is asymptotically stable.

Both of the previous theorems relate to local stability in the vicinity of the origin. Global asymptotic stability is considered next.

Theorem 14.3: The origin is a globally asymptotically stable equilibrium point for the system of equation (*14.18*) if a Lyapunov function $V(\mathbf{x})$ can be found such that (1) $V(\mathbf{x}) > 0$ for all $\mathbf{x} \neq \mathbf{0}$ and $V(\mathbf{0}) = 0$, (2) $\dot{V}(\mathbf{x}) < 0$ for all $\mathbf{x} \neq 0$, and (3) $V(\mathbf{x}) \longrightarrow \infty$ as $\|\mathbf{x}\| \longrightarrow \infty$.

As an aid in understanding these three theorems and their differences, a family of contours of $V(\mathbf{x}) =$ constant is shown in Fig. 14.3 for a two-dimensional state space.

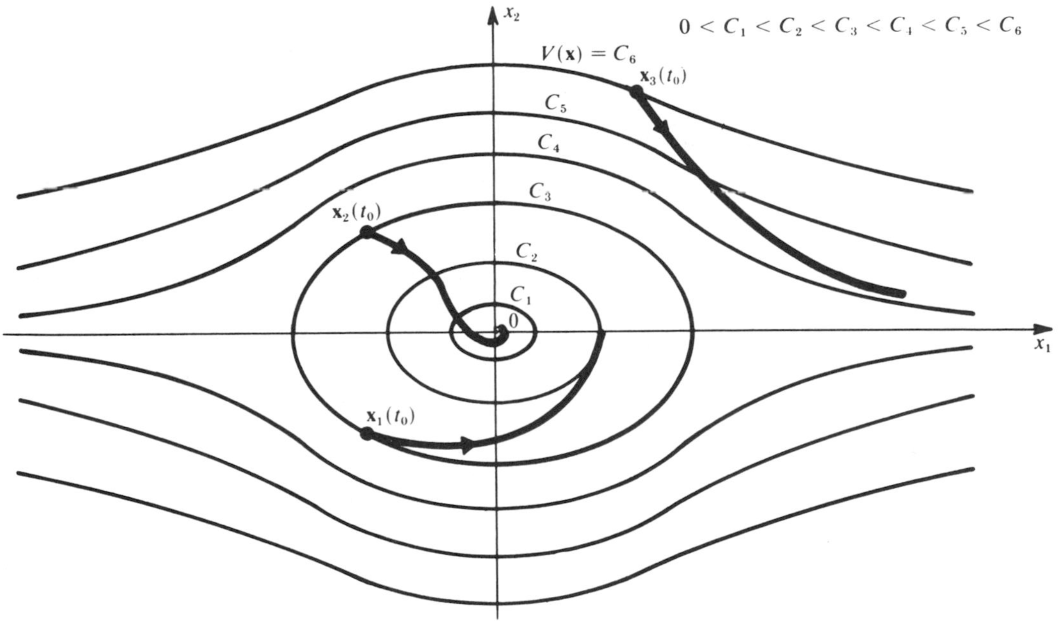

Figure 14.3

These contours can never intersect because $V(\mathbf{x})$ is single valued. They are smooth curves because $V(\mathbf{x})$ and all its partial derivatives are required to be continuous.

For the situation shown, condition (3), Theorem 14.3 is not met. It is possible for $\|\mathbf{x}\| \longrightarrow \infty$ while $V(\mathbf{x})$ remains finite, for example along the contour $V(\mathbf{x}) = C_4$. It could happen that $V(\mathbf{x})$ continually decreases, causing $V(\mathbf{x})$ to decrease from C_6 through C_5 and asymptotically approach C_4, while $\|\mathbf{x}(t)\|$ continues to grow without bound. The trajectory starting at $\mathbf{x}_3(t_0)$ illustrates this. Condition (3) of Theorem 14.3 is necessary to rule out this possibility.

Referring to Fig. 14.3, if $\dot{V}(\mathbf{x}) < 0$ then the origin is asymptotically stable for any initial state sufficiently close to the origin, specifically for any $\mathbf{x}(t_0)$ inside contour C_3. Any initial state inside contour C_3, such as $\mathbf{x}_2(t_0)$, must eventually approach the origin if $\dot{V}(\mathbf{x})$ is always negative for $\mathbf{x} \neq \mathbf{0}$. This is the essence of Theorem 14.2. If the restriction is relaxed so that $\dot{V}(\mathbf{x}) \leq 0$, as in Theorem 14.1, then $V(\mathbf{x})$ can never increase, but might approach a nonzero constant. If this is true and if the initial state is inside contour C_3, the state will remain bounded, but need not approach zero. Initial state $\mathbf{x}_1(t_0)$ illustrates this case and indicates that the origin is stable i.s.L., but not necessarily asymptotically stable.

Unfortunately, the Lyapunov theorems give no indication of how a Lyapunov function might be found. There is no one unique Lyapunov function for a given system. Some are better than others. It might happen that a $V_1(\mathbf{x})$ can be found which indicates stability i.s.L., $V_2(\mathbf{x})$ might indicate asymptotic stability for initial states quite close to the origin, and $V_3(\mathbf{x})$ might indicate asymptotic stability for a much larger region, or even global asymptotic stability. The inability to find a suitable Lyapunov function does not prove instability. If a system is stable in one of the senses mentioned, it is ensured that an appropriate Lyapunov function does exist. Much ingenuity may be required to find it, however.

There is no universally best method of searching for Lyapunov functions. A form for $V(\mathbf{x})$ can be assumed, either as a pure guess or tempered by physical insight and energy-like considerations. Then $\dot{V}(\mathbf{x})$ can be tested, bringing in the system equations $\dot{\mathbf{x}} = \mathbf{f}(\mathbf{x})$ in the process. Another approach is to assume a form for the derivatives of $V(\mathbf{x})$, either $\dot{V}(\mathbf{x})$ or $\nabla_{\mathbf{x}} V(\mathbf{x})$. Then $V(\mathbf{x})$ can be determined by integration and tested to see if it meets the required conditions. Examples of these techniques are presented in the problems.

A more general, time-varying system is now considered:

$$\dot{\mathbf{x}} = \mathbf{f}(\mathbf{x}, t) \tag{14.19}$$

It is assumed that the origin is an equilibrium point, $\mathbf{f}(\mathbf{0}, t) = \mathbf{0}$ for all t. The previous stability theorems are still basically what is needed to conclude stability. However, now the Lyapunov function and its time derivative may be explicit functions of time as well as the state. The definitions of positive and negative definite must be modified to reflect this added generality. Only Theorem 14.3 is generalized.

Theorem 14.4: If a single-valued scalar function $V(\mathbf{x}, t)$ exists, which is continuous and has continuous first partial derivatives and for which

(1) $V(\mathbf{0}, t) = 0$ for all t;
(2) $V(\mathbf{x}, t) \geq \sigma(\|\mathbf{x}\|) > 0$ for all $\mathbf{x} \neq \mathbf{0}$ and for all t, where $\sigma(\cdot)$ is a continuous, nondecreasing scalar function with $\sigma(0) = 0$;
(3) $\dot{V}(\mathbf{x}, t) \leq -\kappa(\|\mathbf{x}\|) < 0$ for all $\mathbf{x} \neq \mathbf{0}$ and for all t, where $\kappa(\cdot)$ is a continuous nondecreasing scalar function with $\kappa(0) = 0$;
(4) $V(\mathbf{x}, t) \leq v(\|\mathbf{x}\|)$ for all $\mathbf{x}$ and t, where $v(\cdot)$ is a continuous nondecreasing scalar function with $v(0) = 0$;
(5) $\sigma(\|\mathbf{x}\|) \longrightarrow \infty$ as $\|\mathbf{x}\| \longrightarrow \infty$;

then $\mathbf{x} = \mathbf{0}$ is uniformly globally asymptotically stable.

The functions σ, κ, and v are all positive definite in the earlier sense, where time did not explicitly appear. $V(\mathbf{x}, t)$ is said to be positive definite if $V(\mathbf{0}, t) = 0$ and if $V(\mathbf{x}, t)$ is always greater than or equal to a time-invariant positive definite function such as σ. Conditions (1) and (2) simply require $V(\mathbf{x}, t)$ to be positive definite. Similarly, condition (3) requires $\dot{V}(\mathbf{x}, t)$ to be negative definite. Condition (5) requires $V(\mathbf{x}, t)$ to become infinite as $\|\mathbf{x}\| \longrightarrow \infty$, as in Theorem 14.3, and condition (4) prevents $V(\mathbf{x}, t)$ from becoming infinite when $\|\mathbf{x}\|$ is finite. The "uniformly" in the conclusion indicates that the asymptotic stability does not depend on any particular initial time t_0.

One final theorem is presented which is sometimes useful in avoiding fruitless searches for Lyapunov functions for unstable systems.

Theorem 14.5: If a scalar function $V(\mathbf{x}, t)$ is continuous, single valued, and has continuous first partial derivatives, and if

(1) $V(\mathbf{0}, t) = 0$ for all t;
(2) $V(\mathbf{x}, t) > 0$ for all $\mathbf{x} \neq 0$ in some region Ω containing the origin;
(3) $\dot{V}(\mathbf{0}, t) = 0$ for all t and $\dot{V}(\mathbf{x}, t) > 0$ for all nonzero $\mathbf{x}$ in Ω and all t;

then the origin is an *unstable* equilibrium point of the system of equation *(14.19)*.

Lyapunov's stability theorems can also be used to investigate the stability of discrete-time systems:

$$\mathbf{x}(k + 1) = \mathbf{f}(\mathbf{x}(k)) \tag{14.20}$$

The origin is assumed to be an equilibrium point, so that $\mathbf{f}(\mathbf{0}) = \mathbf{0}$. The preceding theorems for stability, asymptotic stability, and global asymptotic stability apply provided the time derivative $\dot{V}(\mathbf{x})$ is replaced by the first difference, $\Delta V = V(\mathbf{x}(k + 1)) - V(\mathbf{x}(k))$.

Lyapunov's direct method of stability analysis is a general method of approach, but its successful use requires considerable ingenuity. Beginning with the very general Lyapunov philosophy, various methods of generating Lyapunov functions and easy-to-use results have been developed by restricting the system under consideration in various ways. The methods of Zubov, Lure, Popov, and others fall into this category.

The circle criterion is one popular result of this type which is in essence a generalized frequency domain criterion similar to the Nyquist criterion for linear systems [9, 86].

In addition to stability analysis, Lyapunov's direct method can be used in system design [62, 63, 86]. In particular, the Lyapunov approach can be used to estimate the dominant time constant of linear or nonlinear systems. Control systems can also be designed so as to rapidly correct for disturbances. The approach used is to determine a Lyapunov function which is suitable for the uncontrolled system. Then when the control term $\mathbf{u}(t)$ is considered, $V(\mathbf{x})$ remains unchanged since it does not depend on $\dot{\mathbf{x}}$(or $\mathbf{u}$). $\dot{V}(\mathbf{x})$ does depend upon $\mathbf{u}$ through $\dot{\mathbf{x}}$. The control term $\mathbf{u}(t)$ is selected to make $\dot{V}(\mathbf{x})$ as negative as possible. This ensures that $V(\mathbf{x})$ will approach zero as rapidly as possible. Good approximations to time-optimal control are often obtained in this manner.

ILLUSTRATIVE PROBLEMS

Linear System Stability

14.1 Comment on the stability of the four constant coefficient systems of Problem 9.1, page 244.

a. The eigenvalue is $\lambda = -\alpha$ and the system is asymptotically stable if $\lambda < 0$. If $\lambda = 0$, the system is stable in the sense of Lyapunov. If $\lambda > 0$, it is unstable.
b. The eigenvalue is again $\lambda = -\alpha$ and the results of part a still apply.
c. The eigenvalues of the 2×2 $\mathbf{A}$ matrix satisfy $\lambda^2 + 2\zeta\omega\lambda + \omega^2 = 0$. They are $\lambda = -\zeta\omega \pm \omega\sqrt{\zeta^2 - 1}$. Both eigenvalues are negative for any positive value of ζ and then the system is asymptotically stable. This assumes that ω is a positive constant. If $\zeta = 0$, then $\lambda = \pm j\omega$ and the system is stable i.s.L. For $\zeta < 0$, the system is unstable.
d. The results of part c again apply.

These examples illustrate that the poles of a transfer function are eigenvalues of the matrix $\mathbf{A}$ in the state space representation. This will always be true, but the converse need not be true.

14.2 Is the system of Problem 9.2, page 245, stable? Is it asymptotically stable?

Evaluating $|\mathbf{A} - \mathbf{I}\lambda| = 0$ gives $\lambda^4 + 3\lambda^3 + 3\lambda^2 + \lambda = 0$. Thus one eigenvalue is $\lambda = 0$. The remaining cubic $\lambda^3 + 3\lambda^2 + 3\lambda + 1 = 0$ can be investigated using Routh's criterion. This indicates that there are no right-half-plane eigenvalues. Thus the system is stable, but not asymptotically stable because of the root $\lambda = 0$.

14.3 a. Is the system of Problem 9.3, page 246, asymptotically stable?
b. Is the system of Problem 9.5, page 248, asymptotically stable?

a. All three of the forms given for $\mathbf{A}$ have the same eigenvalues, $\lambda = -3, -1$, and

-6. (These are also poles of the original transfer function.) This system is asymptotically stable.

b. The 6×6 **A** matrix is given in Jordan form, from which it is apparent that the eigenvalues are $0, 0, -3, -3, -3, -1$. This system is unstable because of the double root $\lambda = 0$. With $\mathbf{u} = \mathbf{0}$, $\dot{x}_2 = 0$ or $x_2 =$ constant. Since $\dot{x}_1 = x_2$, $x_1(t)$ is a linear function of time. Thus unless $x_2(0) = 0$, $x_1(t) \longrightarrow \pm\infty$ and as a result $\|\mathbf{x}(t)\| \longrightarrow \infty$.

14.4 Investigate the stability of the continuous-time system and its discrete-time approximation, as described in Problem 10.13, page 274.

The continuous-time system matrix **A** has eigenvalues $\lambda = 0, -1$. Therefore, the system is stable in the sense of Lyapunov but not asymptotically stable. The discrete-time system matrix **A** has eigenvalues $\lambda_1 = 0.368$ and $\lambda_2 = 1$. Since $|\lambda_1| < 1$, but $|\lambda_2| = 1$, the discrete system is stable in the sense of Lyapunov, but not asymptotically stable.

14.5 Comment on the stability of the linear discrete-time systems described in

a. Problem 11.3, page 294,
b. Problem 11.4, page 295,
c. Problem 11.6, page 295,

a. Since the **A** matrix is triangular, its eigenvalues are found by inspection as $\lambda_1 = \lambda_2 = \lambda_3 = 1/2$. Since all $|\lambda_i| < 1$, the system is asymptotically stable. This is verified by the explicit solution found in Problem 11.3, which shows that $\mathbf{x}(k) \longrightarrow \mathbf{0}$ as $k \longrightarrow \infty$.

b. The eigenvalues were given earlier as $\lambda_1 = 1/2$, $\lambda_2 = 1/3$. Since $|\lambda_i| < 1$, the system is asymptotically stable.

c. The eigenvalues are $\lambda_1 = 1$, $\lambda_2 = 1/2$, $\lambda_3 = -1/3$. This system is stable i.s.L., but not asymptotically stable, since $\lambda_1 = 1$.

14.6 Show that if a continuous-time, linear constant system is asymptotically stable, then its adjoint is unstable.

Asymptotic stability of $\dot{\mathbf{x}} = \mathbf{A}\mathbf{x}$ implies that all eigenvalues of **A** have negative real parts. The adjoint equation is $\dot{\mathbf{z}} = -\mathbf{A}^T\mathbf{z}$. The eigenvalues of $-\mathbf{A}^T$ are the roots of $|-\mathbf{A}^T - \mathbf{I}\gamma| = 0$ or $|\mathbf{A}^T - \mathbf{I}(-\gamma)| = \mathbf{0}$. Since $\mathbf{A}^T$ and **A** have the same eigenvalues, the roots γ are the negative of the eigenvalues of **A**. Then **A** having all left-half-plane eigenvalues implies that $-\mathbf{A}^T$ has all right-half-plane eigenvalues, and the adjoint system is unstable.

14.7 Show that if a continuous-time, linear constant system is asymptotically stable, it is also BIBS stable.

The conditions for BIBS stability are given by equations (*14.4*) and (*14.6*). If the system is asymptotically stable, then $\|\mathbf{\Phi}(t, t_0)\| \leq N(t_0)$ for all $t \geq t_0$. Furthermore, using norm inequalities gives

$$\int_{t_0}^{t} \|\mathbf{\Phi}(t, \tau)\mathbf{B}\|\, d\tau \leq \int_{t_0}^{t} \|\mathbf{\Phi}(t, \tau)\| \cdot \|\mathbf{B}\|\, d\tau$$

Since **B** and hence $\|\mathbf{B}\|$ are constant, this term can be taken outside the integral. Also, expressing $\mathbf{\Phi}$ in terms of the Jordan form and modal matrix of **A** (see page 265) leads to

$$\int_{t_0}^{t} \|\mathbf{\Phi}(t, \tau)\mathbf{B}\|\, d\tau \leq \|\mathbf{M}^{-1}\| \cdot \|\mathbf{M}\| \int_{t_0}^{t} \|e^{\mathbf{J}(t-\tau)}\|\, d\tau\, \|\mathbf{B}\|$$

If all the eigenvalues of **A** are distinct, then $\|e^{\mathbf{J}(t-\tau)}\| = e^{\beta_i(t-\tau)}$, where β_i is the largest real part of the eigenvalues of **A**. But β_i is negative since the system is asymptotically stable. In this case the integral is bounded and thus **BIBS** stability follows. If **A** has repeated eigenvalues, then $\|e^{\mathbf{J}(t-\tau)}\|$ can be bounded by $\sqrt{p(t)}\, e^{\beta_i(t-\tau)}$, where $p(t)$ is a polynomial in t. Asymptotic stability ensures that $\beta_i < 0$, and again the integral is bounded. Therefore, for linear constant systems, asymptotic stability implies **BIBS** stability.

14.8 Show that if a continuous-time, linear constant system is asymptotically stable, it is also BIBO stable.

The system $\dot{\mathbf{x}} = \mathbf{Ax} + \mathbf{Bu}$, $\mathbf{y} = \mathbf{Cx} + \mathbf{Du}$ has the output solution

$$\mathbf{y}(t) = \mathbf{C\Phi}(t, t_0)\mathbf{x}(t_0) + \int_{t_0}^{t} \mathbf{C\Phi}(t, \tau)\mathbf{Bu}(\tau)\, d\tau + \mathbf{Du}(t)$$

Since the norm of any constant matrix is bounded, straightforward application of norm inequalities leads to

$$\|\mathbf{y}(t)\| \leq \|\mathbf{C}\|\,\|\mathbf{\Phi}(t, t_0)\|\,\|\mathbf{x}(t_0)\| + \|\mathbf{C}\| \cdot \|\mathbf{B}\| K \int_{t_0}^{t} \|\mathbf{\Phi}(t, \tau)\|\, d\tau + \|\mathbf{D}\| K$$

where $\|\mathbf{u}(t)\| \leq K$ for all t. Asymptotic stability ensures that $\|\mathbf{\Phi}(t, t_0)\|$ is bounded by a decaying exponential and thus $\int_{t_0}^{t} \|\mathbf{\Phi}(t, \tau)\|\, d\tau$ is also bounded for all $t \geq t_0$. It follows that the output is bounded in norm also.

It is also known [9] that *if the system is completely controllable and completely observable*, then the implication can also be reversed. In that case asymptotic stability $\Leftrightarrow$ BIBO stability.

14.9 Consider the stability of a linear constant coefficient system $\dot{\mathbf{x}} = \mathbf{Ax}$, using Lyapunov's direct method.

Assume a quadratic Lyapunov function $V(\mathbf{x}) = \mathbf{x}^T\mathbf{Px}$, with **P** symmetric and positive definite. Then $\dot{V}(\mathbf{x}) = \dot{\mathbf{x}}^T\mathbf{Px} + \mathbf{x}^T\mathbf{P}\dot{\mathbf{x}} = \mathbf{x}^T[\mathbf{A}^T\mathbf{P} + \mathbf{PA}]\mathbf{x}$. If $\mathbf{A}^T\mathbf{P} + \mathbf{PA} = -\mathbf{Q}$ for some positive definite matrix **Q**, then the system is asymptotically stable. Conversely, if a positive definite **Q** is specified, and if the matrix **P** determined by the above equation is positive definite, then asymptotic stability can be concluded. Using this approach, the conditions for a positive definite **P** matrix, in terms of the signs of the principal minors (Sec. 7.7, page 184), lead to conditions for stability in terms of the a_{ij} coefficients. The stability conditions which are obtained in this manner are of the same type as obtained from Routh's criterion.

14.10 Investigate the stability of the system described by

$$\dot{\mathbf{x}} = \begin{bmatrix} -3 & 2 \\ -1 & -1 \end{bmatrix} \mathbf{x} \tag{1}$$

Let $\mathbf{P} = [P_{ij}]$ with $P_{12} = P_{21}$. Then

$$\mathbf{A}^T\mathbf{P} + \mathbf{PA} = \begin{bmatrix} -6P_{11} - 2P_{12} & -4P_{12} - P_{22} + 2P_{11} \\ -4P_{12} - P_{22} + 2P_{11} & 4P_{12} - 2P_{22} \end{bmatrix}$$

Using $\mathbf{Q} = \mathbf{I}_2$, the unit matrix, and solving $\mathbf{A}^T\mathbf{P} + \mathbf{PA} = -\mathbf{Q}$ gives $\mathbf{P} = \begin{bmatrix} 7/40 & -1/40 \\ -1/40 & 18/40 \end{bmatrix}$. The principal minors of **P** are $\Delta_1 = 7/40 > 0$ and $\Delta_2 = |\mathbf{P}| = 5/64 > 0$. Therefore, **P** is positive definite and the system is asymptotically stable.

14.11 Derive conditions for asymptotic stability of the origin for the following systems. Use Lyapunov's direct method and proceed by assuming a suitable form for $\dot{V}(\mathbf{x})$. $V(\mathbf{x})$ is then found by integration.

a. $\dot{x} = ax$;

b. $\dot{x}_1 = x_2,\ \dot{x}_2 = -ax_1 - bx_2$;

c. $\dot{x}_1 = x_2,\ \dot{x}_2 = x_3,\ \dot{x}_3 = -ax_1 - bx_2 - cx_3$.

A form for $\dot{V}$ is assumed, which is at least negative semidefinite. A choice which is often successful is $\dot{V} = -x_n^2$ [98]. Then $V(\mathbf{x}(t)) - V(\mathbf{x}(t_1)) = \int_{t_1}^{t} \dot{V}(\mathbf{x})\,dt$. The lower limit is selected so that $\mathbf{x}(t_1) = \mathbf{0}$ and $V(\mathbf{0}) = 0$. If the $V(\mathbf{x})$ found in this manner is positive definite, then the method is successful.

a. Try $\dot{V} = -x^2$. Assuming $a \neq 0$, this gives $\dot{V} = -x\dot{x}/a$. Then

$$V(x) = -\frac{1}{a}\int_{t_1}^{t} x\dot{x}\,dt = -\frac{1}{a}\int_{0}^{x} x\,dx = -\frac{x^2}{2a}$$

$V(x)$ is positive definite if $a < 0$. Since $\dot{V}$ is negative definite, asymptotic stability results if $a < 0$.

b. Try $\dot{V} = -x_2^2 = (-x_2)x_2$. If $a \neq 0$, then $-x_2 = (\dot{x}_2 + ax_1)/b$ so that $\dot{V} = \dot{x}_2 x_2/b + ax_1x_2/b$. Using $x_2 = \dot{x}_1$ gives

$$V(\mathbf{x}) = (a/b)\int_{0}^{x_1} x_1\,dx_1 + (1/b)\int_{0}^{x_2} x_2\,dx_2 = \frac{ax_1^2}{2b} + \frac{x_2^2}{2b}$$

If $a > 0$ and $b > 0$, $V(\mathbf{x})$ is positive definite. $\dot{V}$ is negative semidefinite, but is never zero on any trajectory of this system except at $\mathbf{x} = \mathbf{0}$. Therefore, $a > 0$ and $b > 0$ ensure asymptotic stability.

c. Try $\dot{V} = -x_3^2$. Then $V(\mathbf{x}) = -\int_{t_1}^{t} x_3^2\,dt = -\int_{t_1}^{t} x_3\dot{x}_2\,dt$. Using integration by parts, $V(\mathbf{x}) = -x_3x_2 + \int x_2\dot{x}_3\,dt$. Using the differential equation to replace $\dot{x}_3$,

$$V(\mathbf{x}) = -x_3x_2 - \int x_2(ax_1 + bx_2 + cx_3)\,dt = -x_3x_2 - \frac{ax_1^2}{2} - b\int x_2^2\,dt - \frac{cx_2^2}{2}$$

The integral term is evaluated, using $\dot{x}_1 = x_2$ and integration by parts:

$$b\int x_2^2\,dt = bx_1x_2 - b\int x_1\dot{x}_2\,dt = bx_1x_2 - b\int x_1x_3\,dt$$

From the differential equation, $-x_1 = (1/a)(\dot{x}_3 + bx_2 + cx_3)$. Hence

$$b\int x_2^2\,dt = bx_1x_2 + \frac{b}{2a}x_3^2 + \frac{b^2}{2a}x_2^2 + (bc/a)\int_{t_2}^{t} x_3^2\,dt$$

which gives

$$V(\mathbf{x}) = -\frac{a}{2}\left(x_1 + \frac{b}{a}x_2\right)^2 - \frac{b}{2a}\left(x_3 + \frac{a}{b}x_2\right)^2 - \frac{c - a/b}{2}x_2^2 - \frac{bc}{a}\int x_3^2\,dt$$

This expression is *not* positive definite. In fact, it can be made negative definite. A new trial function is selected as $V'(\mathbf{x}) = -V(\mathbf{x}) - (bc/a)\int_{t_1}^{t} x_3^2\,dt$. Then

$$\dot{V}'(\mathbf{x}) = -\dot{V}(x) - (bc/a)x_3^2 = -(bc/a - 1)x_3^2$$

$\dot{V}'(\mathbf{x})$ is negative semidefinite if $bc/a > 1$. $V'(\mathbf{x})$ is positive definite if, in addition, $a > 0$, $b > 0$, $c > 0$. Since $\dot{V}'(\mathbf{x})$ can never vanish on a trajectory of this system, the conditions for asymptotic stability are $a > 0$, $b > 0$, $c > 0$, and, $bc - a > 0$.

In all three cases the well-known results of the Routh stability criterion have been determined. The procedures illustrated by these examples can be extended to nonlinear problems [98].

Nonlinear System Stability

14.12 Let the origin be an equilibrium point of a slightly nonlinear system $\dot{\mathbf{x}} = \mathbf{f}(\mathbf{x})$. If this system is linearized about the origin, perhaps using Taylor's series, then $\dot{\mathbf{x}} = \mathbf{A}\mathbf{x} + \mathbf{h}(\mathbf{x})$, where $\mathbf{A}$ is the Jacobian matrix $[\partial \mathbf{f}/\partial \mathbf{x}]$ evaluated at $\mathbf{x} = 0$, and $\mathbf{h}(\mathbf{x})$ represents higher-order terms. Show that if $\|\mathbf{h}(\mathbf{x})\| \leq \alpha \|\mathbf{x}\|$ for some positive constant α, then asymptotic stability of the linear equation $\dot{\mathbf{x}} = \mathbf{A}\mathbf{x}$ implies asymptotic stability of the nonlinear equation as well [29, 62].

Let $\mathbf{\Phi}(t, \tau) = e^{\mathbf{A}(t-\tau)}$ be the transition matrix of the linear equation. Then treating $\mathbf{h}(\mathbf{x})$ as a forcing term, the solution for the nonlinear system can be written as

$$\mathbf{x}(t) = \mathbf{\Phi}(t, t_0)\mathbf{x}(t_0) + \int_{t_0}^{t} \mathbf{\Phi}(t, \tau)\mathbf{h}(\mathbf{x}(\tau))\, d\tau$$

Therefore,

$$\|\mathbf{x}(t)\| \leq \|\mathbf{\Phi}(t, t_0)\| \cdot \|\mathbf{x}(t_0)\| + \int_{t_0}^{t} \|\mathbf{\Phi}(t, \tau)\|\, \|\mathbf{h}(\mathbf{x}(\tau))\|\, d\tau$$

Asymptotic stability of the linear part ensures that $\|\mathbf{\Phi}(t, \tau)\|$ is bounded by a decaying exponential, $\|\mathbf{\Phi}(t, \tau)\| \leq Me^{-k(t-\tau)}$ for all $t \geq \tau$, all $\tau \geq t_0$. Usingth is and the bound on $\|\mathbf{h}(\mathbf{x})\|$ leads to

$$\|\mathbf{x}(t)\| \leq Me^{-k(t-t_0)}\|\mathbf{x}(t_0)\| + \int_{t_0}^{t} \alpha Me^{-k(t-\tau)}\|\mathbf{x}(\tau)\|\, d\tau$$

Multiplying by the positive function e^{kt} leaves the sense of the inequality unchanged, so

$$e^{kt}\|\mathbf{x}(t)\| \leq Me^{kt_0}\|\mathbf{x}(t_0)\| + \int_{t_0}^{t} \alpha Me^{k\tau}\|\mathbf{x}(\tau)\|\, d\tau \qquad (1)$$

Call the right-hand side of equation (*1*) $U(t)$ for convenience. Note that $\dot{U}(t) = \alpha Me^{kt}\|\mathbf{x}(t)\| = \alpha M$ times left-hand side of equation (*1*). Hence

$$\dot{U}/(\alpha M) \leq U \qquad \text{or} \qquad dU/U \leq \alpha M\, dt$$

Integrating both sides gives $\ln (U(t)/C) \leq \alpha M(t - t_0)$, where the integration constant C is the value of U at $t = t_0$, namely $C = Me^{kt_0}\|\mathbf{x}(t_0)\|$. Then

$$U(t) \leq Ce^{\alpha M(t-t_0)} \leq Me^{kt_0}\|\mathbf{x}(t_0)\|e^{\alpha M(t-t_0)}$$

is an explicit upper bound for the right-hand side of equation (*1*). Therefore,

$$e^{kt}\|\mathbf{x}(t)\| \leq Me^{kt_0}\|\mathbf{x}(t_0)\|e^{\alpha M(t-t_0)}$$

or

$$\|\mathbf{x}(t)\| \leq Me^{-(k-\alpha M)(t-t_0)}\|\mathbf{x}(t_0)\|$$

Thus $\|\mathbf{x}(t)\| \longrightarrow 0$ provided the bound on $\|\mathbf{h}(\mathbf{x})\|$ is sufficiently small, i.e., if $\alpha < k/M$. The constant α must satisfy $\alpha \geq \|\mathbf{h}(\mathbf{x})\|/\|\mathbf{x}\|$. Since $\mathbf{h}(\mathbf{x})$ is composed of second- or higher-order terms in components of $\mathbf{x}$, this restriction can be made as small as we please by restricting $\|\mathbf{x}\|$ to be sufficiently small.

Thus asymptotic stability of the linear part of the system implies asymptotic stability of the nonlinear system in a sufficiently small neighborhood of the origin. It can also be shown that if no eigenvalues of the Jacobian matrix have zero real parts, then the nonlinear system is unstable if any one eigenvalue of **A** has a positive real part.

In summary, the stability of the nonlinear system is the same as the linearized portion provided no eigenvalue of **A** is on the imaginary axis. If an eigenvalue is on the imaginary axis, the linearized system gives no definite information about stability.

14.13 Investigate the following nonlinear system for stability:

$$\dot{x}_1 = x_2, \qquad \dot{x}_2 = -g(x_2) - f(x_1) \tag{1}$$

The stability of an *equilibrium point* must be investigated. Since there could be several, it is not really proper to speak of *system stability*. Equilibrium points satisfy $\dot{\mathbf{x}} = 0$, so $x_{2e} = 0$ is required, and then $f(x_{1e}) = -g(0)$. It is assumed that $g(0) = 0$ and that $f(x_1) = 0$ only at $x_1 = 0$. Thus the origin is the only equilibrium point, by assumption.

If x_1 is thought of as a position, then x_2 is a velocity, and the above equations might represent a unit mass connected to a nonlinear spring and damper. The spring force is $f(x_1)$ and the damper force is $g(x_2)$. This analogy suggests trying a Lyapunov function composed of a kinetic energy-like term with x_2^2 and a potential spring energy term (equal to work done by $f(x_1)$):

$$V(\mathbf{x}) = c_1 x_2^2 + c_2 \int_0^{x_1} f(\xi)\, d\xi$$

This term is positive definite if $c_1 > 0$, $c_2 > 0$ and if $f(x_1)$ always has the same sign as x_1, for example, any odd function of x_1. Then

$$\dot{V} = 2c_1 x_2 \dot{x}_2 + c_2 f(x_1)\dot{x}_1$$

Using equation (*1*) gives

$$\dot{V} = 2c_1 x_2[-g(x_2) - f(x_1)] + c_2 f(x_1) x_2$$

Selecting $c_2 = 2c_1$ gives $\dot{V}(\mathbf{x}) = -c_2 x_2 g(x_2)$. $\dot{V}$ is negative semidefinite if $g(x_2)$ always has the same sign as x_2. If this is true, stability i.s.L. is ensured by Theorem 14.1. Actually a slight generalization of Theorem 14.2 is possible. If, instead of $\dot{V}(\mathbf{x})$ being negative definite, $\dot{V}(\mathbf{x})$ can be shown to be always negative *along any trajectory of the system*, asymptotic stability can still be concluded [71].

In this problem, $\dot{V} = 0$ only if $x_2 = 0$, and is negative otherwise. But, if $x_2 \equiv 0$, then $\dot{x}_2 = 0$ also and this requires that $f(x_1) = 0$. By assumption, this means $x_1 = 0$, so $\dot{V} < 0$ for all possible $\mathbf{x}(t)$ trajectories, except at the equilibrium point $\mathbf{x} = \mathbf{0}$. It is concluded that this system is asymptotically stable if $f(x_1)x_1 > 0$ and $g(x_2)x_2 > 0$ for all $\mathbf{x} \neq \mathbf{0}$. Further, since $V(\mathbf{x}) \longrightarrow \infty$ as $\|\mathbf{x}\| \longrightarrow \infty$, the stability is global.

14.14 Use Lyapunov's direct method to study the stability of the origin $\mathbf{x} = 0$ for the system [62] described by

$$\begin{aligned} \dot{x}_1 &= x_2 - a x_1(x_1^2 + x_2^2) \\ \dot{x}_2 &= -x_1 - a x_2(x_1^2 + x_2^2) \end{aligned} \tag{1}$$

A trial Lyapunov function is assumed as $V(\mathbf{x}) = c_1 x_1^2 + c_2 x_2^2$, with c_1 and c_2 unspecified but positive constants. Then $V(\mathbf{x})$ is positive definite and $V(\mathbf{x}) \longrightarrow \infty$ as $\|\mathbf{x}\| \longrightarrow \infty$. The time derivative is

$$\dot{V}(\mathbf{x}) = 2c_1 x_1 \dot{x}_1 + 2c_2 x_2 \dot{x}_2$$

Using equation (*1*), this becomes

$$\dot{V}(\mathbf{x}) = 2c_1x_1[x_2 - ax_1(x_1^2 + x_2^2)] + 2c_2x_2[-x_1 - ax_2(x_1^2 + x_2^2)]$$

If the selection $c_1 = c_2$ is made, then the troublesome x_1x_2 product terms cancel, leaving $\dot{V}(\mathbf{x}) = -2ac_1(x_1^2 + x_2^2)^2$. If the constant a is positive, $\dot{V}(\mathbf{x})$ is negative definite and the origin is globally asymptotically stable by Theorem 14.3.

14.15 Derive conditions which ensure asymptotic stability of the origin for the system in equation (*1*) [98]. Use the integration technique of Problem 14.11 to determine a suitable Lyapunov function:

$$\dot{x}_1 = x_2, \qquad \dot{x}_2 = x_3, \qquad \dot{x}_3 = -(x_1 + cx_2)^n - bx_3 \tag{1}$$

Try $\dot{V}(\mathbf{x}) = -x_3^2$. Then $V(\mathbf{x}) = \int_{t_1}^{t} \dot{V}(\mathbf{x})\,dt = -\int_{t_1}^{t} x_3\dot{x}_2\,dt$. Integration by parts gives

$$V(\mathbf{x}) = -x_3x_2 + \int x_2\dot{x}_3\,dt = -x_3x_2 - \int x_2(x_1 + cx_2)^n\,dt - b\int x_2x_3\,dt$$

But $x_3 = \dot{x}_2$, so $\int_{t_1}^{t} x_2x_3\,dt = \int_0^{x_2} x_2\,dx_2 = x_2^2/2$. Adding and subtracting $\int cx_3(x_1 + cx_2)^n\,dt$ gives

$$V(\mathbf{x}) = -x_3x_2 - \int (x_2 + cx_3)(x_1 + cx_2)^n\,dt + \int cx_3(x_1 + cx_2)^n\,dt - \frac{bx_2^2}{2}$$

From equation (*1*), $x_2 + cx_3 = \dot{x}_1 + c\dot{x}_2$ and $(x_1 + cx_2)^n = -\dot{x}_3 - bx_3$. Therefore,

$$V(\mathbf{x}) = -x_3x_2 - \frac{(x_1 + cx_2)^{n+1}}{n+1} - \frac{bx_2^2}{2} - c\int x_3\dot{x}_3\,dt - bc\int x_3^2\,dt$$

$$= -x_3x_2 - \frac{(x_1 + cx_2)^{n+1}}{n+1} - \frac{bx_2^2}{2} - \frac{cx_3^2}{2} - bc\int x_3^2\,dt$$

This $V(\mathbf{x})$ is not positive definite; in fact, it can be made negative definite. Therefore, a modified function is selected as

$$V'(\mathbf{x}) = -V(\mathbf{x}) - bc\int_{t_1}^{t} x_3^2\,dt = \frac{(x_1 + cx_2)^{n+1}}{n+1} + \frac{bx_2^2}{2} + \frac{cx_3^2}{2} + x_2x_3$$

$$= \frac{(x_1 + cx_2)^{n+1}}{n+1} + \frac{b}{2}\left(x_2 + \frac{x_3}{b}\right)^2 + \frac{(bc - 1)x_3^2}{2b}$$

This is positive definite if $b > 0$, $bc - 1 > 0$ and if $n + 1$ is any even positive integer. Also, $\dot{V}'(\mathbf{x}) = -\dot{V}(\mathbf{x}) - bcx_3^2 = -(bc - 1)x_3^2$. The same conditions ensure that $\dot{V}'(\mathbf{x}) \le 0$ for all $\mathbf{x}$. Since $x_3 \equiv 0$ requires $\dot{x}_3 = 0$ and $\dot{x}_2 = 0$, $x_1 =$ constant, it is seen that $x_3 = 0$ holds along a solution only at the point $\mathbf{x} = \mathbf{0}$. The above conditions thus ensure asymptotic stability.

14.16 Describe a method of generating Lyapunov functions, beginning with an assumed form for the gradient $\nabla_{\mathbf{x}} V$. This method is known as the variable gradient method [86, 106, 107].

Consider a general class of nonlinear, autonomous systems, $\dot{\mathbf{x}} = \mathbf{f}(\mathbf{x})$, for which $\mathbf{f}(\mathbf{0}) = \mathbf{0}$. Then for any candidate Lyapunov function $V(\mathbf{x})$,

$$\frac{dV(\mathbf{x})}{dt} = [\nabla_{\mathbf{x}} V]^T\dot{\mathbf{x}} = [\nabla_{\mathbf{x}} V]^T\mathbf{f}(\mathbf{x}) \tag{1}$$

A general form for the gradient is assumed, such as

$$\nabla_{\mathbf{x}} V = \begin{bmatrix} a_{11}x_1 + a_{12}x_2 + \cdots + a_{1n}x_n \\ a_{21}x_1 + a_{22}x_2 + \cdots + a_{2n}x_n \\ \vdots \qquad\qquad\qquad \vdots \\ a_{n1}x_1 + a_{n2}x_2 + \cdots + a_{nn}x_n \end{bmatrix} \tag{2}$$

The coefficients a_{ij} need not be constants, but may be functions of the components of $\mathbf{x}$. The function $V(\mathbf{x})$ can be determined by evaluating the line integral in state space from $\mathbf{0}$ to a general point $\mathbf{x}$, $V(\mathbf{x}) = \int_0^{\mathbf{x}} (\nabla_{\mathbf{x}} V)^T \, d\mathbf{x}$. The line integral can be made independent of the path of integration if a set of generalized curl conditions is imposed.

Let the ith component of $\nabla_{\mathbf{x}} V$ be called ∇V_i. Then the curl requirements are that $\partial \nabla V_i/\partial x_j = \partial \nabla V_j/\partial x_i$ for $i, j = 1, 2, \ldots, n$. That is, the matrix of second partials $\nabla_{\mathbf{x}}(\nabla_{\mathbf{x}} V)^T = [\partial^2 V/\partial x_i \partial x_j]$ must be symmetric. Satisfying this requirement allows the line integral to be written as the sum of n simple scalar integrals:

$$V(\mathbf{x}) = \int_0^{x_1} \nabla V_1 \, dx_1 \Big|_{x_2 = x_3 = \cdots = x_n = 0} + \int_0^{x_2} \nabla V_2 \, dx_2 \Big|_{\substack{x_1 = x_1 \\ x_3 = x_4 = \cdots = x_n = 0}} + \cdots + \int_0^{x_n} \nabla V_n \, dx_n \Big|_{\substack{x_1 = x_1 \\ x_2 = x_2 \\ \cdots\cdots}} \tag{3}$$

The procedure then consists of assuming $\nabla_{\mathbf{x}} V$ as in equation (2), specifying the coefficients a_{ij} such that (1) $\dot{V}(\mathbf{x})$, equation (1), is at least negative semidefinite and (2) the generalized curl equations are satisfied. Then $V(\mathbf{x})$ is found as in equation (3) and checked for positive definiteness. Usually it will be possible to let many a_{ij} terms equal zero. Generally $a_{nn} = 1$ is chosen at the outset to ensure that $V(\mathbf{x})$ is quadratic in x_n.

14.17 Use the variable gradient technique to investigate the stability of the nonlinear system (see pages 59 and 67 of Reference 71) described by

$$\dot{x}_1 = x_2, \qquad \dot{x}_2 = -f(x_1)x_2 - g(x_1) \tag{1}$$

The gradient is assumed to be of the form

$$\nabla_{\mathbf{x}} V = \begin{bmatrix} \alpha_{11}x_1 + \alpha_{12}x_2 \\ \alpha_{21}x_1 + x_2 \end{bmatrix}$$

The curl equations require $\partial \nabla V_1/\partial x_2 = \partial \nabla V_2/\partial x_1$, or

$$x_1 \frac{\partial \alpha_{11}}{\partial x_2} + \alpha_{12} + x_2 \frac{\partial \alpha_{12}}{\partial x_2} = x_1 \frac{\partial \alpha_{21}}{\partial x_1} + \alpha_{21}$$

Using the assumed gradient and equation (1) gives

$$\dot{V} = (\nabla_{\mathbf{x}} V)^T \begin{bmatrix} x_2 \\ -f(x_1)x_2 - g(x_1) \end{bmatrix}$$
$$= \alpha_{11}x_1x_2 + \alpha_{12}x_2^2 - \alpha_{21}f(x_1)x_1x_2 - f(x_1)x_2^2 - g(x_1)\alpha_{21}x_1 - g(x_1)x_2$$

This expression should be made at least negative semidefinite.

One possible solution begins by setting $\alpha_{21} = 0$. Then the curl equations are satisfied if $\alpha_{12} = 0$ and if α_{11} is not a function of x_2. By setting $\alpha_{11} = g(x_1)/x_1$, we obtain $\dot{V} = -f(x_1)x_2^2$, which is negative semidefinite if $f(x_1) \geq 0$ for all x_1. Then $\nabla_{\mathbf{x}} V = [g(x_1) \quad x_2]^T$ and

$$V = \int_0^{x_1} g(\xi)\, d\xi + \int_0^{x_2} \xi \, d\xi = \int_0^{x_1} g(\xi)\, d\xi + \frac{x_2^2}{2}$$

Thus V is positive definite, and hence a Lyapunov function, if $g(x_1)x_1 > 0$ for all $x_1 \neq 0$.

The following conclusions regarding stability can be drawn:

1. Stable i.s.L. if $g(x_1)x_1 > 0$, $f(x_1) \geq 0$, for all $\mathbf{x} \neq \mathbf{0}$ by Theorem 14.1.
2. Asymptotically stable if, in addition, $f(x_1)$ and $g(x_1) = 0$ only at $x_1 = 0$. This ensures that $\dot{V} \neq 0$ on any solution of equation (*1*) except at $\mathbf{x} = \mathbf{0}$.
3. Globally asymptotically stable if, in addition, $\int_0^{x_1} g(\xi)\, d\xi \longrightarrow \infty$ as $|x_1| \longrightarrow \infty$.

14.18 Use Lyapunov's direct method to investigate the stability of the nonlinear time-varying system [106] $\ddot{x} + a\dot{x} + g(x, t)x = 0$.

The state equations are $\dot{x}_1 = x_2$, $\dot{x}_2 = -ax_2 - g(x_1, t)x_1$.

Assume that $\nabla_{\mathbf{x}} V = \begin{bmatrix} \alpha_{11}x_1 + \alpha_{12}x_2 \\ \alpha_{21}x_1 + x_2 \end{bmatrix}$. Then since $V(\mathbf{x})$ may be an explicit function of time,

$$\dot{V}(\mathbf{x}) = (\nabla_{\mathbf{x}} V)^T \dot{\mathbf{x}} + \partial V/\partial t$$
$$= \alpha_{11}x_1x_2 + \alpha_{12}x_2^2 - a\alpha_{21}x_1x_2 - g(x_1, t)x_1^2\alpha_{21} - ax_2^2 - g(x_1, t)x_1x_2 + \partial V/\partial t$$

In order to remove the x_1x_2 product terms, set $\alpha_{11} = a\alpha_{21} + g(x_1, t)$. The curl equations can then be satisfied if $\alpha_{21} = \alpha_{12} =$ constant. Then

$$V(\mathbf{x}) = \int_0^{x_1} [a\alpha_{21} + g(x_1, t)]x_1\, dx_1 + \int_0^{x_2} [\alpha_{21}x_1 + x_2]\, dx_2 \Big|_{x_1 = \text{constant}}$$
$$= a\alpha_{21}\frac{x_1^2}{2} + \alpha_{21}x_1x_2 + \frac{x_2^2}{2} + \int_0^{x_1} g(x_1, t)x_1\, dx_1$$

If $\int_0^{x_1} g(x_1, t)x_1\, dx_1 > 0$ for all x_1 and t, then $V(\mathbf{x}) > \frac{1}{2}(x_2 + \alpha_{21}x_1)^2 + \frac{1}{2}(a\alpha_{21} - \alpha_{21}^2)x_1^2$. Thus $V(\mathbf{x})$ is positive definite if $a > 0$ and $a\alpha_{21} > \alpha_{21}^2$. This is ensured by selecting $\alpha_{21} = a - \epsilon$, where ϵ is a small positive number. Checking the time derivative,

$$\frac{dV}{dt} = a\alpha_{21}x_1\dot{x}_1 + \alpha_{21}\dot{x}_1x_2 + \alpha_{21}x_1\dot{x}_2 + x_2\dot{x}_2 + g(x_1, t)x_1\dot{x}_1 + \int_0^{x_1} \frac{\partial g(x_1, t)}{\partial t} x_1\, dx_1$$

Using $\alpha_{21} = a - \epsilon$ and the differential equations for $\dot{x}_1$ and $\dot{x}_2$ gives

$$\dot{V} = -\epsilon x_2^2 - (a - \epsilon)g(x_1, t)x_1^2 + \int_0^{x_1} \frac{\partial g(x_1, t)}{\partial t} x_1\, dx_1$$

This expression is negative definite if $g(x_1, t) > 0$ for all x_1, t and if the integral term is sufficiently small. That is, if

$$\int_0^{x_1} \frac{\partial g(x_1, t)}{\partial t} x_1\, dx_1 < ag(x_1, t)x_1^2$$

This will be true, for example, if $\max_{x_1, t}[\partial g(x_1, t)/\partial t] < 2ag(x_1, t)$ for all x_1 and t. Additionally, if $g(x_1, t)$ is bounded for all x_1 and t, then $V(\mathbf{x}, t)$ can be bounded as required by condition (4) of Theorem 14.4. If these conditions are satisfied, the system is uniformly globally asymptotically stable.

14.19 Use the instability Theorem 14.5 to show that the origin is an unstable equilibrium point for van der Pol's equation, $\ddot{x} - \mu(1 - x^2)\dot{x} + x = 0$, with $\mu > 0$ [71, 106].

The state equations are

$$\dot{x}_1 = x_2, \qquad \dot{x}_2 = -x_1 + \mu(1 - x_1^2)x_2 \tag{1}$$

Using the variable gradient technique,

$$\nabla_{\mathbf{x}} V = \begin{bmatrix} \alpha_{11}x_1 + \alpha_{12}x_2 \\ \alpha_{21}x_1 + x_2 \end{bmatrix}$$

and

$$\dot{V} = (\alpha_{11} - 1 + \alpha_{21}\mu)x_1x_2 + (\alpha_{12} + \mu)x_2^2 - \alpha_{21}x_1^2 - \mu x_1^2x_2^2 - \alpha_{21}\mu x_1^3 x_2$$

The curl equations require

$$x_1\frac{\partial \alpha_{11}}{\partial x_2} + \alpha_{12} + x_2\frac{\partial \alpha_{12}}{\partial x_2} = \alpha_{21} + x_1\frac{\partial \alpha_{21}}{\partial x_1}$$

If $\alpha_{11} = 1$, $\alpha_{12} = \alpha_{21} = 0$, the curl equations are satisfied and then

$$\dot{V}(\mathbf{x}) = \mu(1 - x_1^2)x_2^2$$

Integrating,

$$V(\mathbf{x}) = \int_0^{x_1} x_1\,dx_1 + \int_0^{x_2} x_2\,dx_2 = \frac{x_1^2}{2} + \frac{x_2^2}{2}$$

$V(\mathbf{x})$ is positive definite, and since $\dot{V}(\mathbf{x})$ is positive along any solution of equation (*1*) for which $x_1^2 < 1$, the origin is unstable in the sense of Definition 14.1. This does not necessarily mean that $\|\mathbf{x}(t)\|$ will grow without bound. In this particular example, it is well known that $\mathbf{x}(t)$ will approach a finite periodic solution called a *limit cycle* [71].

Other Uses of Lyapunov's Method

14.20 Show how Lyapunov's direct method can be used to estimate the system's speed of response, i.e., the dominant time constant [62, 86].

Assuming that the origin is the equilibrium point, it is desired to obtain an estimate of how rapidly the state will return to the origin after being perturbed away from it. Let $V(\mathbf{x}, t)$ be a Lyapunov function, and define $\eta = -\dot{V}(\mathbf{x}, t)/V(\mathbf{x}, t)$. Then integrating both sides gives

$$\int_{t_0}^{t} \eta\,dt = -\int_{V(\mathbf{x}(t_0),t_0)}^{V(\mathbf{x}(t),t)} \frac{dV}{V}$$

from which

$$\ln\left[\frac{V(\mathbf{x}(t), t)}{V(\mathbf{x}(t_0), t_0)}\right] = -\int_{t_0}^{t} \eta\,dt \qquad \text{or} \qquad V(\mathbf{x}(t), t) = V(\mathbf{x}(t_0), t_0)\,e^{-\int_{t_0}^{t}\eta\,dt}$$

Although η is not generally constant, if $\eta_{\min}$ is defined as the minimum value of η, then

$$V(\mathbf{x}(t), t) \le V(\mathbf{x}(t_0), t_0)e^{-\eta_{\min}(t-t_0)}$$

Since for an asymptotically stable system $\mathbf{x}(t) \longrightarrow \mathbf{0}$ as $V(\mathbf{x}, t) \longrightarrow 0$, it is seen that $1/\eta_{\min}$ can be interpreted as a bound on the system time constant. For a given Lyapunov function, $\eta_{\min}$ provides a figure of merit for the system's response time. When considering several candidate Lyapunov functions, the above considerations indicate that the best Lyapunov function is the one which gives the largest ratio $-\dot{V}/V$.

14.21 Consider the system of equation (*1*) below. The control input $\mathbf{u}(t)$ is limited by $\mathbf{u}^T(t)\mathbf{u}(t) \le 1$ for all time t. Using the results of Problem 14.10, specify the feedback control law which makes $\dot{V}(\mathbf{x})$ as large and negative as possible. This controller will rapidly drive the state back to $\mathbf{0}$ if it is perturbed.

$$\dot{\mathbf{x}} = \begin{bmatrix} -3 & 2 \\ -1 & -1 \end{bmatrix}\mathbf{x} + \begin{bmatrix} 1 & -1 \\ 0 & 1 \end{bmatrix}\begin{bmatrix} u_1(t) \\ u_2(t) \end{bmatrix} \qquad (1)$$

Using the Lyapunov function of Problem 14.10, $V(\mathbf{x}) = \mathbf{x}^T\mathbf{P}\mathbf{x}$ and then $\dot{V}(\mathbf{x}) = \dot{\mathbf{x}}^T\mathbf{P}\mathbf{x} + \mathbf{x}^T\mathbf{P}\dot{\mathbf{x}} = -\mathbf{x}^T\mathbf{x} + 2\mathbf{u}^T\mathbf{B}^T\mathbf{P}\mathbf{x}$. The control $\mathbf{u}(t)$ is selected to minimize $\mathbf{u}^T\mathbf{B}^T\mathbf{P}\mathbf{x}$ subject to the restriction placed upon $\mathbf{u}(t)$. Obviously $\mathbf{u}(t)$ should be parallel to $\mathbf{B}^T\mathbf{P}\mathbf{x}$, but with the opposite sign, and have its largest possible magnitude. That is,

$$\mathbf{u}(t) = -\frac{\mathbf{B}^T\mathbf{P}\mathbf{x}(t)}{\|\mathbf{B}^T\mathbf{P}\mathbf{x}(t)\|}$$

Using the given form for the matrices $\mathbf{B}$ and $\mathbf{P}$, this gives

$$\mathbf{u}(t) = \frac{\begin{bmatrix} -7x_1(t) + x_2(t) \\ 8x_1(t) - 19x_2(t) \end{bmatrix}}{(113x_1^2 - 318x_1x_2 + 362x_2^2)^{1/2}}$$

PROBLEMS

14.22 Comment on the stability of the systems described in

a. Example 10.1, page 262;
b. Problem 10.8, page 270;
c. Figure 10.3, page 275, with $K = 3/16$.

14.23 Investigate the stability of the systems described in conjunction with controllability and observability in

a. Problem 12.2, page 312;
b. Problem 12.17, page 320;
c. Problem 12.18, page 321.

14.24 Consider the time-varying system [29] given by equation (*1*) below and its transition matrix, equation (*2*). Does it necessarily follow that a time-varying system whose eigenvalues are always negative is asymptotically stable? Can stability i.s.L. be concluded?

$$\dot{\mathbf{x}}(t) = \begin{bmatrix} -1 + a\cos^2 t & 1 - a\sin t\cos t \\ -1 - a\sin t\cos t & -1 + a\sin^2 t \end{bmatrix}\mathbf{x}(t) \tag{1}$$

$$\mathbf{\Phi}(t, 0) = \begin{bmatrix} e^{(a-1)t}\cos t & e^{-t}\sin t \\ -e^{(a-1)t}\sin t & e^{-t}\cos t \end{bmatrix} \tag{2}$$

14.25 a. Using precisely the same steps as in Problem 14.11c, find a Lyapunov function and its time derivative for the nonlinear system [45, 71]

$$\dot{x}_1 = x_2, \qquad \dot{x}_2 = x_3, \qquad \dot{x}_3 = -F(x_2)x_3 - ax_2 - bx_1$$

b. Determine sufficient conditions for asymptotic stability.

14.26 If the input of Problem 14.21 has constraints on the individual components, $|u_i(t)| \leq M_i$, what should the control law be if it is desired to rapidly correct for initial state perturbations?

Chapter

15

THE RELATIONSHIP BETWEEN STATE VARIABLE AND TRANSFER FUNCTION DESCRIPTIONS OF SYSTEMS

15.1 INTRODUCTION

Classical control theory makes extensive use of input-output transfer functions to describe physical systems. When multiple inputs and/or outputs must be considered, the transfer matrix is the natural extension of the transfer function. In modern control theory, systems are most often described in terms of state variables. This chapter explores the relationships between these two approaches. Since the usefulness of the transfer function approach is largely restricted to the class of linear constant coefficient systems, the present discussion is similarly restricted. Most of the explanatory material is for continuous-time systems. However, a parallel set of results exists for discrete-time systems. Some examples of this sort are given in Problems 15.7, 15.24, and 15.25.

15.2 TRANSFER MATRICES FROM STATE EQUATIONS

The most general state space description of a linear, constant system with r inputs $\mathbf{u}(t)$, m outputs $\mathbf{y}(t)$, and n state variables $\mathbf{x}(t)$ is given by

$$\dot{\mathbf{x}}(t) = \mathbf{A}\mathbf{x}(t) + \mathbf{B}\mathbf{u}(t) \tag{15.1}$$

$$\mathbf{y}(t) = \mathbf{C}\mathbf{x}(t) + \mathbf{D}\mathbf{u}(t) \tag{15.2}$$

where **A**, **B**, **C**, and **D** are constant matrices of dimensions $n \times n$, $n \times r$, $m \times n$, and $m \times r$, respectively. The Laplace transforms of equations (*15.1*) and (*15.2*) are

$$\begin{aligned} s\mathbf{x}(s) - \mathbf{x}(t=0) &= \mathbf{A}\mathbf{x}(s) + \mathbf{B}\mathbf{u}(s) \\ \mathbf{y}(s) &= \mathbf{C}\mathbf{x}(s) + \mathbf{D}\mathbf{u}(s) \end{aligned} \tag{15.3}$$

As usual when dealing with transfer functions, the initial conditions $\mathbf{x}(t = 0)$ will be ignored. Solving for $\mathbf{x}(s)$ gives

$$\mathbf{x}(s) = [s\mathbf{I}_n - \mathbf{A}]^{-1}\mathbf{B}\mathbf{u}(s)$$

Using this result leads to equation (*15.4*), the input-output relationship for the transformed variables:

$$\mathbf{y}(s) = \{\mathbf{C}[s\mathbf{I}_n - \mathbf{A}]^{-1}\mathbf{B} + \mathbf{D}\}\mathbf{u}(s) \tag{15.4}$$

The $m \times r$ matrix which premultiplies $\mathbf{u}(s)$ is the *transfer matrix* $\mathbf{H}(s)$,

$$\mathbf{H}(s) = \mathbf{C}[s\mathbf{I}_n - \mathbf{A}]^{-1}\mathbf{B} + \mathbf{D} \tag{15.5}$$

The discrete-time equivalent of this result was derived in Problem 11.9. A typical element $H_{ij}(s)$ of $\mathbf{H}(s)$ is the transfer function relating the jth input component u_j to the ith output component y_i, $H_{ij}(s) = y_i(s)/u_j(s)$, with all inputs equal to zero except u_j and all initial conditions being zero.

It is convenient to define $\Delta(s) = |s\mathbf{I}_n - \mathbf{A}|$.† Then $[s\mathbf{I}_n - \mathbf{A}]^{-1} = \text{Adj}\,[s\mathbf{I}_n - \mathbf{A}]/\Delta(s)$ and the transfer matrix can be rewritten as

$$\mathbf{H}(s) = \frac{\mathbf{C}\,\text{Adj}\,[s\mathbf{I}_n - \mathbf{A}]\mathbf{B} + \mathbf{D}\Delta(s)}{\Delta(s)} \tag{15.6}$$

Each element of the adjoint matrix is a polynomial in s of degree less than or equal to $n - 1$. Since $\Delta(s)$ is an nth degree polynomial in s, each element $H_{ij}(s)$ is a ratio of polynomials in s, with the degree of the denominator at least as great as the degree of the numerator. Such an **H** matrix is called a *proper* rational matrix (loosely, at least as many poles as zeros). If $\mathbf{D} = [\mathbf{0}]$, then the numerator of every element in $\mathbf{H}(s)$ will be of degree less than the denominator. In this case, $\mathbf{H}(s)$ is a *strictly proper* rational matrix (loosely, more poles than zeros). Clearly, from equation (*15.6*),

$$\mathbf{D} = \lim_{s \to \infty} \mathbf{H}(s) \tag{15.7}$$

When a state variable description of a system is given, the unique, corresponding transfer matrix is given (equation (*15.5*) or (*15.6*)). If the matrix $\mathbf{H}(s)$ is given, equation (*15.7*) immediately gives the **D** matrix. The determination of $\{\mathbf{A}, \mathbf{B}, \mathbf{C}\}$ from a knowledge of $\mathbf{H}(s)$ is treated in the following sections.

†Recall that in Chapter 7, page 174, $\Delta(\lambda) \triangleq |\mathbf{A} - \mathbf{I}\lambda|$. The above definition is more convenient for this chapter and the next. They differ by the inconsequential factor $(-1)^n$.

Example 15.1

A single-input, single-output system is described in state variable form with

$$\mathbf{A} = \begin{bmatrix} -5 & 1 & 0 \\ 0 & -2 & 1 \\ 0 & 0 & -2 \end{bmatrix}, \quad \mathbf{B} = \begin{bmatrix} 0 \\ 1 \\ 1 \end{bmatrix}, \quad \mathbf{C} = [1 \;\; 0 \;\; 0 \;], \quad D = 0$$

Then

$$[s\mathbf{I} - \mathbf{A}]^{-1} = \frac{\begin{bmatrix} (s+2)^2 & s+2 & 1 \\ 0 & (s+2)(s+5) & (s+5) \\ 0 & 0 & (s+2)(s+5) \end{bmatrix}}{(s+5)(s+2)^2}$$

From equation (*15.5*),

$$H(s) = \frac{s+3}{(s+2)^2(s+5)}$$

This agrees with the results of Example 9.6, where the same problem was worked in reverse order. Note that $\lim_{s\to\infty} H(s) = 0$ as expected, since $D = 0$ in this case. ■

15.3 STATE EQUATIONS FROM TRANSFER MATRICES: REALIZATIONS

Methods of selecting state variables from single-input, single-output transfer functions were presented in Sec. 9.4 and Sec. 11.2, pages 236 and 284. Multivariable systems are now considered.

If a pair of equations (*15.1*) and (*15.2*) can be found which has a specified transfer matrix $\mathbf{H}(s)$, then those equations are called a *realization* of $\mathbf{H}(s)$. For brevity, it is common to refer to the matrices $\{\mathbf{A}, \mathbf{B}, \mathbf{C}, \mathbf{D}\}$ as the realization of $\mathbf{H}(s)$. Specifying $\{\mathbf{A}, \mathbf{B}, \mathbf{C}, \mathbf{D}\}$ is equivalent to giving a prescription for synthesizing a system with a given transfer matrix. Figure 9.3, for example, can be mechanized with physical devices such as operational amplifiers and other standard analog computer equipment. The discrete equivalent in Fig. 9.4 would most often use digital hardware.

Perhaps the simplest method of realizing a given transfer matrix $\mathbf{H}(s)$ is illustrated in Fig. 15.1. Each scalar component $H_{ij}(s)$ is considered individually, and the methods discussed for scalar transfer functions in Sec. 9.4 can be used on each.

Example 15.2

A system with two inputs and two outputs has the transfer matrix

$$\mathbf{H}(s) = \begin{bmatrix} 1/(s+1) & 2/[(s+1)(s+2)] \\ 1/[(s+1)(s+3)] & 1/(s+3) \end{bmatrix}$$

Using the approach suggested in Fig. 15.1, four separate scalar transfer functions are simulated as shown in Fig. 15.2.

Using the state variables x_1 through x_6 as defined in Fig. 15.2, a realization of $\mathbf{H}(s)$ is given by

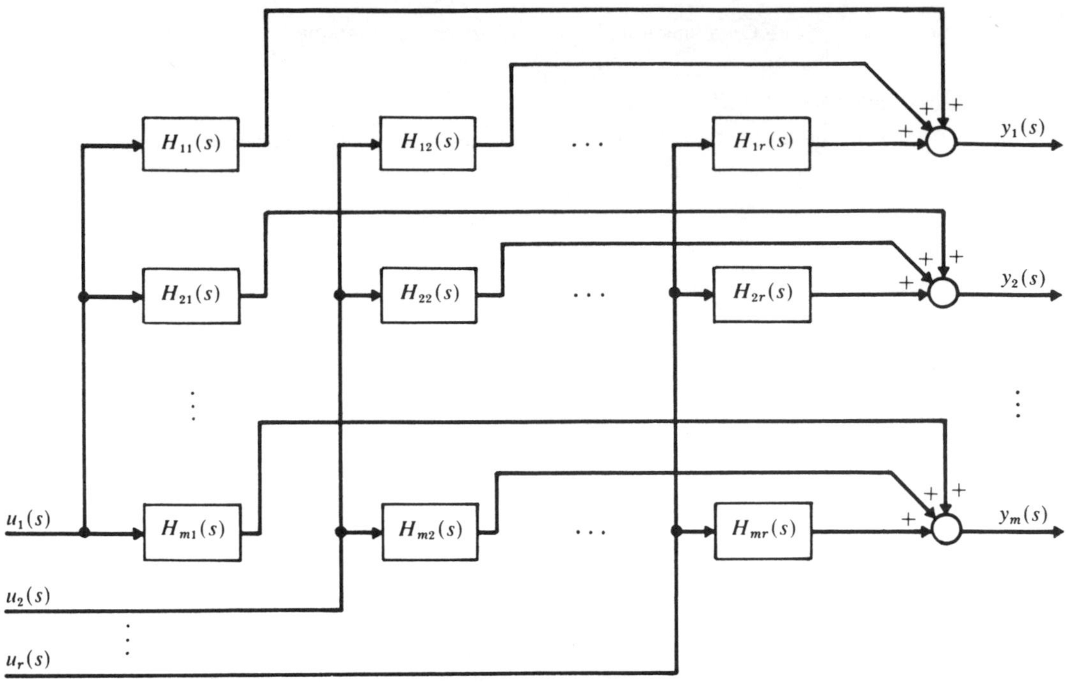

Figure 15.1 Block diagram of $H(s)$ without internal coupling.

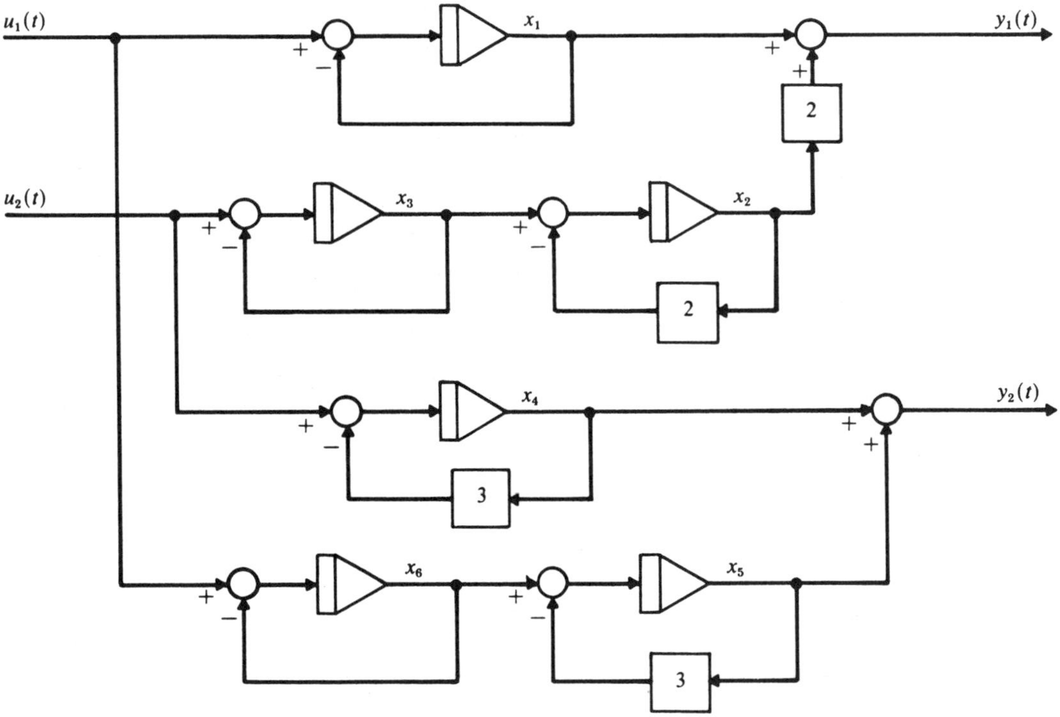

Figure 15.2

$$\mathbf{A}=\begin{bmatrix}-1&0&0&0&0&0\\0&-2&1&0&0&0\\0&0&-1&0&0&0\\0&0&0&-3&0&0\\0&0&0&0&-3&1\\0&0&0&0&0&-1\end{bmatrix},\quad \mathbf{B}=\begin{bmatrix}1&0\\0&0\\0&1\\0&1\\0&0\\1&0\end{bmatrix},\quad \mathbf{C}=\begin{bmatrix}1&0\\2&0\\0&0\\0&1\\0&1\\0&0\end{bmatrix}^T,\quad \mathbf{D}=\begin{bmatrix}0&0\\0&0\end{bmatrix}$$

Application of equation (*15.5*) shows that this sixth-order state representation does have the specified input-output transfer matrix. ■

The method just presented yields a realization for any transfer matrix. The realization of $\mathbf{H}(s)$ is not unique. If one nth order realization of $\mathbf{H}(s)$ can be found, then there are an infinite number of nth order realizations.

Example 15.3

Let $\{\mathbf{A}_1, \mathbf{B}_1, \mathbf{C}_1, \mathbf{D}_1\}$ be a realization of $\mathbf{H}(s)$, where $\mathbf{A}_1$ is an $n \times n$ matrix. Let $\mathbf{M}$ be any constant nonsingular $n \times n$ matrix and define a new state vector $\mathbf{x}_2$ by the transformation (change of basis in the state space Σ) $\mathbf{x}_1 = \mathbf{M}\mathbf{x}_2$. Now $\dot{\mathbf{x}}_1 = \mathbf{A}_1\mathbf{x}_1 + \mathbf{B}_1\mathbf{u}$, $\mathbf{y} = \mathbf{C}_1\mathbf{x}_1 + \mathbf{D}_1\mathbf{u}$ become $\mathbf{M}\dot{\mathbf{x}}_2 = \mathbf{A}_1\mathbf{M}\mathbf{x}_2 + \mathbf{B}_1\mathbf{u}$, $\mathbf{y} = \mathbf{C}_1\mathbf{M}\mathbf{x}_2 + \mathbf{D}_1\mathbf{u}$. Defining $\mathbf{A}_2 = \mathbf{M}^{-1}\mathbf{A}_1\mathbf{M}$, $\mathbf{B}_2 = \mathbf{M}^{-1}\mathbf{B}_1$, $\mathbf{C}_2 = \mathbf{C}_1\mathbf{M}$, and $\mathbf{D}_2 = \mathbf{D}_1$ gives $\dot{\mathbf{x}}_2 = \mathbf{A}_2\mathbf{x}_2 + \mathbf{B}_2\mathbf{u}$, $\mathbf{y} = \mathbf{C}_2\mathbf{x}_2 + \mathbf{D}_2\mathbf{u}$. Then

$$\mathbf{C}_2[s\mathbf{I} - \mathbf{A}_2]^{-1}\mathbf{B}_2 + \mathbf{D}_2 = \mathbf{C}_1\mathbf{M}[s\mathbf{I} - \mathbf{M}^{-1}\mathbf{A}_1\mathbf{M}]^{-1}\mathbf{M}^{-1}\mathbf{B}_1 + \mathbf{D}_1 = \mathbf{C}_1[s\mathbf{I} - \mathbf{A}_1]^{-1}\mathbf{B}_1 + \mathbf{D}_1$$

Thus $\{\mathbf{A}_1, \mathbf{B}_1, \mathbf{C}_1, \mathbf{D}_1\}$ and $\{\mathbf{A}_2, \mathbf{B}_2, \mathbf{C}_2. \mathbf{D}_2\}$ are two different realizations. Equation (*15.7*) indicates that $\mathbf{D}$ is determined only by $\mathbf{H}(s)$. The choice of basis vectors for Σ does not influence $\mathbf{D}$. Hence $\mathbf{D}_1 = \mathbf{D}_2$. ■

This establishes the nonuniqueness of system representations. A more interesting result is that dim (Σ) is not uniquely defined by $\mathbf{H}(s)$.

15.4 DEFINITION AND IMPLICATION OF IRREDUCIBLE REALIZATIONS

The method of Sec. 15.3 is not necessarily an efficient method of deriving state equations. It frequently leads to state equations, and hence a state space, of an unnecessarily high dimension.

Definition 15.1: Of all the possible realizations of $\mathbf{H}(s)$, $\{\mathbf{A}, \mathbf{B}, \mathbf{C}, \mathbf{D}\}$ is said to be an *irreducible* (or minimum) *realization* if the associated state space has the smallest possible dim (Σ).

In the case of a scalar transfer function, the minimum dimension required is equal to the order of the denominator of the transfer function after all common pole-zero cancellations are made. The corresponding result for transfer matrices is not so obvious and is discussed in the next section (see also Problem 15.14).

Whenever a pole-zero pair is cancelled from a transfer function, the system mode associated with the cancelled pole will not be evident in the state equations. Yet, in order to achieve an irreducible realization, these cancellations must be made. What is the implication of irreducible realizations in view of this apparent loss of information about the system?

An irreducible realization is a system, of minimal dimension, which is capable of reproducing the measurable relationships between inputs and outputs. This assumes that the system is originally relaxed, i.e., the initial state vector is zero. For this reason a system and its irreducible realization are said to be *zero state equivalent*. Often, the only knowledge about a system is the information which is obtainable from measurements of inputs and outputs. Transfer functions and matrices can be experimentally determined from these measurements. The irreducible realization does not cause loss of information in this case, because nothing was known about the system's internal structure in the first place. It is not claimed that an irreducible realization is the best decsription of the internal structure of a system.

If the internal structure of a system is known (for example, a circuit diagram), then an irreducible realization may not be the appropriate one to use. State equations can always be written directly from the system's linear graph, as described in Chapter 9. The realization $\{\mathbf{A}, \mathbf{B}, \mathbf{C}, \mathbf{D}\}$ obtained in this manner is the most complete system description, whether it is irreducible or not. Let this realization have $\dim(\Sigma) = n$. The transfer matrix $\mathbf{H}(s)$ can be found using equation (*15.5*). Starting with $\mathbf{H}(s)$, an irreducible realization $\{\mathbf{A}_1, \mathbf{B}_1, \mathbf{C}_1, \mathbf{D}\}$ can be found. If $\mathbf{A}_1$ is $n_1 \times n_1$, with $n_1 < n$, then information about $n - n_1$ modes would be lost by using the irreducible realization (see Problems 15.4 and 15.6). In Example 15.5, the irreducible realization gives no indication that the system is actually unstable. If $n_1 = n$, the system is said to be *completely characterized* by $\mathbf{H}(s)$. In general, transfer matrices, and irreducible realizations of them, describe only that subsystem which is both completely controllable and completely observable (see Problems 15.12 and 15.13). The incompleteness of a transfer matrix description is another reason for preferring state space techniques.

Example 15.4

The input-output equations for a system are

$$\begin{aligned} \dot{y}_1 + 2(y_1 - y_2) &= 4u_1 - u_2 \\ \dot{y}_2 + 3(y_2 - y_1) &= 4u_1 - u_2 \end{aligned} \tag{15.8}$$

Using the simulation diagram of Fig. 15.3, the state equations (*15.9*) and (*15.10*) are obtained:

$$\begin{bmatrix} \dot{x}_1 \\ \dot{x}_2 \end{bmatrix} = \begin{bmatrix} -2 & 2 \\ 3 & -3 \end{bmatrix} \begin{bmatrix} x_1 \\ x_2 \end{bmatrix} + \begin{bmatrix} 4 & -1 \\ 4 & -1 \end{bmatrix} \begin{bmatrix} u_1 \\ u_2 \end{bmatrix} \tag{15.9}$$

$$\begin{bmatrix} y_1 \\ y_2 \end{bmatrix} = \begin{bmatrix} 1 & 0 \\ 0 & 1 \end{bmatrix} \begin{bmatrix} x_1 \\ x_2 \end{bmatrix} \tag{15.10}$$

The state vector has dimension 2. The transfer matrix can be derived directly from equation (*15.8*) by Laplace transforming and a matrix inversion,

$$\mathbf{H}(s) = \begin{bmatrix} 4/s & -1/s \\ 4/s & -1/s \end{bmatrix}$$

Assume now that $\mathbf{H}(s)$ is the only information known about the system. The method of Sec. 15.3 leads to the fourth-order system realization shown in Fig. 15.4.

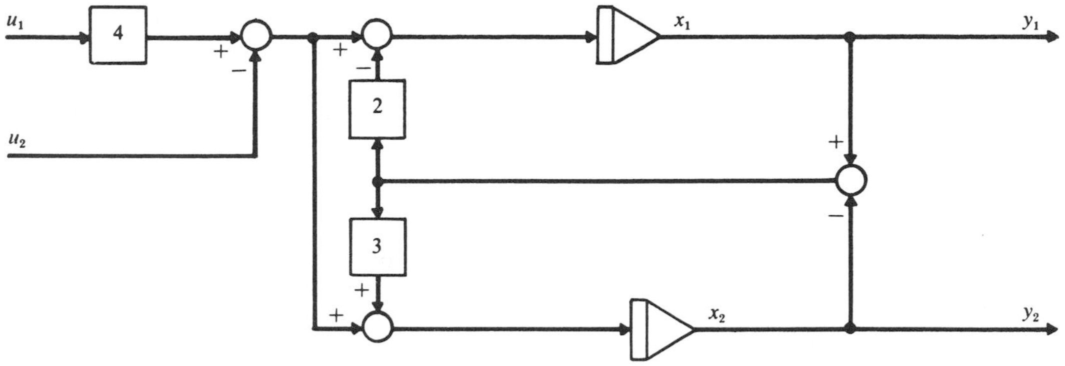

Figure 15.3

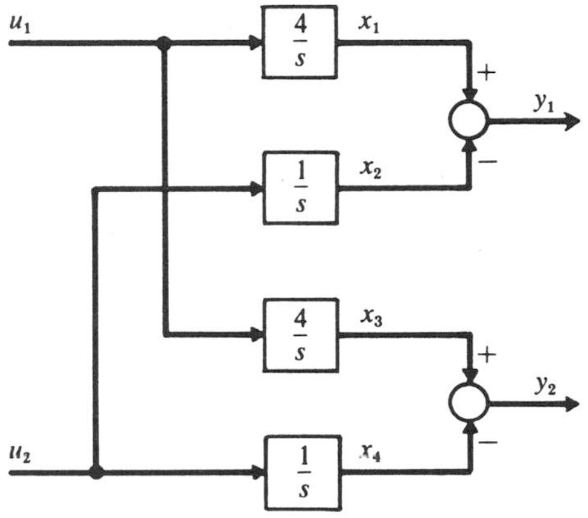

$$\mathbf{A} = [\mathbf{0}_{4\times4}], \quad \mathbf{B} = \begin{bmatrix} 4 & 0 \\ 0 & 1 \\ 4 & 0 \\ 0 & 1 \end{bmatrix}$$

$$\mathbf{C} = \begin{bmatrix} 1 & -1 & 0 & 0 \\ 0 & 0 & 1 & -1 \end{bmatrix}, \quad \mathbf{D} = [\mathbf{0}]$$

Figure 15.4

Two other realizations for $\mathbf{H}(s)$ are shown in Figs. 15.5 and 15.6.

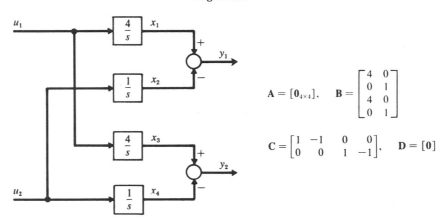

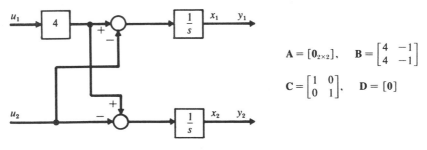

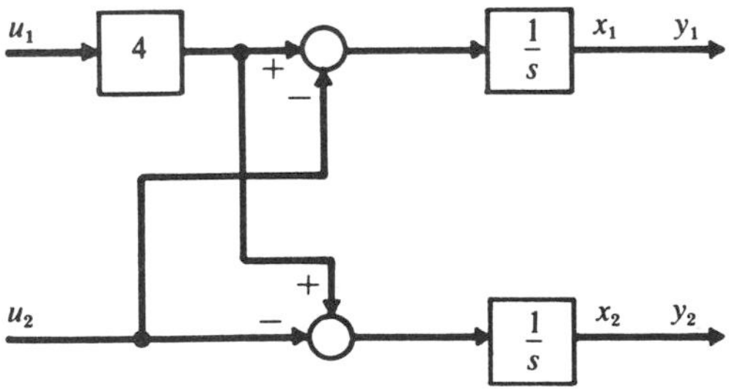

$$\mathbf{A} = [\mathbf{0}_{2\times2}], \quad \mathbf{B} = \begin{bmatrix} 4 & -1 \\ 4 & -1 \end{bmatrix}$$

$$\mathbf{C} = \begin{bmatrix} 1 & 0 \\ 0 & 1 \end{bmatrix}, \quad \mathbf{D} = [\mathbf{0}]$$

Figure 15.5

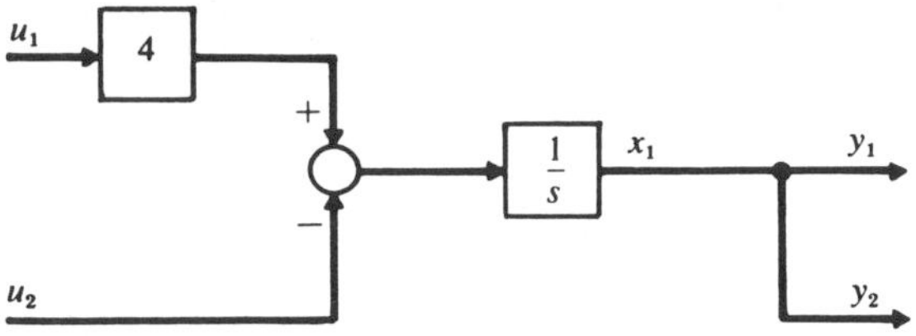

$$\mathbf{A} = 0, \quad \mathbf{B} = [4 \quad -1]$$
$$\mathbf{C} = \begin{bmatrix} 1 \\ 1 \end{bmatrix}, \quad \mathbf{D} = [0]$$

Figure 15.6

None of these three realizations resembles the original system. The one shown in Fig. 15.5 agrees insofar as dim $(\Sigma) = 2$, but the mode corresponding to the pole $(s + 5)$ is not present. The minimum realization, shown in Fig. 15.6, has $n = 1$. The original system cannot be completely controllable and observable. If the original system is decomposed into subsystems as in Problem 12.17, page 320, this is verified. The only mode which is completely controllable and completely observable is the one associated with the eigenvalue (pole) $s = 0$. ■

Consider a scalar transfer function, given in the form of equation (*15.6*). The eigenvalues of the matrix $\mathbf{A}$ are roots of $\Delta(\lambda) = |\lambda\mathbf{I} - \mathbf{A}| = 0$. The poles of $\mathbf{H}(s)$ will be the roots of $\Delta(s) = |s\mathbf{I} - \mathbf{A}| = 0$, *unless* some of the factors of $\Delta(s)$ are cancelled by terms in the numerator of $\mathbf{H}(s)$. Thus the poles of a transfer function will always be eigenvalues of the system matrix $\mathbf{A}$. Eigenvalues of $\mathbf{A}$ need not always be poles of the transfer function because cancellations may occur. If a system is completely characterized by its transfer function (equivalently if $\mathbf{A}$ is associated with an irreducible realization of $\mathbf{H}(s)$), then the poles and the eigenvalues are the same. This will be true if the system is completely controllable and completely observable. A similar relationship exists between the eigenvalues of $\mathbf{A}$ and the roots of the "denominator" of a transfer matrix. This assumes a suitable definition for a matrix "denominator" (see Problem 15.14).

15.5 THE DETERMINATION OF IRREDUCIBLE REALIZATIONS

An irreducible realization for $\mathbf{H}(s)$ can be obtained in two steps. First, obtain a realization by any means whatsoever. Second, decompose this realization into four subsystems which are (1) completely controllable and completely observable, (2) completely controllable but not observable, (3) completely observable but not controllable, and (4) neither controllable nor observable (see Problems 12.14 through 12.17). The subsystem which is completely controllable and completely observable constitutes an irreducible realization of $\mathbf{H}(s)$ [59, 61].

Example 15.5

Suppose that a system realization has been obtained, and it has been transformed into the Jordan canonical form

$$\begin{bmatrix} \dot{q}_1 \\ \dot{q}_2 \\ \dot{q}_3 \end{bmatrix} = \begin{bmatrix} -5 & 0 & 0 \\ 0 & +1 & 0 \\ 0 & 0 & -3 \end{bmatrix} \begin{bmatrix} q_1 \\ q_2 \\ q_3 \end{bmatrix} + \begin{bmatrix} 1 & 0 \\ 0 & 1 \\ 0 & 0 \end{bmatrix} \begin{bmatrix} u_1 \\ u_2 \end{bmatrix}$$

$$\begin{bmatrix} y_1 \\ y_2 \end{bmatrix} = \begin{bmatrix} 1 & 0 & 0 \\ 0 & 0 & 1 \end{bmatrix} \begin{bmatrix} q_1 \\ q_2 \\ q_3 \end{bmatrix} + \begin{bmatrix} 0 & 0 \\ 0 & 1 \end{bmatrix} \begin{bmatrix} u_1 \\ u_2 \end{bmatrix}$$

Because this Jordan form system has distinct eigenvalues, controllability criterion 1 of Sec. 12.3 applies. Since row 3 of $\mathbf{B}_n$ is all zero, mode q_3 is uncontrollable. The second column of $\mathbf{C}_n$ is all zero, so q_2 is an unobservable mode. Modes q_2 and q_3 have no effect on the input-output behavior and can be dropped. An irreducible realization is

$$\dot{q}_1 = -5q_1 + [1 \quad 0]\mathbf{u}$$

$$\mathbf{y} = \begin{bmatrix} 1 \\ 0 \end{bmatrix} q_1 + \begin{bmatrix} 0 & 0 \\ 0 & 1 \end{bmatrix} \mathbf{u}$$

The transfer matrix is $\mathbf{H}(s) = \begin{bmatrix} 1/(s+5) & 0 \\ 0 & 1 \end{bmatrix}$ for both forms of the state equations. ■

Example 15.6

A system realization is found to be

$$\begin{bmatrix} \dot{q}_1 \\ \dot{q}_2 \\ \dot{q}_3 \end{bmatrix} = \begin{bmatrix} -a & 0 & 0 \\ 0 & -a & 0 \\ 0 & 0 & -a \end{bmatrix} \begin{bmatrix} q_1 \\ q_2 \\ q_3 \end{bmatrix} + \begin{bmatrix} 1 & 0 \\ 0 & 1 \\ 1 & 0 \end{bmatrix} \begin{bmatrix} u_1 \\ u_2 \end{bmatrix}$$

$$\mathbf{y} = \begin{bmatrix} 0 & 1 & 1 \\ 1 & 0 & 1 \end{bmatrix} \begin{bmatrix} q_1 \\ q_2 \\ q_3 \end{bmatrix} + [\mathbf{D}]\mathbf{u}$$

Even though the coefficient matrix **A** is diagonal, controllability and observability criteria 1 of Sec. 12.3 cannot be used because of the repeated eigenvalues of **A**. This realization is neither controllable nor observable, and therefore can be reduced. Note that

$$\mathbf{y} = \begin{bmatrix} 0 \\ 1 \end{bmatrix} q_1 + \begin{bmatrix} 1 \\ 0 \end{bmatrix} q_2 + \begin{bmatrix} 1 \\ 1 \end{bmatrix} q_3 + \mathbf{Du} = \begin{bmatrix} 0 \\ 1 \end{bmatrix} (q_1 + q_3) + \begin{bmatrix} 1 \\ 0 \end{bmatrix} (q_2 + q_3) + \mathbf{Du}$$

Because of the linear dependence of the columns of $\mathbf{C}_n$, it is possible to define two new state variables $\tilde{q}_1 = q_1 + q_3$ and $\tilde{q}_2 = q_2 + q_3$. Then $\dot{\tilde{q}}_1 = \dot{q}_1 + \dot{q}_3 = -a(\tilde{q}_1) + 2u_1$ and $\dot{\tilde{q}}_2 = -a\tilde{q}_2 + u_1 + u_2$, or

$$\dot{\tilde{\mathbf{q}}} = \begin{bmatrix} -a & 0 \\ 0 & -a \end{bmatrix} \tilde{\mathbf{q}} + \begin{bmatrix} 2 & 0 \\ 1 & 1 \end{bmatrix} \mathbf{u}$$

$$\mathbf{y} = \begin{bmatrix} 0 & 1 \\ 1 & 0 \end{bmatrix} \tilde{\mathbf{q}} + \mathbf{Du}$$

This system is completely controllable and observable, hence irreducible. The transfer matrix for it and for the original third-order system is

$$\mathbf{H}(s) = \begin{bmatrix} 1/(s+a) & 1/(s+a) \\ 2/(s+a) & 0 \end{bmatrix} + \mathbf{D}$$

■

A direct method of obtaining a Jordan form realization is to expand each $H_{ij}(s)$ element, using partial fractions. Regrouping terms gives a matrix version of the partial fraction expansion of $\mathbf{H}(s)$. A completely controllable realization can be written directly from this.

Example 15.7

The transfer matrix of Example 15.2 is reconsidered. Using partial fractions, it can be written as

$$\mathbf{H}(s) = \begin{bmatrix} \dfrac{1}{s+1} & \dfrac{2}{s+1} - \dfrac{2}{s+2} \\ \dfrac{1/2}{s+1} - \dfrac{1/2}{s+3} & \dfrac{1}{s+3} \end{bmatrix} = \frac{\begin{bmatrix} 1 & 2 \\ \frac{1}{2} & 0 \end{bmatrix}}{s+1} + \frac{\begin{bmatrix} 0 & -2 \\ 0 & 0 \end{bmatrix}}{s+2} + \frac{\begin{bmatrix} 0 & 0 \\ -\frac{1}{2} & 1 \end{bmatrix}}{s+3}$$

$$= \frac{\begin{bmatrix} 1 \\ 0 \end{bmatrix}[1 \quad 2] + \begin{bmatrix} 0 \\ 1 \end{bmatrix}[\frac{1}{2} \quad 0]}{s+1} + \frac{\begin{bmatrix} 1 \\ 0 \end{bmatrix}[0 \quad -2]}{s+2} + \frac{\begin{bmatrix} 0 \\ 1 \end{bmatrix}[-\frac{1}{2} \quad 1]}{s+3}$$

In this case $\lim_{s\to\infty} \mathbf{H}(s) = [\mathbf{0}] = \mathbf{D}$, so that, from equation (*15.5*), $\mathbf{H}(s) = \mathbf{C}[s\mathbf{I} - \mathbf{A}]^{-1}\mathbf{B}$. Comparing this with the expanded form gives

$$\mathbf{H}(s) = \left[\begin{array}{cc|c|c} 1 & 0 & 1 & 0 \\ 0 & 1 & 0 & 1 \end{array}\right] \begin{bmatrix} \dfrac{1}{s+1} & & & \mathbf{0} \\ & \dfrac{1}{s+1} & & \\ & & \dfrac{1}{s+2} & \\ \mathbf{0} & & & \dfrac{1}{s+3} \end{bmatrix} \left[\begin{array}{cc} 1 & 2 \\ \frac{1}{2} & 0 \\ \hline 0 & -2 \\ \hline -\frac{1}{2} & 1 \end{array}\right]$$

In this form the matrices $\mathbf{C}$, $[s\mathbf{I} - \mathbf{A}]^{-1}$, and $\mathbf{B}$ are clearly evident. Since $[s\mathbf{I} - \mathbf{A}]^{-1}$ is diagonal, $\mathbf{A}$ is also diagonal and a realization of $\mathbf{H}(s)$ is given by

$$\mathbf{A} = \operatorname{diag}[-1, \quad -1, \quad -2, \quad -3], \qquad \mathbf{B} = \begin{bmatrix} 1 & 2 \\ \frac{1}{2} & 0 \\ 0 & -2 \\ -\frac{1}{2} & 1 \end{bmatrix}, \qquad \mathbf{C} = \begin{bmatrix} 1 & 0 & 1 & 0 \\ 0 & 1 & 0 & 1 \end{bmatrix}, \qquad \mathbf{D} = [\mathbf{0}]$$

This is an irreducible realization since it is completely controllable and observable. The required state space satisfies dim $(\Sigma) = 4$, and not 6 as in Example 15.2. ■

The method illustrated by Example 15.7 consists of

1. writing $\mathbf{H}(s)$ in a matrix form of partial fraction expansion:

   $$\mathbf{H}(s) = \mathbf{D} + \sum_{i=1}^{k} \frac{\mathbf{N}_i}{s + p_i}$$

   (note that terms like $1/(s + p_i)^m$ are *not* considered here, but will be later);
2. determining the rank r_i of each of the $\mathbf{N}_i$ matrices;
3. factoring each $\mathbf{N}_i$ into the sum of r_i outer products $\mathbf{N}_i = \sum_{j=1}^{r_i} \mathbf{c}_{ij}\rangle\langle\mathbf{b}_{ij}$;

(The double subscripts on the column vectors $\mathbf{c}$ and the row vectors $\mathbf{b}^T$ are used to distinguish which pole they are associated with. This distinction is not always necessary and a single subscript then suffices.)

4. setting

$$\mathbf{A} = \text{diag}\,[-p_1\mathbf{I}_{r_1}, -p_2\mathbf{I}_{r_2}, \ldots, -p_k\mathbf{I}_{r_k}]$$

$$\mathbf{B} = \begin{bmatrix} \mathbf{b}_{11}^T \\ \vdots \\ \mathbf{b}_{1r_1}^T \\ \hline \mathbf{b}_{21}^T \\ \vdots \\ \mathbf{b}_{2r_2}^T \\ \hline \vdots \\ \hline \mathbf{b}_{k1}^T \\ \vdots \\ \mathbf{b}_{kr_k}^T \end{bmatrix}, \qquad \mathbf{C} = [\mathbf{c}_{11} \;\cdots\; \mathbf{c}_{1r_1} \mid \mathbf{c}_{21} \;\cdots\; \mathbf{c}_{2r_2} \mid \cdots \mid \mathbf{c}_{k1} \;\cdots\; \mathbf{c}_{kr_k}]$$

The dimension of $\mathbf{A}$ is $n \times n$, where $n = r_1 + r_2 + \cdots + r_k$. Since $\mathbf{A}$ is diagonal, it is in Jordan form. There is just a single 1×1 Jordan block associated with those poles p_i for which $r_i = 1$. The controllability and observability of these modes is ensured by criteria 1 of Sec. 12.3 since the rows $\mathbf{b}_{i1}^T$ and colunms $\mathbf{c}_{i1}$ are not zero. In general, there will be r_i 1×1 Jordan blocks associated with a given pole p_i. These modes will be both controllable and observable because rank $\mathbf{N}_i = r_i$ implies that the r_i rows $\{\mathbf{b}_{i1}^T, \mathbf{b}_{i2}^T \ldots, \mathbf{b}_{ir_i}^T\}$ are linearly independent and so are the r_i columns $\{\mathbf{c}_{i1}, \mathbf{c}_{i2}, \ldots, \mathbf{c}_{ir_i}\}$ (see Problems 15.10 and 15.11). Since every mode of this realization is both controllable and observable, it is irreducible.

If any one element $H_{ij}(s)$ has a multiple pole, say $1/(s + p_1)^m$, then the matrix partial fraction expansion for $\mathbf{H}(s)$ will also contain this term, plus terms to the $m - 1, m - 2, \ldots, 1$ powers. This requires a modification of the method, which is best explained by example. The essence of the modification amounts to obeying the admonition of Sec. 9.4 against using *unnecessary* integrators when using simulation diagrams. This results in realizations with nondiagonal $\mathbf{A}$ matrices.

Example 15.8

Find an irreducible realization of

$$\mathbf{H}(s) = \begin{bmatrix} \dfrac{1}{(s+2)^3(s+5)} & \dfrac{1}{s+5} \\[2ex] \dfrac{1}{s+2} & 0 \end{bmatrix}$$

Using partial fraction expansion gives

$$\mathbf{H}(s) = \frac{\begin{bmatrix} \frac{1}{3} & 0 \\ 0 & 0 \end{bmatrix}}{(s+2)^3} + \frac{\begin{bmatrix} -\frac{1}{9} & 0 \\ 0 & 0 \end{bmatrix}}{(s+2)^2} + \frac{\begin{bmatrix} \frac{1}{27} & 0 \\ 1 & 0 \end{bmatrix}}{s+2} + \frac{\begin{bmatrix} -\frac{1}{27} & 1 \\ 0 & 0 \end{bmatrix}}{s+5}$$

$$= \frac{\begin{bmatrix} \frac{1}{3} \\ 0 \end{bmatrix}[1 \quad 0]}{(s+2)^3} + \frac{\begin{bmatrix} -\frac{1}{9} \\ 0 \end{bmatrix}[1 \quad 0]}{(s+2)^2} + \frac{\begin{bmatrix} \frac{1}{27} \\ 1 \end{bmatrix}[1 \quad 0]}{s+2} + \frac{\begin{bmatrix} 1 \\ 0 \end{bmatrix}[-\frac{1}{27} \quad 1]}{s+5}$$

There is only one **b** vector associated with the three terms involving $s + 2$, namely $\mathbf{b}_1^T = [1 \quad 0]$. This means that these three terms can be simulated from a series connection of $1/(s + 2)$ terms with a single input $\mathbf{b}_1^T\mathbf{u}$. These terms will also form a single Jordan block. Figure 15.7 gives a simulation diagram from which state equations can be written.

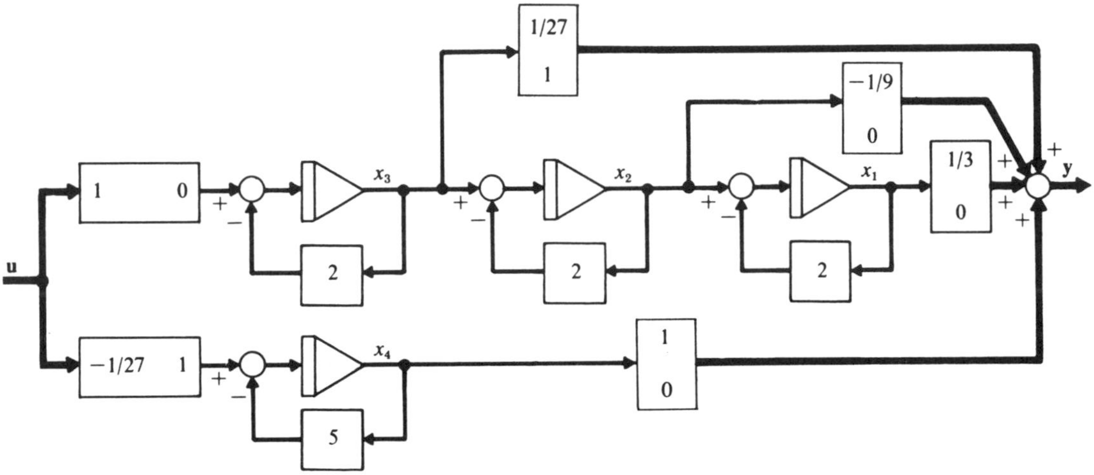

Figure 15.7

Note that

$$x_1(s) = \frac{1}{(s+2)^3}[1 \quad 0]\mathbf{u}(s) \qquad x_2(s) = \frac{1}{(s+2)^2}[1 \quad 0]\mathbf{u}(s)$$

$$x_3(s) = \frac{1}{(s+2)}[1 \quad 0]\mathbf{u}(s)$$

From Fig. 15.7,

$$\begin{bmatrix} \dot{x}_1 \\ \dot{x}_2 \\ \dot{x}_3 \\ \dot{x}_4 \end{bmatrix} = \begin{bmatrix} -2 & 1 & 0 & 0 \\ 0 & -2 & 1 & 0 \\ 0 & 0 & -2 & 0 \\ 0 & 0 & 0 & -5 \end{bmatrix} \begin{bmatrix} x_1 \\ x_2 \\ x_3 \\ x_4 \end{bmatrix} + \begin{bmatrix} 0 & 0 \\ 0 & 0 \\ 1 & 0 \\ -\frac{1}{27} & 1 \end{bmatrix} \mathbf{u}$$

and

$$\mathbf{y} = \begin{bmatrix} \frac{1}{3} \\ 0 \end{bmatrix} x_1 + \begin{bmatrix} -\frac{1}{9} \\ 0 \end{bmatrix} x_2 + \begin{bmatrix} \frac{1}{27} \\ 1 \end{bmatrix} x_3 + \begin{bmatrix} 1 \\ 0 \end{bmatrix} x_4 = \begin{bmatrix} \frac{1}{3} & -\frac{1}{9} & \frac{1}{27} & 1 \\ 0 & 0 & 1 & 0 \end{bmatrix} \mathbf{x}$$

This realization is both completely controllable and completely observable and is therefore irreducible. ■

For each repeated pole the previous method will yield a number of Jordan blocks equal to the number of *linearly independent* $\mathbf{b}_i$ vectors associated with that pole. The realization obtained in this manner will always be completely controllable because one of these $\mathbf{b}_i^T$ vectors will form the last row for each block in $\mathbf{B}$ associated with a Jordan block. (See Problem 15.10.) Observability cannot be ensured in advance. The systematic procedure of Problem 15.7 can also be employed.

Example 15.9

Suppose

$$\mathbf{H}(s) = \frac{\begin{bmatrix}1\\1\\1\end{bmatrix}[1 \quad 0 \quad 0]}{(s+p)^3} + \frac{\begin{bmatrix}0\\1\\0\end{bmatrix}[0 \quad 1 \quad 0]}{(s+p)^3} + \frac{\begin{bmatrix}1\\0\\1\end{bmatrix}[1 \quad 0 \quad 0]}{(s+p)^2} + \frac{\begin{bmatrix}2\\5\\2\end{bmatrix}[0 \quad 0 \quad 1]}{(s+p)^2} + \frac{\begin{bmatrix}2\\2\\3\end{bmatrix}[-1 \quad -1 \quad 0]}{s+p}$$

There are three independent $\mathbf{b}_i^T$ vectors, labeled $\mathbf{b}_1 = \begin{bmatrix}1\\0\\0\end{bmatrix}$, $\mathbf{b}_2 = \begin{bmatrix}0\\1\\0\end{bmatrix}$, and $\mathbf{b}_3 = \begin{bmatrix}0\\0\\1\end{bmatrix}$.

The last $\mathbf{b}^T$ vector can be written as $\begin{bmatrix}-1\\-1\\0\end{bmatrix} = -\mathbf{b}_1 - \mathbf{b}_2$.

Defining $\mathbf{c}_1 = \begin{bmatrix}1\\1\\1\end{bmatrix}$, $\mathbf{c}_2 = \begin{bmatrix}0\\1\\0\end{bmatrix}$, $\mathbf{e}_3 = \begin{bmatrix}1\\0\\1\end{bmatrix}$, $\mathbf{c}_4 = \begin{bmatrix}2\\5\\2\end{bmatrix}$, and $\mathbf{c}_5 = \begin{bmatrix}2\\2\\3\end{bmatrix}$ allows $\mathbf{H}(s)$ to be written as

$$\mathbf{H}(s) = \frac{\mathbf{c}_1\rangle\langle\mathbf{b}_1}{(s+p)^3} + \frac{\mathbf{c}_3\rangle\langle\mathbf{b}_1}{(s+p)^2} - \frac{\mathbf{c}_5\rangle\langle\mathbf{b}_1}{s+p} + \frac{\mathbf{c}_2\rangle\langle\mathbf{b}_2}{(s+p)^3} - \frac{\mathbf{c}_5\rangle\langle\mathbf{b}_2}{s+p} + \frac{\mathbf{c}_4\rangle\langle\mathbf{b}_3}{(s+p)^2}$$

A simulation diagram for this equation is given in Fig. 15.8. The three terms with input $\mathbf{b}_1^T\mathbf{u}$ are linked together in what is called a *Jordan chain*. The three associated state variables will be described by a 3×3 Jordan block. The two terms with input $\mathbf{b}_2^T\mathbf{u}$ form another Jordan chain and hence a Jordan block. Likewise, the term with input $\mathbf{b}_3^T\mathbf{u}$ leads to a third Jordan block. The dimension of a Jordan block is determined by the number of integrators required to simulate the Jordan chain.

The completely controllable realization is given by

$$\mathbf{A} = \left[\begin{array}{ccc|ccc|cc} -p & 1 & 0 & & & & & \\ 0 & -p & 1 & & & \mathbf{0} & & \\ 0 & 0 & -p & & & & & \\ \hline & & & -p & 1 & 0 & & \\ & & & 0 & -p & 1 & & \\ & & & 0 & 0 & -p & & \\ \hline & \mathbf{0} & & & & & -p & 1 \\ & & & & & & 0 & -p \end{array}\right], \quad \mathbf{B} = \left[\begin{array}{c} \mathbf{0} \\ \mathbf{0} \\ \mathbf{b}_1^T \\ \hline \mathbf{0} \\ \mathbf{0} \\ \mathbf{b}_2^T \\ \hline \mathbf{0} \\ \mathbf{b}_3^T \end{array}\right],$$

$$\mathbf{C} = [\mathbf{c}_1 \quad \mathbf{c}_3 \quad -\mathbf{c}_5 \mid \mathbf{c}_2 \quad \mathbf{0} \quad -\mathbf{c}_5 \mid \mathbf{c}_4 \quad \mathbf{0}]$$

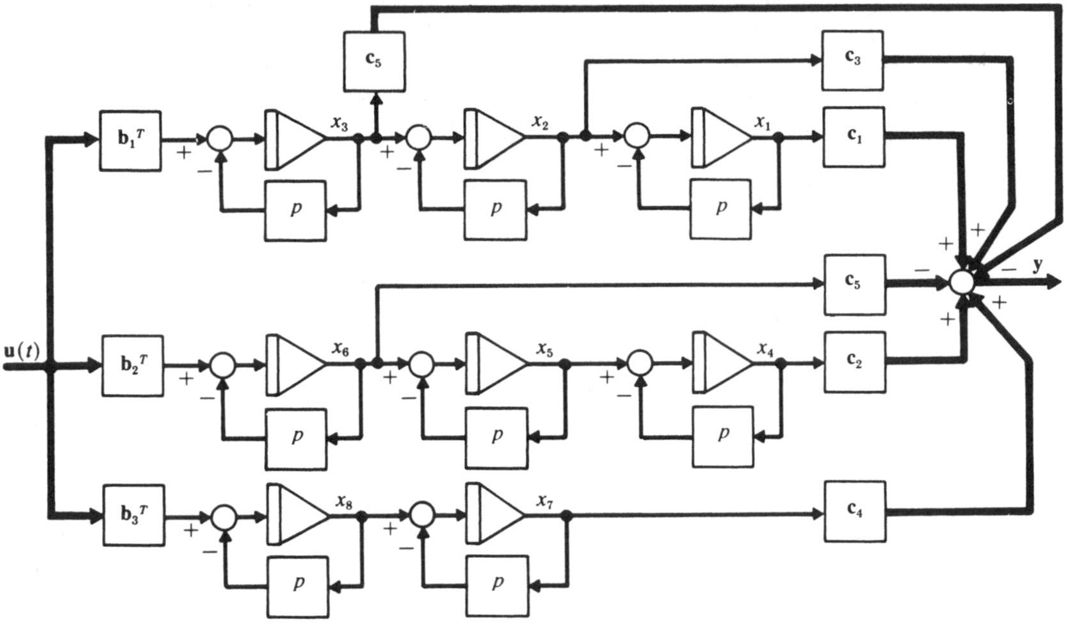

Figure 15.8

It is *not* completely observable because $\{\mathbf{c}_1, \mathbf{c}_2, \mathbf{c}_4\}$ is not a linearly independent set. (See Problem 15.11.) In fact, $\mathbf{c}_4 = 2\mathbf{c}_1 + 3\mathbf{c}_2$. Therefore, this realization can be reduced. ■

Example 15.10

Find an *irreducible* realization for $\mathbf{H}(s)$ of the previous example.

In order to reduce the order of the realization by one, one integrator must be eliminated from Fig. 15.8 without altering the input-output characteristics, that is, $\mathbf{H}(s)$. Using $\mathbf{c}_4 = 2\mathbf{c}_1 + 3\mathbf{c}_2$ allows $\mathbf{H}(s)$ to be rewritten as

$$\mathbf{H}(s) = \frac{\mathbf{c}_1\rangle\langle[\mathbf{b}_1 + 2(s+p)\mathbf{b}_3]}{(s+p)^3} + \frac{\mathbf{c}_3\rangle\langle\mathbf{b}_1}{(s+p)^2} - \frac{\mathbf{c}_5\rangle\langle\mathbf{b}_1}{s+p} + \frac{\mathbf{c}_2\rangle\langle[\mathbf{b}_2 + 3(s+p)\mathbf{b}_3]}{(s+p)^3} - \frac{\mathbf{c}_5\rangle\langle\mathbf{b}_2}{s+p} \tag{15.11}$$

This form indicates that if additional inputs to the first and second Jordan blocks are used, the second-order term in the third Jordan block can be eliminated. However, a new first-order term will apparently be required in the third Jordan block to subtract out all additional unwanted outputs from blocks one and two due to their added inputs. This is made clear by rewriting equation *(15.11)* as

$$\mathbf{H}(s) = \frac{\mathbf{c}_1\rangle\langle[\mathbf{b}_1 + 2(s+p)\mathbf{b}_3]}{(s+p)^3} + \frac{\mathbf{c}_3\rangle\langle[\mathbf{b}_1 + 2(s+p)\mathbf{b}_3]}{(s+p)^2} - \frac{\mathbf{c}_5\rangle\langle\mathbf{b}_1}{s+p} + \frac{\mathbf{c}_2\rangle\langle[\mathbf{b}_2 + 3(s+p)\mathbf{b}_3]}{(s+p)^3} - \frac{\mathbf{c}_5\rangle\langle\mathbf{b}_2}{s+p} - \frac{2\mathbf{c}_3\rangle\langle\mathbf{b}_3}{s+p} \tag{15.12}$$

Figure 15.9 gives the reduced simulation diagram. The new realization is given by

$$\mathbf{A} = \left[\begin{array}{ccc:ccc:c} -p & 1 & 0 & & & & \\ 0 & -p & 1 & & \mathbf{0} & & \\ 0 & 0 & -p & & & & \\ \hdashline & & & -p & 1 & 0 & \\ & \mathbf{0} & & 0 & -p & 1 & \\ & & & 0 & 0 & -p & \\ \hdashline & & & & & & -p \end{array}\right], \qquad \mathbf{B} = \left[\begin{array}{c} \mathbf{0} \\ 2\mathbf{b}_3^T \\ \mathbf{b}_1^T \\ \hdashline \mathbf{0} \\ 3\mathbf{b}_3^T \\ \mathbf{b}_2^T \\ \hdashline \mathbf{b}_3^T \end{array}\right],$$

$$\mathbf{C} = [\mathbf{c}_1 \quad \mathbf{c}_3 \quad -\mathbf{c}_5 \;\vdots\; \mathbf{c}_2 \quad \mathbf{0} \quad -\mathbf{c}_5 \;\vdots\; -2\mathbf{c}_3]$$

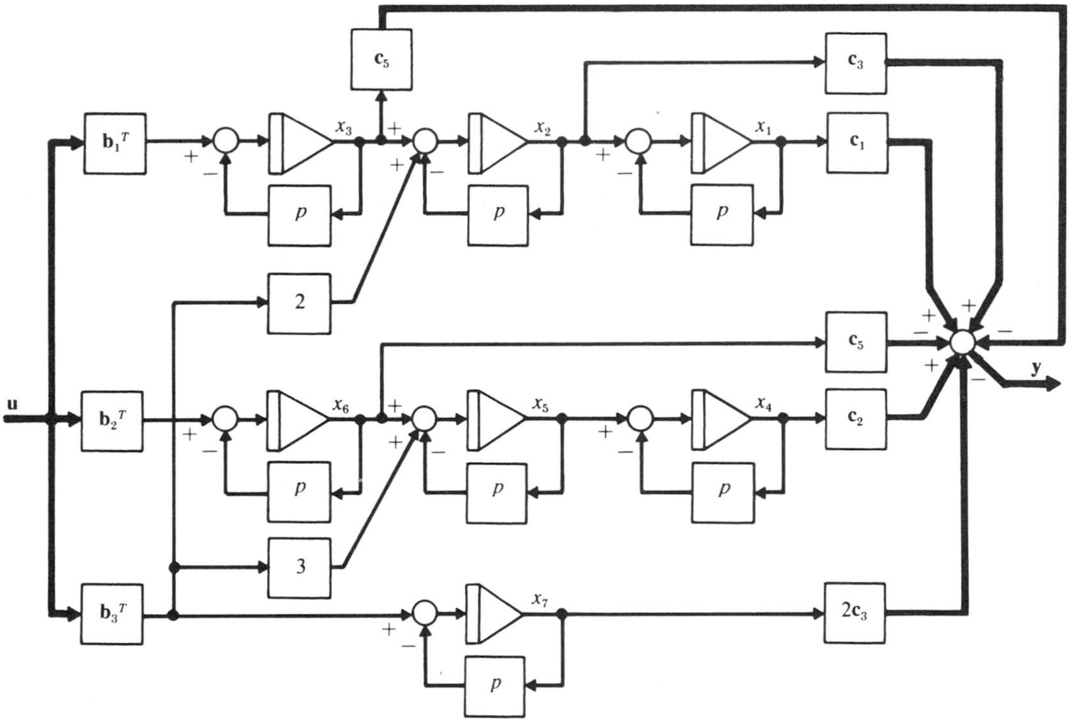

Figure 15.9

This realization is still not observable, and can be further reduced since $-2\mathbf{c}_1 + 2\mathbf{c}_2 = -2\mathbf{c}_3$. This result could have been recognized from equation (*15.12*) and the reduction could have been completed in one step.

Rewriting equation (*15.12*) gives

$$\mathbf{H}(s) = \frac{\mathbf{c}_1\rangle\langle[\mathbf{b}_1 + 2(s+p)\mathbf{b}_3 - 2(s+p)^2\mathbf{b}_3]}{(s+p)^3} + \frac{\mathbf{c}_3\rangle\langle[\mathbf{b}_1 + 2(s+p)\mathbf{b}_3]}{(s+p)^2} - \frac{\mathbf{c}_5\rangle\langle\mathbf{b}_1}{s+p}$$
$$+ \frac{\mathbf{c}_2\rangle\langle[\mathbf{b}_2 + 3(s+p)\mathbf{b}_3 + 2(s+p)^2\mathbf{b}_3]}{(s+p)^3} - \frac{\mathbf{c}_5\rangle\langle\mathbf{b}_2}{s+p}$$

The simulation diagram of Fig. 15.9 can be modified to obtain an irreducible sixth-order realization:

$$\mathbf{A} = \left[\begin{array}{ccc|ccc} -p & 1 & 0 & & & \\ 0 & -p & 1 & & \mathbf{0} & \\ 0 & \mathbf{0} & -p & & & \\ \hline & & & -p & 1 & 0 \\ & \mathbf{0} & & 0 & -p & 1 \\ & & & 0 & 0 & -p \end{array}\right], \qquad \mathbf{B} = \left[\begin{array}{c} -2\mathbf{b}_3^T \\ 2\mathbf{b}_3^T \\ \mathbf{b}_1^T \\ \hline 2\mathbf{b}_3^T \\ 3\mathbf{b}_3^T \\ \mathbf{b}_2^T \end{array}\right],$$

$$\mathbf{C} = [\mathbf{c}_1 \quad \mathbf{c}_3 \quad -\mathbf{c}_5 \mid \mathbf{c}_2 \quad \mathbf{0} \quad -\mathbf{c}_5]$$

■

15.6 CONCLUDING COMMENTS

Linear constant coefficient systems can be described either by state variable techniques or by transfer matrices. If the state equations are known, the unique transfer matrix is given by equation (*15.5*). The reverse process is not unique. Given a transfer matrix, there exist many sets of corresponding state equations. Methods of determining these state equations, called realizations, have been presented. This completes the topic of state variable selection begun in Chapter 9 for continuous-time systems and in Chapter 11 for discrete systems. Also, the results of Problems 15.10 and 15.11 represent generalizations of Criterion 1 of Sec. 12.3.

Emphasis was given to finding the irreducible realization, i.e., the system with the minimum number of state variables, which possesses a specified transfer function. The Jordan form method of determining irreducible realizations is probably the easiest method to understand, but it has two disadvantages:

1. *The elements of* $\mathbf{H}(s)$ *must be factored so that partial fraction expansion can be carried out.* Factoring high-degree polynomials is troublesome.
2. *When* $\mathbf{H}(s)$ *has complex poles, the resulting matrices* **A** *and* **B** *or* **C** *will have complex elements.* This precludes direct simulation with physical devices. The second disadvantage can be overcome, since a similarity transformation can always be used to obtain a realization with all real elements (see Problem 15.9).

An alternative method of obtaining irreducible realizations is provided by the algorithm of B. L. Ho [51, 64]. The method given here is similar to the one discussed by C. T. Chen [17], who also discusses the Ho algorithm. For additional related material see References 19, 29, and 73.

ILLUSTRATIVE PROBLEMS

Transfer Functions and State Equations

15.1 A single-input, single-output system has the transfer function

$$H(s) = \frac{(s+5)(s+\alpha)}{(s+2)(s+3)(s+\alpha)}$$

a. Cancel the common pole, zero pair and write the input-output differential equation for this system. Use a simulation diagram to select state variables.
b. Repeat part a without cancelling the pole, zero pair.
c. Compare the controllability and observability of the two realizations obtained in a and b.

a. The transfer function $y(s)/u(s) = (s+5)/[(s+2)(s+3)]$ implies the differential equation $\ddot{y} + 5\dot{y} + 6y = 5u + \dot{u}$. A simulation diagram is given in Fig. 15.10. Then

$$\begin{bmatrix}\dot{x}_1\\ \dot{x}_2\end{bmatrix} = \begin{bmatrix}0 & 1\\ -6 & -5\end{bmatrix}\begin{bmatrix}x_1\\ x_2\end{bmatrix} + \begin{bmatrix}1\\ 0\end{bmatrix}u, \qquad y = [1 \quad 0]\mathbf{x}$$

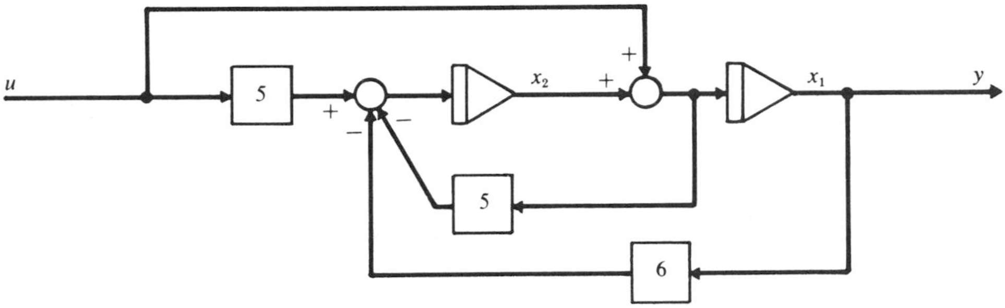

Figure 15.10

b. The differential equation now is $\dddot{y} + (5+\alpha)\ddot{y} + (6+5\alpha)\dot{y} + 6\alpha y = \ddot{u} + (5+\alpha)\dot{u} + 5\alpha u$. From this,

$$y = \int \left\{[u - (5+\alpha)y] + \int \left[(5+\alpha)u - (6+5\alpha)y + \int (5\alpha u - 6\alpha y)\,dt\right] dt\right\} dt$$

A simulation diagram is given in Fig. 15.11. Then

$$\begin{bmatrix}\dot{x}_1\\ \dot{x}_2\\ \dot{x}_3\end{bmatrix} = \begin{bmatrix}-(5+\alpha) & 1 & 0\\ -(6+5\alpha) & 0 & 1\\ -6\alpha & 0 & 0\end{bmatrix}\begin{bmatrix}x_1\\ x_2\\ x_3\end{bmatrix} + \begin{bmatrix}1\\ 5+\alpha\\ 5\alpha\end{bmatrix}u, \qquad y = [1 \quad 0 \quad 0\;]\mathbf{x}$$

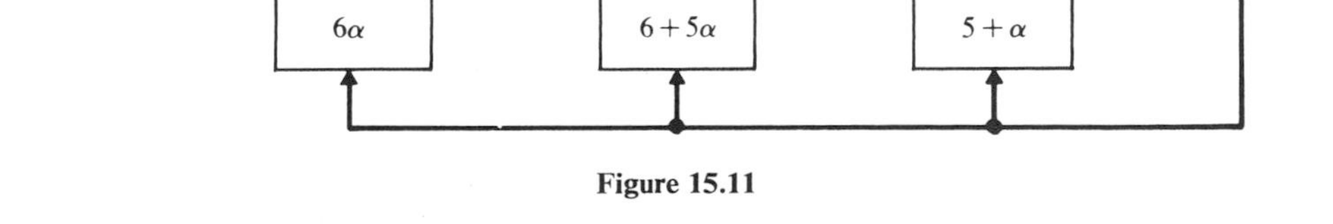

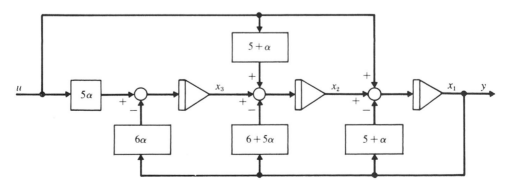

Figure 15.11

c. For system a, $\mathbf{P} = \begin{bmatrix} 1 & 0 \\ 0 & -6 \end{bmatrix}$ has rank 2. $\mathbf{Q}\begin{bmatrix} 1 & 0 \\ 0 & 1 \end{bmatrix}$ has rank 2. System a is completely controllable and observable. For system b, $\mathbf{P} = \begin{bmatrix} 1 & 0 & -6 \\ 5+\alpha & -6 & -6\alpha \\ 5\alpha & -6\alpha & 0 \end{bmatrix}$ has rank 2. $\mathbf{Q} = \begin{bmatrix} 1 & -5-\alpha & \alpha^2+5\alpha+19 \\ 0 & 1 & -5-\alpha \\ 0 & 0 & 1 \end{bmatrix}$ has rank 3. System b is completely observable but not controllable. Other simulation diagrams can be drawn for system b which result in a realization which is completely controllable but not observable. It is not possible to obtain a third-order realization which is both completely controllable and observable.

15.2 Find state equations and the input-output transfer matrix for the system of Fig. 15.12.

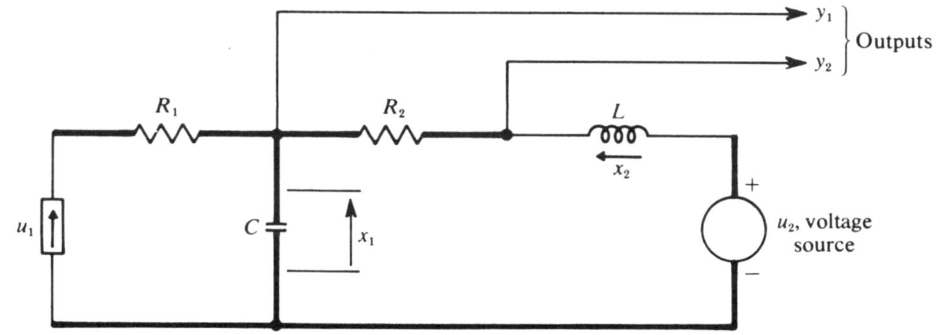

Figure 15.12

The fact that there are multiple inputs and outputs does not change the procedure for selecting state variables from the linear graph, described in Sec. 9.4. The tree to be used is shown in heavy lines. Use the capacitor voltage x_1 and inductor current x_2 as state variables. Then $\dot{x}_1 = x_2/C + u_1/C$, $\dot{x}_2 = -x_1/L - R_2x_2/L + u_2/L$, and $y_1 = x_1$, $y_2 = x_1 + R_2x_2$. Hence

$$\mathbf{A} = \begin{bmatrix} 0 & 1/C \\ -1/L & -R_2/L \end{bmatrix}, \quad \mathbf{B} = \begin{bmatrix} 1/C & 0 \\ 0 & 1/L \end{bmatrix}, \quad \mathbf{C} = \begin{bmatrix} 1 & 0 \\ 1 & R_2 \end{bmatrix}, \quad D = 0$$

Then

$$\mathbf{H}(s) = \mathbf{C}[s\mathbf{I} - \mathbf{A}]^{-1}\mathbf{B} = \frac{\begin{bmatrix} (s+R_2/L)/C & 1/LC \\ s/C & (R_2s+1/C)/L \end{bmatrix}}{s^2 + R_2s/L + 1/LC}$$

15.3 A system with three inputs and two outputs is described by

$$\ddot{y}_1 + 6\dot{y}_1 - \dot{y}_2 + 10y_1 - 6y_2 = \frac{\dot{u}_1}{2} - 2u_1 + \frac{3\dot{u}_2}{2} + 4u_2 - 5\dot{u}_3$$

$$\ddot{y}_2 + 8\dot{y}_2 - 2\dot{y}_1 + 12y_2 - 4y_1 = \dot{u}_1 + 2u_1 + \dot{u}_2 + 2u_2 + \dot{u}_3 + 2u_3$$

Find the transfer matrix $\mathbf{H}(s)$.

A simulation diagram could be drawn, utilizing four integrators. Then equation (*15.5*) could be used to obtain $\mathbf{H}(s)$. Alternatively, the Laplace transforms of the original equations are used here, giving

$$\mathbf{y}(s) = \begin{bmatrix} s^2 + 6s + 10 & -s - 6 \\ -2s - 4 & s^2 + 8s + 12 \end{bmatrix}^{-1} \begin{bmatrix} s/2 - 2 & 3s/2 + 4 & -5 \\ s + 2 & s + 2 & s + 2 \end{bmatrix} \mathbf{u}(s)$$

Therefore,

$$\mathbf{H}(s) = \frac{\begin{bmatrix} s/2 - 1 & 3s/2 + 5 & -4 \\ s + 1 & s + 3 & s \end{bmatrix}}{(s + 2)(s + 4)}$$

15.4 Determine the input-output transfer matrix for a system described by

$$\dot{\mathbf{x}} = \begin{bmatrix} -1 & 1 & 0 \\ 0 & -1 & 0 \\ 0 & 0 & -1 \end{bmatrix} \mathbf{x} + \begin{bmatrix} 2 & 0 \\ 1 & -1 \\ -1 & 1 \end{bmatrix} \mathbf{u}, \qquad \mathbf{y} = \begin{bmatrix} 2 & -1 & 3 \\ 0 & 1 & 0 \\ 3 & -1 & 6 \end{bmatrix} \mathbf{x} + \begin{bmatrix} 1 & 1 \\ 0 & 0 \\ 0 & 0 \end{bmatrix} \mathbf{u}$$

Using equation (*15.5*), the transfer matrix is $\mathbf{H}(s) = \mathbf{C}[s\mathbf{I} - \mathbf{A}]^{-1}\mathbf{B} + \mathbf{D}$. Substituting for **A, B, C,** and **D** and carrying out the multiplication gives

$$\mathbf{H}(s) = \begin{bmatrix} \dfrac{s^2 + 2s + 3}{(s + 1)^2} & \dfrac{s^2 + 6s + 3}{(s + 1)^2} \\[2ex] \dfrac{1}{s + 1} & \dfrac{-1}{s + 1} \\[2ex] \dfrac{-s + 2}{(s + 1)^2} & \dfrac{7s + 4}{(s + 1)^2} \end{bmatrix}$$

15.5 Use the linearized state variable model of the aircraft in Problem 12.7. Find the transfer function matrix which relates the aircraft aileron and rudder deflections to the state variables p, r, β, and ϕ. Draw a block diagram of the relationships contained in the state equations and verify the entries in the transfer function matrix. Use the block diagram to explain in physical terms the mathematical results on controllability (Problem 12.7) and observability (Problem 12.8).

Since the transfer function to *all* the states is desired, set $\mathbf{C} = I$. Then

$$\mathbf{H}(s) = \mathbf{C}[s\mathbf{I} - \mathbf{A}]^{-1}\mathbf{B} = \begin{bmatrix} s + 10 & 0 & 10 & 0 \\ 0 & s + 0.7 & -9 & 0 \\ 0 & 1 & s + 0.7 & 0 \\ 1 & 0 & 0 & s \end{bmatrix}^{-1} \begin{bmatrix} 20 & 2.8 \\ 0 & -3.13 \\ 0 & 0 \\ 0 & 0 \end{bmatrix}$$

$$= \begin{bmatrix} \dfrac{20}{s + 10} & \dfrac{2.8(s - 0.776)(s + 2.176)}{(s + 10)(s^2 + 1.4s + 9.49)} \\[2ex] 0 & \dfrac{-3.13(s + 0.7)}{s^2 + 1.4s + 9.49} \\[2ex] 0 & \dfrac{3.13}{s^2 + 1.4s + 9.49} \\[2ex] \dfrac{20}{s(s + 10)} & \dfrac{2.8(s - 0.776)(s + 2.176)}{s(s + 10)(s^2 + 1.4s + 9.49)} \end{bmatrix}$$

The block diagram of Fig. 15.13 is drawn directly from the state equations. Using elementary block diagram reduction techniques, or Mason's gain formula for a one-step reduction, the transfer functions H_{ij} from each input j to each of the four outputs i can be found. The eight results are of course the entries in the previous matrix.

Figure 15.13 makes it clear why the system is uncontrollable with just δ_a. There is no signal path from this input to either r or β. The other input does feed into all four state variables, either directly or indirectly. The observability results of Problem 12.8 are equally obvious. If ϕ is not an output, it can never be determined since there is no path by which ϕ affects any other state or output. Conversely, if ϕ can be monitored, it is intuitive that its rate, i.e., p, can be deduced. Likewise, the rate of change of p can be deduced, and it contains information about β. This in turn is strongly related to r.

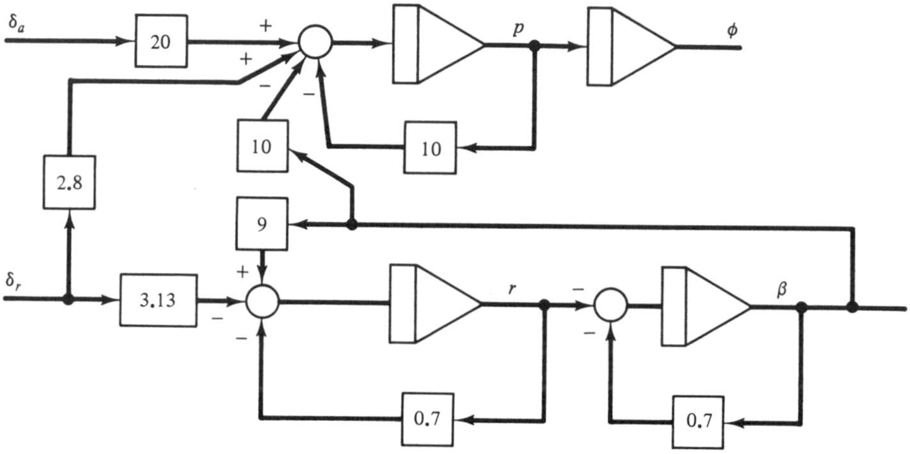

Figure 15.13

Irreducible Realizations

15.6 Find an irreducible realization of $\mathbf{H}(s)$ from Problem 15.4.

Expanding each H_{ij} element yields

$$\mathbf{H}(s) = \begin{bmatrix} 1 + \dfrac{2}{(s+1)^2} & 1 - \dfrac{2}{(s+1)^2} + \dfrac{4}{s+1} \\ \dfrac{1}{s+1} & -\dfrac{1}{s+1} \\ \dfrac{3}{(s+1)^2} - \dfrac{1}{s+1} & \dfrac{-3}{(s+1)^2} + \dfrac{7}{s+1} \end{bmatrix}$$

$$= \frac{\begin{bmatrix} 2 & 2 \\ 0 & 0 \\ 3 & -3 \end{bmatrix}}{(s+1)^2} + \frac{\begin{bmatrix} 0 & 4 \\ 1 & -1 \\ -1 & 7 \end{bmatrix}}{s+1} + \begin{bmatrix} 1 & 1 \\ 0 & 0 \\ 0 & 0 \end{bmatrix}$$

This form indicates that $\mathbf{D} = \begin{bmatrix} 1 & 1 \\ 0 & 0 \\ 0 & 0 \end{bmatrix}$. The remaining two matrices are of rank 1 and 2, respectively. Writing them as outer products gives

$$\mathbf{H}(s) = \frac{\begin{bmatrix} 2 \\ 0 \\ 3 \end{bmatrix}[1 \quad -1]}{(s+1)^2} + \frac{\begin{bmatrix} 0 \\ 1 \\ -1 \end{bmatrix}[1 \quad -1]}{s+1} + \frac{\begin{bmatrix} 4 \\ 0 \\ 6 \end{bmatrix}[0 \quad 1]}{s+1} + \mathbf{D}$$

The vectors $\mathbf{b}_1^T = [1 \quad -1]$, $\mathbf{b}_3^T = [0 \quad 1]$ are linearly independent, and could be used to form the last rows associated with two Jordan blocks $\mathbf{B} = \begin{bmatrix} 0 & 0 \\ 1 & -1 \\ \hdashline 0 & 1 \end{bmatrix}$. However, the corresponding output matrix $\mathbf{C} = \left[\begin{array}{cc:c} 2 & 0 & 4 \\ 0 & 1 & 0 \\ 3 & -1 & 6 \end{array}\right]$ has its first and third columns linearly dependent. This would lead to a realization which is completely controllable, but not observable and hence reducible. Since $\mathbf{c}_3 = 2\mathbf{c}_1$, the first and third terms can be combined:

$$\mathbf{H}(s) = \frac{\mathbf{c}_1\rangle\langle[\mathbf{b}_1 + 2\mathbf{b}_3(s+1)]}{(s+1)^2} + \frac{\mathbf{c}_2\rangle\langle\mathbf{b}_1}{s+1} + \mathbf{D}$$

The corresponding simulation diagram is given in Fig. 15.14.

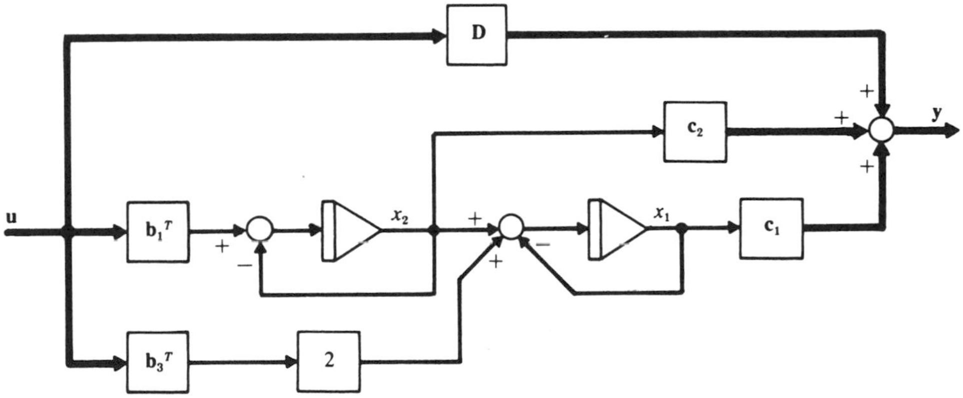

Figure 15.14

The irreducible realization is

$$\begin{bmatrix} \dot{x}_1 \\ \dot{x}_2 \end{bmatrix} = \begin{bmatrix} -1 & 1 \\ 0 & -1 \end{bmatrix}\begin{bmatrix} x_1 \\ x_2 \end{bmatrix} + \begin{bmatrix} 0 & 2 \\ 1 & -1 \end{bmatrix}\begin{bmatrix} u_1 \\ u_2 \end{bmatrix},$$

$$\begin{bmatrix} y_1 \\ y_2 \\ y_3 \end{bmatrix} = \begin{bmatrix} 2 & 0 \\ 0 & 1 \\ 3 & -1 \end{bmatrix}\begin{bmatrix} x_1 \\ x_2 \end{bmatrix} + \begin{bmatrix} 1 & 1 \\ 0 & 0 \\ 0 & 0 \end{bmatrix}\begin{bmatrix} u_1 \\ u_2 \end{bmatrix}$$

This realization is of lesser dimension than the actual system from which the transfer function was obtained. The original system was not completely controllable.

15.7 Find a minimal order state variable realization for the discrete system described by

$$\mathbf{T}(z) = \begin{bmatrix} \dfrac{0.2z}{(z-1)(z-0.5)} & \dfrac{3z(z+0.3)}{(z-1)(z-0.5)} \\ \dfrac{(z+0.8)}{(z-0.5)^2(z-0.7)} & \dfrac{z}{(z-0.5)} \end{bmatrix}$$

A partial fraction expanded form for $\mathbf{T}(z)$ is

$$\mathbf{T}(z) = \frac{1}{(z-0.5)}\begin{bmatrix} -0.2 & 0 \\ -37.5 & 0.5 \end{bmatrix} + \frac{1}{(z-0.5)^2}\begin{bmatrix} 0 & 0 \\ -6.5 & 0 \end{bmatrix}$$
$$+ \frac{1}{(z-1)}\begin{bmatrix} 0.4 & 19.5 \\ 0 & 0 \end{bmatrix} + \frac{1}{(z-0.8)}\begin{bmatrix} 0 & -13.2 \\ 0 & 0 \end{bmatrix}$$
$$+ \frac{1}{(z-0.7)}\begin{bmatrix} 0 & 0 \\ 37.5 & 0 \end{bmatrix} + \begin{bmatrix} 0 & 3 \\ 0 & 1 \end{bmatrix}$$

All of the 2×2 expansion matrices are of rank one except the first, which has rank 2. Therefore, a seventh-order state variable model could be written down immediately as

$$\mathbf{A} = \text{diag}\left[0.5,\quad 0.5,\quad \begin{bmatrix} 0.5 & 1 \\ 0 & 0.5 \end{bmatrix},\quad 1,\quad 0.8,\quad 0.7\right]$$

$$\mathbf{B} = \begin{bmatrix} 1 & 0 \\ 0 & 1 \\ \hline 1 & 0 \\ 0 & 0 \\ \hline 0.4 & 19.5 \\ 0 & 1 \\ 1 & 0 \end{bmatrix} \qquad \mathbf{C} = \left[\begin{array}{cc|cc|ccc} -0.2 & 0 & 0 & 0 & 1 & -13.2 & 0 \\ -37.5 & 0.5 & 0 & -6.5 & 0 & 0 & 37.5 \end{array}\right]$$

$$\mathbf{D} = \begin{bmatrix} 0 & 3 \\ 0 & 1 \end{bmatrix}$$

Note that a 2×2 Jordan block is included for the $(z-0.5)^2$ factor. The first state in this segment has a direct connection to the input but not the output. The reverse is true for the second state in this subsegment. These two states are related, as shown in Fig. 15.15. It can be shown (easily with the right computational aids, otherwise more laboriously) that the above seventh-order system is neither controllable nor observable. The ranks of the controllability and observability matrices are both 5. This indicates that two unneeded states can be deleted. The two states shown in Fig. 15.15, plus the other two uncoupled states associated with $z = 0.5$, must be reduced to just two states. A systematic way of accomplishing this is to assume the segment shown in Fig. 15.16 and select the eight scalar parameters $a, \ldots, h$ so that this segment realization yields the same output as the original fourth-order segment. That is

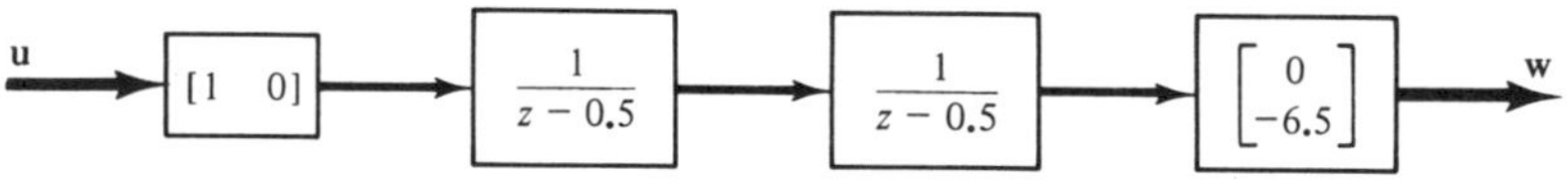

Figure 15.15

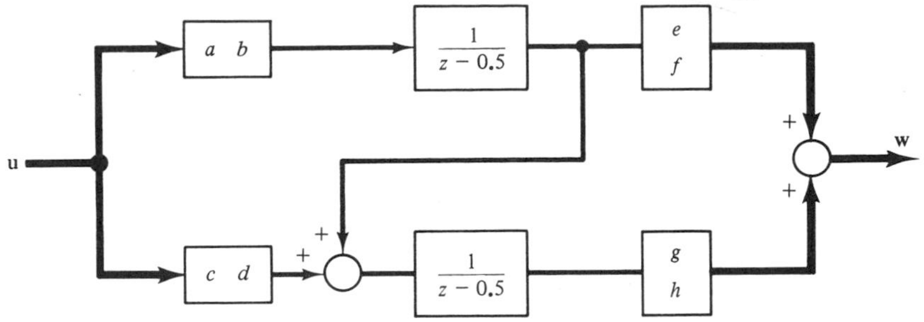

Figure 15.16

$$\mathbf{w} = \left\{ \frac{\begin{bmatrix} e \\ f \end{bmatrix}[a \quad b] + \begin{bmatrix} g \\ h \end{bmatrix}[c \quad d]}{(z-0.5)} + \frac{\begin{bmatrix} g \\ h \end{bmatrix}[a \quad b]}{(z-0.5)^2} \right\} \mathbf{u}$$

$$= \left\{ \frac{\begin{bmatrix} -0.2 & 0 \\ -37.5 & 0.5 \end{bmatrix}}{(z-0.5)} + \frac{\begin{bmatrix} 0 & 0 \\ -6.5 & 0 \end{bmatrix}}{(z-0.5)^2} \right\} \mathbf{u}$$

This requires that $ae + cg = -0.2$, $be + dg = 0$, $af + ch = -37.5$, $bf + dh = 0.5$, $ag = 0$, $bg = 0$, $ah = -6.5$, $bh = 0$. Of the many possibilities, pick $b = c = g = 0$ and $a = 1$. Then solve for $e = -0.2$, $h = -6.5$, $d = -0.076923$, and $f = -37.5$. The 2×2 segment associated with $z = 0.5$ thus has the state equations

$$\begin{bmatrix} x_1(k+1) \\ x_2(k+1) \end{bmatrix} = \begin{bmatrix} 0.5 & 1 \\ 0 & 0.5 \end{bmatrix} \begin{bmatrix} x_1(k) \\ x_2(k) \end{bmatrix} + \begin{bmatrix} 0 & -0.076923 \\ 1 & 0 \end{bmatrix} \mathbf{u}(k)$$

$$\mathbf{w}(k) = \begin{bmatrix} 0 & -0.2 \\ -6.5 & -37.5 \end{bmatrix} \begin{bmatrix} x_1(k) \\ x_2(k) \end{bmatrix}$$

By substituting this second-order segment for the original fourth-order segment, the final fifth-order system realization, which is controllable with index 3 and observable with index 3, is

$$\mathbf{A} = \left[\begin{array}{cc|ccc} 0.5 & 1 & 0 & 0 & 0 \\ 0 & 0.5 & 0 & 0 & 0 \\ \hline 0 & 0 & 1 & 0 & 0 \\ 0 & 0 & 0 & 0.8 & 0 \\ 0 & 0 & 0 & 0 & 0.7 \end{array}\right] \qquad \mathbf{B} = \begin{bmatrix} 0 & -0.07693 \\ 1 & 0 \\ 0.4 & 19.5 \\ 0 & 1 \\ 1 & 0 \end{bmatrix}$$

$$\mathbf{C} = \begin{bmatrix} 0 & -0.2 & 1 & -13.2 & 0 \\ -6.5 & -37.5 & 0 & 0 & 37.5 \end{bmatrix} \qquad \mathbf{D} = \begin{bmatrix} 0 & 3 \\ 0 & 1 \end{bmatrix}$$

15.8 Find an irreducible realization for the transfer matrix

$$\mathbf{H}(s) = \begin{bmatrix} \dfrac{3}{(s^2+2s+10)(s+5)} & \dfrac{2s}{s+5} \\ \dfrac{s+1}{(s^2+2s+10)(s+5)} & \dfrac{s+1}{s+5} \end{bmatrix}$$

Partial fraction expansion gives

$$H_{11}(s) = \frac{\alpha}{s+1+3j} + \frac{\bar{\alpha}}{s+1-3j} + \frac{3/25}{s+5}, \qquad H_{12}(s) = 2 - \frac{10}{s+5}$$

$$H_{21}(s) = \frac{-j\alpha}{s+1+3j} + \frac{-\overline{j\alpha}}{s+1-3j} - \frac{4/25}{s+5}, \qquad H_{22}(s) = 1 - \frac{4}{s+5}$$

where $\alpha = -3/50 + 2j/25$. Let $p_1 = 1 + 3j$. Then

$$\mathbf{H}(s) = \frac{\begin{bmatrix} \alpha & 0 \\ -j\alpha & 0 \end{bmatrix}}{s+p_1} + \frac{\begin{bmatrix} \bar{\alpha} & 0 \\ -\overline{j\alpha} & 0 \end{bmatrix}}{s+\bar{p}_1} + \frac{\begin{bmatrix} 3/25 & -10 \\ -4/25 & -4 \end{bmatrix}}{s+5} + \begin{bmatrix} 0 & 2 \\ 0 & 1 \end{bmatrix}$$

$$= \frac{\begin{bmatrix} \alpha \\ -j\alpha \end{bmatrix}[1 \quad 0]}{s+p_1} + \frac{\begin{bmatrix} \bar{\alpha} \\ -\overline{j\alpha} \end{bmatrix}[1 \quad 0]}{s+\bar{p}_1} + \frac{\begin{bmatrix} 3/25 \\ -4/25 \end{bmatrix}[1 \quad 0]}{s+5} + \frac{\begin{bmatrix} 10 \\ 4 \end{bmatrix}[0 \quad -1]}{s+5} + \mathbf{D}$$

This transfer matrix can be realized with four 1×1 Jordan blocks. Since $\mathbf{c}_3$ and $\mathbf{c}_4$ associated with the pole at -5 are linearly independent, an irreducible realization is given by

$$\mathbf{A} = \text{diag}\,[-p_1, -\bar{p}_1, -5, -5], \qquad \mathbf{B} = \begin{bmatrix} 1 & 0 \\ 1 & 0 \\ 1 & 0 \\ 0 & -1 \end{bmatrix}$$

$$\mathbf{C} = \begin{bmatrix} \alpha & \bar{\alpha} & 3/25 & 10 \\ -j\alpha & -\overline{j\alpha} & -4/25 & 4 \end{bmatrix}, \qquad \mathbf{D} = \begin{bmatrix} 0 & 2 \\ 0 & 1 \end{bmatrix}$$

Because of the complex numbers in **A** and **C**, this realization cannot be directly mechanized. The next problem illustrates a transformation which eliminates the complex numbers.

15.9 Find a suitable transformation (change of basis in Σ) which transforms the realization of Problem 15.8 into one with all real coefficients.

The original state vector is $\mathbf{x}$. A new state vector $\mathbf{x}'$ is sought, satisfying $\mathbf{x} = \mathbf{T}\mathbf{x}'$. Then

$$\dot{\mathbf{x}}' = \mathbf{T}^{-1}\mathbf{A}\mathbf{T}\mathbf{x}' + \mathbf{T}^{-1}\mathbf{B}\mathbf{u}, \qquad \mathbf{y} = \mathbf{C}\mathbf{T}\mathbf{x}' + \mathbf{D}\mathbf{u}$$

and it is required that $\mathbf{A}' = \mathbf{T}^{-1}\mathbf{A}\mathbf{T}$, $\mathbf{B}' = \mathbf{T}^{-1}\mathbf{B}$, and $\mathbf{C}' = \mathbf{C}\mathbf{T}$ all be real.

A tentative transformation is $\mathbf{T} = \left[\begin{array}{cc|c} 1 & j & \mathbf{0} \\ 1 & -j & \\ \hline \multicolumn{2}{c|}{\mathbf{0}} & \mathbf{I}_2 \end{array}\right]$. Then

$$\mathbf{C}' = \begin{bmatrix} \alpha + \bar{\alpha} & j\alpha - j\bar{\alpha} & 3/25 & 10 \\ -j\alpha - \overline{j\alpha} & \alpha + \bar{\alpha} & -4/25 & 4 \end{bmatrix} \quad \text{or} \quad \mathbf{C}' = \begin{bmatrix} -3/25 & -4/25 & 3/25 & 10 \\ 4/25 & -3/25 & -4/25 & 4 \end{bmatrix}$$

The selected transformation does yield a real $\mathbf{C}'$. To evaluate $\mathbf{A}'$ and $\mathbf{B}'$,

$$\mathbf{T}^{-1} = \left[\begin{array}{cc|c} 1/2 & 1/2 & \mathbf{0} \\ -j/2 & j/2 & \\ \hline \multicolumn{2}{c|}{\mathbf{0}} & \mathbf{I}_2 \end{array}\right]$$

is required. Matrix multiplication then gives

$$\mathbf{A}' = \left[\begin{array}{rr|rr} -1 & 3 & 0 & 0 \\ -3 & -1 & 0 & 0 \\ \hline 0 & 0 & -5 & 0 \\ 0 & 0 & 0 & -5 \end{array}\right] \quad \text{and} \quad \mathbf{B}' = \left[\begin{array}{rr} 1 & 0 \\ 0 & 0 \\ \hline 1 & 0 \\ 0 & -1 \end{array}\right]$$

The $\mathbf{D}$ matrix is unaffected by this change of basis, so a real, irreducible realization is $\{\mathbf{A}', \mathbf{B}', \mathbf{C}', \mathbf{D}\}$.

Jordan Form Controllability and Observability Criteria

15.10 A system with n state variables and r inputs is expressed in Jordan form

$$\dot{\mathbf{x}} = \begin{bmatrix} \mathbf{J}_1 & & & & \\ & \mathbf{J}_2 & & & \\ & & \cdot & & \\ & & & \cdot & \\ & & & & \mathbf{J}_p \end{bmatrix} \mathbf{x} + \begin{bmatrix} \mathbf{B}_1 \\ \mathbf{B}_2 \\ \cdot \\ \cdot \\ \cdot \\ \mathbf{B}_p \end{bmatrix} \mathbf{u} \tag{1}$$

Show that the controllability of this system is determined entirely by the last rows $\mathbf{b}_{il}^T$ of each $\mathbf{B}_i$ submatrix. (The subscript l signifies the last row in a given block and is not a fixed integer.) In particular, show that the system is completely controllable if and only if

1. $\{\mathbf{b}_{il}, \mathbf{b}_{jl}, \ldots, \mathbf{b}_{kl}\}$ is a linearly independent set if $\mathbf{J}_i, \mathbf{J}_j, \ldots, \mathbf{J}_k$ are Jordan blocks *with the same eigenvalue* λ_i, and
2. $\mathbf{b}_{pl} \neq \mathbf{0}$ if $\mathbf{J}_p$ is the *only* Jordan block with eigenvalue λ_p.

Note that if all $\mathbf{J}_i$ blocks are 1×1 blocks so that $\mathbf{A}$ is diagonal, the controllability criterion 1 of Sec. 12.3 requires that all $\mathbf{b}^T \neq \mathbf{0}$ for all rows of $\mathbf{B}$. Controllability criterion 2 is used to investigate the more general case. For simplicity, assume there are just three blocks,

$$\mathbf{J}_1 = \begin{bmatrix} \lambda_1 & 1 & 0 \\ 0 & \lambda_1 & 1 \\ 0 & 0 & \lambda \end{bmatrix}, \quad \mathbf{J}_2 = \begin{bmatrix} \lambda_1 & 1 \\ 0 & \lambda_1 \end{bmatrix}, \quad \mathbf{J}_3 = \begin{bmatrix} \lambda_3 & 1 \\ 0 & \lambda_3 \end{bmatrix}$$

with $\mathbf{B}^T = [\mathbf{b}_{11} \quad \mathbf{b}_{12} \quad \mathbf{b}_{13} \mid \mathbf{b}_{21} \quad \mathbf{b}_{22} \mid \mathbf{b}_{31} \quad \mathbf{b}_{32}]$. The controllability matrix is

$$\mathbf{P} = [\mathbf{B} \mid \mathbf{AB} \mid \mathbf{A}^2\mathbf{B} \mid \mathbf{A}^3\mathbf{B} \mid \mathbf{A}^4\mathbf{B} \mid \mathbf{A}^5\mathbf{B} \mid \mathbf{A}^6\mathbf{B}]$$

$$= \left[\begin{array}{c|c|c|c|c|c|c} \mathbf{B}_1 & \mathbf{J}_1\mathbf{B}_1 & \mathbf{J}_1^2\mathbf{B}_1 & \mathbf{J}_1^3\mathbf{B}_1 & \mathbf{J}_1^4\mathbf{B}_1 & \mathbf{J}_1^5\mathbf{B}_1 & \mathbf{J}_1^6\mathbf{B}_1 \\ \hline \mathbf{B}_2 & \mathbf{J}_2\mathbf{B}_2 & \mathbf{J}_2^2\mathbf{B}_2 & \mathbf{J}_2^3\mathbf{B}_2 & \mathbf{J}_2^4\mathbf{B}_2 & \mathbf{J}_2^5\mathbf{B}_2 & \mathbf{J}_2^6\mathbf{B}_2 \\ \hline \mathbf{B}_3 & \mathbf{J}_3\mathbf{B}_3 & \mathbf{J}_3^2\mathbf{B}_3 & \mathbf{J}_3^3\mathbf{B}_3 & \mathbf{J}_3^4\mathbf{B}_3 & \mathbf{J}_3^5\mathbf{B}_3 & \mathbf{J}_3^6\mathbf{B}_3 \end{array}\right]$$

The results of Problem 8.4, page 211, are used for the various powers $\mathbf{J}_i^k$. Then

$$\mathbf{J}_1^k\mathbf{B}_1 = \begin{bmatrix} \lambda_1^k\mathbf{b}_{11}^T + k\lambda_1^{k-1}\mathbf{b}_{12}^T + \frac{1}{2}k(k-1)\lambda_1^{k-2}\mathbf{b}_{13}^T \\ \lambda_1^k\mathbf{b}_{12}^T + k\lambda_1^{k-1}\mathbf{b}_{13}^T \\ \lambda_1^k\mathbf{b}_{13}^T \end{bmatrix}$$

$$\mathbf{J}_2^k\mathbf{B}_2 = \begin{bmatrix} \lambda_1^k\mathbf{b}_{21}^T + k\lambda_1^{k-1}\mathbf{b}_{22}^T \\ \lambda_1^k\mathbf{b}_{22}^T \end{bmatrix}$$

$$\mathbf{J}_3^k\mathbf{B}_3 = \begin{bmatrix} \lambda_3^k\mathbf{b}_{31}^T + k\lambda_3^{k-1}\mathbf{b}_{32}^T \\ \lambda_3^k\mathbf{b}_{32}^T \end{bmatrix}$$

At this point the necessity of condition (2) is obvious since, for example, if $\mathbf{b}_{32} = \mathbf{0}$, the entire seventh row of $\mathbf{P}$ would be zero and rank $\mathbf{P} < n$. To see the necessity of condition (1), let $\mathbf{b}_{22} = \alpha\mathbf{b}_{13}$. Then an elementary row operation ($-\alpha$ times row 3 added to row 5) would make row 5 zero. This would again give rank $\mathbf{P} < n$, so the system would be uncontrollable.

It is tedious but trivial to show that a sequence of elementary column operations can be used to reduce $\mathbf{P}$ to $\mathbf{P}'$. Specifically, subtract λ_1 times each of the first r columns from the corresponding column in the second group of r columns. Then subtract λ_1^2 times column 1 and $2\lambda_1$ times the modified $(r+1)$th column from the $(2r+1)$th column and so on. Continuing this process leads to $\mathbf{P}'$:

$$\mathbf{P}' = \left[\begin{array}{c:c:c:c} \mathbf{b}_{11}^T & \mathbf{b}_{12}^T & \mathbf{b}_{13}^T & \\ \mathbf{b}_{12}^T & \mathbf{b}_{13}^T & \mathbf{0} & \cdots \\ \mathbf{b}_{13}^T & \mathbf{0} & \mathbf{0} & \\ \hdashline \mathbf{b}_{21}^T & \mathbf{b}_{22}^T & \mathbf{0} & \cdots \\ \mathbf{b}_{22}^T & \mathbf{0} & \mathbf{0} & \\ \hdashline \mathbf{b}_{31}^T & (\lambda_3-\lambda_1)\mathbf{b}_{31}^T + \mathbf{b}_{32}^T & (\lambda_3^2-\lambda_1^2)\mathbf{b}_{31}^T + 2(\lambda_3-\lambda_1)\mathbf{b}_{32}^T - 2\lambda_1(\lambda_3-\lambda_1)\mathbf{b}_{31}^T & \cdots \\ \mathbf{b}_{32}^T & (\lambda_3-\lambda_1)\mathbf{b}_{32}^T & (\lambda_3^2-\lambda_1^2)\mathbf{b}_{32}^T - 2\lambda_1(\lambda_3-\lambda_1)\mathbf{b}_{32}^T & \end{array}\right]$$

The last two rows of $\mathbf{P}'$ should be recognized as consisting of $\mathbf{B}_3$, $(\mathbf{J}_3 - \mathbf{I}\lambda_1)\mathbf{B}_3$, $(\mathbf{J}_3 - \mathbf{I}\lambda_1)^2\mathbf{B}_3$, $(\mathbf{J}_3 - \mathbf{I}\lambda_1)^3\mathbf{B}_3, \ldots$. Finally, a series of elementary row operations (row interchanges) gives

$$\mathbf{P}'' = \left[\begin{array}{c:c:c:c:c:c} \mathbf{b}_{13}^T & & & & & \\ \mathbf{b}_{22}^T & & & \mathbf{0} & & \\ \hdashline \mathbf{b}_{12}^T & \mathbf{b}_{13}^T & & & & \\ \mathbf{b}_{21}^T & \mathbf{b}_{22}^T & & & & \\ \hdashline \mathbf{b}_{11}^T & \mathbf{b}_{12}^T & \mathbf{b}_{13}^T & & & \\ \hdashline \mathbf{B}_3 & (\mathbf{J}_2 - \mathbf{I}\lambda_1)\mathbf{B}_3 & (\mathbf{J}_3 - \mathbf{I}\lambda_1)^2\mathbf{B}_3 & (\mathbf{J}_3 - \mathbf{I}\lambda_1)^3\mathbf{B}_3 & (\mathbf{J}_3 - \mathbf{I}\lambda_1)^4\mathbf{B}_3 & \cdots \end{array}\right]$$

By definition, the rank of $\begin{bmatrix} \mathbf{b}_{13}^T \\ \mathbf{b}_{22}^T \end{bmatrix}$ is 2 if and only if $\mathbf{b}_{13}$ and $\mathbf{b}_{22}$ are linearly independent. If they are, then a 2×2 nonzero determinant can be formed by deleting columns. Linear independence requires $\mathbf{b}_{13} \neq \mathbf{0}$. It is easy to show that if any one element of $\mathbf{b}_{32}$ is nonzero, then a nonzero 2×2 determinant can be obtained from $[(\mathbf{J}_3 - \mathbf{I}\lambda_1)^3\mathbf{B}_3 \mid (\mathbf{J}_3 - \mathbf{I}\lambda_1)^4\mathbf{B}_3]$. Thus an $n \times n$ lower block triangular matrix can be formed from $\mathbf{P}''$, whose determinant is nonzero. Rank $\mathbf{P}''$ = rank $\mathbf{P} = n$ implies complete controllability. The procedure used for this example quickly becomes unwieldy, but the result generalizes to any number of Jordan blocks [18].

15.11 If the output equation for the system in Problem 15.10 is

$$\mathbf{y} = [\mathbf{C}_1 \mid \mathbf{C}_2 \mid \cdots \mid \mathbf{C}_p]\mathbf{x} + \mathbf{D}\mathbf{u}$$

show that observability depends only on the first columns $\mathbf{c}_{i1}$ of each $\mathbf{C}_i$ block. In particular, show that the system is completely observable if and only if

1. $\{\mathbf{c}_{i1}, \mathbf{c}_{j1}, \ldots, \mathbf{c}_{k1}\}$ is a linearly independent set if $\mathbf{J}_i, \mathbf{J}_j, \ldots, \mathbf{J}_k$ are Jordan blocks *with the same eigenvalue* λ_i, and
2. $\mathbf{c}_{p1} \neq \mathbf{0}$ if $\mathbf{J}_p$ is the *only* Jordan block with eigenvalue λ_p.

Employ the same seventh-order system of Problem 15.10 for simplicity, and use $\mathbf{C} = [\mathbf{c}_{11} \quad \mathbf{c}_{12} \quad \mathbf{c}_{13} \mid \mathbf{c}_{21} \quad \mathbf{c}_{22} \mid \mathbf{c}_{31} \quad \mathbf{c}_{32}]$. The observability matrix is

$$\mathbf{Q} = [\bar{\mathbf{C}}^T \mid \bar{\mathbf{A}}^T\bar{\mathbf{C}}^T \mid (\bar{\mathbf{A}}^T)^2\bar{\mathbf{C}}^T \mid \cdots \mid (\bar{\mathbf{A}}^T)^6\bar{\mathbf{C}}^T]$$

Using the same kind of elementary column and row operations as in Problem 15.10, $\mathbf{Q}$ can be reduced to $\mathbf{Q}''$, a matrix of a similar form to $\mathbf{P}''$. Because now $\bar{\mathbf{A}}^T$ is used instead of $\mathbf{A}$, the *first* rows of $\bar{\mathbf{C}}_i^T$ play the role of the *last* rows of $\mathbf{B}_i$. Thus the conclusions regarding the leading columns of $\mathbf{C}_i$ follow directly from the results of Problem 15.10.

Relation Between Irreducibility and Controllability, Observability

15.12 Show that an irreducible realization of $\mathbf{H}(s)$ must be completely controllable and observable. Show that a completely controllable and observable realization cannot be reduced. Assume that $\mathbf{D} = [0]$.

Every $m \times r$ transfer matrix can be expanded into an infinite series $\mathbf{H}(s) = \mathbf{H}_1/s + \mathbf{H}_2/s^2 + \mathbf{H}_3/s^3 + \cdots$, where $\mathbf{H}_i$ are $m \times r$ constant matrices. Every realization satisfies $\mathbf{H}(s) = \mathbf{C}\mathbf{\Phi}(s)\mathbf{B}$, and $\mathbf{\Phi}(s) = \mathcal{L}\{e^{\mathbf{A}t}\} = \mathbf{I}/s + \mathbf{A}/s^2 + \mathbf{A}^2/s^3 + \cdots$. Therefore, for every realization, $\mathbf{H}_i = \mathbf{C}\mathbf{A}^{i-1}\mathbf{B}$.

Let n be the smallest integer such that $\mathbf{H}_j$ can be written as a combination of $\mathbf{H}_1$ through $\mathbf{H}_n$, for $j > n$. Then $\mathbf{H}(s) = \beta_1(s)\mathbf{H}_1 + \beta_2(s)\mathbf{H}_2 + \cdots + \beta_n(s)\mathbf{H}_n$. If dim $(\Sigma) = n$ so that $\mathbf{A}$ is $n \times n$, the Cayley-Hamilton theorem gives $\mathbf{\Phi}(s) = \alpha_0\mathbf{I} + \alpha_1\mathbf{A} + \cdots + \alpha_{n-1}\mathbf{A}^{n-1}$. $\mathbf{A}$ cannot be less than $n \times n$, otherwise $\mathbf{A}^{n-1}$ could be expressed as a linear combination of lower powers of $\mathbf{A}$. This implies that $\mathbf{H}_n$ can be expressed as a combination of $\mathbf{H}_1$ through $\mathbf{H}_{n-1}$ and contradicts the manner in which n was chosen. Thus minimum dim $(\Sigma) = n$.

Suppose $\mathbf{A}$ is $p \times p$, with $p > n$. Then

$$\mathbf{H}(s) = \beta_1\mathbf{C}\mathbf{B} + \beta_2\mathbf{C}\mathbf{A}\mathbf{B} + \cdots + \beta_n\mathbf{C}\mathbf{A}^{n-1}\mathbf{B}$$

must equal

$$\mathbf{H}(s) = \mathbf{C}[\alpha_0\mathbf{I} + \alpha_1\mathbf{A} + \cdots + \alpha_{p-1}\mathbf{A}^{p-1}]\mathbf{B}$$

This requires that for $i \geq n$, either $\mathbf{A}^i\mathbf{B}$ is a linear combination of $\mathbf{B}, \mathbf{A}\mathbf{B}, \ldots, \mathbf{A}^{n-1}\mathbf{B}$, or $\mathbf{C}\mathbf{A}^i$ is a linear combination of $\mathbf{C}, \mathbf{C}\mathbf{A}, \ldots, \mathbf{C}\mathbf{A}^{n-1}$. The first possibility means the system is uncontrollable. The second means the system is unobservable. A realization can have dim $(\Sigma) > n$ only if it is either not controllable or not observable. Conversely, if $\mathbf{A}$ is $n \times n$ and both controllable and observable, then $\mathbf{H}(s)$ will contain terms up to and including $\mathbf{H}_n$, and thus cannot be reduced.

15.13 Give a geometrical interpretation of the relationship between irreducible realizations of transfer matrices and controllability and observability.

Assume that $\mathbf{D} = [\mathbf{0}]$ since $\mathbf{D}$ is not influenced by the dimension of the state space Σ. Then $\mathbf{H}(s) = \mathbf{C}\mathbf{\Phi}(s)\mathbf{B}$. In the time domain the mapping from the input space $\mathcal{U}$ to the output space $\mathcal{Y}$ is given by

$$\mathbf{y}(t) = \mathbf{C}\int_0^t e^{\mathbf{A}(t-\tau)}\mathbf{B}\mathbf{u}(\tau)\,d\tau$$

Define $\mathcal{A}_1 : \mathcal{U} \longrightarrow \Sigma$ by $\mathbf{x} = \int_0^t e^{-\mathbf{A}\tau}\mathbf{B}\mathbf{u}(\tau)\,d\tau$ and $\mathcal{A}_2 : \Sigma \longrightarrow \mathcal{Y}$ by $\mathbf{y} = \mathbf{C}e^{\mathbf{A}t}\mathbf{x}$. Then $\mathbf{y} = \mathcal{A}_2(\mathcal{A}_1(\mathbf{u}))$. The state space can be decomposed into $\Sigma = \mathfrak{R}(\mathcal{A}_1) \oplus \mathfrak{N}(\mathcal{A}_1^*)$ or alternatively into $\Sigma = \mathfrak{R}(\mathcal{A}_2^*) \oplus \mathfrak{N}(\mathcal{A}_2)$. Thus $\dim(\Sigma) = \dim(\mathfrak{R}(\mathcal{A}_1) + \dim(\mathfrak{N}(\mathcal{A}_1^*) = \dim(\mathfrak{R}(\mathcal{A}_2^*)) + \dim(\mathfrak{N}(\mathcal{A}_2))$. No matter which input $\mathbf{u}(t)$ is used, $\mathcal{A}_1(\mathbf{u}) \in \mathfrak{R}(\mathcal{A}_1)$, so the component of $\mathbf{x} \in \mathfrak{N}(\mathcal{A}_1^*)$ will be zero. Since $\mathcal{A}_2(\mathbf{0}) = \mathbf{0}$, no part of the output $\mathbf{y}$ due to any input $\mathbf{u}$ will depend upon states in $\mathfrak{N}(\mathcal{A}_1^*)$. Thus $\dim(\Sigma)$ can be reduced without affecting the input-output relationship by requiring $\dim(\mathfrak{N}(\mathcal{A}_1^*)) = 0$, that is, $\mathfrak{N}(\mathcal{A}_1^*) = \{\mathbf{0}\}$. This is the condition for complete controllability. Likewise, any state $\mathbf{x} \in \mathfrak{N}(\mathcal{A}_2)$ contributes nothing to the output $\mathbf{y}$.

Making $\dim(\mathfrak{N}(\mathcal{A}_2) = 0$, that is, $\mathfrak{N}(\mathcal{A}_2) = \{\mathbf{0}\}$, reduces the dimension of Σ without affecting the input-output relationship. But $\mathfrak{N}(\mathcal{A}_2) = \{\mathbf{0}\}$ is the condition for complete observability. If the realization is completely controllable and observable, then $\dim(\Sigma) = \dim(\mathfrak{R}(\mathcal{A}_1)) = \dim(\mathfrak{R}(\mathcal{A}_2^*)) = n$, the dimension of the irreducible realization. A suggestive sketch is given in Fig. 15.17.

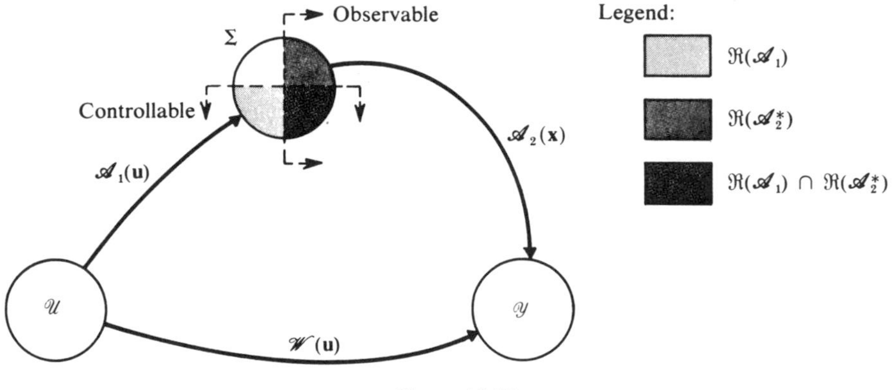

Figure 15.17

15.14 For a scalar transfer function, the order of the irreducible realization is the order of the denominator *after* all common poles and zeros are cancelled. Chen [17] defines the "denominator" of a rational transfer matrix as the lowest common denominator of all minors of all orders, after all possible cancellations have been made. Using this definition, the order of the irreducible realization is again equal to the order of the "denominator." Use this to verify that the irreducible realization for $\mathbf{H}(s)$ of Example 15.2 has $n = 4$ as found in Example 15.7.

$\mathbf{H}(s) = \begin{bmatrix} \dfrac{1}{s+1} & \dfrac{2}{(s+1)(s+2)} \\ \dfrac{1}{(s+1)(s+3)} & \dfrac{1}{s+3} \end{bmatrix}$. The first-order minors are just the H_{ij} elements, and the lowest common denominator of these four terms is $(s+1) \times (s+2)(s+3)$. There is just one second-order minor in this problem, namely

$$|\mathbf{H}(s)| = \frac{1}{(s+1)(s+3)} - \frac{2}{(s+1)^2(s+2)(s+3)} = \frac{s(s+3)}{(s+1)^2(s+2)(s+3)}$$

The lowest common denominator of all first- and second-order minors is $(s+1)^2 \times (s+2)(s+3)$. It has the order $n = 4$. Furthermore, the poles at $s = -1, -1, -2$, and -3 are the eigenvalues of the 4×4 irreducible $\mathbf{A}$ matrix of Example 15.7.

PROBLEMS

15.15 Find the state equations and the input-output transfer function for the circuit of Fig. 15.18.

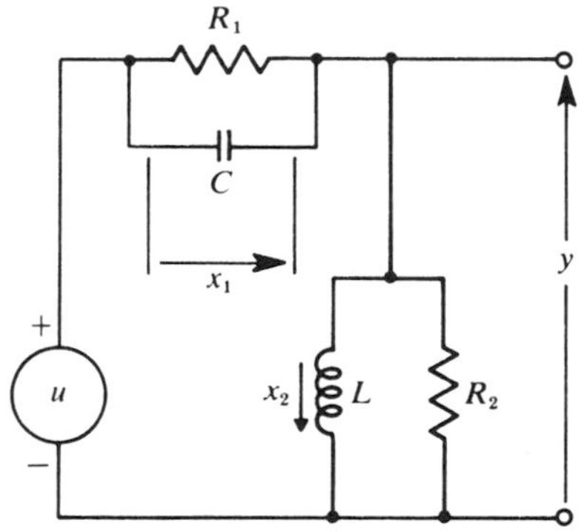

Figure 15.18

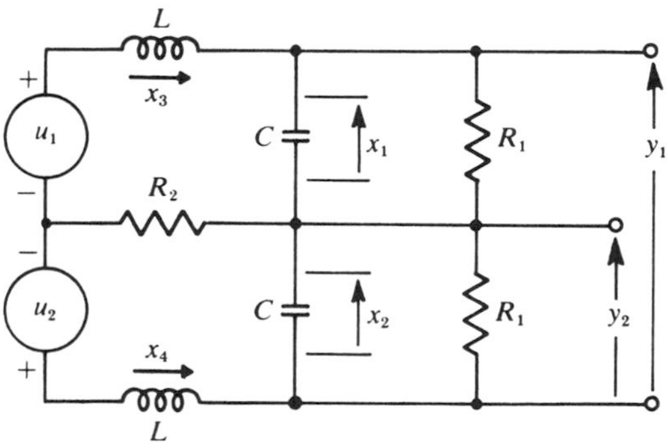

Figure 15.19

15.16 Find the state equations for the circuit of Fig. 15.19.

15.17 Find the transfer matrix for the system given in Problem 12.17, page 320.

15.18 Interchange the current source and voltage source of Fig. 15.12, and repeat Problem 15.2.

15.19 Find an irreducible realization for $\mathbf{H}(s) = \begin{bmatrix} \dfrac{1}{(s+1)} & \dfrac{1}{(s+1)(s+2)} & \dfrac{s}{(s+1)} \\ 0 & \dfrac{1}{(s+1)(s+3)} & \dfrac{(s+1)}{(s+3)} \end{bmatrix}$.

15.20 Which of the following system realizations $\dot{\mathbf{x}} = \mathbf{Ax} + \mathbf{Bu}$, $\mathbf{y} = \mathbf{Cx} + \mathbf{Du}$ are irreducible?

a. $\mathbf{A} = \begin{bmatrix} -6 & 1 & 0 & 0 \\ 0 & -6 & 0 & 0 \\ 0 & 0 & -6 & 0 \\ 0 & 0 & 0 & 6 \end{bmatrix}$, $\mathbf{B} = \begin{bmatrix} 1 & 1 & 1 \\ 1 & 1 & 1 \\ 2 & 2 & 2 \\ 0 & 1 & 0 \end{bmatrix}$

$\mathbf{C} = \begin{bmatrix} 3 & 1 & 4 & 0 \\ 0 & 1 & 1 & 1 \end{bmatrix}$, $\mathbf{D} = [0]$

b. Same $\mathbf{A}$ as in a, $\mathbf{B} = \begin{bmatrix} 0 & 0 & 0 \\ 1 & 1 & 1 \\ 2 & 0 & 2 \\ 0 & 1 & 0 \end{bmatrix}$, $\mathbf{C} = \begin{bmatrix} 3 & 1 & 4 & 0 \\ 0 & 1 & 0 & 1 \end{bmatrix}$, $\mathbf{D} = [0]$.

c. Same $\mathbf{A}$ as in a, same $\mathbf{B}$ as in b, same $\mathbf{C}$ as in a, $\mathbf{D} = [0]$.

15.21 Find an irreducible realization of $\mathbf{H}(s) = \begin{bmatrix} \dfrac{s}{s+1} & \dfrac{1}{(s+1)(s+2)} & \dfrac{1}{s+3} \\ \dfrac{-1}{s+1} & \dfrac{1}{(s+1)(s+2)} & \dfrac{1}{s} \end{bmatrix}$. See page 220 of Reference 17.

15.22 Find an irreducible realization of $\mathbf{H}(s) = \begin{bmatrix} \dfrac{-s}{(s+1)^2} & \dfrac{1}{s+1} \\ \dfrac{2s+1}{s(s+1)} & \dfrac{1}{s+1} \end{bmatrix}$.

15.23 Find an irreducible realization of

$$\mathbf{H}(s) = \frac{1}{s^4}\begin{bmatrix} s^3 - s^2 + 1 & 1 & -s^3 + s^2 - 2 \\ 1.5s + 1 & s + 1 & -1.5s - 2 \\ s^3 - 9s^2 - s + 1 & -s^2 + 1 & s^3 - s - 2 \end{bmatrix}$$

See pages 244 and 249 of Reference 17.

15.24 A discrete-time system's matrices are

$\mathbf{A} = \begin{bmatrix} 1 & 0 & 0 \\ 0 & 0.5 & 0.8 \\ 0 & -0.8 & 0.5 \end{bmatrix}$, $\mathbf{B} = \begin{bmatrix} 1 & 0 \\ 0 & 1 \\ 1 & 1 \end{bmatrix}$

$\mathbf{C} = \begin{bmatrix} 0.5 & -1 & 0 \\ 0.5 & 1 & 1 \end{bmatrix}$, $\mathbf{D} = [0]$

Find the input-output transfer function matrix $\mathbf{T}(z)$.

15.25 Find a minimal realization for the following system. Be sure that the final system matrices **A**, **B**, **C**, and **D** are all real so that they can be synthesized with real hardware.

$$\mathbf{T}(z) = \begin{bmatrix} \dfrac{z}{(z^2 - z + 0.5)(z-1)(z-0.3679)} \\ \dfrac{(z+0.2)}{(z-1)(z-0.3679)} \end{bmatrix}$$

Hint: Use partial fraction expansion on all real poles, but leave the complex conjugate pair in the form of a second-order segment.

Chapter

16

DESIGN OF LINEAR FEEDBACK CONTROL SYSTEMS

16.1 INTRODUCTION

A discussion of open-loop versus closed-loop control was presented in Chapter 1. Important advantages of feedback were discussed at that time. Chapter 2 presented a review of the analysis and design of single-input, single-output feedback systems using classical control techniques. It will be recalled that a fundamental method of classical design consists of forcing the dominant closed-loop poles to be suitably located in the s-plane or the Z-plane. Just what constitutes a suitable location depends upon the design specifications regarding relative stability, response times, accuracy, and so on.

This chapter considers the design of feedback compensators for linear, constant coefficient multivariable systems. One of the fundamental design objectives is, again, the achievement of suitable pole locations in order to ensure satisfactory transient response. This problem is analyzed, first under the assumption that *all* state variables can be used in forming feedback signals. Output feedback, i.e., incomplete state feedback, is also considered.

An additional design objective which cannot arise in single-input, single-output systems is the achievement of a decoupled or noninteracting system. This means that each input component affects just one output component, or possibly some prescribed subset of output components.

16.2 STATE FEEDBACK AND OUTPUT FEEDBACK

The state equations that have been discussed extensively in the past seven chapters will be considered to constitute the open-loop system. Most of what is to be discussed in this chapter applies equally well to continuous-time systems

$$\begin{aligned} \dot{\mathbf{x}} &= \mathbf{A}\mathbf{x} + \mathbf{B}\mathbf{u} \\ \mathbf{y} &= \mathbf{C}\mathbf{x} + \mathbf{D}\mathbf{u} \end{aligned} \tag{16.1}$$

or to discrete-time systems

$$\begin{aligned} \mathbf{x}(k+1) &= \mathbf{A}\mathbf{x}(k) + \mathbf{B}\mathbf{u}(k) \\ \mathbf{y}(k) &= \mathbf{C}\mathbf{x}(k) + \mathbf{D}\mathbf{u}(k) \end{aligned} \tag{16.2}$$

The system matrices $\{\mathbf{A}, \mathbf{B}, \mathbf{C}, \mathbf{D}\}$ have different meanings in the two cases, and of course the locations of "good" poles will differ between the s-plane and the Z-plane. However, if s_1 is a good pole location in the s-plane, then its image $z_1 = \exp(s_1 T)$ will inherit the same good features in the Z-plane. In any event, the methods and procedures to be developed will look the same for both types of systems in terms of the four system matrices.

It is assumed that these matrices are specified and cannot be altered by the designer to improve performance. It will be consistently assumed that the state vector $\mathbf{x}$ is $n \times 1$, the input vector $\mathbf{u}$ is $r \times 1$, and the output or measurement vector $\mathbf{y}$ is $m \times 1$. Thus $\mathbf{A}$ is $n \times n$, $\mathbf{B}$ is $n \times r$, $\mathbf{C}$ is $m \times n$, and $\mathbf{D}$ is $m \times r$. If the system performance is to be altered, it must be accomplished by some form of signal manipulation outside of the given open-loop system. Two commonly used possibilities are shown in Figs. 16.1 and 16.2 for continuous-time systems. These arrangements are referred to as state variable feedback and output feedback, respectively.

The feedback gain matrices $\mathbf{K}$ and $\mathbf{K}'$ are $r \times n$ and $r \times m$ respectively and are assumed constant. The inputs $\mathbf{v}$ and $\mathbf{v}'$ are assumed to be $l \times 1$ vectors. Thus the feed-forward matrices $\mathbf{F}$ and $\mathbf{F}'$ are also assumed to be constant and are of dimension $r \times l$.

It could be justifiably argued that state variable feedback is only of academic interest because, by definition, the outputs are the only signals which are accessible. State variable feedback seems to violate our dictum about using only signals external to the given open-loop system. In spite of this objection, state feedback is considered for the following reasons:

1. The state $\mathbf{x}$ contains all pertinent information about the system. It is of interest to determine what can be accomplished by using feedback in this ideal limiting case.
2. There are instances for which the state variables are all measurable, i.e., outputs. This will be the case if $\mathbf{C} = \mathbf{I}_n$ and $\mathbf{D} = [\mathbf{0}]$.
3. Several optimal control laws (see Chapter 17) take the form of a state feedback control. Anticipating this result, it is worthwhile to have an understanding of the effects of state feedback.

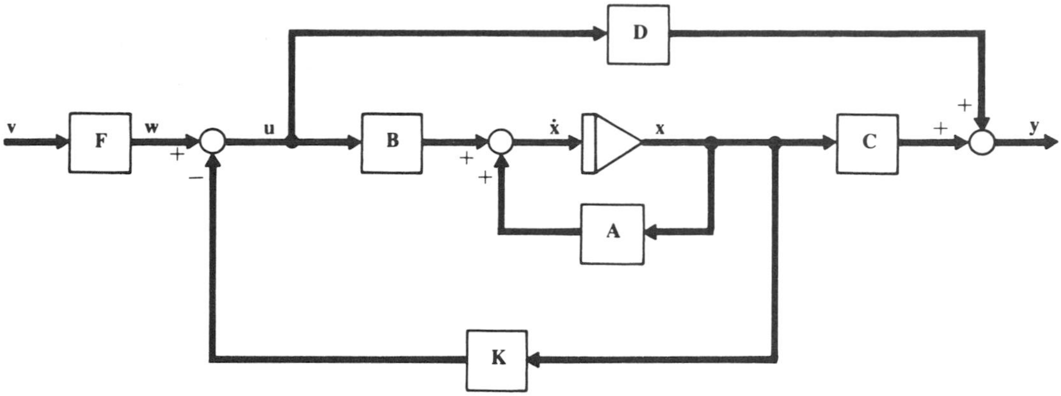

Figure 16.1 State variable feedback system.

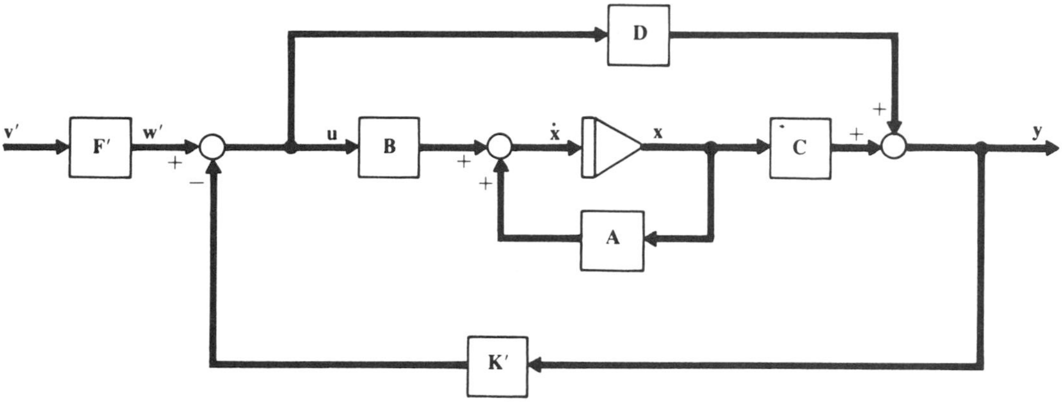

Figure 16.2 Output feedback system.

4. There are effective means available for estimating or reconstructing the state variables from the available inputs and outputs (see Section. 16.7).

The equations which describe the state feedback problem are equation (*16.1*) or (*16.2*), plus the relation

$$\mathbf{u}(t) = \mathbf{Fv}(t) - \mathbf{Kx}(t) \quad \text{or} \quad \mathbf{u}(k) = \mathbf{Fv}(k) - \mathbf{Kx}(k) \tag{16.3}$$

Combining gives

$$\dot{\mathbf{x}} = [\mathbf{A} - \mathbf{BK}]\mathbf{x} + [\mathbf{BF}]\mathbf{v} \quad \text{or} \quad \mathbf{x}(k+1) = [\mathbf{A} - \mathbf{BK}]\mathbf{x}(k) + [\mathbf{BF}]\mathbf{v}(k) \tag{16.4}$$

and

$$\mathbf{y} = [\mathbf{C} - \mathbf{DK}]\mathbf{x} + [\mathbf{DF}]\mathbf{v} \quad \text{or} \quad \mathbf{y}(k) = [\mathbf{C} - \mathbf{DK}]\mathbf{x}(k) + [\mathbf{DF}]\mathbf{v}(k) \tag{16.5}$$

Equations (*16.4*) and (*16.5*) are of the same form as equations (*16.1*) and (*16.2*). Considering $\{\mathbf{A}, \mathbf{B}, \mathbf{C}, \mathbf{D}\}$ as fixed system elements, the question is, "What changes in overall system characteristics can be achieved by choice of **K** and **F**?" *Stability* of the state feedback system depends on the eigenvalues of $[\mathbf{A} - \mathbf{BK}]$. *Controllability* depends on the pair $\{[\mathbf{A} - \mathbf{BK}], \mathbf{BF}\}$. *Observability* depends on the pair $\{[\mathbf{A} - \mathbf{BK}], [\mathbf{C} - \mathbf{DK}]\}$. The effect of feedback on these properties is investigated in Sec. 16.3.

The output feedback system is described by equation (*16.1*) or (*16.2*) plus $\mathbf{u}(t) = \mathbf{F}'\mathbf{v}' - \mathbf{K}'\mathbf{y}$. Therefore, for continuous-time systems

$$\mathbf{y}(t) = \mathbf{Cx} + \mathbf{DF}'\mathbf{v}' - \mathbf{DK}'\mathbf{y}$$

or

$$\mathbf{y}(t) = [\mathbf{I}_m + \mathbf{DK}']^{-1}\{\mathbf{Cx} + \mathbf{DF}'\mathbf{v}'\} \tag{16.6}$$

Using this gives $\dot{\mathbf{x}} = \{\mathbf{A} - \mathbf{BK}'[\mathbf{I}_m + \mathbf{DK}']^{-1}\mathbf{C}\}\mathbf{x} + \mathbf{B}\{\mathbf{F}' - \mathbf{K}'[\mathbf{I}_m + \mathbf{DK}']^{-1}\mathbf{DF}'\}\mathbf{v}'$. The matrix inversion identity $\mathbf{I}_r - \mathbf{K}'[\mathbf{I}_m + \mathbf{DK}']^{-1}\mathbf{D} \equiv [\mathbf{I}_r + \mathbf{K}'\mathbf{D}]^{-1}$ simplifies this to

$$\dot{\mathbf{x}} = \{\mathbf{A} - \mathbf{BK}'[\mathbf{I}_m + \mathbf{DK}']^{-1}\mathbf{C}\}\mathbf{x} + \mathbf{B}[\mathbf{I}_r + \mathbf{K}'\mathbf{D}]^{-1}\mathbf{F}'\mathbf{v}' \tag{16.7}$$

Again, equations (*16.7*) and (*16.6*) are of the same form as equation (*16.1*). The properties of stability, controllability, and observability are now determined by $\{\mathbf{A} - \mathbf{BK}'[\mathbf{I}_m + \mathbf{DK}']^{-1}\mathbf{C}\}$, $\mathbf{B}[\mathbf{I}_r + \mathbf{K}'\mathbf{D}]^{-1}\mathbf{F}'$, and $[\mathbf{I}_m + \mathbf{DK}']^{-1}\mathbf{C}$. Exactly the same kind of results applies to the discrete-time system.

16.3 THE EFFECT OF FEEDBACK ON SYSTEM PROPERTIES

The closed-loop systems, obtained by using either state feedback or output feedback, are described by four new system matrices. The reason for adding feedback is to impove the system characteristics in some sense. The effect of feedback on the properties of controllability, observability, and stability should be understood.

Controllability

Let **P** be the controllability matrix of Chapter 12. The open-loop system has $\mathbf{P} = [\mathbf{B} \mid \mathbf{AB} \mid \mathbf{A}^2\mathbf{B} \mid \cdots \mid \mathbf{A}^{n-1}\mathbf{B}]$. With state feedback the controllability matrix becomes

$$\tilde{\mathbf{P}} = [\mathbf{BF} \mid (\mathbf{A} - \mathbf{BK})\mathbf{BF} \mid (\mathbf{A} - \mathbf{BK})^2\mathbf{BF} \mid \cdots \mid (\mathbf{A} - \mathbf{BK})^{n-1}\mathbf{BF}]$$

If the feed-forward matrix **F** satisfies rank $(\mathbf{F}) = r$, then **F** will not affect the rank of $\tilde{\mathbf{P}}$. Physically, this means that there are as many independent input components **v** after adding feedback as there were in the input **u** without feedback. Assuming this is true, and letting $\mathbf{F} = \mathbf{I}_r$ for convenience, a series of elementary column operations can be used to reduce $\tilde{\mathbf{P}}$ to **P**. For example, the columns of **BKB** are linear combinations of the columns of $\mathbf{B}$. Elementary operations can reduce these, as well as all other extra terms in $\tilde{\mathbf{P}}$, to $\mathbf{0}$. Therefore, rank $(\tilde{\mathbf{P}})$ = rank $(\mathbf{P})$ for any gain matrix $\mathbf{K}$. Thus state feedback does not alter the controllability of the open-loop system.

Since $\tilde{\mathbf{P}}$ and $\mathbf{P}$ have the same rank for *any* $\mathbf{K}$, including the special case $\mathbf{K} = \mathbf{K}'[\mathbf{I}_m + \mathbf{D}\mathbf{K}']^{-1}\mathbf{C}$, system controllability is also unaltered when output feedback is used. This assumes that $\mathbf{I}_r + \mathbf{K}'\mathbf{D}$ is nonsingular and rank $(\mathbf{F}') = r$.

Observability

The open-loop observability matrix is $\mathbf{Q} = [\bar{\mathbf{C}}^T \mid \bar{\mathbf{A}}^T\bar{\mathbf{C}}^T \mid \cdots \mid (\bar{\mathbf{A}}^{n-1})^T\bar{\mathbf{C}}^T]$. When state feedback is used, observability is obviously lost if $\mathbf{C} = \mathbf{D}\mathbf{K}$. This set of simultaneous linear equations has a solution $\mathbf{K}$ if rank $[\mathbf{D}]$ = rank $[\mathbf{D} \mid \mathbf{C}]$. This is just one illustration of the general result: state feedback can cause a loss of observability. When output feedback is used, observability of the open- and closed-loop systems is the same.

Assume, without loss of generality, that $\mathbf{D} = [\mathbf{0}]$. Then the output feedback observability matrix is

$$\tilde{\mathbf{Q}} = [\bar{\mathbf{C}}^T \mid (\bar{\mathbf{A}} - \overline{\mathbf{B}\mathbf{K}'\mathbf{C}})^T\bar{\mathbf{C}}^T \mid \cdots \mid (\bar{\mathbf{A}} - \overline{\mathbf{B}\mathbf{K}'\mathbf{C}})^{n-1\,T}\bar{\mathbf{C}}^T]$$

A series of elementary column operations, precisely like those used on $\tilde{\mathbf{P}}$, can be used to reduce $\tilde{\mathbf{Q}}$ to $\mathbf{Q}$. This proves that rank $(\tilde{\mathbf{Q}})$ = rank $(\mathbf{Q})$. Observability is preserved when output feedback is used, regardless of $\mathbf{K}'$. The reason why system observability is invariant for output feedback and not for state feedback is the presence of the matrix $\mathbf{C}$ in the $\mathbf{A} - \mathbf{B}\mathbf{K}'\mathbf{C}$ terms. That is, columns of $\bar{\mathbf{C}}^T\bar{\mathbf{K}}'^T\bar{\mathbf{B}}^T\bar{\mathbf{C}}^T$ can be shown to be linearly related to columns of $\bar{\mathbf{C}}^T$, whereas columns of $\bar{\mathbf{K}}^T\bar{\mathbf{B}}^T\bar{\mathbf{C}}^T$ need not be.

Stability

Stability of linear systems depends entirely on the location of the eigenvalues in the complex plane. Both state and output feedback can alter the open-loop eigenvalues. The degree of freedom one has in specifying closed-loop eigenvalue locations by choice of $\mathbf{K}$ or $\mathbf{K}'$ is the crux of the pole assignment problems of the following sections.

16.4 POLE ASSIGNMENT USING STATE FEEDBACK [11]

Equation (*16.4*) indicates that the eigenvalues of the closed-loop state feedback system are roots of

$$\Delta'(\lambda) \triangleq |\lambda\mathbf{I} - \mathbf{A} + \mathbf{B}\mathbf{K}| = 0 \qquad (16.8)$$

It has been proven [25, 116] that if (and only if) the open-loop system $(\mathbf{A}, \mathbf{B})$ is completely controllable, then any set of desired closed-loop eigenvalues $\Gamma = \{\lambda_1, \lambda_2, \ldots, \lambda_n\}$ can be achieved using a constant state feedback matrix $\mathbf{K}$. In order to synthesize the system with real hardware, all elements of **K** must be real. This will be the case if, for each complex $\lambda_i \in \Gamma$, $\bar{\lambda}_i$ is also assigned to Γ.

A method of finding a matrix **K** which yields any specified set of eigenvalues Γ is now given [11]. It is applicable to any system described by equation (*16.1*) or (*16.2*). The method given for finding **K** constitutes a constructive proof that controllability is sufficient for arbitrary pole placement. Proof of the necessity may be found in the references.

The problem is to determine **K** such that equation (*16.8*) is satisfied for n specified values $\lambda_i \in \Gamma$. Equation (*16.8*) is rewritten as

$$\Delta'(\lambda) = |(\lambda\mathbf{I}_n - \mathbf{A})[\mathbf{I}_n + (\lambda\mathbf{I}_n - \mathbf{A})^{-1}\mathbf{BK}]| = |\lambda\mathbf{I}_n - \mathbf{A}| \cdot |\mathbf{I}_n + (\lambda\mathbf{I}_n - \mathbf{A})^{-1}\mathbf{BK}|$$

The open-loop characteristic polynomial is $|\lambda\mathbf{I}_n - \mathbf{A}| \triangleq \Delta(\lambda)$. Since $(\lambda\mathbf{I}_n - \mathbf{A})^{-1}$ is of the same form as the Laplace transform of the open-loop transition matrix, this term is denoted by $\mathbf{\Phi}(\lambda)$. The fact that the Z-transform of a discrete-time system transition matrix is

$$\mathbf{\Phi}(z) = z[\mathbf{I}z - \mathbf{A}]^{-1}$$

and thus has an extra z in it, should not be a matter of concern. The use of the symbol $\mathbf{\Phi}$ is really an arbitrary choice. Also, need for the existence of the inverse in the definition of $\mathbf{\Phi}$ is more apparent than real, as will be seen later. The open- and closed-loop characteristic polynomials are related by

$$\Delta'(\lambda) = \Delta(\lambda)|\mathbf{I}_n + \mathbf{\Phi}(\lambda)\mathbf{BK}| = \Delta(\lambda)|\mathbf{I}_r + \mathbf{K\Phi}(\lambda)\mathbf{B}| \tag{16.9}$$

The second form of equation (*16.9*) is obtained by using the determinant identity of Problems 3.5, page 83, and 7.18, page 192. The matrix **K** must be selected so that $\Delta'(\lambda_i) = 0$ for each $\lambda_i \in \Gamma$. This will be accomplished by forcing the $r \times r$ determinant to vanish. If any desired λ_i is also a root of $\Delta(\lambda)$, the following procedure is still valid (see Problem 16.2). A sufficient condition for the determinant of $\mathbf{I}_r + \mathbf{K\Phi}(\lambda_i)\mathbf{B}$ to be zero is if any one row or column is zero. More general conditions under which a determinant will be zero are well known, such as the existence of any linear dependencies among the rows or columns. Columns will be forced to zero here because of the simplicity that results. Define the jth column of $\mathbf{I}_r$ as $\mathbf{e}_j$ and define $\mathbf{\Psi}(\lambda_i) = \mathbf{\Phi}(\lambda_i)\mathbf{B}$,· with the jth column being $\boldsymbol{\psi}_j$. Then λ_i is a root of $\Delta'(\lambda)$ if **K** is selected to satisfy $\mathbf{e}_j + \mathbf{K}\boldsymbol{\psi}_j(\lambda_i) = \mathbf{0}$, since this forces column j to be zero. Thus

$$\mathbf{K}\boldsymbol{\psi}_j(\lambda_i) = -\mathbf{e}_j \tag{16.10}$$

This equation by itself is not sufficient for determining **K**. However, if an independent equation of this type can be found for every $\lambda_i \in \Gamma$, then **K** can be determined. Controllability of (**A**, **B**) is sufficient to guarantee that rank $\mathbf{\Psi}(\lambda_i) = r$ for each λ_i [17]. If all the desired λ_i are distinct, it will always be possible to find n linearly independent columns $\boldsymbol{\psi}_{j_1}(\lambda_1), \boldsymbol{\psi}_{j_2}(\lambda_2), \ldots, \boldsymbol{\psi}_{j_n}(\lambda_n)$ from the columns of the $n \times nr$ matrix

$[\Psi(\lambda_1) \quad \Psi(\lambda_2) \quad \cdots \quad \Psi(\lambda_n)]$. Then

$$\mathbf{K}[\psi_{j_1}(\lambda_1) \quad \psi_{j_2}(\lambda_2) \quad \cdots \quad \psi_{j_n}(\lambda_n)] = -[\mathbf{e}_{j_1} \quad \mathbf{e}_{j_2} \quad \cdots \quad \mathbf{e}_{j_n}]$$

or

$$\mathbf{K} = -[\mathbf{e}_{j_1} \quad \mathbf{e}_{j_2} \quad \cdots \quad \mathbf{e}_{j_n}][\psi_{j_1}(\lambda_1) \quad \psi_{j_2}(\lambda_2) \quad \cdots \quad \psi_{j_n}(\lambda_n)]^{-1} \tag{16.11}$$

Example 16.1

Let the system of equation (*16.1*) have $\mathbf{A} = \begin{bmatrix} 0 & 2 \\ 0 & 3 \end{bmatrix}$, $\mathbf{B} = \begin{bmatrix} 0 \\ 1 \end{bmatrix}$.

The open-loop system is unstable. Controllability is easily verified. Then

$$\Phi(\lambda) = \frac{\begin{bmatrix} \lambda - 3 & 2 \\ 0 & \lambda \end{bmatrix}}{\lambda(\lambda - 3)}, \qquad \Psi(\lambda) = \psi_1(\lambda) = \frac{\begin{bmatrix} 2 \\ \lambda \end{bmatrix}}{\lambda(\lambda - 3)}$$

If the desired poles are $\lambda_1 = -3$, $\lambda_2 = -4$, then $\psi_1(\lambda_1) = [1/9 \quad -1/6]^T$ and $\psi_1(\lambda_2) = [1/14 \quad -1/7]^T$ are linearly independent. Equation (*16.11*) gives $\mathbf{K} = -[1 \quad 1]\begin{bmatrix} 36 & 18 \\ -42 & -28 \end{bmatrix} = [6 \quad 10]$. This feedback gain matrix gives closed-loop eigenvalues at $\lambda = -3$ and -4. ■

Example 16.2

The system is modified so that $\mathbf{B} = \begin{bmatrix} 1 & 0 \\ 0 & 1 \end{bmatrix}$ and the desired roots are now $\lambda_1 = -3$, $\lambda_2 = -5$. Then $\psi_1(\lambda_1) = [-1/3 \quad 0]^T$, $\psi_1(\lambda_2) = [-1/5 \quad 0]^T$, $\psi_2(\lambda_1) = [1/9 \quad -1/6]^T$, and $\psi_2(\lambda_2) = [1/20 \quad -1/8]^T$.

Using $\psi_1(\lambda_1)$ and $\psi_2(\lambda_2)$ as the independent columns gives $\mathbf{K} = -\begin{bmatrix} 1 & 0 \\ 0 & 1 \end{bmatrix}\begin{bmatrix} -3 & -6/5 \\ 0 & -8 \end{bmatrix} = \begin{bmatrix} 3 & 6/5 \\ 0 & 8 \end{bmatrix}$. If, instead, $\psi_2(\lambda_1)$ and $\psi_1(\lambda_2)$ are used, then $\mathbf{K} = -\begin{bmatrix} 0 & 1 \\ 1 & 0 \end{bmatrix}\begin{bmatrix} 0 & -6 \\ -5 & -10/3 \end{bmatrix} = \begin{bmatrix} 5 & 10/3 \\ 0 & 6 \end{bmatrix}$.

It is easily verified that both of these solutions yield $\lambda_1 = -3$, $\lambda_2 = -5$. The gain matrix is not unique and the remaining freedom of choice may be useful in meeting other system specifications. ■

When repeated poles are desired, the previous technique may or may not yield n linearly independent $\psi_j(\lambda_i)$ columns. In Example 16.2 a double root could be specified, but in Example 16.1 only one ψ_j column exists. For repeated roots a modification is required. If λ_i is desired to have algebraic multiplicity m_i, then $\left.\frac{d^v\Delta'(\lambda)}{d\lambda^v}\right|_{\lambda=\lambda_i} = 0$ for $v = 1, 2, \ldots, m_i - 1$. Recalling the rule for differentiating a determinant, Sec. 3.11, page 80,

$$\begin{aligned} \frac{d\Delta'(\lambda)}{d\lambda} = \left\{\frac{d\Delta(\lambda)}{d\lambda}\right\}|\mathbf{I}_r + \mathbf{K}\Phi(\lambda)\mathbf{B}| &+ \left|\mathbf{K}\frac{d\psi_1}{d\lambda} \;\vdots\; \mathbf{e}_2 + \mathbf{K}\psi_2 \;\vdots\; \cdots \;\vdots\; \mathbf{e}_r + \mathbf{K}\psi_r \;\vdots\right|\Delta(\lambda) \\ &+ \left|\mathbf{e}_1 + \mathbf{K}\psi_1 \;\vdots\; \mathbf{K}\frac{d\psi_2}{d\lambda} \;\vdots\; \cdots \;\vdots\; \mathbf{e}_r + \mathbf{K}\psi_r\right|\Delta(\lambda) + \cdots \\ &+ \left|\mathbf{e}_1 + \mathbf{K}\psi_1 \;\vdots\; \mathbf{e}_2 + \mathbf{K}\psi_2 \;\vdots\; \cdots \;\vdots\; \mathbf{K}\frac{d\psi_r}{d\lambda}\right|\Delta(\lambda) \end{aligned} \tag{16.12}$$

Suppose $\mathbf{K}\boldsymbol{\psi}_j(\lambda_i) = -\mathbf{e}_j$. Then every term in equation (*16.12*) has one zero column, except the determinant containing $\mathbf{K}\, d\boldsymbol{\psi}_j/d\lambda$. Thus an additional independent equation for forcing $\Delta'(\lambda_i) = 0$ is

$$\mathbf{K}\frac{d\boldsymbol{\psi}_j}{d\lambda}\bigg|_{\lambda=\lambda_i} = \mathbf{0} \tag{16.13}$$

If necessary, higher-order derivatives can also be used. A total of n linearly independent columns, either $\boldsymbol{\psi}_{j_i}(\lambda_i)$ or derivatives, can always be found in this way. The procedure is analogous to finding n linearly independent eigenvectors and generalized eigenvectors from the adjoint and differentiated adjoint matrices (Sec. 7.4). When all n values of λ_i are the same, differentiation will always be required (assuming $r < n$). In general, a certain amount of a priori ambiguity will exist, just as it does in elementary treatments of generalized eigenvectors. The similarity to the eigenvector-generalized eigenvector problem is not just a coincidence, as will be seen in the next section.

Example 16.3

Find a **K** matrix which yields $\lambda_1 = \lambda_2 = -1$ for the system of Example 16.1.

Two equations are required to determine **K**, one like equation (*16.10*) and one like (*16.13*):

$$\frac{d\boldsymbol{\psi}_1}{d\lambda} = \frac{[-4\lambda + 6 \quad -\lambda^2]^T}{(\lambda^2 - 3\lambda)^2}$$

so that

$$\mathbf{K} = -[1 \quad 0]\left[\boldsymbol{\psi}_1(-1) \quad \frac{d\boldsymbol{\psi}_1}{d\lambda}(-1)\right]^{-1} = [1/2 \quad 5]$$ ■

Let the $n \times n$ matrix formed from the columns of $\boldsymbol{\psi}_{j_i}(\lambda_i)$ (and possibly their derivatives) be denoted as **G**. Note that **G** must be nonsingular, and must have exactly one column associated with each $\lambda_i \in \Gamma$. Define $\mathfrak{J}$ as the $r \times n$ matrix whose columns are $\mathbf{e}_{j_i}$(or a zero column for each derivative column in **G**). Then n equations like (*16.10*) and (*16.13*) can be combined into

$$\mathbf{KG} = -\mathfrak{J} \tag{16.14}$$

The notation has been chosen purposefully to point out the similarity to classical single-input, output techniques such as root locus, which makes extensive use of the equation $KG = -1$. The state feedback gain matrix for specified pole placement is

$$\mathbf{K} = -\mathfrak{J}\mathbf{G}^{-1} \tag{16.15}$$

16.5 EIGENVECTORS AND STATE VARIABLE FEEDBACK

If equation (*16.8*) is true, then there exists at least one nonzero vector $\boldsymbol{\psi}_i$, such that

$$(\lambda_i\mathbf{I} - \mathbf{A} + \mathbf{BK})\boldsymbol{\psi}_i = \mathbf{0}$$

Rearranged, this says that

$$(\mathbf{A} - \mathbf{BK})\boldsymbol{\psi}_i = \lambda_i\boldsymbol{\psi}_i$$

This makes it clear that $\boldsymbol{\psi}_i$ is an eigenvector of the closed-loop system matrix $(\mathbf{A} - \mathbf{BK})$ associated with the closed-loop eigenvalue (pole) λ_i. Rewriting equation (*16.8*) in yet another way gives

$$(\lambda_i\mathbf{I} - \mathbf{A})\boldsymbol{\psi}_i = -\mathbf{BK}\boldsymbol{\psi}_i$$

or

$$[(\lambda_i\mathbf{I} - \mathbf{A}) \mid \mathbf{B}]\begin{bmatrix} \boldsymbol{\psi}_i \\ \mathbf{K}\boldsymbol{\psi}_i \end{bmatrix} = [\mathbf{0}] \tag{16.16}$$

If the system described by the pair $\{\mathbf{A}, \mathbf{B}\}$ is controllable, it is known that the coefficient matrix in the above homogeneous equation has rank n for any value of λ_i [25, 116]. If equation (*16.8*) is to be true for each specified eigenvalue λ_i in Γ, then equation (*16.16*) must also be true. The solution of homogeneous equations was discussed in Chapters 5 and 7. Let the maximal set of linearly independent solution vectors for a given eigenvalue be called $\mathbf{U}(\lambda_i)$.

The columns of $\mathbf{U}$ constitute a basis for the null space of $[(\lambda_i\mathbf{I} - \mathbf{A}) \mid \mathbf{B}]$. Note that the inverse of $\lambda_i\mathbf{I} - \mathbf{A}$ need not exist in order to find these solutions.

From equation (*16.16*) it is clear that the unknown vectors that have been solved for are partitioned into two parts. Therefore, the matrix $\mathbf{U}$ is partitioned accordingly as

$$\mathbf{U}(\lambda_i) = \begin{bmatrix} \boldsymbol{\psi}_1 & \boldsymbol{\psi}_2 & \cdots & \boldsymbol{\psi}_r \\ \hline \mathbf{f}_1 & \mathbf{f}_2 & \cdots & \mathbf{f}_r \end{bmatrix} = \begin{bmatrix} \boldsymbol{\Psi}(\lambda_i) \\ \hline \boldsymbol{\mathfrak{F}}(\lambda_i) \end{bmatrix}$$

where the substitution $\mathbf{f} = \mathbf{K}\boldsymbol{\psi}$ has been made. Collectively, all of these relations are

$$\mathbf{K}[\boldsymbol{\Psi}(\lambda_1) \quad \boldsymbol{\Psi}(\lambda_2) \quad \cdots \quad \boldsymbol{\Psi}(\lambda_n)] = [\boldsymbol{\mathfrak{F}}(\lambda_1) \quad \boldsymbol{\mathfrak{F}}(\lambda_2) \quad \cdots \quad \boldsymbol{\mathfrak{F}}(\lambda_n)] \tag{16.17}$$

Equation (*16.17*) cannot be solved directly for $\mathbf{K}$ because it is generally overdetermined and thus has inconsistent equations. However, if the system is controllable a nonsingular $n \times n$ matrix of $\boldsymbol{\psi}_i(\lambda_j)$'s can be found by selecting n linearly independent columns from both sides of equation (*16.17*) and deleting the remaining columns. In this selection process, one column must be selected for each specified eigenvalue λ_i. Let the n columns from the left-hand side be called $\mathbf{G}$ and the corresponding columns from the right side be called $\boldsymbol{\mathfrak{G}}$. If this is done, the result can be solved for the feedback gain matrix $\mathbf{K}$,

$$\mathbf{K} = \boldsymbol{\mathfrak{G}}\mathbf{G}^{-1} \tag{16.18}$$

Comparing this with equation (*16.15*) shows that the method of Sec. 16.4 is essentially the same. It is now clear that the vectors which were arbitrarily selected, subject only to the requirement of a nonsingular $\mathbf{G}$ matrix, are actually eigenvectors of the closed-loop system. The exception is in the cases where differentiation of $\boldsymbol{\psi}_i$ vectors was required. The matrix $\mathfrak{J}$ corresponds to $-\boldsymbol{\mathfrak{G}}$ of this section.

Knowing that $\boldsymbol{\psi}_i$ are closed-loop eigenvectors may help guide the choice of which columns to retain. Actually, any linear combination of columns from a given partition of $\boldsymbol{\Psi}$ can be used as long as the same linear combination of $\mathbf{f}$'s is used. It is because of this extra flexibility that the $\mathcal{G}$ matrix need not be made up of the simple $\mathbf{e}_i$ columns of $\mathfrak{J}$. It seems worthwhile to try to make **G** be as nearly orthogonal as possible because this improves its invertibility robustness and tends to give less interaction between modes of the closed-loop system. This usually improves system sensitivity.

This alternate view of pole placement is nicely suited to machine computation. It eliminates the need to find analytic inverses. (This really wasn't a requirement of the previous approach either, except in cases like Problem 16.2 where a limit has to be found [33, 34]). It also makes it clear when differentiation will be required. Actually, a method of avoiding differentiation will be developed. Suppose it is desired that λ_i be a pth order root of the characteristic equation. If the dimension of the null space, that is the rank of $\boldsymbol{\Psi}(\lambda_i)$, is p or larger, then the above procedure will suffice. If this rank is less than p, then it is not possible to specify p eigenvectors, but the desired p eigenvalues can still be achieved. The excess will be associated with generalized eigenvectors, that is

$$(\mathbf{A} - \mathbf{BK})\boldsymbol{\psi}_g = \lambda_i \boldsymbol{\psi}_g + \boldsymbol{\psi}_i$$

or *(16.19)*

$$[(\lambda_i \mathbf{I} - \mathbf{A}) \mid \mathbf{B}]\begin{bmatrix} \boldsymbol{\psi}_g \\ \mathbf{K}\boldsymbol{\psi}_g \end{bmatrix} = \boldsymbol{\psi}_i$$

This is the standard formulation of a generalized eigenvector problem and, as was shown in Chapter 7, one method of solution involves differentiation with respect to λ. That is the method used in Sec. 16.4 whenever it was not possible to find the required number of eigenvectors. As was shown in Chapter 7, there are ways of finding generalized eigenvectors which do not involve differentiation.

Example 16.4

Repeat Example 16.3, but without analytically finding $\boldsymbol{\psi}(\lambda)$ and differentiating it. Find all solutions of equation (*16.16*) which here specializes to

$$\begin{bmatrix} -1 & -2 & \mid & 0 \\ 0 & -4 & \mid & 1 \end{bmatrix}\begin{bmatrix} \boldsymbol{\psi} \\ \mathbf{K}\boldsymbol{\psi} \end{bmatrix} = \mathbf{0}$$

It is found that all nontrivial solutions are proportional to

$$[\boldsymbol{\psi}^T \mid (\mathbf{K}\boldsymbol{\psi})^T]^T = [0.5 \quad -0.25 \mid -1]^T$$

Obviously only one eigenvector $\boldsymbol{\psi} = [0.5 \quad -0.25]^T$ can be selected from this one-dimensional space, so a generalized eigenvector is needed. It is found from

$$\begin{bmatrix} -1 & -2 & \mid & 0 \\ 0 & -4 & \mid & 1 \end{bmatrix}\begin{bmatrix} \boldsymbol{\psi}_g \\ \mathbf{K}\boldsymbol{\psi}_g \end{bmatrix} = \begin{bmatrix} 0.5 \\ -0.25 \end{bmatrix}$$

A solution is $[-0.625 \quad 0.0625 \mid 0]^T$. Therefore, the two equations for finding **K** are

$$\mathbf{K}\begin{bmatrix} 0.5 & -0.625 \\ -0.25 & 0.0625 \end{bmatrix} = [-1 \quad 0]$$

so that matrix inversion or its equivalent gives $\mathbf{K} = [0.5 \quad 5]$. ■

Outline of Pole Placement Algorithm (Eigenvalue-Eigenvector Assignment)

Given **A**, **B**, and the desired set of eigenvalues Γ

I. *For each* $\lambda_i \in \Gamma$

1. Form $[(\lambda_i \mathbf{I} - \mathbf{A} \mid \mathbf{B}]$.
2. Find the null space basis set **U** by solving equation (*16.16*).
3. Partition **U**, using the top n rows as the $n \times r$ matrix $\mathbf{\Psi}(\lambda_i)$.
4. Use the remaining r rows of **U** as the $r \times r$ matrix $\mathfrak{F}(\lambda_i)$.

II. *Form the composite matrices*

$$\mathbf{\Omega} = [\mathbf{\Psi}(\lambda_1) \mid \mathbf{\Psi}(\lambda_2) \mid \cdots \mid \mathbf{\Psi}(\lambda_n)];\ n \times nr, \text{ rank } r \text{ if controllable}$$

and

$$\mathbf{\Lambda} = [\mathfrak{F}(\lambda_1) \mid \mathfrak{F}(\lambda_2) \mid \cdots \mid \mathfrak{F}(\lambda_n)];\ r \times nr$$

so that equation (*16.17*) can be compactly written as $\mathbf{K\Omega} = \mathbf{\Lambda}$.

III. *Select n linearly independent columns of* $\mathbf{\Omega}$ *to from the* $n \times n$ *matrix* **G**. One column (or any linear combinations of columns) must be selected from each $\mathbf{\Psi}(\lambda_i)$ partition. The selected columns will be closed-loop eigenvectors. As a preliminary screening out of linearly dependent columns, the inner products should be checked. This can be done by forming $\bar{\mathbf{\Omega}}^T\mathbf{\Omega}$. The i,j element, normalized by the square root of the i,i and j,j elements, is the cosine of the angle between columns i and j. If the result is 1 for any pair of selected columns, they are linearly dependent and the selection must be modified accordingly.

IV. *Use the same column numbers selected in step III to form the* $r \times n$ *matrix* $\mathfrak{G}$ *from* $\mathbf{\Lambda}$.

V. *Solve* $\mathbf{KG} = \mathfrak{G}$ *for the* $r \times n$ *gain matrix* **K**. It may be more convenient with some equation solver packages to solve $\mathbf{G}^T\mathbf{K}^T = \mathfrak{G}^T$ instead. Note that passing the pairwise inner product test in step III is necessary but not sufficient for the existence of $\mathbf{G}^{-1}$. Therefore, it may be necessary to return to step III and modify the eigenvector column selections.

VI. *General Comments*

1. A more sophisticated automatic column selection process could be incorporated. A modified Gram-Schmidt orthogonalization process could be used on $\mathbf{\Omega}$ to guarantee the selection of an independent set. However, making the selection interactive allows the user to interject judgment regarding the desirability of certain closed-loop eigenvectors.
2. If a certain eigenvalue is specified more times than allowed by eigenvector considerations, then one or more generalized eigenvectors will need to be found, using equation (*19.19*) in place of equation (*16.16*) in step I.2 above.

3. A version of this algorithm has been implemented for real eigenvalues, in Microsoft FORTRAN [83] running under CP/M [118]. Problems of up to $n = 20$, $nr = 40$ fit into a 64K memory system. Subroutine LINEQ [81] (with corrections) was used for step I.2. The algorithm outlined here will solve the complex eigenvalue case if a more powerful FORTRAN dialect having complex arithmetic capabilities is used.
4. The single-input problem ($r = 1$) can be solved by simpler methods which deal directly with the coefficients of the characteristic polynomial. A program which does this (and more) is STVARFDBK [81].

Example 16.5

A model [1] of the lateral dynamics of an F-8 aircraft at a particular set of flight condition was given in Problem 12.7. A discrete approximation for this system is obtained using the method of Problem 10.11, with $T = 0.2$ and retaining terms through thirtieth order.

The Approximate Discrete Transition Matrix Phi

$$\begin{bmatrix} 1.3533533E-01 & 9.8391518E-02 & -7.2121400E-01 & 0.0000000E+00 \\ 0.0000000E+00 & 7.1751231E-01 & 1.4726298E+00 & 0.0000000E+00 \\ 0.0000000E+00 & -1.6362552E-01 & 7.1751231E-01 & 0.0000000E+00 \\ 8.6466469E-02 & 7.8584068E-03 & -1.0389241E-01 & 1.0000000E+00 \end{bmatrix}$$

The Approximate Input Matrix **B1** = Gamma

$$\begin{bmatrix} 1.7293293E+00 & 2.1750930E-01 \\ 0.0000000E+00 & -5.5092329E-01 \\ 0.0000000E+00 & 5.5393357E-02 \\ 2.2706707E-01 & 3.0423844E-02 \end{bmatrix}$$

Design a state feedback controller which will provide closed-loop Z-plane poles corresponding to s-plane poles at $s = -2, -5, -8$, and -10. Using $z = e^{Ts}$, this means that the desired Z-plane poles are

$\lambda_i = 0.6703, 0.3679, 0.2019$, and 0.1353.

Solving equation (*16.16*) for each λ_i in turn gives

$$\mathbf{U}(\lambda_1) = \begin{bmatrix} -1.0000000E+00 & 0.0000000E+00 \\ 0.0000000E+00 & 4.4247019E-01 \\ 0.0000000E+00 & 3.5991812E-01 \\ 4.7545883E-01 & 1.0787997E-02 \\ \hdashline 3.0937374E-01 & 8.4780902E-04 \\ 0.0000000E+00 & -1.0000000E+00 \end{bmatrix}$$

$$\mathbf{U}(\lambda_2) = \begin{bmatrix} -1.0000000E+00 & 0.0000000E+00 \\ 0.0000000E+00 & 7.5502163E-01 \\ 0.0000000E+00 & 1.9485569E-01 \\ 1.8517032E-01 & 5.8909208E-03 \\ \hdashline 1.3450530E-01 & 8.7471344E-02 \\ 0.0000000E+00 & -1.0000000E+00 \end{bmatrix}$$

$$\mathbf{U}(\lambda_3) = \begin{bmatrix} -1.0000000\text{E}+00 & 0.0000000\text{E}+00 \\ 0.0000000\text{E}+00 & 7.2149646\text{E}-01 \\ 0.0000000\text{E}+00 & 1.2148339\text{E}-01 \\ 1.1934122\text{E}-01 & 3.6517035\text{E}-03 \\ \hdashline 3.8512699\text{E}-02 & 1.1616344\text{E}-01 \\ 0.0000000\text{E}+00 & -1.0000000\text{E}+00 \end{bmatrix}$$

$$\mathbf{U}(\lambda_4) = \begin{bmatrix} -1.0000000\text{E}+00 & 0.0000000\text{E}+00 \\ 0.0000000\text{E}+00 & 6.9379878\text{E}-01 \\ 0.0000000\text{E}+00 & 9.9803299\text{E}-02 \\ 1.0003469\text{E}-01 & 2.9885261\text{E}-03 \\ \hdashline 0.0000000\text{E}+00 & 1.2362902\text{E}-01 \\ 0.0000000\text{E}+00 & -1.0000000\text{E}+00 \end{bmatrix}$$

The portions above the partition lines are $\boldsymbol{\Psi}(\lambda_i)$ and the lower portions are $\mathfrak{F}(\lambda_i)$. In order to find a suitable gain, some combination of columns (1 or 2) and (3 or 4) and (5 or 6) and (7 or 8) must be selected. For example, if columns 1, 4, 6, and 7 are used, equation (*16.18*) gives

The Feedback Gain Matrix

$$\mathbf{K} = \begin{bmatrix} 8.2435042\text{E}-02 & 2.4582298\text{E}-01 & -5.2851838\text{E}-01 & 8.2406455\text{E}-01 \\ 0.0000000\text{E}+00 & -1.5015275\text{E}+00 & 6.8607545\text{E}-01 & 0.0000000\text{E}+00 \end{bmatrix}$$

As a check of the result, the closed-loop system matrix $\mathbf{A} - \mathbf{BK}$ is formed. The eigenvalues of this closed-loop system matrix are given below, and verify that the desired pole locations have been achieved.

The Resulting Closed-Loop Eigenvalues Are:

Real Part	Imaginary Part
1.3529983E − 01	0.0000000E + 00
2.0190017E − 01	0.0000000E + 00
3.6790001E − 01	0.0000000E + 00
6.7029989E − 01	0.0000000E + 00

If columns 2, 4, 5, and 7 are selected, the following alternative gain matrix is obtained. Notice that the two sets of results have totally different **K** matrices and closed-loop eigenvectors, although the closed-loop eigenvalues will be the same.

The Feedback Gain Matrix

$$\mathbf{K} = \begin{bmatrix} 1.9954935\text{E}-01 & 1.6860582\text{E}-01 & -2.6471341\text{E}-01 & 1.9948014\text{E}+00 \\ 0.0000000\text{E}+00 & -8.8968951\text{E}-01 & -1.6846579\text{E}+00 & 0.0000000\text{E}+00 \end{bmatrix} \quad \blacksquare$$

Before concluding the discussion of state variable feedback, assume that the gain matrix **K** is factored into $\mathbf{K} = \mathbf{F}_1\mathbf{K}_1$, and that $\mathbf{K}_1$ is left in the feedback path and $\mathbf{F}_1$ is placed in the forward path, as shown in Fig. 16.3. There is no change in the characteristic equation and hence the poles. The numerator of the closed-loop transfer function can differ greatly, however. If $\mathbf{F}_1$ is just a scalar, then only the gains of the

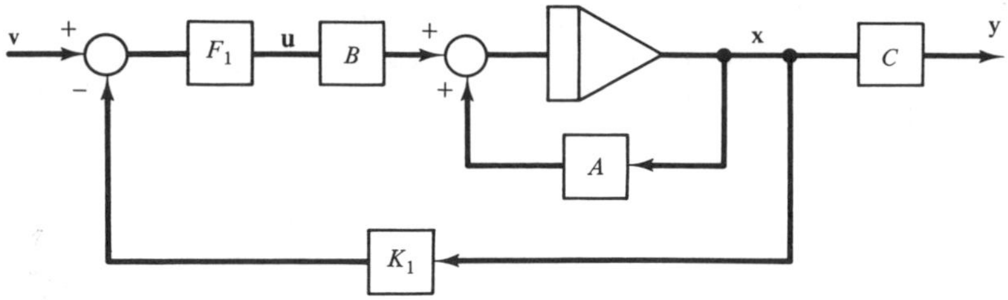

Figure 16.3

transfer function elements change. If $\mathbf{F}_1$ is a full $r \times r$ matrix, the zeros of the system can change as well.

Example 16.6

Show that if

$$\mathbf{K} = \begin{bmatrix} 2 & 0 \\ -4.5 & 9 \end{bmatrix}$$

is used in a state feedback controller for the second-order system of Example 15.6, the closed-loop poles are at -5 and -10. Then consider two optional implementations,

$$\mathbf{K} = \begin{bmatrix} 2 & 0 \\ 1 & 1 \end{bmatrix}\begin{bmatrix} 1 & 0 \\ -5.5 & 9 \end{bmatrix} = \begin{bmatrix} 1 & 1 \\ -1 & 1 \end{bmatrix}\begin{bmatrix} 3.25 & -4.5 \\ -1.25 & 4.5 \end{bmatrix}$$

and find the closed-loop transfer functions.

If the control law is $\mathbf{u} = -\mathbf{Kx} + \mathbf{F}_1\mathbf{v}$, the expression for the closed-loop transfer function is

$$\mathbf{H}(s) = \mathbf{C}[s\mathbf{I} - \mathbf{A} + \mathbf{BK}]^{-1}\mathbf{BF}_1 \quad \text{and} \quad \mathbf{A} - \mathbf{BK} = \begin{bmatrix} -5 & 0 \\ 2.5 & -10 \end{bmatrix}$$

For the three values of $\mathbf{F}_1$, namely $\mathbf{I}$, $\begin{bmatrix} 2 & 0 \\ 1 & 1 \end{bmatrix}$, and $\begin{bmatrix} 1 & 1 \\ -1 & 1 \end{bmatrix}$, the resultant transfer functions are

$$\mathbf{H}(s) = \begin{bmatrix} \dfrac{1}{s+5} & \dfrac{1}{s+10} \\ \dfrac{2}{s+5} & 0 \end{bmatrix}, \begin{bmatrix} \dfrac{3(s+\frac{25}{3})}{(s+5)(s+10)} & \dfrac{1}{s+10} \\ \dfrac{4}{s+5} & 0 \end{bmatrix} \text{ and } \begin{bmatrix} \dfrac{5}{(s+5)(s+10)} & \dfrac{2(s+\frac{15}{2})}{(s+5)(s+10)} \\ \dfrac{2}{s+5} & \dfrac{2}{s+5} \end{bmatrix}$$

■

16.6 POLE ASSIGNMENT USING OUTPUT FEEDBACK [11]

When output feedback is used, the closed-loop eigenvalues are roots of

$$|\lambda\mathbf{I}_n - \mathbf{A} + \mathbf{BK}'[\mathbf{I}_m + \mathbf{DK}']^{-1}\mathbf{C}| = 0 \tag{16.20}$$

Frequently it is assumed that either $\mathbf{D} = \mathbf{0}$ or the inputs $\mathbf{u}$ can be measured and

$\mathbf{y} - \mathbf{Du}$ can be used as the output. This effectively means that $\mathbf{D}$ could be assumed zero. Here $\mathbf{D}$ is left general, except that it is assumed $(\mathbf{I}_m + \mathbf{DK}')^{-1}$ exists.

Using manipulations like those in Sec. 16.4, equation (*16.20*) becomes

$$\Delta_0'(\lambda) = \Delta(\lambda)|\mathbf{I}_r + \mathbf{K}'[\mathbf{I}_m + \mathbf{DK}']^{-1}\mathbf{C}\boldsymbol{\Phi}(\lambda)\mathbf{B}| \tag{16.21}$$

where $\Delta_0'(\lambda)$ signifies the closed-loop characteristic polynomial with output feedback. The matrix $\mathbf{C}\boldsymbol{\Phi}(\lambda)\mathbf{B}$ plays a role in the concept of output controllability [17]. Define this matrix as $\boldsymbol{\Psi}'(\lambda)$. The number of linearly independent columns $\boldsymbol{\psi}_j'(\lambda_i)$ (and their derivatives if λ_i is repeated) that can be found from the $m \times nr$ matrix $[\boldsymbol{\Psi}'(\lambda_i) \quad \boldsymbol{\Psi}'(\lambda_2) \quad \cdots \quad \boldsymbol{\Psi}'(\lambda_n)]$ will never exceed rank $(\mathbf{C})$. It has been shown [26] that if $(\mathbf{A}, \mathbf{B})$ is completely controllable and if $\mathbf{C}$ has full rank m ($m \leq n$ is assumed), then m of the n eigenvalues of the closed-loop system can be arbitrarily specified to within any degree of accuracy (but not always exactly). Complex eigenvalues must be specified in conjugate pairs. Assume that an $m \times m$ nonsingular matrix can be found from columns $\boldsymbol{\psi}_{j_i}'(\lambda_i)$ (and possibly their derivatives). Call this matrix $\mathbf{G}'$. Then

$$\mathbf{K}'[\mathbf{I}_m + \mathbf{DK}']^{-1}\mathbf{G}' = -\mathfrak{F}' \tag{16.22}$$

where the $r \times m$ matrix $\mathfrak{F}'$ is defined as before, with columns formed from columns of $\mathbf{I}_r$ (or zero columns if derivatives of $\boldsymbol{\psi}_j$ are required). Then

$$\mathbf{K}'[\mathbf{I}_m + \mathbf{DK}']^{-1} = -\mathfrak{F}'(\mathbf{G}')^{-1}$$

so that

$$\mathbf{K}' - -\mathfrak{F}'(\mathbf{G}')^{-1}[\mathbf{I}_m + \mathbf{DK}']$$

or

$$\mathbf{K}' = -[\mathbf{I}_r + \mathfrak{F}'(\mathbf{G}')^{-1}\mathbf{D}]^{-1}\mathfrak{F}'(\mathbf{G}')^{-1} \tag{16.23}$$

Example 16.7

Specify an output feedback matrix K' so that the controllable system $\mathbf{A} = \begin{bmatrix} 0 & 1 \\ -3 & -4 \end{bmatrix}$, $\mathbf{B} = \begin{bmatrix} 1 & 0 \\ 0 & 1 \end{bmatrix}$, $\mathbf{C} = [1 \quad 1]$, $\mathbf{D} = [0]$ will have $\lambda_1 = -5$ as a closed-loop eigenvalue:

$$\boldsymbol{\psi}'(\lambda) = \frac{[1 \quad 1]}{\lambda + 3}$$

Since rank $(\mathbf{C}) = 1$ and $\boldsymbol{\psi}_1'(\lambda_1) = \boldsymbol{\psi}_2'(\lambda_1)$, there is only one independent column, $\mathbf{G}' = -1/2$. Using a column of $\mathbf{I}_2$ gives $\mathfrak{F}' = \begin{bmatrix} 1 \\ 0 \end{bmatrix}$, so that $\mathbf{K}' = -\mathfrak{F}'(\mathbf{G}')^{-1} = \begin{bmatrix} 2 \\ 0 \end{bmatrix}$. Using equation (*16.20*), it is easy to verify that this given $\Delta_0'(\lambda) = (\lambda + 1)(\lambda + 5)$. ■

Example 16.8

An unstable but completely controllable system is described by

$$\mathbf{A} = \begin{bmatrix} -2 & 1 & 0 \\ 0 & -2 & 0 \\ 0 & 0 & 4 \end{bmatrix}, \quad \mathbf{B} = \begin{bmatrix} 0 & 0 \\ 0 & 1 \\ 1 & 0 \end{bmatrix}, \quad \mathbf{C} = \begin{bmatrix} 0 & 0 & 1 \\ 1 & 0 & 0 \end{bmatrix}, \quad \mathbf{D} = \begin{bmatrix} 1 & 0 \\ 0 & 0 \end{bmatrix}$$

Find an output feedback matrix which gives closed-loop eigenvalues at $\lambda_1 = -5, \lambda_2 = -6$.

For this case, $\mathbf{\Psi}'(\lambda) = \begin{bmatrix} 1/(\lambda - 4) & 0 \\ 0 & 1/(\lambda + 2)^2 \end{bmatrix}$. The columns are used to form the nonsingular matrix $\mathbf{G}' = \begin{bmatrix} -1/9 & 0 \\ 0 & 1/16 \end{bmatrix}$ and the corresponding $\mathfrak{F}' = \begin{bmatrix} 1 & 0 \\ 0 & 1 \end{bmatrix}$. Then from equation (*16.23*),

$$\mathbf{K}' = -\left\{\mathbf{I}_2 + \begin{bmatrix} -9 & 0 \\ 0 & 16 \end{bmatrix}\begin{bmatrix} 1 & 0 \\ 0 & 0 \end{bmatrix}\right\}^{-1}\begin{bmatrix} -9 & 0 \\ 0 & 16 \end{bmatrix} = \begin{bmatrix} -9/8 & 0 \\ 0 & -16 \end{bmatrix}$$

Using this result in equation (*16.20*) leads to

$$\Delta_0'(\lambda) = (\lambda + 5)(\lambda + 6)(\lambda - 2)$$

The two desired roots have been achieved, but the system is still unstable due to the root $\lambda_3 = 2$. ■

The preceding method of analyzing output feedback is satisfactory for explanatory purposes and for application to low-order systems whose $\mathbf{\Phi}(\lambda)$ can easily be found in closed form. It is now shown how to take better advantage of the algorithmic results of Sec. 16.5. First a feedback gain $\mathbf{K}_*$ will be found which will multiply $\mathbf{y} - \mathbf{Du}$. If $\mathbf{D} = \mathbf{0}$, $\mathbf{K}_*$ will be the desired output gain $\mathbf{K}'$. If $\mathbf{D}$ is not zero, an additional step will be needed at the end of the procedure.

Equation (*16.20*) becomes

$$|\lambda\mathbf{I} - \mathbf{A} + \mathbf{BK}_*\mathbf{C}| = 0$$

which implies that one or more nonzero vectors $\boldsymbol{\psi}$ exist such that

$$[\lambda\mathbf{I} - \mathbf{A} \mid \mathbf{B}]\begin{bmatrix} \boldsymbol{\psi} \\ \mathbf{K}_*\mathbf{C}\boldsymbol{\psi} \end{bmatrix} = \mathbf{0} \tag{16.24}$$

Except for the presence of $\mathbf{C}$ this is again equation (*16.16*). All independent nontrivial solutions must found, and then $\mathbf{U}(\lambda_i)$, $\mathbf{\Psi}(\lambda_i)$ and $\mathbf{\Omega}$ are formed as in the previous algorithm, without change. Likewise, the lower partition of $\mathbf{U}(\lambda_i)$ is used to form $\mathfrak{F}'(\lambda_i)$. The meaning of these lower partition columns is now different, namely $\mathbf{f}_i - \mathbf{K}_*\mathbf{C}\boldsymbol{\psi}_i$. These are used to form $\mathbf{\Lambda}'$ essentially as before. One extra operation is now required because of the presence of the $\mathbf{C}$ matrix. Define $\mathbf{\Omega}' \triangleq \mathbf{C}\mathbf{\Omega}$. Then the counterpart to equation (*16.17*) is $\mathbf{K}_*\mathbf{\Omega}' = \mathbf{\Lambda}'$. Now $\mathbf{\Omega}'$ is an $m \times nr$ matrix whose rank is m if $\mathbf{C}$ is full rank. Therefore, at most m linearly independent columns of $\mathbf{\Omega}'$ can be selected in step III of the algorithm to form $\mathbf{G}'$. The corresponding columns of $\mathbf{\Lambda}'$ are used to form $\mathfrak{F}'$. It is clear that at most m poles can be arbitrarily placed using output feedback. The solution for $\mathbf{K}_*$ is found as in step V, $\mathbf{K}_* = \mathfrak{F}'(\mathbf{G}')^{-1}$

If $\mathbf{D}$ is not zero, then

$$\mathbf{u} = -\mathbf{K}_*(\mathbf{y} - \mathbf{D}\mathbf{u}) + \mathbf{F}'\mathbf{v}'$$

can be solved to give

$$\mathbf{u} = -[\mathbf{I} - \mathbf{K}_*\mathbf{D}]^{-1}\mathbf{K}_*\mathbf{y} + [\mathbf{I} - \mathbf{K}_*\mathbf{D}]^{-1}\mathbf{F}'\mathbf{v}'$$

The output feedback gain matrix is

$$\mathbf{K}' = [\mathbf{I} - \mathbf{K}_*\mathbf{D}]^{-1}\mathbf{K}_* \tag{16.25}$$

A simple examination shows that this is exactly the same as given by equation (*16.23*). Note that the multiplier of the external input $\mathbf{v}'$ is also modified. This common term can be placed in the forward loop, as shown in Fig. 16.4.

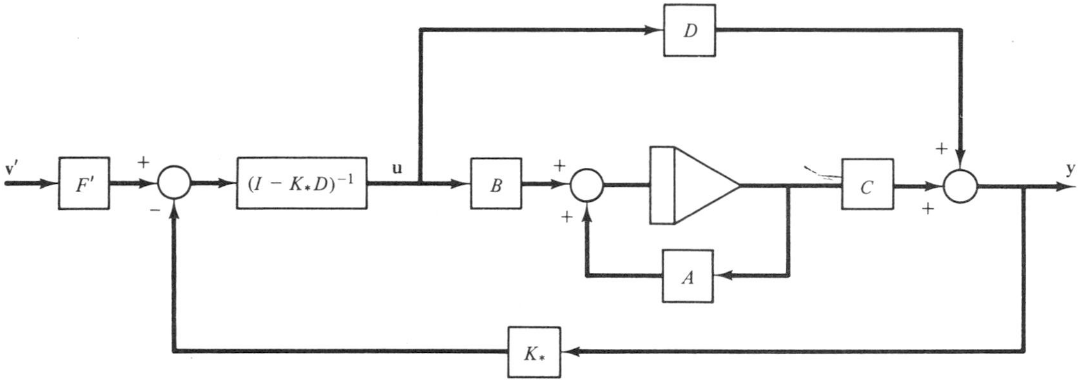

Figure 16.4

16.7 RECONSTRUCTING THE STATE FROM AVAILABLE OUTPUTS

A system described by equation (*16.1*) is considered first. For simplicity it is now assumed that $\mathbf{D} = [\mathbf{0}]$. If $\mathbf{D} \neq [\mathbf{0}]$, an equivalent output $\mathbf{y}' = \mathbf{y} - \mathbf{D}\mathbf{u}$ can be used in place of $\mathbf{y}$ in what follows. It is desired to obtain a good estimate of the state $\mathbf{x}(t)$, given a knowledge of the output $\mathbf{y}(t)$, the input $\mathbf{u}(t)$, and the system matrices $\mathbf{A}, \mathbf{B}, \mathbf{C}, \mathbf{D}$. This problem is referred to as the *state reconstruction problem*. If $\mathbf{C}$ is square and nonsingular, then $\mathbf{x}(t) = \mathbf{C}^{-1}\mathbf{y}(t)$. In general, this trivial result will not be applicable.

Another dynamic system, called an *observer* [74], is to be constructed. Its input will depend on $\mathbf{y}$ and $\mathbf{u}$ and its state (output) should be a good approximation to $\mathbf{x}(t)$. The form of the observer is selected as

$$\dot{\hat{\mathbf{x}}} = \mathbf{A}_c\hat{\mathbf{x}} + \mathbf{B}_c\mathbf{y} + \mathbf{z} \tag{16.26}$$

where $\hat{\mathbf{x}}$ is the $n \times 1$ vector approximation to $\mathbf{x}$.* $\mathbf{A}_c$ and $\mathbf{B}_c$ are $n \times n$ and $n \times m$ matrices and $\mathbf{z}$ is an $n \times 1$ vector, to be determined. Defining the estimation error as $\mathbf{e} = \mathbf{x} - \hat{\mathbf{x}}$ and using equation (*16.1*) gives

$$\dot{\mathbf{e}} = \mathbf{A}\mathbf{x} - \mathbf{A}_c\hat{\mathbf{x}} - \mathbf{B}_c\mathbf{y} + \mathbf{B}\mathbf{u} - \mathbf{z} \tag{16.27}$$

By selecting $\mathbf{z} = \mathbf{B}\mathbf{u}$ and using $\mathbf{y} = \mathbf{C}\mathbf{x}$, this reduces to

$$\dot{\mathbf{e}} = (\mathbf{A} - \mathbf{B}_c\mathbf{C})\mathbf{x} - \mathbf{A}_c\hat{\mathbf{x}}$$

Selecting $\mathbf{A}_c = \mathbf{A} - \mathbf{B}_c\mathbf{C}$ gives

$$\dot{\mathbf{e}} = \mathbf{A}_c\mathbf{e} \tag{16.28}$$

If the eigenvalues of $\mathbf{A}_c$ all have negative real parts, then an asymptotically stable error equation results. This indicates that $\mathbf{e}(t) \longrightarrow \mathbf{0}$, or $\hat{\mathbf{x}}(t) \longrightarrow \mathbf{x}(t)$ as $t \longrightarrow \infty$. The term $\mathbf{B}_c$ of the observer is still unspecified. If the original system (*16.1*) is completely observable, then it is always possible to find a $\mathbf{B}_c$ which will yield any set of desired eigenvalues for $\mathbf{A}_c$. Thus it is possible to control the rate at which $\hat{\mathbf{x}} \longrightarrow \mathbf{x}$. This is shown by using manipulations very similar to those in Sec. 16.4 [11]:

$$\begin{aligned} \Delta_c(\lambda) &= |\lambda\mathbf{I}_n - \mathbf{A}_c| = |\lambda\mathbf{I}_n - \mathbf{A} + \mathbf{B}_c\mathbf{C}| \\ &= |\lambda\mathbf{I}_n - \mathbf{A}| \cdot |\mathbf{I}_n + \mathbf{\Phi}(\lambda)\mathbf{B}_c\mathbf{C}| = \Delta(\lambda)|\mathbf{I}_m + \mathbf{C}\mathbf{\Phi}(\lambda)\mathbf{B}_c| = 0 \end{aligned} \tag{16.29}$$

Complete observability guarantees that n linearly independent rows can be selected from the rows of $\mathbf{C}\mathbf{\Phi}(\lambda)$ (or its derivatives if necessary when repeated eigenvalues are desired). Therefore, equation (*16.29*) can be made zero for n specified eigenvalues λ_k by defining the ith row of $\mathbf{C}\mathbf{\Phi}(\lambda)$ as $\boldsymbol{\psi}_i^T$ and requiring

$$\boldsymbol{\psi}_i^T(\lambda_k)\mathbf{B}_c = -\mathbf{e}_i^T \qquad \text{or} \qquad \left.\frac{d\boldsymbol{\psi}_i^T}{d\lambda}\right|_{\lambda=\lambda_k}\mathbf{B}_c = \mathbf{0}^T \tag{16.30}$$

where $\mathbf{e}_i^T$ is the ith row of $\mathbf{I}_m$ and $\mathbf{0}^T$ is the $1 \times m$ zero row vector. Using one equation like (*16.30*) for each desired eigenvalue, defining the $n \times n$ nonsingular matrix $\mathbf{G}_c$, and letting $\mathfrak{F}_c$ be the $n \times m$ matrix whose rows are either $\mathbf{e}_i^T$ or $\mathbf{0}^T$, leads to

$$\mathbf{B}_c = -\mathbf{G}_c^{-1}\mathfrak{F}_c \tag{16.31}$$

Since a square matrix and its transpose have the same determinant, equation (*16.29*) can also be written as

$$|\lambda\mathbf{I} - \mathbf{A}^T + \mathbf{C}^T\mathbf{B}_c^T| = 0$$

This is exactly the same form as equation (*16.8*), but with $\mathbf{A}$, $\mathbf{B}$, and $\mathbf{K}$ replaced by $\mathbf{A}^T$, $\mathbf{C}^T$, and $\mathbf{B}_c^T$. The observer design problem is the dual of the pole placement

*The circumflex ^ will be used in this chapter to indicate an estimate for a quantity, e.g., $\hat{\mathbf{x}}$ is an estimate of $\mathbf{x}$. This should not be confused with the notation for a unit vector introduced in Chapter 4.

problem using full state feedback. The discussion and algorithm of Sec. 16.5 thus apply in their entirety. As before, there will be times when generalized eigenvectors will be required in order to obtain multiple poles.

Example 16.9

Design an observer for the system in Example 16.1, such that the observer eigenvalues are $\lambda_1 = \lambda_2 = -8$. Use $\mathbf{C} = [1 \quad 0]$.

Using $\mathbf{\Phi}(\lambda)$ from Example 16.1 gives $\mathbf{C\Phi}(\lambda) = \{1/\lambda \quad 2/[\lambda(\lambda - 3)]\}$. One row of $\mathbf{G}_c$ is $\mathbf{C\Phi}(-8) = [-1/8 \quad 1/44]$. A second independent row can be obtained from $\frac{d}{d\lambda}[\mathbf{C\Phi}(\lambda)]\Big|_{\lambda=-8}$ $= \begin{bmatrix} -\dfrac{1}{\lambda^2} & \dfrac{-2(2\lambda - 3)}{\lambda^2(\lambda - 3)^2} \end{bmatrix}_{\lambda=-8}$ This requires that $\mathfrak{J}_c = \begin{bmatrix} 1 \\ 0 \end{bmatrix}$, and then equation (*16.31*) gives

$$\mathbf{B}_c = -\begin{bmatrix} -1/8 & 1/44 \\ -1/64 & 19/3872 \end{bmatrix}^{-1} \begin{bmatrix} 1 \\ 0 \end{bmatrix} = \begin{bmatrix} 19 \\ 121/2 \end{bmatrix}$$

Alternatively, $[-8\mathbf{I} - \mathbf{A}^T \mid \mathbf{C}^T]\begin{bmatrix} \boldsymbol{\psi} \\ \mathbf{B}_c^T\boldsymbol{\psi} \end{bmatrix} = \mathbf{0}$ has the solution

$$\mathbf{U}_1(-8) = \begin{bmatrix} -0.125 \\ 0.0227273 \\ \hline -1. \end{bmatrix} = \begin{bmatrix} \boldsymbol{\psi}_1 \\ \hline \mathbf{f}_1 \end{bmatrix}$$

A generalized eigenvector is needed to find a second independent column.

$$[-8\mathbf{I} - \mathbf{A}^T \mid \mathbf{C}^T]\begin{bmatrix} \boldsymbol{\psi}_2 \\ \mathbf{B}_c^T\boldsymbol{\psi}_2 \end{bmatrix} = \boldsymbol{\psi}_1$$

is solved to find $\mathbf{U}_2(-8) = [0.015625 \quad -0.004907 \mid 0]^T$ so $\boldsymbol{\psi}_2 = [0.015625 \quad -0.004907]^T$ and $\mathbf{f}_2 = 0$. Using these as columns of $\mathbf{G}$ and $\mathcal{G}$ respectively gives $\mathbf{B}_c^T\mathbf{G} = \mathcal{G}$, which gives the same answer as before. It is easily verified that $|\lambda\mathbf{I} - \mathbf{A} + \mathbf{B}_c\mathbf{C}| = (\lambda + 8)^2$. Thus the observer, described by equation (*16.32*), has the desired eigenvalues:

$$\begin{aligned} \dot{\hat{\mathbf{x}}} &= (\mathbf{A} - \mathbf{B}_c\mathbf{C})\hat{\mathbf{x}} + \mathbf{B}_c\mathbf{y} + \mathbf{z} \\ &= \begin{bmatrix} -19 & 2 \\ -121/2 & 3 \end{bmatrix}\hat{\mathbf{x}} + \begin{bmatrix} 19 \\ 121/2 \end{bmatrix}\mathbf{y} + \begin{bmatrix} 0 \\ 1 \end{bmatrix}\mathbf{u} \end{aligned} \qquad (16.32)$$

■

For the discrete-time system of equation (*16.2*) the observer form is postulated as

$$\hat{\mathbf{x}}(k + 1) = \mathbf{A}_c\hat{\mathbf{x}}(k) + \mathbf{B}_c\mathbf{y}(k) + \mathbf{z}(k)$$

The estimation error $\mathbf{e}(k) = \mathbf{x}(k) - \hat{\mathbf{x}}(k)$ will satisfy the homogeneous difference equation

$$\mathbf{e}(k + 1) = (\mathbf{A} - \mathbf{B}_c\mathbf{C})\mathbf{e}(\mathbf{k}) = \mathbf{A}_c\mathbf{e}(k)$$

provided that $\mathbf{z}(k)$ is selected as $(\mathbf{B} - \mathbf{B}_c\mathbf{D})\mathbf{u}(k)$. Just as in the continuous-time case, the error $\mathbf{e}(k)$ will decay towards zero if $\mathbf{A}_c$ is stable. This means that all its eigenvalues must be inside the unit circle. The decay rate depends on the location of these eigenvalues. If the original system of equation (*16.2*) is observable, it is possible to force the n eigenvalues to any desired locations by proper choice of $\mathbf{B}_c$. The mechanics of doing this are exactly the same as in the continuous-time case.

The observers described above are used to estimate $\mathbf{x}$. It a constant state feedback matrix $\mathbf{K}$ is then used, with $\hat{\mathbf{x}}$ as input instead of $\mathbf{x}$, the system shown in Fig. 16.5 is

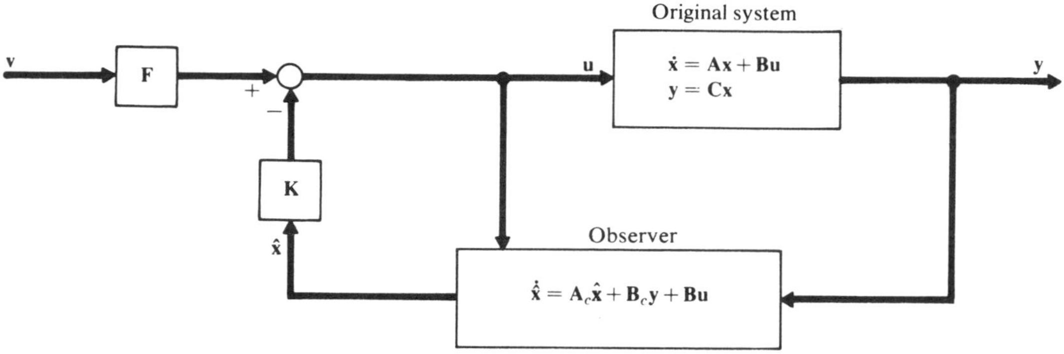

Figure 16.5

obtained. The composite system is of order $2n$. By proper selection of $\mathbf{K}$, n of the closed-loop eigenvalues can be specified as in Secs. 16.4. and 16.5. By proper selection of $\mathbf{B}_c$, the remaining n eigenvalues of the observer can be specified. This represents a *separation principle.* That is, a feedback system with the desired poles can be designed, proceeding as if all states were measurable. Then a separate design of the observer can be used to provide the desired observer poles. To see this, note that

$$\dot{\mathbf{x}} = \mathbf{Ax} + \mathbf{Bu}, \qquad \dot{\hat{\mathbf{x}}} = \mathbf{A}_c\hat{\mathbf{x}} + \mathbf{B}_c\mathbf{Cx} + \mathbf{Bu}, \qquad \mathbf{u} = \mathbf{Fv} - \mathbf{K}\hat{\mathbf{x}}$$

can be combined into

$$\begin{bmatrix} \dot{\mathbf{x}} \\ \dot{\hat{\mathbf{x}}} \end{bmatrix} = \begin{bmatrix} \mathbf{A} & -\mathbf{BK} \\ \mathbf{B}_c\mathbf{C} & \mathbf{A}_c - \mathbf{BK} \end{bmatrix} \begin{bmatrix} \mathbf{x} \\ \hat{\mathbf{x}} \end{bmatrix} + \begin{bmatrix} \mathbf{BF} \\ \mathbf{BF} \end{bmatrix} \mathbf{v}$$

The $2n$ closed-loop eigenvalues are roots of

$$\Delta_T(\lambda) = \begin{vmatrix} \mathbf{I}_n\lambda - \mathbf{A} & \mathbf{BK} \\ -\mathbf{B}_c\mathbf{C} & \mathbf{I}_n\lambda - \mathbf{A}_c + \mathbf{BK} \end{vmatrix} = 0$$

This can be reduced to a block triangular form by a series of elementary operations. Subtracting rows $i\,(1 \le i \le n)$ from rows $n + i$, and then adding columns $n + j$ $(1 \le j \le n)$ to columns j gives

$$\Delta_T(\lambda) = \begin{vmatrix} \mathbf{I}_n\lambda - \mathbf{A} & \mathbf{BK} \\ -\mathbf{I}_n\lambda + \mathbf{A} - \mathbf{B}_c\mathbf{C} & \mathbf{I}_n\lambda - \mathbf{A}_c \end{vmatrix} = \begin{vmatrix} \mathbf{I}_n\lambda - \mathbf{A} + \mathbf{BK} & \mathbf{BK} \\ \mathbf{0} & \mathbf{I}_n\lambda - \mathbf{A}_c \end{vmatrix}$$
$$= |\mathbf{I}_n\lambda - \mathbf{A} + \mathbf{BK}| \cdot |\mathbf{I}_n\lambda - \mathbf{A}_c| = \Delta'(\lambda) \cdot \Delta_c(\lambda)$$

This verifies that the two sets of n eigenvalues can be specified separately, provided $(\mathbf{A}, \mathbf{B}, \mathbf{C})$ is completely controllable and completely observable. Experience indicates that a good design results if the continuous-time observer poles are selected to be a little farther to the left in the s-plane than the desired closed-loop state feedback poles. The discrete-time observer poles should be somewhat nearer the Z-plane origin so that the transients die out faster than the dominant system modes.

The observers described in this section are called full state or *identity observers* [74], since the total state vector is reconstructed. Because information about some of the states is directly obtainable from $\mathbf{y}$, the above scheme contains a certain amount of redundancy. A special case of the so-called "reduced order observer" is considered here, where the measurements of $\mathbf{y}$ are easily related to some m component subset of the state $\mathbf{x}$ and perhaps the input $\mathbf{u}$. It is assumed that $\mathbf{C} = [\mathbf{I}_m \mid 0]$. (See Problem 16.12 to see how general or special this case really is.) The discrete-time system of equation (*16.2*) is treated here. Continuous-time systems can be treated similarly with obvious minor changes.

Partition the state vector into the m measured components and the n-m remaining components. Then the state equations become

$$\mathbf{x}_1(k+1) = \mathbf{A}_{11}\mathbf{x}_1(k) + \mathbf{A}_{12}\mathbf{x}_2(k) + \mathbf{B}_1\mathbf{u}(k)$$

$$\mathbf{x}_2(k+1) = \mathbf{A}_{21}\mathbf{x}_1(k) + \mathbf{A}_{22}\mathbf{x}_2(k) + \mathbf{B}_2\mathbf{u}(k)$$

and

$$\mathbf{y}(k) = \mathbf{x}_1(k) + \mathbf{D}\mathbf{u}(k)$$

Since only the noise-free case is being discussed, $\mathbf{y}$ and $\mathbf{u}$ are assumed known exactly at all times. The third equation plays only the role of giving $\mathbf{x}_1$ if $\mathbf{D}$ is nonzero. Therefore $\mathbf{x}_1$ can be treated as known quantity at all times. Define two more "known" quantities

$$\begin{aligned} \mathbf{y}_r(k) &\triangleq \mathbf{x}_1(k+1) - \mathbf{A}_{11}\mathbf{x}_1(k) - \mathbf{B}_1\mathbf{u}(k) \\ \mathbf{v}_r(k) &\triangleq \mathbf{A}_{21}\mathbf{x}_1(k) + \mathbf{B}_2\mathbf{u}(k) \end{aligned} \tag{16.33}$$

Then the new equivalent dynamic equation is

$$\mathbf{x}_2(k+1) = \mathbf{A}_{22}\mathbf{x}_2(k) + \mathbf{v}_r(k)$$

and the new equivalent measurement equation is

$$\mathbf{y}_r(k) = \mathbf{A}_{12}\mathbf{x}_2(k)$$

A linear observer form is postulated as

$$\hat{\mathbf{x}}_2(k+1) = \mathbf{A}_r\hat{\mathbf{x}}_2(k) + \mathbf{B}_r\mathbf{y}_r(k) + \mathbf{z}_r(k)$$

The error $\mathbf{e}(k) \triangleq \mathbf{x}_2(k) - \hat{\mathbf{x}}_2(k)$ satisfies

$$\mathbf{e}(k+1) = \mathbf{A}_{22}\mathbf{x}_2(k) + \mathbf{v}_r(k) - \mathbf{A}_r\hat{\mathbf{x}}_2 - \mathbf{B}_r\mathbf{y}_r(k) - \mathbf{z}_r(k)$$

If the selections $z_r(k) = \mathbf{v}_r(k)$ and $\mathbf{A}_r = \mathbf{A}_{22} - \mathbf{B}_r\mathbf{A}_{12}$ are made, then

$$\mathbf{e}(k+1) = (\mathbf{A}_{22} - \mathbf{B}_r\mathbf{A}_{12})\mathbf{e}(k)$$

The convergence rate of $\mathbf{e}(k)$ to zero can be controlled by proper choice of $\mathbf{B}_r$, just as in the case of the full state observer. Figure 16.6 gives the reduced order observer in block diagram form.

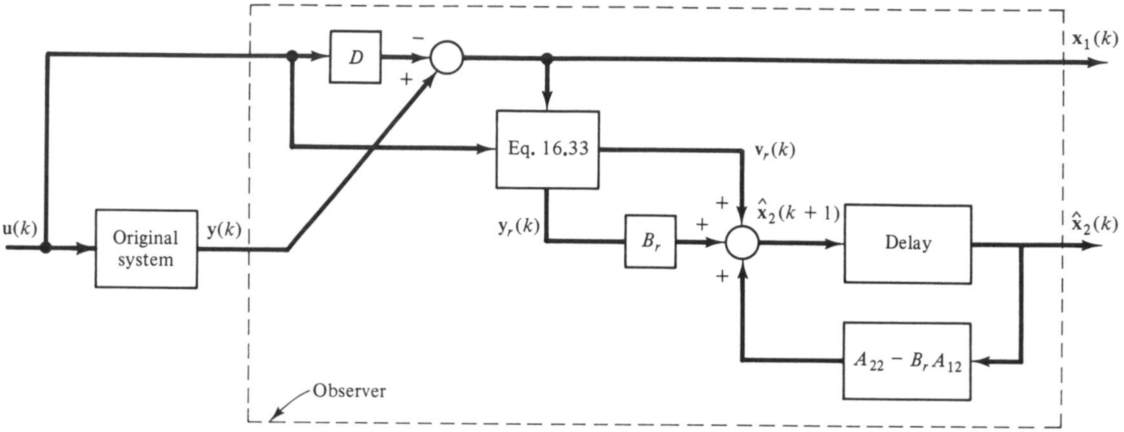

Figure 16.6

16.8 DESIGN OF DECOUPLED OR NONINTERACTING SYSTEMS [35, 84]

A system with an equal number m of inputs and outputs is considered in this section. If the $m \times m$ transfer matrix $\mathbf{H}(s)$ is diagonal and nonsingular, the system is said to be *decoupled* because each input affects one and only one output. When the number of inputs and outputs are not equal, $\mathbf{H}(s)$ is not square, and thus cannot be diagonal. Various other kinds of decoupling and partial decoupling have been defined where $\mathbf{H}(s)$ is triangular or block diagonal and so on [85, 102].

The problem of reducing a system with m inputs and m outputs to decoupled form, using a state feedback control law $\mathbf{u} = -\mathbf{K}_d\mathbf{x} + \mathbf{F}_d\mathbf{v}$, is considered. Assuming that $\mathbf{D} = [\mathbf{0}]$, the transfer matrix for the state feedback system of equations (*16.4*) and (*16.5*) is

$$\mathbf{H}(s) = \mathbf{C}[s\mathbf{I} - \mathbf{A} + \mathbf{B}\mathbf{K}_d]^{-1}\mathbf{B}\mathbf{F}_d \qquad (16.34)$$

The decoupling problem is that of selecting matrices $\mathbf{F}_d(m \times m)$ and $\mathbf{K}_d(m \times n)$ so that $\mathbf{H}(s)$ is diagonal and nonsingular. Consider the inverse transform of equation (*16.34*) and the Cayley-Hamilton theorem applied to $e^{[\mathbf{A}-\mathbf{B}\mathbf{K}_d]t}$. This leads to an alternative statement of decoupling. The matrices

$$\mathbf{C}[\mathbf{A} - \mathbf{B}\mathbf{K}_d]^j\mathbf{B}\mathbf{F}_d, \qquad j = 0, 1, \ldots, n-1 \tag{16.35}$$

must all be diagonal if the system is decoupled. Let the ith row of $\mathbf{C}$ be $\mathbf{c}_i$ and define a set of m integers by

$$d_i = \min_j \{j | \mathbf{c}_i\mathbf{A}^j\mathbf{B} \neq \mathbf{0}, j = 0, 1, \ldots, n-1\} \tag{16.36}$$

or

$$d_i = n - 1 \qquad \text{if } \mathbf{c}_i\mathbf{A}^j\mathbf{B} = \mathbf{0} \text{ for all } j$$

The original system can be decoupled using state feedback [35] if and only if the following $m \times m$ matrix is nonsingular:

$$\mathbf{N} = \begin{bmatrix} \mathbf{c}_1\mathbf{A}^{d_1}\mathbf{B} \\ \mathbf{c}_2\mathbf{A}^{d_2}\mathbf{B} \\ \vdots \\ \mathbf{c}_m\mathbf{A}^{d_m}\mathbf{B} \end{bmatrix}$$

In particular, one set of decoupling matrices is

$$\mathbf{F}_d = \mathbf{N}^{-1} \quad \text{and} \quad \mathbf{K}_d = \mathbf{N}^{-1}\begin{bmatrix} \mathbf{c}_1\mathbf{A}^{d_1+1} \\ \vdots \\ \mathbf{c}_m\mathbf{A}^{d_m+1} \end{bmatrix} \tag{16.37}$$

Example 16.10

Determine whether the open-loop system of Example 16.8 can be decoupled using state feedback. Assume now that $\mathbf{D} = [\mathbf{0}]$.

Since $\mathbf{c}_1\mathbf{A}^0\mathbf{B} = [1 \quad 0] \neq \mathbf{0}$, set $d_1 = 0$. Also, $\mathbf{c}_2\mathbf{A}^0\mathbf{B} = \mathbf{0}$, but $\mathbf{c}_2\mathbf{A}\mathbf{B} = [0 \quad 1] \neq \mathbf{0}$, so that $d_2 = 1$. Therefore, $\mathbf{N} = \begin{bmatrix} \mathbf{c}_1\mathbf{B} \\ \mathbf{c}_2\mathbf{A}\mathbf{B} \end{bmatrix} = \begin{bmatrix} 1 & 0 \\ 0 & 1 \end{bmatrix}$ is nonsingular and decoupling is possible. The decoupling matrices are $\mathbf{F}_d = \mathbf{N}^{-1} = \mathbf{I}$ and $\mathbf{K}_d = \mathbf{N}^{-1}\begin{bmatrix} \mathbf{c}_1\mathbf{A} \\ \mathbf{c}_2\mathbf{A}^2 \end{bmatrix} = \begin{bmatrix} 0 & 4 & 4 \\ 4 & -4 & 0 \end{bmatrix}$. Using these, the decoupled transfer matrix is obtained from equation (*16.34*):

$$\mathbf{H}(s) = \begin{bmatrix} 1/s & 0 \\ 0 & 1/s^2 \end{bmatrix}$$

■

Results of the previous example are typical in that all the poles of the decoupled system are at the origin. This is always the result if equation (*16.37*) is used. In fact, the general result is $\mathbf{H}(s) = \text{diag}\,[s^{-d_1-1} \quad s^{-d_2-1} \quad \cdots \quad s^{-d_m-1}]$. This system is said to be *integrator decoupled.* The performance of such a decoupled system would usually not be acceptable. Two questions naturally arise. First, can decoupling be accomplished by using only the available outputs? The answer to this is yes, provided dynamic feedback compensators (observers for example) of sufficiently high dimension are allowed [54]. Second, can other state feedback matrices or output feedback matrices be used to prespecify closed-loop pole locations while preserving the decoupled nature of the system? The answer to this question is yes in some cases, no in others. At least $m + \sum_{i=1}^{m} d_i$ poles can be prespecified, but not necessarily all of the poles [36].

Example 16.11

Consider the decoupled system found in Example 16.10. Find a constant feedback matrix which moves the closed-loop poles from zero to $\{-5, -5, -10\}$. Is the resultant system still decoupled?

The decoupled system of Example 16.10 is treated as the open-loop system. That is,

$$\tilde{\mathbf{A}} = \mathbf{A} - \mathbf{BK}_d = \begin{bmatrix} -2 & 1 & 0 \\ -4 & 2 & 0 \\ 0 & 0 & 0 \end{bmatrix}, \qquad \tilde{\mathbf{B}} = \mathbf{BF}_d = \begin{bmatrix} 0 & 0 \\ 0 & 1 \\ 1 & 0 \end{bmatrix}, \qquad \tilde{\mathbf{C}} = \mathbf{C} = \begin{bmatrix} 0 & 0 & 1 \\ 1 & 0 & 0 \end{bmatrix}$$

Since $\mathbf{F}_d$ is nonsingular, this system is still completely controllable (Sec. 16.3). Thus the specified eigenvalues can be achieved using a constant state feedback matrix $\tilde{\mathbf{K}}$.

Let $\tilde{\Psi}(\lambda) = [\lambda\mathbf{I} - \tilde{\mathbf{A}}]^{-1}\tilde{\mathbf{B}} = \begin{bmatrix} 0 & 1/\lambda^2 \\ 0 & (\lambda + 2)/\lambda^2 \\ 1/\lambda & 0 \end{bmatrix}$. Selecting

$$\tilde{\mathbf{G}} = [\psi_2(-10) \quad \psi_2(-5) \quad \psi_1(-5)] = \begin{bmatrix} 1/100 & 1/25 & 0 \\ -8/100 & -3/25 & 0 \\ 0 & 0 & -1/5 \end{bmatrix}, \qquad \tilde{\mathcal{G}} = \begin{bmatrix} 0 & 0 & 1 \\ 1 & 1 & 0 \end{bmatrix}$$

equation (*16.15*) gives $\tilde{\mathbf{K}} = \begin{bmatrix} 0 & 0 & 5 \\ 20 & 15 & 0 \end{bmatrix}$.

The combined system, using $\mathbf{K}_d$ and $\mathbf{F}_d$ to achieve decoupling and $\tilde{\mathbf{K}}$ to achieve pole placement, is shown in Fig. 16.7. In this case the resulting system is still decoupled and has closed-loop poles, as specified, at $-5, -5, -10$. The closed-loop transfer function is

$$\mathbf{H}(s) = \mathbf{C}[s\mathbf{I} - \mathbf{A} + \mathbf{BK}_d + \mathbf{BF}_d\tilde{\mathbf{K}}]^{-1}\mathbf{BF}_d = \begin{bmatrix} 1/(s+5) & 0 \\ 0 & 1/[(s+5)(s+10)] \end{bmatrix}$$ ■

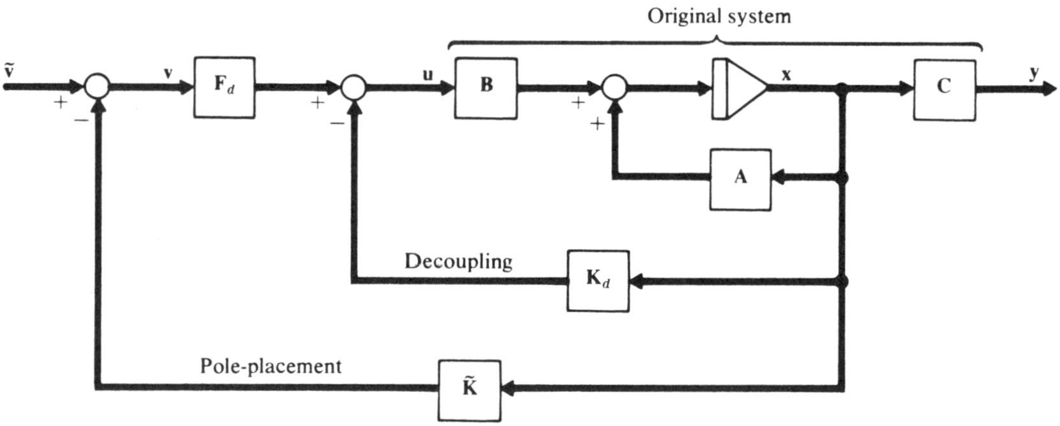

Figure 16.7

ILLUSTRATIVE PROBLEMS

State Feedback

16.1 A system is described by $\mathbf{A} = \begin{bmatrix} -2 & 1 & 0 \\ 0 & -2 & 0 \\ 0 & 0 & 4 \end{bmatrix}$, $\mathbf{B} = \begin{bmatrix} 0 & 0 \\ 0 & 1 \\ 1 & 0 \end{bmatrix}$. Find a constant state feedback matrix $\mathbf{K}$ which yields closed-loop poles $\Gamma = \{-2, -3, -4\}$.

Inverting $(\lambda\mathbf{I} - \mathbf{A})$ leads to

$$\mathbf{\Phi}(\lambda)\mathbf{B} = \begin{bmatrix} 0 & 1/(\lambda+2)^2 \\ 0 & 1/(\lambda+2) \\ 1/(\lambda-4) & 0 \end{bmatrix} \triangleq [\boldsymbol{\psi}_1(\lambda) \quad \boldsymbol{\psi}_2(\lambda)]$$

A nonsingular $\mathbf{G}$ matrix can be obtained as

$$\mathbf{G} = [\boldsymbol{\psi}_1(-2) \quad \boldsymbol{\psi}_2(-3) \quad \boldsymbol{\psi}_2(-4)] = \begin{bmatrix} 0 & 1 & \frac{1}{4} \\ 0 & -1 & -\frac{1}{2} \\ -\frac{1}{6} & 0 & 0 \end{bmatrix}$$

Corresponding to this choice, $\mathfrak{F} = [\mathbf{e}_1 \quad \mathbf{e}_2 \quad \mathbf{e}_2] = \begin{bmatrix} 1 & 0 & 0 \\ 0 & 1 & 1 \end{bmatrix}$. Then equation *(16.15)* gives $\mathbf{K} = \begin{bmatrix} 0 & 0 & 6 \\ 2 & 3 & 0 \end{bmatrix}$.

16.2 Repeat Problem 16.1 if $\Gamma = \{-2, -2, -20\}$.

Since two of the open-loop roots are equal to -2, it is not possible to immediately select three independent and finite columns from $\boldsymbol{\psi}_j(\lambda_i)$. However, a limiting process can be used. Since $\boldsymbol{\psi}_2(-2)$ has infinite components, define $\alpha = 1/(\lambda + 2)$ and let

$$\mathbf{G} = [\boldsymbol{\psi}_1(-2) \quad \boldsymbol{\psi}_2(\lambda) \quad \boldsymbol{\psi}_2(-20)] = \begin{bmatrix} 0 & \alpha^2 & 1/(18)^2 \\ 0 & \alpha & -1/18 \\ -1/6 & 0 & 0 \end{bmatrix}$$

Corresponding to this choice,

$$\mathfrak{F} = [\mathbf{e}_1 \quad \mathbf{e}_2 \quad \mathbf{e}_2], \quad \mathbf{G}^{-1} = \frac{6(18)^2}{\alpha^2(18 + 1/\alpha)} \begin{bmatrix} 0 & 0 & -\alpha^2/18 - \alpha/(18)^2 \\ 1/[6(18)] & 1/[6(18)^2] & 0 \\ \alpha/6 & -\alpha^2/6 & 0 \end{bmatrix}$$

As $\lambda \longrightarrow -2, \alpha \longrightarrow \infty$, but $\lim_{\lambda \to -2} \mathbf{G}^{-1} = \begin{bmatrix} 0 & 0 & -6 \\ 0 & 0 & 0 \\ 0 & -18 & 0 \end{bmatrix}$. Using this in equation *(16.15)* gives $\mathbf{K} = \begin{bmatrix} 0 & 0 & 6 \\ 0 & 18 & 0 \end{bmatrix}$.

16.3 Repeat Problem 16.1 if $\Gamma = \{-3, -2 + j, -2 - j\}$.

In this case select $\mathbf{G} = [\boldsymbol{\psi}_1(-3) \quad \boldsymbol{\psi}_2(-2 + j) \quad \boldsymbol{\psi}_2(-2 - j)]$. This gives the non-singular, complex matrix

$$\mathbf{G} = \begin{bmatrix} 0 & -1 & -1 \\ 0 & -j & j \\ -1/7 & 0 & 0 \end{bmatrix}, \quad \text{with} \quad \mathbf{G}^{-1} = \frac{7}{2j}\begin{bmatrix} 0 & 0 & -2j \\ -j/7 & -1/7 & 0 \\ -j/7 & 1/7 & 0 \end{bmatrix}$$

Then $\mathbf{K} = -\begin{bmatrix} 1 & 0 & 0 \\ 0 & 1 & 1 \end{bmatrix}\mathbf{G}^{-1} = \begin{bmatrix} 0 & 0 & 7 \\ 1 & 0 & 0 \end{bmatrix}$. Note that even though $\mathbf{G}$ and $\mathbf{G}^{-1}$ are complex, $\mathbf{K}$ is real since the complex eigenvalue and its conjugate were both in Γ.

16.4 Repeat Problem 16.1 if $\Gamma = \{-3, -3, -3\}$.

Since $\mathbf{\Psi}(\lambda) = \mathbf{\Phi}(\lambda)\mathbf{B}$ has only two columns, and there are three repeated eigenvalues, a differentiated column must be used in order to obtain a nonsingular 3×3 $\mathbf{G}$ matrix. Column 1 cannot be used for this purpose since $d\boldsymbol{\psi}_1/d\lambda$ and $\boldsymbol{\psi}_1$ will be constant multiples of each other regardless of the numerical value of λ. Therefore, select $\mathbf{G} = \left[\boldsymbol{\psi}_1(-3) \quad \boldsymbol{\psi}_2(-3) \quad \left.\dfrac{d\boldsymbol{\psi}_2}{d\lambda}\right|_{\lambda=-3}\right]$. This requires that $\mathfrak{F} = \begin{bmatrix} 1 & 0 & 0 \\ 0 & 1 & 0 \end{bmatrix}$, so that $\mathbf{K} = -\begin{bmatrix} 1 & 0 & 0 \\ 0 & 1 & 0 \end{bmatrix}\begin{bmatrix} 0 & 1 & 2 \\ 0 & -1 & -1 \\ -1/7 & 0 & 0 \end{bmatrix}^{-1} = \begin{bmatrix} 0 & 0 & 7 \\ 1 & 2 & 0 \end{bmatrix}$.

16.5 A system is described by equation (*16.1*), with $\mathbf{A} = \begin{bmatrix} -2 & 1 & 0 \\ 0 & -2 & 1 \\ 0 & 0 & -2 \end{bmatrix}$, $\mathbf{B} = \begin{bmatrix} 0 \\ 0 \\ 1 \end{bmatrix}$. Find a constant state feedback matrix $\mathbf{K}$ which gives closed-loop eigenvalues $\lambda_1 = \lambda_2 = \lambda_3 = -1$.

Form $[\lambda\mathbf{I} - \mathbf{A}]$ and its inverse, $\mathbf{\Phi}(\lambda) = \begin{bmatrix} 1/(\lambda+2) & 1/(\lambda+2)^2 & 1/(\lambda+2)^3 \\ 0 & 1/(\lambda+2) & 1/(\lambda+2)^2 \\ 0 & 0 & 1/(\lambda+2) \end{bmatrix}$.

Then

$$\mathbf{\Psi}(\lambda) = \mathbf{\Phi}(\lambda)\mathbf{B} = \begin{bmatrix} 1/(\lambda+2)^3 \\ 1/(\lambda+2)^2 \\ 1/(\lambda+2) \end{bmatrix}$$

In order to obtain the required three linearly independent columns, it is necessary to use $d\mathbf{\Psi}/d\lambda = [-3/(\lambda+2)^4 \quad -2/(\lambda+2)^3 \quad -1/(\lambda+2)^2]^T$ and $d^2\mathbf{\Psi}/d\lambda^2$. When $\lambda = -1$ is used, the numerical result is $\mathbf{G} = \begin{bmatrix} 1 & -3 & 12 \\ 1 & -2 & 6 \\ 1 & -1 & 2 \end{bmatrix}$. Correspondingly, $\mathfrak{F} = [1 \quad 0 \quad 0]$, so that $\mathbf{K} = -[1 \quad 0 \quad 0]\mathbf{G}^{-1} = [-1 \quad 3 \quad -3]$.

Output Feedback

16.6 Specify a constant output feedback matrix so that a system with $\mathbf{A} = \begin{bmatrix} -1 & 0 \\ 0 & -4 \end{bmatrix}$, $\mathbf{B} = \begin{bmatrix} 0 & 1 \\ 1 & 0 \end{bmatrix}$, $\mathbf{C} = [0 \quad 1]$, and $\mathbf{D} = [0 \quad 0]$ will have a closed-loop eigenvalue at -10.

Compute $\Phi(\lambda) = \begin{bmatrix} 1/(\lambda+1) & 0 \\ 0 & 1/(\lambda+4) \end{bmatrix}$ and then form $C\Phi(\lambda)B = [1/(\lambda+4) \quad 0]$. Column 1 must be used, so $G' = -1/6$. The corresponding 2×1 $\mathfrak{J}'$ matrix is just column 1 of I_2. Equation (*16.22*) then becomes $K'(-1/6) = -\begin{bmatrix} 1 \\ 0 \end{bmatrix}$ or $K' = [6 \quad 0]^T$. As a check, this result is substituted into equation (*16.20*) to give $\Delta_0'(\lambda) = \begin{vmatrix} \lambda+1 & 0 \\ 1 & \lambda+10 \end{vmatrix}$, indicating that the closed-loop eigenvalues are $\lambda_1 = -1$ and $\lambda_2 = -10$.

16.7 Another output measurement is added to the system of Problem 16.6, so that now $C = \begin{bmatrix} 1 & 1 \\ 0 & 1 \end{bmatrix}$. Is it possible to use constant output feedback so that both λ_1 and λ_2 equal -10 for the closed-loop system? If yes, find the gain matrix K'.

Since rank $C = 2 = n$, both eigenvalues can be prescribed. Now $C\Phi(\lambda)B = \begin{bmatrix} 1/(\lambda+4) & 1/(\lambda+1) \\ 1/(\lambda+4) & 0 \end{bmatrix}$. Using $\lambda = -10$ and choosing these columns to form G' reduces equation (*16.22*) to $K'\begin{bmatrix} -1/6 & -1/9 \\ -1/6 & 0 \end{bmatrix} = -\begin{bmatrix} 1 & 0 \\ 0 & 1 \end{bmatrix}$ or $K' = \begin{bmatrix} 0 & 6 \\ 9 & -9 \end{bmatrix}$.

16.8 A system [26] is described by equation (*16.1*) with $A = \begin{bmatrix} 0 & 1 & 0 \\ 0 & 0 & 1 \\ 1 & 0 & 0 \end{bmatrix}$, $B = \begin{bmatrix} 0 \\ 1 \\ 0 \end{bmatrix}$, $C = \begin{bmatrix} 1 & 0 & 0 \\ 1 & 1 & 0 \end{bmatrix}$, $D = \begin{bmatrix} 0 \\ 0 \end{bmatrix}$. Since this system is completely controllable and rank $C = m = 2$, two closed-loop eigenvalues can be made arbitrarily close to any specified values by using output feedback. Find K' so that $\lambda_1 = 1$, $\lambda_2 = \epsilon$, where ϵ is arbitrarily close to zero.

As before, form the matrices

$$\Phi(\lambda) = \frac{\begin{bmatrix} \lambda^2 & \lambda & 1 \\ 1 & \lambda^2 & \lambda \\ \lambda & 1 & \lambda^2 \end{bmatrix}}{\lambda^3 - 1} \quad \text{and} \quad C\Phi(\lambda)B = \frac{\begin{bmatrix} \lambda \\ \lambda + \lambda^2 \end{bmatrix}}{\lambda^3 - 1}$$

If $\lambda = 0$, this reduces to a zero vector, and this is what prevents $\lambda = 0$ from being achieved exactly. However, setting $\lambda_2 = \epsilon$ and then using a limiting process allows λ_2 to approach 0. The value $\lambda_1 = 1$ causes unbounded components, so another limiting process must be used. Let $\alpha = \lambda_1^3 - 1$ and select $G' = \begin{bmatrix} 1/\alpha & \epsilon/(\epsilon^3 - 1) \\ 2/\alpha & (\epsilon + \epsilon^2)/(\epsilon^3 - 1) \end{bmatrix}$. Then

$$K' = -[1 \quad 1]G'^{-1} = -\frac{[\alpha(1+\epsilon) - 2(\epsilon^2 - 1/\epsilon) \quad -\alpha + \epsilon^2 - 1/\epsilon]}{\epsilon - 1}$$

Then, as $\alpha \longrightarrow 0$ and $\epsilon \longrightarrow 0$, $K' \longrightarrow [2/\epsilon \quad -1/\epsilon]$. This result indicates that $\lambda_2 = 0$ can be achieved only by using infinite feedback gains. This is analogous to the well-known result for single-input, output systems. An infinite gain is required to cause a closed-loop pole to coincide with an open-loop zero.

Observers

16.9 A system described by equation (*16.1*) has $\mathbf{A} = \begin{bmatrix} -2 & -2 & 0 \\ 0 & 0 & 1 \\ 0 & -3 & -4 \end{bmatrix}$, $\mathbf{B} = \begin{bmatrix} 1 & 0 \\ 0 & 0 \\ 0 & 1 \end{bmatrix}$, $\mathbf{C} = \begin{bmatrix} 1 & 0 & 1 \\ 0 & 1 & 0 \end{bmatrix}$, $\mathbf{D} = [\mathbf{0}]$. Design an observer whose output settles to the actual state vector within one second.

It should be verified that the system is completely observable. The specification can be translated into desired eigenvalue locations. The response of the error equation (*16.28*) will be made up of terms involving $e^{\lambda_i t}$, where λ_i is an eigenvalue to be selected for $\mathbf{A}_c$. By considering, for example, e^{-4} as being a close enough approximation to zero, all eigenvalues of $\mathbf{A}_c$ must have real parts more negative than -4. Select $\lambda_1 = \lambda_2 = -5$ and $\lambda_3 = -6$. For this system (see Problem 8.20),

$$\mathbf{\Phi}(\lambda) = (\lambda\mathbf{I} - \mathbf{A})^{-1} = \begin{bmatrix} \dfrac{1}{\lambda+2} & \dfrac{-2(\lambda+4)}{(\lambda+1)(\lambda+2)(\lambda+3)} & \dfrac{-2}{(\lambda+1)(\lambda+2)(\lambda+3)} \\ 0 & \dfrac{\lambda+4}{(\lambda+1)(\lambda+3)} & \dfrac{1}{(\lambda+1)(\lambda+3)} \\ 0 & \dfrac{-3}{(\lambda+1)(\lambda+3)} & \dfrac{\lambda}{(\lambda+1)(\lambda+3)} \end{bmatrix}$$

and

$$\mathbf{C\Phi}(\lambda) = \begin{bmatrix} \dfrac{1}{\lambda+2} & \dfrac{-5\lambda-14}{(\lambda+1)(\lambda+2)(\lambda+3)} & \dfrac{\lambda^2+2\lambda-2}{(\lambda+1)(\lambda+2)(\lambda+3)} \\ 0 & \dfrac{\lambda+4}{(\lambda+1)(\lambda+3)} & \dfrac{1}{(\lambda+1)(\lambda+3)} \end{bmatrix}$$

Using rows 1 and 2 with $\lambda = -5$ and row 2 again with $\lambda = -6$ gives

$$\mathbf{G}_c = \begin{bmatrix} -1/3 & -11/24 & -13/24 \\ 0 & -1/8 & 1/8 \\ 0 & -2/15 & 1/15 \end{bmatrix} \quad \text{and} \quad \mathfrak{F}_c = \begin{bmatrix} 1 & 0 \\ 0 & 1 \\ 0 & 1 \end{bmatrix}$$

Using these in equation (*16.31*) gives $\mathbf{B}_c = \begin{bmatrix} 3 & -8 \\ 0 & 7 \\ 0 & -1 \end{bmatrix}$, and then $\mathbf{A}_c = \begin{bmatrix} -5 & 6 & -3 \\ 0 & -7 & 1 \\ 0 & -2 & -4 \end{bmatrix}$.
The observer is defined by $\dot{\hat{\mathbf{x}}} = \mathbf{A}_c\hat{\mathbf{x}} + \mathbf{B}_c\mathbf{y} + \mathbf{Bu}$, where $\mathbf{u}$ and $\mathbf{y}$ are the input and output of the system being observed.

16.10 Consider the open-loop system described by the matrices **A**, **B**, **C**, and **D** of Example 16.8. Design an observer with eigenvalues of -8, -8, and -10.

Since the output is $\mathbf{y} = \mathbf{Cx} + \mathbf{Du}$, the observer equations are modified to take into account the nonzero **D** matrix. Let $\dot{\hat{\mathbf{x}}} = \mathbf{A}_c\hat{\mathbf{x}} + \mathbf{B}_c\mathbf{y} + \mathbf{z} = \mathbf{A}_c\hat{\mathbf{x}} + \mathbf{B}_c\mathbf{Cx} + \mathbf{B}_c\mathbf{Du} + \mathbf{z}$. It is seen that setting $\mathbf{z} = (\mathbf{B} - \mathbf{B}_c\mathbf{D})\mathbf{u}$ again leads to the error equation (*16.28*). Therefore, the observer matrices $\mathbf{A}_c$ and $\mathbf{B}_c$ are designed as in Sec. 6.7, but the observer input is modified.

$$\mathbf{\Phi}(\lambda) = (\lambda\mathbf{I} - \mathbf{A})^{-1} = \begin{bmatrix} 1/(\lambda+2) & 1/(\lambda+2)^2 & 0 \\ 0 & 1/(\lambda+2) & 0 \\ 0 & 0 & 1/(\lambda-4) \end{bmatrix}$$

$$\mathbf{C\Phi} = \begin{bmatrix} 0 & 0 & 1/(\lambda-4) \\ 1/(\lambda+2) & 1/(\lambda+2)^2 & 0 \end{bmatrix}$$

Evaluating rows 1 and 2 with $\lambda = -8$, and row 2 again with $\lambda = -10$ yields

$$\mathbf{G}_c = \begin{bmatrix} 0 & 0 & -1/12 \\ -1/6 & 1/36 & 0 \\ -1/8 & 1/64 & 0 \end{bmatrix}$$

and the corresponding matrix is $\mathfrak{J}_c = \begin{bmatrix} 1 & 0 \\ 0 & 1 \\ 0 & 1 \end{bmatrix}$. Equation (*16.31*) gives $\mathbf{B}_c = \begin{bmatrix} 0 & 14 \\ 0 & 48 \\ 12 & 0 \end{bmatrix}$.
Then using $\mathbf{A}_c = \mathbf{A} - \mathbf{B}_c\mathbf{C}$ and $\mathbf{z} = (\mathbf{B} - \mathbf{B}_c\mathbf{D})\mathbf{u}$ gives the observer shown in Fig. 16.8.

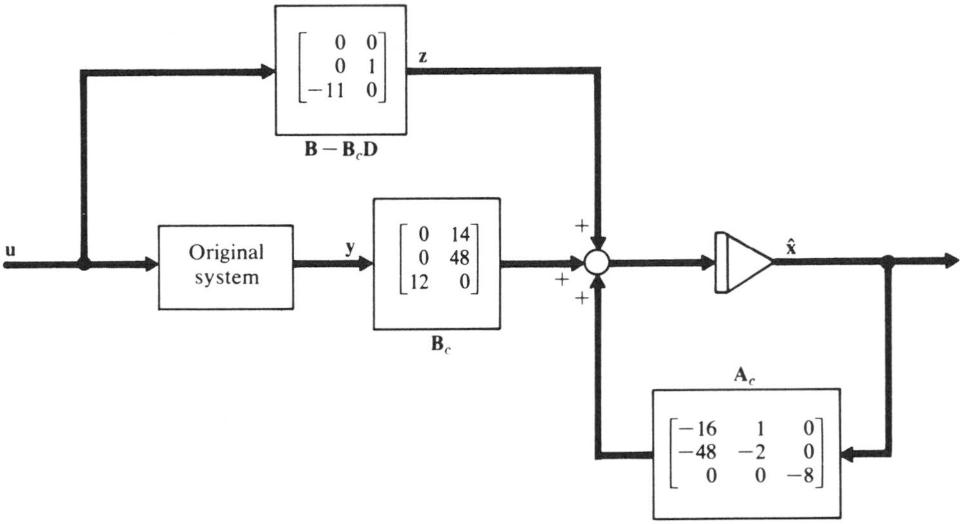

Figure 16.8

16.11 Consider the open-loop unstable system of Example 16.8. Design a closed-loop system with dynamic feedback, which has eigenvalues -5, -6, and -6. The eigenvalues added by the feedback compensation should have real parts more negative than -7.

It is first assumed that all state variables are available for feedback. A state feedback matrix $\mathbf{K}$ is selected to satisfy the specification $\lambda_i \in \{-5, -6, -6\}$.

$$\mathbf{\Phi}(\lambda)\mathbf{B} = \begin{bmatrix} 0 & 1/(\lambda + 2)^2 \\ 0 & 1/(\lambda + 2) \\ 1/(\lambda - 4) & 0 \end{bmatrix} \triangleq \mathbf{\Psi}(\lambda)$$

Selecting $\mathbf{G} = [\boldsymbol{\psi}_2(-5) \quad \boldsymbol{\psi}_2(-6) \quad \boldsymbol{\psi}_1(-6)]$ and $\mathfrak{J} = \begin{bmatrix} 0 & 0 & 1 \\ 1 & 1 & 0 \end{bmatrix}$ gives, from equation (*16.15*), $\mathbf{K} = -\begin{bmatrix} 0 & 0 & 1 \\ 1 & 1 & 0 \end{bmatrix} \begin{bmatrix} 1/9 & 1/16 & 0 \\ -1/3 & -1/4 & 0 \\ 0 & 0 & -1/10 \end{bmatrix}^{-1} = \begin{bmatrix} 0 & 0 & 10 \\ 12 & 7 & 0 \end{bmatrix}$. This result cannot be used directly since all state variables are not available. However, an observer was designed for this system in Problem 16.10. Since that observer meets the specification, it can be used to provide an estimate of $\mathbf{x}$. Then $\mathbf{K}$ is applied to that estimate, as shown in Fig. 16.9.

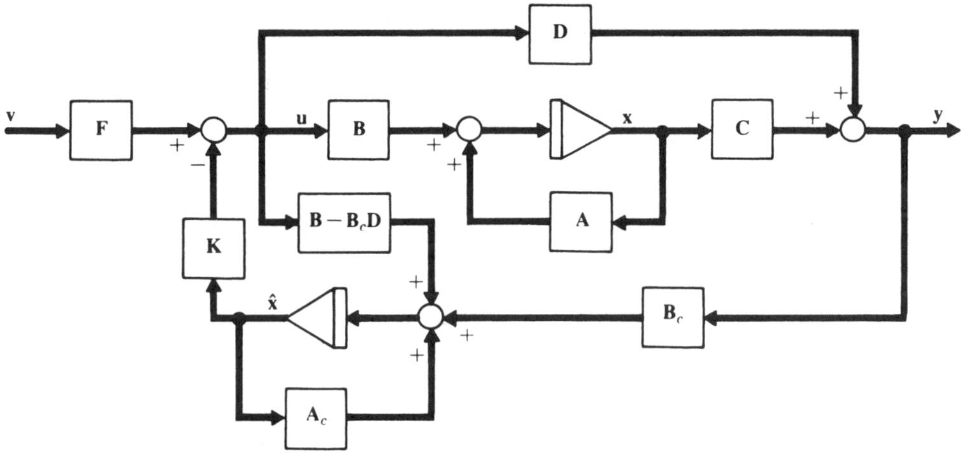

Figure 16.9

16.12 When can a system of the general form of equation (*16.1*) or (*16.2*) be transformed so that $\mathbf{C} = [\mathbf{I} \mid \mathbf{0}]$?

From Sec. 3.10 it is clear that the first requirement is that **C** must have full rank m. Assume that this is so and that the first m columns of **C** are linearly independent. Then by a series of row operations (equivalent to premultiplication by a nonsingular transformation matrix $\mathbf{T}_1$) **C** can be brought into row-reduced-echelon form. That is

$$\mathbf{T}_1\mathbf{y} = \mathbf{T}_1\mathbf{C}\mathbf{x} + \mathbf{T}_1\mathbf{D}\mathbf{u}$$

or

$$\mathbf{y}' = [\mathbf{I} \mid \mathbf{C}']\mathbf{x} + \mathbf{D}'\mathbf{u}$$

To further reduce **C** (i.e., to eliminate $\mathbf{C}'$), column operations are required. These are accomplished by postmultiplying by a nonsingular transformation matrix $\mathbf{T}_2$. The only way this can be accomplished is to redefine the state vector according to

$$\mathbf{x} = \mathbf{T}_2\mathbf{x}'$$

Then $\mathbf{y}' = [\mathbf{I} \mid \mathbf{0}]\mathbf{x}' + \mathbf{D}'\mathbf{u}$. The dynamics equation must similarly be transformed into $\mathbf{x}'$ variables

$$\begin{aligned}\mathbf{x}'(k+1) &= \mathbf{T}_2^{-1}\mathbf{A}\mathbf{T}_2\mathbf{x}'(k) + \mathbf{T}_2^{-1}\mathbf{B}\mathbf{u}(k)\\ &= \mathbf{A}'\mathbf{x}'(k) + \mathbf{B}'\mathbf{u}(k)\end{aligned}$$

16.13 A third-order discrete-time system with two inputs and one output is described by

$$\mathbf{A} = \begin{bmatrix} 0 & -0.5 & -0.1 \\ -0.1 & -0.7 & 0.2 \\ 0.8 & -0.8 & 1 \end{bmatrix} \quad \mathbf{B} = \begin{bmatrix} 1 & -1 \\ -1 & 1 \\ 0 & 3 \end{bmatrix} \quad \mathbf{C} = [1 \quad 0 \quad 0]^T$$

Design a reduced-order observer to estimate x_2 and x_3. Place its poles at $z = 0.1$ and 0.2.

Using the algorithm of Sec. 16.5 with the equivalent **A** and **B** matrices selected as

$$\mathbf{A}_{22}^T = \begin{bmatrix} -0.7 & -0.8 \\ 0.2 & 1 \end{bmatrix} \quad \text{and} \quad \mathbf{A}_{12}^T = \begin{bmatrix} -0.5 \\ -0.1 \end{bmatrix}$$

gives a gain matrix **K** which is

$$\mathbf{B}_r^T = [0.391608 \quad -1.95804]^T$$

Using this to form $\mathbf{A}_r = \mathbf{A} - \mathbf{B}_r\mathbf{A}_{12}$ gives the observer the result shown in Fig. 16.10.

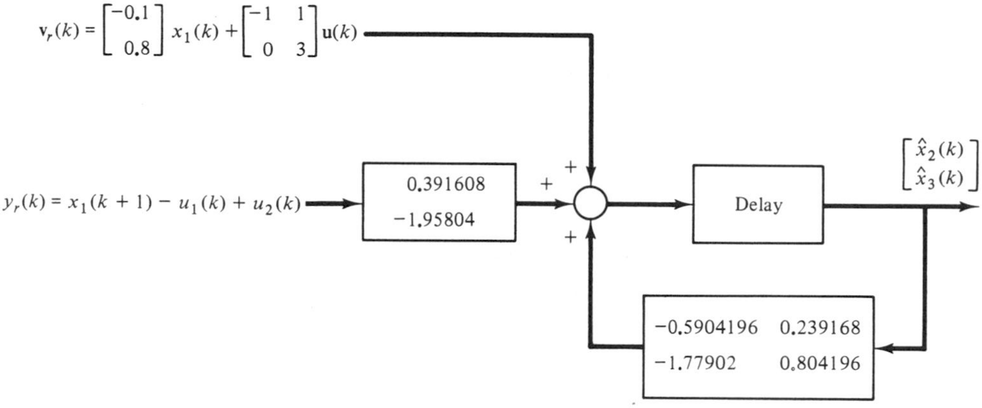

Figure 16.10

Decoupling

16.14 Use state feedback to decouple the system [36] described by

$$\mathbf{A} = \begin{bmatrix} 1 & 1 & 0 \\ 0 & 1 & 0 \\ 0 & 0 & 1 \end{bmatrix}, \quad \mathbf{B} = \begin{bmatrix} 0 & 1 \\ 1 & 0 \\ 1 & 0 \end{bmatrix}, \quad \mathbf{C} = \begin{bmatrix} 1 & 1 & -1 \\ 0 & 1 & 0 \end{bmatrix}, \quad \mathbf{D} = [\mathbf{0}]$$

Using equation (*16.36*), $\mathbf{c}_1\mathbf{B} = [0 \quad 1] \neq \mathbf{0}$, so $d_1 = 0$. $\mathbf{c}_2\mathbf{B} = [1 \quad 0] \neq \mathbf{0}$, so $d_2 = 0$. Then $\mathbf{N} = \begin{bmatrix} 0 & 1 \\ 1 & 0 \end{bmatrix}$ is nonsingular and the system can be decoupled. $\mathbf{F}_d = \mathbf{N}^{-1} = \begin{bmatrix} 0 & 1 \\ 1 & 0 \end{bmatrix}$ and $\mathbf{K}_d = \begin{bmatrix} 0 & 1 \\ 1 & 0 \end{bmatrix}\begin{bmatrix} \mathbf{c}_1\mathbf{A} \\ \mathbf{c}_2\mathbf{A} \end{bmatrix} = \begin{bmatrix} 0 & 1 & 0 \\ 1 & 2 & -1 \end{bmatrix}$. Using these results in equation (*16.34*) gives the integrator decoupled result, $\mathbf{H}(s) = \begin{bmatrix} 1/s & 0 \\ 0 & 1/s \end{bmatrix}$.

Note that the original system is not completely controllable, and thus the decoupled system is not controllable either. Any attempt to specify all three eigenvalues using the method of Sec. 16.4 or 16.5 will fail. To see this, let $\tilde{\mathbf{A}} = \mathbf{A} - \mathbf{B}\mathbf{K}_d$, $\tilde{\mathbf{B}} = \mathbf{B}\mathbf{F}_d$. Then $\boldsymbol{\Psi}(\lambda) = \boldsymbol{\Phi}(\lambda)\tilde{\mathbf{B}} = \begin{bmatrix} 1/\lambda & 0 \\ 0 & 1/\lambda \\ 0 & 1/\lambda \end{bmatrix}$, so that only two independent columns

$\psi_j(\lambda_i)$ can be selected, regardless of the values of λ_i. Derivatives of $\psi_j(\lambda_i)$ do not produce additional independent columns.

16.15 Can the system of Example 15.8 be decoupled using state feedback?

The irreducible realization found in Example 15.8 is used. The product $\mathbf{c}_1\mathbf{B} = [0 \quad 1]$ is not zero, so $d_1 = 0$. Likewise, $\mathbf{c}_2\mathbf{B} = [1 \quad 0]$, so $d_2 = 0$. Then $\mathbf{N} = \begin{bmatrix} 0 & 1 \\ 1 & 0 \end{bmatrix}$ is nonsingular, so the system can be decoupled.

The integrator decoupled system is obtained by using $\mathbf{F}_d = \mathbf{N}^{-1} = \begin{bmatrix} 0 & 1 \\ 1 & 0 \end{bmatrix}$, and $\mathbf{K}_d = \mathbf{N}^{-1}\begin{bmatrix} \mathbf{c}_1\mathbf{A} \\ \mathbf{c}_2\mathbf{A} \end{bmatrix} = \begin{bmatrix} 0 & 0 & -2 & 0 \\ -2/3 & 5/9 & -5/27 & -5 \end{bmatrix}$. Equation (*16.34*) gives the decoupled closed-loop transfer function $\mathbf{H}(s) = \begin{bmatrix} 1/s & 0 \\ 0 & 1/s \end{bmatrix}$.

16.16 Consider the open-loop, decoupled system described by $\tilde{\mathbf{A}}$, $\tilde{\mathbf{B}}$, and $\tilde{\mathbf{C}}$ of Example 16.11 and with $\mathbf{D} = \begin{bmatrix} 1 & 0 \\ 0 & 0 \end{bmatrix}$. Find an output feedback matrix which places closed-loop poles at $\lambda_1 = -5$, $\lambda_2 = -10$. Is the resultant system still decoupled? Is it stable?

Let the output control law be $\mathbf{u} = -\mathbf{K}'\mathbf{y} + \mathbf{v}'$, that is, $\mathbf{F}' = \mathbf{I}$. Using the results of Example 16.11, $\mathbf{\Psi}'(\lambda) = \mathbf{C}[\lambda\mathbf{I} - \mathbf{A}]^{-1}\mathbf{B} = \begin{bmatrix} 1/\lambda & 0 \\ 0 & 1/\lambda^2 \end{bmatrix}$. Selecting

$$\mathbf{G}' = [\boldsymbol{\psi}_1(-5) \quad \boldsymbol{\psi}_2(-10)] = \begin{bmatrix} -1/5 & 0 \\ 0 & 1/100 \end{bmatrix}, \qquad \mathfrak{F}' = \begin{bmatrix} 1 & 0 \\ 0 & 1 \end{bmatrix}$$

and using equation (*16.23*) gives the output feedback matrix

$$\mathbf{K}' = -\left\{\mathbf{I}_2 + \begin{bmatrix} -5 & 0 \\ 0 & 100 \end{bmatrix}\begin{bmatrix} 1 & 0 \\ 0 & 0 \end{bmatrix}\right\}^{-1}\begin{bmatrix} -5 & 0 \\ 0 & 100 \end{bmatrix} = \begin{bmatrix} -5/4 & 0 \\ 0 & -100 \end{bmatrix}$$

When this feedback is used, the characteristic equation (*16.20*) is

$$\Delta_0'(\lambda) = |\lambda\mathbf{I}_3 - \tilde{\mathbf{A}} + \tilde{\mathbf{B}}\mathbf{K}'[\mathbf{I}_2 + \tilde{\mathbf{D}}\mathbf{K}']^{-1}\tilde{\mathbf{C}}| = \left|\begin{bmatrix} \lambda + 2 & -1 & 0 \\ -96 & \lambda - 2 & 0 \\ 0 & 0 & \lambda + 5 \end{bmatrix}\right|$$
$$= (\lambda + 5)(\lambda^2 - 100) = 0$$

Therefore, the eigenvalues are at -5, -10, and $+10$ and the system is unstable. Using equations (*16.6*) and (*16.7*), and the fact that $\mathbf{F}' = \mathbf{I}_2$, leads to

$$\mathbf{H}(s) = [\mathbf{I}_2 + \mathbf{D}\mathbf{K}']^{-1}\{\tilde{\mathbf{C}}\{s\mathbf{I}_3 - \tilde{\mathbf{A}} + \tilde{\mathbf{B}}\mathbf{K}'[\mathbf{I}_2 + \mathbf{D}\mathbf{K}']^{-1}\tilde{\mathbf{C}}\}^{-1}\tilde{\mathbf{B}}[\mathbf{I}_2 + \mathbf{K}'\mathbf{D}]^{-1} + \mathbf{D}\}$$
$$= \begin{bmatrix} -4(s+1)/(s+5) & 0 \\ 0 & 1/(s^2 - 100) \end{bmatrix}$$

The system is still uncoupled. If $\mathbf{D} = [\mathbf{0}]$, as in Example 16.10, then $\mathbf{K}' = \begin{bmatrix} 5 & 0 \\ 0 & -100 \end{bmatrix}$ and $\mathbf{H}(s) = \begin{bmatrix} 1/(s+5) & 0 \\ 0 & 1/(s^2 - 100) \end{bmatrix}$.

16.17 Assume that a particular system can be decoupled by use of state feedback. Assume that the system is also completely observable. Show that the system can be decoupled by using an observer to estimate the state and then by using the estimated state with the decoupling matrices.

Let the system be described by $\dot{\mathbf{x}} = \mathbf{Ax} + \mathbf{Bu}$, $\mathbf{y} = \mathbf{Cx}$, and the decoupling state feedback is $\mathbf{u} = \mathbf{F}_d\mathbf{v} - \mathbf{K}_d\mathbf{x}$. Let the observer be described by $\dot{\hat{\mathbf{x}}} = \mathbf{A}_c\hat{\mathbf{x}} + \mathbf{B}_c\mathbf{y} + \mathbf{Bu}$. It is to be shown that $\mathbf{x}$ can be replaced by $\hat{\mathbf{x}}$ in the control law without altering the decoupling. Writing $\hat{\mathbf{x}} = \mathbf{x} - \mathbf{e}$ and using $\mathbf{u} = \mathbf{F}_d\mathbf{v} - \mathbf{K}_d\hat{\mathbf{x}} = \mathbf{F}_d\mathbf{v} - \mathbf{K}_d\mathbf{x} + \mathbf{K}_d\mathbf{e}$ gives

$$\begin{bmatrix} \dot{\mathbf{x}} \\ \dot{\mathbf{e}} \end{bmatrix} = \begin{bmatrix} \mathbf{A} - \mathbf{BK}_d & \mathbf{BK}_d \\ \mathbf{0} & \mathbf{A}_c \end{bmatrix} \begin{bmatrix} \mathbf{x} \\ \mathbf{e} \end{bmatrix} + \begin{bmatrix} \mathbf{BF}_d \\ \mathbf{0} \end{bmatrix} \mathbf{v}$$

Using Laplace transforms gives

$$\begin{bmatrix} \mathbf{x}(s) \\ \mathbf{e}(s) \end{bmatrix} = \begin{bmatrix} s\mathbf{I} - \mathbf{A} + \mathbf{BK}_d & -\mathbf{BK}_d \\ \mathbf{0} & s\mathbf{I} - \mathbf{A}_c \end{bmatrix}^{-1} \begin{bmatrix} \mathbf{BF}_d \\ \mathbf{0} \end{bmatrix} \mathbf{v}(s)$$

Only the upper-left block of the inverse is needed in order to solve for $\mathbf{x}(s)$. Using results of Sec. 3.9, this reduces to $\mathbf{x}(s) = [s\mathbf{I} - \mathbf{A} + \mathbf{BK}_d]^{-1}\mathbf{BF}_d\mathbf{v}(s)$ and the output is $\mathbf{y} = \mathbf{C}[s\mathbf{I} - \mathbf{A} + \mathbf{BK}_d]^{-1}\mathbf{BF}_d\mathbf{v}(s)$. Comparing this with equation (*16.34*) shows that the system is still decoupled.

16.18 Figure 16.11 shows a system described by $\{\mathbf{A}, \mathbf{B}, \mathbf{C}\}$, an observer described by $\{\mathbf{A}_c, \mathbf{B}_c\}$, decoupling matrices $\{\mathbf{K}_d, \mathbf{F}_d\}$, and a pole placement feedback matrix $\tilde{\mathbf{K}}$. Let the system be the one in Example 16.8, but with $\mathbf{D} = [0]$. The decoupling matrices are calculated in Example 16.10, and the pole placement matrix is the one found in Example 16.11. Show that using the output and an observer yields the same decoupled system transfer matrix as was obtained in Example 16.11.

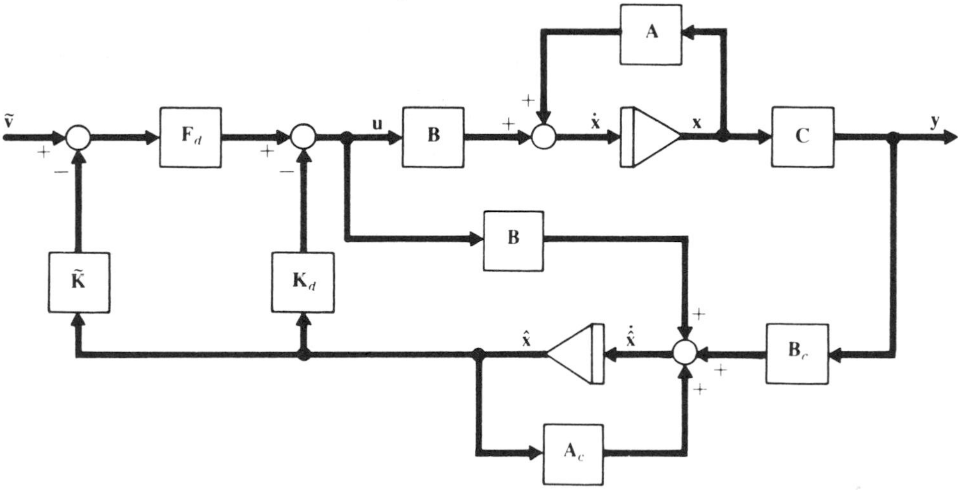

Figure 16.11

The dynamic equations are $\dot{\mathbf{x}} = \mathbf{Ax} + \mathbf{Bu}$ and $\dot{\hat{\mathbf{x}}} = \mathbf{A}_c\hat{\mathbf{x}} + \mathbf{B}_c\mathbf{Cx} + \mathbf{Bu}$. The control law is $\mathbf{u} = \mathbf{F}_d\tilde{\mathbf{v}} - (\mathbf{F}_d\tilde{\mathbf{K}} + \mathbf{K}_d)\hat{\mathbf{x}}$. Using $\mathbf{e} = \mathbf{x} - \hat{\mathbf{x}}$ to eliminate $\hat{\mathbf{x}}$ leads to

$$\begin{bmatrix} \dot{\mathbf{x}} \\ \dot{\mathbf{e}} \end{bmatrix} = \begin{bmatrix} \mathbf{A} - \mathbf{B}(\mathbf{F}_d\tilde{\mathbf{K}} + \mathbf{K}_d) & \mathbf{B}(\mathbf{F}_d\tilde{\mathbf{K}} + \mathbf{K}_d) \\ \mathbf{0} & \mathbf{A}_c \end{bmatrix} \begin{bmatrix} \mathbf{x} \\ \mathbf{e} \end{bmatrix} + \begin{bmatrix} \mathbf{BF}_d \\ \mathbf{0} \end{bmatrix} \tilde{\mathbf{v}}$$

Taking the Laplace transform and solving for $\mathbf{x}(s)$ gives

$$\mathbf{y}(s) = \mathbf{Cx}(s) = \mathbf{C}[s\mathbf{I} - \mathbf{A} + \mathbf{B}(\mathbf{F}_d\tilde{\mathbf{K}} + \mathbf{K}_d)]^{-1}\mathbf{BF}_d\tilde{\mathbf{v}}$$

This is the same result as in Example 16.11 and gives

$$\mathbf{H}(s) = \begin{bmatrix} 1/(s+5) & 0 \\ 0 & 1/[(s+5)(s+10)] \end{bmatrix}$$

16.19 Discuss the design techniques of this chapter as contrasted to classical techniques.

In the classical methods, reviewed in Chapter 2, there are usually just one input and one output. Feedback is always output feedback, but perhaps it involves dynamic compensation (lead networks, lead-lag networks, etc.). The design is often developed (at the introductory level) as a trial and error process. The goal is to obtain satisfactory locations for the dominant poles, to ensure acceptable overshoot, settling time, natural frequency, etc. In addition, sensitivity to parameter variations and disturbances is to be reduced and steady-state accuracy must be acceptable. See Reference 82 for a treatment of classical problems using the state variable approach. See also Reference 76.

In this chapter a systematic means of locating *all* poles is presented, provided the system is completely controllable and observable. Multiple inputs and outputs can be treated, but a computer will be required in analyzing most realistic applications. When the feedback is restricted to output signals, observers (or reduced dimensional observers or other dynamic compensators) will be required. This is analogous to using dynamic network compensators in classical design. Sensitivity reduction [24, 90] has not been considered here because of space limitations. However, equations (*16.9*) and (*16.21*) are related to the concept of return difference and the reduction of sensitivity [60].

16.20 What can be said about pole placement by means of state feedback if a system is not completely controllable?

A system can always be decomposed into a controllable subsystem and an uncontrollable subsystem. The poles of the controllable subsystem can be relocated as desired. The poles of the uncontrollable part are unaffected by state feedback. Therefore, an uncontrollable, unstable system cannot be stabilized by state feedback if the unstable eigenvalues (poles) are associated with the uncontrollable part.

PROBLEMS

16.21 Assuming that all state variables can be measured, find a state feedback matrix **K** for the system of Problem 16.9. The closed-loop poles are to be placed at $\lambda_i = -3, -3, -4$.

16.22 The open-loop system of Problem 16.9 is compensated by adding feedback consisting of the observer designed in Problem 16.9 followed by the constant matrix **K** of Problem 16.21. Verify that the 6th-order closed-loop system has eigenvalues at $\{-3, -3, -4, -5, -5, -6\}$.

16.23 Repeat Example 16.7, except use $\boldsymbol{\psi}_2'(\lambda_1)$ instead of $\boldsymbol{\psi}_1'(\lambda_1)$. This means that $\mathfrak{F}' = \begin{bmatrix} 0 \\ 1 \end{bmatrix}$ instead of $\begin{bmatrix} 1 \\ 0 \end{bmatrix}$.

16.24 What should the constant output feedback matrix $\mathbf{K}'$ be if the system in Example 16.7 is to have a closed-loop eigenvalue at -20?

16.25 If Problem 16.6 is reconsidered, with $\mathbf{C} = \begin{bmatrix} 1 & 0 \\ 0 & 1 \end{bmatrix}$, find $\mathbf{K}'$ so that $\lambda_1 = \lambda_2 = -10$.

16.26 Repeat Problem 16.8 with the desired eigenvalues $\lambda_1 = \epsilon$, $\lambda_2 = -\epsilon$.

16.27 Let $\mathbf{A} = \begin{bmatrix} 0 & 1 & 0 \\ 0 & 0 & 1 \\ 1 & 0 & 0 \end{bmatrix}$, $\mathbf{B} = \begin{bmatrix} 0 & 1 \\ 1 & 0 \\ 0 & 0 \end{bmatrix}$, $\mathbf{C} = \begin{bmatrix} 1 & 0 & 0 \\ 0 & 0 & 1 \end{bmatrix}$. Can this system be decoupled?

16.28 Can the system described by $\mathbf{A} = \begin{bmatrix} 1 & 0 & 0 \\ 0 & 0 & 1 \\ 0 & 1 & 0 \end{bmatrix}$, $\mathbf{B} = \begin{bmatrix} 1 & 0 \\ 1 & 0 \\ 1 & 1 \end{bmatrix}$, $\mathbf{C} = \begin{bmatrix} 1 & 0 & 0 \\ 0 & 1 & 0 \end{bmatrix}$ be decoupled?

16.29 Find the matrices $\mathbf{F}_d$ and $\mathbf{K}_d$ such that state feedback reduces the system of Problem 16.9 to integrator decoupled form. Find the decoupled transfer matrix also.

16.30 Find a state feedback gain matrix which will give the following discrete-time system closed-loop poles at $z = 0$ and 0.5.

$$\mathbf{x}(k+1) = \begin{bmatrix} 0 & 1 \\ 1 & 1 \end{bmatrix} \mathbf{x}(k) + \begin{bmatrix} 0 \\ 1 \end{bmatrix} \mathbf{u}(k)$$

16.31 Repeat Problem 16.30, but force the poles to be at $z = 0.3 \pm j\,0.5$.

16.32 A discrete-time system has

$$\mathbf{A} = \begin{bmatrix} 0.65 & -0.15 \\ -0.15 & 0.65 \end{bmatrix} \quad \text{and} \quad \mathbf{B} = \begin{bmatrix} 1 & 1 \\ -1 & 1 \end{bmatrix}$$

Find two different feedback gain matrices which will both give closed-loop poles at $z = 0$ and 0.2.

16.33 A system is described by

$$\dot{y}_1 + 3y_1 - y_2 = u_1$$

$$\ddot{y}_2 + 2(\dot{y}_1 + \dot{y}_2 - \dot{y}_3) + 4(y_2 - y_1) = u_2 + 5u_1$$

$$\ddot{y}_3 + 6\dot{y}_3 - 2\dot{y}_1 + y_3 = u_2$$

Put the system into state variable form, with the first three states being y_1, y_2, and y_3 respectively. Design a state feedback controller which will give closed-loop poles at -3, -4, -4, -5, and -5. Simulate the system for various test inputs, assuming all states are available for use. Then design a full state observer, with poles at -6, -7, -8, -9, and -10. Again on a simulation, compare the transient response of the cascaded observer, state feedback system with full state feedback results.

16.34 Design a reduced-order observer to estimate only the nonmeasured states of Problem 16.33. Use this observer, along with the previous state feedback controller, and compare transient behavior with that obtained in the previous problem.

16.35 Using the continuous system model of Problem 12.7, show that the closed-loop poles will be at $\lambda_i = -2, -5, -8$, and -10 if any of the following state feedback matrices are used. In each case the columns selected for use in equation (*16.17*) are shown. Compare these results with the discrete equivalent of Example 16.5.

The Feedback Gain Matrix Using Columns 1, 3, 6, 8

$$\mathbf{K} = \begin{bmatrix} -1.5000001E-01 & 7.4249178E-01 & -3.1340556E+00 & 5.0000000E-01 \\ 0.0000000E+00 & -5.3035126E+00 & 1.8814680E+01 & 0.0000000E+00 \end{bmatrix}$$

The Feedback Gain Matrix Using Columns 2, 4, 5, 7

$$\mathbf{K} = \begin{bmatrix} 4.0000007E-01 & 2.5047928E-01 & -3.4747612E-01 & 4.0000005E+00 \\ 0.0000000E+00 & -1.7891376E+00 & -1.0894562E+00 & 0.0000000E+00 \end{bmatrix}$$

The Feedback Gain Matrix Using Columns 1, 4, 6, 7

$$\mathbf{K} = \begin{bmatrix} 9.9999994E-02 & 5.1884985E-01 & -1.5014696E+00 & 1.0000000E+00 \\ 0.0000000E+00 & -3.7060702E+00 & 7.1533542E+00 & 0.0000000E+00 \end{bmatrix}$$

Chapter

17

AN INTRODUCTION TO OPTIMAL CONTROL THEORY

17.1 INTRODUCTION

The word "optimal" intuitively means doing a job in the best of all possible ways. Before a more definitive meaning can be attached, the "job" must be defined, a mathematical scale must be established for indicating what "best" means, and the "possible" alternatives must be spelled out. Without these qualifiers, a claim that a system is optimal is really meaningless. A crude, inaccurate system might be optimal because it is inexpensive and easy to fabricate and its performance is adequate. Conversely, a very precise system could be nonoptimal because it is too expensive or too heavy.

An introductory account of dynamic programming [7] and the minimum principle [95] approaches to optimal control problems is given in this chapter. These and other optimization techniques are discussed from various points of view and at various levels in References 14, 16, 65, 70, 96, 100, and 115. A linear programming approach has already been given in Problem 5.17. The minimum norm solutions of Problems 4.25 and 6.13 illustrate still another approach, which is generalized in Problem 17.1.

17.2 STATEMENT OF THE OPTIMAL CONTROL PROBLEM

A mathematical statement of the optimal control problem consists of:

1. A description of the system to be controlled.
2. A description of system constraints and possible alternatives.

3. A description of the task to be accomplished.
4. A statement of the criterion for judging optimal performance.

The dynamic systems to be considered are described in state variable form by one of the forms in equation (*17.1*). It is assumed that all states are available as output measurements, i.e., $\mathbf{y}(t) = \mathbf{x}(t)$. If this is not the case, the state of an *observable* linear system can be estimated by using an observer or a Kalman filter.

$$\dot{\mathbf{x}} = \mathbf{f}(\mathbf{x}(t), \mathbf{u}(t), t) \qquad \text{or} \qquad \mathbf{x}(k+1) = \mathbf{f}(\mathbf{x}(k), \mathbf{u}(k)) \tag{17.1}$$

Constraints will sometimes exist on allowable values of the state variables. However, in this chapter only control variable constraints are considered. The set of admissible controls U is a subset of the r-dimensional input space $\mathcal{U}^r$ of Chapter 9, $U \subseteq \mathcal{U}^r$. For example, U could be defined as the set of all piecewise continuous vectors $\mathbf{u}(t) \in \mathcal{U}^r$ satisfying $|\mathbf{u}_i(t)| \leq M$ or $||\mathbf{u}(t)|| \leq M$ for all t, and for some positive constant M. If there are no constraints, $U = \mathcal{U}^r$.

The task to be performed often takes the form of additional boundary conditions on equation (*17.1*). An example might be to transfer the state from a known initial state $\mathbf{x}(t_0)$ a specified final state $\mathbf{x}(t_f) = \mathbf{x}_d$ at a specified time t_f, or the minimum possible t_f. The task might be to transfer the state to a specified region of state space, called a *target set*, rather than to a specified point. Often, the task to be performed is implicitly specified by the performance criterion.

The most general continuous or discrete general performance criteria to be considered are

$$J = S(\mathbf{x}(t_f), t_f) + \int_{t_0}^{t_f} L(\mathbf{x}(t), \mathbf{u}(t), t)\, dt$$

or

$$J = S(\mathbf{x}(N)) + \sum_{k=0}^{N-1} L(\mathbf{x}(k), \mathbf{u}(k)) \tag{17.2}$$

S and L are real, scalar-valued functions of the indicated arguments. S is the cost or penalty associated with the error in the stopping or terminal state at time t_f (or N). L is the cost or loss function associated with the transient state errors and control effort. The functions S and L must be selected by the system designer to put more or less emphasis on terminal accuracy, transient behavior, and the expended control effort in the total cost function J.

Example 17.1

Set $S = 0$, $L = 1$. Then $J = \int_{t_0}^{t_f} dt = t_f - t_0$. This is the minimum time problem. ■

Example 17.2

Set $S = 0$, $L = \mathbf{u}^T\mathbf{u}$. Then $J = \int_{t_0}^{t_f} \mathbf{u}^T\mathbf{u}\, dt$ is a measure of the control effort expended. In many cases this term can be interpreted as control energy. This is called the least-effort problem. ■

Example 17.3

Set $S = [\mathbf{x}(t_f) - \mathbf{x}_d]^T[\mathbf{x}(t_f) - x_d]$ and $L = 0$. Then minimizing J is equivalent to minimizing the square of the norm of the error between the final state $\mathbf{x}(t_f)$ and a desired final state $\mathbf{x}_d$. This is the minimum terminal error problem. ■

Example 17.4

Set $S = 0$, $L = [\mathbf{x}(t) - \eta(t)]^T[\mathbf{x}(t) - \eta(t)]$. Then minimizing J is equivalent to minimizing the integral of the norm squared of the transient error between the actual state trajectory $\mathbf{x}(t)$ and a desired trajectory $\eta(t)$. ■

Example 17.5

A general quadratic criterion which gives a weighted trade-off between the previous three criteria uses $S = [\mathbf{x}(t_f) - \mathbf{x}_d]^T\mathbf{M}[x(t_f) - \mathbf{x}_d]$ and $L = [\mathbf{x}(t) - \eta(t)]^T\mathbf{Q}[\mathbf{x}(t) - \eta(t)] + \mathbf{u}(t)^T\mathbf{R}\mathbf{u}(t)$. The $n \times n$ weighting matrices $\mathbf{M}$ and Q are assumed to be positive semidefinite to ensure a well-defined finite minimum for J. The $r \times r$ matrix $\mathbf{R}$ is assumed to be positive definite because $\mathbf{R}^{-1}$ will be required in future manipulations. Values for $\mathbf{M}$, $\mathbf{Q}$, and $\mathbf{R}$ should be selected to give the desired trade-offs among terminal error, transient error, and control effort. ■

The discrete-time versions of the above examples require obvious minor modifications. Other performance criteria would be appropriate in specific cases.

The *optimal control problem* is now stated as:

From among all admissible control functions (or sequences) $\mathbf{u} \in U$, find that one which minimizes J of equation (*17.2*) subject to the dynamic system constraints of equation (*17.1*) and all initial and terminal boundary conditions that may be specified.

If the control is determined as a function of the initial state and other given system parameters, the control is said to be open-loop. If the control is determined as a function of the current state, then it is a closed-loop or feedback control law. Examples of both types will be given in the sequel.

The importance of the property of controllability should be evident. If the system is completely controllable, there is at least one control which will transfer any initial state to any desired final state. If the system is not controllable, it is not meaningful to search for the optimal control. However, controllability does not guarantee that a solution exists for every optimal control problem. Whenever the admissible controls are restricted to the set U, certain final states may not be attainable. Even though the system is completely controllable, the required control may not belong to U (see Problem 17.27).

17.3 DYNAMIC PROGRAMMING

General Introduction to the Principle of Optimality

Dynamic programming provides an efficient means for sequential decision-making. Its basis is R. Bellman's principle of optimality, "An optimal policy has the property that whatever the initial state and the initial decision are, the remaining decisions must constitute an optimal policy with regard to the state resulting from the first decision." As used here, a "decision" is a choice of control at a particular time and the "policy" is the entire control sequence (or function).

Consider the nodes in Fig. 17.1 as "states," in a general sense. A "decision" is

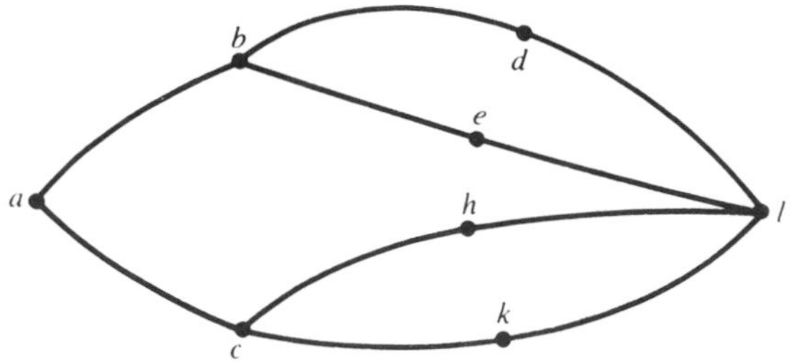

Figure 17.1

the choice of alternative paths leaving a given node. The goal is to move from state a to state l with minimum cost. A cost is associated with each segment of the line graph. Define J_{ab} as the cost between a and b. J_{bd} is the cost between b and d, etc. For path a, b, d, l the total cost is $J = J_{ab} + J_{bd} + J_{dl}$, and the optimal path (policy) is defined by

$$\min J = \min [J_{ab} + J_{bd} + J_{dl}, J_{ab} + J_{be} + J_{el}, J_{ac} + J_{ch} + J_{hl}, J_{ac} + J_{ck} + J_{kl}] \tag{17.3}$$

If the initial state is a and if the initial decision is to go to b, then the path from b to l must certainly be selected optimally if the overall path from a to l is to be optimum. If the initial decision is to go to c, then the path from c to l must then be selected optimally.

Let g_b and g_c be the minimum costs from b and c, respectively, to l. Then $g_b = \min[J_{bd} + J_{dl}, J_{be} + J_{el}]$ and $g_c = \min[J_{ch} + J_{hl}, J_{ck} + J_{kl}]$. The principle of optimality allows equation (*17.3*) to be written as

$$g_a \triangleq \min J = \min [J_{ab} + g_b, J_{ac} + g_c] \tag{17.4}$$

The key feature is that the quantity to be minimized consists of two parts:

1. The part directly attributable to the current decision, such as costs J_{ab} and J_{ac}.
2. The part representing the minimum value of all future costs, starting with the state which results from the first decision.

The principle of optimality replaces a choice between all alternatives (equation (*17.3*)) by a sequence of decisions between fewer alternatives (find g_b, g_c, and then g_a from equation (*17.4*)). Dynamic programming allows us to concentrate on a sequence of current decisions rather than being concerned about all decisions simultaneously.

Division of cost into the two parts, current and future, is typical, but these parts do not necessarily appear as a sum. A simple example illustrates the sequential nature of the method.

Example 17.6

Given N numbers $x_1, x_2, \ldots, x_N$, find the smallest one.

Rather than consider all N numbers simultaneously, define g_k as the minimum of x_k through x_N. Then $g_N = x_N$, $g_{N-1} = \min\{x_{N-1}, g_N\}$, $g_{N-2} = \min\{x_{N-2}, g_{N-1}\}, \ldots,$ $g_k = \min\{x_k, g_{k+1}\}$. Continuing to choose between two alternatives eventually leads to $g_1 = \min\{x_1, g_2\} = \min\{x_1, x_2, \ldots, x_N\}$.

The desired result g_1 need not be unique, since more than one number may have the same smallest value. The recursive nature of the formula $g_k = \min\{x_k, g_{k+1}\}$ is typical of all discrete dynamic programming solutions. ■

Example 17.7

Use dynamic programming to find the minimum cost route between node a and node l of Fig. 17.2. The line segments can only be traversed from left to right, and the travel costs are shown beside each segment.

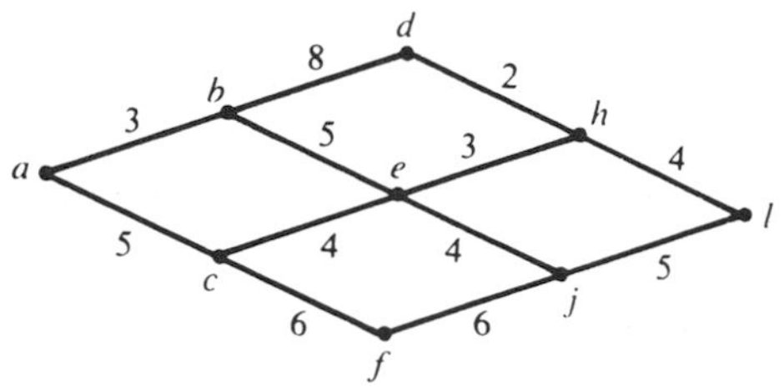

Figure 17.2 Simple routing problem.

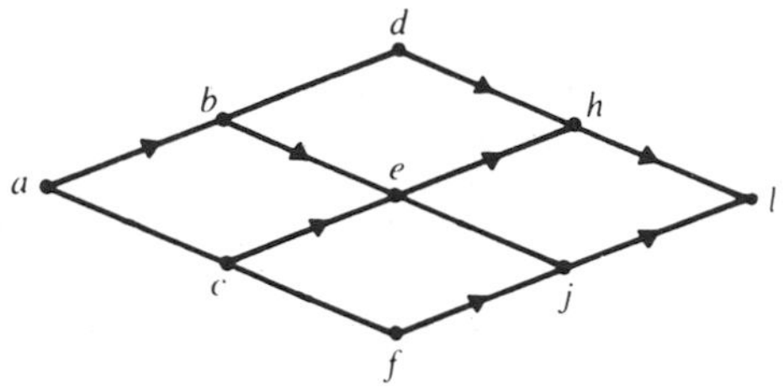

Figure 17.3 The optimal paths.

Define g_k as the minimum cost from a general node k to l. Obviously, $g_l = 0$. Since there is only one admissible path from h to l, $g_h = 4$. Similarly, $g_j = 5$, $g_d = 2 + g_h = 6$, and $g_f = 6 + g_j = 11$. The first point where a decision must be made is at node e. $g_e - \min\{3 + g_h, 4 + g_j\} - \min\{3 + 4, 4 + 5\} - 7$. The best route from e to l passes through h. Continuing, it is found that $g_b = \min\{8 + g_d, 5 + g_e\} = 12$, $g_c = \min\{4 + g_e, 6 + g_f\} = 11$, and $g_a = \min\{3 + g_b, 5 + g_c\} = 15$. The minimum cost is 15, and the best path is a, b, e, h, l.

The above procedure has actually answered the sequence of questions: if past decisions cause the route to be at point k, what is the best decision to make, starting at that point? The answers to these questions must be stored for future use and are shown by the arrows in Fig. 17.3. The minimum cost path from any node to l is now obvious. For example, if the path starts at c, then c, e, h, l is cheapest. From f, the cheapest path is f, j, l. ■

Application to Discrete-Time Optimal Control

To make the transition from the simple graph to the more general control problems, the following analogies are made.The graph of Fig. 17.2 or17.3is a plotof possible of states (nodes) at discrete-time points t_k. Point a is at t_0, b and c at $t_1, \ldots, l$ at t_4. The choice of possible directions, say up or down from e, is analogous to the set of admissible controls. Selecting a control $\mathbf{u}(k)$ is analogous to selecting a direction of departure from a given node $\mathbf{x}(k)$. The line segments connecting nodes play the same role as the difference equation (*17.1*), since both determine the next node $\mathbf{x}(k+1)$ to be encountered.

Example 17.8

Consider the scalar system $x(k+1) = x(k) + u(k)$ with boundary conditions $x(0) = 0$ and $x(3) = 3$. Find the controls $u(0)$, $u(1)$, and $u(2)$ which minimize $J = \sum_{k=0}^{2} \{u(k)^2 + \Delta t_k^2\}$. This performance criterion is the sum of the squares of three hypotenuses in the tx plane. This is a form of a minimum distance problem and the optimal sequence of points $x(k)$ will lie on a straight line in the tx plane. Verifying this obvious result will serve to illustrate the dynamic programming procedure.

Let $g(x(k))$ be the minumum cost from $x(k)$ to the terminal point. Since $x(3)$ is the terminal point, $g(x(3)) = 0$. Then

$$g(x(2)) = \min_{u(2)} \{\text{cost from } x(2) \text{ to the terminal point}\} = \min_{u(2)} \{u(2)^2 + \Delta t_2^2 + g(x(3))\}$$

This minimization is not without restriction since $x(3)$ must equal 3. Using $x(3) = x(2) + u(2)$, it is found that $u(2) = 3 - x(2)$. For any $x(2)$, there is a uniquely required $u(2)$. This is analogous to the decisions at points h and j of Example 17.7.

For convenience, let $\Delta t_k = 1$. Then $g(x(2)) = [3 - x(2)]^2 + 1$. Moving back one stage, $g(x(1)) = \min_{u(1)} \{u(1)^2 + 1 + g(x(2))\}$. Using $x(2) = x(1) + u(1)$ leads to

$$g(x(1)) = \min_{u(1)} \{u(1)^2 + 2 + [3 - x(1) - u(1)]^2\}$$

There are no restrictions on $u(1)$, so the minimization can be accomplished by setting $\dfrac{\partial}{\partial u(1)}\{u(1)^2 + 2 + [3 - x(1) - u(1)]^2\} = 0$. This gives $u(1) = [3 - x(1)]/2$, and then

$$g(x(1)) = \left[3 - x(1) - \frac{3 - x(1)}{2}\right]^2 + 2 + \left[\frac{3 - x(1)}{2}\right]^2$$

This is used along with $x(1) = x(0) + u(0)$ in $g(x(0)) = \min_{u(0)} \{u(0)^2 + 1 + g(x(1))\}$. Minimizing again gives $u(0) = [3 - x(0)]/3$. Since $x(0) = 0$ is given, $u(0) = 1$. The difference equation gives $x(1) = 1$. This is used in the previously computed expression to give $u(1) = 1$. Then $x(2) = x(1) + u(1) = 2$ and $u(2) = 3 - x(2) = 1$. Finally, $x(3) = x(2) + u(2) = 3$ as required. Note that the sequence $x(0), x(1), x(2), x(3)$ lies on a straight line in tx space, as expected. ■

Discrete dynamic programming solutions of optimal control problems usually consist of two stage-by-stage passes through the time stages. First, a *backward* pass answers the questions, "What is the minimum cost if the problem is started at time t_k with state $\mathbf{x}(k)$, and what is the optimal control as a function of $\mathbf{x}(k)$?" If $\mathbf{u}(k)$ is unrestricted and the criterion function is simple enough, the optimal $\mathbf{u}(k)$ can be obtained in explicit equation form in terms of $\mathbf{x}(k)$. This can be done by setting the gradient with respect to $\mathbf{u}(k)$ equal to zero as in Example 17.8. When this is not possible, a discrete set of grid points for the state and control variables would be used. A computer search routine would be used to find the optimal $\mathbf{u}(k)$, in tabular form, for each discrete value of $\mathbf{x}(k)$. This is a generalization of the method in Example 17.7.

The backward pass is completed when time t_0 is reached. Since $\mathbf{x}(0)$ is known, $\mathbf{u}(0)$ can be found in terms of that specific state. The second pass is in the *forward* direction. The system difference equation uses $\mathbf{x}(0)$ and $\mathbf{u}(0)$ to obtain $\mathbf{x}(1)$. The previously computed function or table is used to determine the value of $\mathbf{u}(1)$ associated with $\mathbf{x}(1)$. Then, in turn, $\mathbf{u}(1)$ gives $\mathbf{x}(2)$ and $\mathbf{x}(2)$ gives $\mathbf{u}(2)$, and so on.

The preceding verbal description is now expressed in equation form for the discrete versions of equations (*17.1*) and (*17.2*). Define $g(\mathbf{x}(k))$ as the minimum cost of the process, starting at t_k, $\mathbf{x}(k)$. Obviously, $g(\mathbf{x}(N)) = S(\mathbf{x}(N))$. Then the principle of optimality gives

$$g(\mathbf{x}(N-1)) = \min_{\mathbf{u}(N-1)} \{L(\mathbf{x}(N-1), \mathbf{u}(N-1)) + g(\mathbf{x}(N))\}$$

Equation (*17.1*) is used to eliminate $\mathbf{x}(N)$. The minimization with respect to $\mathbf{u}(N-1)$ is carried out by setting the gradient with respect to $\mathbf{u}(N-1)$ equal to zero (*if* there are no restrictions on $\mathbf{u}$), or by a computer search routine. In either case the optimal $\mathbf{u}(N-1)$ must be stored for each possible $\mathbf{x}(N-1)$. Also, $g(\mathbf{x}(N-1))$ must be stored for use in the next stage. This continues, stage-by-stage, with a typical step being

$$\begin{aligned} g(\mathbf{x}(k)) &= \min_{\mathbf{u}(k)} \{L(\mathbf{x}(k), \mathbf{u}(k)) + g(\mathbf{x}(k+1))\} \\ &= \min_{\mathbf{u}(k)} \{L(\mathbf{x}(k), \mathbf{u}(k)) + g(\mathbf{f}(\mathbf{x}(k), \mathbf{u}(k)))\} \end{aligned} \qquad (17.5)$$

Equation (*17.5*) is an extremely powerful and general result. It is the key to the solution of many discrete-time optimal control problems. This nonlinear difference equation has the boundary condition $g(\mathbf{x}(N)) = S(\mathbf{x}(N))$. The solution of equation (*17.5*) yields the optimal control $\mathbf{u}^*(k)$† at each time step, and also the optimal trajectory $\mathbf{x}^*(k)$. Certain special cases are easier to solve. If the dynamic system is linear and if the cost function is quadratic in the state and/or control variables (the so-called L-Q problem), then the solution to equation (*17.5*) reduces to a set of recursive matrix difference equations. (See Problem 17.4.) These equations take the same form as the Kalman filter equations (*1*) to (*3*) of Problem 11.9. The L-Q solution shows that the optimal control consists of a linear state variable feedback form

$$\mathbf{u}^*(k) = \mathbf{F}(k)\mathbf{V}(k) - \mathbf{G}(k)\mathbf{x}(k) \qquad (17.6)$$

The time-variable gain matrices $\mathbf{F}(k)$ and $\mathbf{G}(k)$ and the external input $\mathbf{V}(k)$ are determined as in Problem 17.4. The L-Q class of problems is well understood and easily solved, and as a result most practical applications of optimal control theory have been of this variety.

For constant coefficient systems whose operating time is very long compared

†The * on a vector, such as $\mathbf{u}^*(k)$, indicates the optimal vector. It should not be confused with the notation for an adjoint transformation, such as $\mathcal{A}^*$.

with system time constants, it is often justifiable to use only the steady-state constant gains with little or no degradation in optimality. This is because the gains, solved in the reversed-time direction, approach constant values in a few time constants from the terminal time. For the bulk of the operating period they remain constant. Starting at the initial time, the controller drives the state to its final value (often zero) in a similarly short period, long before the variable gains would be encountered.

The advangate of the constant gain version is ease of implementation. Note that when **F** and **G** are constant, equation (*17.6*) is of the same form as the pole-placement controller of equation (*16.3*). The methods of finding these gains and the philosophies which underlie them are quite different, however. Some comparisons are made in Problems 17.6 and 17.11.

The more general case of equation (*17.5*) can still be solved (in principle at least) by using a tabular computational approach. That is, a discretized grid of possible $\mathbf{x}(k)$ and $\mathbf{u}(k)$ values is determined at each time point. The results of the backward-in-time pass through this grid will consist of a table of optimal $\mathbf{u}^*(k)$ values for each possible $\mathbf{x}(k)$ value. The storage requirements quickly become excessive for all but the lowest order systems.

Application to Continuous-Time Optimal Control

Dynamic programming applies in a similar way to continuous-time systems. The cost of operating the system from a general time and state t, $\mathbf{x}(t)$ to the terminal time and state t_f, $\mathbf{x}(t_f)$ is defined as

$$g(\mathbf{x}(t), t_f - t) \triangleq \min_{\mathbf{u}(t)} \left\{ S(\mathbf{x}(t_f), t_f) + \int_t^{t_f} L(\mathbf{x}(t), \mathbf{u}(t), t)\, dt \right\}$$

Breaking the integral into two segments gives

$$g(\mathbf{x}(t), t_f - t) = \min_{\mathbf{u}(t)} \left\{ S(\mathbf{x}(t_f), t_f) + \int_{t+\delta t}^{t_f} L(\mathbf{x}(t), \mathbf{u}(t), t)\, dt + \int_t^{t+\delta t} L(\mathbf{x}(t), \mathbf{u}(t), t)\, dt \right\} \tag{17.7}$$

The principle of optimality states that if the total cost is to be minimum, then the cost from $t + \delta t$, $\mathbf{x}(t + \delta t)$, must also be minimum. The first two terms on the right side of equation (*17.7*) must therefore be equal to $g(\mathbf{x}(t + \delta t), t_f - t - \delta t)$, so that

$$g(\mathbf{x}(t), t_f - t) = \min_{\mathbf{u}(t)} \left\{ g(\mathbf{x}(t + \delta t), t_f - t - \delta t) + \int_t^{t+\delta t} L(\mathbf{x}(t), \mathbf{u}(t), t)\, dt \right\} \tag{17.8}$$

For δt sufficiently small,

$$\int_t^{t+\delta t} L(\mathbf{x}(t), \mathbf{u}(t), t)\, dt \simeq L(\mathbf{x}(t), \mathbf{u}(t), t)\, \delta t \qquad \text{and} \qquad \mathbf{x}(t + \delta t) \simeq \mathbf{x}(t) + \dot{\mathbf{x}}\, \delta t$$

Taylor series expansion gives

$$g(\mathbf{x}(t + \delta t), t_f - t - \delta t) \simeq g(\mathbf{x}(t), t_f - t) + [\nabla_{\mathbf{x}} g]^T \dot{\mathbf{x}}\, \delta t - \frac{\partial g}{\partial t_r}\, \delta t$$

where $t_r \triangleq t_f - t$ is the time remaining. Using these in equation *(17.8)* gives

$$g(\mathbf{x}(t), t_r) = \min_{\mathbf{u}(t)} \left\{ g(\mathbf{x}(t), t_r) + [\nabla_{\mathbf{x}} g]^T \dot{\mathbf{x}}\, \delta t - \frac{\partial g}{\partial t_r}\, \delta t + L\, \delta t \right\} \tag{17.9}$$

By definition, $g(\mathbf{x}(t), t_r)$ is a function only of the current state and the time remaining (and not a function of $\mathbf{u}(t)$). Therefore, equation *(17.9)* leads to the *Hamilton-Jacobi-Bellman* (H.J.B.) partial differential equation [58]:

$$\frac{\partial g}{\partial t_r} = \min_{\mathbf{u}(t)} \{ L(\mathbf{x}(t), \mathbf{u}(t), t) + [\nabla_{\mathbf{x}} g]^T \dot{\mathbf{x}} \} \tag{17.10}$$

The boundary condition is $g(\mathbf{x}(t), t_r)|_{t_r=0} = S(\mathbf{x}(t_f), t_f)$. Equation *(17.10)* is the continuous-time counterpart of equation *(17.5)*. It is the major key to solving continuous-time optimal control problems. Whether it can be solved or not, and with what difficulty level, depends upon the class of systems, cost functions, admissible controls, and state constraints. The optimal control is the one which minimizes the right side of equation *(17.10)*. If there are no restrictions on $\mathbf{u}(t)$, i.e., the admissible control set $\mathbf{U}$ is the entire space $\mathcal{U}^r$, then the minimum can be found by differentiating with respect to $\mathbf{u}(t)$ and setting the resultant gradient vector to zero. This gives the necessary condition for optimality. (A review of Sec. 13.6 might be worthwhile here).

$$\frac{\partial L}{\partial \mathbf{u}} + \frac{\partial \mathbf{f}}{\partial \mathbf{u}} \nabla_{\mathbf{x}} g = \mathbf{0}$$

If the system is linear, $\dot{\mathbf{x}} = \mathbf{A}\mathbf{x} + \mathbf{B}\mathbf{u}$ and $\partial \mathbf{f}/\partial \mathbf{u} = \mathbf{B}^T(t)$. If the loss function L is quadratic, i.e., L of Example 17.5, then $\partial L/\partial \mathbf{u} = 2\mathbf{R}\mathbf{u}(t)$ so that the optimal $\mathbf{u}(t)$ for the continuous version of the *L-Q* problem is given by

$$\mathbf{u}^*(t) = -\tfrac{1}{2}\mathbf{R}^{-1}\mathbf{B}^T \nabla_{\mathbf{x}} g(\mathbf{x}(t), t_r) \tag{17.11}$$

The Hamilton-Jacobi-Bellman equation is solved in Problem 17.7 for the linear system, quadratic cost case. It is shown there that $\nabla_{\mathbf{x}} g$ is a linear function of $\mathbf{x}(t)$. Therefore, equation (17.11) constitutes a closed-loop control law. In fact, it is a linear but time-varying state feedback control law. In the L-Q case, solution of the H.J.B. partial differential equation reduces to solving an ordinary (but still nonlinear) matrix differential equation called the Riccati equation. The time-varying state feedback gain matrix depends on the solution of this Riccati equation. For simple low-order cases, complete analytical solutions are possible (see Problem 17.8). In problems of higher order, numerical solutions of the Riccati equation are employed. Just as in the discrete-time case, it is sometimes justifiable to implement only the steady-state version of the solution. This is attractive because then the controller is just a matrix of constant feedback gains.

17.4 PONTRYAGIN'S MINIMUM PRINCIPLE

The continuous-time versions of equations (*17.1*) and (*17.2*) are considered. If the set of admissible controls is unrestricted, the calculus of variations [42] can be used to derive necessary conditions which characterize the optimal solution. When the admissible control set is bounded in some way, unrestricted variations in $\mathbf{u}(t)$ are not allowed. This situation is analogous to the boundary points for the function shown in Fig. 13.3 page 329. If the minimum occurs on the boundary, then it is not necessarily true that the first variation (analogous to the slope) vanishes at that point. Pontryagin's minimum principle [95] is an extension of the methods and results of variational calculus to the case of bounded control variables. The minimum principle provides a set of necessary conditions for optimality (see Problems 17.12 and 17.13).

The *Hamiltonian* is defined as the scalar function

$$\mathcal{H}(\mathbf{x}, \mathbf{u}, \mathbf{p}, t) \triangleq L(\mathbf{x}, \mathbf{u}, t) + \mathbf{p}^T(t)\mathbf{f}(\mathbf{x}, \mathbf{u}, t) \tag{17.12}$$

where $\mathbf{p}(t)$ is the $n \times 1$ *costate vector* and satisfies

$$\dot{\mathbf{p}} = -\left[\frac{\partial \mathbf{f}}{\partial \mathbf{x}}\right]^T \mathbf{p} - \nabla_{\mathbf{x}} L(\mathbf{x}, \mathbf{u}, t) \tag{17.13}$$

The minimum principle states that the optimal control $\mathbf{u}^*(t)$ is that member of the admissible control set U which minimizes $\mathcal{H}$ at every time. If $\mathbf{u}(t)$ has r components, then minimizing $\mathcal{H}$ gives r algebraic equations which allow the determination of $\mathbf{u}^*(t)$ in terms of the still unknown $\mathbf{p}(t)$ and $\mathbf{x}(t)$. Then $\mathbf{u}(t)$ can be eliminated from equations (*17.1*) and (*17.13*). These equations can also be written in canonical form as

$$\dot{\mathbf{x}} = \frac{\partial \mathcal{H}}{\partial \mathbf{p}}, \qquad \dot{\mathbf{p}} = -\frac{\partial \mathcal{H}}{\partial \mathbf{x}} \tag{17.14}$$

Equation (*17.14*) consists of $2n$ first-order differential equations, so $2n$ boundary conditions are required for solution. The initial conditions $\mathbf{x}(t_0) = \mathbf{x}_0$ give n of them. The remaining n conditions will apply at the final time t_f. Their exact nature depends on the particular problem. $\mathbf{x}(t_f)$ will be specified directly in some cases. If $\mathbf{x}(t_f)$ is free, then $\mathbf{p}(t_f) = \nabla_{\mathbf{x}} S(\mathbf{x}(t), t)|_{tf}$. Combinations of these two kinds of conditions, as well as others, can arise. If the final time t_f is not specified, an additional equation is required to determine it (see Problem 17.13).

Equation (17.14), along with n boundary conditions at t_0 and n more at t_f, constitutes a two-point boundary value problem. Linear two-point boundary value problems are easily solved, at least in principle. Nonlinear two-point boundary value problems are generally difficult to solve, even numerically. Much of the effort in optimal control theory has been devoted to the development of efficient algorithms for computer solution of these problems (Problem 17.14).

Example 17.9

A simplified model of the linear motion of an automobile is $\dot{x} = u$, where $x(t)$ is the vehicle velocity and $u(t)$ is the acceleration or deceleration. The car is initially moving at x_0 ft/sec. Find the optimal $u(t)$ which brings the velocity $x(t_f)$ to zero in *minimum time* t_f. Assume that acceleration and braking limitations require $|u(t)| \leq M$ for all t.

The minimum time performance criterion is $J = \int_0^{t_f} 1\,dt$, so the Hamiltonian is $\mathcal{H} = 1 + p(t)u(t)$. In order to minimize $\mathcal{H}$, it is obvious that $u(t) = -M$ if $p(t) > 0$ and $u(t) = M$ if $p(t) < 0$. That is, $u^*(t) = -M \operatorname{sign}(p(t))$. The optimal control has been found as a function of the unknown $p(t)$. The differential equation for p is $\dot{p} = -\partial\mathcal{H}/\partial x = 0$, so $p(t)$ is constant. The value of this constant must be determined from boundary conditions. (Actually only the sign of the constant is needed for this problem.) The available boundary conditions are $x(0) = x_0$, $x(t_f) = 0$. The form of the solution for $x(t)$ is $x(t) = x_0 + \int_0^t u(\tau)\,d\tau$, but $u(t)$ is a constant, either $+M$ or $-M$. Therefore, $x(t_f) = x_0 \pm Mt_f = 0$. Clearly, if $x_0 > 0$, then $u = -M$, maximum deceleration. If $x_0 < 0$, then $u = M$, maximum acceleration. That is, $u^*(t) = -M \operatorname{sign}(x_0)$. The minimum stopping time is $t_f = |x_0|/M$. Note that $u^*(t)$ is expressed in terms of $x(t_0)$, so this represents an open-loop control law. This is a typical result of using the minimum principle. ■

Example 17.10

Consider the linear, constant system $\dot{\mathbf{x}} = \mathbf{A}\mathbf{x} + \mathbf{B}\mathbf{u}$ with $\mathbf{u}(t)$ unrestricted. Find $\mathbf{u}(t)$ which minimizes a trade-off between terminal error and control effort,

$$J = [\mathbf{x}(t_f) - \mathbf{x}_d]^T[\mathbf{x}(t_f) - \mathbf{x}_d] + \int_0^{t_f} \mathbf{u}^T(t)\mathbf{u}(t)\,dt$$

Time t_f is fixed.

The Hamiltonian is $\mathcal{H} = \mathbf{u}^T\mathbf{u} + \mathbf{p}^T[\mathbf{A}\mathbf{x} + \mathbf{B}\mathbf{u}]$. $\mathcal{H}$ is minimized by setting

$$\frac{\partial\mathcal{H}}{\partial\mathbf{u}} = \mathbf{0} = 2\mathbf{u} + \mathbf{B}^T\mathbf{p} \quad \text{or} \quad \mathbf{u}^*(t) = -\tfrac{1}{2}\mathbf{B}^T\mathbf{p}(t)$$

This is not yet a useful answer, since $\mathbf{p}(t)$ is unknown. However, $\dot{\mathbf{p}} = -\partial\mathcal{H}/\partial\mathbf{x} = -\mathbf{A}^T\mathbf{p}$. Combining this with the original system equation gives, after eliminating $\mathbf{u}(t)$,

$$\begin{bmatrix} \dot{\mathbf{x}} \\ \dot{\mathbf{p}} \end{bmatrix} = \begin{bmatrix} \mathbf{A} & -\tfrac{1}{2}\mathbf{B}\mathbf{B}^T \\ \mathbf{0} & -\mathbf{A}^T \end{bmatrix} \begin{bmatrix} \mathbf{x} \\ \mathbf{p} \end{bmatrix} \tag{17.15}$$

The $2n$ boundary conditions are $\mathbf{x}(0) = \mathbf{x}_0$ and $\mathbf{p}(t_f) = \nabla_{\mathbf{x}} S|_{t_f} = 2[\mathbf{x}(t_f) - \mathbf{x}_d]$. This linear two-point boundary value problem is treated in Problem 17.15. ■

17.5 THE SEPARATION THEOREM

In previous sections of this chapter it has been assumed that the entire state vector $\mathbf{x}$ can be measured and used in forming the feedback control signal. In Chapter 16 it was shown that if the system is *linear* and if the control law is *linear*, then a *linear* observer can be used to estimate $\mathbf{x}$ from the available outputs $\mathbf{y}$ without changing the closed-loop poles which the controller is designed to give.

Another version of the separation theorem is more commonly stated in conjunction with optimal control problems. It states that *if*

a. the system models are linear (both the dynamics of equation (*9.11*) or (*9.13*) and the output equation of equation (*9.12*) or (*9.14*)) and
b. all measurement errors and disturbances have Gaussian probability density functions, and
c. the cost function of equation (*17.2*) is quadratic,

then the expected value of the cost function J is minimized by

1. designing an optimal controller based on the assumption that all states are available,
2. designing an optimal estimator to provide an estimate $\hat{\mathbf{x}}$ of $\mathbf{x}$,
3. using $\hat{\mathbf{x}}$ in place of $\mathbf{x}$ in the control law of step 1.

These two versions of the separation principle are closely related because of the following facts: The optimal control law for a linear system with a quadratic cost function is a linear control law. The optimal estimator for a linear system subjected to Gaussian noise is a linear estimator of the same form as the linear observer of Chapter 16. While version one is true for *any* linear controller and linear estimator, regardless of how they were designed, and without any Gaussian assumptions, all that is guaranteed is that the desired closed-loop poles are still achieved. Version two guarantees the optimality of J in a stochastic sense, but demands somewhat stricter assumptions. The class of problems that meet these restrictions is referred to as L-Q-G problems (linear, quadratic, Gaussian). Most successful applications of optimal control theory so far have been members of this class.

The implication of the separation theorem is that the assumption that the full state is available can be relaxed. The optimal controllers in this chapter can be cascaded with an observer (as in Chapter 16) or a Kalman filter (Problem 11.11) to obtain the overall design. This practice is generally not valid outside the L-Q-G class, although it is still sometimes used as a rather effective suboptimal scheme.

Intuitively, one might expect that if control inputs could be tailored in some way, certain system modes would be more strongly excited, making their states easier to estimate. This idea is the basis for "probing"-type control signals. Likewise, it seems intuitive that if one or more states are only known to within some level of uncertainty, but the control system behaves as if it were exactly known, then overly bold or decisive control actions might incorrectly be taken at times. A more cautious controller which somehow takes into account its knowledge of the uncertainty in the state estimates might give better overall performance. The notions of probing and caution come into play in those cases where the separation theorem does not apply. This is still a subject of active research. To give any more than the above intuitive definitions would go beyond the scope of this book.

17.6 CONCLUDING COMMENTS

A comparison of equations (*17.10*) and (*17.12*) shows that the Hamiltonian of the minimum principle also appears in the H.J.B. equation. The costate vector $\mathbf{p}$ plays the role of $\nabla_{\mathbf{x}} g$, the sensitivity of the minimum cost with respect to changes in the state. Detailed consideration of the relationship between the two methods, as well as other topics, may be found in more advanced texts on optimal control theory.

Analytical solutions to optimization problems are possible when low-order, constant coefficient linear systems with quadratic performance criteria are considered. This is true whether dynamic programming or the minimum principle is used. These simple solutions can usually be expressed in either open- or closed-loop form (see Problem 17.16).

If the set of admissible controls is constrained, then the minimization required in the H.J.B. equation is more difficult. Even for linear systems, the minimum principle can be expected to yield a nonlinear two-point boundary value problem. In this case, and also for high-order or time-varying or nonlinear systems or for nonquadratic performance criteria, a computer will be required for solution.

In the context of computer solutions, the biggest limitation on dynamic programming is the "curse of dimensionality" [7]. For a system with n state variables, Fig. 17.2 must be replaced with the equivalent of an $n + 1$ dimensional grid. Discretizing each state variable into 100 possible values gives $(100)^n$ grid points at each time point. Values of $g(\mathbf{x}(k))$ and $\mathbf{u}(k)$ must be stored for each grid point. The fast access storage capacity of even the largest computer can be exhausted. In its favor, dynamic programming yields a globally optimum solution in feedback form.

The major limitation of the minimum principle is the necessity of numerically solving a nonlinear two-point boundary value problem. Various iterative methods have been proposed, but none is foolproof. If the system's state equations are stable, the costate equations will be unstable (see Problem 14.6). Thus small errors in boundary conditions at either t_0 or t_f can propagate into large errors at the other time limit, regardless of whether the equations are integrated forward or backward in time. Solutions resulting from the minimum principle are normally of the open-loop type. They may represent local, rather than global, optimums.

The necessity of computer solutions has limited real-time implementation of optimal control theory in the past. Still, off-line optimal solutions have provided valuable standards of comparison for evaluating easier-to-implement suboptimal control schemes.

The control problems for which the optimal solutions are the most feasible to implement are of the L-Q or L-Q-G variety. Off-line solution of the Riccati equation, or its discrete counterpart, allows optimal feedback control matrices to be precomputed. Often, the constant steady-state version of the feedback matrices gives sufficiently good control [60, 67]. The advantages of linear system control can be extended to nonlinear systems which are slightly perturbed from nominal conditions (see Problem 13.3).

Digital controllers have been used for many years in environmental control systems, food, fiber, and chemical processing industries, manufacturing, aerospace, power generation, and other industries. Many of these controllers have been simple on-off programmed logic devices. In many cases they were direct replacements for the older, mechanical relay controllers. As computer technology continues to advance, it seems certain that optimal controllers will be used more extensively.

One aspect of these applications that has not yet been discussed has to do with unknown or uncertain system dynamics. There are many worthwhile references which deal with these subjects in depth [46, 48, 55, 108]. The material in this book is more than adequate preparation for the deterministic portions of these subjects. A knowledge of random process theory [87] is also important. As a hint as to how things go, if a model structure is assumed, then the least-squares techniques of Chapters 5, 11, and 13 can be used to estimate the unknown model parameters from measurements of inputs and outputs. This is the essence of some commonly used system identification schemes. If the parameters are estimated on-line and if the model is concurrently used in a pole-placement controller of Chapter 16, or an optimal controller of this chapter, the resultant system is called a self-tuning controller. Adaptive controllers are similar in that they also attempt to overcome unknown or varying system dynamics while achieving adequate control. The future will bring many exciting applications of control theory in robotics and other intelligent or learning systems. At present, however, this book must end here.

ILLUSTRATIVE PROBLEMS

Linear, Time-Varying Minimum Norm Problem

17.1 Consider the general linear time-varying system $\dot{\mathbf{x}}(t) = \mathbf{A}(t)\mathbf{x}(t) + \mathbf{B}(t)\mathbf{u}(t)$. Find the control which transfers the state from $\mathbf{x}(t_0)$ to $\mathbf{x}_d$ at a fixed time t_f while minimizing the control effort, $J = \int_{t_0}^{t_f} \mathbf{u}^T(t)\mathbf{u}(t)\, dt$.

Since at the final time t_f, $\mathbf{x}(t_f)$ must equal $\mathbf{x}_d$, the solution to the state equation must satisfy

$$\mathbf{x}_d - \mathbf{\Phi}(t_f, t_0)\mathbf{x}(t_0) = \int_{t_0}^{t_f} \mathbf{\Phi}(t_f, \tau)\mathbf{B}(\tau)\mathbf{u}(\tau)\, d\tau \tag{1}$$

The right-hand side of equation (*1*) defines a linear operator $\mathcal{A}(\mathbf{u})$ such that $\mathcal{A}: \mathcal{U} \longrightarrow \Sigma$. The left-hand side is a fixed vector in the state space Σ. The input function $\mathbf{u}_{[t_0, t_f]}$ is an element of the infinite dimensional input function space $\mathcal{U}$. The performance criterion is the norm squared of an element in $\mathcal{U}$. The problem is therefore one of finding the minimum norm solution to a linear operator equation.

Results of Problem 6.13b are directly applicable, $\mathbf{u}^*(t) = \mathcal{A}^*(\mathcal{A}\mathcal{A}^*)^{-1}[\mathbf{x}_d - \mathbf{\Phi}(t_f, t_0)\, \mathbf{x}(t_0)]$. The adjoint operator $\mathcal{A}^*$ is defined by $\langle \mathcal{A}(\mathbf{u}), \mathbf{v}\rangle_\Sigma = \langle \mathbf{u}, \mathcal{A}^*\mathbf{v}\rangle_{\mathcal{U}}$.

Assuming $\mathbf{\Phi}$, $\mathbf{B}$, and $\mathbf{u}$ are real, the results of Sec. 12.5 give $\mathcal{A}^* = \mathbf{B}^T(t)\mathbf{\Phi}^T(t_f, t)$. The optimal control is

$$\mathbf{u}^*(t) = \mathbf{B}^T(t)\mathbf{\Phi}^T(t_f, t)\left[\int_{t_0}^{t_f} \mathbf{\Phi}(t_f, \tau)\mathbf{B}(\tau)\mathbf{B}^T(\tau)\mathbf{\Phi}^T(t_f, \tau)\, d\tau\right]^{-1}[\mathbf{x}_d - \mathbf{\Phi}(t_f, t_0)\mathbf{x}(t_0)]$$

Note that the indicated inverse exists if the system is controllable. By making use of $\mathbf{\Phi}(t_f, \tau) = \mathbf{\Phi}(t_f, t_0)\mathbf{\Phi}(t_0, \tau)$, the solution can be expressed in alternative but equivalent forms.

Other problems can be solved in the same way whenever the performance criterion J can be interpreted as some more general norm of a function $\mathbf{u}(t) \in \mathcal{U}$ [96].

Discrete Dynamic Programming

17.2 Find the minimum cost path which starts at point a and ends at any one of the points h, j, k, l of Fig. 17.4. Travel costs are shown beside each path segment and toll charges are shown by each node.

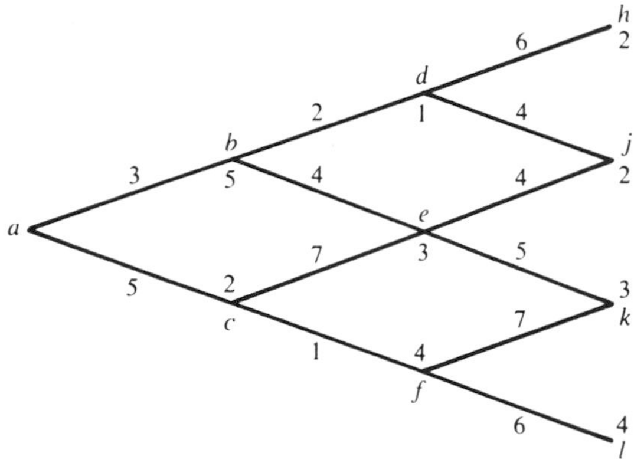

Figure 17.4

Let g_α be the minimum cost from node α to any one of the four possible termination points.

Last stage: $g_h = 2 \quad g_k = 3$

$g_j = 2 \quad g_l = 4$

Next stage:

$g_d = 1 + \min[6 + g_h, 4 + g_j]$

$= 1 + \min[6 + 2, 4 + 2] = 7,$ from d to j

$g_e = 3 + \min[4 + g_j, 5 + g_k] = 9,$ from e to j

$g_f = 4 + \min[7 + g_k, 6 + g_l] = 14,$ from f to k or l

Next stage:

$g_b = 5 + \min[2 + g_d, 4 + g_e] = 14,$ from b to d

$g_c = 2 + \min[7 + g_e, 1 + g_f] = 17,$ from c to f

Initial stage: $g_a = 0 + \min[3 + g_b, 5 + g_c] = 17,$ from a to b

The minimum total cost is $g_a = 17$ and is achieved on path a, b, d, j.

17.3 A two-stage discrete-time system is described by $x(k + 1) = x(k) + u(k)$ with $x(0) = 10$. Use dynamic programming to find $u(0)$ and $u(1)$ which minimize $J = [x(2) - 20]^2 + \sum_{k=0}^{1} \{x^2(k) + u^2(k)\}$. There are no restrictions on $u(k)$.

$g[x(k)]$ is defined as the minimum cost from state $x(k)$ at time k to some final state $x(2)$. Then $g[x(2)] = [x(2) - 20]^2$ and

$$g[x(1)] = \min_{u(1)} \{x^2(1) + u(1)^2 + g[x(2)]\}$$
$$= \min_{u(1)} \{x^2(1) + u^2(1) + [x(1) + u(1) - 20]^2\}$$

Differentiating with respect to $u(1)$ and equating to zero gives $u(1) = [20 - x(1)]/2$, so that $g[x(1)] = x^2(1) + [20 - x(1)]^2/2$. Using this in $g[x(0)] = \min_{u(0)}\{x^2(0) + u^2(0) + g[x(1)]\}$ and setting $\partial\{\ \}/\partial u(0) = 0$ gives $u(0) = [20 - 3x(0)]/5$. This completes the backward pass. Since it is known that $x(0) = 10$, $u(0) = -2$. This, plus the difference equation, yields $x(1) = 8$. Using this in the expression found for $u(1)$ gives $u(1) = 6$. Finally, $x(2) = x(1) + u(1) = 14$. The minimum cost is $g[x(0)] = 240$.

17.4 A system is described by $\mathbf{x}(k+1) = \mathbf{A}\mathbf{x}(k) + \mathbf{B}\mathbf{u}(k)$ with $\mathbf{x}(0)$ known. Find the control sequence $\mathbf{u}(k)$ which minimizes

$$J = \tfrac{1}{2}\mathbf{x}(N)^T\mathbf{M}\mathbf{x}(N) + \tfrac{1}{2}\sum_{k=0}^{N-1}\{\mathbf{x}^T(k)\mathbf{Q}\mathbf{x}(k) + \mathbf{u}^T(k)\mathbf{R}\mathbf{u}(k)\}$$

$\mathbf{M}$ and $\mathbf{Q}$ are symmetric positive semidefinite $n \times n$ matrices and $\mathbf{R}$ is a symmetric positive definite $r \times r$ matrix. No restrictions are placed on $\mathbf{u}(k)$.

Let $g[\mathbf{x}(k)]$ = min cost from k, $\mathbf{x}(k)$ to N, $\mathbf{x}(N)$. Equation (*17.5*) gives

$$g[\mathbf{x}(k)] = \min_{\mathbf{u}(k)} \{\tfrac{1}{2}\mathbf{x}^T(k)\mathbf{Q}\mathbf{x}(k) + \tfrac{1}{2}\mathbf{u}^T(k)\mathbf{R}\mathbf{u}(k) + g[\mathbf{x}(k+1)]\} \tag{1}$$

This is a difference equation and the boundary condition is $g[\mathbf{x}(N)] = \tfrac{1}{2}\mathbf{x}(N)^T\mathbf{M}\mathbf{x}(N)$. This equation is solved by assuming a solution

$$g[\mathbf{x}(k)] = \tfrac{1}{2}\mathbf{x}(k)^T\mathbf{W}(N-k)\mathbf{x}(k) + \mathbf{x}(k)^T\mathbf{V}(N-k) + Z(N-k) \tag{2}$$

where $\mathbf{W}$, $\mathbf{V}$, and Z are an $n \times n$ matrix, an $n \times 1$ vector, and a scalar, respectively. They will be selected so as to force equation (*2*) to satisfy equation (*1*). Equation (*1*) becomes

$$\begin{aligned}
&\tfrac{1}{2}\mathbf{x}(k)^T\mathbf{W}(N-k)\mathbf{x}(k) + \mathbf{x}(k)^T\mathbf{V}(N-k) + Z(N-k)\\
&\quad = \min_{\mathbf{u}(k)} \{\tfrac{1}{2}\mathbf{x}^T(k)\mathbf{Q}\mathbf{x}(k) + \tfrac{1}{2}\mathbf{u}^T(k)\mathbf{R}\mathbf{u}(k) + \tfrac{1}{2}\mathbf{x}(k+1)^T\mathbf{W}(N-k-1)\mathbf{x}(k+1)\\
&\qquad + \mathbf{x}(k+1)^T\mathbf{V}(N-k-1) + Z(N-k-1)\}
\end{aligned} \tag{3}$$

Substituting $\mathbf{x}(k+1) = \mathbf{A}\mathbf{x}(k) + \mathbf{B}\mathbf{u}(k)$ and regrouping, the right-hand side of equation (*3*) becomes

$$\begin{aligned}
\text{R.H.S.} = \min_{\mathbf{u}(k)} \{&\tfrac{1}{2}\mathbf{x}(k)^T[\mathbf{Q} + \mathbf{A}^T\mathbf{W}(N-k-1)\mathbf{A}]\mathbf{x}(k)\\
&+ \tfrac{1}{2}\mathbf{u}(k)^T[\mathbf{R} + \mathbf{B}^T\mathbf{W}(N-k-1)\mathbf{B}]\mathbf{u}(k)\\
&+ \mathbf{u}(k)^T[\mathbf{B}^T\mathbf{V}(N-k-1) + \mathbf{B}^T\mathbf{W}(N-k-1)\mathbf{A}\mathbf{x}(k)]\\
&+ \mathbf{x}(k)^T\mathbf{A}^T\mathbf{V}(N-k-1) + Z(N-k-1)\}
\end{aligned}$$

Since there are no restrictions on $\mathbf{u}(k)$, the minimizing $\mathbf{u}(k)$ is found by setting $\partial\{\ \}/\partial\mathbf{u}(k) = \mathbf{0}$. This yields

$$\mathbf{u}^*(k) = -[\mathbf{R} + \mathbf{B}^T\mathbf{W}(N-k-1)\mathbf{B}]^{-1}\mathbf{B}^T[\mathbf{V}(N-k-1) + \mathbf{W}(N-k-1)\mathbf{A}\mathbf{x}(k)] \tag{4}$$

After rearranging and simplifying, this gives

$$\begin{aligned}\text{R.H.S.}&\\ &= \tfrac{1}{2}\mathbf{x}(k)^T[\mathbf{Q} + \mathbf{A}^T\mathbf{W}(N-k-1)\mathbf{A} - \mathbf{A}^T\mathbf{W}(N-k-1)\mathbf{B}\mathbf{U}\mathbf{B}^T\mathbf{W}(N-k-1)\mathbf{A}]\mathbf{x}(k)\\ &\quad + \mathbf{x}(k)^T[\mathbf{A}^T\mathbf{V}(N-k-1) - \mathbf{A}^T\mathbf{W}(N-k-1)\mathbf{B}\mathbf{U}\mathbf{B}^T\mathbf{V}(N-k-1)]\\ &\quad + [Z(N-k-1) - \tfrac{1}{2}\mathbf{V}(N-k-1)^T\mathbf{B}\mathbf{U}\mathbf{B}^T\mathbf{V}(N-k-1)]\end{aligned}$$

where $\mathbf{U} \triangleq [\mathbf{R} + \mathbf{B}^T\mathbf{W}(N-k-1)\mathbf{B}]^{-1}$. Equating the left-hand and right-hand sides, the assumed form for $g[\mathbf{x}(k)]$ can be forced to be a solution *for all* $\mathbf{x}(k)$ by requiring that the quadratic terms, the linear terms, and the terms not involving $\mathbf{x}$ all balance individually. This requires

$$\mathbf{W}(N-k) = \mathbf{Q} + \mathbf{A}^T\mathbf{W}(N-k-1)\mathbf{A} - \mathbf{A}^T\mathbf{W}(N-k-1)\mathbf{B}\mathbf{U}\mathbf{B}^T\mathbf{W}(N-k-1)\mathbf{A} \quad (5)$$

$$\mathbf{V}(N-k) = \mathbf{A}^T\mathbf{V}(N-k-1) - \mathbf{A}^T\mathbf{W}(N-k-1)\mathbf{B}\mathbf{U}\mathbf{B}^T\mathbf{V}(N-k-1) \quad (6)$$

$$Z(N-k) = Z(N-k-1) - \tfrac{1}{2}\mathbf{V}(N-k-1)^T\mathbf{B}\mathbf{U}\mathbf{B}^T\mathbf{V}(N-k-1) \quad (7)$$

The boundary conditions are $\mathbf{W}(N-N) = \mathbf{M}$, $\mathbf{V}(N-N) = \mathbf{0}$, and $Z(N-N) = 0$. A computer solution, backward in time, easily gives $\mathbf{W}(N-k)$. Normally *(5)* is solved first. Then its solution $\mathbf{W}(N-k)$ is used as a known coefficient matrix while solving *(6)* for $\mathbf{V}(N-k)$. Then $\mathbf{V}$ acts as a known forcing function in *(7)*. Actually, *(7)* never needs to be solved if the only interest is in finding the optimal control. $Z(N-k)$ is only needed if $J_{\min}$ must be calculated. For the cost function considered here (the so-called regulator problem), *(6)* is a homogeneous equation with zero initial conditions, so $\mathbf{V}(N-k)$ is zero for all stages. Therefore, *(6)* is not needed either.

Equation *(5)* remains as the principal result. It is interesting to rewrite *(5)* as follows: Let $k' = N - k$ be a backward running time index and let $\mathbf{W}(N-k) = \mathbf{M}(k')$. Then *(5)* can be replaced by

$$\mathbf{M}(k') = \mathbf{A}^T\mathbf{P}(k'-1)\mathbf{A} + \mathbf{Q} \quad (8)$$

$$\mathbf{K}(k') = \mathbf{M}(k')\mathbf{B}[\mathbf{B}^T\mathbf{M}(k')\mathbf{B} + \mathbf{R}]^{-1} \quad (9)$$

$$\mathbf{P}(k') = [\mathbf{I} - \mathbf{K}(k')\mathbf{B}^T]\mathbf{M}(k') \quad (10)$$

These are exactly the same as *(1)*, *(2)*, and *(3)* of Problem 11.11 (the Kalman filter algorithm), except that $\mathbf{B}^T$ replaces $\mathbf{C}$, $\mathbf{A}^T$ replaces $\mathbf{\Phi}$, and k' replaces k. The optimal mean square estimator problem and the optimal regulator problem are duals of each other. The initial condition on *(8)* is $\mathbf{M}(0) = \mathbf{M}$. The optimal feedback control is given by

$$\mathbf{u}^*(k) = -\mathbf{K}^T(k'-1)\mathbf{A}\mathbf{x}(k)$$

so that the feedback gain matrix is $\mathbf{G}(k) = \mathbf{K}^T(k'-1)\mathbf{A}$. This duality has a practical significance. If a computer program is available to compute the Kalman gain matrix $\mathbf{K}(k)$ (equation *(2)* of Problem 11.11), then the same algorithm can be used to find the control gains $\mathbf{G}(k)$. Just make the interchanges with $\mathbf{B}$, $\mathbf{C}$, $\mathbf{A}$, and $\mathbf{\Phi}$ as mentioned above and remember that time is reversed.

17.5 Find the optimal control sequence for the linear system of Problem 17.4, but with a more general cost function.

$$J = \tfrac{1}{2}[\mathbf{x}(N) - \mathbf{x}_d]^T\mathbf{M}[\mathbf{x}(N) - \mathbf{x}_d] + \tfrac{1}{2}\sum_{k=0}^{N-1}\{[\mathbf{x}(k) - \boldsymbol{\eta}(k)]^T\mathbf{Q}[\mathbf{x}(k) - \boldsymbol{\eta}(k)] + \mathbf{u}(k)^T\mathbf{R}\mathbf{u}(k)\} \quad (1)$$

This cost function attempts to make $\mathbf{x}(k)$ follow the specified sequence $\boldsymbol{\eta}(k)$. This is called the optimal tracking problem.

Most of the solution details are the same as Problem 17.4 and will not be repeated. Expanding the quadratic terms in J shows that there are four additional terms to deal with, due to $\mathbf{x}_d(k)$ and $\boldsymbol{\eta}(k)$. Two of these terms become forcing functions on equations (6) and (7), modified here as (6′) and (7′).

$$\mathbf{V}(N-k) = \mathbf{A}^T\mathbf{V}(N-k-1) - \mathbf{A}^T\mathbf{W}(N-k-1)\mathbf{BUB}^T\mathbf{V}(N-k-1) - \mathbf{Q}\boldsymbol{\eta}(k) \quad (6')$$

$$Z(N-k) = Z(N-k-1) - \tfrac{1}{2}\mathbf{V}^T(N-k-1)\mathbf{BUB}^T\mathbf{V}(N-k-1) + \tfrac{1}{2}\boldsymbol{\eta}(k)^T\mathbf{Q}\boldsymbol{\eta}(k) \quad (7')$$

Equation (5)—or its expanded counterparts (8), (9), and (10)—remains unchanged. The other two additional terms come into the boundary conditions, which are

$$\mathbf{W}(N-N) = \mathbf{M} \quad \text{(unchanged)}$$

$$\mathbf{V}(N-N) = -\mathbf{M}\mathbf{x}_d$$

$$Z(N-N) = \tfrac{1}{2}\mathbf{x}_d^T\mathbf{M}\mathbf{x}_d$$

A form of (6′) which is consistent with the notation introduced in (8), (9), and (10) of the previous problem is

$$\mathbf{V}(k') = \mathbf{A}^T[\mathbf{I} - \mathbf{K}^T(k'-1)\mathbf{B}^T]\mathbf{V}(k'-1) - \mathbf{Q}\boldsymbol{\eta}(k) \quad (6'')$$

The control law is

$$\mathbf{u}^*(k) = -\mathbf{K}^T(k'-1)\mathbf{A}\mathbf{x}(k) - \mathbf{U}(k'-1)\mathbf{B}^T\mathbf{V}(k'-1) \quad (2)$$

To demonstrate the use of the algorithm, Problem 17.3 is recalculated. For this case $\mathbf{A} = \mathbf{B} = \mathbf{M} = \mathbf{Q} = \mathbf{R} = 1$ and $\boldsymbol{\eta}(k) = 0$. The final time is $N = 2$. The initial conditions at $k = 2$ (or equivalently $k' = 0$) are $\mathbf{M}(0) = 1$ and $\mathbf{V}(0) = -20$. The calculations are tabulated from left to right in the order performed. The table also indicates the correct starting points and time sequencing relationships among the variables. The extra quantities $\mathbf{U}$ and $(\mathbf{I} - \mathbf{KB}^T)$ are given for convenience since they are both used more than once per line of table entries in the optimal tracking problem.

k	$k'=N-k$	$\mathbf{u}(k)=-\mathbf{K}^T\mathbf{A}\mathbf{x}(k)-\mathbf{UB}^T\mathbf{V}$	$\mathbf{V}$	$\mathbf{M}$	$\mathbf{U}=[\mathbf{B}^T\mathbf{MB}+\mathbf{R}]^{-1}$	$\mathbf{K}=\mathbf{MBU}$	$(\mathbf{I}-\mathbf{KB}^T)$	$\mathbf{P}$
2	0	—	−20	1	1/2	1/2	1/2	1/2
1	1	$\mathbf{u}(1)=-1/2\mathbf{x}(1)+10$	−10	3/2	2/5	3/5	2/5	3/5
0	2	$\mathbf{u}(0)=-3/5\mathbf{x}(0)+4$	−4	8/5	—	—	—	—

Once $k = 0$ is reached on the backward pass, the knowledge that $\mathbf{x}(0) = 10$ is used to give $\mathbf{u}(0) = -3/5(10) + 4 = -2$. This control gives $\mathbf{x}(1) = 8$ and then $\mathbf{u}(1) = -1/2(8) + 10 = 6$. The final state is $\mathbf{x}(2) = \mathbf{x}(1) + \mathbf{u}(1) = 14$. These all agree with Problem 17.3.

The general mechanization of the optimal controller for the optimal tracker is given in Fig. 17.5. This also applies to the optimal regulator of Problem 17.4 as special case with $\mathbf{V}(k') = \mathbf{0}$.

17.6 In order to compare quadratic optimal controllers with earlier methods, the dc motor system of Example 2.5 and Problems 2.22 through 2.24 is reconsidered. Using a sample

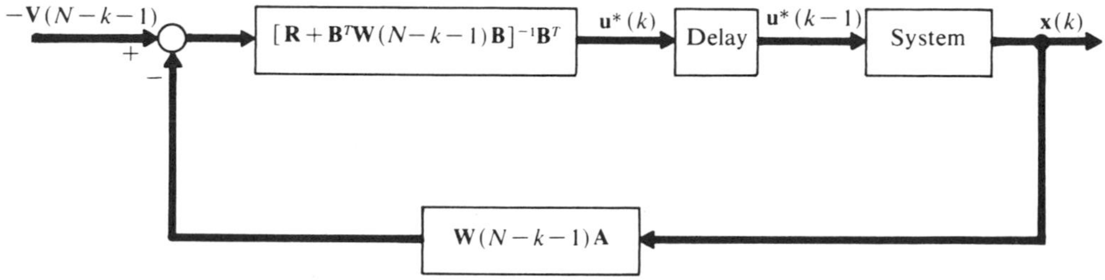

Figure 17.5

time of $T = 1$ and a motor gain of $K = 0.5$ allows the following observable canonical form state equations to be written for the open-loop system.

$$\mathbf{x}(k+1) = \begin{bmatrix} 1.6065 & 1 \\ -0.6065 & 0 \end{bmatrix} \mathbf{x}(k) + \begin{bmatrix} 0.1065 \\ 0.0902 \end{bmatrix} E(k)$$

The first component $\mathbf{x}_1(k)$ is the motor shaft position angle, and its measurement is assumed available for use. It can be seen that the second state can be determined as $\mathbf{x}_2(k) = -0.6065\mathbf{x}_1(k-1) - 0.0902E(k-1)$. Therefore, it is justifiable to assume that both states are available for use. Design a controller which minimizes the cost function J of Problem 17.5.

Equations *(8)*, *(9)*, and *(10)* of Problem 17.4, plus equations *(2)* and *(6″)* of Problem 17.5, were solved with $N = 15$ steps of $T = 1$ second each. The first group of cases studied all had $\boldsymbol{\eta}(k) = \mathbf{0}$ and $\mathbf{x}_d = \mathbf{0}$ (i.e., the regulator problem). For each case the steady-state feedback gain matrix is listed, along with the closed-loop eigenvalue locations. In each case steady state was reached in about seven steps, a little over three motor time constants.

Case	M	Q	R	$\mathbf{G}_{ss} = \mathbf{K}^T\mathbf{A}$	Closed-loop Eigenvalues
1	**I**	**I**	1	[2.106 1.9160]	$0.6057 \pm j0.1756$
2	0	**I**	1	[2.1060 1.9160]	$0.6057 \pm j0.1756$
3	20**I**	**I**	1	[2.1060 1.9160]	$0.6057 \pm j0.1756$
4	**I**	**I**	10	[0.8078 0.7747]	0.6304 , 0.8201 (real)
5	**I**	**I**	0.1	[5.9290 4.6290]	$0.2787 \pm j0.3062$

From this it is seen that the final results are independent of the initial **M**. For relatively smaller **R** values, larger gains and a faster, more responsive system is obtained. Larger **R** values give slower, smoother response and smaller gains. Smaller motor inputs will be required. The function of the **R** weighting term is to prevent large control signals from being called for. The **Q** weighting term is intended to get the state near to $\boldsymbol{\eta}(k)$.

Another group of cases with $\boldsymbol{\eta}(k) = [1. \quad -0.6065]^T$ was run, and for simplicity **M** was set to zero and $\mathbf{R} = 1$ was held fixed. In addition to the steady-state feedback gains, the steady-state values of the external inputs $\mathbf{v} = -\mathbf{U}\mathbf{B}^T\mathbf{V}$ are given, along with the scalar closed-loop transfer function $\mathbf{H}(z)$ from $\mathbf{v}(z)$ to $\mathbf{x}_1(z)$ (normalized as described below). The eigenvalues are, of course, the denominator roots.

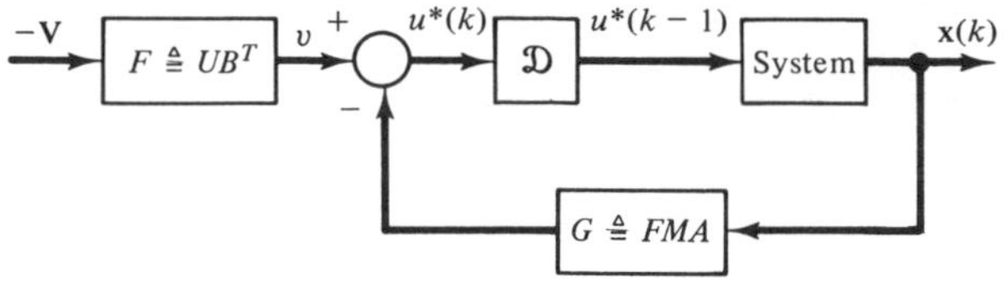

(a) Move F

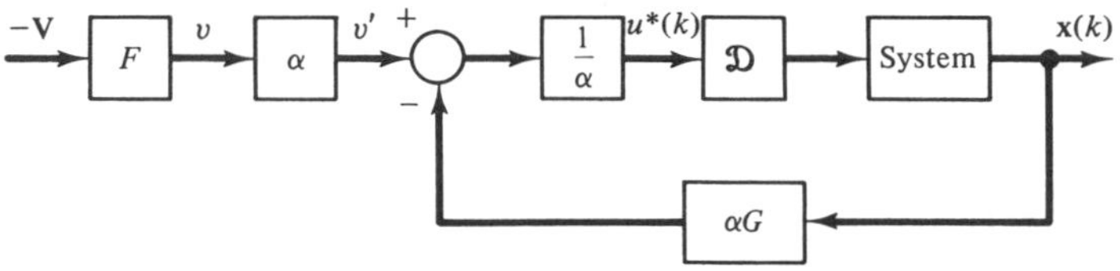

(b) Pick normalizing $\alpha = \dfrac{1}{v_{ss}}$

Figure 17.6

Case	Q	Gain	v	Transfer Function
6	**I**	[same as case 1]	0.9439	$0.3971(z - 0.5324)/[(z - 0.6057 \pm j0.1756)]$
7	100**I**	[7.1525 5.6840]	3.7052	$1.2744(z - 0.4281)/[(z - 0.166 \pm j0.182)]$
8	10000**I**	[8.3299 6.4773]	4.4013	$1.4714(z - 0.4116)/[(z - 0.0068)(z - 0.1283)]$

The increase in **Q** is equivalent to a relative decrease in **R**. The previous observation remains true: as **Q** gets bigger relative to **R** the system response gets faster, but at the expense of larger control effort. The role of the external input needs to be clarified. This input **V** is a two-component vector. By shifting gains around, an equivalent normalized scalar input system can be obtained from Fig. 17.5. The two steps used in shifting and normalizing are shown in Fig. 17.6. The transfer functions given above are from v' to **x**. The final value theorem shows that each transfer function has a steady-state gain of one, meaning that there will be no steady state error in the position $\mathbf{x}_1$ after a unit input at v'. This means that in steady state **Gx** must equal v.

The desired state $\boldsymbol{\eta}(k)$ above happens to be a point of equilibrium where x_1 and x_2 are in balance when the input E is zero. Two more cases are solved where this is not true. x_1 is selected as unity and x_2 is chosen arbitrarily. In case 9 the errors in both states are equally weighted, and in case 10 the error in x_2 is ignored so η_2 could have been any value without changing the results. (This is not to say that the solution does not depend on Q_{22}, however.)

Case	Q	Gain	v	Transfer Function
9	100**I**	[7.1525 5.684]	1.06	$1.2766(z - 0.4260)/[z - 0.1650 \pm j0.1883]$
10	Diag[100, 0]	[7.7430 5.696]	4.288	$1.3384(z - 0.3698)/[z - 0.1340 \pm j0.3060]$

We see from the above results that the closed-loop system poles can be moved around by changing the relative importance of **Q** and **R**. If any one of the above sets of eigenvalues had been selected a priori, then the pole-placement methods of Chapter 16 would give the same feedback gain. Sometimes it may be more natural to select pole locations based on their relation to a desired time response. Pole placement would easily give the gains. Sometimes it may seem more natural to express the design goals in terms of a cost function to be minimized. In either case, if extremes are requested, the magnitudes of the resultant control signals may become excessive or even ridiculous. It is not possible to make a turbogenerator set, which normally takes minutes to come up to rated speed, respond with a time constant in the microsecond range just because "the math in Chapter 16 says you can put the poles anywhere you want." Neither can state variable feedback transform a light pleasure aircraft into a high-performance military fighter. The fallacy is due to the validity of the linear models breaking down and due to state or control limits being exceeded. The linear models are intended for use over a limited range of values of the signals. Conductors tend to vaporize when their rated loads are drastically exceeded. Wings or control surfaces tend to get ripped off under excessive loads. Similar limitations exist in other situations.

The step responses for the the designs of cases 6, 7, 8, and 10 are shown in Fig. 17.7 for comparison with the Chapter 2 results. Perhaps case 7 gives the best response of any considered here or in Chapter 2. Further parametric studies may uncover a better response between case 6 and case 7.

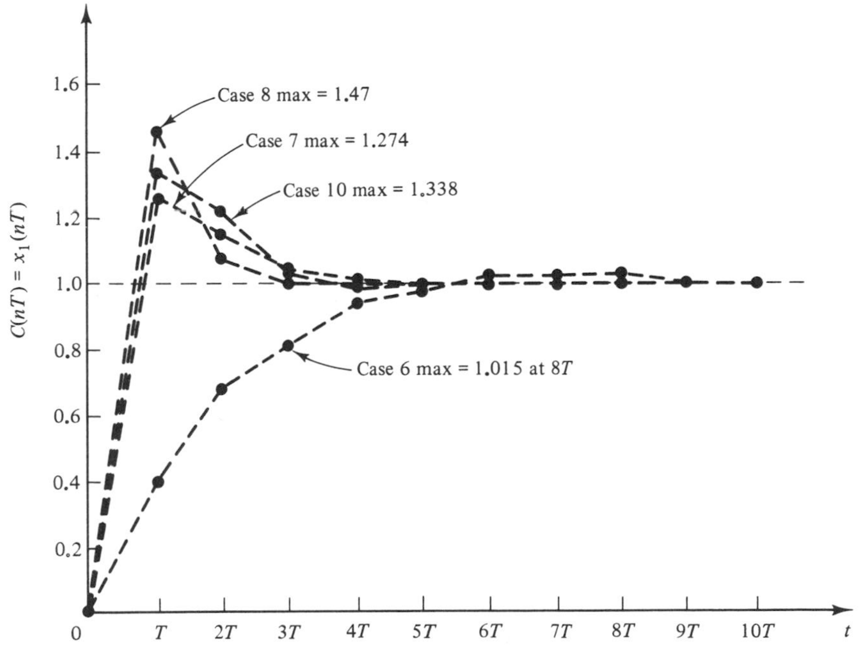

Figure 17.7 (Compare with Figs. 2.25, 2.26, and 2.28.)

Continuous Dynamic Programming

17.7 Consider the linear system $\dot{\mathbf{x}} = \mathbf{A}\mathbf{x} + \mathbf{B}\mathbf{u}$. Find the optimal control which minimizes the quadratic performance criterion of Example 17.5. There are no restrictions on $\mathbf{u}(t)$, and t_f is fixed.

Equation (*17.9*) specializes to

$$\frac{\partial g}{\partial t_r} = \min_{u(t)} \{[\mathbf{x}(t) - \boldsymbol{\eta}(t)]^T\mathbf{Q}[\mathbf{x}(t) - \boldsymbol{\eta}(t)] + \mathbf{u}(t)^T\mathbf{R}\mathbf{u}(t) + (\nabla_{\mathbf{x}} g)^T[\mathbf{A}\mathbf{x} + \mathbf{B}\mathbf{u}]\}$$

with boundary conditions

$$g[\mathbf{x}(t), t_r]|_{t_r=0} = [\mathbf{x}(t_f) - \mathbf{x}_d]^T\mathbf{M}[\mathbf{x}(t_f) - \mathbf{x}_d]$$

Since $\mathbf{u}(t)$ is unrestricted, taking the derivative with respect to $\mathbf{u}(t)$ and setting it equal to zero gives $\mathbf{u}^*(t) = -\frac{1}{2}\mathbf{R}^{-1}\mathbf{B}^T\nabla_{\mathbf{x}} g[\mathbf{x}(t), t_r]$. The H.J.B. equation then reduces to

$$\frac{\partial g}{\partial t_r} = [\mathbf{x}(t) - \boldsymbol{\eta}(t)]^T\mathbf{Q}[\mathbf{x}(t) - \boldsymbol{\eta}(t)] + (\nabla_{\mathbf{x}} g)^T\mathbf{A}\mathbf{x}(t) - \tfrac{1}{4}(\nabla_{\mathbf{x}} g)^T\mathbf{B}\mathbf{R}^{-1}\mathbf{B}^T\,\nabla_{\mathbf{x}} g$$

This nonlinear partial differential equation can be solved [58] by assuming a solution $g[\mathbf{x}(t), t_r] = \mathbf{x}^T(t)\mathbf{W}(t_r)\mathbf{x}(t) + \mathbf{x}^T(t)\mathbf{V}(t_r) + Z(t_r)$, where $\mathbf{W}$ is an unknown symmetric $n \times n$ matrix, $\mathbf{V}$ is an unknown $n \times 1$ vector, and Z is an unknown scalar. Differentiating the assumed answer gives

$$\frac{\partial g}{\partial t_r} = \mathbf{x}^T(t)\frac{d\mathbf{W}}{dt_r}\mathbf{x}(t) + \mathbf{x}^T(t)\frac{d\mathbf{V}}{dt_r} + \frac{dZ}{dt_r} \qquad \text{and} \qquad \nabla_{\mathbf{x}} g = 2\mathbf{W}(t_r)\mathbf{x}(t) + \mathbf{V}(t_r)$$

Using these, the H.J.B. equation becomes

$$\begin{aligned}\mathbf{x}^T(t)\frac{d\mathbf{W}}{dt_r}\mathbf{x}(t) + \mathbf{x}^T(t)\frac{d\mathbf{V}}{dt_r} + \frac{dZ}{dt_r} &= \mathbf{x}^T\{\mathbf{Q} + 2\mathbf{W}\mathbf{A} - \mathbf{W}\mathbf{B}\mathbf{R}^{-1}\mathbf{B}^T\mathbf{W}\}\mathbf{x} \\ &\quad + \mathbf{x}^T\{-2\mathbf{Q}\boldsymbol{\eta} + \mathbf{A}^T\mathbf{V} - \mathbf{W}\mathbf{B}\mathbf{R}^{-1}\mathbf{B}^T\mathbf{V}\} \\ &\quad + \{\boldsymbol{\eta}^T\mathbf{Q}\boldsymbol{\eta} - \tfrac{1}{4}\mathbf{V}^T\mathbf{B}\mathbf{R}^{-1}\mathbf{B}^T\mathbf{V}\}\end{aligned}$$

In order for the assumed form to actually be a solution for all $\mathbf{x}(t)$, the quadratic terms in $\mathbf{x}$, the linear terms, and the terms not involving $\mathbf{x}$ must balance individually. Therefore,

$$\frac{dZ}{dt_r} = \boldsymbol{\eta}(t)^T\mathbf{Q}\boldsymbol{\eta}(t) - \tfrac{1}{4}\mathbf{V}^T(t_r)\mathbf{B}\mathbf{R}^{-1}\mathbf{B}^T\mathbf{V}(t_r) \qquad \text{(terms not involving } \mathbf{x}(t)) \qquad (1)$$

The linear terms in $\mathbf{x}(t)$ require

$$\frac{d\mathbf{V}}{dt_r} = -2\mathbf{Q}\boldsymbol{\eta}(t) + \mathbf{A}^T\mathbf{V}(t_r) - \mathbf{W}(t_r)\mathbf{B}\mathbf{R}^{-1}\mathbf{B}^T\mathbf{V}(t_r) \qquad (2)$$

Each matrix involved in the quadratic terms is symmetric except $\mathbf{x}^T\{2\mathbf{W}\mathbf{A}\}\mathbf{x}$, which is rewritten as $\mathbf{x}^T\{\mathbf{W}\mathbf{A} + \mathbf{A}^T\mathbf{W}\}\mathbf{x} + \mathbf{x}^T\{\mathbf{W}\mathbf{A} - \mathbf{A}^T\mathbf{W}\}\mathbf{x}$. The second term is always zero since the matrix is skew-symmetric. All quadratic terms in the H.J.B. equation can be combined into the form $\mathbf{x}^T\mathbf{F}\mathbf{x} = \mathbf{0}$. Since every matrix term in $\mathbf{F}$ is now symmetric, it can be concluded that $\mathbf{F} = \mathbf{0}$, or

$$\frac{d\mathbf{W}}{dt_r} = \mathbf{Q} + \mathbf{W}\mathbf{A} + \mathbf{A}^T\mathbf{W} - \mathbf{W}\mathbf{B}\mathbf{R}^{-1}\mathbf{B}^T\mathbf{W} \qquad (3)$$

The boundary conditions for the three sets of differential equations are $\mathbf{W}(t_r = 0) = \mathbf{M}$, $\mathbf{V}(t_r = 0) = -2\mathbf{M}\mathbf{x}_d$, and $Z(t_r = 0) = \mathbf{x}_d^T\mathbf{M}\mathbf{x}_d$.

Equation (*3*) is known as the *Riccati equation*. It can be solved first, and the result can then be used in solving equation (*2*), after which, equation (*1*) can be integrated.

The optimal feedback control system is shown in Fig. 17.8.

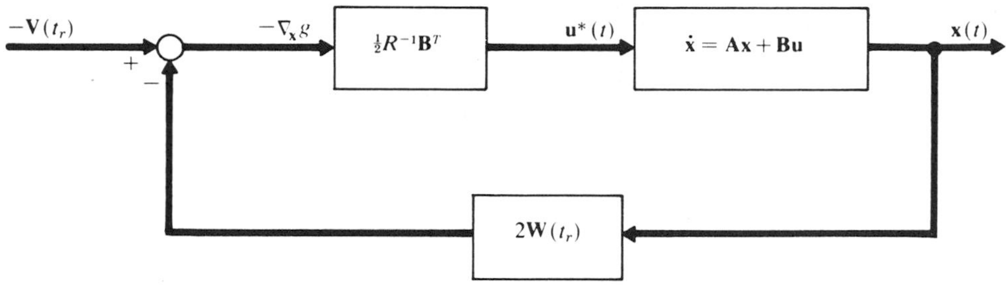

Figure 17.8

17.8 How can the nonlinear Riccati equation (equation (*3*), Problem 17.7) be solved?

It is always possible to solve it by numerical integration, backward in time from $t = t_f(t_r = 0)$ to $t = t_0(t_r = t_f - t_0)$. It can also be solved analytically, since a change of variables reduces it to a pair of *linear* matrix equations. Let $\mathbf{W}(t_r) = \mathbf{E}(t_r)\mathbf{F}^{-1}(t_r)$. Then the Riccati equation becomes

$$\frac{d\mathbf{E}}{dt_r}\mathbf{F}^{-1} - \mathbf{EF}^{-1}\frac{d\mathbf{F}}{dt_r}\mathbf{F}^{-1} = \mathbf{Q} + \mathbf{EF}^{-1}\mathbf{A} + \mathbf{A}^T\mathbf{EF}^{-1} - \mathbf{EF}^{-1}\mathbf{BR}^{-1}\mathbf{B}^T\mathbf{EF}^{-1}$$

Postmultiply by $\mathbf{F}$ and require $d\mathbf{E}/dt_r = \mathbf{A}^T\mathbf{E} + \mathbf{QF}$. Then it must also be true that $-\mathbf{EF}^{-1}(d\mathbf{F}/dt_r) = \mathbf{EF}^{-1}\mathbf{AF} - \mathbf{EF}^{-1}\mathbf{BR}^{-1}\mathbf{B}^T\mathbf{E}$. Premultiplying by $(\mathbf{EF}^{-1})^{-1}$ gives two sets of linear equations $\begin{bmatrix} \frac{d\mathbf{E}}{dt_r} \\ \frac{d\mathbf{F}}{dt_r} \end{bmatrix} = \left[\begin{array}{c|c} \mathbf{A}^T & \mathbf{Q} \\ \hline \mathbf{BR}^{-1}\mathbf{B}^T & -\mathbf{A} \end{array}\right]\begin{bmatrix} \mathbf{E} \\ \hline \mathbf{F} \end{bmatrix}$. The boundary conditions are $\mathbf{E}(t_r)\mathbf{F}^{-1}(t_r)|_{t_r=0} = \mathbf{M}$, or $\mathbf{E}(t_r = 0) = \mathbf{M}$, $\mathbf{F}(t_r = 0) = \mathbf{I}_n$. When $\mathbf{A}$, $\mathbf{B}$, $\mathbf{Q}$, and $\mathbf{R}$ are constant, this linear equation can be solved, at least in principle, in terms of a matrix exponential. After finding $\mathbf{E}(t_r)$ and $\mathbf{F}(t_r)$, the desired result is $\mathbf{W}(t_r) = \mathbf{E}(t_r)\mathbf{F}(t_r)^{-1}$.

17.9 Find the optimal feedback control law for the unstable scalar system $\dot{x} = x + u$ which minimizes $J = Mx(t_f)^2 + \int_0^{t_f} u(t)^2\, dt$. There are no restrictions on $u(t)$, and t_f is fixed.

Using dynamic programming and the results of Problems 17.7 and 17.8, with $A = 1$, $B = 1$, $x_d = 0$, $Q = 0$, $R = 1$, the optimal control is $u^*(t) = -\frac{1}{2}\nabla_x g[x(t), t_r]$, where $\nabla_x g = 2W(t_r)x(t) + V(t_r)$.

The Riccati equation for $W(t_r)$ is $dW/dt_r = 2W - W^2$, with $W(t_r = 0) = M$.

The equation for $V(t_r)$ is $dV/dt_r = [1 - W(t_r)]V$, but since $V(t_r = 0) = 0$, $V(t_r) \equiv 0$ for all t_r.

Using $W(t_r) = E(t_r)/F(t_r)$ for this scalar case leads to $dE/dt_r = E$, $dF/dt_r = E - F$, with $E(t_r = 0) = M$, $F(t_r = 0) = 1$. Solving gives $E(t_r) = e^{t_r}M$ and $F(t_r) = e^{-t_r} + (e^{t_r} - e^{-t_r})M/2$.

Since $W(t_r) = E(t_r)/F(t_r)$, it is seen that $W \longrightarrow 2$ for large t_f. This in turn gives $u^*(t) = -2x(t)$ and the stable closed-loop system $\dot{x} = -x(t)$.

The general feedback control system is given in Fig. 17.9.

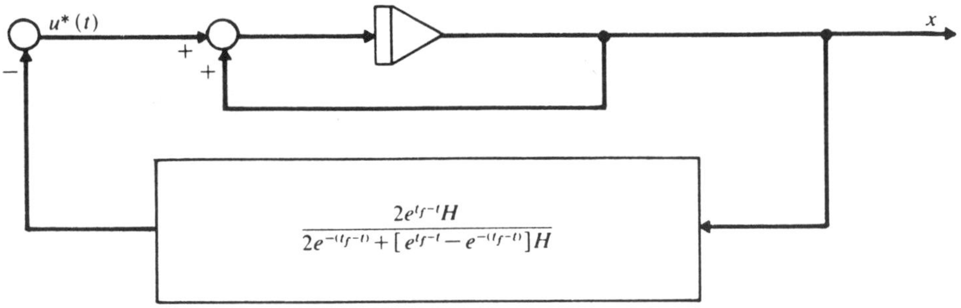

Figure 17.9

17.10 Consider the scalar control system $\dot{x} = Ax + u$. Investigate the constant state feedback control system which minimizes

$$J = x(t_f)^2 M + \int_0^{t_f} \{Qx^2 + u^2\}\, dt \qquad \text{as } t_f \longrightarrow \infty$$

What is $u^*(t)$ if $Q = 0$? What if $Q = -A^2$?

The Riccati equation is $dW/dt_r = Q + 2WA - W^2$. In general, the steady-state solution can be found by numerically integrating until steady state is reached, or by setting the derivative equal to zero and solving a nonlinear algebraic equation for W. In this scalar case, setting $dW/dt_r = 0$ gives a simple quadratic equation for the steady-state values of W. Its solutions are $W = A \pm \sqrt{A^2 + Q}$.

Since $V(t_r) = 0$, $u^*(t) = -W(t_r)x(t)$, the steady-state feedback system satisfies $\dot{x} = Ax - [A \pm \sqrt{A^2 + Q}]x = \mp\sqrt{A^2 + Q}x$.

For a stable system, the plus sign in W must be selected, giving $\dot{x} = -\sqrt{A^2 + Q}x$. If $Q = 0$, $\dot{x} = -|A|\mathbf{x}$. If the original system was stable, $-|A| = A$, so this implies that $u^* = 0$. But $Q = 0$ means that it does not matter what $x(t)$ is, so the optimal control strategy is to do nothing. This gives a zero value to the integral part of J. Also, since $t_f \longrightarrow \infty$, $x(t_f) \longrightarrow 0$ for a stable system. If the original system is unstable, the feedback system is stabilized. If $Q = -A^2$, then $W = A$ in steady state, so $u^*(t) = -Ax$. This gives $\dot{x} = 0$, and the integral term in J is once again identically zero.

17.11 A linearized aircraft model was introduced in Problem 12.7 where its controllability and observability properties were examined. Its transfer functions were obtained in Problem 15.5 and pole-placement feedback controllers for it were found in Problem 16.34. Design a controller using the steady-state optimal regulator approach. For convenience the system models are repeated here.

$$\mathbf{A} = \begin{bmatrix} -10 & 0 & -10 & 0 \\ 0 & -0.7 & 9 & 0 \\ 0 & -1 & -0.7 & 0 \\ 1 & 0 & 0 & 0 \end{bmatrix} \qquad \mathbf{B} = \begin{bmatrix} 20 & 2.8 \\ 0 & -3.13 \\ 0 & 0 \\ 0 & 0 \end{bmatrix}$$

The solution involves selecting the appropriate weighting matrices $\mathbf{M}$, $\mathbf{Q}$, and $\mathbf{R}$ and then solving the Riccati equation (*3*) in Problem 17.7 until steady state is reached.

The controller is found by combining equation (*17.11*) with the results for $\nabla_{\mathbf{x}} g$ from Problem 17.7, and the result is

$$\mathbf{u}^*(t) = -(1/2)\mathbf{R}^{-1}\mathbf{B}^T[2\mathbf{W}\mathbf{x}(t) + \mathbf{V}]$$

as shown in Fig. 17.8. Two issues remain. What method is to be used to solve the Riccati equation? There are methods of solving the algebraic Riccati that results when $d\mathbf{W}/dt_r$ is set to zero [2]. The matrix case is much more challenging than the scalar example of Problem 17.10. The alternative approach is used here. Numerical integration is used until $\mathbf{W}$ effectively becomes constant. For an asymptotically stable constant coefficient system, a reasonable estimate is that this will occur within four time constants. The time constant is estimated as the reciprocal of the smallest nonzero real part of the eigenvalues of $\mathbf{A}$. The eigenvalues of $\mathbf{A}$ are 0, $-0.7 \pm 3j$, and -10, so the time constant is estimated as $1/0.7 = 1.4$ sec. The integration stepsize ΔT must not be too large, or poor accuracy will result. If it is too small, excessive computer time is required. A method which seems satisfactory is to select ΔT as about $1/(20|\lambda_{\max}|)$, which means that the total number of integration steps might be on the order of $T/\Delta T = 80|\lambda_{\max}|/|\lambda_{\min}|$, which can easily exceed 1000. These are just estimates. The magnitudes of the changes in $\mathbf{W}$ must be monitored to determine when all components have essentially stopped changing.

The second issue is selection of the weighting matrices. For small problems with only a few parameters it may be feasible to parametrically examine the range of possibilities. For most problems a more focused approach is desirable. The expanded quadratic will contain terms of the form $x_i^2 Q_{ii} + u_i^2 R_{ii}$. Similar treatment of the final value terms can be done, but here $M = 0$ is selected because it will have no effect on steady-state answers. If x_i is a position variable with a magnitude of thousands of feet, and if u_i is an angle of say 0.01 radian, it is clear that u_i will have no effect on J unless $R_{ii} \gg Q_{ii}$. The point is that scaling units and variable magnitudes are important, as well as the subjective choice of the importance of keeping u_i small compared to keeping x_i small. If all variables in the quadratic cost function are intended to be equally important, then one method is to estimate the maximum possible or allowable values of each variable and use these estimates to select the weights. For the airplane model, suppose that structural limits and prevention of pilot blackout require that the roll rate p be < 300 deg/sec and the yaw rate r be < 18 deg/sec. The maximum side-slip angle is estimated as $\beta < 15$ deg and the maximum roll angle as $\phi < 180$ deg. Likewise, there are maximums for the control surface deflection angles, and the hypothetical values assumed here are $\delta_a < 40$ deg and $\delta_r < 10$ deg. Then setting each term in J to unity when the variables are at their limits gives

$$\mathbf{Q} = \text{Diag}\,[0.0011,\ 0.308,\ 0.44,\ 0.003]$$

$$\mathbf{R} = \text{Diag}\,[0.0626,\ 1]$$

The relative magnitudes are all that matter. The above components have been scaled so the largest entry is unity. Considerable rounding off is probably justified because of the gross approximations involved in estimating the maximums. Using the above values for $\mathbf{Q}$ and $\mathbf{R}$, the approximate steady-state feedback gain matrix and the closed-loop eigenvalues it yields are

$$\mathbf{G} = \begin{bmatrix} 0.0358 & 0.0197 & -0.0355 & 0.1942 \\ 0.0001 & -0.3072 & -0.1191 & -0.0001 \end{bmatrix}$$

$$\lambda_i = \{-0.376,\ -1.181 \pm 2.898j,\ -10.340\}$$

The resultant system has moved the open-loop pole from the origin to -0.376, which means that the perturbations caused by any step disturbance will decay back to zero. The settling time and damping ratio of the dominant complex poles have been improved considerably and the remaining nondominant pole is not much changed.

If the maximum value of each state variable is multiplied by its two gain values, the resultant commands for δ_a and δ_r can be estimated. These values are valid if only one state variable is at its maximum and all others are zero.

Cause		x_1	x_2	x_3	x_4
Commands	δ_a	10	0.3	0.5	35
	δ_r	0.03	5.5	1.7	0.01

This shows that no maximum command is exceeded, unlike the results that can occur if arbitrarily selected **Q** or **R** values are used, or if extreme requests are made of a pole-placement design. In this particular example it is clear that δ_a is primarily controlled by the roll rate x_1 and roll angle x_4, and δ_r is primarily controlled by yaw rate x_2 and side-slip angle x_3 as expected from the physics of the problem.

A series of other sets of Q and R have been analyzed for comparison. With $\mathbf{Q} = \mathbf{I}$ and $\mathbf{R} = \mathbf{I}$,

$$\mathbf{G} = \begin{bmatrix} 0.6587 & 0.0768 & -0.2610 & 0.9909 \\ 0.0802 & -0.7184 & -0.2743 & 0.0729 \end{bmatrix}$$

$$\lambda_i = \{-0.889,\ -1.82 \pm j2.6,\ -22.5\}$$

The dominant complex poles are about the same as the earlier case and the dominant real pole has an improved settling time, but the major difference is the shift in the non-dominant pole, which will not be reflected much in system behavior. The required gains are much higher here and commanded deflection angles could easily exceed their limits.

With $\mathbf{Q} = 10\mathbf{I}$ and $\mathbf{R} = 0.1\mathbf{I}$ the resultant eigenvalues are all real $\{-0.995, -1.27, -30.7$, and $-202.2\}$ and the gains are so high as to be ridiculous. Several are on the order of 9 or 10!

With $\mathbf{Q} = \text{Diag}\,[0.01, 0.1, 0.1, 0.01]$ and $\mathbf{R} = \text{diag}\,[0.1, 1]$ the results are

$$\mathbf{G} = \begin{bmatrix} 0.1160 & 0.0275 & -0.0952 & 0.2956 \\ 0.0012 & -0.1148 & -0.0559 & 0.0004 \end{bmatrix}$$

$$\lambda_i = \{-0.5,\ -0.8798 \pm 2.965j,\ -11.823\}$$

The dominant poles are less damped and have a poorer settling time than the first case in spite of the fact that gains are typically higher and command limits could be exceeded.

The Minimum Principle

17.12 A system is described by $\dot{\mathbf{x}} = \mathbf{f}(\mathbf{x}, \mathbf{u}, t)$, with $\mathbf{x}(t_0)$ given. Find the necessary conditions which $\mathbf{x}(t)$ and $\mathbf{u}(t)$ must satisfy if they are to minimize

$$J = S(\mathbf{x}(t_f), t_f) + \int_{t_0}^{t_f} L(\mathbf{x}(t), \mathbf{u}(t), t)\, dt$$

The admissible controls must satisfy $\mathbf{u}(t) \in U$. Assume t_f is fixed.

Let $\mathbf{x}^*(t)$, $\mathbf{u}^*(t)$ be the optimal quantities and let $\mathbf{x}(t)$ be arbitrary and let $\mathbf{u}(t)$ be arbitrary but admissible. Let $\mathbf{p}(t)$ be an $n \times 1$ vector of Lagrange multipliers (also called the *costate* or *adjoint* variables). Adjoin the differential constraints to J and call the result J':

$$J' = S(\mathbf{x}(t_f), t_f) + \int_{t_0}^{t_f} \{L(\mathbf{x}, \mathbf{u}, t) + \mathbf{p}^T(t)[\mathbf{f}(\mathbf{x}, \mathbf{u}, t) - \dot{\mathbf{x}}]\}\, dt$$

Since $\mathbf{x}^*$, $\mathbf{u}^*$ are optimal, $\Delta J = J'(\mathbf{x}, \mathbf{u}) - J'(\mathbf{x}^*, \mathbf{u}^*) \geq 0$ for $\mathbf{x}$ arbitrary and $\mathbf{u} \in U$.

Let $\mathbf{x} = \mathbf{x}^* + \delta\mathbf{x}$. Using Taylor series expansion gives

$$J'(\mathbf{x}^* + \delta\mathbf{x}, \mathbf{u}) = S(\mathbf{x}^*(t_f), t_f) + [\nabla_{\mathbf{x}} S]^T|_{t_f}\, \delta\mathbf{x}(t_f) + \int_{t_0}^{t_f} \Big\{L(\mathbf{x}^*, \mathbf{u}, t) + [\nabla_{\mathbf{x}} L(\mathbf{x}^*, \mathbf{u}, t)]^T\, \delta\mathbf{x} + \mathbf{p}^T\Big[\mathbf{f}(\mathbf{x}^*, \mathbf{u}, t) + \frac{\partial \mathbf{f}}{\partial \mathbf{x}}\, \delta\mathbf{x} - \dot{\mathbf{x}}^* - \delta\dot{\mathbf{x}}\Big]\Big\}\, dt + \text{higher-order terms}$$

Therefore,

$$\Delta J = (\nabla_{\mathbf{x}} S)^T|_{t_f}\, \delta\mathbf{x}(t_f) + \int_{t_0}^{t_f} \{L(\mathbf{x}^*, \mathbf{u}, t) + \mathbf{p}^T\mathbf{f}(\mathbf{x}^*, \mathbf{u}, t) - L(\mathbf{x}^*, \mathbf{u}^*, t) - \mathbf{p}^T\mathbf{f}(\mathbf{x}^*, \mathbf{u}^*, t)\}\, dt + \int_{t_0}^{t_f} \Big\{[\nabla_{\mathbf{x}} L(\mathbf{x}^*, \mathbf{u}, t)]^T\, \delta\mathbf{x} + \mathbf{p}^T \frac{\partial \mathbf{f}}{\partial \mathbf{x}}\, \delta\mathbf{x} - \mathbf{p}^T\, \delta\dot{\mathbf{x}}\Big\}\, dt + \text{higher-order terms}$$

When the higher-order terms are dropped, the result is called δJ, the *first variation* of J. The condition for optimality is that $\Delta J \geq 0$ for arbitrary $\delta\mathbf{x}(t)$ and $\mathbf{u}(t) \in U$. If $\mathbf{x}$, $\mathbf{u}$ are sufficiently close to $\mathbf{x}^*$, $\mathbf{u}^*$, then the sign of ΔJ is the same as the sign of δJ. By defining the Hamiltonian as $\mathcal{H}(\mathbf{x}^*, \mathbf{u}^*, \mathbf{p}, t) = L(\mathbf{x}^*, \mathbf{u}^*, t) + \mathbf{p}(t)^T\mathbf{f}(\mathbf{x}^*, \mathbf{u}^*, t)$ and using integration by parts, $\int_{t_0}^{t_f} \mathbf{p}^T\, \delta\dot{\mathbf{x}}\, dt = \mathbf{p}^T\delta\mathbf{x}\Big|_{t_0}^{t_f} - \int_{t_0}^{t_f} \dot{\mathbf{p}}^T\, \delta\mathbf{x}\, dt$, we obtain

$$\partial J = [\nabla_{\mathbf{x}} S - \mathbf{p}]_{t_f}^T\, \delta\mathbf{x}(t_f) + \int_{t_0}^{t_f} [\mathcal{H}(\mathbf{x}^*, \mathbf{u}, \mathbf{p}, t) - \mathcal{H}(\mathbf{x}^*, \mathbf{u}^*, \mathbf{p}, t)]\, dt + \int_{t_0}^{t_f} \Big\{\Big[[\nabla_{\mathbf{x}} L(\mathbf{x}^*, \mathbf{u}, t)]^T + \mathbf{p}^T \frac{\partial \mathbf{f}}{\partial \mathbf{x}} + \dot{\mathbf{p}}^T\Big]\, \delta\mathbf{x}\Big\}\, dt$$

In order for this to be nonnegative for arbitrary $\delta\mathbf{x}$, the coefficient of $\delta\mathbf{x}(t)$ inside the integral must vanish on the optimal trajectory,

$$\dot{\mathbf{p}} = -\Big[\frac{\partial \mathbf{f}}{\partial \mathbf{x}}\Big]^T \mathbf{p} - \nabla_{\mathbf{x}} L$$

The term involving $\delta\mathbf{x}(t_f)$ must also vanish. If $\mathbf{x}(t_f)$ is fixed, then $\delta\mathbf{x}(t_f) = \mathbf{0}$. If $\mathbf{x}(t_f)$ is not fixed, then $\delta\mathbf{x}(t_f) \neq \mathbf{0}$, so $\mathbf{p}(t_f) = \nabla_{\mathbf{x}} S|_{t_f}$. The remaining integral term must be nonnegative for any admissible $\mathbf{u}$, including a $\mathbf{u}$ which equals $\mathbf{u}^*$ for all except an infinitesimal time interval. Therefore, it is concluded that

$$\mathcal{H}(\mathbf{x}^*, \mathbf{u}, \mathbf{p}, t) \geq \mathcal{H}(\mathbf{x}^*, \mathbf{u}^*, \mathbf{p}, t) \qquad \text{for all } t, \text{ all } \mathbf{u} \in U$$

This constitutes one version of the minimum principle, since $\mathbf{u}^*(t)$ is that $\mathbf{u}$ which minimizes $\mathcal{H}$. If there are no restrictions on $\mathbf{u}$, this reduces to $\partial\mathcal{H}/\partial\mathbf{u} = \partial L/\partial\mathbf{u} + \mathbf{p}^T(\partial\mathbf{f}/\partial\mathbf{u}) = \mathbf{0}$.

17.13 What modifications are necessary in the previous problem if t_f is not fixed in advance, but must be determined as part of the solution?

In forming ΔJ, variations in the final time must now be considered. Let t_f^* be the optimal stopping time and $t_f = t_f^* + \delta t_f$ is the perturbed time. Referring to Fig. 17.10

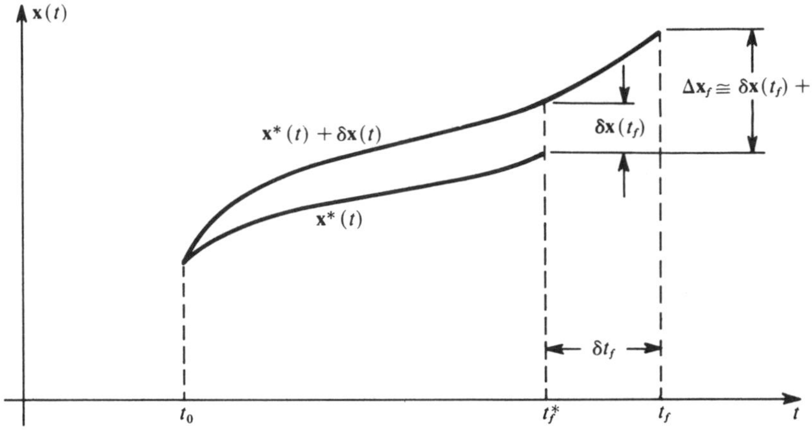

Figure 17.10

it is apparent that two different variations in $\mathbf{x}$ at the final time must be considered. $\delta\mathbf{x}(t_f)$ is the variation used in Problem 17.12 and $\Delta\mathbf{x}(t_f)$ is the total variation.

$$\begin{aligned}\Delta J &= J'(\mathbf{x}^* + \delta\mathbf{x}, \mathbf{u}, t_f^* + \delta t_f) - J'(\mathbf{x}^*, \mathbf{u}^*, t_f^*)\\ &= S(\mathbf{x}^*(t_f^*) + \Delta\mathbf{x}_f, t_f^* + \delta t_f) - S(\mathbf{x}^*(t_f^*), t_f^*)\\ &\quad + \int_{t_0}^{t_f^*+\delta t_f} \{L(\mathbf{x}^* + \delta\mathbf{x}, \mathbf{u}, t) + \mathbf{p}^T[\mathbf{f}(\mathbf{x}^* + \partial\mathbf{x}, \mathbf{u}, t) - \dot{\mathbf{x}}^* - \delta\dot{\mathbf{x}}]\}\, dt\\ &\quad - \int_{t_0}^{t_f^*} \{L(\mathbf{x}^*, \mathbf{u}^*, t) + \mathbf{p}^T[\mathbf{f}(\mathbf{x}^*, \mathbf{u}^*, t) - \dot{\mathbf{x}}^*]\}\, dt\end{aligned}$$

In computing δJ, the following relation is used:

$$\int_{t_0}^{t_f^*+\delta t_f} \{\ \}\, dt = \int_{t_0}^{t_f^*} \{\ \}\, dt + \int_{t_f^*}^{t_f^*+\delta t_f} \{\ \}\, dt \simeq \int_{t_0}^{t_f^*} \{\ \}\, dt + \delta t_f L$$

Then

$$\begin{aligned}\delta J &= (\nabla_{\mathbf{x}} S)^T|_{t_f} \Delta\mathbf{x}_f - \mathbf{p}^T(t_f)\,\delta\mathbf{x}(t_f) + \left[\frac{\partial S}{\partial t} + L(\mathbf{x}^*, \mathbf{u}, t)\right]\bigg|_{t_f^*} \delta t_f\\ &\quad + \int_{t_0}^{t_f^*} [\mathcal{H}(\mathbf{x}^*, \mathbf{u}, t) - \mathcal{H}(\mathbf{x}^*, \mathbf{u}^*, t)]\, dt\\ &\quad + \int_{t_0}^{t_f} \left\{\left[[\nabla_{\mathbf{x}} L(\mathbf{x}^*, \mathbf{u}, t)]^T + \mathbf{p}^T \frac{\partial \mathbf{f}}{\partial \mathbf{x}} + \dot{\mathbf{p}}^T\right] \delta\mathbf{x}\right\}\, dt\end{aligned}$$

From this, conclusions regarding the two integral terms are the same as in Problem 17.12. The only changes are in the boundary terms. If $\mathbf{x}^*(t_f)$ is free, then using $\delta\mathbf{x}(t_f) = \Delta\mathbf{x}_f - \dot{\mathbf{x}}^*(t_f)\delta t_f$ leads to boundary terms $[(\nabla_{\mathbf{x}} S)^T|_{t_f^*} - \mathbf{p}^T(t_f)]\Delta\mathbf{x}_f + [\partial S/\partial t) + L(\mathbf{x}^*, \mathbf{u}, t) + \mathbf{p}^T\dot{\mathbf{x}}^*]|_{t_f}\delta t_f$. The conclusion is that $\mathbf{p}(t_f) = \nabla_{\mathbf{x}} S|_{t_f^*}$ as before, and the additional scalar equation required for determining t_f^* is $\partial S/\partial t + \mathcal{H}|_{t_f} = 0$.

If the final state had been restricted in some way, say to lie on a surface $\psi(\mathbf{x}(t_f), t_f)$, then $\Delta \mathbf{x}_f$ and δt_f are interrelated. Their coefficients cannot then be separately set equal to zero, but instead, additional restrictions must be imposed [14, 95].

17.14 Develop an iterative method of solving the two-point boundary value problem which results when the minimum principle is applied to:

$$\dot{\mathbf{x}} = \mathbf{f}(\mathbf{x}, \mathbf{u}), \qquad \mathbf{x}(t_0) \text{ known}, \qquad t_f \text{ fixed},$$

$$\text{minimize } J = \tfrac{1}{2}[\mathbf{x}(t_f) - \mathbf{x}_d]^T\mathbf{M}[\mathbf{x}(t_f) - \mathbf{x}_d] + \int_{t_0}^{t_f} L\, dt$$

Minimizing $\mathcal{H}$ allows $\mathbf{u}^*(t)$ to be found as a function of $\mathbf{p}(t)$. Then the two-point boundary value problem of equation (*17.14*) can be written as $\dot{\mathbf{x}} = \mathbf{f}(\mathbf{x}, \mathbf{p})$, $\dot{\mathbf{p}} = \mathbf{h}(\mathbf{x}, \mathbf{p})$; $\mathbf{x}(t_0) = \mathbf{x}_0$, $\mathbf{p}(t_f) = \mathbf{M}[\mathbf{x}(t_f) - \mathbf{x}_d]$. If $\mathbf{p}(t_0)$ were known, the equations for $\mathbf{x}$ and $\mathbf{p}$ could be solved by mumerical integration.

One way to proceed is to assume a $\mathbf{p}(t_0)^{(0)}$, numerical integrate to find $\mathbf{x}(t_f)^{(0)}$ and $\mathbf{p}(t_f)^{(0)}$. Then the terminal conditions can be checked, and the error can be used to estimate a new $\mathbf{p}(t_0)^{(1)}$. Figure 17.11 is used to illustrate the correction procedure [65]. Since $\mathbf{p}(t_f)^{(0)}$ and $\mathbf{x}(t_f)^{(0)}$ are determined by $\mathbf{p}(t_0)^{(0)}$, an unknown functional relation exists among these variables, as implied by the graph. The "slope" of the function multiplied by $\Delta\mathbf{p}(t_0)$ is set equal to the error in the terminal conditions. That is, a New-

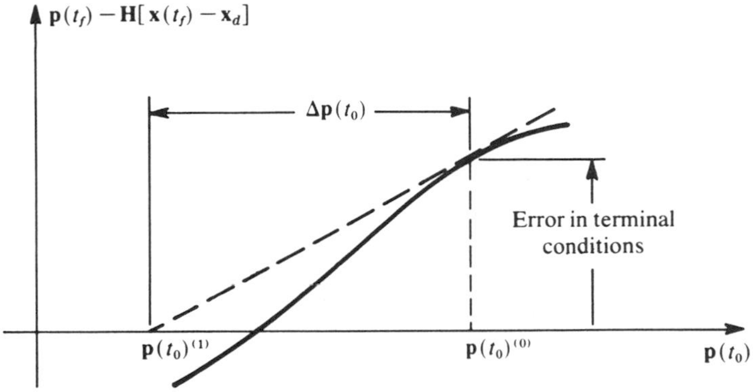

Figure 17.11

ton-Raphson correction scheme is used (see Problem 13.6).

The new estimate is

$$\mathbf{p}(t_0)^{(1)} = \mathbf{p}(t_0)^{(0)} - \Delta\mathbf{p}(t_0)$$

where

$$\left[\frac{\partial \mathbf{p}(t_f)}{\partial \mathbf{p}(t_0)} - \mathbf{M}\frac{\partial \mathbf{x}(t_f)}{\partial \mathbf{p}(t_0)}\right]^{(0)} \Delta\mathbf{p}(t_0) = \mathbf{p}(t_f)^{(0)} - \mathbf{M}[\mathbf{x}(t_f)^{(0)} - \mathbf{x}_d]$$

The two $n \times n$ sensitivity matrices $\mathbf{S}_\mathbf{p}(t_f) \triangleq \partial\mathbf{p}(t_f)/\partial\mathbf{p}(t_0)$ and $\mathbf{S}_\mathbf{x}(t_f) \triangleq \partial\mathbf{x}(t_f)/\partial\mathbf{p}(t_0)$ can be found by solving two sets of $n \times n$ matrix equations, obtained from the differential equations for $\mathbf{x}$ and $\mathbf{p}$ by interchanging $\partial/\partial\mathbf{p}(t_0)$ and d/dt. They are

$$\dot{\mathbf{S}}_\mathbf{x} = \frac{\partial \mathbf{f}}{\partial \mathbf{x}}\mathbf{S}_\mathbf{x} + \frac{\partial \mathbf{f}}{\partial \mathbf{p}}\mathbf{S}_\mathbf{p}, \qquad \mathbf{S}_\mathbf{x}(0) = [\mathbf{0}]$$

$$\dot{\mathbf{S}}_\mathbf{p} = \frac{\partial \mathbf{h}}{\partial \mathbf{x}}\mathbf{S}_\mathbf{x} + \frac{\partial \mathbf{h}}{\partial \mathbf{p}}\mathbf{S}_\mathbf{p}, \qquad \mathbf{S}_\mathbf{p}(0) = \mathbf{I}_n$$

These can be integrated along with the $\mathbf{x}$ and $\mathbf{p}$ equations. Only the terminal values are needed in the correction scheme, which generalizes to

$$\mathbf{p}(t_0)^{(k+1)} = \mathbf{p}(t_0)^{(k)} - [\mathbf{S}_\mathbf{p}(t_f)^{(k)} - \mathbf{M}\mathbf{S}_\mathbf{x}(t_f)^{(k)}]^{-1}\{\mathbf{p}(t_f)^{(k)} - \mathbf{M}[\mathbf{x}(t_f)^{(k)} - \mathbf{x}_d]\}$$

Success of the method depends on a good initial estimate for $\mathbf{p}(t_0)$ as well as the characteristics of the functions $\mathbf{f}$ and $\mathbf{h}$ and their derivatives.

17.15 Assuming that matrices $\mathbf{A}$ and $\mathbf{B}$ and vectors $\mathbf{x}(t_0) = \mathbf{x}_0$ and $\mathbf{x}_d$ are given, find the solution for the optimal control in Example 17.10.

When $\mathbf{A}$ and $\mathbf{B}$ are constant, the form of the solution for equation (*17.15*) is

$$\begin{bmatrix} \mathbf{x}(t) \\ \mathbf{p}(t) \end{bmatrix} = e^{\begin{bmatrix} \mathbf{A} & -1/2\mathbf{B}\mathbf{B}^T \\ \mathbf{0} & -\mathbf{A}^T \end{bmatrix}(t-t_0)} \begin{bmatrix} \mathbf{x}(t_0) \\ \mathbf{p}(t_0) \end{bmatrix}$$

For convenience, the $2n \times 2n$ exponential matrix is written in partitioned form as

$$\begin{bmatrix} \boldsymbol{\phi}_{11}(t, t_0) & \boldsymbol{\phi}_{12}(t, t_0) \\ \boldsymbol{\phi}_{21}(t, t_0) & \boldsymbol{\phi}_{22}(t, t_0) \end{bmatrix}$$

This may be computed using the methods of Chapter 8. Actually $\boldsymbol{\phi}_{21}(t, t_0)$ will be the $n \times n$ null matrix for this problem, but the more general form is treated. At the final time,

$$\mathbf{x}(t_f) = \boldsymbol{\phi}_{11}(t_f, t_0)\mathbf{x}_0 + \boldsymbol{\phi}_{12}(t_f, t_0)\mathbf{p}(t_0)$$

$$\mathbf{p}(t_f) = \boldsymbol{\phi}_{21}(t_f, t_0)\mathbf{x}_0 + \boldsymbol{\phi}_{22}(t_f, t_0)\mathbf{p}(t_0)$$

Using the boundary condition $\mathbf{p}(t_f) = 2[\mathbf{x}(t_f) - \mathbf{x}_d]$ gives

$$2[\boldsymbol{\phi}_{11}(t_f, t_0)\mathbf{x}_0 + \boldsymbol{\phi}_{12}(t_f, t_0)\mathbf{p}(t_0) - \mathbf{x}_d] = \boldsymbol{\phi}_{21}(t_f, t_0)\dot{\mathbf{x}}_0 + \boldsymbol{\phi}_{22}(t_f, t_0)\mathbf{p}(t_0)$$

The unknown $\mathbf{p}(t_0)$ can now be found:

$$\mathbf{p}(t_0) = [\boldsymbol{\phi}_{22}(t_f, t_0) - 2\boldsymbol{\phi}_{12}(t_f, t_0)]^{-1}\{[2\boldsymbol{\phi}_{11}(t_f, t_0) - \boldsymbol{\phi}_{21}(t_f, t_0)]\mathbf{x}_0 - 2\mathbf{x}_d\}$$

With $\mathbf{p}(t_0)$ known, $\mathbf{p}(t)$ and $\mathbf{u}^*(t)$ are given by

$$\mathbf{p}(t) = \boldsymbol{\phi}_{21}(t, t_0)\mathbf{x}_0 + \boldsymbol{\phi}_{22}(t, t_0)\mathbf{p}(t_0)$$

$$\mathbf{u}^*(t) = -\tfrac{1}{2}\mathbf{B}^T\mathbf{p}(t)$$

The optimal control law is open loop since $\mathbf{u}^*(t)$ is expressed as a function of $\mathbf{x}_0$ and $\mathbf{x}_d$.

17.16 Convert the control law of Problem 17.15 to a closed-loop control law.

Instead of writing $\begin{bmatrix} \mathbf{x}(t_f) \\ \mathbf{p}(t_f) \end{bmatrix} = \boldsymbol{\Phi}(t_f, t_0)\begin{bmatrix} \mathbf{x}_0 \\ \mathbf{p}(t_0) \end{bmatrix}$, use $\begin{bmatrix} \mathbf{x}(t_f) \\ \mathbf{p}(t_f) \end{bmatrix} = \boldsymbol{\Phi}(t_f, t)\begin{bmatrix} \mathbf{x}(t) \\ \mathbf{p}(t) \end{bmatrix}$, for a general time t.

Repeating much of Problem 17.15, but solving for $\mathbf{p}(t)$ instead of $\mathbf{p}(t_0)$, gives

$$\mathbf{p}(t) = [\boldsymbol{\phi}_{22}(t_f, t) - 2\boldsymbol{\phi}_{12}(t_f, t)]^{-1}\{[2\boldsymbol{\phi}_{11}(t_f, t) - \boldsymbol{\phi}_{21}(t_f, t)]\mathbf{x}(t) - 2\mathbf{x}_d\}$$

Once again, $\mathbf{u}^*(t) = -\frac{1}{2}\mathbf{B}^T\mathbf{p}(t)$. The feedback control law is illustrated in Fig. 17.12.

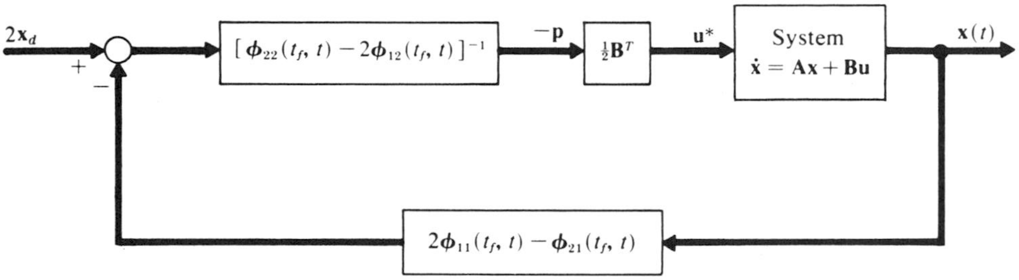

Figure 17.12

17.17 Use the minimum principle to find the input voltage $u(t)$ which charges the capacitor of Fig. 17.13 from x_0 at $t = 0$ to x_d at a fixed t_f while minimizing the energy dissipated in R. There are no restrictions on $u(t)$.

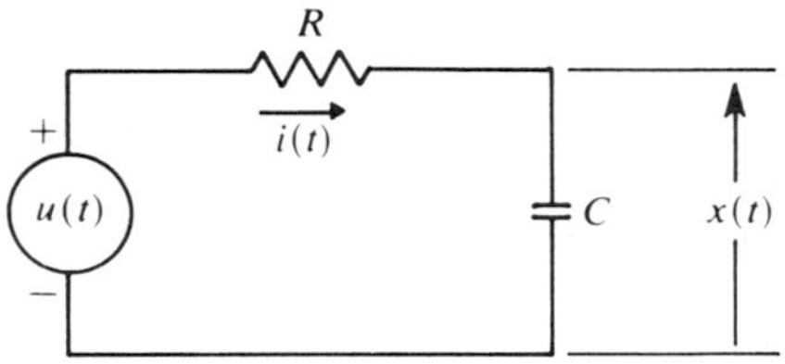

Figure 17.13

The state equation is $\dot{x} = -x/RC + u/RC$. The energy dissipated is

$$J = \int_0^{t_f} i^2 R\, dt = \int_0^{t_f} \frac{[u(t) - x(t)]^2}{R}\, dt$$

The Hamiltonian is $\mathcal{H} = (u - x)^2/R + p(-x/RC + u/RC)$. Minimize $\mathcal{H}$ by setting $\partial\mathcal{H}/\partial u = 0$, or $2(u - x)/R + p/RC = 0$. Therefore, $u^*(t) = -p(t)/2C + x(t)$.

The costate equation is $\dot{p} = -\partial\mathcal{H}/\partial x = 2(u - x)/R + p/RC$.

Eliminating $u^*(t)$, and simplifying, the two-point boundary value problem is $\dot{x} = -p/(2RC^2)$, $\dot{p} = 0$, with $x(0) = x_0$, $x(t_f) = x_d$. This implies that $p(t) = \text{constant}$, α. Therefore, $x(t) = x_0 - \alpha t/(2RC^2)$. From the terminal boundary condition, $\alpha = -2RC^2(x_d - x_0)/t_f$. Therefore, $u^*(t) = RC(x_d - x_0)/t_f + x(t)$. But $x(t) = x_0(1 - t/t_f) + x_d t/t_f$, so $u^*(t) = x_0 + (x_d - x_0)(RC + t)/t_f$.

17.18 A spin-stabilized satellite is wobbling slightly, with components of angular velocity due to the wobble being $x_1(t)$ and $x_2(t)$. The state equations are (see Problems 9.11, page 253, and 10.6, page 269).

$$\begin{bmatrix} \dot{x}_1 \\ \dot{x}_2 \end{bmatrix} = \begin{bmatrix} 0 & -\Omega \\ \Omega & 0 \end{bmatrix} \begin{bmatrix} x_1 \\ x_2 \end{bmatrix} + \begin{bmatrix} u_1(t) \\ u_2(t) \end{bmatrix}$$

Find the control $\mathbf{u}^*(t)$ which drives $\mathbf{x}(t_f)$ to $\mathbf{0}$ at a fixed t_f while minimizing the control energy $J = \int_0^{t_f} \mathbf{u}^T\mathbf{u}\, dt$. There are no restrictions on $\mathbf{u}(t)$.

The Hamiltonian is $\mathcal{H} = \mathbf{u}^T\mathbf{u} + \mathbf{p}^T[\mathbf{A}\mathbf{x} + \mathbf{u}]$. It is minimized by $\mathbf{u}^*(t) = -\frac{1}{2}\mathbf{p}(t)$, where the costate vector satisfies $\dot{\mathbf{p}} = -\partial\mathcal{H}/\partial\mathbf{x} = -\mathbf{A}^T\mathbf{p}$. Note that since $\mathbf{A} = \begin{bmatrix} 0 & -\Omega \\ \Omega & 0 \end{bmatrix}$, we have $-\mathbf{A}^T = \mathbf{A}$.

The two-point boundary value problem is

$$\begin{bmatrix} \dot{\mathbf{x}} \\ \hline \dot{\mathbf{p}} \end{bmatrix} = \left[\begin{array}{c|c} \mathbf{A} & -\frac{1}{2}\mathbf{I}_2 \\ \hline \mathbf{0} & \mathbf{A} \end{array}\right] \begin{bmatrix} \mathbf{x} \\ \hline \mathbf{p} \end{bmatrix} \qquad \text{with } \mathbf{x}(0) = \mathbf{x}_0,\ \mathbf{x}(t_f) = \mathbf{0}$$

From Problem 8.17, the 4×4 transition matrix for this system is

$$\mathbf{\Phi}(t, 0) = \left[\begin{array}{c|c} e^{\mathbf{A}t} & -\dfrac{t}{2}e^{\mathbf{A}t} \\ \hline \mathbf{0} & e^{\mathbf{A}t} \end{array}\right], \qquad \text{where } e^{\mathbf{A}t} = \begin{bmatrix} \cos\Omega t & -\sin\Omega t \\ \sin\Omega t & \cos\Omega t \end{bmatrix}$$

Using the terminal boundary condition gives $\mathbf{0} = e^{\mathbf{A}t_f}\mathbf{x}_0 - (t_f/2)e^{\mathbf{A}t_f}\mathbf{p}(0)$, from which $\mathbf{p}(0) = (2/t_f)\mathbf{x}_0$. Using $\mathbf{p}(0)$ gives $\mathbf{p}(t) = e^{\mathbf{A}t}\mathbf{p}(0)$, so $\mathbf{u}^*(t) = -(1/t_f)e^{\mathbf{A}t}\mathbf{x}_0$. When this control is used, the state satisfies $\mathbf{x}^*(t) = (1 - t/t_f)e^{\mathbf{A}t}\mathbf{x}_0$ and the minimum cost is

$$J = \frac{1}{t_f^2}\int_0^{t_f} \mathbf{x}_0^T[e^{\mathbf{A}t}]^T e^{\mathbf{A}t}\mathbf{x}_0\, dt = \mathbf{x}_0^T\mathbf{x}_0/t_f \qquad \text{since } [e^{\mathbf{A}t}]^T = [e^{\mathbf{A}t}]^{-1}$$

17.19 Consider a more general version of Example 17.9. Let the position of the automobile be $x_1(t)$ and the velocity be $x_2(t)$. The admissible controls must satisfy $|u(t)| \leq 1$ for all t. Find the $u^*(t)$ which drives the position and velocity to zero simultaneously, in minimum time.

The system equations are $\dot{x}_1 = x_2$, $\dot{x}_2 = u$. The Hamiltonian is $\mathcal{H} = 1 + p_1\dot{x}_1 + p_2\dot{x}_2 = 1 + p_1x_2 + p_2u$. $\mathcal{H}$ is minimized by selecting $\mathbf{u}^*(t) = -\text{sign}\,[p_2(t)]$.

The equations for $\mathbf{p}(t)$ are $\dot{p}_1 = -\partial\mathcal{H}/\partial x_1 = 0$ and $\dot{p}_2 = -\partial\mathcal{H}/\partial x_2 = -p_1$. Therefore, $p_1(t) = $ constant, $p_1(0)$, and $p_2(t) = p_2(0) - p_1(0)t$. This indicates that $p_2(t)$ is a linear function of t and changes sign at most once (ignoring the exceptional case where $p_1(0) = p_2(0) = 0$). Therefore, $u^*(t)$ changes sign at most once.

Rather than attempting to determine $p_1(0)$ and $p_2(0)$, the behavior of the system is investigated for $u(t) = +1$ and $u(t) = -1$. With $u = +1$, $x_2(t) = x_2(0) + t$ and $x_1(t) = x_1(0) + x_2(0)t + t^2/2$, or $x_1(t) = \frac{1}{2}x_2^2(t) + x_1(0) - \frac{1}{2}x_2^2(0)$. In the x_1x_2 plane, this represents a family of parabolas open toward the positive x_1 axis. Similarly with $u = -1$, $x_1(t) = -\frac{1}{2}x_2^2(t) + x_1(0) + \frac{1}{2}x_2^2(0)$. This is a family of parabolas open toward the negative x_1 axis. Just one member of each family passes through the specified terminal point $x_1 = x_2 = 0$.

Figure 17.14 shows the portion of these two families that are of interest. Segments of the two parabolas through the origin form the switching curve. For any initial state above this curve, $u = -1$ is used until the state intersects the switching curve. Then $u = +1$ is used to reach the origin. For initial states below the switching curve, $u = +1$ is used first, then $u = -1$.

This problem is an example of bang-bang control. It illustrates that the fastest method of coming to a red light and stopping is to use maximum acceleration until the last possible moment, and then use full braking to stop (hopefully) at the intersection. This example also illustrates the remarks in Sec. 17.1. This control is certainly not optimal in terms of tire wear or the number of traffic tickets received.

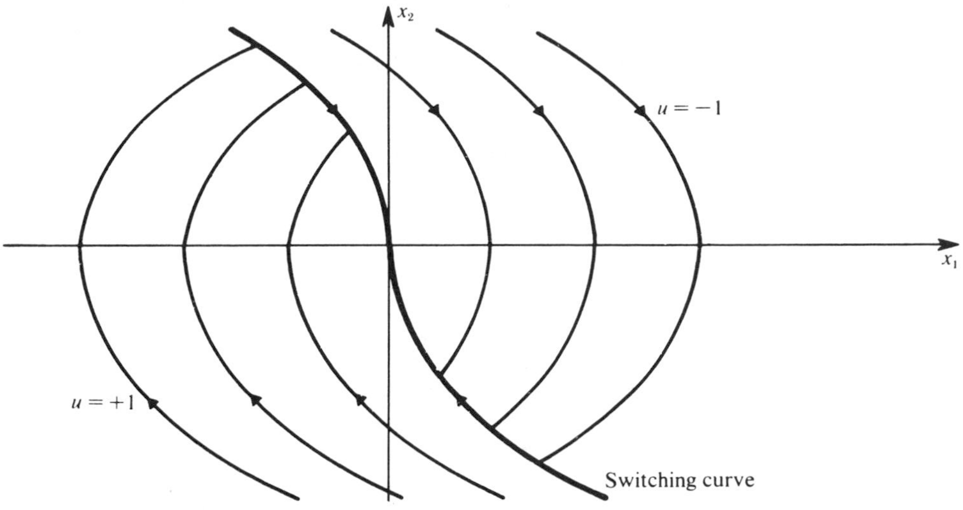

Figure 17.14

PROBLEMS

Dynamic Programming

17.20 Use dynamic programming to find the path which moves left to right from point a to point z of Fig. 17.15 while minimizing the sum of the costs on each path traveled.

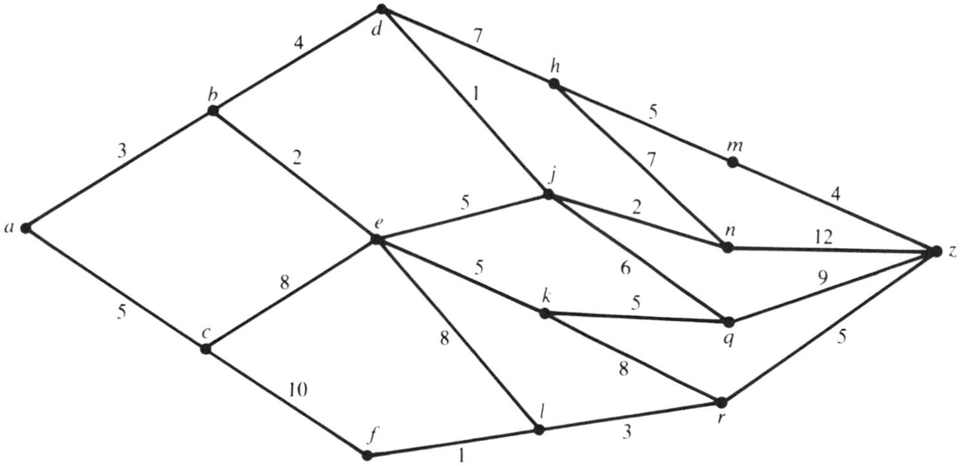

Figure 17.15

17.21 A student has four hours available to study for four exams. He will earn the scores shown in Table 17.1 for various study times. Use dynamic programming to find the optimal allocation of time in order to maximize the sum of his four scores. Consider only integer numbers of hours.

TABLE 17.1 TEST SCORES

Study Hours	Course No.			
	1	2	3	4
0	20	40	40	80
1	45	45	52	91
2	65	57	62	95
3	75	61	71	97
4	83	69	78	98

17.22 Find the control sequence which minimizes $J = [x(2) + 2]^2 + \sum_{k=0}^{1} u(k)^2$ for the scalar system $x(k + 1) = \frac{1}{2}x(k) + u(k)$, $x(0) = 10$, no restrictions on $u(k)$. Also find the resulting sequence $x(k)$ and the minimum value of J.

17.23 The same system and initial conditions of Problem 17.22 are considered. Find the optimal control and state sequences which minimize $J = \sum_{k=0}^{4} u(k)^2$ and which give $x(5) = 0$.

17.24 A scalar system is described by $x(k + 1) = \frac{1}{2}x(k) + 2u(k)$, with $x(0) = x_0$. Find $x^*(k)$ and $u^*(k)$ which minimize $J = \sum_{k=1}^{2} \{x^2(k) + u^2(k - 1)\}$.

17.25 An unstable discrete-time system is described by

$$\mathbf{x}(k + 1) = \begin{bmatrix} 1 & 2 \\ -2 & -3 \end{bmatrix}\mathbf{x}(k) + \begin{bmatrix} 1 \\ 0 \end{bmatrix}u(k) \qquad (1)$$

$$y(k) = [0 \quad 1]\mathbf{x}(k) \qquad (2)$$

Find the steady-state feedback gain matrix for the optimal regulator problem, assuming all states are available for use. Also determine the closed-loop eigenvalues that result.

Use the following weights

(*a*) $\mathbf{Q} = \mathbf{I}$, $\mathbf{R} = 1$

(*b*) $\mathbf{Q} = \mathbf{I}$, $\mathbf{R} = 10$

(*c*) $\mathbf{Q} = 10\mathbf{I}$, $\mathbf{R} = 1$

(*d*) $\mathbf{Q} = 1000\mathbf{I}$, $\mathbf{R} = 1$

17.26 The system of Problem 17.25 is now assumed to have additive white noises $\mathbf{w}(k)$ on equation (*1*) and $\mathbf{v}(k)$ on equation (*2*). The noise covariance matrices are

$$\mathbf{Q} = \begin{bmatrix} 9 & 0 \\ 0 & 4 \end{bmatrix} \quad \text{and} \quad \mathbf{R} = 16 \text{ respectively.}$$

a. A Kalman filter (recursive least-squares estimator) is to be used rather than a deterministic observer. Find the steady-state Kalman gain K and the pole locations of the filter's dynamics (i.e., equivalent to the observer poles).

b. The above filter is used along with the controllers found in Problem 17.25 as indicated by the separation theorem. How does the presence of the filter and the use of estimated states rather than actual states affect the overall system performance in each case? Remember that the separation theorem guarantees that this approach is the best that can be done. But it does not guarantee that the results will be what the controller asked for or that the system will even work well.

17.27 Repeat Problem 17.26 if the measurement noise covariance is reduced to $\mathbf{R} = 0.1$.

17.28 Find the optimal control sequence and the resulting state sequence for the system

$$\mathbf{x}(k+1) = \begin{bmatrix} 1.0 & 0.0 \\ -0.5 & 0.5 \end{bmatrix}\mathbf{x}(k) + \begin{bmatrix} 1 \\ -1 \end{bmatrix}u(k)$$

with initial conditions $\mathbf{x}(0) = [4 \quad -2]^T$. The optimal control is to minimize the cost function J defined in Problem 17.5, with final time $N = 5$ and

$$\mathbf{M} = \text{Diag}\,[1000, 1000], \qquad \mathbf{Q} = \text{Diag}\,[10, 10], \qquad \mathbf{R} = 1, \qquad \mathbf{x}_d = [8 \quad -7]^T,$$

$$\boldsymbol{\eta} = [-1 \quad 1]^T$$

Note the conflict being requested between $\mathbf{x}_d$ and $\boldsymbol{\eta}$. In general, fixed terminal conditions $\mathbf{x}(N) = \mathbf{x}_d$ can be closely approximated by using a sufficiently large $\mathbf{M}$ final weighting matrix.

17.29 Use the discrete-time approximate model of the airplane in Example 16.5 and design a steady-state optimal regulator with

$$\mathbf{Q} = \text{Diag}\,[0.1, 1.0, 1.0, 0.1] \quad \text{and} \quad \mathbf{R} = \mathbf{I}$$

17.30 A commonly used simplification of the previous system is that $\delta_a = 4\delta_r$. This allows the simpler single-input analysis to be carried out, using as the input matrix $\mathbf{B}$ four times column 1 plus column 2 of the original $\mathbf{B}$ matrix. Using this simplification, find the feedback gains for generating the equivalent control δ_r if

(*a*) $\mathbf{Q} = \mathbf{I}$, $\mathbf{R} = 1$

(*b*) $\mathbf{Q} = \mathbf{I}$, $\mathbf{R} = 0.1$

(*c*) $\mathbf{Q} = \mathbf{I}$, $\mathbf{R} = 25$

(*d*) $\mathbf{Q} = \text{Diag}\,[0.1, 1.0, 1.0, 0.1]$, $\mathbf{R} = 1$

The Minimum Principle

17.31 Use the minimum principle to find the optimal control $u^*(t)$ and the corresponding $x^*(t)$ for the system $\dot{x} = u(t)$ with $x(0) = 0$, $x(1) = 1$ and no restrictions on $u(t)$. The performance criterion is $J = \int_0^1 (x^2 + u^2)\,dt$.

17.32 Use the minimum principle to solve the optimization problem: $\dot{x}_1 = x_2$, $\dot{x}_2 = u$, $x_1(0) = 1$, $x_2(0) = 1$, $x_1(2) = 0$, $x_2(2) = 0$, minimize $J = \frac{1}{2}\int_0^2 u^2(t)\,dt$ with $u(t)$ unrestricted.

17.33 Solve the following two-point boundary value problem on the interval $0 \le t \le 1$: $\dot{x} = -2x - \frac{1}{2}p$, $\dot{p} = -2x + 2p$, $x(0) = 10$, $p(1) = 4x(1)$.

17.34 The equations of motion for a rocket flying in a vertical plane under the influence of constant gravity g and constant thrust T can be written as $\dot{x}_1 = x_3$, $\dot{x}_2 = x_4$, $\dot{x}_3 = (T/m)\cos u(t)$, $\dot{x}_4 = (T/m)\sin u(t) - g$, where x_1 and x_2 are horizontal and vertical position components, x_3 and x_4 are horizontal and vertical veclocity components, and the control $u(t)$ is the thrust angle measured from the horizontal. The rocket is to be flown to a specified terminal altitude with zero vertical velocity at t_f and maximum horizontal velocity. Use the minimum principle to establish that the optimal thrust angle follows the linear-tangent steering law.

$$\tan u^*(t) = \alpha t - \beta \qquad \text{where } \alpha \text{ and } \beta \text{ are constants}$$

(*Hint:* Minimize $J = -x_3(t_f)$.)

17.35 Consider a system which is nonlinear with respect to $\mathbf{x}$, but linear with respect to $\mathbf{u}$, i.e., $\dot{\mathbf{x}} = \mathbf{f}(\mathbf{x}, t) + \mathbf{Bu}$. In each of the following cases, use the minimum principle to find the form of $\mathbf{u}^*(t)$ in terms of $\mathbf{p}(t)$.

a. $J = \int_{t_0}^{t_f} \mathbf{u}^T\mathbf{u}\, dt$, $\mathbf{x}(t_f) = \mathbf{0}$, t_f fixed, $\mathbf{u}^T(t)\mathbf{u}(t) \leq 1$ for all t.

b. Drive $\mathbf{x}(t_f)$ to zero in minimum time, $\mathbf{u}^T(t)u(t) \leq 1$ for all t.

c. Same as b except each component of $\mathbf{u}$ satisfies $|u_i(t)| \leq 1$ for all t.

d. $J = \int_{t_0}^{t_f} \Sigma |u_i(t)|\, dt$, with $|u_i(t)| \leq 1$ for all t.

17.36 A scalar system is described by $\dot{x} = x + u$ with $x(0) = 2$. The admissible controls must satisfy $|u(t)| \leq 1$. Use the minimum principle to find $u^*(t)$ which drives $x(t)$ to zero in minimum time.

17.37 Analyze an equivalent single-input model of the aircraft in Problem 17.11 by assuming that $\delta_a = 4\delta_r$. Use $\mathbf{Q} = \text{Diag}\,[0.1, 1.0, 1.0, 0.1]$ and $\mathbf{R} = 1$.

Appendix A

SOFTWARE AND COMPUTATIONAL AIDS

The first edition of this text was designed so that all the problems, with perhaps a dozen exceptions, could be worked with pencil and paper. All the principles and techniques could be demonstrated in fairly simple fashion by this approach. The current edition retains almost all of these simple illustrative examples, but adds to them some others which require computational assistance for efficient solution. Even if the reader does not have access to these kinds of aids, it is still worthwhile to study the numerical results that have been presented. Certain trade-offs, trends, and other interesting results thus become apparent.

A second approach to the computationally oriented problems is to use existing packages to the extent available, without becoming involved in the details of how the algorithms work. This is somewhat dangerous because most software packages have inherent limitations and pitfalls that might manifest themselves at inopportune times. Furthermore, a major thrust of this book has been to explain how and why certain methods and techniques work. That is, enough material has been presented on most topics to allow the development of code.

A more in-depth approach would be to view the development of operating code as the ultimate test of an understanding of a given method. In a tightly scheduled tour through this book, as in a university course, there will be insufficient time to take this approach with each topic. A couple such major projects, plus the development of several of the more minor "tools" along the way, might be feasible. The more powerful programmable calculators should not be overlooked in this regard. They are probably capable of doing all the computations required in this book, but perhaps less conveniently. If new routines are to be developed, the numerical results found here will be helpful in calibrating and checking them out.

Most large corporations and university departments interested in control theory now have some offering of software support, either "homegrown" or acquired. For a partial look at what is available, see References 3, 5, 27 (Appendix B), 32, 38, 49, 50, and 81.

The software used in creating and solving the problems in this book is of two types. In the minority of cases, some larger packages, primarily LSAP [50], were run on a VAX 11/780 system with graphics terminals. Most of the programs used were specifically developed to fit on a microcomputer with only 64K of RAM. A major source for the code developed was Melsa (with Jones in the second edition) [81]. The programs found in this reference were written years ago and intended for use in a batch mode with card input. Practically all of these have been ported to a modern microcomputer for interactive use, with some modifications and the correction of a few minor, but annoying and hard to find errors. The chapter-by-chapter listing of major programs used will refer to Melsa's original names [81], although various modifications have been made. In addition to the modified Melsa programs, a number of personally developed programs were used. It is worth noting that the FORTRAN source code for all these microcomputer programs and supporting subroutines, plus some other routines not needed in this book, fill just about half of one 630K floppy disk.

Chapter	*Major Programs Used and Related Comments*
1	ZTIME computers inverse Z-transform time sequences by long division of transfer function polynomials. See Problem 1.16.
2	LSAP [49]. Root locus, frequency response, time response, conversion from s-domain to Z-domain, partial fraction expansion, plus miscellaneous. RTLOC modified Melsa root locus, line printer plots. NYQUIST is a modified version of Melsa's FRESP for doing Bode and Nyquist analyses. PRFEXP Melsa's partial fraction expansion program results differ from LSAP's when applied in Z-plane. See Example 11.6.
3	The usual matrix utility routines. BASMAT (Melsa) for inverting a matrix, among other things used later. Melsa's SIMEQ (simultaneous equation solver). LINEQ (Melsa) more general package for solving linear equations. HERMIT (Melsa) gives Hermite normal form, i.e., row-reduced echelon form. CHOLSKY does Cholesky decomposition.
4	GRAMS does the Gram-Schmidt expansion (GSE). Also, something like HERMIT or LINEQ is needed to test rank.
5	GRAMS, SOLVE (based on LINEQ), BATCHLS a batch least-squares program. SVD [38] is a singular value decomposition routine. This is a more reliable method for finding rank.
6	No additional new programs.
7	BASMAT (eigenvalues), SOLVE or LINEQ, etc., for eigenvectors. PROOT (Melsa) routine for factoring polynomials was the only really unacceptable subroutine found in Melsa [81]. It was an order of magnitude

slower than POLRT on a micro, so it is not used in any of these programs. SVD is also useful.

8 BASMAT finds exponential matrix (nonrepeated root case only). By proper interpretation it also gives $\mathbf{A}^k$.

9 No new programs.

10 TRESP (Melsa) uses Runge-Kutta integration to find time response to state equations and to plot results on line printer. Original version restricted to one input-output. Effort is under way to modify for multiple input-outputs and give menu-driven selection of the more common input signals. APPROX finds approximate discrete-time system from a continuous model. See Problem 10.11.

11 LSAP is useful for converting from *s*- to *Z*-plane. Various partial fraction and/or factoring programs useful. KALMAN is a linear discrete Kalman filter program with many options. Can be viewed as a recursive least-squares program.

12 CBILITY forms the controllability/observability matrices and determines their ranks as well as the controllability/observability index.

13 NONLIN is a Marquardt-type algorithm for solving simultaneous nonlinear equations.

14 No new programs.

15 STVARFBK (modified Melsa) computes input-output transfer functions from state equations. It also checks for observability, computes feedback gains for specified closed-loop poles, etc. It is set up for one input-one output, but repeated use can give matrix transfer functions. This feature could easily be generalized for multiple input-output.

16 STVARFBK does the single-input state feedback design. FULLST does the multiple-input state feedback pole-placement design. See Sec. 16.5. It can also do output feedback design and observer design. FULLST currently will only do real poles. This is not an inherent limitation of the method.

17 CONTRL solves the discrete quadratic optimal tracking problem of Problem 17.5 (regulator and steady-state cases are subsets). RICATI solves the continuous-time Riccati equation to steady state as discussed in Problem 17.11. It also has a handy option of computing the closed-loop eigenvalues for an arbitrary feedback gain matrix. ZTIME is useful here, as in earlier chapters, for finding discrete-time responses by long division of the *Z*-transfer function. The *Z*-transform of interest may be obtained using the capabilities of STVARFBK to get from state equations to transfer functions.

REFERENCES AND SUGGESTIONS FOR FURTHER STUDY

1. Alag, G. and H. Kaufman: "An Implementable Digital Adaptive Flight Controller Designed Using Stabilized Single-Stage Algorithms," *IEEE Transactions on Automatic Control*, Vol. AC-22, No. 5, October 1977, pp. 780–788.
2. Anderson, B. D. O. and J. B. Moore: *Linear Optimal Control*, Prentice-Hall, Englewood Cliffs, N. J., 1971.
3. Armstrong, E. S.: *ORACLS, A Design System for Multivariable Control*, Marcel Dekker, New York, 1980.
4. Aseltine, J. A.: *Transform Method in Linear Systems Analysis*, McGraw-Hill, New York, 1958.
5. Astrom, K. J.: "Computer Aided Modeling, Analysis and Design of Control Systems—A Perspective," *IEEE Control Systems Magazine*, Vol. 3, No. 2, May 1983, pp. 4–16.
6. Bellman, R.: *Matrix Analysis*, McGraw-Hill, New York, 1960.
7. Bellman, R. and S. E. Dreyfus: *Applied Dynamic Programming*, Princeton University Press, Princeton, New Jersey, 1962.
8. Bellman, R. and R. Kalaba: *Quasilinearization and Nonlinear Boundary Value Problems*, Elsevier Press, New York, 1965.
9. Brockett, R. W.: "The Status of Stability Theory for Deterministic Systems," *IEEE Transactions on Automatic Control*, Vol. AC-11, No. 3, July 1966, pp. 596-606.
10. Brogan, W. L.: "Optimal Control Theory Applied to Systems Described by Partial Differential Equations," *Advances in Control Systems*, Vol. 6, C. T. Leondes, Ed., Academic Press, New York, 1968.
11. Brogan, W. L.: "Applications of a Determinant Identity to Pole-Placement and Observer Problems," *IEEE Transactions on Automatic Control*, Vol. AC-19, October 1974, pp. 612–614.

12. Brown, K. M. and S. D. Conte: "The Solution of Simultaneous Nonlinear Equations," *Proceedings of the Association of Computing Machinery,* National Meeting, 1967, pp. 111–114.

13. Brown, K. M. and J. E. Dennis: "Derivative-Free Analogues of the Levenberg-Marquardt and Gauss Algorithms for Nonlinear Least Squares," *Numerische Math.*, Vol. 18, 1972, pp. 289–297.

14. Bryson, A. E. and Y. C. Ho: *Applied Optimal Control*, Blaisdell (*now* Xerox), Waltham, Mass., 1969.

15. Cannon, R. H., Jr.: *Dynamics of Physical Systems*, McGraw-Hill, New York, 1967.

16. Canon, M. D., C. D. Cullum, Jr., and E. Polak: *Theory of Optimal Control and Mathematical Programming*, McGraw-Hill, New York, 1970.

17. Chen, C. T.: *Introduction to Linear System Theory*, Holt, Rinehart and Winston, New York, 1970.

18. Chen, C. T. and C. A. Desoer: "A Proof of Controllability of Jordan Form State Equations," *IEEE Transactions on Automatic Control*, Vol. AC-13, No. 2, April 1968, pp. 195-196.

19. Chen, C. T. and D. P. Mital: "A Simplified Irreducible Realization Algorithm," *IEEE Transactions on Automatic Control*, Vol. AC-17, No. 4, August 1972, pp. 535–537.

20. Chestnut, H.: "A Systems Approach to the Economic Use of Computers for Controlling Systems in Industry," *General Electric Report* No. 70-C-089, Schenectady, N.Y., February 1970.

21. Churchill, R. V.: *Introduction to Complex Variables and Applications*, 2nd Ed., McGraw-Hill, New York, 1960.

22. Coddington, E. A. and N. Levinson: *Theory of Ordinary Differential Equations*, McGraw-Hill, New York, 1955.

23. Courant, R. and D. Hilbert: *Methods of Mathematical Physics*, Vol. 1, Interscience: Wiley, New York, 1953.

24. Cruz, J. B., Jr., and W. R. Perkins: "A New Approach to the Sensitivity Problem in Multivariable Feedback System Design," *IEEE Transactions on Automatic Control*, Vol. AC-9, No. 3, July 1964, pp. 216–223.

25. Davison, E. J.: "On Pole Assignment in Multivariable Linear Systems," *IEEE Transactions on Automatic Control*, Vol. AC-13, No. 6. December 1968, pp. 747-748.

26. Davison, E. J.: "On Pole Assignment in Linear Systems with Incomplete State Feedback," *IEEE Transactions on Automatic Control*, Vol. AC-15, No. 3, June 1970, pp. 348–351.

27. D'Azzo, J. J. and C. H Houpis: *Linear Control System Analysis and Design*, 2nd Ed., McGraw-Hill, New York, 1981.

28. De Russo, P. M., R. J. Roy and C. M. Close: *State Variables for Engineers*, Wiley, New York, 1965.

29. Desoer, C. A.: *Notes for a Second Course on Linear Systems*, Van Nostrand Reinhold, New York, 1970.

30. Dorf, R. C.: *Modern Control Systems*, Addison-Wesley, Reading, Mass., 1967.

31. Elgerd, O. I.: *Control Systems Theory*, McGraw-Hill, New York, 1967.

32. Emami-Naeini, A. and G. F. Franklin: "Interactive Computer-Aided Design of Control Systems," *IEEE Control Systems Magazine*, Vol. 1, No. 4, December 1981, pp. 31–36.
33. Fahmy, M. M. and J. O'Reilly: "On Eigenstructure Assignment in Linear Multivariable Systems," *IEEE Transactions on Automatic Control*, Vol. AC-27, No. 3, June 1982, pp. 690–693.
34. Fahmy, M. M. and J. O'Reilly: "Eigenstructure Assignment in Multivariable Systems—A Parametric Solution ,"*IEEE Transactions on Automatic Control*, Vol. AC-28, No. 10, October 1983, pp. 990–994.
35. Falb, P. L. and W. A. Wolovich: "Decoupling in the Design and Synthesis of Multivariable Control Systems," *IEEE Transactions on Automatic Control*, Vol. AC-12, No. 6, December 1967, pp. 651–659.
36. Falb, P. L. and W. A. Wolovich: "On the Decoupling of Multivariable Systems," *Proc. 1967 Joint Automatic Control Conference*, pp. 791–796.
37. Fletcher, R. and M. J. D. Powell: *Computer Journal*, Vol. 6, 1963, pp. 163–168.
38. Forsythe, G. E., M. A. Malcolm and C. Moler: *Computer Methods for Mathematical Computations*, Prentice-Hall, Englewood Cliffs, N. J., 1977.
39. Franklin, G. F. and J. D. Powell: *Digital Control of Dynamic Systems*, Addison-Wesley, Reading, Mass.: 1980.
40. Friedman, B.: *Principles and Techniques of Applied Mathematics*, Wiley, New York, 1956.
41. Gass, S. I.: *Linear Programming, Methods and Applications*, 2nd Ed., McGraw-Hill, New York, 1964.
42. Gelfand, I. M. and S. V. Fomin: *Calculus of Variations*, Prentice-Hall, Englewood Cliffs, New Jersey, 1963 (Translated from Russian edition).
43. Gilbert, E. G.: "Controllability and Observability in Multivariable Control Systems," *Journ. Soc. Ind. Appl. Math—Control Series*, Series A, Vol. 1, No. 2, 1963, pp. 128–151.
44. Goldstein, H.: *Classical Mechanics*, Addison-Wesley, Reading, Mass., 1959.
45. Goldwyn, R. M. and K. S. Narendra: "Stability of Certain Nonlinear Differential Equations," *IEEE Transactions on Automatic Control*, Vol. AC-8, No. 4, October 1963, pp. 381–382.
46. Goodwin, G. C. and R. L. Payne: *Dynamic System Identification—Experiment Design and Data Analysis*, Academic Press, New York, 1977.
47. Halmos, P. R.: *Finite Dimensional Vector Spaces*, 2nd Ed., D. Van Nostrand, Princeton, New Jersey, 1958.
48. Harris, C. J. and S. A. Billings: *Self-Tuning and Adaptive Control*, Peter Peregrinus on behalf of the Institution of Electrical Engineers (United Kingdom), 1981.
49. Herget, C. J. and A. J. Laub, editors: "Computer Aided Design of Control Systems Special Issue," *IEEE Control Systems Magazine* Vol. 2, No. 4, December 1982,
50. Herget, C. J. and T. P. Weis: "Linear System Analysis Program (LSAP) Users Manual," *Lawrence Livermore Laboratory*, University of California Report UCID-30184, October 1980.
51. Ho, B. L. and R. E. Kalman: "Effective Construction of Linear State-Variable Models from Input/Output Functions, "*Proc. Third Allerton Conf.*, 1965, pp. 449–459.

52. Hoffman, K., and R. Kunze: *Linear Algebra*, Prentice-Hall, Englewood Cliffs, N. J., 1961.

53. Hovanessian, S. A. and L. A. Pipes: *Digital Computer Methods in Engineering*, McGraw-Hill, New York, 1969.

54. Howze, J. W. and J. B. Pearson: "Decoupling and Arbitrary Pole Placement in Linear Systems Using Output Feedback," *IEEE Transactions on Automatic Control*, Vol. AC-15. No. 6, December 1970, pp. 660–663.

55. Hsia, T. C.: *System Identification*, Lexington Books, Lexington, Mass., 1977.

56. Jeger, M. and B. Eckman: *Vector Geometry and Linear Algebra*, Interscience: Wiley, New York, 1966

57. Kalman, R. E.: "A New Approach to Linear Filtering and Prediction Problems," *Trans. of the ASME, Journ. of Basic Engineering*, Vol. 82, 1960, pp. 35–45.

58. Kalman, R. E.: "The Theory of Optimal Control and the Calculus of Variations," *Mathematical Optimization Techniques*, University of California Press, Los Angeles, California, 1963.

59. Kalman, R. E.: "Mathematical Description of Linear Dynamical Systems," *Journ. Soc. Ind. Appl. Math—Control Series*, Series A, Vol. 1, No. 2, 1963, pp. 152–192.

60. Kalman, R. E.: "When is a Linear Control System Optimal?" *Trans. of the ASME, Journ. of Basic Engineering*, Vol. 86D, March 1964, pp. 51–60.

61. Kalman, R. E.: "Irreducible Realizations and the Degree of a Rational Matrix," *Journ. Soc. Ind. Appl. Math—Control Series*, Series A, Vol. 13, 1965, pp. 520–544.

62. Kalman, R. E. and J. E. Bertram: "Control System Analysis and Design Via the 'Second Method' of Lyapunov. I. Continuous-Time Systems," *Trans. ASME Journal of Basic Engineering*, Vol. 82D, June 1960, pp. 371-393.

63. Kalman, R. E. and J. E. Bertram: "Control System Analysis and Design Via the 'Second Method' of Lyapunov. II. Discrete-Time Systems," *Trans. ASME Journal of Basic Engineering*, Vol. 82D, June 1960, pp. 394–400.

64. Kalman, R. E. and N. DeClaris: *Aspects of Network and System Theory*, Holt, Rinehart and Winston, New York, 1971, pp. 385–407.

65. Kirk, D. E.: *Optimal Control Theory*, Prentice-Hall, Englewood Cliffs, New Jersey, 1970.

66. Kolmogorov, A. N. and S. V. Formin: *Elements of the Theory of Functions and Functional Analysis*, Vol. 1 (1957) and Vol. 2 (1961), Graylock Press, Albany, New York (Translated from Russian editions).

67. Kreindler, E. and A. Jameson: "Optimality of Linear Control Systems," *IEEE Transactions on Automatic Control*, Vol. AC-17, No. 3, June 1972, pp. 349–351.

68. Kunz, K. S.: *Numerical Analysis*, McGraw-Hill, New York, 1957.

69. Kuo, B. C.: *Analysis and Synthesis of Sampled-Data Control Systems*, Prentice-Hall, Englewood Cliffs, New Jersey, 1963.

70. Lapidus, L. and R. Luus: *Optimal Control of Engineering Processes*, Blaisdell, Waltham, Mass., 1967.

71. LaSalle, J. and S. Lefschetz: *Stability by Liapunov's Direct Method, with Applications*, Academic Press. New York, 1961.

72. Lee, E. A.: LCAP2 "LCAP2-Linear Controls Analysis Program," *IEEE Control Systems Magazine,* Vol. 2, No. 4, December 1982, pp. 15–18.

73. Leondes, C. T. and L. M. Novak: "Optimal Minimal-Order Observers for Discrete-Time Systems—A United Theory," *Automatica*, Vol. 8, No. 4, July 1972, pp. 379–387.

74. Luenberger, D. G.: "An Introduction to Observers," *IEEE Transactions on Automatic Control*, Vol. AC-16, No. 6, December 1971, pp. 596–602.

75. MacCamy, R. C. and V. J. Mizel: *Linear Analysis and Differential Equations*, Macmillan, New York, 1969.

76. MacFarlane, A. G. J.: "A Survery of Some Recent Results in Linear Multivariable Feedback Theory," *Automatica*, Vol. 8, No. 4, July 1972, pp. 455–492.

77. Marquardt, D. W.: "An Algorithm for Least Squares Estimation of Nonlinear Parameters," *SIAM Journal of Applied Mathematics*, Vol. 11, 1963, pp. 431–441.

78. Martens, H. R. and D. R. Allen: *Introduction to Systems Theory*, Charles E. Merrill Publishing Co., Columbus, Ohio, 1969.

79. McGill, R. and P. Kenneth: "Solution of Variational Problems by Means of a Generalized Newton-Raphson Operator," *AIAA Journal*, Vol. 2, 1964, pp. 1761–1766.

80. Meditch, J. S.: *Stochastic Optimal Linear Estimation and Control*, McGraw-Hill, New York, 1969.

81. Melsa, J. L. and S. K. Jones: *Computer Programs for Computational Assistance in the Study of Linear Control Theory*, 2nd Ed., McGraw-Hill, New York, 1973.

82. Melsa, J. L. and D. G. Schultz: *Linear Control Systems*, McGraw-Hill, New York, 1969.

83. *Microsoft Fortran*-80 *Reference Manual*, Copyright Microsoft, 1979, 1980.

84. Morse, A. S. and W. M. Wonham: "Status of Noninteracting Control," *IEEE Transactions on Automatic Control*, Vol. AC-16, No. 6, December 1971, pp. 568–581.

85. Morse, A. S., and W. M. Wonham: "Triangular Decoupling of Linear Multivariable Systems," *IEEE Transactions on Automatic Control*, Vol. AC-15, No. 4, August 1970, pp. 447–449.

86. Ogata, K.: *State Space Analysis of Control Systems*, Prentice-Hall, Englewood Cliffs, New Jersey, 1967.

87. Papoulis, A.: *Probability, Random Variables and Stochastic Processes*, McGraw-Hill, New York, 1965, pp. 344–346.

88. Pearson, J. B. and C. Y. Ding: "Compensator Design for Multivariable Linear Systems," *IEEE Transactions on Automatic Control*, Vol. AC-14, No. 2, April 1969, pp. 130–134.

89. Penrose, R.: "A Generalized Inverse for Matrices," *Proceedings of the Cambridge Philosophical Society*, Vol. 51, Part 3, 1955, pp. 406–413.

90. Perkins, W. R. and J. B. Cruz, Jr.: "Feedback Properties of Linear Regulators," *IEEE Transactions on Automatic Control*, Vol. AC-16, No. 6, December 1971, pp. 659–664.

91. Perlis, S.: *Theory of Matrices*, Addison-Wesley, Reading, Mass., 1958.

92. Pio, R. L.: "Symbolic Representation of Coordinate Transformations," *IEEE Transactions on Aerospace and Navigational Electronics*, Vol. ANE-11, No. 2, June 1964, pp. 128–134.

93. Pio, R. L.: "Euler Angle Transformations," *IEEE Transactions on Automatic Control*, Vol. AC-11, No. 4, October 1966, pp. 707–715.

94. Pipes, L. A.: *Applied Mathematics for Engineers and Physicists*, 3rd Ed., McGraw-Hill, New York, 1971.

95. Pontryagin, L. S., V. G. Boltyanskii, R. V. Gamkrelidze and E. F. Mishchenko: *The Mathematical Theory of Optimal Processes*, Interscience: Wiley, New York, 1962 (Translated from Russian edition).

96. Porter, W. A.: *Modern Foundations of Systems Engineering*, Macmillan, New York, 1966.

97. Raven, E. A.: "A Minimum Realization Method," *IEEE Control Systems Magazine*, Vol. 1, No. 3, September 1981, pp. 14–20.

98. Reiss, R. and G. Geiss: "The Construction of Liapunov Functions," *IEEE Transactions on Automatic Control*, Vol. AC-8, No. 4, October 1963, pp. 382–383.

99. Rubin, W. B.: "A Simple Method for Finding the Jordan Form of a Matrix," *IEEE Transactions on Automatic Control*, Vol. AC-17, No. 1, February 1972, pp. 145–146.

100. Sage, A. P.: *Optimum Systems Control*, Prentice-Hall, Englewood Cliffs, New Jersey, 1968.

101. *SAS User's Guide*, 1979 Ed., SAS Institute, Box 8000, Cary, N.C. 27511.

102. Sato, S. M. and P. V. Lopresti: "On the Generalization of State Variable Decoupling Theory," *IEEE Transactions on Automatic Control*, Vol. AC-16, No. 2, April 1971, pp. 133–139.

103. Saucedo, R. and E. Schiring: *Introduction to Continuous and Digital Control Systems*, Macmillan, New York, 1968.

104. Savas, E. S.: "Feedback Controls on Urban Pollution," *IEEE Spectrum*, July 1969, pp. 77–81.

105. Schied, F.: *Theory and Problems of Numerical Analysis*, Schaum's Outline Series, p. 355, McGraw-Hill, New York, 1968.

106. Schultz, D. G.: "The Variable Gradient Method of Generating Liapunov Functions, with Applications to Control Systems," Ph.D. Dissertation, Purdue University, April 1962.

107. Schultz, D. G. and J. E. Gibson: "The Variable Gradient Method for Generating Liapunov Functions," *AIEE Trans.*, Part II, Vol. 81, Sept. 1962, pp. 203–210.

108. Schweppe, F. C.: *Uncertain Dynamic Systems*, Prentice-Hall, Engelwood Cliffs, N. J., 1973.

109. Shearer, J. L., A. T. Murphy and H. H. Richardson: *Introduction to System Dynamics*, Addison-Wesley, Reading, Mass., 1967.

110. Simonnard, M.: *Linear Programming*, Prentice-Hall, Englewood Cliffs, New Jersey, 1966.

111. Taylor, A. E. and R. W. Mann: *Advanced Calculus*, Xerox, Boston, Mass., 1972.

112. Taylor, A. E.: *Functional Analysis*, Wiley, New York, 1958.

113. Waespy, C. M.: "An Application of Linear Programming to Minimum Fuel Optimal Control," Engineering Report No. 67-162, University of California, Los Angeles, December 1976.

114. Westlake, J.: *A Handbook of Matrix Inversion and Solution of Linear Equations*, p. 13, Wiley, New York, 1968.

115. Wilde, D. J. and C. S. Beightler: *Foundations of Optimization*, Prentice-Hall, Englewood Cliffs, New Jersey, 1967.

116. Wonham, W. M.: "On-Pole Assignment in Multi-input, Controllable Linear Systems," *IEEE Transactions on Automatic Control*, Vol. AC-12, No. 6, December 1967, pp. 660–665.

117. Zadeh, L. A. and C. A. Desoer: *Linear System Theory, The State Space Approach*, McGraw-Hill, New York, 1963.

118. Zaks, R.: *The CP/M Handbook*, Sybex, 1980. (CP/M is the registered trademark of Digital Research, Pacific Grove, Calif.)

ANSWERS TO PROBLEMS

CHAPTER 1

1.21 $Q = \dfrac{A}{\rho g}\dot{P}$

1.22 $v(t) = v(t_0) + \int_{t_0}^{t} Q(\tau)\,d\tau$

1.23 $v_0 = L(df_3/dt) + Rf_3$

1.24 $y(s)/u(s) = [Cs + 1/R_1]/[Cs + (1/R_1 + 1/R_2)]$

1.25 See Reference 104 for a general discussion.

1.26 $Y(z) = \sum_{n=0}^{\infty} e^{-0.1nT}e^{-nsT} = \dfrac{1}{1 - z^{-1}e^{-0.2}} = \dfrac{z}{z - e^{-0.2}}$

1.27 $H(z) = \dfrac{10z^2(z+5)}{(z-2)(z-4)(z-8)}$

Poles at $z = 2, 4, 8$

Zeros at 0., 0., -5

Stable

1.28 $y(nT) = 15.625 - 0.2916(0.2)^n + 3(0.4)^n - 17.333(0.8)^n$

CHAPTER 2

2.25 $3 < K < 9$

2.26 $K = 20$, $\omega = 2$ rad/sec

2.27 With $K = \sqrt{327{,}680} \cong 572$, the s row of Routh's array is zero. The auxiliary equation then is $44.2s^2 + 2288 = 0$, indicating the cross-over frequency is $\omega = 7.19$ rad/sec.

2.28 a. Type 0, $K_p = 20$ db $= 10$, $K_v = 0$, $K_a = 0$

b. Type 1, $K_p = \infty$, $K_v = -20$ db $= 0.1$, $K_a = 0$

c. Type 2, $K_p = \infty$, $K_v = \infty$, $K_a = -60$ db $= 0.001$

2.30 Stable for all K, but very low stability margins for small values of K.

2.31 For $K > 7.5$ there is one clockwise and one counterclockwise encirclement. Therefore, $N = 0$ and system is stable. There are two unstable roots for $0 < K < 7.5$.

2.32 Gain margin $\cong 1.76$, i.e., $K > 8800$ causes system instability. Phase margin $\cong 33°$. $\omega \cong 15$ rad/sec for a $-180°$ phase angle.

2.33 $K = 9.65$, $\omega = 11.8$ rad/sec (*Hint*: Dominant roots at $-\alpha \pm j\omega$; $e^{-\alpha} = 0.0015$.)

2.34 Nonminimum phase

2.35 Direct realization: $y(t_k) = -0.5y(t_{k-1}) + E(t_{k-1}) - 0.5E(t_{k-2})$

Parallel realization: $y_1(t_k) = E(t_{k-1})$

$y_2(t_k) = -0.5y_2(t_{k-1}) + E(t_{k-1})$

$y(t_k) = -y_1(t_k) + 2y_2(t_k)$

Cascade realization: $E_1(t_k) = E(t_{k-1})$

$y(t_k) = E_1(t_k) - 0.5[E_1(t_{k-1}) + y(t_{k-1})]$

2.36 $C(nT) \cong 13.3333 - 20(0.5)^n + 6.6667(-0.5)^n$

2.37 a. $C(z) = 1 + (\alpha + \beta - a - b)z^{-1} + (ab + \alpha^2 + \alpha\beta + \beta^2 - a\alpha - b\alpha - a\beta - b\beta)z^{-2} + \cdots$

Therefore, $C(0) = 1$, $C(T) = (\alpha + \beta - a - b)$ and $C(2T) = ab + \alpha^2 + \alpha\beta + \beta^2 - a\alpha - b\alpha - a\beta - b\beta$.

b. $$C(nT) = \frac{ab}{\alpha\beta}\delta_{0n} + \left[\frac{\alpha^2 - a\alpha - b\alpha + ab}{\alpha(\alpha - \beta)}\right]\alpha^n + \left[\frac{\beta^2 - a\beta - b\beta + ab}{\beta(\beta - \alpha)}\right]\beta^n$$

where $\delta_{0n} = \begin{cases} 0 & \text{if } n \neq 0 \\ 1 & \text{if } n = 0 \end{cases}$

c. $C(nT) \longrightarrow 0$ as $n \longrightarrow \infty$

2.38 a. $$\frac{C(z)}{R(z)} = \frac{0.004845K(z + 0.9672)}{z^2 - (1.9239 - 0.00484K)z + (0.90484 + 0.004686K)}$$

b.

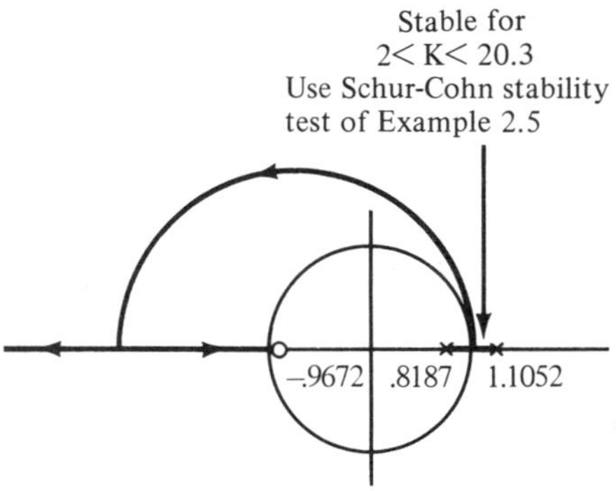

Figure A2.38

2.39 Let [ ]* indicate the Z-transform of whatever is inside [ ].

$$E(z) = \frac{[RG_1]^*}{1 + [G_1G_2H_1]^* + [G_1G_2G_3H_2]^*}$$

$$C(s) = \frac{[RG_1]^*G_2G_3G_5/[1 + G_5H_3]}{1 + [G_1G_2H_1]^* + [G_1G_2G_3H_2]^*} + \frac{RG_4G_5}{1 + G_5H_3}$$

$C(z)$ is the same except put an * on $[G_2G_3G_5/[1 + G_5H_3]]$ and $[RG_4G_5/[1 + G_5H_3]]$

2.40 $G_c(z) = \dfrac{Az^2 + Bz + D}{(1 + \tau/T)z^2 - (2\tau/T + 1)z + \tau/T}$

where $A = K_D/T + K_P(1 + \tau/T) + K_I(T + \tau)$

$B = -2(K_D/T + K_P\tau/T) - K_P - K_I\tau$

$D = K_D/T + K_P\tau/T$

For simplicity, let $\tau \to 0$. Then one obtains an algorithm of the form

$$E_1(t_k) = E_1(t_{k-1}) + \alpha E(t_k) - \beta E(t_{k-1}) + \gamma E(t_{k-2})$$

CHAPTER 3

3.27 $i_1 = \dfrac{(h_{22}R_L + 1)}{\Delta}v_1,\ i_2 = \dfrac{h_{21}}{\Delta}v_1,\ v_2 = \dfrac{-h_{21}R_L}{\Delta}v_1$, where $\Delta = (h_{11}h_{22} - h_{12}h_{21})R_L + h_{11}$

3.28 $v_1 = \dfrac{1}{\Delta}[h_{11}(1 + h_{22}R_L) - h_{12}h_{21}R_L]i_1,\ i_2 = \dfrac{h_{21}i_1}{\Delta},\ v_2 = -\dfrac{1}{\Delta}h_{21}R_Li_1,$

where $\Delta = 1 + h_{22}R_L$

3.29 $|\mathbf{A}| = -468$

3.30 $\mathbf{A}^{-1} = \dfrac{1}{17}\begin{bmatrix} 3 & 4 & -3 \\ 7 & -19 & 10 \\ -2 & 3 & 2 \end{bmatrix},\ \mathbf{B}^{-1} = \dfrac{1}{58}\begin{bmatrix} 59 & 2 & 3 & 4 & 5 \\ 1 & 60 & 3 & 4 & 5 \\ -4 & -8 & 46 & -16 & -20 \\ -2 & -4 & -6 & 50 & -10 \\ -8 & -16 & -24 & -32 & 18 \end{bmatrix}$

$$\mathbf{C}^{-1} = \left[\begin{array}{cc|cc|cc} \frac{1}{5} & \frac{-2}{5} & 0 & 0 & 0 & 0 \\ \frac{2}{5} & \frac{1}{5} & 0 & 0 & 0 & 0 \\ \hline 0 & 0 & 0 & 1 & 0 & 0 \\ 0 & 0 & \frac{1}{7} & \frac{-5}{7} & 0 & 0 \\ \hline 0 & 0 & 0 & 0 & \frac{-3}{55} & \frac{8}{55} \\ 0 & 0 & 0 & 0 & \frac{8}{55} & \frac{-3}{55} \end{array}\right]$$

3.31 $\mathbf{A}(s) = \begin{bmatrix} \dfrac{1}{s} & \dfrac{1}{s^2} \\ \dfrac{1}{s+a} & \dfrac{b\beta}{s^2+\beta^2} \\ \dfrac{2}{s^3} & \dfrac{s+1}{(s+1)^2+\beta^2} \end{bmatrix},\quad \mathbf{B}(s) = \begin{bmatrix} \dfrac{s}{s^2-\beta^2} & \dfrac{\beta}{s^2-\beta^2} \\ \dfrac{1}{(s+a)^2} & \dfrac{s}{s^2+\beta^2} \end{bmatrix}$

3.32 $\mathbf{A}(t) = [t^4 \quad t\cos\beta t],\quad \mathbf{B}(t) = \begin{bmatrix} e^{-t} - e^{-2t} & -e^{-t} + 2e^{-2t} \\ 0 & \delta(t) \end{bmatrix}$

3.33 a. $\mathbf{S} = \begin{bmatrix} 2 & -0.5 & 1 \\ 0 & 2.78388 & 1.61645 \\ 0 & 0 & 2.32101 \end{bmatrix}$ b. Not possible, $\mathbf{A}$ not positive definite.

c. $\mathbf{S} = \begin{bmatrix} 4 & 1 & 0.25 & -0.25 & 0.75 \\ 0 & 3 & 1.25 & 0.75 & 0.91666 \\ 0 & 0 & 4.83477 & 0.64636 & 0.40505 \\ 0 & 0 & 0 & 3.15551 & 2.41267 \\ 0 & 0 & 0 & 0 & 3.10035 \end{bmatrix}$

CHAPTER 4

4.26 a. $|\mathbf{G}| = 16$

b. $\hat{\mathbf{v}}_1 = \frac{1}{\sqrt{14}}[1 \quad 2 \quad 3]^T$, $\hat{\mathbf{v}}_2 = \frac{1}{\sqrt{35}}[1 \quad -5 \quad 3]^T$, $\hat{\mathbf{v}}_3 = \frac{1}{\sqrt{10}}[-3 \quad 0 \quad 1]^T$

4.27 $\mathbf{r}_1 = \left[\frac{5}{4} \quad \frac{1}{4} \quad -\frac{1}{4}\right]^T$, $\mathbf{r}_2 = \left[-\frac{1}{4} \quad -\frac{1}{4} \quad \frac{1}{4}\right]^T$, $\mathbf{r}_3 = [-3 \quad 0 \quad 1]^T$

4.28 $\mathbf{z} = \frac{5}{\sqrt{14}}\hat{\mathbf{v}}_1 - \frac{23}{\sqrt{35}}\hat{\mathbf{v}}_2 - \frac{21}{\sqrt{10}}\hat{\mathbf{v}}_3$

4.29 $\mathbf{z} = \frac{37}{4}\mathbf{x}_1 - \frac{13}{4}\mathbf{x}_2 - 21\mathbf{x}_3$

4.30 $\mathbf{r} = \frac{1}{10}[3 \quad -1 \quad 0]^T$, $\mathbf{r}_2 = \frac{1}{70}[-7 \quad 19 \quad -10]^T$, $\mathbf{r}_3 = \frac{1}{7}[0 \quad -1 \quad 2]^T$

4.31 $\mathbf{r}_1 = \left[0 \quad \frac{1}{2} \quad \frac{1}{2}\right]^T$, $\mathbf{r}_2 = \left[0 \quad -\frac{1}{2} \quad \frac{1}{2}\right]^T$, $\mathbf{r}_3 = [1 \quad 0 \quad -1]^T$, $\mathbf{z} = 2\mathbf{x}_1 - \mathbf{x}_2 + 5\mathbf{x}_3$

4.33

$$\mathbf{G} = \begin{bmatrix} 2.5784502E+03 & 6.3518805E+02 & -1.1518515E+02 & 3.3449918E+02 \\ 6.3518805E+02 & 6.2262292E+02 & 6.9457092E+01 & 8.6028557E+01 \\ -1.1518515E+02 & 6.9457092E+01 & 2.9569002E+01 & -1.9252985E+01 \\ 3.3449918E+02 & 8.6028557E+01 & -1.9252985E+01 & 1.1678545E+02 \end{bmatrix}$$

$|\mathbf{G}| = 3.12183 \times 10^8$

The orthonormal basis vectors $\mathbf{v}_i$ are

$$\begin{bmatrix} 8.7438685E-01 \\ 2.5207549E-01 \\ 2.9540095E-02 \\ -4.1356131E-01 \end{bmatrix}, \begin{bmatrix} -1.4671828E-01 \\ 8.4976387E-01 \\ 4.4605288E-01 \\ 2.3960790E-01 \end{bmatrix}, \begin{bmatrix} -3.4713072E-01 \\ 4.2903343E-01 \\ -6.5270102E-01 \\ -5.1904905E-01 \end{bmatrix}, \begin{bmatrix} 3.0564949E-01 \\ 1.7403936E-01 \\ -6.1167425E-01 \\ 7.0862055E-01 \end{bmatrix}$$

Then $\mathbf{x} = -0.6804674\mathbf{v}_1 + 1.269\mathbf{v}_2 - 2.10276\mathbf{v}_3 + 27.774197\mathbf{v}_4$. Note that the sum of the squares of the components of $\mathbf{x}$ in the original and the new coordinates is 1375.065.

4.34 The Grammian is

$$\begin{bmatrix} 4.0000000E+00 & 4.0008001E+00 & -7.9900002E+00 \\ 4.0008001E+00 & 4.0016012E+00 & -7.9916024E+00 \\ -7.9900002E+00 & -7.9916024E+00 & 1.5960100E+01 \end{bmatrix}$$

Its determinant is not zero, but very small, because the three vectors are very nearly colinear. To proceed with these poorly conditioned vectors as a basis set would produce very unreliable results.

4.35 An orthonormal basis for the subspace is, from Gram-Schmidt,

$$\begin{bmatrix} 7.3127246E-01 \\ 3.6563623E-01 \\ -3.6563623E-01 \\ 7.3127240E-02 \\ 4.3876347E-01 \end{bmatrix}, \begin{bmatrix} -1.1037266E-01 \\ -2.8796402E-01 \\ -7.9833186E-01 \\ -4.9211112E-01 \\ -1.5933463E-01 \end{bmatrix}, \begin{bmatrix} 5.2741766E-01 \\ 1.2584069E-01 \\ 2.0743863E-01 \\ -2.8101894E-01 \\ -7.6419383E-01 \end{bmatrix}$$

The components of $\mathbf{x}_1$ and $\mathbf{x}_2$, expressed in these coordinates, are

$$\begin{bmatrix} 1.2431632E+00 \\ -1.8481143E+00 \\ -1.8451571E-01 \end{bmatrix}, \begin{bmatrix} 4.8995252E+00 \\ 3.3418725E+00 \\ -1.1238300E+01 \end{bmatrix}$$

4.36 Dimension is 2.

4.37 $\|\mathbf{x}_n\| = \dfrac{5}{\sqrt{14}}, \quad x_1 = \dfrac{-5}{7}, \quad x_2 = \dfrac{-15}{14}, \quad x_3 = \dfrac{5}{14}$

4.38 $\mathbf{y}_p = \begin{bmatrix} \frac{1}{2} & 2 & \frac{9}{2} \end{bmatrix}^T$

4.39 The sine and/or cosine functions form an orthogonal basis set, $\{\mathbf{v}_i, i = 1, \infty\}$. The reciprocal basis $\{\mathbf{r}_i\}$ differs only by normalizing constants. The inner product is that defined in Problem 4.20 and the Fourier coefficients can be thought of as components of an infinite dimensional vector.

4.40 $\mathbf{A}$ must be an $n \times n$ real matrix which satisfies $\mathbf{x}^T\mathbf{A}\mathbf{x} > 0$ for all $\mathbf{x} \neq \mathbf{0}$. This last condition is the definition of a positive definite matrix.

4.42 $\langle \mathbf{f}, \mathbf{g} \rangle = \int_a^b \bar{\mathbf{f}}^T(\tau)\mathbf{g}(\tau)\, d\tau, \ \|\mathbf{f}\| = \langle \mathbf{f}, \mathbf{f} \rangle^{1/2}$

4.45 Yes. Use the columns of $\mathbf{A}$ as vectors $\mathbf{a}_i$. If necessary, add additional vectors orthogonal to these to get a full set; use Gram-Schmidt to find the set of $\mathbf{v}_i$ vectors. Then $\mathbf{T} \triangleq \begin{bmatrix} \mathbf{v}_1^T \\ \mathbf{v}_2^T \\ \vdots \end{bmatrix}$. Then $\mathbf{T}\mathbf{T}^T = \mathbf{I} = \mathbf{T}^T\mathbf{T}$ and $\mathbf{TA}$ = upper triangular.

CHAPTER 5

5.23 $\mathbf{x} = \dfrac{1}{27}[40 \quad -49 \quad 48]^T$

5.24 $\mathbf{x}_2(0) = \phi_{22}^{-1}(T)[\mathbf{x}_2(T) - \phi_{21}(T)\mathbf{x}_1(0)]$, where $\mathbf{\Phi}(T)$ is partitioned into $\left[\begin{array}{c|c} \phi_{11}(T) & \phi_{12}(T) \\ \hline \phi_{21}(T) & \phi_{22}(T) \end{array}\right]$.

5.26 $\mathbf{x} = [-4\alpha \quad 0 \quad \alpha]^T$, α an arbitrary scalar

5.27 $\mathbf{x}_1 = [-1 \quad 0 \quad 1]^T$, $\mathbf{x}_2 = [-1 \quad 1 \quad 0]^T$ and any linear combination of these.

5.28 Yes, since $r_A = 1$, $\mathbf{x} = [-2\alpha \quad \alpha]^T$, α arbitrary.

5.29 The degeneracy of **A** is 3.; the null space basis is

$$\begin{bmatrix} 0.0000000E+00 \\ -1.0000000E+00 \\ 1.1764708E-01 \\ 4.1176471E-01 \\ 0.0000000E+00 \end{bmatrix}, \begin{bmatrix} -1.0000000E+00 \\ 0.0000000E+00 \\ -7.6470608E-01 \\ 8.2352948E-01 \\ 0.0000000E+00 \end{bmatrix}, \begin{bmatrix} 0.0000000E+00 \\ 0.0000000E+00 \\ 6.4705873E-01 \\ 7.6470590E-01 \\ -1.0000000E+00 \end{bmatrix}$$

5.30 $Z_{\max} = 14$, $x_1 = 0$, $x_2 = 19$, $x_3 = x_4 = x_5 = 0$, $x_6 = 7$

5.31 $x_1 = 0.16317$, $x_2 = 0.06385$, $x_3 = 0.12904$, $x_4 = 0$, $x_5 = 0$, $x_6 = 178.9$, $x_7 = 2.2402$, $x_8 = 0$, $Z_{\max} = 3.052$ are approximate answers from computer solution.

5.32 a. $x_1 = \frac{13}{5}$, $x_2 = \frac{1}{5}$ b. $x_1 = \frac{84}{35}$, $x_2 = \frac{14}{35}$

5.33 $x = 2$. Column space is one dimensional with basis $[2 \quad 1]^T$. $\mathbf{y} - \mathbf{A}x$ is perpendicular to this line, i.e., $\mathbf{A}x$ is the orthogonal projection of $\mathbf{y}$ onto this line.

5.35 $a = 1$, $b = 3$

5.36 $a = \frac{32}{37}$, $b = \frac{107}{37}$, $\|\mathbf{e}\|^2 = \frac{8}{37}$

5.37 $a = \frac{104}{109}$, $b = \frac{323}{109}$

5.38 $i(0) \cong 16.1$

5.39 $\mathbf{a} = [2.6286 \quad 0.082145 \quad -0.003572]^T$, $\|\mathbf{y}_e\| = 0.4326$, and the estimated final GPA is 3.057, which is probably more realistic even though the residual error is larger here.

5.40 $C = 5.3704$, $\alpha = 1.077$, estimated mile time is 5 min, 22 sec.

5.41 After processing first four equations $\mathbf{x}_4 = [-0.497 \quad -0.022 \quad 0.450]^T$, and after five $\mathbf{x}_5 = [-0.382 \quad -0.137 \quad 0.565]^T$. Roughly these same values were obtained using several initial estimates for $\mathbf{x}$, as long as the first P is very large. Disagreement with Example 5.6 after $k = 4$ is not surprising since unreliable results will be given by the recursive equations whenever **A** is not full rank. The system is not observable, as will be discussed in Chapter 12.

CHAPTER 6

6.25 a. $\mathcal{A}(\alpha\mathbf{x}_1 + \beta\mathbf{x}_2) = \alpha\mathcal{A}(\mathbf{x}_1) + \beta\mathcal{A}(\mathbf{x}_2)$ because of the properties of the inner product.
b. $\mathfrak{D}(\mathcal{A}) = \mathfrak{X}$. Codomain is the real line, $\mathcal{R}^1$. $\mathfrak{R}(\mathcal{A}) = \mathcal{R}^1$ also, so $\mathcal{A}$ is onto. $\mathfrak{N}(\mathcal{A}) = \{\mathbf{x} \mid \mathbf{x}$ orthogonal to $\mathbf{b}\}$. $\mathcal{A}$ is not one-to-one since $\mathfrak{N}(\mathcal{A}) \neq \{\mathbf{0}\}$.

6.26 $\mathbf{A} = \begin{bmatrix} 0 & -1 & 2 \\ -3 & 2 & 2 \\ 0 & 2 & 3 \end{bmatrix}$

6.27 $\mathbf{A}' = \begin{bmatrix} 1 & 0 & 0 \\ 0 & 2 & 0 \\ 0 & 0 & 3 \end{bmatrix} = \mathbf{B}^{-1}\mathbf{A}\mathbf{B}$

6.28 No, it depends upon the fact that $\mathbf{B}^{-1} = \mathbf{B}^T$.

6.29 $\mathbf{A}^{-1} = \frac{1}{\sin^2 \alpha}\begin{bmatrix} \hat{\mathbf{u}}^T - \hat{\mathbf{v}}^T \cos\alpha \\ \hat{\mathbf{v}}^T - \hat{\mathbf{u}}^T \cos\alpha \\ \mathbf{w}^T \end{bmatrix}$

6.30 $\mathbf{T}_{BE} = \begin{bmatrix} 0 & \omega_z & -\omega_y \\ -\omega_z & 0 & \omega_x \\ \omega_y & -\omega_x & 0 \end{bmatrix} \mathbf{T}_{BE}$, where $\boldsymbol{\omega} = [\omega_x \quad \omega_y \quad \omega_z]^T$ is the angular velocity of the body coordinates with respect to the inertial coordinates, which could be measured with body-mounted rate gyros. (*Hint*: Let $\mathbf{x}_I = [1 \quad 0 \quad 0]$ be a fixed inertial vector. The same vector expressed in body coordinates is $\mathbf{x}_B$, the first column of $\mathbf{T}_{BE}$. Then use Euler's formula for differentiating a vector in rotating coordinates,

$$\left.\frac{d\mathbf{x}}{dt}\right|_I = \left.\frac{d\mathbf{x}}{dt}\right|_B + \boldsymbol{\omega} \times \mathbf{x}_B$$

Repeat for two other vectors $\mathbf{y}_I$ and $\mathbf{z}_I$, and use the result of Problem 4.6 for the cross product.)

6.31 $\mathbf{y}_r = \frac{1}{7}[6 \quad 19 \quad 32]^T$, $\mathbf{y}_p = \frac{1}{14}[34 \quad 5 \quad 53]^T$

6.34 a. *Hint*: Result of Problem 7.34 will be helpful.
b. *Hint*: Results of Problem 7.22 will be helpful.

6.35 Ignoring the boundary terms, the formal adjoint equation is $\mathbf{y}(k-1) = \mathbf{A}_k^T\mathbf{y}(k)$. Note that the time index k on the adjoint equation runs backward.

6.36 Introduction of a suitable weighting matrix can be handled by defining a weighted inner product, $\langle \mathbf{x}_1, \mathbf{x}_2 \rangle_Q = \mathbf{x}_1^T\mathbf{Q}\mathbf{x}_2$. This new inner product modifies the meaning of $\mathcal{A}^*$, since $\langle \mathbf{y}, \mathcal{A}(\mathbf{x}) \rangle_Q = \langle \mathcal{A}^*(\mathbf{y}), \mathbf{x} \rangle_Q$. With this definition for $\mathcal{A}^*$, the previous results still apply.

CHAPTER 7

7.36 $\lambda_i = 1, -2, 3$. $\mathbf{x}_i = [-1 \quad 1 \quad 1]^T$, $[11 \quad 1 \quad -14]^T$, and $[1 \quad 1 \quad 1]^T$. $\mathbf{J} = \text{diag}\,[1, \quad -2, \quad 3]$

7.37 $\lambda_1 = \lambda_2 = 2$, $m = 2$, $q = 2$, $\mathbf{x}_1 = [1 \quad 0]^T$, $\mathbf{x}_2 = [0 \quad 1]^T$

7.38 $|\mathbf{A} - \mathbf{I}\lambda| = 0 \Rightarrow \lambda^3(2 - \lambda) = 0$. Therefore, $\lambda = 0$ is an eigenvalue with algebraic multiplicity 3 and index $k = 2$. Rank $(\mathbf{A} - \mathbf{I}\lambda)|_{\lambda=0} = 2$. Therefore, $q = 2$, so there are two eigenvectors $\mathbf{x}_1, \mathbf{x}_3$ and one generalized eigenvector associated with $\lambda = 0$, and they are solutions of $\mathbf{A}^2\mathbf{x} = \mathbf{0}$. $\lambda_4 = 2$ has the eigenvector $\mathbf{x}_4$. These are shown as columns of

$$\mathbf{M} = \begin{bmatrix} 0 & 0 & 1 & 1 \\ 1 & 0 & 0 & 0 \\ 0 & 0 & -2 & 0 \\ 0 & 1 & 0 & 0 \end{bmatrix}$$

7.39 $\lambda_1 = 3 + j, \lambda_2 = 3 - j, \lambda_3 = 6, \mathbf{x}_1 = [6 - 2j \quad 2 - 4j \quad 0]^T, \mathbf{x}_2 = \bar{\mathbf{x}}_1, \mathbf{x}_3 = [0 \quad 0 \quad 1]^T$, $\mathbf{J} = \text{diag}\,[3 + j, 3 - j, 6]$

7.40 No. Although both matrices have $\lambda_1 = \lambda_2 = 2$, the Jordan form for $\mathbf{A}$ is $\mathbf{J} = \begin{bmatrix} 2 & 1 \\ 0 & 2 \end{bmatrix}$. Since $\mathbf{B}$ is already in Jordan form and since $\mathbf{J} \neq \mathbf{B}$, they are not similar.

7.41 $\lambda_1 = 15$, $\mathbf{x}_1 = [1 \quad 1 \quad -1]^T$, $\lambda_2 = 9$, $\mathbf{x}_2 = [1 \quad -2 \quad -1]^T$, $\lambda_3 = 3$, $\mathbf{x}_3 = [1 \quad 0 \quad 1]^T$

7.42 $\lambda_1 = 5.049$, $\lambda_2 = 0.643$, $\lambda_3 = 0.308$, $\mathbf{x}_1 = [1 \quad 0.802 \quad 0.445]^T$, $\mathbf{x}_2 = [1 \quad -0.555 \quad -1.247]^T$, $\mathbf{x}_3 = [1 \quad -2.247 \quad 1.802]^T$

7.43 a. *Hint*: Set $\lambda = 0$ in the definition of $\Delta(\lambda)$ and in $\Delta'(\lambda)$.
b. Use results of Problem 7.11 and note: (i) the only factor in $\Delta(\lambda)$ which involves all the diagonal elements a_{ii}, and which gives rise to all λ^{n-1} terms, is of the form $(a_{11} - \lambda)(a_{22} - \lambda) \cdots (a_{nn} - \lambda)$, and (ii) the form of the coefficient of λ^{n-1}, and (iii) $c_i' = (-1)^n c_i$.

7.44 a. Negative definite, b. positive semidefinite, c. negative semidefinite, d. positive definite, e. indefinite

7.45 $\mathbf{x} = (1/\sqrt{2})[-1 \quad 1]^T$ (an eigenvector), $Q_{\max} = 4 = \lambda_{\max}$ = eigenvalue associated with $\mathbf{x}$.

7.46 From equation (7.4), $\mathbf{E}_i = \mathbf{x}_i\rangle\langle\mathbf{r}_i$. This is a projection since $\mathbf{E}_i\mathbf{E}_i = \mathbf{E}_i$ and $\mathbf{E}_i\mathbf{E}_j = 0$ with $i \neq j$.

7.47 Similar to the preceding problem except the eigenspaces are now mutually orthogonal.

CHAPTER 8

8.24 $\mathbf{A}^{-1} = \frac{1}{6}[-\mathbf{A}^2 + 2\mathbf{A} + 3\mathbf{I}] = \frac{1}{6}\begin{bmatrix} -2 & 4 & 0 \\ 2 & 2 & 0 \\ 3 & -3 & 3 \end{bmatrix}$

8.25 $\begin{bmatrix} e^{-t} & \frac{1}{4}(e^t - e^{-t}) \\ 0 & e^t \end{bmatrix}$

8.26 $\begin{bmatrix} e^{-3t} & 2te^{-3t} \\ 0 & e^{-3t} \end{bmatrix}$

8.28 $\begin{bmatrix} \frac{1}{2}(e^{7t} + e^{-t}) & \frac{5}{8}(e^{7t} - e^{-t}) \\ \frac{2}{5}(e^{7t} - e^{-t}) & \frac{1}{2}(e^{7t} + e^{-t}) \end{bmatrix}$

8.29 $\frac{1}{3}\begin{bmatrix} e^{-t} + 2e^{-4t} & e^{-t} - e^{-4t} \\ 2(e^{-t} - e^{-4t}) & 2e^{-t} + e^{-4t} \end{bmatrix}$

8.30 $\begin{bmatrix} (\frac{1}{2})^k & -k(\frac{1}{2})^k & k(2-k)(\frac{1}{2})^{k-1} \\ 0 & (\frac{1}{2})^k & k(\frac{1}{2})^{k-2} \\ 0 & 0 & (\frac{1}{2})^k \end{bmatrix}$

8.31 $\begin{bmatrix} -3e^{-t} + 4e^{-2t} & -6(e^{-t} - e^{-2t}) & 0 \\ 2(e^{-t} - e^{-2t}) & 4e^{-t} - 3e^{-2t} & 0 \\ 0 & 0 & e^{-3t} \end{bmatrix}$

8.32 $\begin{bmatrix} \alpha_0 & \alpha_1 & \alpha_2 \\ \alpha_2 & \alpha_0 - 3\alpha_2 & \alpha_1 + 3\alpha_2 \\ \alpha_1 + 3\alpha_2 & -3\alpha_1 - 8\alpha_2 & \alpha_0 + 3\alpha_1 + 6\alpha_2 \end{bmatrix}$ with $\begin{cases} \alpha_0 = e^t - te^t + \frac{1}{2}t^2e^t \\ \alpha_1 = te^t - t^2e^t \\ \alpha_2 = \frac{1}{2}t^2e^t \end{cases}$

8.33

$$e^{\mathbf{A}t} = \begin{bmatrix} 0 & 0 & 0 & 0 \\ 0 & 0 & 0 & 0 \\ 0 & 0 & 0 & 0 \\ 0.1 & 0.1054 & -0.0738 & 0.1 \end{bmatrix} + e^{-t}\begin{bmatrix} 1 & 0.1047 & 0.9739 & 0 \\ 0 & 0 & 0 & 0 \\ 0 & 0 & 0 & 0 \\ -0.1 & -0.0105 & -0.0974 & 0 \end{bmatrix}$$

$$+ e^{-0.7t}\sin(3t)\begin{bmatrix} 0 & 0.3246 & -0.3142 & 0 \\ 0 & 0 & 3 & 0 \\ 0 & -0.333 & 0 & 0 \\ 0 & -0.0571 & -0.2847 & 0 \end{bmatrix} + e^{-0.7t}\cos(3t)\begin{bmatrix} 0 & -0.1047 & -0.9739 & 0 \\ 0 & 1 & 0 & 0 \\ 0 & 0 & 1 & 0 \\ 0 & -0.0949 & 0.1712 & 0 \end{bmatrix}$$

8.34

$$\mathbf{A}^k = (0.5)^k\begin{bmatrix} 0 & 4 & -4 \\ 0 & 2 & -2 \\ 0 & 1 & -1 \end{bmatrix} + \begin{bmatrix} 0 & -1 & 2 \\ 0 & -1 & 2 \\ 0 & -1 & 2 \end{bmatrix}$$

CHAPTER 9

9.13 $\mathbf{A}_1 = \begin{bmatrix} 0 & 1 \\ -b & -a \end{bmatrix}$, $\mathbf{B}_1 = \begin{bmatrix} 0 \\ 1 \end{bmatrix}$, $\mathbf{C}_1 = [1 \quad 0]$, $D_1 = 0$

$\mathbf{A}_2 = \begin{bmatrix} -a & 1 \\ -b & 0 \end{bmatrix}$, $\mathbf{B}_2 = \mathbf{B}_1$, $\mathbf{C}_2 = \mathbf{C}_1$, $D_2 = D_1$

9.14 Let $x_1 = y_1$, $x_2 = y_2$, $x_3 = \dot{y}_2$, $\mathbf{x} = [x_1 \quad x_2 \quad x_3]^T$, $\mathbf{y} = [y_1 \quad y_2]^T$, $\mathbf{u} = [u_1 \quad u_2]^T$,

$$\mathbf{A} = \begin{bmatrix} -3 & -3 & 0 \\ 0 & 0 & 1 \\ 0 & -3 & -4 \end{bmatrix}, \quad \mathbf{B} = \begin{bmatrix} 1 & 0 \\ 0 & 0 \\ 0 & 1 \end{bmatrix}, \quad \mathbf{C} = \begin{bmatrix} 1 & 0 & 0 \\ 0 & 1 & 0 \end{bmatrix}, \quad \mathbf{D} = \begin{bmatrix} 0 & 0 \\ 0 & 0 \end{bmatrix}$$

9.15 Of the many possible answers, one is obtained by setting $x_1 = y_1$, $x_2 = \dot{y}_1 - u_2$, $x_3 = y_2 - 2u_1$. Then $\mathbf{A} = \begin{bmatrix} 0 & 1 & 0 \\ -2 & -3 & 2 \\ 3 & 0 & -3 \end{bmatrix}$, $\mathbf{B} = \begin{bmatrix} 0 & 1 \\ 5 & -3 \\ -6 & 1 \end{bmatrix}$, $\mathbf{C} = \begin{bmatrix} 1 & 0 & 0 \\ 0 & 0 & 1 \end{bmatrix}$, $\mathbf{D} = \begin{bmatrix} 0 & 0 \\ 2 & 0 \end{bmatrix}$

9.16

$$\begin{bmatrix} \dot{x}_1 \\ \dot{x}_2 \\ \dot{x}_3 \end{bmatrix} = \left[\begin{array}{cc|c} -1 & 1 & 0 \\ 0 & -1 & 0 \\ \hline 0 & 0 & -6 \end{array}\right] \left[\begin{array}{c} x_1 \\ x_2 \\ \hline x_3 \end{array}\right] + \left[\begin{array}{c} 0 \\ 1 \\ \hline 1 \end{array}\right] u \quad \text{and} \quad y = [1/5 \quad -1/25 \quad 1/25]\mathbf{x}$$

9.17 With x_1 = voltage across C_1, x_2 = current through L_1, x_3 = current through L_2,

$$u = i_s, \quad \begin{bmatrix} \dot{x}_1 \\ \dot{x}_2 \\ \dot{x}_3 \end{bmatrix} = \begin{bmatrix} 0 & 1/(C_1 + C_2) & -1/(C_1 + C_2) \\ -1/L_1 & -R/L_1 & 0 \\ 1/L_2 & 0 & 0 \end{bmatrix} \begin{bmatrix} x_1 \\ x_2 \\ x_3 \end{bmatrix} + \begin{bmatrix} 0 \\ R/L_1 \\ 0 \end{bmatrix} u$$

9.18 x_1, x_2 = voltage across C_1, C_2. x_3 = current through L. Then

$$\begin{bmatrix} \dot{x}_1 \\ \dot{x}_2 \\ \dot{x}_3 \end{bmatrix} = \begin{bmatrix} \dfrac{-1}{C_1(R_1 + R_3)} & 0 & \dfrac{R_1}{C_1(R_1 + R_3)} \\ \dfrac{NK}{R_4 C_2} & -\dfrac{1}{R_4 C_2} & 0 \\ \dfrac{-R_1}{L(R_1 + R_3)} & 0 & \dfrac{-(R_1R_2 + R_1R_3 + R_2R_3)}{L(R_1 + R_3)} \end{bmatrix} \begin{bmatrix} x_1 \\ x_2 \\ x_3 \end{bmatrix} + \begin{bmatrix} 0 \\ -\dfrac{NK}{R_4 C_2} \\ \dfrac{1}{L} \end{bmatrix} v_s$$

$y = [0 \quad 1 \quad 0]\mathbf{x}$

9.19

$$\dot{x}_1 = \frac{1}{C_1}\left[-\frac{x_1}{R_1 + R_2} - \frac{R_1}{R_1 + R_2}x_3 + \frac{u_1(t)}{R_1 + R_2}\right]$$

$$\dot{x}_2 = \frac{-1}{C_2}u_2(t)$$

$$\dot{x}_3 = \frac{1}{L}\left[\frac{R_1 x_1}{R_1 + R_2} - \frac{(R_1R_2 + R_1R_3 + R_2R_3)x_3}{R_1 + R_2} + \frac{R_2}{R_1 + R_2}u_1(t) - R_3 u_2(t)\right]$$

CHAPTER 10

10.21 $y_1(t) = +\frac{9}{2}te^{-3t} + \frac{25}{4}e^{-3t} - \frac{21}{4}e^{-t}$; $y_2(t) = \frac{7}{2}e^{-t} - \frac{3}{2}e^{-3t}$

10.22 $y_1(t) = 11e^{-2t} - 3e^{-3t} - 7e^{-t}$; $y_2(t) = \frac{7}{2}e^{-t} - \frac{3}{2}e^{-3t}$

10.23 $\mathbf{x}(t) = \begin{bmatrix} (10 - 9t)e^{-t} \\ (1 - 9t)e^{-t} \end{bmatrix}$

10.24 $\mathbf{x}(t) = \begin{bmatrix} \frac{83}{9}e^{-t} - \frac{25}{3}te^{-t} + \frac{7}{9}e^{2t} \\ \frac{8}{9}e^{-t} - \frac{25}{3}te^{-t} + \frac{1}{9}e^{2t} \end{bmatrix}$

10.25 $\mathbf{x}(t_f) = \mathbf{0}$. Note that $\mathbf{u}(t) = -\frac{1}{t_f}\mathbf{\Phi}(t, 0)\mathbf{x}(0)$.

10.27 $\mathbf{\Phi}(t, 0) = e^{-t/2}\begin{bmatrix} \cos\frac{3}{2}t - \frac{1}{3}\sin\frac{3}{2}t & -\frac{5}{3}\sin\frac{3}{2}t \\ \frac{2}{3}\sin\frac{3}{2}t & \cos\frac{3}{2}t + \frac{1}{3}\sin\frac{3}{2}t \end{bmatrix}$

CHAPTER 11

11.14 $x_1(k) = y_1(k)$, $x_2(k) = y_1(k+1) + 10y_1(k) - 2u_1(k) - y_2(k)$, $x_3(k) = y_2(k)$. Then

$$\mathbf{x}(k+1) = \begin{bmatrix} -10 & 1 & 1 \\ -3 & 0 & -2 \\ 4 & 0 & -4 \end{bmatrix}\mathbf{x}(k) + \begin{bmatrix} 2 & 0 \\ 1 & 0 \\ -1 & 2 \end{bmatrix}\mathbf{u}(k)$$

$$\mathbf{y}(k) = \begin{bmatrix} 1 & 0 & 0 \\ 0 & 0 & 1 \end{bmatrix}\mathbf{x}(k)$$

11.15 Let $x_1(k) = y(k)$, $x_2(k) = y(k+1)$. Then

$$\mathbf{x}(k+1) = \begin{bmatrix} 0 & 1 \\ -2 & -3 \end{bmatrix}\mathbf{x}(k) + \begin{bmatrix} 0 \\ 1 \end{bmatrix}u(k), \quad y(k) = [1 \quad 0]\mathbf{x}(k)$$

Let $x_1(k) = y(k)$, $x_2(k) = y(k+1) + 3y(k)$. Then

$$\mathbf{x}(k+1) = \begin{bmatrix} -3 & 1 \\ -2 & 0 \end{bmatrix}\mathbf{x}(k) + \begin{bmatrix} 0 \\ 1 \end{bmatrix}u(k), \quad y(k) = [1 \quad 0]\mathbf{x}(k)$$

11.16 $x_1(k) = 2 - 5k + 5k(3 - k)$, $x_2(k) = 5 + 10k$, $x_3(k) = 10$

11.17 $\mathbf{x}(k) = \begin{bmatrix} 0 \\ 10 \end{bmatrix} + \sum_{j=1}^{k} (0.632)(0.368)^{k-j}\begin{bmatrix} 1 \\ -1 \end{bmatrix}$

11.18 $\mathbf{x}(k) = \begin{bmatrix} 1 & 0 \\ -1 & 1 \end{bmatrix}\mathbf{q}(k)$ gives $\mathbf{q}(k+1) = \begin{bmatrix} 1 & 0 \\ 0 & 1/2 \end{bmatrix}\mathbf{q}(k) + \begin{bmatrix} 1 \\ 0 \end{bmatrix}u(k)$

$\mathbf{y}(k) = [4 \quad 1]\mathbf{q}(k)$

11.19 No. The system matrix $\mathbf{A}$ is singular and consequently $\mathbf{\Phi}^{-1}$ does not exist.

11.20 $\dot{\mathbf{x}} = \begin{bmatrix} 0 & 1 \\ 0 & -1/\tau \end{bmatrix}\mathbf{x} + \begin{bmatrix} 0 \\ K/\tau \end{bmatrix}u(t)$

$$\mathbf{x}(k+1) = \begin{bmatrix} 1 & \tau(1 - e^{-\Delta t/\tau}) \\ 0 & e^{-\Delta t/\tau} \end{bmatrix}\mathbf{x}(k) + \begin{bmatrix} \Delta t - \tau(1 - e^{-\Delta t/\tau}) \\ 1 - e^{-\Delta t/\tau} \end{bmatrix}Ku(k)$$

11.21 $\mathbf{x}(k+1) = \begin{bmatrix} 1 - K\Delta t + K\tau(1 - e^{-\Delta t/\tau}) & \tau(1 - e^{-\Delta t/\tau}) \\ -K(1 - e^{-\Delta t/\tau}) & e^{-\Delta t/\tau} \end{bmatrix}\mathbf{x}(k) + \begin{bmatrix} \Delta t - \tau(1 - e^{-\Delta t/\tau}) \\ 1 - e^{-\Delta t/\tau} \end{bmatrix}Kr(k)$

$\theta(k) = [1 \quad 0]\mathbf{x}(k)$

11.22

$$\mathbf{A} = \begin{bmatrix} 0.3666 & 0.1010 & 0.0110 \\ 0 & 0.6703 & 0.1341 \\ 0 & 0 & 0.6703 \end{bmatrix} \qquad \mathbf{B} = \begin{bmatrix} 0.0136 \\ 0.1802 \\ 0.1648 \end{bmatrix}$$

11.23 One of many possible answers, obtained by using a parallel realization, is

$$\mathbf{x}(k+1) = \begin{bmatrix} 0.04979 & 0 & 0 \\ 0 & 0.22313 & 0 \\ 0 & 0 & 0.60653 \end{bmatrix} \mathbf{x}(k) + \begin{bmatrix} 0.2121 \\ 0.1934 \\ 0.0649 \end{bmatrix} u(k)$$

$$y(k) = [0.04979 \quad -0.22313 \quad 0.60653]\mathbf{x}(k)$$

CHAPTER 12

12.18 System is completely controllable and completely observable.

12.19 a. Completely controllable, b. completely controllable, c. not completely controllable.

12.20 a. Not controllable, not observable. This system will reappear in Example 15.2. b. Not controllable; it is observable. This system is a special case of Problem 15.1b with $\alpha = 1$. c. Both controllable and observable. See also Problems 15.8 and 15.9.

12.21 Yes. When put into state variable form, the system is completely controllable.

12.22 Completely controllable and completely observable for all finite values of k_1 and b_2 except when $k_1 b_2 = 1$.

12.23 Controllable, unless $R_3 C_2 = R_2 C_1$. Observable unless $R_1 = 0$. In these cases pole-zero cancellation occurs.

12.24 a. Uncontrollable and unobservable. Two identical *RLC* systems in parallel.
b. Not completely controllable, due to pole-zero cancellation. Completely observable.

12.26 *Hint:* Use the semigroup property of transition matrices.

CHAPTER 13

13.20 $Q \approx Q_n + (c/2)\sqrt{\rho g/h_n}\,\delta h$, where $h = h_n + \delta h$ and $Q_n = c\sqrt{\rho g h_n}$

13.21 a. $\theta = \cos^{-1}\left[\dfrac{\langle(\mathbf{r}_1 - \mathbf{x}), (\mathbf{r}_2 - \mathbf{x})\rangle}{\|\mathbf{r}_1 - \mathbf{x}\|\,\|\mathbf{r}_2 - \mathbf{x}\|}\right]$

b. $\delta\theta = [\nabla_{\mathbf{x}}\theta]_n^T \delta\mathbf{x}$. Letting $\mathbf{r}_1 - \mathbf{x} = \mathbf{u}$ and $\mathbf{r}_2 - \mathbf{x} = \mathbf{v}$, the three components of $\nabla_{\mathbf{x}}\theta$ are given by

$$\frac{\partial\theta}{\partial x_i} = \frac{(u_i + v_i)\,\|\mathbf{u}\|\,\|\mathbf{v}\| - (v_i\,\|\mathbf{u}\|^2 + u_i\,\|\mathbf{v}\|^2)\cos\theta}{\|\mathbf{u}\|^2\,\|\mathbf{v}\|^2 \sin\theta}$$

13.22 $x = -0.5466$ is obtained after 4 iterations from $x^{(0)} = 0$ and after 5 iterations from $x^{(0)} = -2$. The other two roots are $-1.2267 \pm j1.4677$.

13.23 $x^{(1)}(t) = \left[2 - \dfrac{1 + 2\tanh 2}{\tanh 2}\right] e^{-3t \sinh 2/\cosh^2 2} + \dfrac{1 + 2\tanh 2}{\tanh 2}$. For comparison purposes, the exact solution of the original nonlinear equation is $x(t) = \sinh^{-1}(3t + \sinh 2)$.

13.24 No. The gradient $\nabla_{\mathbf{x}} f = \mathbf{p}$ is well defined and is never zero, regardless of $\mathbf{x}$, except in the trivial case. This partially explains why the extreme values always occur on constraint boundaries, as discussed in Sec. 5.5.

CHAPTER 14

14.22 a. Eigenvalues are $\lambda_1 = -4$, $\lambda_2 = 2$. Every term in $\mathbf{\Phi} \longrightarrow \infty$, as e^{2t} when $t \longrightarrow \infty$. Therefore, $\|\mathbf{\Phi}(t, 0)\| \longrightarrow \infty$ and the system is unstable.

b. Eigenvalues are $\lambda = -1, -2$, and -3. $\|\Phi(t, 0)\| \longrightarrow 0$ as $t \longrightarrow \infty$ and the system is globally asymptotically stable.

c. Eigenvalues are $\lambda = -1/4$ and $-3/4$. The system is asymptotically stable.

14.23 a. $\lambda_1 = 1$, $\lambda_2 = 1/2$. Stable i.s.L. but not asymptotically stable; observable but not controllable.

b. $\lambda_1 = -1$, $\lambda_2 = -3$, $\lambda_3 = -5$. Asymptotically stable, uncontrollable and unobservable.

c. $\lambda = 1 \pm \sqrt{21}$. System is unstable, completely controllable and observable.

These results illustrate that stability, controllability and observability are independent system properties. One property does not imply or require any of the others.

14.24 The eigenvalues for this system are roots of $\lambda^2 + (2 - a)\lambda + (2 - a) = 0$. They are independent of time and are in the left-half plane if $a < 2$. But $\|\Phi(t, 0)\|$ obviously goes to infinity if $a > 1$, meaning that the system is unstable. For time-varying systems left-half-plane eigenvalues for all times *do not* ensure stability.

14.25 a. $$V'(\mathbf{x}) = \frac{1}{2b}(ax_2 + bx_1)^2 + \frac{1}{2ab}(ax_3 + bx_2)^2 + \int_0^{x_2} \left[F(x_2) - \frac{b}{a}\right] x_2\, dx_2$$

$$\dot{V}'(\mathbf{x}) = -\frac{a}{b}\left[F(x_2) - \frac{b}{a}\right] x_3^2$$

b. Asymptotically stable if $a > 0$, $b > 0$, and $F(x_2) > b/a$ for all x_2.

14.26 $u_1(t) = -M_1 \operatorname{sign}[7x_1(t) - x_2(t)]$
$u_2(t) = -M_2 \operatorname{sign}[-8x_1(t) + 19x_2(t)]$

CHAPTER 15

15.15 $$\dot{\mathbf{x}} = \begin{bmatrix} -1/R_1C - 1/R_2C & -1/C \\ 1/L & 0 \end{bmatrix}\mathbf{x} + \begin{bmatrix} -1/R_2C \\ 1/L \end{bmatrix}u, \quad y = [1 \quad 0]\mathbf{x} + u,$$

$$H(s) = \frac{s(s + 1/R_1C)}{s^2 + (1/R_1C + 1/R_2C)s + 1/LC}$$

15.16 $$\dot{\mathbf{x}} = \begin{bmatrix} -1/CR_1 & 0 & 1/C & 0 \\ 0 & -1/CR_1 & 0 & -1/C \\ -1/L & 0 & -R_2/L & -R_2/L \\ 0 & 1/L & -R_2/L & -R_2/L \end{bmatrix}\mathbf{x} + \begin{bmatrix} 0 & 0 \\ 0 & 0 \\ 1/L & 0 \\ 0 & 1/L \end{bmatrix}\begin{bmatrix} u_1 \\ u_2 \end{bmatrix}, \quad \mathbf{y} = \begin{bmatrix} 1 & 1 & 0 & 0 \\ 0 & 1 & 0 & 0 \end{bmatrix}\mathbf{x}$$

15.17 $$\mathbf{H}(s) = \begin{bmatrix} 0 & 0 \\ 1/(s + 3) & 0 \end{bmatrix}$$

15.18 First-order state equation $\dot{x}_1 = [-x_1/R_1 + u_1/R_1 + u_2]/C$, $y_1 = x_1$, $y_2 = x_1 + u_2R_2$, $$\mathbf{H}(s) = \frac{\begin{bmatrix} 1/R_1 & 1 \\ 1/R_1 & 1 + R_2C(s + 1/R_1C) \end{bmatrix}}{(s + 1/R_1C)C}$$

15.19 $\mathbf{D} = \begin{bmatrix} 0 & 0 & 1 \\ 0 & 0 & 1 \end{bmatrix}$. One possible choice for $\{\mathbf{A}, \mathbf{B}, \mathbf{C}\}$ is

$$\mathbf{A} = \begin{bmatrix} -1 & 0 & 0 & 0 \\ 0 & -1 & 0 & 0 \\ 0 & 0 & -2 & 0 \\ 0 & 0 & 0 & -3 \end{bmatrix}, \quad \mathbf{B} = \begin{bmatrix} 1 & 1 & -1 \\ 0 & 1 & 0 \\ 0 & 1 & 0 \\ 0 & 1 & 4 \end{bmatrix}, \quad \mathbf{C} = \begin{bmatrix} 1 & 0 & -1 & 0 \\ 0 & \frac{1}{2} & 0 & -\frac{1}{2} \end{bmatrix}$$

15.20 a. is reducible since it is completely observable but not controllable.
b. is reducible since it is completely controllable but not observable.
c. is irreducible. It is completely controllable and observable.

15.21 $\mathbf{D} = \begin{bmatrix} 1 & 0 & 0 \\ 0 & 0 & 0 \end{bmatrix}$. One possible solution for {**A**, **B**, **C**} is

$$\mathbf{A} = \begin{bmatrix} -1 & 0 & 0 & 0 \\ 0 & -2 & 0 & 0 \\ 0 & 0 & -3 & 0 \\ 0 & 0 & 0 & 0 \end{bmatrix}, \quad \mathbf{B} = \begin{bmatrix} -1 & 1 & 0 \\ 0 & 1 & 0 \\ 0 & 0 & 1 \\ 0 & 0 & 1 \end{bmatrix}, \quad \mathbf{C} = \begin{bmatrix} 1 & -1 & 1 & 0 \\ 1 & -1 & 0 & 1 \end{bmatrix}$$

15.22 $\mathbf{D} = \begin{bmatrix} 0 & 0 \\ 0 & 0 \end{bmatrix}$. One possible solution for {**A**, **B**, **C**} is

$$\mathbf{A} = \left[\begin{array}{cc|c|c} -1 & 1 & 0 & 0 \\ 0 & -1 & 0 & 0 \\ \hline 0 & 0 & -1 & 0 \\ \hline 0 & 0 & 0 & 0 \end{array}\right], \quad \mathbf{B} = \left[\begin{array}{cc} 0 & 0 \\ 1 & 0 \\ \hline 1 & 1 \\ \hline 1 & 0 \end{array}\right], \quad \mathbf{C} = \left[\begin{array}{cc|c|c} 1 & -2 & 1 & 0 \\ 0 & 0 & 1 & 1 \end{array}\right]$$

Another solution has the same **A**, but $\mathbf{B} = \left[\begin{array}{cc} -1 & 1 \\ 1 & 0 \\ \hline 1 & 1 \\ \hline 1 & 0 \end{array}\right]$ and $\mathbf{C} = \left[\begin{array}{cc|c|c} 1 & 0 & 0 & 0 \\ 0 & 0 & 1 & 1 \end{array}\right]$. A third solution is given by Chen (see page 251 of Reference 17).

15.23 One eighth-order irreducible realization is

$$\dot{\mathbf{x}} = \left[\begin{array}{cccc|ccc|c} 0 & 1 & 0 & 0 & & & & \\ 0 & 0 & 1 & 0 & & \mathbf{0} & & \\ 0 & 0 & 0 & 1 & & & & \\ 0 & 0 & 0 & 0 & & & & \\ \hline & & & & 0 & 1 & 0 & \\ & \mathbf{0} & & & 0 & 0 & 1 & \\ & & & & 0 & 0 & 0 & \\ \hline & & & & & & & 0 \end{array}\right]\mathbf{x} + \left[\begin{array}{ccc} 0 & 0 & 0 \\ -1 & 0 & 1 \\ 0 & 0 & 0 \\ 1 & 1 & -2 \\ \hline 0 & 0 & 0 \\ 2 & 0 & -2 \\ 1 & 0 & 1 \\ \hline 1 & 0 & -1 \end{array}\right]\mathbf{u}$$

$$\mathbf{y} = \left[\begin{array}{cccc|cccc|c} 1 & 0 & 0 & 0 & 0 & 0 & 0 & & 1 \\ 1 & 1 & 0 & 0 & \tfrac{1}{2} & 0 & 0 & & 1 \\ 1 & 0 & -1 & 0 & -1 & -5 & 1 & & 10 \end{array}\right]\mathbf{x}$$

See Reference 17 for a detailed solution.

15.24

$$\mathbf{T}(z) = \begin{bmatrix} \dfrac{0.5z^2 - 1.3z + 1.245}{(z-1)} & -(z+0.3) \\ \dfrac{1.5z^2 - 1.2z + 0.145}{(z-1)} & 2(z-0.5) \end{bmatrix} \frac{1}{(z^2 - z + 0.89)}$$

15.25 One possible answer is

$$\mathbf{A} = \begin{bmatrix} 1 & 0 & 0 & 0 \\ 0 & 0.3679 & 0 & 0 \\ 0 & 0 & 1 & 1 \\ 0 & 0 & -0.5 & 0 \end{bmatrix} \quad \mathbf{B} = \begin{bmatrix} 1 \\ 1 \\ -0.9878 \\ -1.3761 \end{bmatrix} \quad \mathbf{D} = [0]$$

$$\mathbf{C} = \begin{bmatrix} 3.1641 & -2.1762 & 1 & 0 \\ 1.8984 & -0.8984 & 0 & 0 \end{bmatrix}$$

CHAPTER 16

16.21 $\mathbf{K} = \begin{bmatrix} 1 & -5 & -1 \\ 0 & 9 & 3 \end{bmatrix}$. $\left(\textit{Hint:} \text{ Define } \alpha = \dfrac{1}{\lambda + 3} \text{ and let } \lambda \longrightarrow -3 \text{ after finding } \mathbf{G}^{-1}.\right)$

16.23 $\mathbf{K}' = [0 \quad 2]^T$, but $\Delta_0'(\lambda) = (\lambda + 1)(\lambda + 5)$ as before.

16.24 One solution is $\mathbf{K}' = [17 \quad 0]^T$, another is $\mathbf{K}' = [0 \quad 17]^T$.

16.25 One solution is $\mathbf{K}' = \begin{bmatrix} 0 & 6 \\ 9 & 0 \end{bmatrix}$. For another solution using a slightly different method, see [26].

16.26 $\mathbf{K}' = [-1/\epsilon^2 \quad 1/\epsilon^2]$

16.27 No, because $\mathbf{N} = \begin{bmatrix} 0 & 1 \\ 0 & 1 \end{bmatrix}$ is singular [36].

16.28 No, $\mathbf{N} = \begin{bmatrix} 1 & 0 \\ 1 & 0 \end{bmatrix}$ is singular [36].

16.29 $\mathbf{F}_d = \begin{bmatrix} 1 & -1 \\ 0 & 1 \end{bmatrix}, \quad \mathbf{K}_d = \begin{bmatrix} -2 & -2 & 0 \\ 0 & -3 & -4 \end{bmatrix}, \quad \mathbf{H}(s) = \begin{bmatrix} 1/s & 0 \\ 0 & 1/s^2 \end{bmatrix}$

16.30 $\mathbf{K} = [1 \quad 0.5]$

16.31 $\mathbf{K} = [1.34 \quad 0.4]$

16.32 Two choices are

$$\mathbf{K} = \begin{bmatrix} 0.4 & -0.3 \\ 0.25 & 0.15 \end{bmatrix} \text{ and } \begin{bmatrix} 0.3 & -0.4 \\ 0.15 & 0.25 \end{bmatrix}$$

16.33 Using the nested integrator method (Example 9.5) gives

$$\mathbf{A} = \begin{bmatrix} -3 & 1 & 0 & 0 & 0 \\ -2 & -2 & 2 & 1 & 0 \\ 2 & 0 & -6 & 0 & 1 \\ 4 & -4 & 0 & 0 & 0 \\ 0 & 0 & -1 & 0 & 0 \end{bmatrix} \quad \mathbf{B} = \begin{bmatrix} 1 & 0 \\ 0 & 0 \\ 0 & 0 \\ 5 & 1 \\ 0 & 1 \end{bmatrix} \quad \mathbf{C} = \begin{bmatrix} 1 & 0 & 0 & 0 & 0 \\ 0 & 1 & 0 & 0 & 0 \\ 0 & 0 & 1 & 0 & 0 \end{bmatrix} \quad \mathbf{D} = [0]$$

One choice for $\mathbf{K}$ is

$$\begin{bmatrix} -2.1251 & -67.1349 & -26.9825 & -12.045 & -15.107 \\ 4.8752 & 171.368 & 70.0198 & 34.4558 & 40.8934 \end{bmatrix}$$

A choice for the full state observer is

$$\mathbf{B}_c = \begin{bmatrix} 6.3127 & 1.0443 & -0.0136 \\ -35.2083 & 13.0842 & 1.5313 \\ -43.6735 & -2.2322 & 9.6030 \\ -288.0065 & 49.2646 & -2.4314 \\ -367.7839 & -17.7946 & 59.789 \end{bmatrix}$$

CHAPTER 17

17.20 $g_a = 21$ is minimum cost. Optimal path is a, b, e, l, r, z.

17.21 2 hr on course 1, score = 65; 0 hr on course 2, score = 40; 1 hr on course 3, score = 52; 1 hr on course 4, score = 91. Total max. score = 248.

17.22 $u^*(0) = -1$, $u^*(1) = -2$, $x^*(0) = 10$, $x^*(1) = 4$, $x^*(2) = 0$, $J^* = 9$. For comparison, if no control is used, $u(k) = 0$, $x(0) = 10$, $x(1) = 5$, $x(2) = 2.5$, and $J = 20.25$.

17.23

$k =$	0	1	2	3	4	5
$u^*(k)$	−0.0147	−0.0293	−0.0587	−0.1173	−0.2346	—
$x^*(k)$	10.0	4.985	2.463	1.173	0.469	0.0

17.24 $u^*(0) = -21x_0/104$, $u^*(1) = -x_0/52$, $x^*(0) = x_0$, $x^*(1) = 5x_0/52$, $x^*(2) = x_0/104$

17.25 The gain matrices and eigenvalues are:

a. $[-1.620 \quad -1.959]$, $0.190 \pm j0.148$
b. $[-1.199 \quad -1.413]$, $-0.4 \pm j0.26$
c. $[-1.755 \quad -2.136]$, -0.033, -0.0212 (both real)
d. $[-1.775 \quad -2.162]$, 0.0044, -0.229 (both real)

17.26 a. $\mathbf{K}_{ss}^T = [-0.5202 \quad 0.8588]$. Filter eigenvalues are $0.2319 \pm j0.2956$.

b. Based on the filter pole locations, the filter will be much slower than controllers c and d and a little slower than a. Therefore, the filter will have a major impact on system response, maybe even be the dominant effect. In case b the filter is somewhat faster than the controller, and the total system will act more nearly like the full state controller.

17.27 a. $\mathbf{K}_{ss}^T = [-0.4561 \quad 0.9977]$. Eigenvalues are $0.0404 \pm j0.0257$.

b. Because of the lower noise level, the filter response is much faster. The effect on the control loop will be essentially negligible and the controller will work about as well as if it had the actual state to work with rather than estimate. Stated differently, with this low noise level the estimated states will agree very closely with the true values.

17.28 On backward pass find

k	k'	$\mathbf{G}^T$	$v = -\mathbf{FV}$
5	0	—	—
4	1	[0.74963 −0.24988]	7.4963
3	2	[0.72159 −0.23296]	−0.5684
2	3	[0.72140 −0.23305]	−0.9369
1	4	[0.72140 −0.23305]	−0.9537
0	5	[0.72140 −0.23305]	−0.9544

Then on the forward pass, starting with the given $\mathbf{x}(0)$, find

k	$\mathbf{x}(k)^T$	$u(k) = -\mathbf{G}(k)\mathbf{x}(k) + v(k)$
0	[4 −2]	−4.3061
1	[−0.3061 1.3061]	−0.4284
2	[−0.7345 1.2345]	−0.1192
3	[−0.8538 1.1038]	0.3048
4	[−0.5490 0.6740]	8.0762
5	[7.5273 −7.4648]	—

17.29

$$\mathbf{G} = \begin{bmatrix} 0.0422 & 0.0487 & -0.1039 & 0.2606 \\ 0.0075 & -0.4757 & -0.7322 & 0.0019 \end{bmatrix}$$

$\lambda_i = \{0.102, 0.898, 0.6068 \pm j0.3517\}$

17.30 The feedback matrix $\mathbf{G}$ and the resulting eigenvalues are:

a. $[0.0227 \quad -0.1097 \quad -0.1507 \quad 0.1120]$, $\{0.0227, 0.8304, 0.699 \pm j0.455\}$
b. $[0.0271 \quad -0.0032 \quad -0.1099 \quad 0.1255]$, $\{0.0025, 0.8196, 0.721 \pm j0.479\}$
c. $[0.0274 \quad -0.0034 \quad -0.1113 \quad 0.1265]$, $\{4.34 \times 10^{-4}, 0.818, 0.721 \pm j0.4818\}$
d. $[0.0208 \quad 0.0004 \quad -0.0833 \quad 0.0152]$, $\{0.0045, 0.8437, 0.7197 \pm j0.4818\}$

For reference the open-loop eigenvalues are $\{1.0, 0.135, 0.7175 \pm j0.4907\}$

17.31 $u^*(t) = \dfrac{\cosh t}{\sinh (1)}, \; x^*(t) = \dfrac{\sinh t}{\sinh (1)}$

17.32 $u^*(t) = -7/2 + 3t, \; x_1^*(t) = 1 + t - 7t^2/4 + t^3/2, \; x_2^*(t) = 1 - 7t/2 + 3t^2/2$

17.33 $x(t) \simeq 10 \cosh \sqrt{5}\, t - 10.065 \sinh \sqrt{5}\, t, \; p(t) \simeq 5.001 \cosh \sqrt{5}\, t - 4.462 \sinh \sqrt{5}\, t$

17.35 a. $u^*(t) = -\frac{1}{2}\mathbf{B}^T\mathbf{p}(t)$ if $\|\frac{1}{2}\mathbf{B}^T\mathbf{p}(t)\| \le 1$; $\mathbf{u}^*(t) = -\mathbf{B}^T\mathbf{p}(t)/\|\mathbf{B}^T\mathbf{p}(t)\|$ otherwise. Linear control with saturation.

b. $u^*(t) = -\mathbf{B}^T\mathbf{p}(t)/\|\mathbf{B}^T\mathbf{p}(t)\|$ for all t unless $\mathbf{B}^T\mathbf{p}(t) \equiv \mathbf{0}$ on a finite time interval (singular control case). Optimal control is always on the boundary of U except in the case of singular control.

c. Each component satisfies $u_i^*(t) = -\text{sign}\,[\mathbf{B}^T\mathbf{p}(t)]_i$, where $[\mathbf{B}^T\mathbf{p}(t)]_i$ is the ith component of $\mathbf{B}^T\mathbf{p}(t)$. This is the so-called bang-bang control and is valid provided $[\mathbf{B}^T\mathbf{p}(t)]_i \neq \mathbf{0}$. If $[\mathbf{B}^T\mathbf{p}(t)]_i \equiv 0$ on a finite interval, we have singular control.

d. $u_i^*(t) = 0$ when $|[\mathbf{B}^T\mathbf{p}(t)]_i| \le 1$; $u_i^*(t) = -\text{sign}\,[\mathbf{B}^T\mathbf{p}(t)]_i$ otherwise.

17.36 No solution exists. It is not possible to reach the origin with this unstable system and the bounded admissible controls unless $|x(0)| < 1$. The minimum principle provides only necessary conditions. It does not guarantee existence of a solution.

17.37 $\mathbf{G} = [0.2215 \quad -0.3728 \quad 0.2190 \quad 0.3144]$, $\lambda_i = \{-0.921, -28.15, -0.9165 \pm j2.88\}$

INDEX